D0084950

EARTHSCAPE

A PHYSICAL GEOGRAPHY

EARTHSCAPE

A PHYSICAL GEOGRAPHY

WILLIAM M. MARSH

University of Michigan, Flint

JOHN WILEY & SONS

New York *Chichester* *Brisbane* *Toronto* *Singapore*

Cover photo by Tom Bean
(Paria Canyon Wilderness Area, Arizona-Utah)

Chapter opening illustrations by James Turner

Library of Congress Cataloging-in-Publication Data
Marsh, William M.
Earthscape: a physical geography.
Bibliography: p.
1. Physical geography—Text-books—1945-
I. Title.
GB54.5.M367 1987 910'.02 86-28209
ISBN 0-471-85055-1
Printed in the United States of America
10 9 8 7 6 5 4 3 2 1

Introduction to the Student

The quests of the *Starship Enterprise* are fictional expressions of a rich tradition of geographic exploration in the Western world. Several centuries before the voyages of the *Starship Enterprise*, exploration was focused on our own solar system. The manned landing on the moon in 1969, the unmanned landings on Mars in 1975, and the satellite voyages to the outer planets of our solar system in the 1970s and 1980s are also expressions of this great geographic tradition. Only 100 years before the planetary missions, the thrust of American geographic exploration was in the American West where in the 1870s John Wesley Powell traversed the Grand Canyon and F. V. Hayden explored the Yellowstone region. And a hundred years before that Captain James Cook explored the Pacific discovering Antarctica, making the first Western visit to Hawaii, and charting the coasts of New Zealand and Australia.

Quite literally, we are a people preoccupied with learning about what is where, about finding out how things are arranged geographically in the world, the solar system, and the Universe. What motivates individuals and nations to pursue geographic exploration varies—political expansion, economic gain, individual glory, and intellectual curiosity—but the results are often the same. In any case, exploration and geographic discovery in the nineteenth and twentieth centuries have consistently led to mapping projects, and mapping has led to questions about the way things are distributed, what influences distributions, and why the patterns of certain distributions correlate with one another. These questions in turn lead to questions about the forces and processes that control the land masses, climates, vegetation, water features, minerals, human populations, agriculture, diseases, and so on.

Although our drive to explore continues to push the geographic adventure to greater and greater distances from earth, at the same time the job of mapping and analyzing the geographic features of earth continues. In fact, the work of earth-centered geographers grows with each decade as the earth changes, new problems arise (such as the extent and effects of drought in Central Africa), and world population continues to grow and spread.

The efforts of geographers have been advanced tremendously in this century with new and more sophisticated mapping techniques. Prior to 1930 or so, mapping had to be done on foot with surveying instruments. The pace was slow even in easy terrain, and in areas of rugged terrain, mapping was often crude at best. After 1930 aircraft made aerial photography possible, and today we have the capacity to remap all of the United States and Canada each decade at a scale large enough to show individual houses. With the development of radar in World War II, aerial mapping could be done in cloud-shrouded regions because radar can "see" through clouds, something that photography cannot do. In addition, airborne scanning apparatuses were also developed, enabling geographers to detect such things as surface temperatures in land and water, moisture differences in soil, and variations in land use and crop types. And with the advent of earth-orbiting satellites carrying earth-viewing scanners, the earth's atmosphere, oceans, lands, and vegetation can be monitored almost continuously. We see one of the benefits of these modern mapping systems in daily weather reporting and forecasting.

With the types and amounts of geographic data and information increasing each decade, the need to understand how things are interrelated geographically over vast reaches of the earth is also growing. From the standpoint of survival, it has become clear that the world can no longer be treated as a geographically compartmentalized sphere: Witness that the same ocean currents that periodically warm the coast of Peru and influence the fish catch there also cause weather changes that affect the magnitude and frequency of destructive storms in the western United States. Increased sulphur dioxide emissions from power plants in the United States produce acidic rainwater, which has altered the chemical balance of lakes in another country, Canada, causing many of them to become ecologically sterile. And air pollution from the industrialized world as a whole may be causing large-scale climatic changes that could reduce food production and cause famine in many Third World countries.

But despite the glaring need to understand broadly based earth problems, we in the United States as a society are actually less geographically aware than we were twenty or thirty years ago. Paradoxically, this comes at a time when the opportunities and the need for advancing geographic knowledge are greatest. Our perspectives on the global and international problems of our day—the border wars, mineral depletion issues, regional famine, and international migration—are often frighteningly simple and geographically naive. Not surprisingly, the solutions we propose for these problems often fail to recognize the broad foundation of scientific knowledge we have available to us.

As national and international problems expand in scope and complexity, the need to illuminate the larger picture and illustrate the interconnectedness of the features and processes over the earth's surface is greater than ever. Herein lies one of the most important roles of physical geography in modern college and university education, one that this book is designed to promote.

Traditional geography books focused on telling us what is where: oceans, rivers, minerals, mountains, countries, cities, and so on. In modern geography we are more concerned with figuring out why things are distributed as they are. In physical geography we are concerned with understanding patterns of climate, soil, river systems, vegetation, landforms, and land use—what we call the geographic landscape. This book is aimed at illustrating how the geographic landscape is formed, how it functions, and how it changes under both natural and human influences. To achieve this we have to understand the nature of the *forces* that drive change on the earth's surface. Any change that takes place on the earth is driven by energy, such as radiation from the sun, heat from the earth's interior, and chemical compounds from plants. This energy drives the movement of the atmosphere and the oceans, the exchange of water between air and land, and the growth of plants. How the driving forces are distributed over the earth and what factors influence their distributions are important questions in physical geography. In order to understand the distribution of energy over the earth's surface, it is important to realize that it is arranged in great *systems*. These systems are characterized by energy flows and material among different parts of the environment, such as the flow of solar radiation to the earth's surface, the flow of heat between the soil surface and the lower atmosphere, and the flow of heated ocean currents into cold regions of the oceans.

Although the driving forces of energy are at the root of landscape change, the actual instruments of change are surface *processes*. Processes are characterized by movements or changes in earth substances such as air, water, and soil. Processes are the agents of change on earth; they are the tools of the geosystems in a manner of speaking. Understanding the mechanisms of earth surface processes is an important concern to geographers, but equally important is understanding how these processes vary over time, as with the seasons, and over geographic space, as from the Pacific Coast to the interior of North America.

From the question of how processes function we go to the question of the *work* that is accomplished by them. We wish to know not only how much work takes place when water runs over a barren farmfield or waves crash into a shore, but also how the resultant changes are physically manifested in the landscape. These physical manifestations are the features we see in the landscape: forests, mountains, beaches, and so on, as well as human artifacts in the form of cities, farms, and reservoirs—the ingredients of landscape.

The landscape was once described as "the excited skin of the earth," meaning that it is vibrant with never-ending change. It is amazing to discover how remarkably different many landscapes were as little as 100 years ago. For example, since the nineteenth century the great forests of many tropical areas have converted to grasslands, stream valleys in many urban areas have been eradicated as cities have grown over them, and extensive areas of harsh, windy desert, such as the Imperial Valley of California, have been transformed into manicured landscapes of agricultural green.

Contemporary landscapes are changing at faster and faster rates, largely because of the increased influences of growing human populations and expanding technology. This is a monumental issue for humans because the landscape is our habitat and its condition is central to our survival. Geographers are chiefly concerned with the causes and forms of landscape change, but as members of society they must also be concerned with the directions of change and with the quality of change when weighed against human values. As you read this book, consider how you stand on various issues and ask yourself what basis you have for your values in different geographic contexts; as a member of a community, region, nation, continent, planet.

ACKNOWLEDGMENTS

A long list of editors, reviewers, and contributors provided a share of the resources necessary to the preparation of this book. The contributors are

Jeff Dozier
James Turner
George Kish
Bruce Decker
John Street
Walter A. Schroeder
Tim F. Ball
Geoffrey A. J. Scott
Bruce D. Marsh
Charles B. Belt, Jr.
Chester Wilson
Richard Hill-Rowley
John M. Grossa, Jr.

Special recognition goes to my old friend and colleague, Jeff Dozier, of the University of California-Santa Barbara, for his input more than 10 years ago in shaping Chapters 3, 4, 5, 6, and 8. Special recognition also goes to my friend and colleague, James Turner, of Richardson-Verdoorn, Inc., who created and rendered the opening drawing for each chapter. The remaining contributors, by no means less significant, are thanked for their thoughtful words in notes and appendixes.

Many people reviewed the manuscript at different stages of its development. Several early reviewers were anonymous but their remarks were useful, and they are thanked for their help. Later reviewers include: Anthony Davis, University of Toronto; Ian Campbell, University of Alberta; Nigel Allen, Louisiana State University; David Greenland, University of Colorado; David Miller, SUNY-Cortland; and Robert Howard, California State University, Northridge. My thanks to each.

Among the many editors who worked on this project, Katie Vignery, Carol Verburg, Ron Pullins, John Hendry, and Elizabeth Meder are recognized for their guidance and patience.

A very special thanks goes to my wife, Nina L. Marsh, for her unfailing support and assistance in this project.

W.M.M.
Flint, Michigan
1987

Introduction to the Instructor

This book builds on its predecessor, *Landscape*, published in 1981, but it is different in many ways. It is written in a more colloquial style, with more attention given to certain topics such as weather and climate, and ocean circulation. *Earthscape* also introduces an element from old-time physical geography—physiography, which is used as a summary chapter to the book. The quality of the graphics in *Landscape* has been retained and expanded somewhat in *Earthscape*.

The balance of topics (coverage) in *Earthscape* is designed to meet the needs of the one-semester physical geography course. Nine chapters deal with weather and climate, seven with earth materials, crustal processes, and landforms, four with vegetation and soil, three with water, and one with physiography. The general organizational scheme starts with weather and climate (Chapters 1–9), then goes to vegetation and soils (Chapters 10–14), followed by landforms (Chapters 15–24). The chapters on water are presented in the weather and climate section (Chapter 6) and in the landforms section (Chapters 17 and 18). Water is also discussed in the chapters on vegetation and soils and in the final chapter on the physiography of North America.

The twenty-four chapters in *Earthscape* are written so that they can be rearranged to fit several different course organization schemes. For courses that begin with earth materials, crustal processes, and landforms instead of weather and climate, the appropriate sequence of chapters should be as follows:

1. Planet Earth and Its Atmosphere
2. The Forces and Systems on the Earth's Surface

15. Structure and Composition of the Solid Earth

16. Crustal Mechanics, Rock Structures, and Related Landforms
17. Infiltration and Groundwater
18. Runoff, Streamflow, and Flooding
19. Weathering, Slope Processes, and Landforms
20. The Work of Streams in Shaping the Land
21. The Formation of Shores and Coastlines
22. Glaciation, Glacial Landforms, and Landscapes
23. Airflow and the Work of Wind on the Land

13. Soil Materials and Properties
14. Soil Formation, Classification, and Problems

3. The Flow of Radiation to and from the Earth
4. The Heating of Land, Water, and Air
5. The Circulation of the Atmosphere and Oceans
6. Atmospheric Moisture, Precipitation, and Weather
7. The Hydrologic System over Land and Water
8. Climates of the World
9. Climate Change: Then and Now

10. Basic Plants and Vegetation of the World
11. Plant Processes in the Ecosystems
12. Patterns and Processes of Vegetation Distributions

24. The Physiography of the United States and Canada

Another variation that may be more appropriate for some courses is one that places weathering, soils, and water together:

.
.
.

15. Structure and Composition of the Solid Earth
16. Crustal Mechanics, Rock Structures, and Related Landforms

19. Weathering, Slope Processes, and Landforms

13. Soil Materials and Properties
14. Soil Formation, Classification, and Problems

17. Infiltration and Groundwater
18. Runoff, Streamflow, and Flooding

20. The Work of Streams in Shaping the Land
21. The Formation of Shores and Coastlines
22. Glaciation, Glacial Landforms, and Landscapes
23. Airflow and the Work of Wind on the Land
24. The Physiography of the United States and Canada

Five appendices are also presented in *Earthscape*. The first two, "Maps and Map Reading" and "Remote Sensing and Image Interpretation," are introductions to the materials and techniques in these areas. Appendix C presents supplementary climatic maps for the United States and Canada. Appendixes D and E, Soil Tables and Units of Measurements and Conversions, are for reference purposes.

Contents

APPENDIXES

EARTHSCAPE

A PHYSICAL GEOGRAPHY

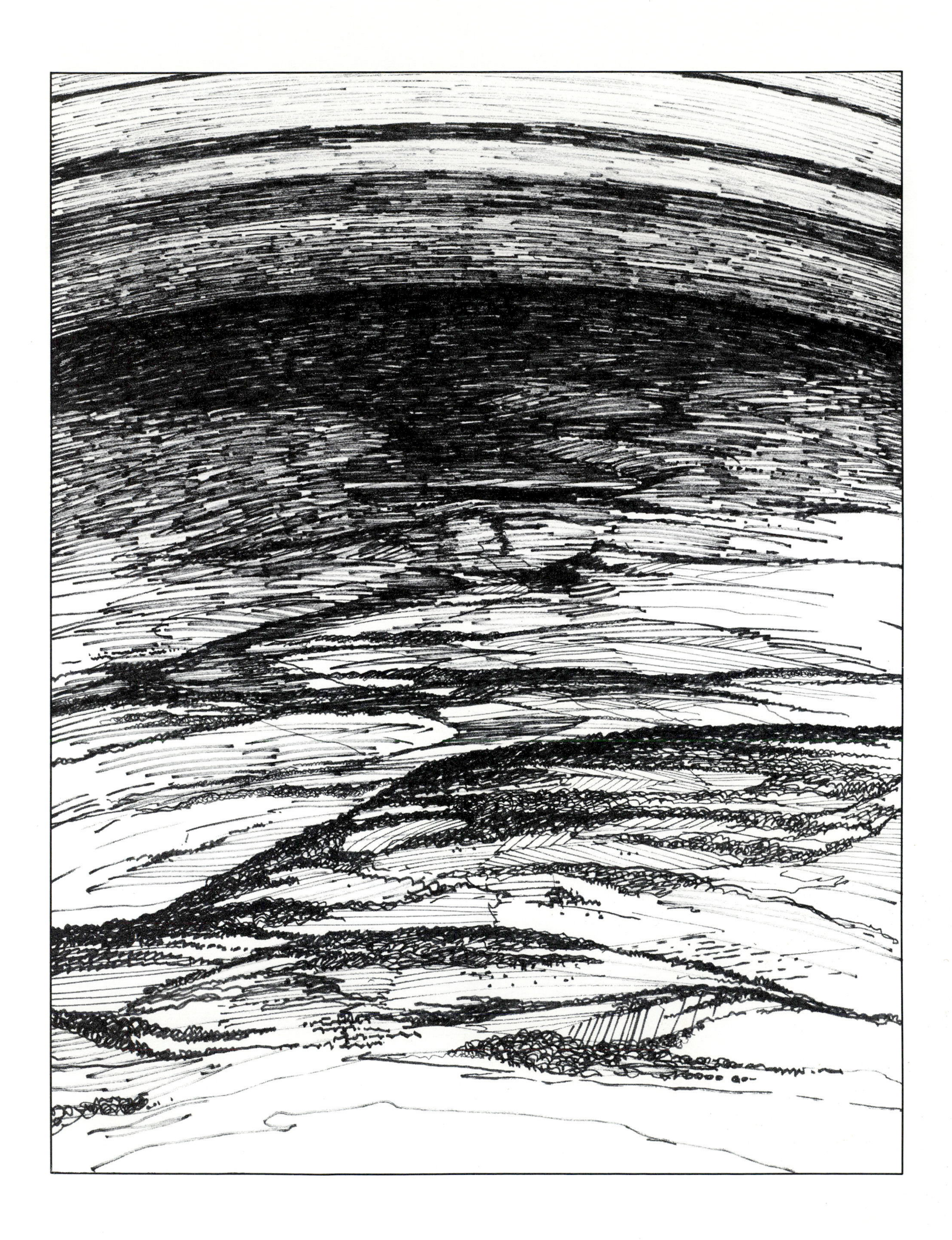

Chapter 1

Planet Earth and Its Atmosphere

Geographic curiosity is one of the deeply rooted passions of Western culture and modern science. The size and shape of the earth, for example, have been a source of inquiry for centuries. Modern science continues to work on these and related questions because they are important in furthering our understanding of earth, its atmosphere, and climates. What controls the seasons? What is the nature of the great envelope of gas that surrounds earth? How does the earth's diverse surface composition influence the atmosphere?

Almost every culture we know about has created a story about the nature of the earth and its origins. Why? Certainly not because such stories are necessary for human survival. The more likely answer is that they are somehow necessary for human intellect. The Eskimo (or Inuit), various groups of American Indians, and, of course, the ancient Greeks each had their own stories. The Greeks envisioned the earth as a great organism whose organs were thought to be the volcanoes, the rivers, the seas, and so on.

Western culture also has its stories about the earth. In the past several centuries another story has unfolded built mainly on information supplied by exploration and scientific inquiry. From this information we have shaped hypotheses, built theories about the origin of the earth and its inhabitants, and discovered new questions for investigation.

Scientific investigation in any form depends on measurement; we know about the earth's size, features, and motion, for example, because we have measured them in some way. Measurement and observation are essential to our culture; they are part of our system of norms or standards of proper behavior. When we have a question about nature, we are inclined to pursue it by observing and measuring things for ourselves rather than just wondering about it or seeking answers in dreams or divine revelations.

Much of our technology today is devoted to building devices, such as cameras, weather observation aircraft, orbiting satellites, surveying instruments, and computers for making measurements, storing the data, and analyzing the results. Our story about the earth has been built and refined over the centuries from repeated observations and measurements. The chapters to the story are the scientific theories, such as the theory of plate tectonics and the theory of evolution. Once written, however, the story is subject to constant revision and repair as new information is brought to light on the different theories. Thus, we must bear in mind that the science we read today represents only the best approximation of a portrait of the earth to date.

THE PHYSICAL CHARACTER OF EARTH

Although we have done pretty well in measuring our planet's physical features, we have not done so well in figuring out its origins. This is not surprising, for the measurements needed to pursue questions of events that happened billions of years ago, such as the formation of the planet's early rocks, are difficult to make. One model of the earth's formation as a planet holds that in the first stage of development, debris (both gaseous and solid) from a stellar explosion gathered (condensed) about a nucleus of some sort. As the mass of debris grew, its gravitational force also grew, compressing the mass into a denser body. As this process took place, the various chemical substances that made up the debris began to rearrange themselves according to their density into a gravitationally stable configuration, meaning that the heavier substances were pulled by gravity farther toward the interior of the young planet than the lighter substances. This entailed movement of huge volumes of material, and we reason that this was possible because some of the internal substances had turned liquid under the high temperatures of the heat from compression within the planet.

The densest matter, such as iron and nickel, migrated toward the center of the mass while the lighter substances, such as silicon and nitrogen, were displaced outward. In this way, the major zones or *spheres* of the planet were formed: the *lithosphere* (core and mantle), with the highest densities, in the center; the *hydrosphere* over the lithosphere; and the *atmosphere*, where densities are lowest, forming the outer envelope. When life developed on the planet, a fourth sphere was formed—the *biosphere*, where matter achieved densities intermediate between water and air (Fig. 1.1).

The biosphere is the most tenuous sphere of all. Here all but a tiny part of life is housed in a zone only

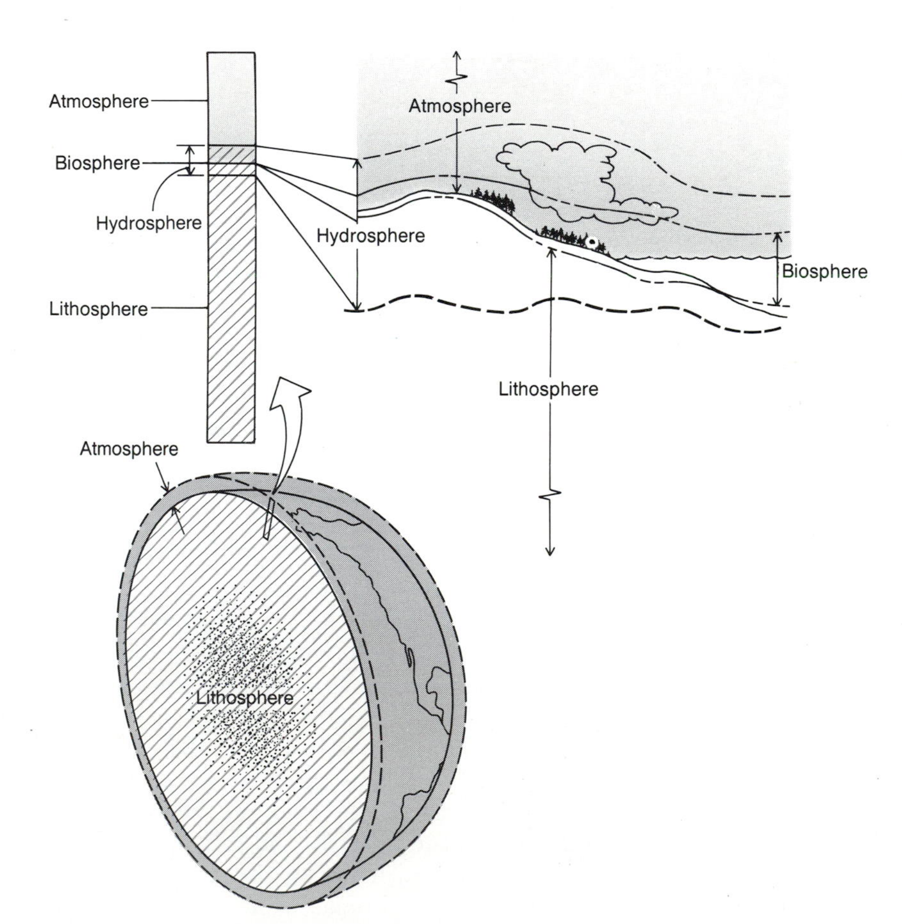

Figure 1.1 The principal spheres of the earth. At the land surface all four spheres merge, forming the landscape.

meters deep. In earth-scale perspective, the biosphere is hardly more than a fine membrane lying over the oceans and lithosphere and under the atmosphere. Above the biosphere, the atmosphere extends outward thousands of meters, forming a transitional layer or medium between the zone of the sun and space on one hand and the earth's surface on the other.

The atmosphere is both a vital regulator of the quantity of solar radiation and an important medium in the redistribution of heat across the earth's surface. In physical geography it is important to understand how light and heat get to and from the lower atmosphere because this energy drives most of the important processes that shape the landscape, such as wind, rain, plant growth, and the flow of glaciers. Before we examine why these processes vary as they do over the earth, we must first examine the nature of the earth as a physical body and the atmosphere that surrounds it.

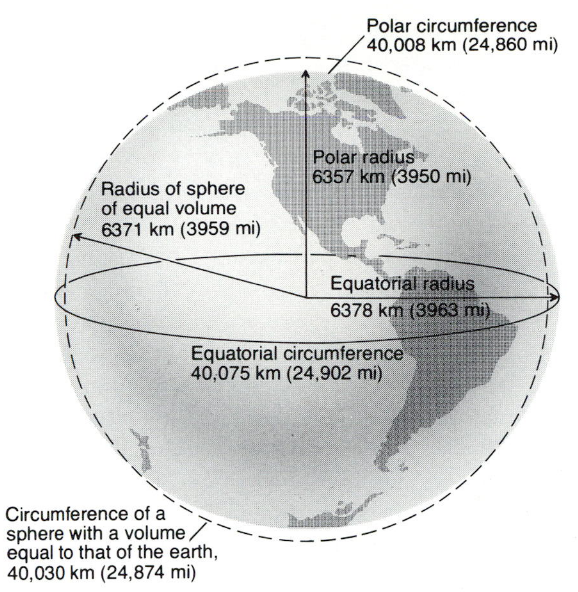

Figure 1.2 Important dimensions of the earth.

The Size and Shape of Earth

Since our objective is to explain the broad differences in climate, vegetation, and related landscape features, we begin by identifying the basic forces controlling the processes on the earth's surface. The most basic of these forces is solar energy. To understand physical geography, then, we need to know about the factors that influence the global distribution of solar energy. To begin with, we can examine the earth's shape, size, motion, and relationship to the sun. These explain north-south variations in solar energy receipt, the seasons, and the duration of day and night, as well as certain aspects of climatic and weather processes.

Our planet has two types of motion: it *rotates*, or spins on its axis; and it *revolves*, or travels in an orbit through space around the sun. Any rotating body produces a centrifugal effect; that is, matter in or on the body tends to be thrown to the outside. In the case of the earth, rotation causes a slight equatorial bulge which produces a slight departure in the earth's shape from that of a true sphere. The term we give to the true shape of the earth is *geoid*.

The shape of earth can be described using two circles. One, the *equator*, is drawn east-west directly around the middle of the planet. The other, which we will call the polar circle, is drawn north-south through the poles (Fig. 1.2). The radius of each circle is the distance from the center of the earth to a point on the circle at the earth's surface.

The shape of the equator is very close to a true circle, with a radius of 6378 km (3963 miles). The polar circle, however, has more the shape of an ellipse rather than a true circle, with a radius of 6357 km (3950 miles) —21 km less than the equatorial radius. Map makers must take the 21 km difference into account when they construct large-scale maps; for studies of climate, however, it is not severe enough to be important. Therefore, we will consider the earth a true sphere with a radius of 6371 km (3959 miles) (Fig. 1.2).

Determining Location on Earth

Throughout this book, one of our main objectives will be to describe the distributions of geographic features, events, and activities on the earth's surface. But the surface of the earth, with an area of 510 million square kilometers (197 million square miles), is so vast that our job would be impossible without a geographic reference system. The system universally used is the *global coordinate system*. This system consists of a network, or grid, of lines running east-west and north over the globe. On a flat surface, such as that used for laying out the streets and property lines in most towns, constructing the coordinate system simply involves building a square or rectangular grid of lines intersecting at right angles. On a curved surface, the job is somewhat more difficult. For the earth, constructing a coordinate system involves measuring angles on the equatorial and polar circles. We can best describe this system by showing how the lines are drawn.

Parallels and Latitude The east-west lines are called parallels. They are all drawn parallel to the equator and each is a circle, but no two are the same size because the circumferences of the earth grow smaller from the equator to the poles (Fig. 1.3*a*). The first step in constructing parallels is to bisect the globe along the polar circle, that is, north-south. Next, a protractor is placed on the plane of bisection, with the base of the protractor aligned with the equator (Fig. 1.3*b*). Starting at the equator, angles northward to the pole are ticked off; the procedure is repeated toward the South Pole. The angles are then numbered, beginning with 0 degrees at the equator and ending with 90 degrees at each of the poles. The point at the edge of the circle (the earth's surface) marking a 45 degree angle is where the 45° parallel is drawn. This is the reason why parallels are numbered in degrees—because they represent angles between the equator and one pole drawn from the center of the earth.

Parallels are used to measure latitude. The *latitude* of any place on the earth's surface is a reference to location north or south of the equator, and it is given in degrees, minutes, and seconds (symbolized °, ′, ″, respectively). There are 360 degrees in a full circle; each degree is divided into 60 minutes; and each minute is divided into 60 seconds. In this book our considerations of angles are generally not precise enough to justify their measurement to the second, so we will generally use just degrees and minutes. The distance represented on a sphere by a degree of latitude is always the same. On the earth this distance is 111 km (69 miles), and a second (1/3600 of a degree) of latitude is a mere 31 meters. If we consider the true, ellipsoidal shape of the earth, some variation is represented by the distance of a degree of latitude—about 1 km between the equator and the poles.

Only one parallel, the equator, traces the full circumference of the earth, and because of this feature, it qualifies as a *great circle*. A great circle is defined as the perimeter of any plane that passes through the center of the earth. All of the other parallels are *small circles* because the planes they define do not pass through the center of the earth; hence, their perimeters represent less than the earth's full circumference. (One drawn around the pole at latitude 89°59′59″ would be very small indeed.)

The concept of a great circle is not limited to parallels, for any number of great circles can be drawn in any direction on the globe. Great circles are important in navigation because the shortest distance between any two points on the earth's surface follows a great circle route. We should note, however, that, although the great circle principle is employed in the global coordinate system, there are no lines in the grid named great circles per se.

Meridians and Longitude The north-south lines of the global coordinate system are called *meridians*. They are constructed in basically the same fashion as the parallels (Fig. 1.4), with one important difference: they do not run parallel to each other but converge at the poles. In constructing parallels, there is a problem as to where to start the system because, unlike the equator for parallels, there is no geometrically convenient point at which to place zero. According to international agreement, an arbitrary starting point was specified that coincides with the Royal Observ-

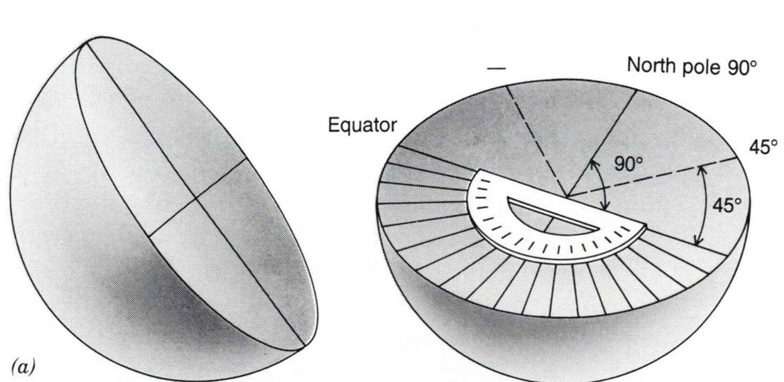

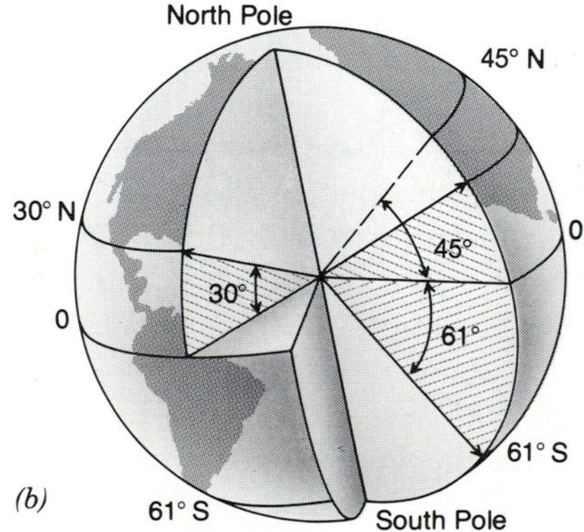

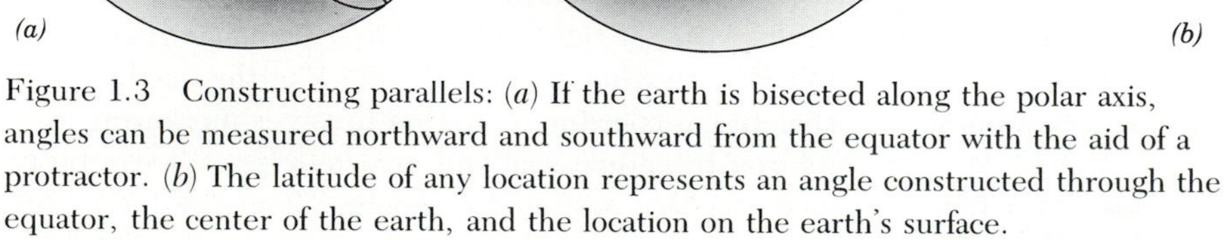
Figure 1.3 Constructing parallels: (*a*) If the earth is bisected along the polar axis, angles can be measured northward and southward from the equator with the aid of a protractor. (*b*) The latitude of any location represents an angle constructed through the equator, the center of the earth, and the location on the earth's surface.

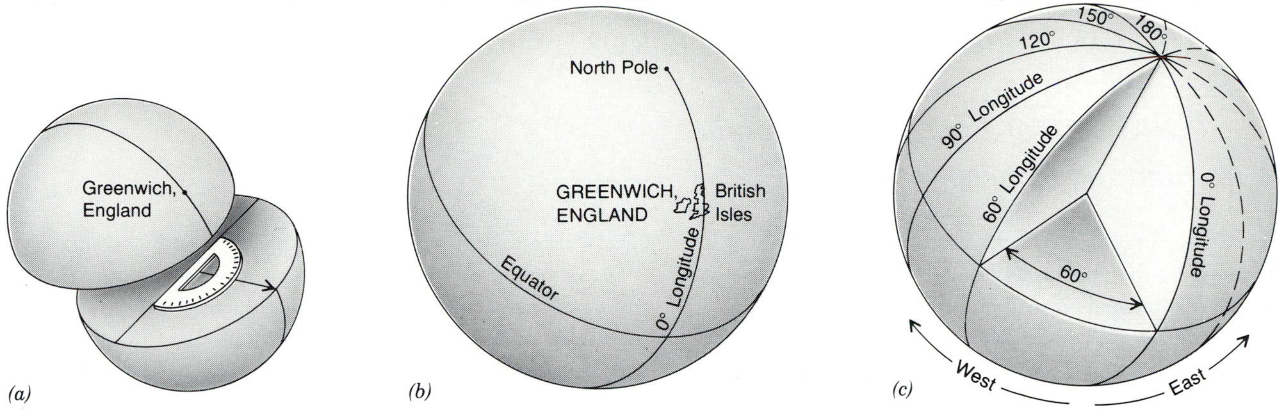

Figure 1.4 Constructing meridians: (*a*) Meridians are constructed on the perimeter of the equatorial plane, with the protractor fixed on the earth's center. (*b*) Since there is no geometrically convenient place to begin, the 0 meridian is fixed on Greenwich, England, and is called the prime meridian. (*c*) Every meridian is half of a great circle. Longitude is given as degrees east or west of the Greenwich Meridian, which is 0° longitude.

atory at Greenwich, England (Fig. 1.4*a*). A north-south line drawn through this point to the North Pole and the South Pole is the *Greenwich* (or Prime) *Meridian* (Fig. 1.4*b*), and it is labeled 0 degrees longitude. From the Greenwich Meridian all meridians westward to 180 degrees are designated west longitude, and those eastward to 180 degrees are designated east longitude.

Because the 0° meridian does not completely encircle the earth, but only half of it, the maximum range of longitude is from 0° to 180°W and 0° to 180°E (Fig. 1.4*c*). Both 180°E and 180°W are halfway around the globe, in opposite directions from the Greenwich Meridian. Thus, they are really the same meridian, and for this reason, the direction indication for 180° is omitted. Because meridians converge at the poles, the *length* (distance) represented by a degree of longitude varies from 111 km at the equator, to 96 km at 30° latitude, to 56 km at 60° latitude, to 0° km at 90° latitude. The development of the global coordinate system and its relationship to measurement systems and concepts about earth is a fascinating story which geographer George Kish outlines for us in Note 1.

Zones of Latitude

The earth's climates are arranged roughly into several great belts of latitude. Much of our discussion in the following chapters uses this framework. Three broad zones of latitude can be defined in both hemispheres: the *high latitudes,* which cover the upper 23.5 degrees of latitude, between 66.5° and 90°; the *middle latitudes,* which extend from 66.5 to 23.5 degrees latitude; and the *low latitudes*, which lie between 23.5 degrees and the equator. Latitude 23.5 degrees is significant because it marks the highest latitude that receives direct solar radiation, that is, where the sun's rays hit the earth's surface perpendicularly (at a 90 degree angle). (More will be said about sun angles later.) In the Northern Hemisphere the parallel at 23.5 degrees is called the *Tropic of Cancer*, and in the Southern Hemisphere, it is called the *Tropic of Capricorn* (Fig. 1.5). Latitude 66.5° marks the *Arctic* and *Antarctic* circles, above which are the only locations on earth to experience day-long (24 hours) light and day-long dark each year.

Used correctly, the term *tropics* refers to the Tropic of Capricorn and the Tropic of Cancer. It follows that the zone between 23.5° south latitude and 23.5° north latitude should be the *intertropical zone*, and indeed many scientists do follow this convention. However, the term *intertropical* has declined in usage, and today *tropics* or *tropical zone* seems to be the preferred term for this zone. The *equatorial zone* is the middle belt of the earth, extending 10° latitude or so north and south of the equator (Fig. 1.5).

Within the broad belt of the middle latitudes, three additional zones are often designated, though their locations are somewhat arbitrary. Just above the tropics is the *subtropical zone* which extends from 23.5 degrees to 35 degrees latitude. Much of the American South, the Mediterranean region, Australia, and southern China lie in this zone. The *midlatitude zone* is the center belt, and, though its limits are somewhat

THE VIEW FROM THE FIELD / NOTE 1

The Origin of Longitude, Latitude, and the Concept of a Meter

Measuring the size of the earth and establishing accurately the geographic location of a place on its surface have been two great challenges faced by geographers for over 2000 years.

It was in the third century before Christ that Erastosthenes, a Greek geographer working in Alexandria in Egypt, first calculated the size of the earth. By measuring the differences in the angle of observation of the sun from two different locations—Alexandria in northern Egypt and Syene, located 500 miles to the south—and by using his sound knowledge of geometry, he produced a remarkably accurate result, close to 40,000 kilometers (24,000 miles).

To reference the exact location of a place—that is, how far north and south and east and west—was a much more difficult task. Measurements of latitude were undertaken at a very early time. Before the beginning of the Christian era, they were based on the elevation of the sun above the horizon at noon at specified times in the year. Longitude, on the other hand, required solving two problems: first, the accurate placing of a line of departure, or starting point, for the first or zero meridian; and second, the accurate measurement of time. Changes in daily time with the rotation of the earth proved to be a convenient way to measure distances east and west of any point on the earth's surface. This is based on the fact that the circumference of the earth, 360 degrees, equals 24 hours, the time for one complete revolution of the earth. Thus, 15 degrees of longitude, east-west distance in the terrestrial grid, equals one hour's difference between two meridians 15 degrees apart.

In 1530 a Dutch astronomer-mathematician, Gemma Frisius, suggested that establishing local noon time at a given point, then moving an hour's distance east or west from that point, would give an exact measure of longitude. But it was not until more than a century later, in the mid-1600s, that an accurate clock, operated by a pendulum and thus transportable, became available. Combined with the availability, for the first time, of a telescope, to make accurate astronomical observations, the way was open for the measurement of longitudes.

Galileo, an Italian astronomer who first opened our eyes to the heavens, discovered the satellites or "moons" of the planet Jupiter in 1610. His work was further refined by another Italian, Cassini, who made accurate measurements of the *appearance* or *emersion*, and *disappearance* or *immersion*, of the satellites of Jupiter, thus establishing a more accurate "celestial clock" than any available before. He published his data in 1668.

The following year, Cassini was invited to the newly established Paris Observatory, the first of its kind. Working with French scientists, he was able to measure the longitude as well as the latitude of forty-three locations scattered all over the globe, and in 1696 he completed the first world map based on accurately determined locations on the terrestrial grid.

During the 1600s and 1700s a controversy arose over the exact shape of the earth: was it a perfect sphere or not? Isaac Newton, working on a theoretical base, declared that the earth was in fact a spheroid, that is, a sphere flattened at the poles. To test Newton's hypothesis, and at the same time to measure the circumference of the earth, French scientists measured arcs of the meridian (a slice of a circle) at midlatitude (halfway between the equator and the North Pole) in France as well as arcs of the meridian at the equator in present-day Ecuador and at the Arctic circle in Sweden. These measurements re-

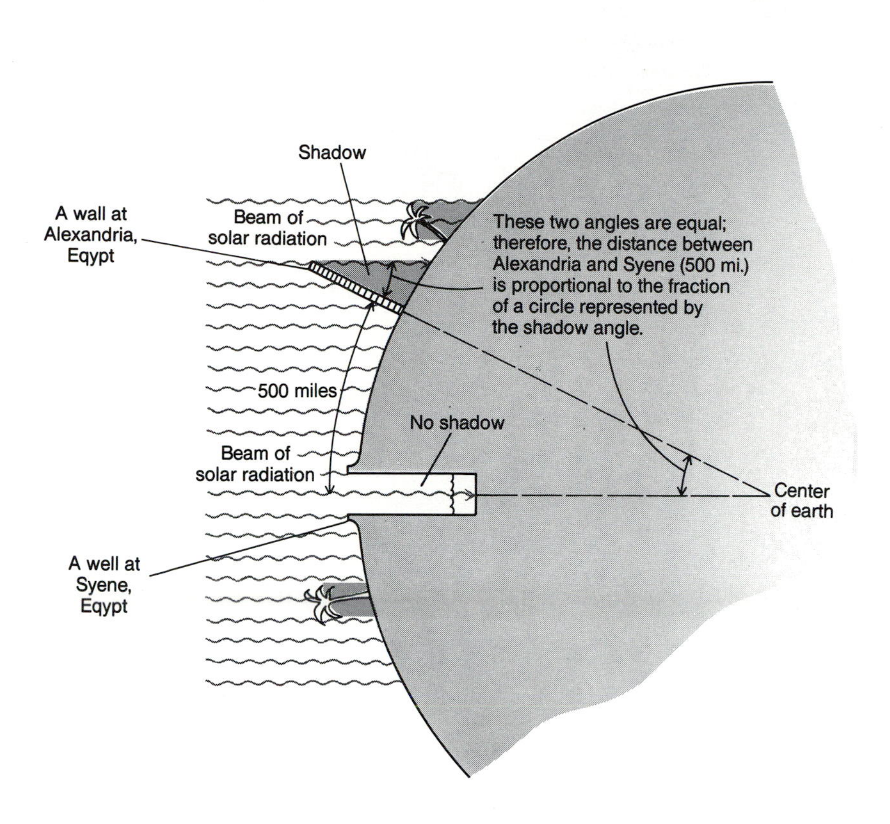

vealed that an arc of the meridian is longer near the poles than near the equator, proving Newton's Theorem.

After completing these great surveys, in the second half of the eighteenth century, French scientists called for the establishment of a new standard of linear length, which they named the metric system. Its standard unit, the meter, was defined as one forty-millionth the circumference of the earth. After a careful remeasurement of the arc of the meridian in France, in 1798–99, the French National Assembly on December 10, 1799, defined the meter as one forty-millionth of a great circle meridian and made the metric system the basis of all measurements. A platinum rod measuring one meter in length was deposited in a special vault outside Paris; it remained the world standard for well over a century and a half.

The search for a more exact definition of a meter continued, however. In 1960 the meter was redefined as 1,650,763 wavelengths of the light emitted by krypton 86, a rare atmospheric gas. And in 1983, the General Conference on Weights and Measures, a worldwide scientific body, redefined the meter once more, this time as the distance light travels in 1/299,792,458th of a second. This is based on the established, constant speed of light, which can now be measured with this degree of accuracy by atomic clocks.

George Kish
University of Michigan

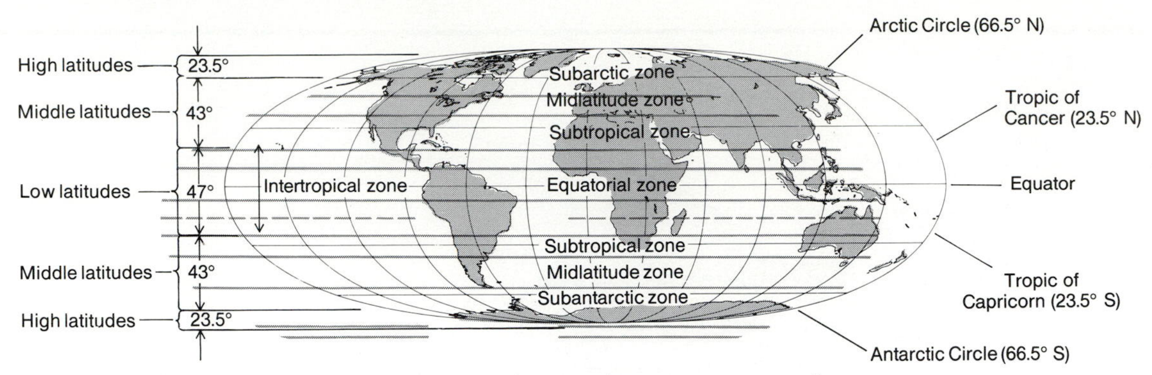

Figure 1.5 Zones of latitude: In this map projection, the area shown in proportional to the true areas covered. The areas of the zones as a percentage of the earth's total area are as follows:

- Equatorial (0–10°) 17%
- Tropical (0–23.5°) 40% (includes Equatorial)
- Subtropical (23.5–35°) 18%
- Midlatitude (35–55°) 24%
- Subarctic (55–66.5°) 10%
- Arctic (66.5 to 90°) 8%

arbitrary, 35 to 55 degrees is usually given for it. Peking, Tokyo, London, Paris, Berlin, Moscow, New York, Philadelphia, Chicago, Montreal, Buenos Aires, and Sydney lie in this zone. The *subarctic zone* lies between 55 degrees and the Arctic and Antarctic circles. The bulk of the land in this zone is occupied by the USSR, Canada, and the United States. Beyond 66.5°N and S are the *arctic* and *antarctic zones*, the uppermost part of which (generally given as above 75 degrees latitude) is the *polar zone*.

Global Distributions of Land and Water

It is important to know about the global distribution of land and water because it has a strong influence on climatic patterns. The continents and oceans contrast sharply in their thermal properties, and these differences are reflected in the characteristics of the envelopes of air that lie over them. Large land masses are associated with a *continental regime*, characterized by wide fluctuations in seasonal and daily tempera-

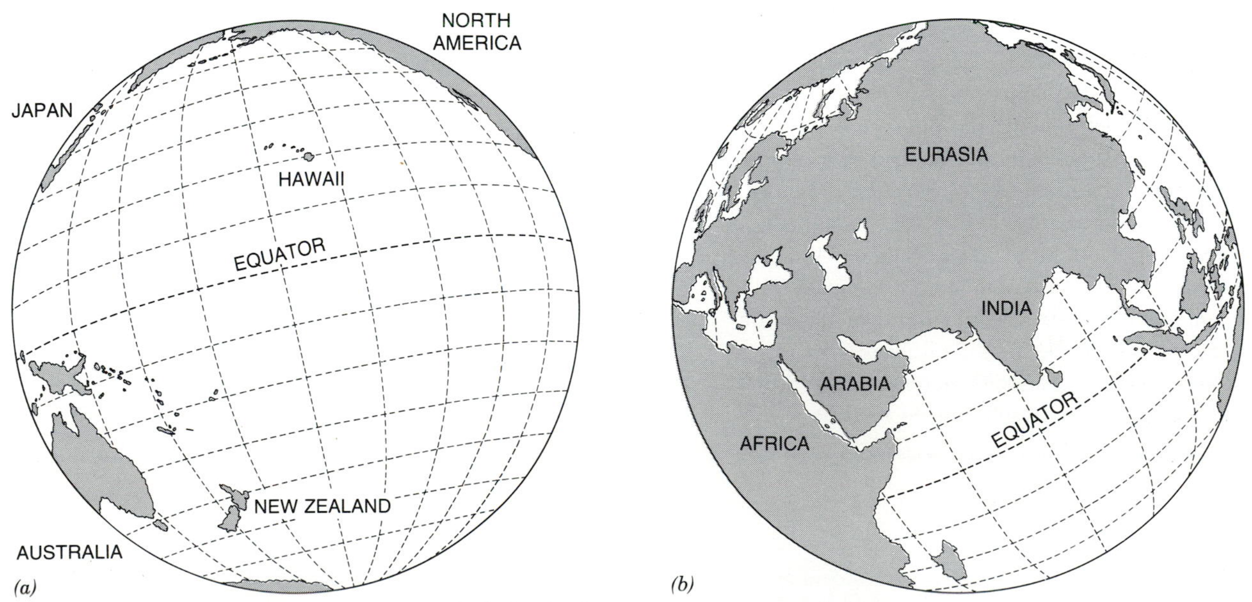

Figure 1.6 The hemispheres of (*a*) water and (*b*) land. The strongest continental effect on earth is produced by the Eurasia land mass.

tures and a tendency to be dry. Great bodies of water have a *maritime regime*, with moderate fluctuations in seasonal and daily temperatures and a tendency to be moist. Which of these regimes has the greatest influence on climate in any given part of the world depends on the amount of surface area taken up by land and water and the prevailing patterns of atmospheric circulation between the two. (The latter topic is taken up in Chapter 5.)

Not surprisingly, then, the distribution of land and water on the earth is an important factor in physical geography. We tend to think of our planet in terms of the land masses we live on, but, in fact, water covers a little more than 70 percent of the earth. Water is especially abundant in the Southern Hemisphere, where it constitutes about 80 percent of the surface area. In the middle latitude and subarctic zone of the Southern Hemisphere, water forms a belt around the globe that is interrupted only by New Zealand and the southern portions of South America and Australia (Fig. 1.6*a*).

This contrasts sharply with the Northern Hemisphere where the middle latitude and subarctic zones contain the bulk of the North American and Eurasian land masses. The largest, of course, is Eurasia which constitutes 37 percent of the earth's land area and develops the strongest continental climatic regime on the globe (Fig. 1.6*b*). Another contrast is found in the polar zones, the Southern Hemisphere's being covered by land (Antarctica) and the Northern Hemisphere's being covered by water and sea ice (the Arctic Ocean).

On balance, then, what can one deduce about world climate from the distribution of land and water alone? Our first hypothesis would probably be that the oceans should have a significant influence on climate for the earth as a whole. Second, we would propose that the Southern Hemisphere should generally favor the maritime-type regime. Third, the middle latitude and subarctic zones of the Southern Hemisphere should be overwhelmingly maritime, whereas the same zone in the Northern Hemisphere should show a contrasting pattern of continental and maritime with the alternate masses of land and water. Fourth, the polar zone of the Northern Hemisphere should be somewhat more moderate thermally than its counterpart in the Southern Hemisphere. Indeed, this is what we find.

EARTH-SUN RELATIONS

We mentioned earlier that the earth in space has two principal motions, both of which are very important in the distribution of solar radiation: its *revolution* in an orbit around the sun and its *rotation* on its axis. If we could observe these motions from the North Pole, both would appear counterclockwise. That is, a point on the surface of the earth rotates from west to east, and the path of the earth in its orbit around the sun is in the same direction.

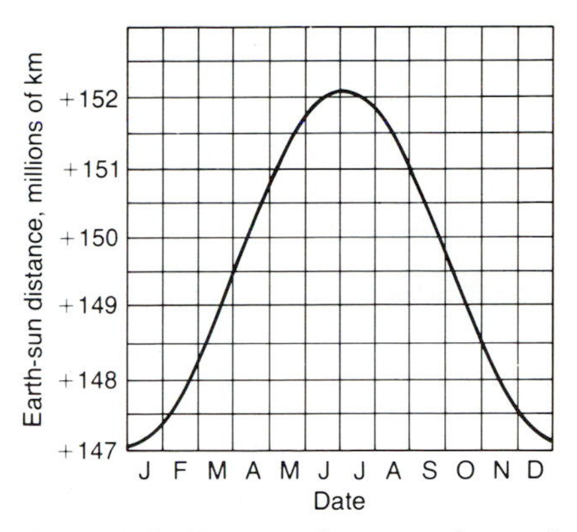

Figure 1.7 Distance between the earth and the sun, in millions of kilometers.

The period of revolution is called the *year* and is equal to 365.242 solar days. A *solar day* is the length of time it takes a single point on the earth's surface to make one complete rotation with respect to the sun. Although we consider it to be a constant, the length of day varies during the year. The earth's orbit is slightly elliptical, rather than circular, and because of this the earth moves (revolves) faster at one end of its orbit and slower at the other. Therefore, the length of the solar day is actually an *average* of 24 hours. In addition, the apparent rotation speed—measured by sighting the sun—averages slightly slower than the true rotation speed. As a result, the rotation period of a solar day is slightly longer than that of a day measured relative to heavenly bodies beyond our solar system. An earth day measured by the position of distant stars is called a *sidereal day*.

As the earth moves in its elliptical orbit, its distance from the sun varies (Fig. 1.7). The mean distance from the earth to the sun is 149.6 million kilometers (93 million miles). *Perihelion*, the shortest distance, is 147 million kilometers (81.5 million miles) and occurs on January 3. *Aphelion*, the longest distance, is 152 million kilometers (94.5 million miles) and occurs on July 5 (Fig. 1.8). You might expect that the earth would be warmest when it is closest to the sun; but in the Northern Hemisphere, the opposite is true. Winter in the Northern Hemisphere coincides with perihelion and summer with aphelion.

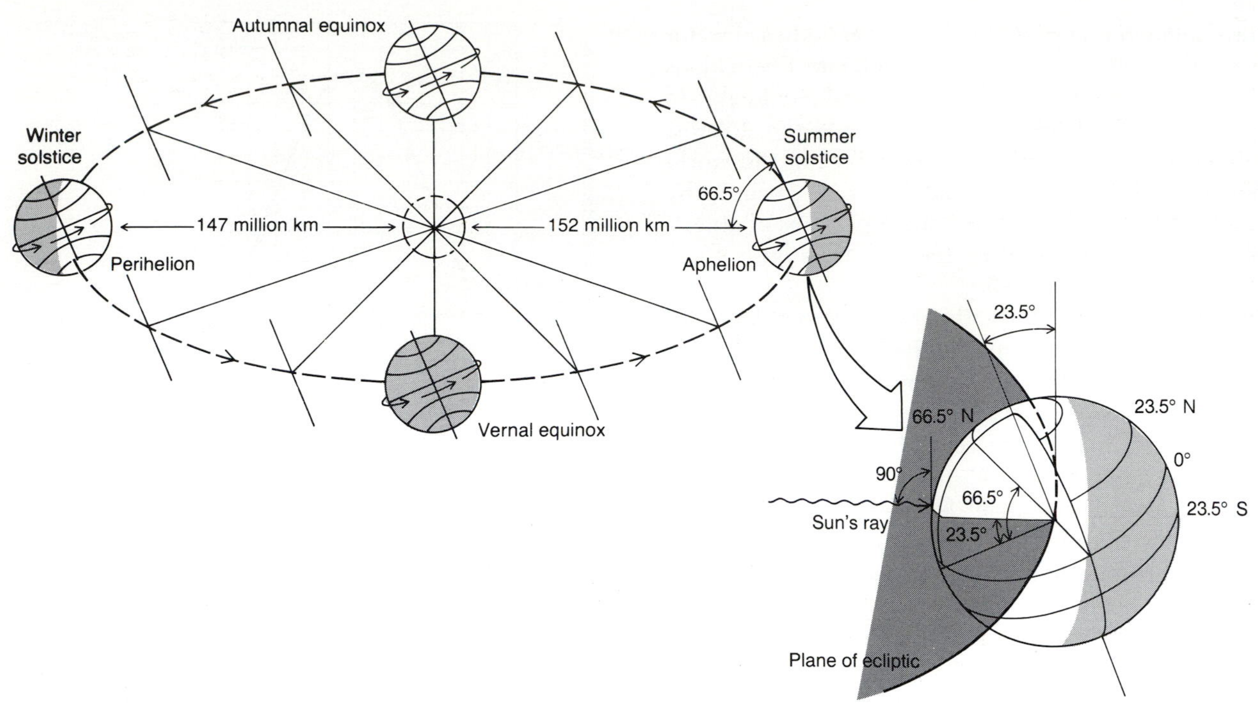

Figure 1.8 The revolution of the earth and the seasons. Note that the angle of inclination of the earth's axis is the same in all seasons.

The Seasons

We recognize the seasons as the rhythmical oscillations of climate and landscape that occur with a certain amount of predictability year after year. In different parts of the world, the seasons have different characteristics. March may be wet or dry; July may be cold or warm, light or dark, or fair or stormy, for example. Throughout most parts of the world, however, the processes of life are closely related to seasonal patterns. In turn, the seasons everywhere are related to changes in the amount of energy the earth is receiving from the sun. What is the cause of these changes?

We have already seen that the reason is not the elliptical shape of the earth's orbit. The variation in distance between earth and sun has only a minor effect (about plus or minus 3.5 percent) on the seasonal variations in the receipt of solar energy. The true explanation of the seasons becomes clear when we look at Figure 1.8. As the earth revolves around the sun, its axis is not perpendicular to the plane of revolution, called *plane of the ecliptic;* instead, it inclines at an angle of 66 degrees 33 minutes, or 23°27′ off the vertical.[1] Because the orientation of the axis remains constant throughout revolution, the tilt does not favor the same hemisphere throughout the year. During one part of the orbit, the Southern Hemisphere is pointed toward the sun, and during the opposite part, the Northern Hemisphere is pointed toward the sun. (See the positions marking summer solstice and winter solstice in Fig. 1.8.)

When one hemisphere is pointed toward the sun, the sun angle in that hemisphere is greater than it is at comparable latitudes in the other hemisphere. *Sun angle* is the angle at which a ray of solar radiation hits the earth's atmosphere and surface; in the inset in Figure 1.8, it is shown as a 90° angle at 23.5°N latitude. This angle (90°) is the greatest sun angle possible on earth; if you were standing at that location, the noon sun would be directly overhead. The lowest sun angle is 0 degrees, and this occurs when the sun's rays travel parallel to the earth's surface. Standing on the earth's surface, you would see the sun on the horizon with only its upper half showing. When the earth is at one "end" or the other of its orbit, and the full 23.5 degrees tilt of the axis is aligned with the sun, all sun angles over the earth are either increased or decreased by 23.5 degrees. (More will be said about sun angles in Chapter 3.)

[1]For our purposes we can round these numbers off to 66.5° and 23.5°.

Exactly half of the planet is illuminated by the sun at all times. But as Figure 1.8 makes clear, the geographic position of the illuminated sector changes as the attitude of the earth's axis changes with revolution about the sun. On June 20–22, when the axial tilt brings the Northern Hemisphere toward the sun by a full 23.5°—what we described above as one "end" of the orbit—the North Pole lies entirely within the illuminated sector and the South Pole entirely within the dark sector. The boundary separating the two sectors, called the *circle of illumination*, falls 23.5° of latitude below the poles (Fig. 1.8). This latitude marks the Arctic and Antarctic circles, and this date, June 20–22, is the *summer solstice* for the Northern Hemisphere. The circle of illumination at this time reaches beyond the North Pole, so that light washes the whole arctic region from morning to morning. The same date in the Southern Hemisphere is the *winter solstice*. December 20–22 represents the same geometry in reverse and is also called the *solstice*, though the winter and summer designations are also reversed. *Solstice* means furthest point, or point of culmination, when the trend toward longer or shorter days reverses.

The word *equinox* is Latin for "equal night." The equinox dates, March 20–22 and September 20–22, are the intermediate points in the earth's orbit between the solstices. On these days the circle of illumination falls across both poles. Thus, as the earth rotates, every geographic location spends exactly half of the day in the dark sector and half in the light sector. Solar radiation is thus broadcast to each hemisphere in equal amounts.

So far we have discussed solar radiation as if nothing affected its reaching the earth's surface but the planet's position in space. In fact, before this radiation can reach the earth's surface, it must pass through the atmosphere.

THE NATURE OF AIR AND THE ATMOSPHERE

Our atmosphere is a layer of air held near the earth's surface by gravity. Nitrogen and oxygen make up over 99 percent of the atmosphere by volume (Table 1.1). The other fraction of 1 percent consists of a mixture of gases referred to as *minor constituents*. Despite their low quantities, some of the minor constituents are important. Ozone absorbs ultraviolet radiation, shielding us from potentially lethal exposures of radiation. Carbon dioxide is a good absorber of the longwave radiation emitted from the earth's surface; along with water vapor, it performs the important function of slowing the outflow of energy from the atmosphere.

Water vapor varies from 0 to 4 percent in the lower atmosphere (the *troposphere*), depending on the air temperature and the availability of moisture from sources such as the oceans. The warmer the air, the more vapor it can hold. In addition, moist air can retain more heat than dry air. Thus, vapor has an important influence on air temperature. We can see this when we compare the moderate day/night temperature ranges in humid climates with the more extreme ranges in dry climates.

TABLE 1.1 COMPOSITION OF THE ATMOSPHERE

Gas	% by Volume in Troposphere	Notes
Nitrogen (N_2)	78.084	
Oxygen (O_2)	20.946	Has developed principally with the evolution of plant life in past two billion years.
Aragon (A)	0.934	
Carbon dioxide (CO_2)	0.033	Has increased since 19th century with burning of fossil fuels; absorbs longwave radiation.
Neon (Ne)	0.00182	
Helium (He)	0.000524	
Methane (CH_4)	0.00016	
Krypton (Kr)	0.00014	
Hydrogen (H_2)	0.00005	
Nitrous oxide (N_2O)	0.000035	Absorbs longwave radiation.
Important Variable Gases		
Water vapor (H_2O)	0–4	Absorbs longwave radiation.
Ozone (O_3)	0–.000007	Absorbs ultraviolet radiation in upper atmosphere.

Air Pressure

Being a mixture of gases, the atmosphere behaves as a fluid. As in any fluid, the deeper one is in the atmosphere, the greater the pressure from the overlying fluid. For this reason, air pressure is greatest at sea level and decreases with altitude. When you gain altitude rapidly in a car or an airplane, the unpleasant feeling in your ears is caused by the pressure differential between the environment and the air in the Eustachian tubes in your ears. Similarly, when you dive to the bottom of a swimming pool, you can feel the increased pressure from the increased depth of the overlying water. For our purposes, however, there are two major differences between the atmosphere and the swimming pool: (1) since water is so much denser than air, greater pressure changes in the pool take place over a relatively small depth; and (2) since air is compressible (whereas water is not), its density is not constant throughout the atmosphere. As a result of the gravitation force acting on it, the bulk of the atmosphere is compressed into a very thin envelope over the earth's surface.

Air pressure is normally measured in *millibars* (abbreviated mb). We can define *pressure* as a force exerted on a surface of some specified area, which in the case of air, is normally one square meter. The basic unit of force in the metric system is the *newton*. Because a newton per square meter is a very small pressure, pressure is often measured in bars instead. (One bar is 100,000 newtons per square meter.) A millibar is 0.001 bar, or 100 newtons per square meter. Normal sea-level pressure is 1013.25 mb, or just a little more than a bar (Fig. 1.9*a*). At 5500 m (18,000 feet) pressure is only about 500 mb; thus, half of the atmosphere is below this altitude. Figure 1.9*b* shows how pressure varies with altitude.

Mapping Air Pressure

At any geographic scale, from local to global, heating of the earth's surface and the lower atmosphere often results in areas of low air pressure, whereas cooling of the surface and the lower atmosphere usually results in areas of high air pressure. The reason for the low pressure is that as air is heated, it decreases in density and becomes buoyant relative to the surrounding air; thus, it rises, displacing cooler air above it. The reason for high pressure is basically the opposite; the air grows denser and heavier as it cools, and sinks downward. At scales in between the local and global, what we call the *synoptic* scale (which would include areas the size of North America), high- and low-pressure areas can also result from complications of fluid flow in the atmosphere.

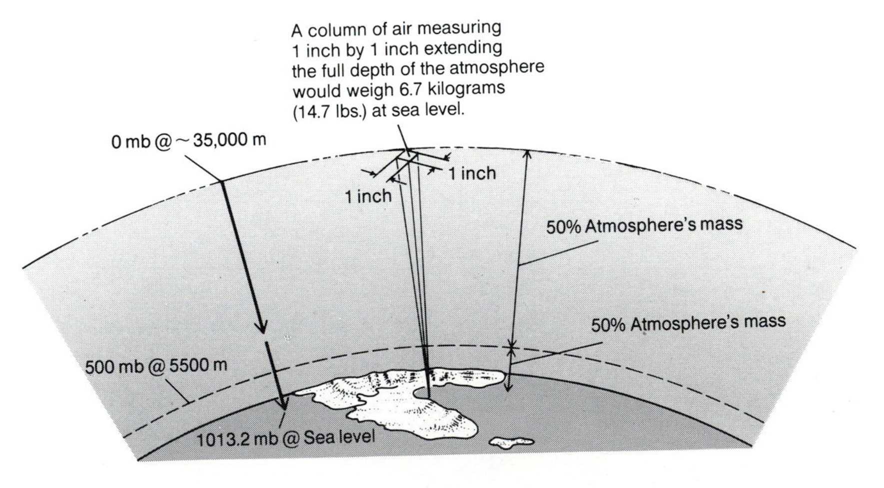

Figure 1.9 (*a*) Some facts about the distribution of mass and pressure in the atmosphere. (*b*) The change in atmospheric pressure with altitude above sea level. In the polar regions, pressures tend to be somewhat lower; in the equatorial regions, they are somewhat higher at given altitudes. (*c*) Adjustments in pressure readings to the sea level datum.

Mapping the distribution of atmospheric pressure is essential to weather forecasting and the study of climates. Air pressure in any one place is easy to measure using several types of *barometers.* Constructing a barometric map can be difficult, however, especially over the land, because variations in elevation distort the readings. Consequently, a common datum plane must be used, and the one in worldwide use is sea level. All barometric readings are converted to their sea-level equivalents: readings above sea level are adjusted upward, and readings at places below sea level, such as Death Valley, California, are reduced. Once this adjustment is made, any pressure differences that appear on maps can be attributed to atmospheric processes such as heating, cooling, and mixing (Fig. 1.9*c*).

The type of map used to depict air pressure is called an *isobaric* map. It consists of a series of lines, called *isobars,* that connect points of equal (sea-level) pressure. The isobaric interval (the number of millibars separating successive isobars) is 4 millibars on synoptic scale maps. The isobars are usually shown with surface wind direction depicted in small arrows (Fig. 1.10*a*). Upper atmospheric pressure and winds are also mapped by weather agencies such as the U.S. National Weather Service of the National Oceanic and Atmospheric Administration (NOAA) (Fig. 1.10*b*). These data are usually printed together with surface pressure maps. The isobaric maps are so important to weather forecasting that a complete set of new maps for North America is printed every 6 hours.

Thermal Structure of the Atmosphere

The generalized picture of the atmosphere's thermal structure—that is, its temperature—is more complex than that of pressure. Instead of declining steadily with altitude above the surface, temperature reverses trend at certain levels (altitudes). These reverses mark the boundaries of the various subdivisions used by meteorologists. The principal subdivisions and their corresponding temperatures and altitudes are shown in Figure 1.11.

On the basis of thermal structure, we can divide the atmosphere into four zones. The *troposphere* extends from the earth's surface to an average altitude of about 12 km, ranging from 8 km over the poles to 18 km over the equator. This layer is characterized by temperatures that decrease with altitude at a rate of about 6.5°C per km to a bitter cold temperature of −60°C or so at the troposphere's upper boundary. The troposphere is the most dynamic part of the atmosphere, where rapid mixing, cloud formation, and precipitation take place.

Above the troposphere is the *stratosphere,* which extends to an altitude of about 50 km. In contrast to the troposphere, little vertical mixing takes place in

Height, meters
15,000
14,000
13,000
12,000
11,000
10,000
9000
8000
7000
6000
5000
4000
3000
2000
1000
0
12,000 m
High-flying jetliner
8848 m
Mt. Everest
6193 m
Mt. McKinley
4420 m
Mt. Whitney
Mexico City
Empire State Building
Dead Sea Depression
200 400 600 800 1000
Pressure, millibars
Height, feet
50,000
45,000
40,000
35,000
30,000
25,000
20,000
15,000
10,000
5000
0

(b)

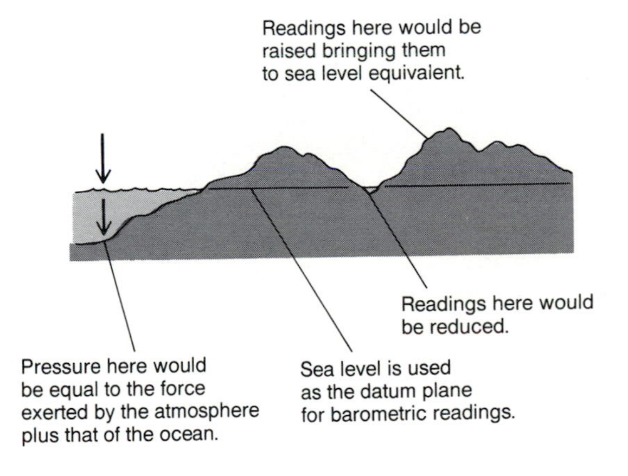

(c)

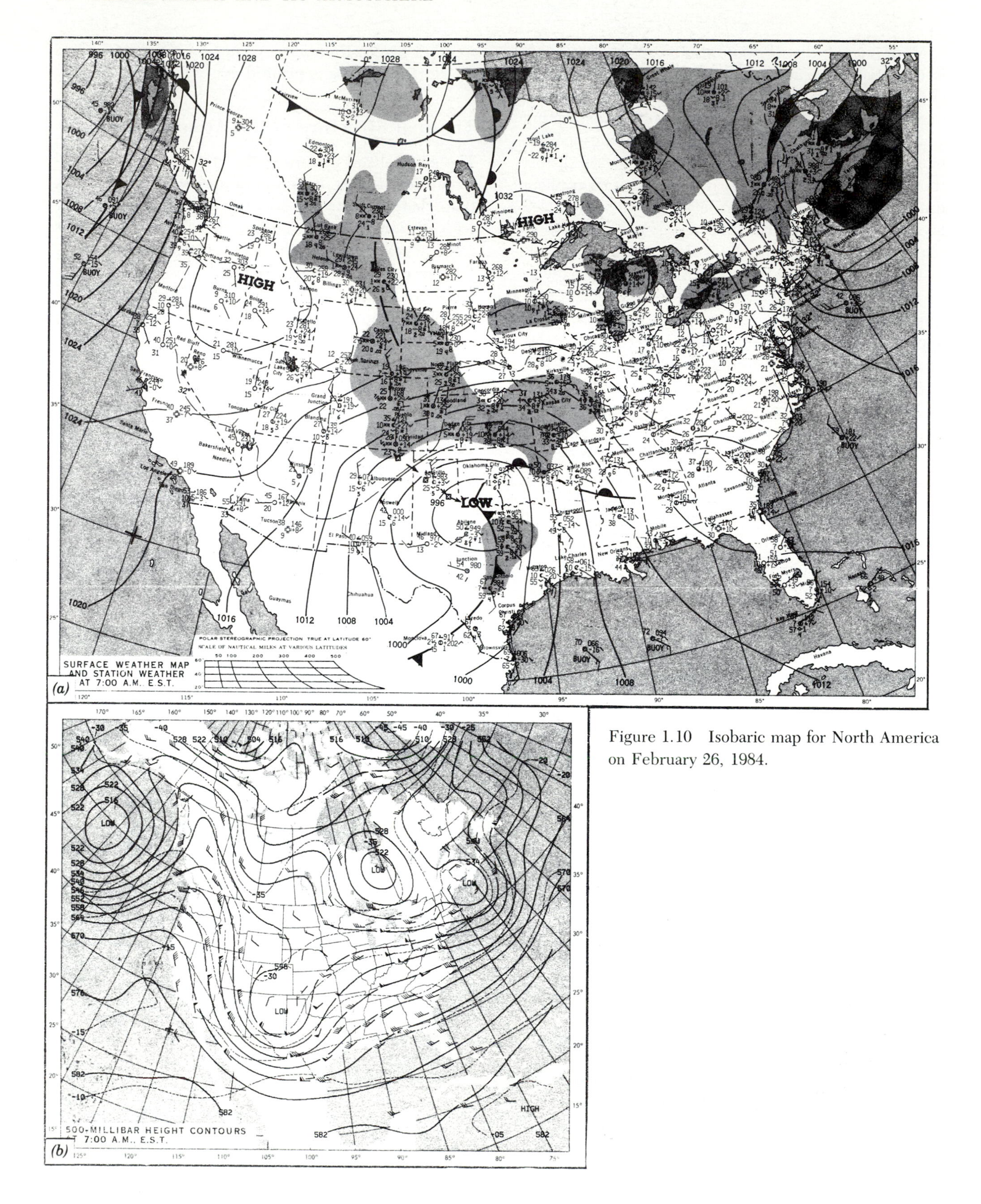

Figure 1.10 Isobaric map for North America on February 26, 1984.

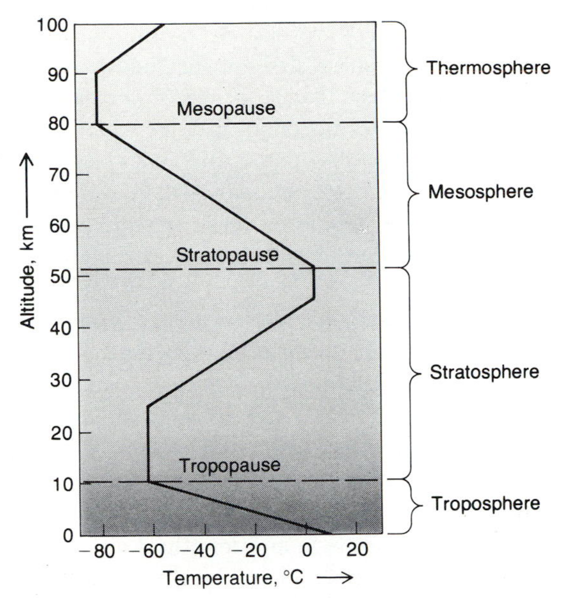

Figure 1.11 Temperature profile of the atmosphere, with subdivisions. This change in temperature with altitude is considered to be typical of temperature conduction in the middle latitudes and is called the U.S. Standard Atmosphere. The boundaries between the subdivisions are termed pauses, meaning change.

the stratosphere. Temperatures hold roughly constant in the first 15 km and then increase over the next 20 km. At the top of the stratosphere, temperatures are almost as high as they are at the earth's surface. This is because a layer of ozone, concentrated around 50 km altitude, absorbs ultraviolet radiation from the sun which would otherwise penetrate to the earth's surface. The *mesosphere* extends from 50 km to about 90 km. Beyond it is the *thermosphere*, where there are a few air molecules and temperatures are above 1500°C.

The Troposphere: The Zone of Weather and Life

The troposphere is the lowermost division of the atmosphere and therefore most important with respect to earth surface processes. In addition to being the zone of active weather, it is also the only part of the atmosphere that is directly influenced by the earth's surface. This influence includes heat and moisture exchanges with the surface as well as frictional effects on the movement of air induced by surface irregularities such as mountains. The lower 1 to 2 km of the troposphere is often referred to as the *frictional layer* or *boundary layer* of the atmosphere, where winds are slowed and reoriented as they slide over sea and land.

Because all parts of the troposphere are accessible by aircraft, our scientific understanding of the troposphere has increased tremendously in the past 40 years. In addition to monitoring large-scale weather events such as the development of hurricanes and frontal systems, we can also study the thermal and aerodynamic mechanisms of the upper troposphere. Such research has been rewarded with meaningful discoveries, among them the importance of upper atmospheric circulation in the development of regional weather patterns. In the past decade satellites have also come into wide use in monitoring major weather systems in the troposphere.

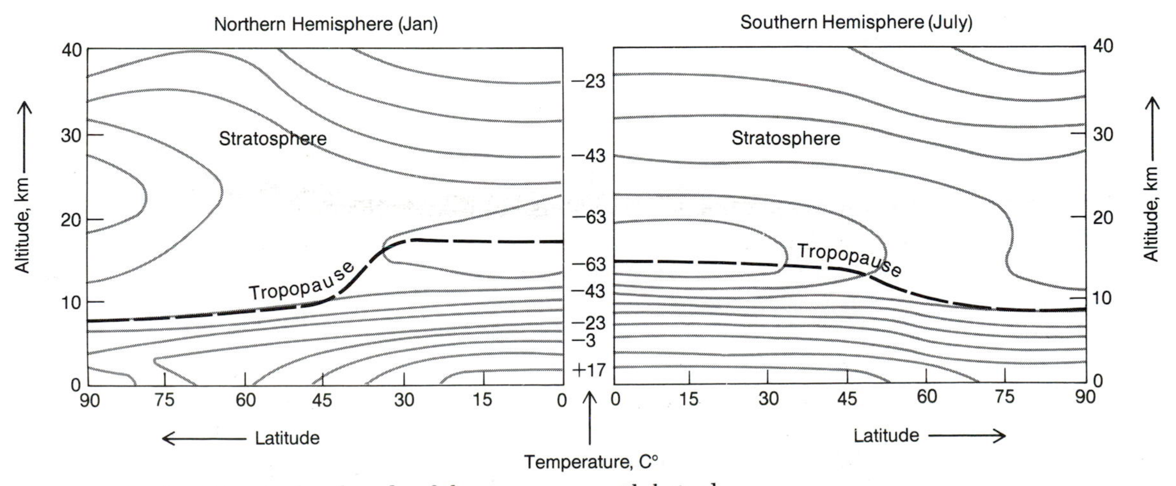

Figure 1.12 Variation in the altitude of the tropopause with latitude.

The *tropopause*, the upper boundary of the troposphere, is marked by the beginning of a major reversal, or *inversion*, in the vertical cooling trend of the lower atmosphere. The altitude of the tropopause is greatest between 35°N latitude and 40°S latitude, where, because of the centrifugal effect of earth rotation and the air pushed aloft by intensive solar heating in the intertropics, the tropopause reaches 16 to 18 km above the surface (Fig. 1.12). Poleward of this zone in both hemispheres the tropopause declines over the next 15 degrees of latitude to an elevation of 8 to 10 km where it remains throughout the polar zones.

What this tells us is that the zone of active weather is much deeper with a greater mass of gas—and higher amounts of energy in the lower latitudes. Weather processes such as thunderstorms near the equator, for example, commonly reach as high as 20 km (12 miles) altitude. As we will see in later chapters, much of this is explained by the high concentration of solar energy in the intertropical zone, which produces a great heat reservoir around the middle of the planet.

Summary

The earth's shape deviates slightly from that of a true sphere, but for most geographical problems it can be treated as a sphere. Locations on earth are measured according to the global coordinate system, which consists of a network of intersecting lines called parallels and meridians. Parallels are east-west-running lines which parallel the equator. Except for the equator, all parallels encircle less than a fill circumference of the earth. Meridians are north-south lines which run from pole to pole. Each meridian forms half of a great circle. Locations on earth are measured in degrees, minutes, and seconds longitude (east or west of the Greenwich Meridian) and latitude (north or south of the equator).

As a planet, the earth has two principal motions, revolution and rotation. Both are important in the distribution of solar radiation over the earth's surface. Although the distance between earth and sun varies by about 5 million km over the year, this has little influence on solar receipt and the seasons. The best explanation for the seasons is the influence of the earth's axial tilt on sun angle and length of day. The solstices mark the extremes in sun angle and length of day over the planet and correspond to winter and summer in both hemispheres.

Air is a highly compressible mixture of gases that exert an average pressure on the earth's surface at sea level of 1013.2 millibars. Unlike atmospheric pressure, which declines steadily with altitude to the top of the atmosphere, atmospheric temperature is characterized by inversions that are used to define the subdivisions of the atmosphere. The troposphere is the zone of active weather, characterized by vertical mixing and precipitation. Isobaric maps are used to depict the distribution of pressure at different levels in the troposphere, and they are essential to modern weather forecasting.

Concepts and Terms for Review

- perspectives on earth
- earth spheres
- geoid
- global coordinate system
  - parallels and meridians
  - longitude and latitude
  - degrees, minutes, seconds
  - concept of a meter
- zones of latitude
  - high, middle, low
  - equatorial zone and intertropical zone
  - tropical and subtropical zones
  - arctic and subarctic zones
- global distribution of land and water
  - Northern and Southern hemispheres
  - continental thermal regime
  - maritime thermal regime
- earth-sun relations
  - rotation and revolution
  - solstice and equinox
  - season and sun angle
- atmospheric composition
  - nitrogen and oxygen
  - minor constituents
- atmospheric pressure
  - millibar
  - sea-level pressure
  - isobaric map
- atmospheric thermal structure
  - troposphere
  - stratosphere
  - mesosphere
- troposphere
  - boundary layer
  - tropopause

Sources and References

Bates, M. (1960) *The Forest and the Sea.* New York: Random House.

Davidson, D. (1978) *Science for Physical Geographers.* London: Arnold.

Eiseley, L. (1961) *Darwin's Century.* New York: Anchor Press.

Hare, F. K. (1971) "Future Climates and Future Environments." *Bulletin American Meteorological Society,* 52: pp. 451–456.

Jellicoe, G., and Jellicoe, S. (1975) *The Landscape of Man.* New York: Viking Press.

Marsh, W. M., and Dozier, J. (1981) *Landscape: An Introduction to Physical Geography.* Reading, MA: Addison-Wesley.

Morner, Nils-Arel. (1984) "Terrestrial, Solar and Galactic Origin of the Earth's Geophysical Variables." *Geografiska Annaler,* 66:1–2, pp. 1–9.

National Weather Service. (1978) *Operations of the National Weather Service.* Silver Spring, MD: NOAA, U.S. Dept. of Commerce.

NOAA, NASA, and U.S. Air Force. (1976) *U.S. Standard Atmosphere.* Wash., D.C.: U.S. Government Printing Office.

Popper, K. R. (1968) *The Logic of Scientific Discovery.* New York: Harper and Row.

Chapter 2

The Forces and Systems on the Earth's Surface

The earth's surface is the focus of the geographer's scientific curiosity. This is where different forces, such as solar radiation, heat, and wind, interact to shape the thin layer where all life resides. What drives these forces? How are they distributed and interconnected geographically? What measures can we use to describe conditions at the earth's surface which are meaningful in geographic analysis?

Not all the world looks at nature the way we do; our notion of nature and its workings may look quite odd to people in societies different from our own. For one thing, we see nature in a much larger geographic framework than do many traditional people; the nature we have come to know extends the world over and, in many respects, beyond to neighboring planets in the solar system. For another thing, we believe in a one-nature concept of earth. As much as the thought intrigues us, we are reluctant to accept the idea that beyond some range of mountains, or in some remote corner of the ocean, there are different natures operating, natures that are not governed by the same principles and forces as the nature we know.

The one-nature concept is an extremely important foundation block of modern science. It enables us to apply certain principles, such as the law of gravity, to the entire planet. It also encourages us to perceive nature as an interconnected whole, as a great network laced together by flows of air, water, energy, food, and organisms. We call these networks *systems.* No people on earth uses the concept of systems more than we do: communication systems, railroad systems, sewer systems, ecosystems, wind systems—the basic wiring and plumbing of the planet's geography.

Loosely defined, a *system* is a set of components (some sort of entities such as cities, human brains, or hills and valleys) linked together by a flow of something (energy, ideas, or water, for example). The flow may be one way or two ways or more, but what is most critical is that the linkage signifies a relationship of some kind. On the basis of such relationships, *cause and effect* associations are implied among the components of a system.

Discovering and explaining the nature of relations is the basic job of science. In physical geography the focus is on the systems and processes that shape the earth's surface, and one of geography's abiding interests is in the geographic patterns of systems and how different systems relate to one another spatially. The story usually begins with the natural systems and forces that drive them and then goes into human systems because human factors are also important in shaping the landscape.

This chapter opens with a look at the flow of energy at the earth's surface. We will see that the processes that change the earth's surface are driven by energy and that the flow of this energy to and from the earth's surface can be described as a system. From this system, others—such as the hydrologic system and ecosystems—are built, each driven by energy on or near the earth's surface. To understand systems scientifically, we must learn about the principles that govern them and how energy is measured.

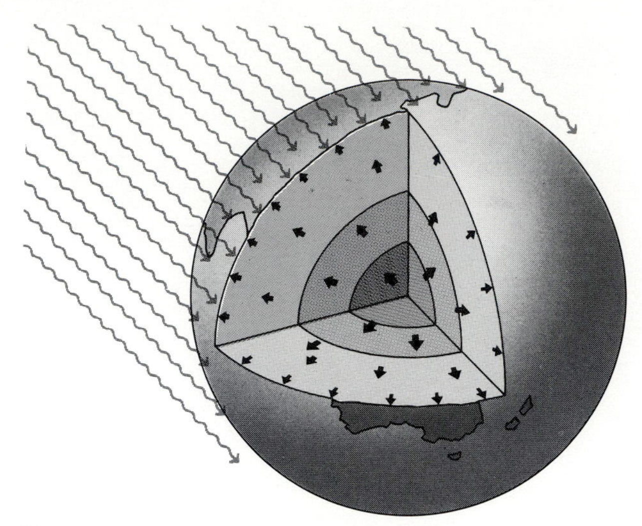

Figure 2.1 A cross-sectional view of the earth, showing the gross pattern of energy flow from endogenous and exogenous sources. The earth's surface is the critical boundary layer toward which energy flows from both sources.

ENERGY AND PROCESSES ON THE EARTH'S SURFACE

Compared to our neighbor the moon, the earth's surface is very lively. We do not, for instance, find active volcanoes or large earthquakes on the moon because the moon has apparently lost much of its internal heat. This heat is necessary to drive these and related processes, which on earth, even after billions of years, continue to help shape the planet's surface.

But there are two other conditions that also account for the earth's active surface: (1) the occurrence of large concentrations of heat and mechanical energy in the presence of (2) highly mobile substances, namely, air, water, and living organisms. When these substances are set into motion by this energy, the resultant movements, which we call *processes*, can effect change in the earth's surface. Because the distribution of energy is not uniform over the earth, the processes that it drives and the changes that the processes effect are not uniformly distributed either. This is an important point, for it begins to help us understand why the earth's surface is so diverse geographically.

Origin and Destination of Earth Energy

The energy that drives the earth's processes comes from two sources. The primary source is the sun, which supplies energy only in the form of radiation. A large share of this energy is represented by sunlight. The secondary source is the earth's interior, which provides energy in the form of heat and movement in rocks. The geosciences have termed these energy sources *exogenous* (external origin) and *endogenous* (internal origin). Figure 2.1 shows the overall flow of external and internal energy in and around a cross-section of the earth.

As the diagram suggests, the earth's surface is a critical boundary layer for energy. This thin layer is made up of air, water, plants, animals, soil, and rock. When energy encounters these materials, it usually changes in some way. It may change form, as when plants transform light energy into biochemical energy by photosynthesis; or it may be absorbed and stored in the landscape, as when heat is stored in soil after a hot summer day. Some forms of energy penetrate the surface materials and flow through them, whereas other forms are unable to penetrate them at all. Which result takes place depends both on the form of energy coming to the earth's surface and on the substances it encounters. For example, much of the solar radiation received by earth can penetrate kilometers of air and meters of seawater, but it is unable to penetrate more than a few millimeters of quartz sand particles on a beach.

For energy that reaches earth in the form of heat, the surface layer plays the important role of regulating its rate of flow in and out of earth environments. This process is similar to the way the different materials in a house (glass, wood, concrete, plaster) regulate heat

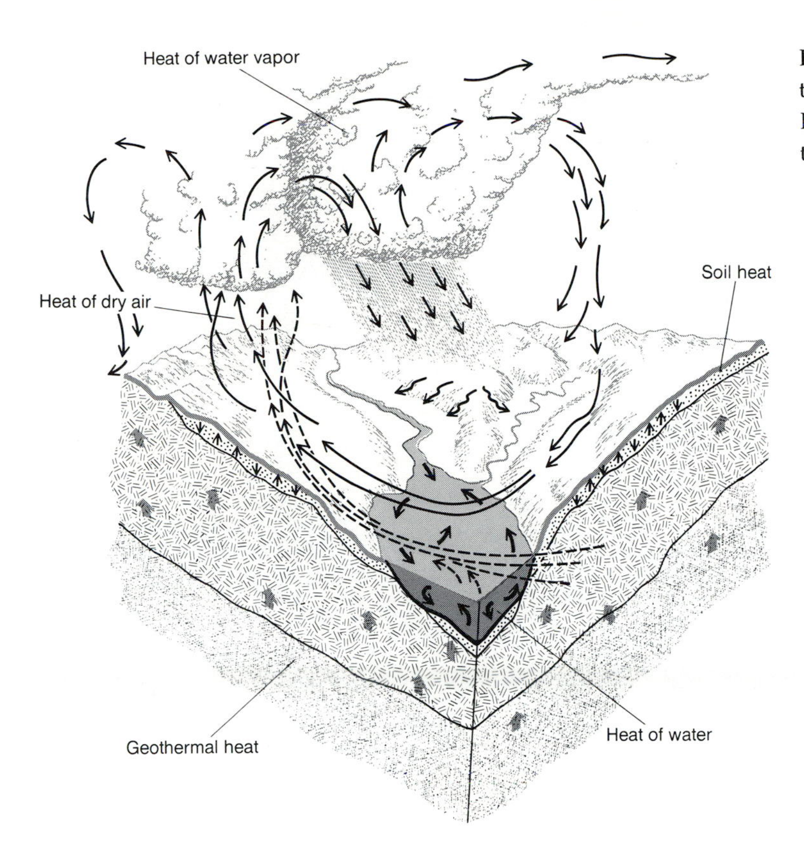

Figure 2.2 A schematic illustration showing the nature of heat flow in air, water, and rock. Flow is very slow in rock and fast in air. Water tends to be intermediate.

loss. In moving air, for example, heat flow near the earth's surface is rapid, whereas in rock and soil under it, it is very slow, mainly because these substances lack mixing motion. The heat flow in water would be intermediate between air and immobile materials. If we could see heat and if we could watch it diffuse throughout the surface layer, we might be fascinated by its fast-flowing, almost bursting motion in air and its slower motion in ocean water. In contrast, heat flow in rock, soil, and still water (deep in the oceans) would appear constrained, with movement limited to small volumes of material over long periods of time (Fig. 2.2).

Effects of Energy at the Surface

In all materials, heat has the capacity to initiate movement. This movement can actually perform work, that is, cause physical changes in and on the earth's surface. For instance, heated air can generate wind. As the wind slides over open water, some of its momentum is transferred to the water surface and waves are formed. As the waves move into shallow water, they rub on the bottom, thereby exerting force on sand and pebbles along the shore. If this force exceeds the strength of the gravitational and molecular forces that hold the sand and pebbles in place, they are moved, work is affected, and the earth's surface is changed.

The total flow of energy toward the earth's surface amounts to an equivalent of 172,032,000,000 megawatts of power per second (1 megawatt = 1 million watts). Except for the small amount that comes from the earth's interior, only 0.002 percent of the total (32 million megawatts), all of this energy emanates from the sun (Fig. 2.3). But not all of the solar energy that reaches the atmosphere gets to the earth's surface; about half of it is either reflected from the atmosphere back into space or is absorbed by the atmosphere itself.

The total power available every second to drive the processes of the landscape, atmosphere, and oceans is the sum of: (1) the earth's heat, at 32 million megawatts; (2) the solar radiation absorbed by the atmosphere itself, at 30.1 billion megawatts; and (3) the solar radiation absorbed by the earth's surface, at 81.7 billion megawatts. Thus, the skin of the earth is powered by a grand total of 111.8 billion megawatts of energy each second. This energy permits photosyn-

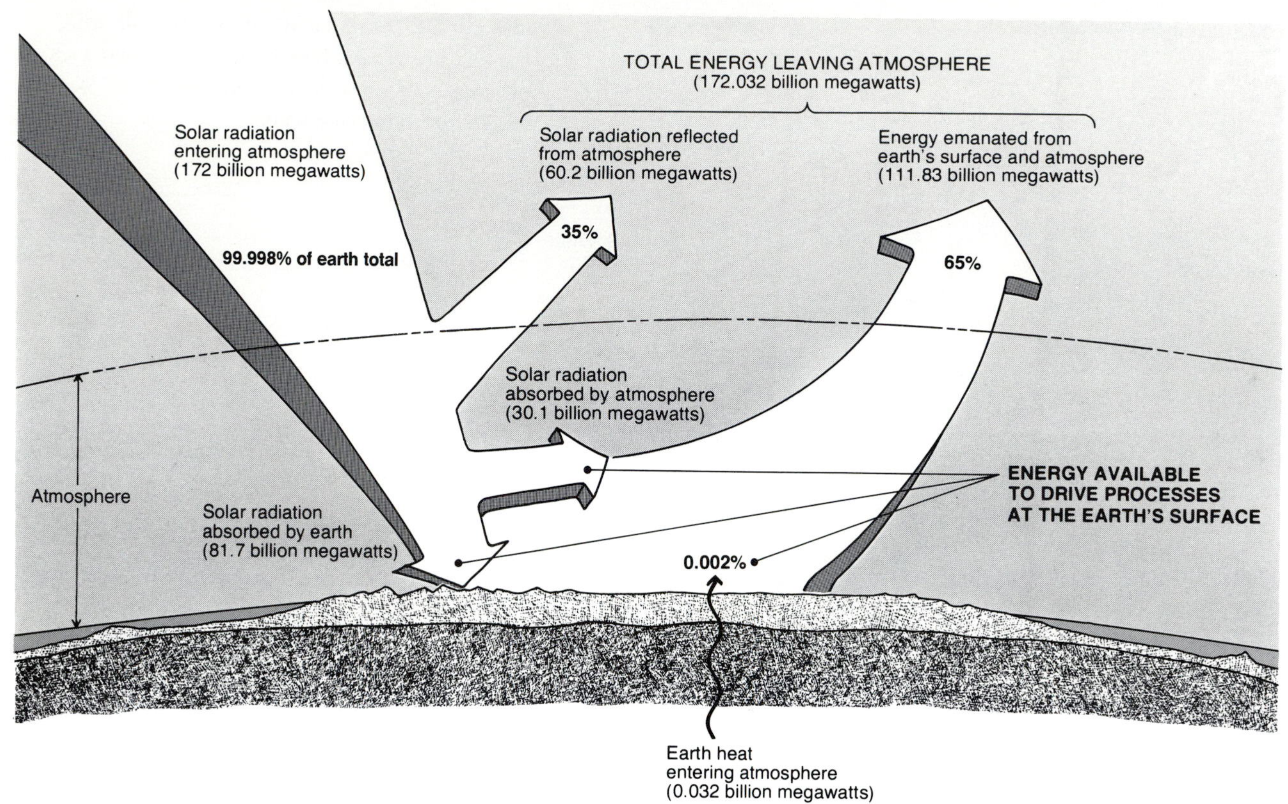

Figure 2.3 Energy to and from the earth's surface in megawatts per second. All but 0.0002 percent comes from the sun.

thesis and plant growth, induces winds which generate waves and erode soil, and evaporates water which becomes rainfall on the land and sea.

THE ENERGY-BALANCE CONCEPT

For the earth's surface as a whole, the sun can be considered the sole source of energy (since the amount of endogenous energy is tiny by comparison). This energy heats the land and water and since all objects that contain heat can also emit heat, the earth's surface emits heat. The rate at which the surface emits heat is governed by its temperature; the higher the temperature, the more heat emitted. If the amount of solar energy received by the earth's surface over some period of time is exactly the same as the amount emitted, then the surface temperature holds steady and is termed an *equilibrium surface temperature*. The energy system is described as being in an *energy balance*. Imbalances occur when the inflow exceeds the outflow or vice versa, and are reflected by rising or falling surface temperatures.

The energy-balance concept is similar to an accounting problem. To begin with, we can imagine money as energy and a piece of landscape as a financial account. The account is an active one, with energy being constantly deposited (as radiation from the sun) and withdrawn (heat passing from the land to the atmosphere and radiation passing from the atmosphere back into space) from the basic energy reserve. When inflows exceed outflows, the reserve grows and we call the system *positively balanced*. When outflows exceed inflows, the reserve dwindles, and the system is said to be *negatively balanced*. Precisely between these states is a state of equilibrium, in which the flows are balanced—deposits equal withdrawals. At that point, the system maintains what physicists term a *steady state:* The reserve, though continuously undergoing energy loss and replacement, is unchanging in its total amount of energy from moment to moment.

Daily and seasonal changes in weather and sun position may cause short-term energy imbalances for periods of a few minutes to a few months. But the atmosphere, soil, and oceans have the capacity to store some of the energy produced during periods when the account is positively balanced, and these reserves usually make up for the deficits in periods of negative balance. Thus, for a given moment, over the vast surface and atmosphere of the entire earth, the energy balance holds steady. If this were not the case, the earth as a whole would get hotter or cooler.

Some Governing Principles

The operation of any energy system, whether as large as the earth's or as small as an ant hill's, is subject to certain physical principles. First, the *conservation-of-energy principles* tells us that there can be no absolute loss of energy within any system. Energy may change form, as from light (a radiant form) to heat, or it may be stored, but it cannot be created or destroyed—and so eventually the ledger must balance. In other words, all energy that flows into the system, whether the system is your body or the atmosphere, must be accounted for in (1) outflows of energy, (2) storage of energy, or (3) work performed.

Second, because the energy system is *dynamic* (rather than static), it is capable of undergoing spontaneous adjustments to variations in the energy flow. All adjustments represent a trend toward a new state of equilibrium. When the inflow of energy increases and the internal energy reserve grows, the outflow of energy tends to become more rapid. When the outflow adjusts to a level equal to that of the inflow, the system has attained a new level of equilibrium. In reality, natural systems rarely attain equilibrium, but are usually described as being in a state of continuous adjustment with periods of near equilibrium. Note 2 offers a practical example of the possible variations in the balance of the energy system of a greenhouse.

Energy Systems in the Landscape

The greenhouse is an example or model of an energy system that is similar in many respects to the earth's atmosphere. Reference to the earth's "greenhouse effect" is based mainly on the role of two atmospheric gases, carbon dioxide and water vapor, in absorbing radiation emanating from the earth's surface just as the roof of the greenhouse intercepts the energy emanating from inside the greenhouse. When the atmosphere heats up with absorption of this radiation, it in turn becomes a radiating body, releasing its energy toward space or back to earth (Fig. 2.4*a*).

This system, like all energy systems, is made up of

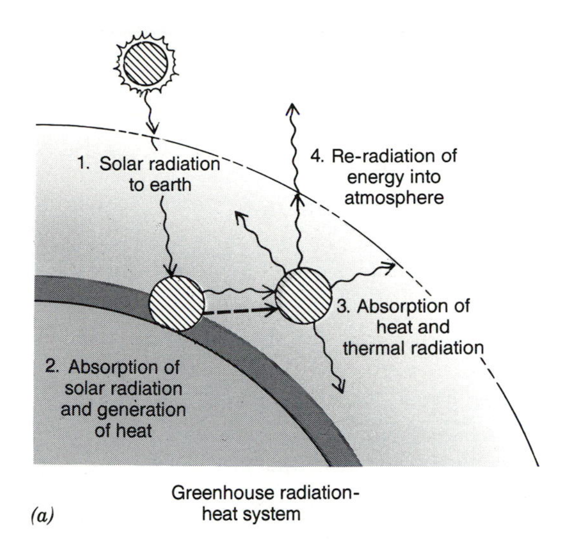

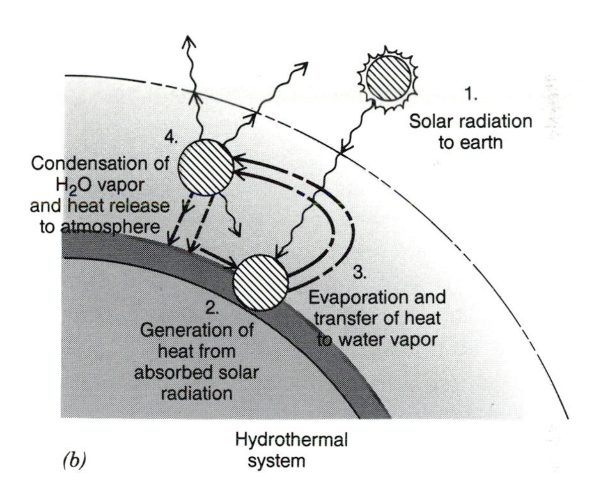

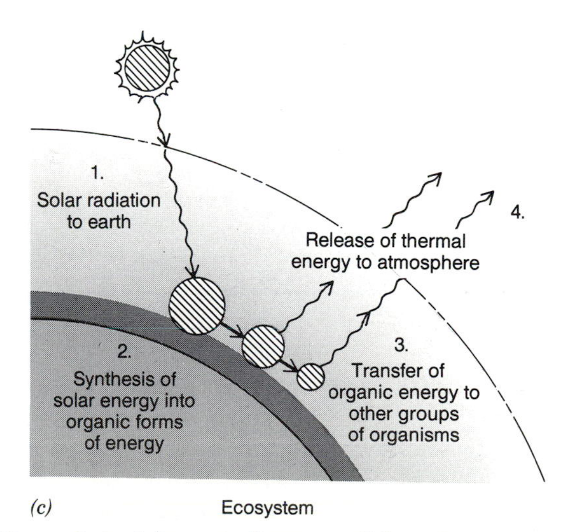

Figure 2.4 Schematic diagrams of three important energy systems; (*a*) greenhouse radiation-heat system, (*b*) hydrothermal system, and (*c*) ecosystem.

THE VIEW FROM THE FIELD / NOTE 2

Energy Balance and the Geography of the Greenhouse

Let us use a greenhouse to illustrate the variations in the energy balance. On a sunny day, solar radiation is transmitted through the glass roof of a greenhouse and is absorbed by the plants and other objects on the inside. As these materials take on energy, their temperatures rise, and they begin to emit energy themselves, which in turn heats the air inside the greenhouse. As with a closed car left in the sunshine, this energy (heat and thermal radiation) cannot pass out of the greenhouse as easily as the solar radiation passed in. The balance of energy flow is positive because the greenhouse environment is gaining more energy than it is releasing. One measure of this condition is a steadily rising temperature.

Because extremely high temperatures injure many plants, most greenhouses are designed to maintain an energy equilibrium throughout much of the day. Therefore, as the temperature rises near midday, vents in the roof are opened to let hot air out. The vents are continually adjusted throughout the day as the inflow of solar radiation varies with the passage of clouds and with changes in the sun's angle of elevation.

The inflow of solar radiation ceases altogether at night, of course, but the greenhouse still continues to release some of the heat gained during the day. The balance of energy flow is now negative because more energy is being lost than gained. Vents are normally closed at night to conserve this energy.

Several additional means could be employed to alter the energy balance of the greenhouse. We could paint some of the panels in the glass roof to reduce the inflow of energy. Or, we could change the ventilation system to increase or decrease the outflow of heated air. Or we could draw off some of the heat and use it for other purposes, such as heating water or growing plants. In any event, the sum total represented by energy losses, conversion to storage, and work accomplished in the greenhouse must always equal the total amount of energy received in the first place.

But there is more to the climate of a greenhouse than its energy balance. The energy is part of a flow system which has directional components, called *vectors*, that result in spatial variations in the concentrations of energy. These spatial patterns of energy change with the time of the day as the sun moves across the sky, with cloud patterns, with the growth of plants, and so on. As a result, some parts of the greenhouse are brighter and warmer than others, that is, somewhat different climatically from place to place. This in turn affects the growth rates of plants and the observant gardener arranges his/her plants to take best advantage of these conditions, thereby creating a special greenhouse geography. In other words, the landscape of the greenhouse, like that of earth, is not a random arrangement of plants, soil, and related artifacts formed by a great energy system, but a response to spatial variations *within* the energy system.

specific physical components (in this case, water, soil, and vegetation on the earth's surface and carbon dioxide and water vapor in the atmosphere) linked together by flows of energy. Each link or set in the system is characterized by a conversion of energy—for example, radiation into heat or heat into radiation, or energy into work as in the expansion of heated air.

Many other energy systems also operate on or near the earth's surface, and in addition to having the characteristics mentioned above, they also have important geographic attributes. The hydrologic system, for instance, involves the cycle of moisture between the atmosphere and the earth's surface, and is highly effective in redistributing heat over the earth (Fig.

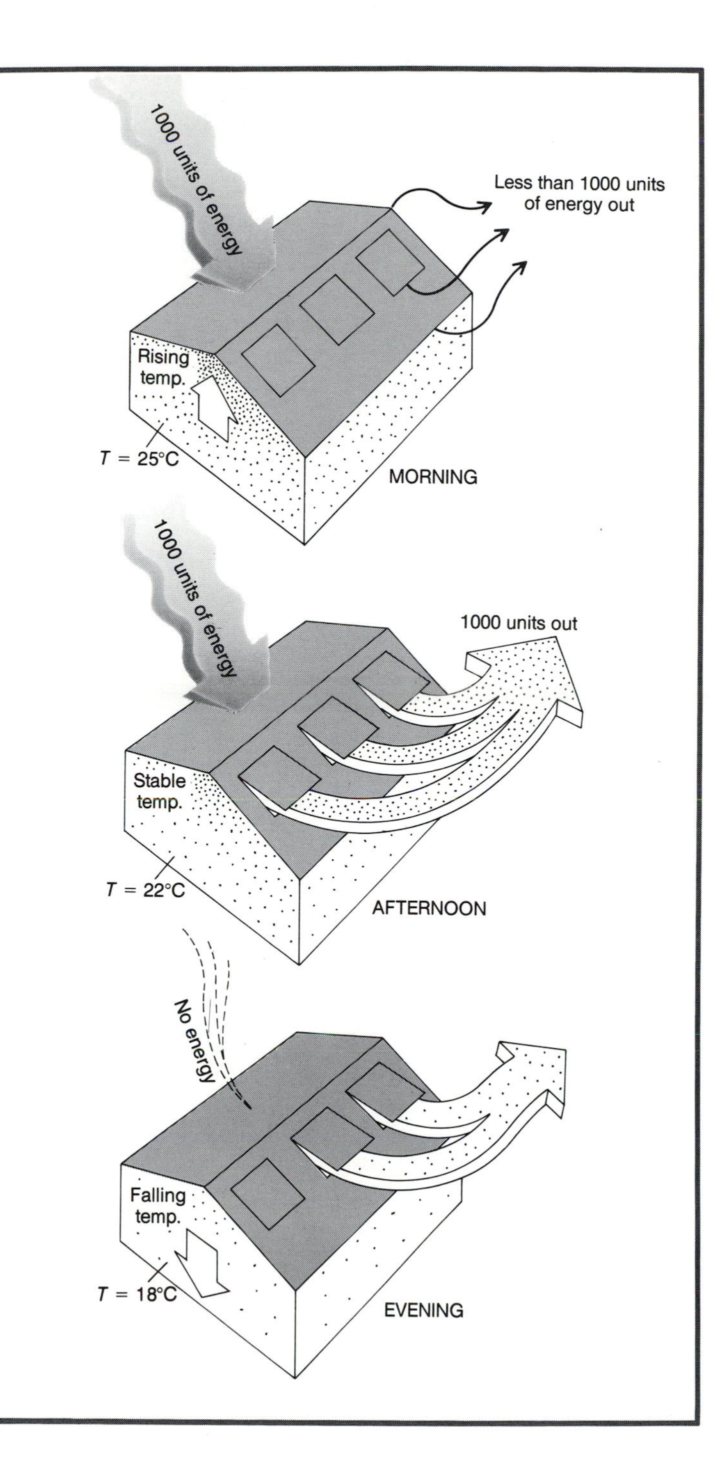

2.4*b*). The pattern of redistribution is uneven geographically, however, owing to variations in wind patterns and in the availability of water and heat, so that some areas receive or give up much larger amounts of heat (and water) than others.

Another energy system is the ecosystem, a group of organisms linked together in food chains. The flow of energy through an ecosystem begins with the conversion of solar energy by plants into chemical energy which is stored in the plant tissues. Microflora and herbivores then consume the plants, carnivores consume the herbivores, and certain carnivores consume other carnivores. In this way, energy in the form of organic compounds is passed from one set of organisms to another. With each set, however, a large part of the energy is used in respiration, that is, in the maintenance of the organism. Most of this energy is in turn given up to the atmosphere as heat and thermal radiation (Fig. 2.4*c*). Because ecosystems are built on plants, their geographic character is closely related to the distribution of vegetation and its growth rates. Thus, ecosystems in tropical wet environments, where the resources for plant growth are most abundant, are the most active on earth; where one or more of the basic plant resources is limited, however, energy flow in ecosystems is much lower.

ENERGY, HEAT, AND TEMPERATURE

It should be clear by now that we cannot go much further in the discussion without defining energy, heat, and the units used to measure them. *Energy is the ability to do work*, such as the transportation of water by ocean currents and the erosion of soil by wind. Work involves the transfer of energy from one body to another or the conversion of energy from one form to another and is defined as the product of force over distance. Energy is found in various *forms* in the environment, for example, *radiation* in the atmosphere, *heat* in the soil, *motion* (*kinetic* energy) in the oceans, or in various stored or *potential* forms, for instance, organic matter in the soil or a mass of water raised above a sea level.

In seeking explanations for phenomena in the landscape, whether it be the behavior of a glacier or the influence of drought on agriculture in the Corn Belt, geographers are frequently led to questions about energy. These questions usually center on energy available to drive certain essential processes, such as the evaporation of water from the soil or the melting of a mountain snowpack, and call for information in two forms: (1) The amount of energy held in some material, such as heat in air, water, or soil; and (2) the rate of energy flow from, through, or into some material. The rate of energy flow to or through a given area of space, such as a square meter, is referred to as energy flux, radiant-flux density, or irradiance. In this book we will use the term *energy flux*.

In the case of heat, we can measure both the heat content of a substance and the rate of heat flow in the substance. In the case of radiation, however, we can measure only its rate of flow. Solar radiation passes through the atmosphere so quickly that it makes no sense to measure its content in the atmosphere.

Energy Flux

To measure the flux of energy we must agree on some units of measurement. The flow of anything, whether it be water, traffic, or energy, must be referenced in terms of the *space* it takes place in and the *time* it takes some quantity to pass by. In the case of solar radiation coming through the atmosphere, we ask: "How many units of energy are passing through an area of some size, say, a window, in a minute of time?" Or, "How many units are being received by a surface, say, a square foot of ground per minute"? Or, conversely, "How many units of energy are being released by a surface of some area over some amount of time?"

The units used to represent a quantity of energy are usually joules, calories, or BTUs (British Thermal Units). According to the Systeme International (SI), which specifies the preferred units of measurement based on an international consensus of scientists, the most appropriate unit is a *joule* (J) for energy, a square meter for area (m^2), and a second (s) for time. A joule is equal to one unit of force, a newton, applied over a length of one meter. In terms of heat, 4186 joules are needed to raise the temperature of one kilogram of water one degree C when the water is at a temperature of 14.5°C.

The *calorie* (cal) is also widely used as an energy unit; one calorie represents the amount of heat energy required to raise the temperature of one gram of water 1°C (from 14.5°C to 15.5°C). (A calorie of heat is not to be confused with a nutrition calorie, which is used in diet books and on food packages and is equivalent to 1000 calories, or a kilocalorie.) A joule is equal to about one-fourth calorie. One calorie, or gram calorie as it is often called, is equal to 4.186 joules. When calories are used, the appropriate unit for area is a square centimeter (cm^2), and for time, the minute (min). The flux of energy is always given in energy units (joules or calories) per unit of area (m^2 or cm^2) per unit of time (second or minute). See Appendix E for more complete definitions of energy, work, and force.

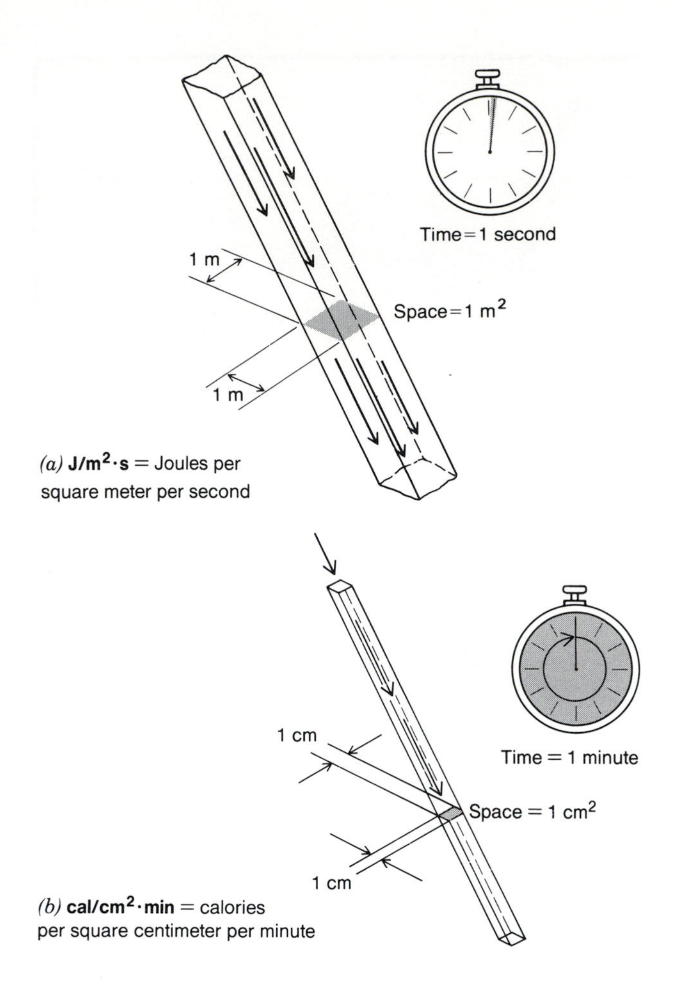

(a) **J/m²·s** = Joules per square meter per second

(b) **cal/cm²·min** = calories per square centimeter per minute

Temperature

The molecules of all earth substances—air, water, rock, and organisms in all places—are in a state of continual vibration. When energy is added to a substance, its molecules begin to vibrate more rapidly, heat is generated, and its temperature rises. As the vibration of molecules increases, a substance will expand; therefore, for a given substance, changes in volume can be used as a measure of changes in heat. Herein lies the principle of a thermometer. In a liquid-filled thermometer, for example, a heat-sensitive liquid—usually mercury—is placed in a narrow tube, and when the substance expands and contracts, it rises and falls in the tube. The tube is scaled numerically so that a temperature can be read from the fluid level.

Several different temperature scales are in use today. The Fahrenheit scale, which is the most arbitrary

TABLE 2.1 KEY TEMPERATURES ON THE CENTIGRADE, FAHRENHEIT, AND KELVIN SCALES

	°C	°F	°K
Absolute zero	−273.15	−459.67	0
Normal freezing point of H_2O[a]	0	32	273.15
Normal boiling point of H_2O[a]	100	212	373.15

[a]At sea level.

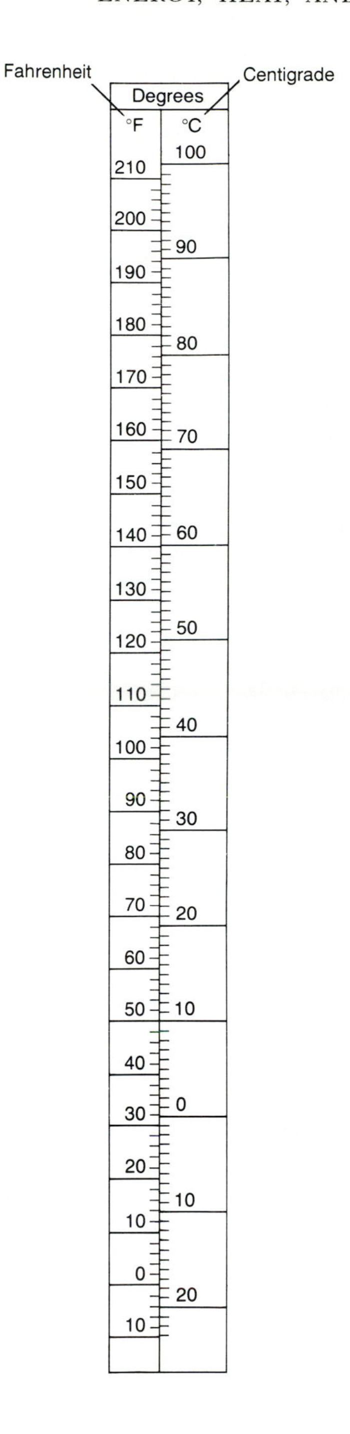

one, sets 32 degrees as the freezing temperature of water and 212 degrees as its boiling temperature. The centigrade, or Celsius, scale sets 0 degrees and 100 degrees for these same two temperatures (see Table 2.1). The Kelvin scale is based on *absolute zero* (0°K), the state at which there is no molecular vibration in a substance and hence no heat, which is equivalent to −273.15°C. The increment or unit of the Kelvin scale is the same as that of the Celsius.

Until recently, official atmospheric temperatures in the United States were recorded in Fahrenheit units. Today, however, scientists prefer Celsius units. Thus, it is often necessary to convert from one scale to the other. The following formulas can be used for conversion:

$$^\circ C = (F^\circ - 32)/18$$
$$^\circ F = (1.8 \times C^\circ) + 32.$$

The chart next to Table 2.1 provides a graphic comparison of the two temperature scales.

Temperature and Heat

We should understand at the outset that temperature and heat are not the same. Temperature is a measure of the heat content of a substance as represented by the average motion (vibration) of each of its molecules, whereas heat is the actual energy content of the substance, defined as the total motion of all its molecules. To interpret heat and temperature properly, you must know about the thermal properties of the substances you find at the earth's surface. In evaluating the temperature levels of air and soil, or moist and dry soil, for example, after a day of sunshine it is necessary to understand that these substances have markedly different thermal properties.

A thermal property called *specific heat* determines how a substance's temperature responds to changes in heat energy. The temperature of a gram of water, for example, increases very slowly with the addition of heat compared to that of a gram of air because water has a much higher specific heat than air. Strictly speaking, therefore, temperature readings taken at different places are of greatest utility in geographic analysis when they represent a single component of

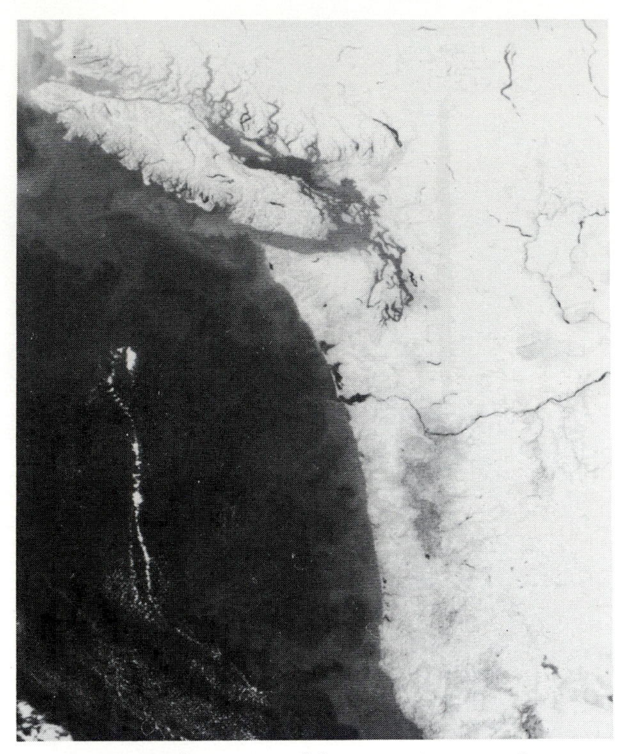

Figure 2.5 In this image of the West Coast, cooler temperatures are represented by light tones and warmer temperatures by dark tones. In addition to the thermal contrast between land and water, a cool ocean current can be seen along the shore.

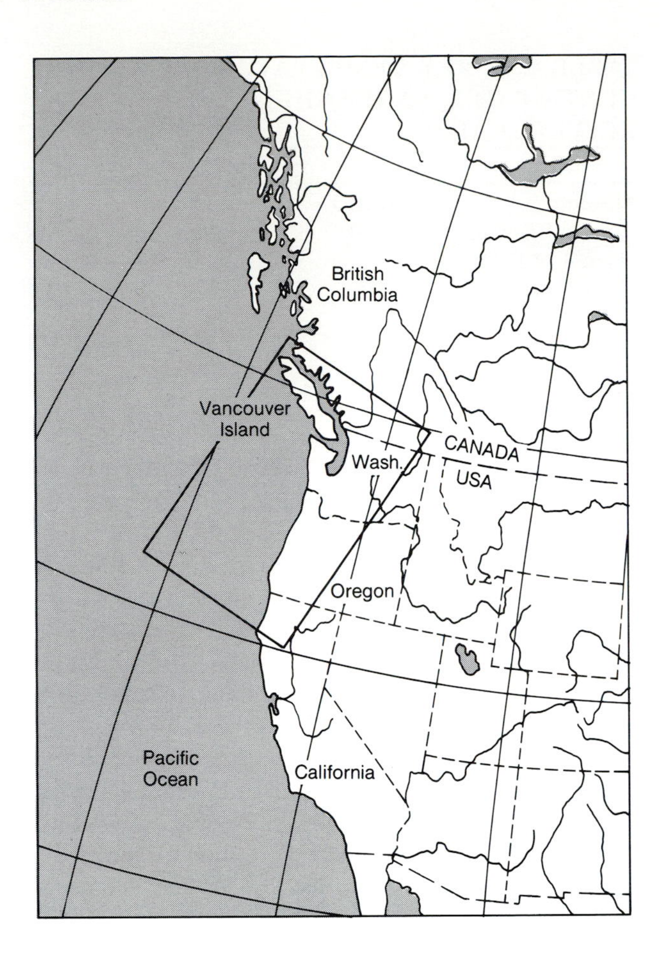

the environment, such as air or water, or similar substances such as soils of comparable compositions. Accordingly, data and maps of earth temperatures must specify whether the readings represent air, the soil surface, or the surface of the oceans if they are to give an accurate portrayal of distribution of heat on the earth's surface (Fig. 2.5).

Monitoring the Earth's Temperature

Air is the only component of the earth's surface environment whose temperature is monitored by virtually all nations of the world on a regular basis. There are several reasons for this. Besides being easy to measure and record, air temperature is one of the two most important elements of climate (the other being precipitation), and humans and their activities are acutely influenced by the temperature of air near the earth's surface. In addition, air temperature is a good choice to represent the temperature conditions of the planet's surface because air easily reaches thermal equilibrium with the surfaces under it. Therefore, when we examine near-surface air temperatures over broad geographic regions, we are getting a fairly accurate portrayal of thermal conditions at the earth's surface.

Although the air temperatures at thousands of locations on the continents have been monitored on a regular basis for decades, comparatively few data have been collected over the oceans. However, with the widespread use of airborne and satellite sensing systems in recent years, we have greatly improved our understanding of the thermal conditions of the oceans and the air above them (Fig. 2.5). The results are revealing that the oceans are very important controls on global climate, including the climate of the continents. Indeed, some researchers in North America now argue that Pacific Ocean temperatures are the best indicator of winter weather trends across the middle zone of the continent.

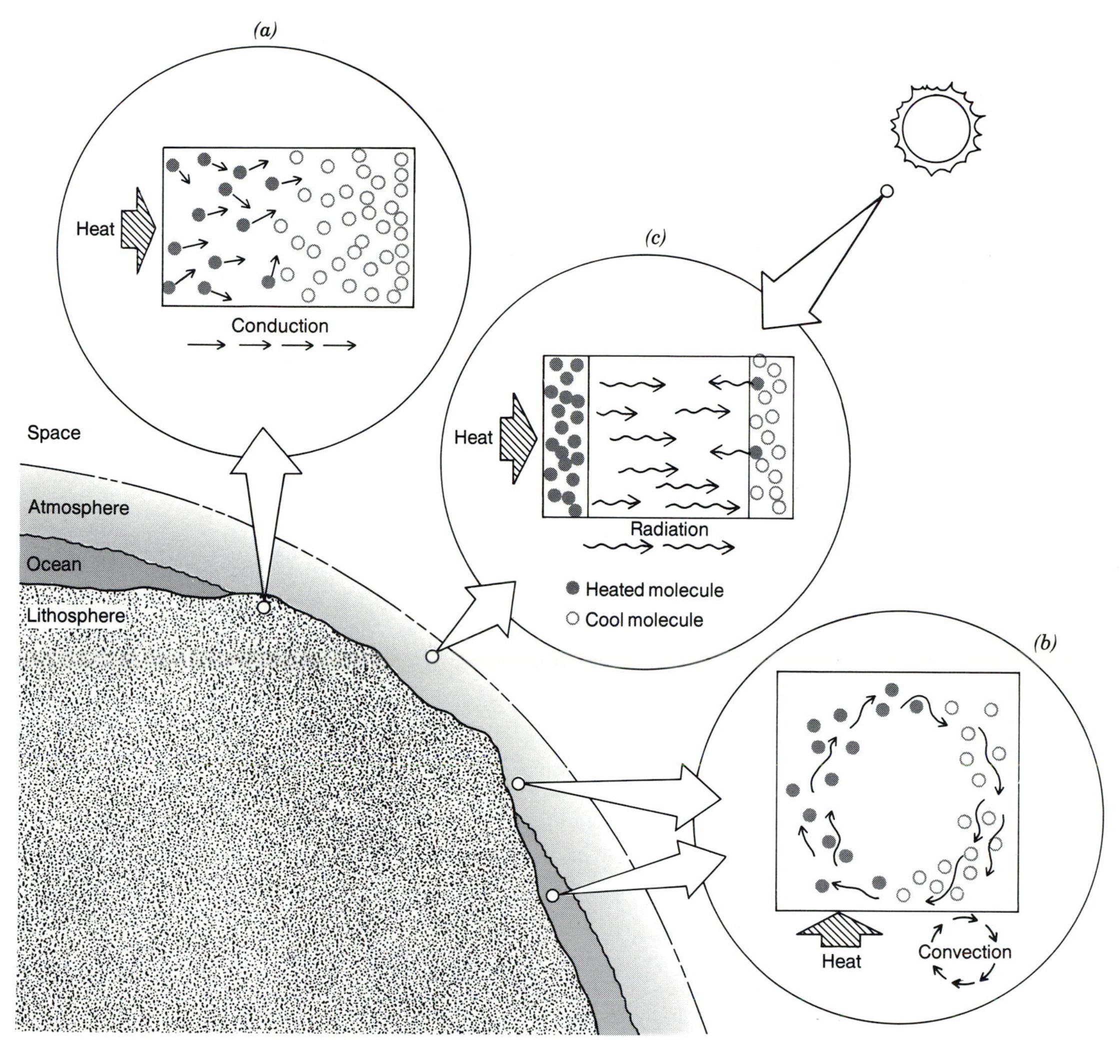

Figure 2.6 Schematic diagrams illustrating the character of (*a*) conduction, (*b*) convection, and (*c*) radiation and the environments where they are prevalent.

Heat Transfer

The redistribution of heat from one medium to another, such as from soil to air, and from one place to another, such as from the Gulf of Mexico to the middle of North America with the movement of air masses, is essential to an understanding of the weather and climate of earth. Three mechanisms are involved in the transfer of heat: conduction, convection, and radiation.

Conduction, the process of heat transfer at the molecular level, takes place when rapidly moving molecules collide with one another. The resultant heat flow is always from a warmer body or part of a body to a colder, which is true for all three mechanisms. Conduction is illustrated in the heating of a skillet placed on a hot stove; heat is transferred but no displacement of the substance itself takes place (Fig. 2.6*a*).

Heat transfer by *convection* takes place when a fluid (either gas or liquid) flows into contact with a solid body or another fluid body of a different temperature. The fluid either gains or loses heat at the contact surface and is then displaced and mixed with cooler or warmer parts of the fluid, thereby effecting heat transfer (Fig. 2.6*b*). In the study of weather and climate, the term *convection* is used to describe mainly the vertical component of mixing in the atmosphere, whereas the term *advection* is used to describe the horizontal component where air in contact with the surface moves great distances in the transfer of energy.

Heat transfer by *radiation* involves the propagation

of energy in electromagnetic waves. This energy is transformed into heat when it is absorbed by a substance (air, water, rock, etc.). Unlike conduction and convection, this mechanism does not depend on a medium such as air or water to transmit the energy. It can take place in the absence of a connecting medium, such as in interstellar space, or within certain substances, such as air and clean water which are transparent to certain forms of radiation. Opaque substances, which soil and dark rocks are to sunlight, do not transmit electromagnetic waves but instead absorb them (Fig. 2.6*c*).

All three mechanisms of heat transfer are active in the atmosphere, but convection and radiation are decidedly the most effective. In the oceans, convection is the principal mechanism, whereas in soil and surface rock, conduction is the principal mechanism of heat transfer. Of the three, conduction is by far the slowest mechanism of heat transfer.

Thermal Character of Earth as a Planet

As we learn more about conditions on neighboring planetary bodies, we can better appreciate the unique character of thermal conditions on the earth's surface. The equilibrium surface temperature for earth is about 15°C (58°F). Differences in surface temperatures from day to night vary widely over the earth, but a day/night variation of 5 to 10°C could be considered representative for the earth as a whole. These temperatures are modest compared to those on Mars, for example, where the equilibrium surface temperature is close to −50°C and day/night differences are typically 50 degrees or more. These thermal contrasts help explain some of the major environmental differences between earth and the other terrestrial planets, among them the state of water.

Virtually all the water on Mars (held in polar ice caps and underground) is frozen, whereas on earth, water exists abundantly in all three states: solid, liquid, and vapor. Moreover, the thermal regime of the earth's surface with its relatively strong variations from region to region and season to season drives the rapid exchange of water between the atmosphere and the surface via the processes of evaporation, condensation, and precipitation. A global thermal regime only 25°C colder would produce ice over most of the oceans, permafrost over the bulk of the land masses, and greatly reduce the vapor content of the atmosphere; whereas a 25°C warmer air surface temperature would put a great deal more vapor in the atmosphere and melt the polar ice caps which in turn would raise the oceans by 65 meters (210 feet). A major change in atmospheric water vapor would dramatically alter world climate, both in terms of temperature and moisture, and a rise in sea level of 65 meters would flood most of the world's major cities.

Summary

Radiation from the sun is the principal source of energy for the earth's surface. This energy drives the essential processes of the atmosphere, soil, oceans, and biota. In order for the earth to maintain an energy balance, however, it must give up as much energy as it receives; the measure of the earth's energy balance is its equilibrium surface temperature.

In tracing the flow of energy on earth, we find that it moves through various systems, changing form and geographic distribution. The conservation-of-energy principle tells us that energy can neither be created nor destroyed; thus, all that enters the realm of earth can be accounted for in some form at some place at a given time.

Energy can be measured in various units, but joules and calories are preferred in science. Temperature is an indicator of the heat content of a substance, and air temperature at the earth's surface is a good measure of the thermal conditions of earth. Heat transfer takes place by conduction, convection, and radiation, and these mechanisms are essential to the redistribution of heat on earth. The thermal regime of earth may be unique among the terrestrial planets in the solar system, for it allows water to exist in all three states and promotes the rapid exchange of water between the atmosphere and the planet surface.

Concepts and Terms for Review

concept of a system
  components
  linkage
  cause and effect associations
energy and processes
sources of earth energy
  exogenous
  endogenous
energy balance
  equilibrium temperature
  steady state
conservation-of-energy-principle
energy systems
  greenhouse
  hydrologic
  ecosystems
energy terminology
  work
  energy forms
  flux
  joules
  calories
temperature
  Celsius
  Fahrenheit
  Kelvin
  absolute zero
heat
  specific heat
  conduction
  convection and advection
  radiation
earth temperature
  monitoring
  surface temperature
  relative to Mars

Sources and References

Chorley, R. J., and Kennedy, B. A. (1971) *Physical Geography: A Systems Approach*. New Jersey: Prentice Hall.

Gribbin, John R. *Future Weather, 1983: Carbon Dioxide, Climate, and the Greenhouse Effect*. New York: Penguin Books.

Hubert, L. F., and Lehr, P. E. (1967) *Weather Satellites*. Waltham, Mass.: Blaisdell.

Iribarne, J. V., and Godson, W. L. (1973) *Atmospheric Thermodynamics*. Hingham, Ma.: Reidel.

Palmén, E., and Newton, C. W. (1969) *Atmospheric Circulation Systems*. New York: Academic Press.

Perry, A. H., and Walker, J. M. (1977) *The Ocean Atmosphere System*. London: Longman.

Ras, P. K., et al. (1972) "Global Sea-Surface Temperature Distribution Determined from an Environmental Satellite." *Monthly Weather Review*, Vol. 100, pp. 101–114.

Seidel, Stephen, and Keyes, Dale. (1983) *Can We Delay a Greenhouse Warming?* Washington, D.C.: Environmental Protection Agency.

Chapter 3

The Flow of Radiation to and from Earth

We cannot begin to understand the character of the earth's surface and the processes that change it without first examining solar radiation. What is its makeup and rate of flow to earth? How is it influenced by different surface and atmospheric conditions? How is it distributed over the earth, and what are the means by which this energy is released from the earth's surface and atmosphere?

Religion, mythology, and literature are rich with allusions to the theme of light and dark. The ancient Zoroastrian religion, which is considered to be a forerunner of Judaism, Christianity, and Islam, portrayed light as good and dark as evil. Not surprisingly, fire was, and still is, an important symbol of this religion. Ra, the sun god, was a cornerstone in the religious mythology of the ancient Egyptians. In *Genesis*, God's first creation is light; the King James version of the Bible describes it this way: "And God said, Let there be light; and there was light. And God saw the light, and it was good . . ." Conversely, another name for Satan is the "prince of darkness."

If we could uncover the origins of these ideas, we might find that they grew out of a need to explain complex natural and human phenomena, in this case, the relationship between light and life. That light induced plant growth and heated the ground and the air was no mystery to the ancients, but the mechanisms involved were a great unknown. The manner in which solar energy related to larger scale phenomena such as the wind systems and ocean currents was also largely a mystery to the ancients.

We now understand that practically all of what we see in the landscape—both *features* such as forests and stream valleys and *processes* such as evaporation of water, the flow of air, and the motion of water waves—is somehow tied to solar energy. If we are interested in learning why these features are arranged geographically as they are and why these processes affect change as they do on the earth, it is necessary to learn about solar energy. We must learn not only about its physical properties and the principles governing its behavior, but also about its geographic distribution over the planet. Furthermore, we should understand how solar energy is disposed of at the earth's surface and in the atmosphere, for we know that energy does not build up on earth over the years.

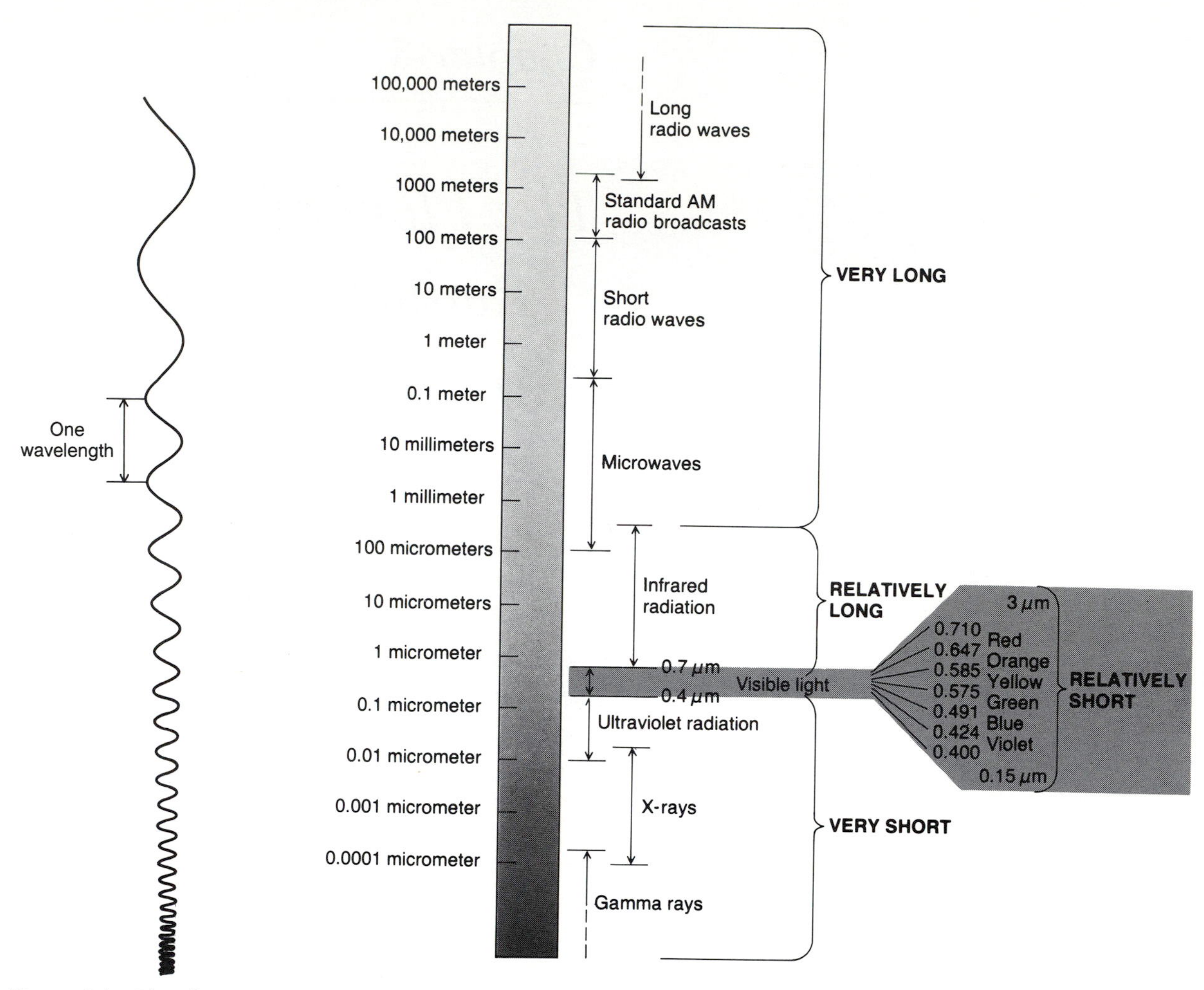

Figure 3.1 The electromagnetic spectrum, including the terms for the various types of radiation.

This leads us to the concept of the earth's energy balance. There are two essential parts to the energy system of the earth's surface and atmosphere: the radiation balance and the heat balance. In this chapter we examine the radiation balance; first, solar radiation, its physical characteristics, rate of flow to the earth, and distribution over the earth; and second, the radiation earth gives up, its characteristics, and its role in heating the atmosphere.

SOLAR AND EARTH RADIATION

The sun produces a massive amount of energy, which it broadcasts into space in the form of radiation. Only a minute fraction of this radiation (about one two-billionth of the total) is intercepted by earth, where it lights and heats the earth's atmosphere and surface. At the top of the atmosphere, when the earth is at its mean distance from the sun (149.6 million km), incoming solar radiation has a strength of 1372 joules per m^2 per second (J/m^2 · s) or 1.97 calories per cm^2 per minute (cal/cm^2 · min). Because the flow of this radiation is believed to be very steady, this value is called the *solar constant.*[1]

Much, but not all, solar radiation that enters the realm of the earth, is absorbed by the atmosphere, the land, and the oceans. To maintain an energy balance, the earth must ultimately return this energy to space. But before it does so, the energy enters the

[1]This is the latest value as determined by satellite radiometers and is 19 joules greater than the value previously accepted.

earth's air, water, and biological systems where it drives these systems, enabling them to perform work.

The Nature of Radiation

Many varieties of radiant energy (also known as *electromagnetic radiation*) make up the energy that pours from the sun. We classify these according to a scheme called the *electromagnetic spectrum*. Because radiant energy travels in the form of electromagnetic waves, it makes sense to define the various types of radiation in the spectrum according to a wavelength. A *wavelength* is the distance from the crest of one wave to the next, and the unit of measurement in which it is commonly expressed is the micrometer (or micron), equal to one millionth of a meter and abbreviated μm.

For our purposes we can subdivide the electromagnetic spectrum as shown in Figure 3.1. The key subdivisions are as follows: (1) very short wavelengths, less than 0.15 micrometer, which include gamma rays, X-rays, and some ultraviolet radiation; (2) relatively short wavelengths, between 0.15 and 3.0–4.0 micrometers, which include ultraviolet, visible light, and near infrared radiation; and (3) relatively long wavelengths, between 3.0–4.0 and 100 micrometers, which is infrared radiation; and (4) very long wavelengths, greater than 100 micrometers, which include radio waves, television waves, and microwaves. Categories 2 and 3 include most radiation coming from the sun and returned by earth, and for ease of reference are termed *shortwave* and *longwave*, respectively.

Some Governing Principles

At what rate is radiation produced by a radiating body, and what controls the type of radiation produced? The rate at which a body emits radiation is a function of its temperature; the higher the temperature, the higher the rate of emission. More precisely, the rate of emission increases at an increasing rate with temperature; as temperature gets higher, the rate itself gets higher even faster.

The sun, with a surface temperature of 5800 degrees K (6073 degrees C), produces a vastly greater amount of radiant energy than does the earth, whose equilibrium surface temperature is 15°C. Earth, however, produces more than Mars whose surface temperature is below zero degrees C, and Mars produces more than some frigid asteroid lying in the shadow of a planet.

Solar radiation reaches its peak intensity in the visible, or "light," part on the electromagnetic spectrum, whereas radiation emitted by earth reaches its peak intensity in the infrared portion of the spectrum. Thus, the sun emits not only more energy than does earth, but also a much greater proportion of shorter wavelengths. These, as well as several other important facts, are illustrated in Figure 3.2. The left-hand curves represent the energy emitted by the sun and received by the earth's atmosphere, and the right-hand curve, the energy emitted by the earth. The region from 0.4 to 0.7 micrometer on the spectrum is the visible, or light, portion.

The bulk of solar radiation is concentrated around 0.48 micrometer, and the bulk of earth radiation, at much longer wavelengths, around 10 micrometers. Although it receives energy as relatively shortwave radiation, the earth returns most radiation to space as relatively long waves in the infrared portion of the spectrum. This important fact is described by *Wien's law*, which in this context states that the wavelength of maximum-intensity radiation grows longer as the absolute temperature of the radiating body decreases (Fig. 3.3).

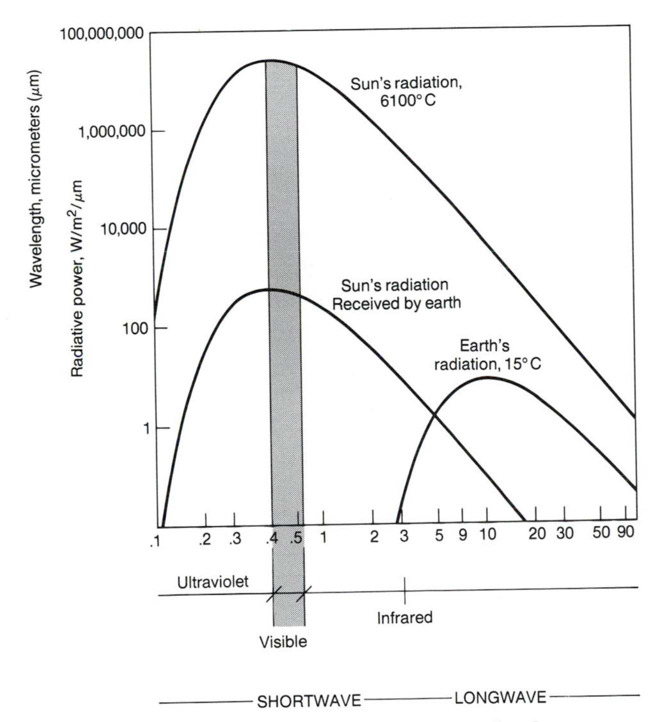

Figure 3.2 The distribution of intensities of radiation produced by the sun and the earth. The vertical axis represents the intensity of output; the horizontal axis represents micrometers, representing wavelength. Solar radiation is concentrated around 0.5 micrometers, between the ultraviolet and infrared wavelengths, whereas earth radiation is entirely infrared.

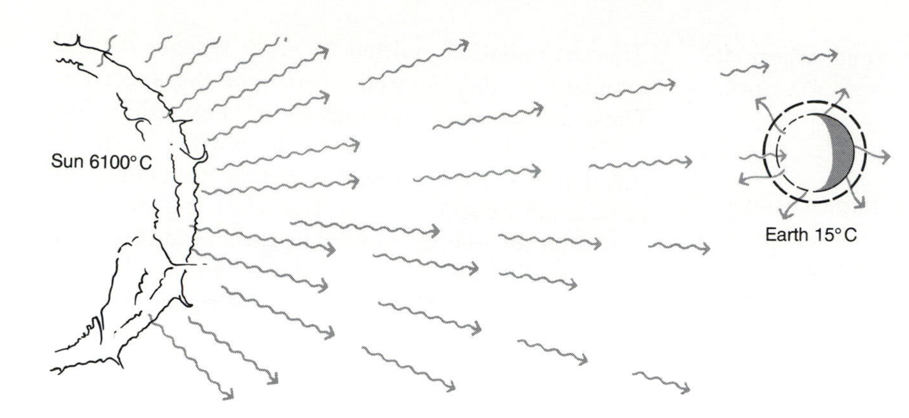

Figure 3.3 A schematic diagram illustrating the difference in solar and earth radiation in relation to the surface temperature of each body.

This difference is an extremely important factor in the earth's energy balance because the atmosphere is less transparent to longwave radiation than it is to shortwave radiation. Therefore, the return flow of longwave radiation from the earth's surface to the top of the atmosphere is less efficient than the inflow of shortwave. As a result, the atmosphere produces a greenhouse effect for the earth, trapping energy as it absorbs longwave radiation. So effective is the greenhouse effect that without it the earth's equilibrium surface temperature would be 33°C lower than it is, −18°C instead of +15°C!

RECEIPT OF SOLAR RADIATION

If we measure the average amount of solar radiation getting to the earth's surface—no matter where we are on the planet—we would find that it is much less than the solar constant. To explain this phenomenon, we would have to examine three important factors: (1) absorption and reflection by the atmosphere; (2) the directness of the angle at which the radiation strikes the ground; and (3) the daily duration of daylight. To explain how much of that radiation that reaches the ground is actually absorbed by it, a fourth factor, surface reflection, or *albedo*, would also have to be considered.

Absorption and Reflection by the Atmosphere

In most places the atmosphere is not really clear to solar radiation. Clouds get in the way as do particles of dirt and molecules of various gases. On any given day these factors can reduce the amount of incoming shortwave radiation by as much as 75 percent. This reduction begins in the upper atmosphere, where ultraviolet radiation is *absorbed* by ozone, resulting in a 3-percent decline in the strength of the radiation beam. Farther into the atmosphere, additional absorption takes place. Infrared radiation, which constitutes a small percentage of solar radiation, is absorbed by molecules of carbon dioxide and water vapor, and virtually none reaches the earth's surface directly from the sun. Absorption also takes place when visible wavelengths are absorbed directly by particles of water and dirt. For the earth as a whole, atmosphere absorption totals 15 to 20 percent on the average (Fig. 3.4).

The beam is reduced even more by the *scattering* (reflection) of radiation off gas molecules, dust particles, and ice or water droplets in clouds. Scattering is divided into two types: *sky scattering*, which involves dust and gas molecules, and *cloud scattering*, or *reflection* which involves ice and water particles. In both types some radiation is scattered toward the earth, and some (often termed *backscattered radiation*) is scattered toward space. Moreover, certain wavelengths are scattered more than others. The scattering of blue and violet wavelengths is greater than that of the longer red wavelengths, and this gives the sky its blue color (Fig. 3.4). Together, scattering by particles, air, and clouds reduces the flow of solar radiation by an average of 31 percent worldwide. On any given day, however, this figure may be two or three times greater for cloudy and heavily polluted areas or as little as 5 to 10 percent for areas with cloudless, clean air.

Beam and Diffuse Radiation

The solar radiation that reaches the earth's surface arrives in two forms: *beam* and *diffuse*. Beam radiation is directed shortwave radiation whose direct line of flow through the atmosphere has not been broken up

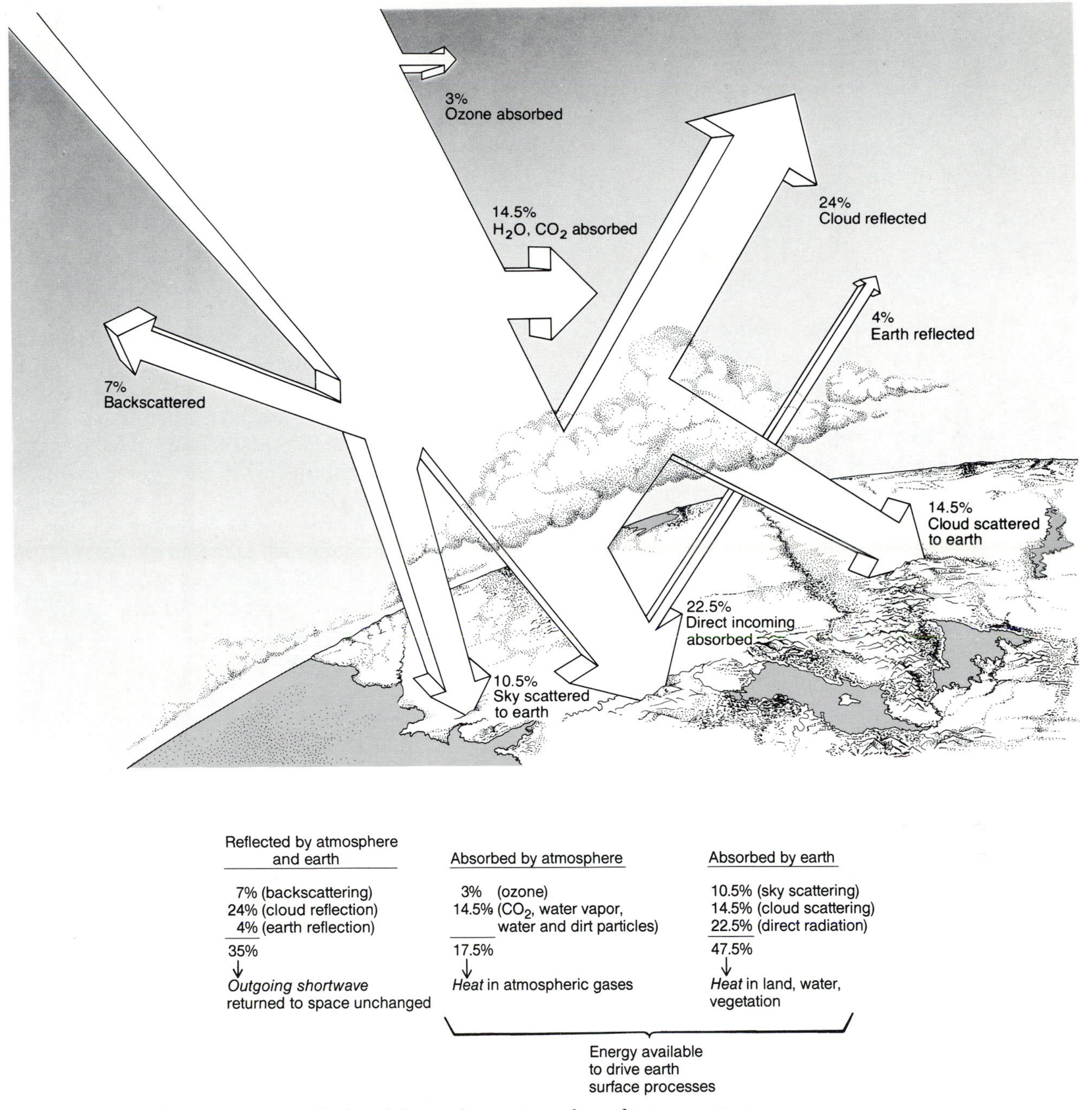

Figure 3.4 The key processes in the breakdown of incoming solar radiation: scattering, reflection, and absorption are illustrated in the diagram and summarized in the accompanying table. The values given here are representative of the earth as a whole; bear in mind that seasonal and geographical variations are quite pronounced.

by scattering. Although beam radiation changes orientation slightly by bending, or *refracting*, as it passes through the atmosphere, it is essentially unidirectional and creates strong shadows on the back sides of opaque objects. Diffuse radiation moves in multiple directions as rays of light ricochet among air molecules and particles of dust, water, and ice. As experienced photographers know, diffuse light does not produce good shadows, and it is less influenced by sun angle than is beam light. The incidence of diffused radiation is greatest in cloudy regions of the world such as the equatorial belt and the maritime polar and subpolar zones.

The Effects of Sun Angle

The angle at which the radiation beam passes through the atmosphere and strikes the earth is called *sun angle*. Sun angle is very important for two reasons: First, if it is low (i.e., small), the beam of radiation slices through the atmosphere obliquely, taking a long route to the surface. This increases the loss of energy because the longer the route, the greater the amount of scattering and absorption (Fig. 3.5*a*). Second, sun angle is important because we are interested in the radiation incident on the earth's surface. The more direct the angle of incidence, the greater the intensity of bombardment of the surface by radiation.

Consider in Figure 3.5*b* the beam of sunlight of 2 m^2 illuminating an area of the panel 2 m, so that the sun angle is 90 degrees. The 2-m^2 surface is intercepting the entire 2-m^2 beam. Now let us tilt the panel back, so that the sun angle is only 30 degrees. The beam of sunlight is still 2m^2 but now it is spread over an area twice as large (4 m^2). Thus, the density of radiation on it would be half as great as when the sun angle was 90 degrees. This illustrates the influence of sun angle on incident radiation. The *incident radiation* value for any surface is equal to the energy of the beam divided by the area it illuminates.

Let us apply this principle to the landscape. Instead of tilting a hypothetical surface, however, let us use natural slopes in the land. In the Northern Hemisphere poleward of the tropics, the sun always shines from the southern part of the sky; therefore, slopes that face south receive more direct radiation than do their northfacing counterparts (Fig. 3.6*a*). In spring it is not uncommon to find patches of snow on north-facing slopes and barren ground on southfacing slopes. This results from the fact that the sun's radiation heats the southfacing slope, while its northfacing counterpart is in shadow much of each day (Fig. 3.6*b*).

In hilly terrain, sun angle has a strong influence on ground-level climate which can often be traced to differences in soil, moisture, and plant conditions. For example, soil-moisture evaporation rates may be so much greater on southfacing slopes that the growth rates of plants such as grasses are measurably slower

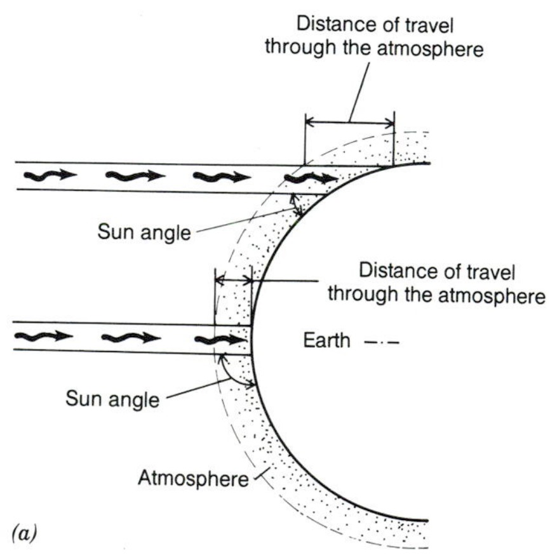

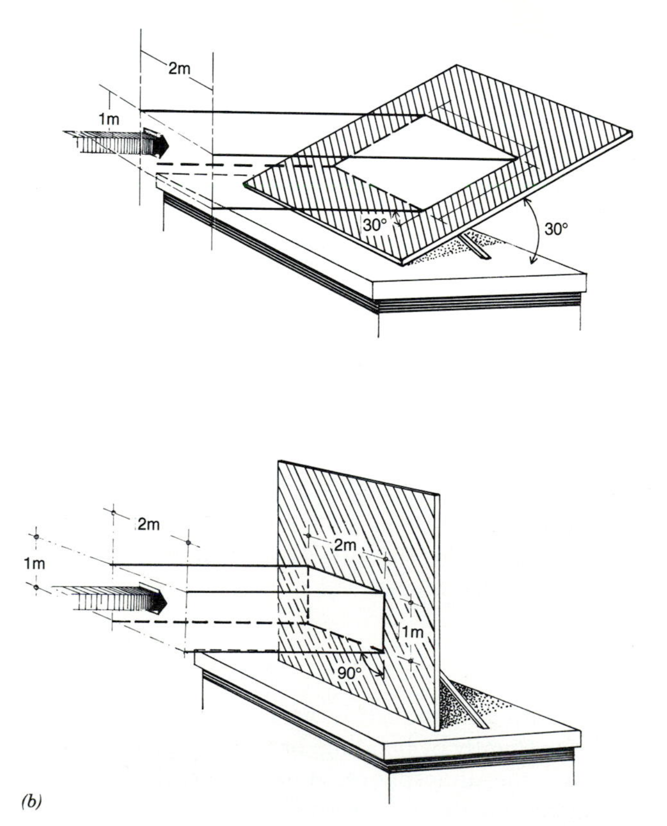

Figure 3.5 The influence of the earth's curvature (*a*) on sun angle and the distance of travel of the solar beam through the atmosphere; (*b*) the influence of sun angle on the intensity of radiation. Where the beam strikes the surface directly, radiation is most intense because the beam irradiates an area proportional to its cross-sectional area. If the surface is tilted 60°, only half of the beam of radiation strikes the surface, thereby reducing the unit area intensity by 50 percent.

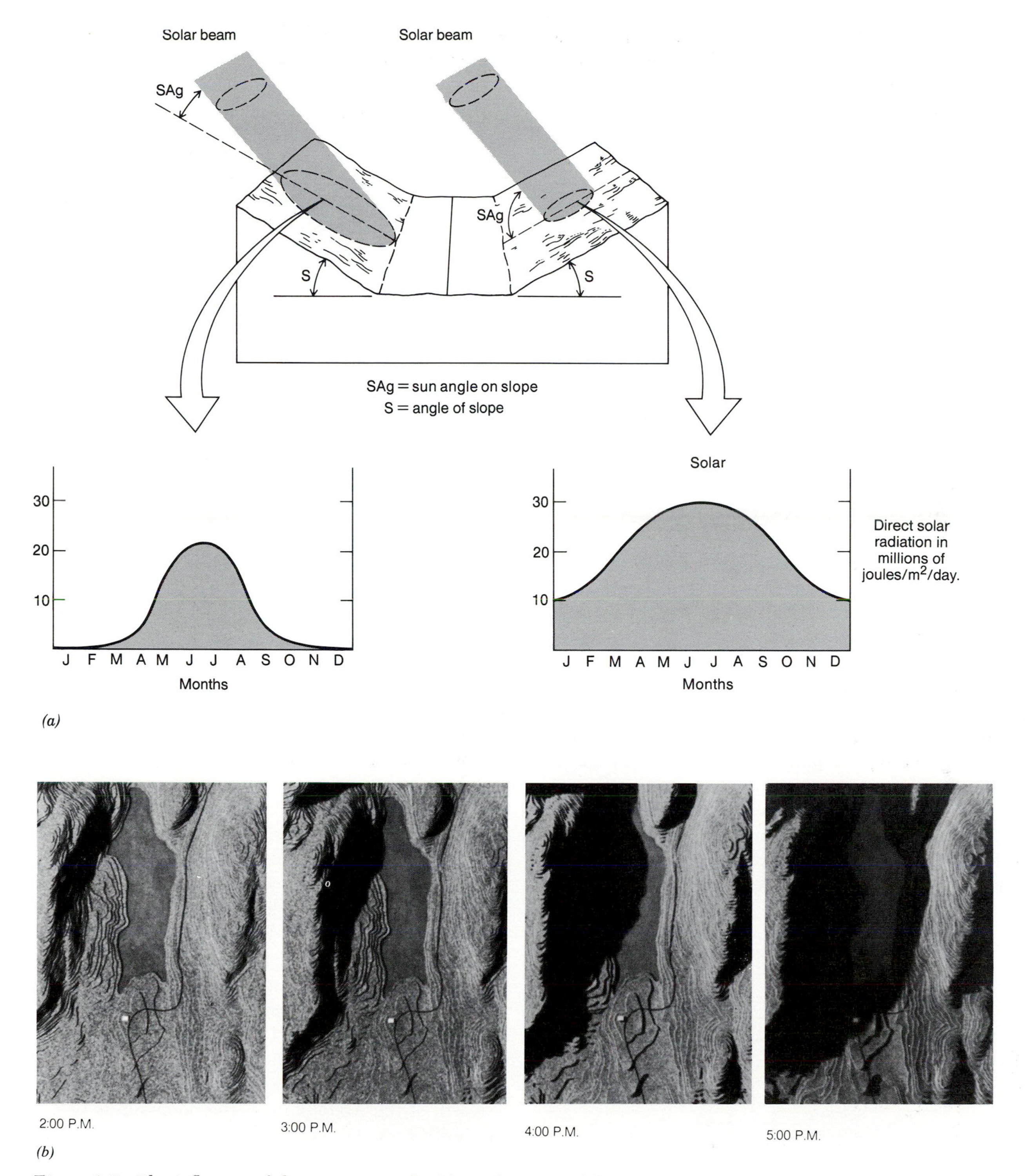

Figure 3.6 The influence of slope on sun angle: (*a*) Total receipt of direct solar radiation may be two times higher on 30° southfacing slopes than on 30° northfacing slopes at latitude 50°N; (*b*) the afternoon patterns of shadow and sunlight in Jordon Pond Valley, Acadia National Park, Maine, on August 1 based on a simulation model.

than are those of the same plants on the northfacing slope. In mountainous areas, even the plant types may be radically different on north- and southfacing slopes (see Fig. 12.6).

Sun Declination, Solstices, and Equinoxes

Now let us step back and examine the larger question of seasonal and latitudinal variations in sun angle. At any moment in the earth's revolution about the sun, there is always one, and only one, latitude at which the noon sun angle is vertical to the earth's surface, that is, directly overhead to someone standing there. This latitude is called the *declination* of the sun.

The declination of the sun changes every day by an average of one-quarter of a degree of latitude. In the course of one annual revolution of the earth, it migrates back and forth across the tropics. Its northernmost position is 23.5 degrees N (or 23°27′) (the Tropic of Cancer) on June 20–22; in the Northern Hemisphere this date is called the *summer solstice* (see also Chapter 1). The southernmost position is 23.5 degrees S (the Tropic of Capricorn) on December 20–22, which is called the *winter solstice* for the Northern Hemisphere. At two dates midway between the solstices, the sun is vertical at the equator. These dates—when the declination of the sun is 0 degrees—are the *equinoxes*, the *vernal equinox* occurring on March 20–22 and the *autumnal equinox* on September 20–22 in the Northern Hemisphere. These four dates may vary by a day in either direction because our calendar has to be adjusted every 4 years to bring it back in phase with the period of the earth's revolution. The graph in Figure 3.7 shows the declination of the sun for the entire year. The dates of the solstices and equinoxes are opposite for the Southern Hemisphere.

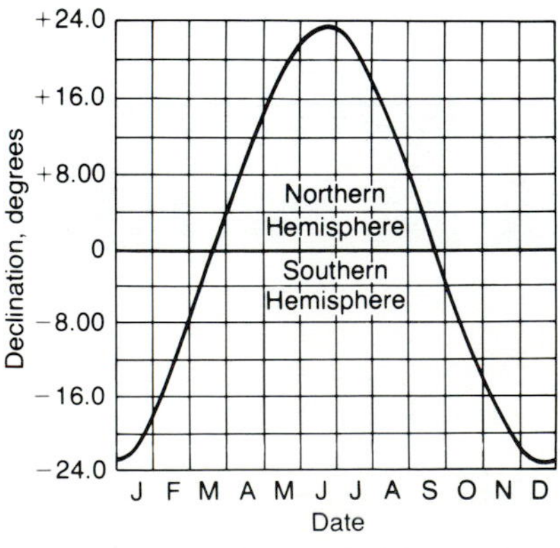

Figure 3.7 Declination of the sun, in degrees.

Calculation of Sun Angle

Calculation of the noon sun angle for any given latitude and declination is simplified by considering the solar *zenith angle* (Z). The *zenith angle* is the angle formed between a vertical line and the sun at high noon. If the sun is directly overhead, the zenith angle is 0 degrees; when the sun is at the horizon, the zenith angle is 90 degrees. The *sun angle*, which is the angle between noon sun and the earth's horizon, is thus equal to 90 degrees $-Z$. Figure 3.8 depicts the relationship among zenith angle, latitude, and declination for a variety of possible situations: Zenith angle is equal to the difference between the latitude of the place in question and the declination of the sun. For the location 22 degrees north latitude, the zenith angle is 30 degrees on March 1 (equal to the 22 degrees of latitude north of the equator plus the 8 degrees of latitude of the sun south of the equator, as read from the graph in Fig. 3.7).

Global Variations in Sun Angle

Now let us consider sun-angle variation with season and latitude. At the equator, the sun angle is never less than 66.5 degrees because the declination of the sun is never farther away than the Tropic of Cancer and Tropic of Capricorn. Thus, the sun angle at the equator varies only 23.5 degrees, from 90 degrees to 66.5 degrees over the year.

In the midlatitudes, the variation in the sun angle is twice as great as at the equator. At 40 degrees N, for example, the sun angle is 73.5 degrees on June 20–22 (which is about 10 degrees greater than it is at the equator on this date), and 26.5 degrees on December 20–22. Thus, the annual variation in sun angle for this midlatitude location is 47 degrees. At the Arctic Circle (66.5 degrees N), the sun angle is nearly 47 degrees on June 20–22 and 0 degrees on December 20–22, and the noon sun is on the horizon.

North of the Arctic Circle, the noon sun is below the horizon on December 20–22, and at the North Pole itself, the sun angle is minus 23.5 degrees on this date, or effectively 0 degrees. The highest annual sun angle at the poles is + 23.5 degrees, which occurs on June 20–22 in the north and on December 20–22 in the south. The effective variation in sun angle at the poles—assuming that any negative angle is 0—is

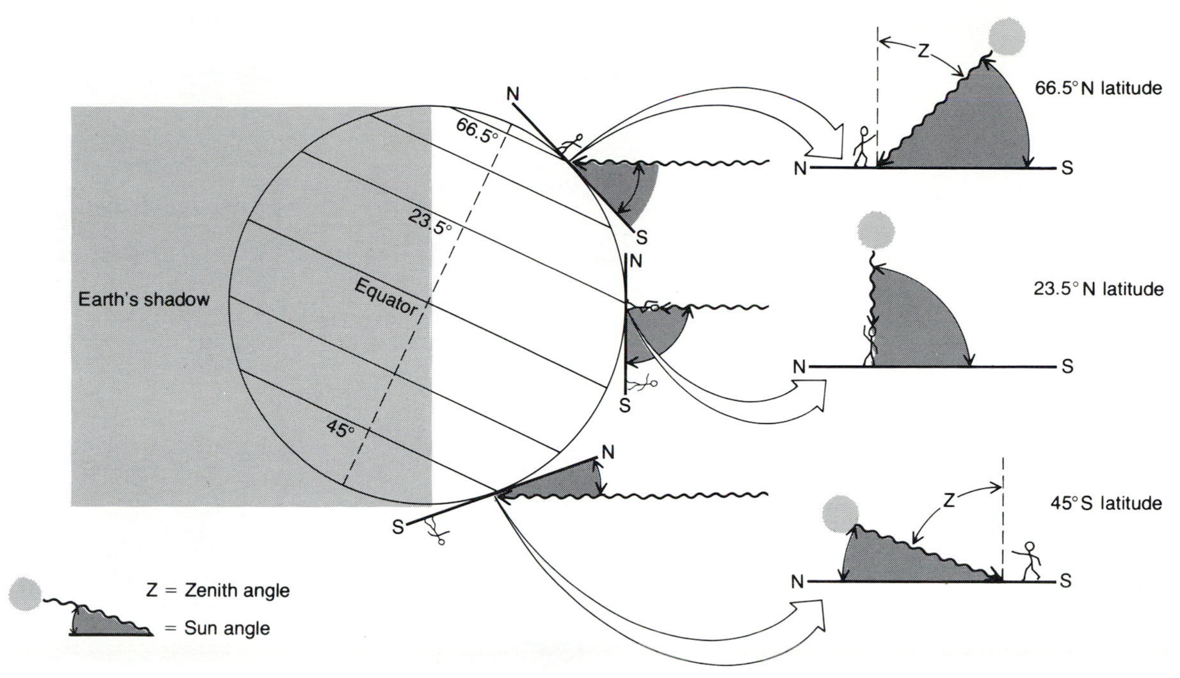

Figure 3.8 The relationship among zenith angle, latitude, and sun angle on June 22. Note that the sun is in the southern sky for the observer at 45°N, but in the northern sky for the observer at 45°S.

therefore 23.5 degrees annually. Thus, in considering the world as a whole, the annual variation in sun angle is greatest for the midlatitudes, and, accordingly, we would expect the climates there to be more seasonal in character than those in the polar and equatorial zones.

OTHER EFFECTS ON SOLAR RADIATION

The influence of the atmosphere and sun angle is of paramount importance in our examination of the receipt of solar radiation by earth. But two other factors must also be taken into account: the duration of sunlight and the reflective capacity of earth materials.

Variations in the Length of Day

Like the variations in sun angle, variations in day length are also related to latitude and solar declination. Figure 3.9 shows the portions of the earth illuminated by the sun at three separate positions: summer solstice, winter solstice, and equinox. Note that the *circle of illumination* which separates the lighted portion from the unlighted portion is a great circle and as such always divides the earth in half. The equator is also a great circle, and because two great circles always bisect each other, the equator is always half in the light and half in the dark, regardless of the date and declination. Thus, any point on the equator will be 12 hours in light and 12 hours in darkness as the earth makes a complete rotation. But all of the other parallels of latitude are small circles. Therefore, except when the earth is in one of the equinox positions, the circle of illumination intersects these latitudes unequally, and a point on the surface of the earth will not spend equal time in light and darkness as the earth rotates. Furthermore, on June 20–22 all latitudes north of the Arctic Circle are in darkness for 24 hours. The situation is reversed on December 20–22.

On any day, regardless of its length, the daily receipt of incoming shortwave radiation should, barring interference from clouds, approximate a bell-shaped curve corresponding to the rise and fall in the sun's altitude. As the day progresses and the sun rises toward its high-noon position, the intensity of incoming radiation increases; likewise, radiation decreases as the sun sinks from noon to dusk (Fig. 3.10). The curve

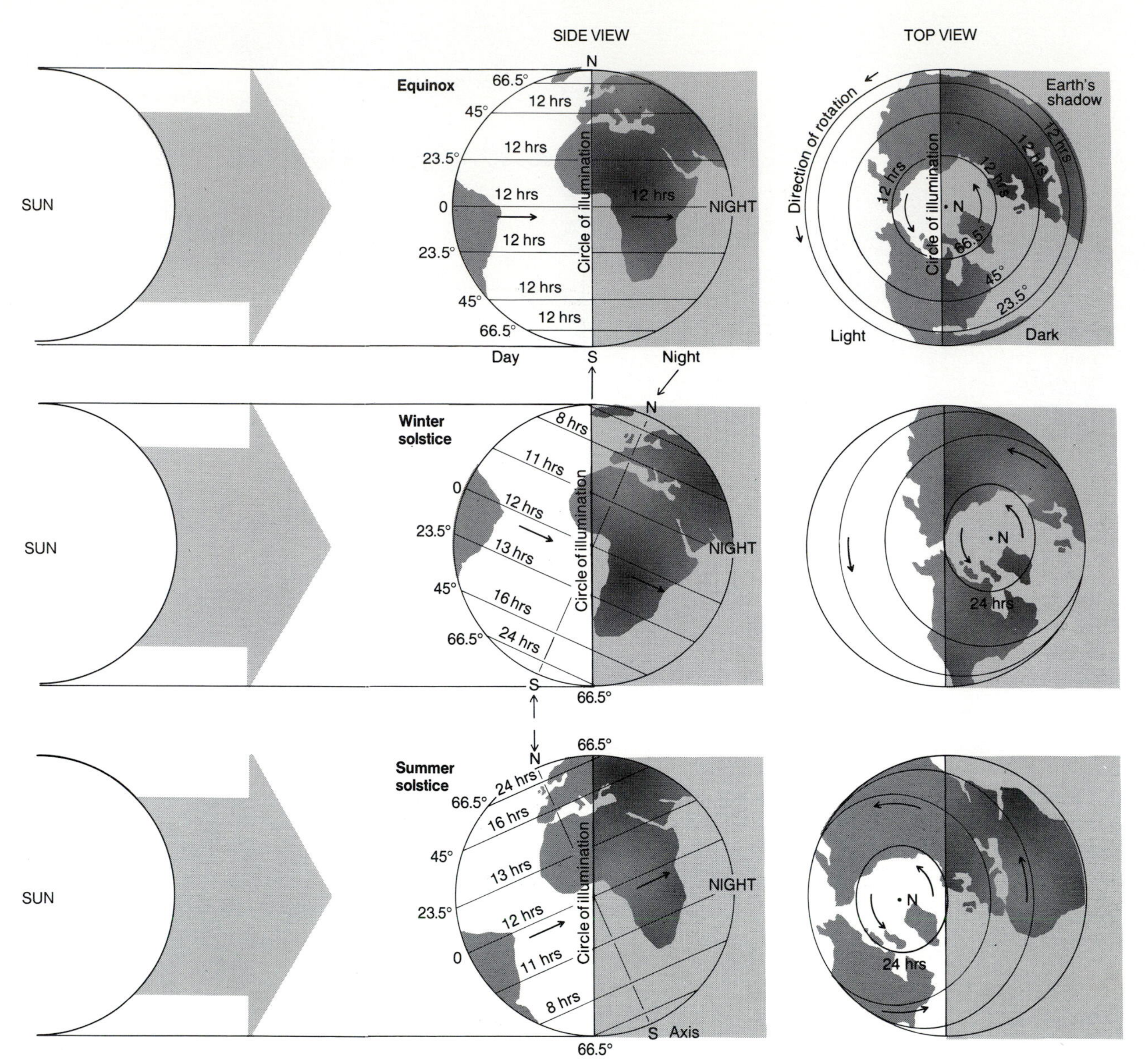

Figure 3.9 Schematic illustrations showing the position of the circle of illumination in relation to the earth's axis (side view) during the equinoxes and solstices. The hours give the approximate time in a day spent in the light and dark sectors at the latitudes shown. The top view shows the relationship of the circle of illumination to the North Pole looking down on the earth from above the North Pole.

expands and contracts with changes of season (i.e., length of day), and the magnitude of annual variation increases with distance poleward of the equator. In addition, the curve heightens with length of day because sun angle increases as days grow longer (Fig. 3.10).

Surface Reflection

For incoming shortwave radiation reaching the landscape, part is absorbed and part is reflected back into the atmosphere. The capacity of the earth's surface to reflect shortwave radiation is highly variable from

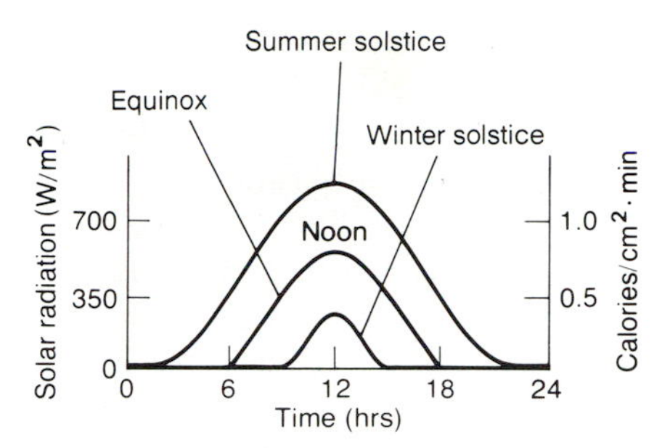

Figure 3.10 Idealized curves for incoming solar radiation for a midlatitude location representing the solstices and the equinoxes.

place to place, in turn making radiation absorption equally variable.

Surface reflectance is termed *albedo* and is defined as the percentage of the incoming shortwave radiation at ground level that is not absorbed by the surface. We can designate incoming shortwave radiation as Si and the amount of shortwave radiation reflected (outgoing) as So. Thus, Albedo = So/Si × 100.[2]

[2]The So/Si value, a decimal, is multiplied times 100 to convert it into a percent, the standard expression for albedo.

Albedo is influenced by many characteristics of a surface, including texture, water content, and orientation with respect to the sun. Because earth materials are so varied in these respects, some surfaces are much better reflectors (and therefore much poorer absorbers) of solar radiation than others. Generally, dark-toned and rough-textured materials, such as a soil with high organic content or the canopy of a pine forest, have lower albedos than do light-toned materials, such as snow and concrete. The actual values range from as high as 90 percent (over fresh snow) to 10 percent or less (over moist, black soil).

Albedo also varies with sun angle. This is especially pronounced on water; at angles above 70 degrees or so, the albedo of water is only 5 to 10 percent, whereas at angles below 20 degrees, it can be as great as 90 percent. Figure 3.11 presents an example of the wide differences in albedo which can be found in high mountain terrain. With this sort of range, it is easy to appreciate the extreme variations in surface heating which can be produced within small areas on sunny days.

Finally, there are seasonal changes in radiation to consider. At midlatitude sites, such as the midcontinental United States, components of incoming and

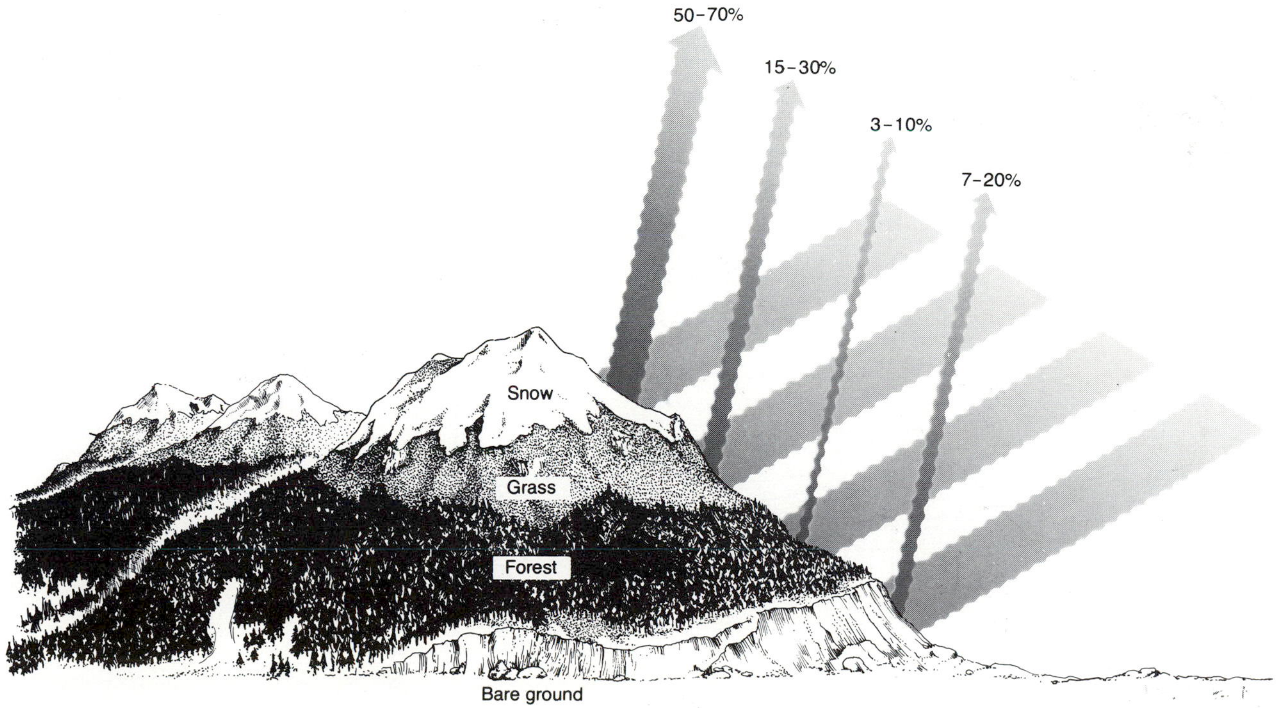

Figure 3.11 Four types of mountain surfaces: snow, grass, forest, and bare ground. With equal amounts of incoming radiation on each surface, major differences in the radiation balance are attributable to differences in albedo.

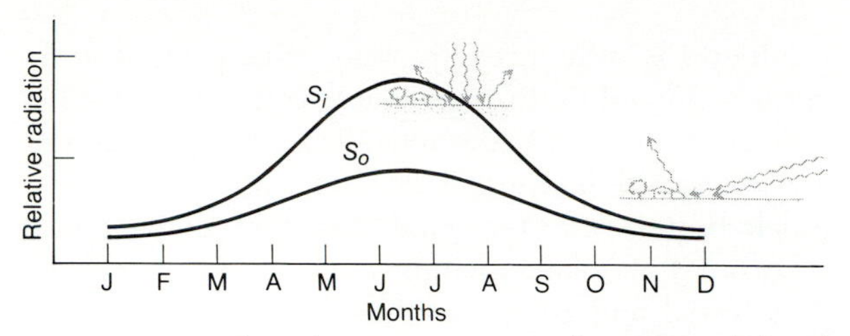

Figure 3.12 The relative proportion of incoming (S_i) and reflected (S_o) shortwave radiation over a midlatitude year. The S_i is weak in winter, and as much as 50 to 70 percent of it may be reflected back into the atmosphere.

outgoing shortwave radiation change relative to each other over the seasons. Because of high sun angles in summer, both total incoming radiation and the proportion of it absorbed are usually greater than in any other season of the year. In winter, the presence of snow increases the proportion of shortwave radiation reflected, and the incoming shortwave radiation is also less; therefore, the net shortwave radiation is far less in winter (Fig. 3.12).

THE SHORTWAVE RADIATION BALANCE

As shortwave incoming radiation passes through the atmosphere, part of it is absorbed, part of it is reflected back into space, part of it is scattered across the sky and to the ground, and part of it is directly transmitted to the surface. Clouds, which are made up of water (including ice) particles, are highly effective in reflecting and scattering solar radiation. The radiation reaching the ground—about 50 percent of the total on the average for the earth—is either reflected or absorbed, depending on the nature of the surface it encounters (Fig. 3.4).

Since no surface is a perfect reflector (that is, has a 100-percent albedo), some shortwave radiation is always being absorbed by the ground during the daylight hours; the balance between incoming and outgoing thus favors incoming and creates a positive net shortwave flux between dawn and dusk. At night incoming shortwave ceases, and since outgoing shortwave is merely reversed incoming shortwave, outgoing must also cease, creating a zero net flux. At no time is net shortwave radiation negative.

For the earth as a whole, net shortwave radiation is strongly positive, with about one of every two calories of solar radiation entering the atmosphere actually being absorbed by the earth's surface. In addition, recall that 15 to 20 percent is also absorbed by the atmosphere. From a geographic standpoint, however, regional and seasonal variations in net shortwave radiation are probably more meaningful than the average global value.

Global Variations in Solar Radiation

In the absence of cloud cover (or airborne dirt), the amount of incoming solar radiation possible at any given latitude on the earth's surface is a product of

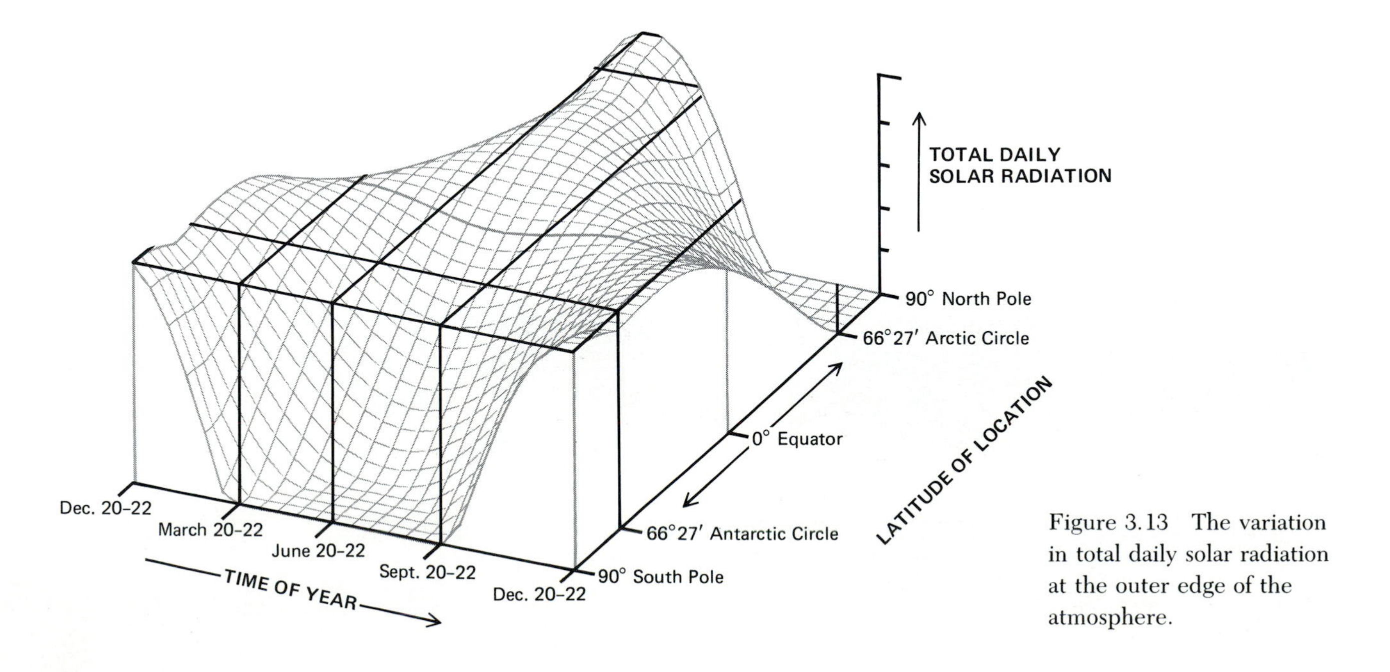

Figure 3.13 The variation in total daily solar radiation at the outer edge of the atmosphere.

sun angle and duration of daylight. Figure 3.13 shows relative distribution of incoming solar radiation by latitude and month disregarding the effects of the atmosphere. The world map in Figure 3.14 shows the average annual solar energy received at the earth's surface. In comparing the two, notice that one shows a smooth change in incoming solar radiation with latitude, while the other (Fig. 3.13) shows an uneven pattern of change between 60 degrees N latitude and 60 degrees S latitude. This difference is attributed to the various influences of the atmosphere on incoming solar radiation, especially cloud cover.

The cloud-cover effect is most evident in the equatorial and subtropical zones. The heavy cloud cover of the equatorial zone, particularly over land masses, reduces the total annual receipt of solar radiation in some areas by as much as 50 percent. In contrast, the subtropical deserts, which are cloudless for all but a few days per year, experience very little reduction in solar radiation. The influence of cloud cover (coupled with sun angle and length of day) can also be seen in the data presented in Table 3.1, which gives incoming shortwave radiation at the surface as a percentage of solar radiation received at the top of the atmosphere for eight locations in the Northern Hemisphere.

TABLE 3.1 PERCENTAGE SOLAR RADIATION REACHING THE EARTH'S SURFACE

Location	Latitude	Si at Surface[a]
San Juan, Puerto Rico	18.5°N	64%
Honolulu, Hawaii	21.3°N	64%
New Orleans, La.	29.9°N	46%
Albuquerque, N.M.	35.0°N	66%
Columbia, Mo.	40.0°N	58%
Karlsruhe, Germany	49.1°N	49%
Pavlovsk, USSR	59.7°N	38%
Fairbanks, Alaska	64.8°N	50%

Source: From H. E. Landsberg, *Physical Climatology*, 2d ed. (DuBois, Pa.: Gray Printing Co., 1960), 446 pp.

[a]Based on annual averages compiled before 1960.

The influence of cloud cover is also evident in Figure 3.15. This graph shows the change in annual av-

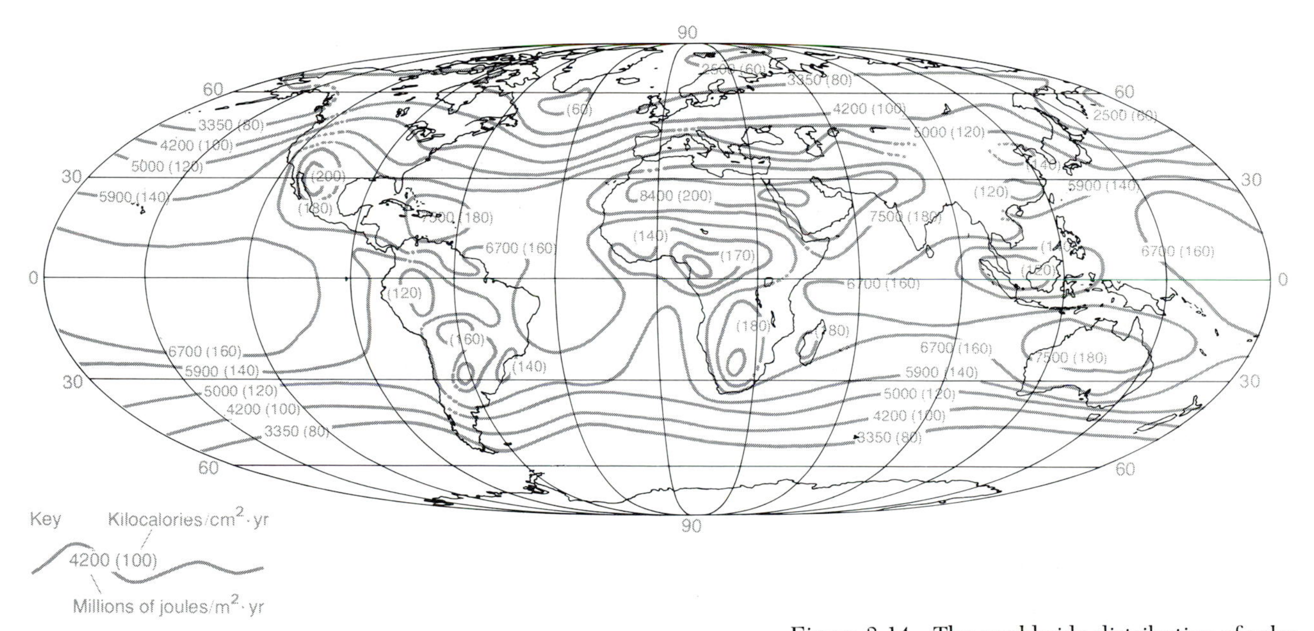

Figure 3.14 The worldwide distribution of solar radiation in millions of joules per year and kilocalories per square centimeter per year. Note that the values in the equatorial zone are lower than those in the subtropics, mainly because of differences in cloud cover.

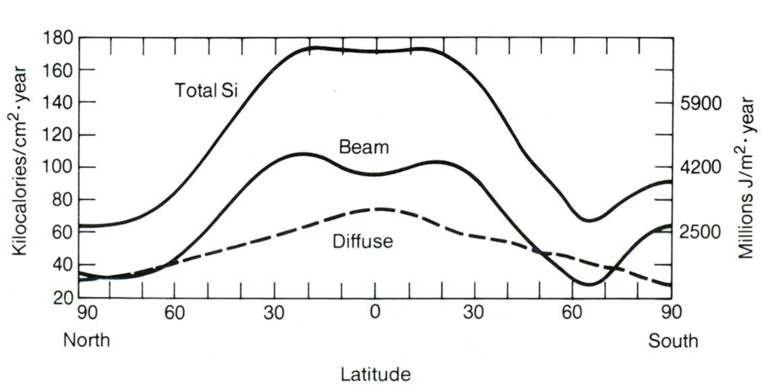

Figure 3.15 Variations in beam and diffuse solar radiation with latitude based on worldwide average values.

erage beam and diffuse radiation with latitude north and south of the equator. Notice that beam radiation values dip in the 20 degrees belt along the equator and in the belts along the Arctic and Antarctic circles. In addition, diffuse radiation is much greater than the beam along the Antarctic Circle where heavy cloud covers develop over the extensive areas of ocean.

Solar Radiation in Summary

The receipt of solar radiation by the earth's surface and lower atmosphere is the first major consideration in the study of world climates. The flux of incoming solar radiation varies with latitude, season, and atmospheric conditions. The declination of the sun migrates between the Tropic of Cancer on June 20–22 to the Tropic of Capricorn on December 20–22 in response to the earth's annual revolution about the sun. As the declination changes, the sun angles and duration of daylight change for every location of earth. Coupled with the influence of cloud cover, a global pattern of solar radiation emerges (measured at the base of the atmosphere), in which the highest values occur not in the equatorial zone but in the subtropical deserts. At ground level itself, solar radiation is influenced by the slope and albedo of the terrain it encounters. Slopes inclined toward the sun and ground with low albedo induce the highest absorption of energy per unit area of surface.

LONGWAVE RADIATION

Barring the minute contributions from endogenous (geothermal) sources, all primary heating of the earth's surface is caused by the absorption of solar radiation. In order to maintain an energy balance, the earth must in turn emit this energy back into space. Using Wien's law, we can reason that this energy must be emitted as longwave radiation, for earth surfaces are much cooler than those of the sun.

We also know that the rate at which longwave radiation is emitted from a surface is a function of its temperature. Therefore, outgoing longwave radiation reaches its highest value during the day when the surface temperature is highest (Fig. 3.16). This peak will usually lag 2 to 4 hours after the midday peak of incoming shortwave radiation, owing to the time taken to heat the soil (or surface water). Although both incoming and outgoing shortwave radiation becomes zero after sunset, outgoing longwave radiation continues all night.

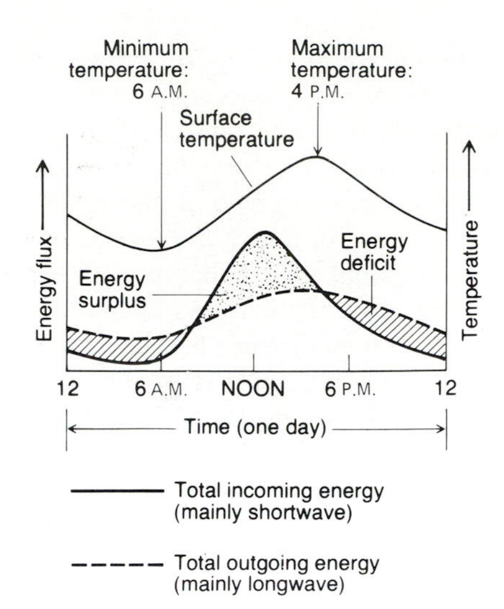

Figure 3.16 Peak period of daily outgoing longwave radiation.

Longwave Radiation in the Atmosphere

Once in the air, longwave radiation is subject to absorption by water droplets, water vapor, and carbon dioxide, and from these materials it may be reradiated back toward the surface. If the lower atmosphere is highly humid, as it often is when there is a low cloud ceiling, the longwave outflow is slow, and nighttime temperatures will typically remain relatively high. Within this atmospheric greenhouse, longwave radiation is absorbed and then reradiated back to the surface, where it is reabsorbed to produce secondary heating of the ground. Added to this reradiated energy is longwave radiation produced from the absorption of solar radiation in the atmosphere.

Not uncommonly, the energy brought to the surface over a 24-hour period by incoming longwave radiation more than doubles that brought to the earth's surface by incoming shortwave radiation. This is possible because much of the longwave radiation present in the lower atmosphere is recycled several times between atmosphere and earth before it is finally released into space. Thus, in calculating the net radiation for all waves, both *outgoing* and *incoming* longwave radiation must be considered.

With each cycle of longwave radiation, the amount of energy decreases as some of it is converted to heat and some of it is lost in skyward radiation. In addition, some longwave radiation is lost directly into space because carbon dioxide and water are simply ineffective in absorbing it. This is particularly so for wavelengths between 8 and 11 micrometers, for which the lower atmosphere presents a "window" to outgoing

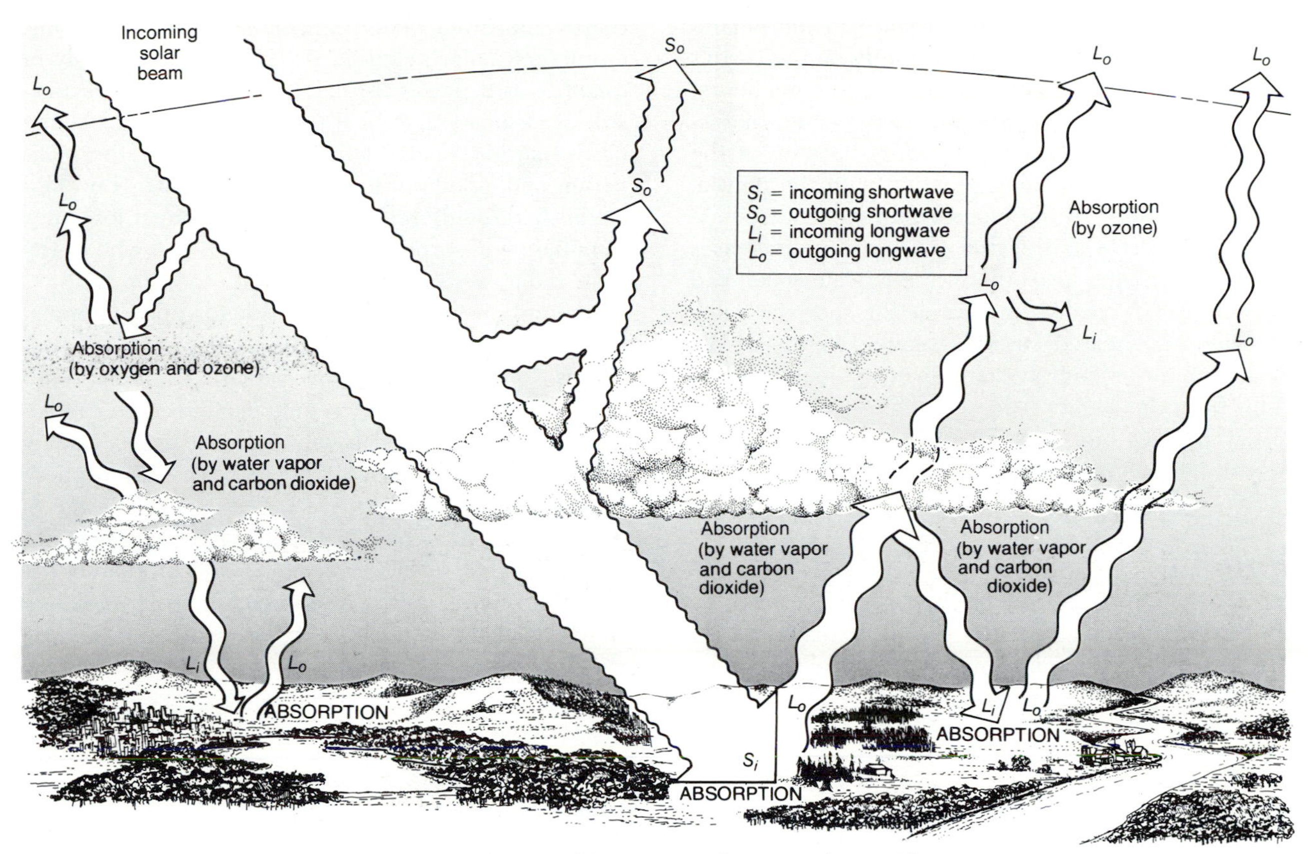

Figure 3.17 The general pattern of flow of short- and longwave radiation in the earth's atmosphere. Critical to the flow scheme are the points of radiation absorption. For incoming shortwave radiation (S_i) it is the earth's surface: this absorption is the primary source of earth surface heat. For outgoing longwave radiation (L_o), it is water vapor and carbon dioxide. These gases in turn emit longwave radiation toward the ground (L_i) and toward the sky (L_o). The L_i may be absorbed by surface materials, and the energy in turn reradiates into the atmosphere as L_o. After several iterations of this L_i/L_o cycle, the total energy received by the earth's surface as longwave radiation often exceeds that received as shortwave radiation.

longwave radiation. Higher up, however, some of this radiation (between 9 and 10 micrometers) is absorbed by ozone. The pattern of long- and shortwave radiation flow is summarized schematically in Figure 3.17.

The Net Radiation Balance

Taking both shortwave and longwave radiation into account, we can determine the total radiation balance for any geographic area over some period of time. This balance of all radiation is called the *net radiation*, and it tells us what the status of the radiation flux is for an area in terms of radiant energy gained and lost. In its simplest form, the net radiation can be thought of as the balance between net shortwave radiation and net longwave radiation. It can also be given as the balance between all incoming radiation and outgoing radiation (both longwave and shortwave in both cases):

$$\text{Net Radiation} = \text{Total Incoming Radiation} - \text{Total Outgoing Radiation}$$

If a surface is losing radiant energy, its net radiation will be negative; if it is gaining radiant energy, its net radiation will be positive.

Both negative and positive balance in the net radiation are occurring almost constantly on the earth's surface. For short periods of time, say, several hours, the net radiation rarely achieves a zero value, that is, a perfect balance. This is so mainly because of the daily rise and fall of solar radiation with the diurnal cycle of the sun. In the morning the net radiation is typically positive as the flux of shortwave increases and the landscape heats up; in the late afternoon and evening it is typically negative as the shortwave flux declines and heated surfaces emit a comparatively large amount of longwave. But over the long run, the shifts from the negative flows of night to positive flow of day balance out, and the earth thus maintains an overall balance of radiation flow.

Summary

Energy is received from the sun as a steady flow of radiation called the solar constant. Most of this radiation is shortwave, ranging from 0.15 to 3.0–4.0 micrometers. Solar radiation is reduced substantially in quantity as it passes through the atmosphere as a result of absorption, reflection, and scattering. At the earth's surface, it is further influenced by the inclination and albedo of the receiving surface. Low sun angles and highly reflective materials affect low rates of radiation absorption. For the earth as a whole, the solar radiation absorbed varies with location, season, sky conditions, topography, and landscape materials.

The energy absorbed by the earth's surface materials is reradiated as longwave radiation, and in a manner consistent with the greenhouse effect, this radiation may be absorbed by atmospheric gases, mainly water vapor and carbon dioxide. The atmosphere in turn reradiates longwave radiation itself, some of which is directed back to earth with the remainder moving higher in the atmosphere and ultimately beyond into space. In the end, incoming solar radiation must equal the outgoing longwave radiation leaving the atmosphere.

Concepts and Terms for Review

- solar constant
- electromagnetic spectrum
  - wavelength
  - ultraviolet radiation
  - visible radiation
  - infrared radiation
- shortwave and longwave radiation
- Wien's law
- atmosphere's influence on radiation
  - absorption
  - reflection and scattering
- beam and diffuse radiation
- sun angle and incidence radiation
- sun declination and seasons
- global variations in sun angle
- length of day
  - circle of illumination
- albedo
- shortwave radiation balance
- global variations in solar radiation
  - cloud cover influences
- longwave radiation from earth
  - absorption and reradiation
- greenhouse effect
- net radiation of both shortwave and longwave

Sources and References

Barry, R.G., and Chorley, R.J. (1982) *Atmosphere, Weather, and Climate,* 4th ed. New York: Methuen.

Cartledge, O. (1973) "Solar Radiation and Climate in a Subtropical Region." *Nature,* Vol. 242, pp. 11–12.

Climate Research Board. (1979) *Carbon Dioxide and Climate: A Scientific Assessment.* Washington, D.C. National Academy of Sciences.

Coulson, K.L. (1975) *Solar and Terrestrial Radiation—Methods and Measurements.* New York: Academic Press.

Goody, R. (1964) *Atmospheric Radiation.* London: Clarendon Press.

Liou, K.-N. (1980) *An Introduction to Atmospheric Radiation.* New York: Academic Press.

Sellers, William D. (1965) *Physical Climatology.* Chicago: University of Chicago Press. A comprehensive description of energy-balance principles and methods of calculating components.

White, O.R., ed. (1977) *The Solar Output and Its Variation.* Colorado: Colorado Associated University Press.

Chapter 4

The Heating of Land, Water, and Air

The topic of this chapter is heat, both the kind that can be measured with a thermometer and the kind that cannot. We know that all materials do not absorb radiation equally; likewise, all materials do not heat equally when they absorb energy. How does heat move in earth materials? Can it be stored in the landscape? Is heat readily converted into other energy forms? How deep does heat penetrate in the soil? How is it redistributed in air and water?

The modern landscape is made up of natural and human phenomena woven together into all sorts of complex relationships. To understand these phenomena and their interrelationships, it is helpful to simplify and streamline the way we think about them. One device that helps in this process is the scientific model. Models are a means of structuring our thought processes. They are a conceptual architecture of sorts, which enables one to identify critical relationships, measure change, and test questions.

Of the many types of models used in science, the analogue model is one of the most popular. We use analogies in everyday conversation, and the basic logic is the same: Analogue models borrow the structure of a known condition, situation, or system to help us think through, analyze, and explain a problem or phenomenon under study. The camera is an analogue model of the human eye. The energy balance of a warm-blooded animal is an analogue model of the earth's energy balance.

As in most analogues, there are differences between the model and the real world. In the case of the energy balance, one of the most important differences is that the principal source of body energy is endogenous heat, whereas for earth, it is exogenous radiation. But many principles and processes are the same; in particular, the skin of both bodies is a critical boundary across which huge amounts of energy are constantly flowing. Furthermore, in the thin layer of air that envelops the skin, different climates are created from the different combinations of heat, moisture, and radiation flows. In fact, our sense of physical comfort is a good measure of the status of that climate. When we feel cold, the heat loss is too great and we turn to various ways of bringing the thermal climate to a warmer balance. This may entail stepping into sunshine to increase heat production on the skin or adding clothing to reduce the rate of heat loss. How we modify body climate helps us to understand some aspects of earth climate and how it is modified by vegetation, water, land use features, and so on.

Our objective in this chapter will be to complete our examination of the energy balance of the earth's surface by adding heat flows to the radiation flows we discussed in the last chapter. Recall that radiation (both shortwave and longwave) is virtually the sole source of energy for heat on or near the earth's surface. This heat is generated when radiation is absorbed by air, water, and soil. As these materials heat up, they in turn generate flows of radiation and heat.

THE ESSENTIAL ENERGY BALANCE

Thrust your bare arm into brilliant sunshine, and three things happen: (1) the surface temperature of your arm rises, (2) the outflow of longwave radiation from your skin increases, and (3) heat is transferred from the skin to the air above it and the tissue below it. The same things happen as the sun rises on a lawn or soil surface in the morning. Very simply what we are observing is the conservation of energy principle in action: a large quantity of solar radiant energy comes to the surface and is absorbed, a portion of it is converted into longwave radiation which leaves the surface, and the remainder is transformed into heat energy.

We know intuitively that heating takes place because the air and soil warm up as the day progresses. Where does the heat go? It goes into storage in soil, air, and water. Why do these materials not grow warmer and warmer over days, months, and years? Because the net radiation reverses (that is, becomes negative) at night or during the winter, and heat is transferred out of the soil, air, and water, converted into longwave radiation, and lost into the atmosphere.

The Basic Energy-Balance Model

There are two basic versions of the energy-balance model: (1) positive net radiation, with heat flows from the surface *into* soil, air, and water; and (2) negative net radiation, with heat flows *from* soil, air, and water to the surface. Using the ground (or water) surface as our plane of reference, there are just three heat flows to consider. We will define them briefly here and discuss them in more detail later.

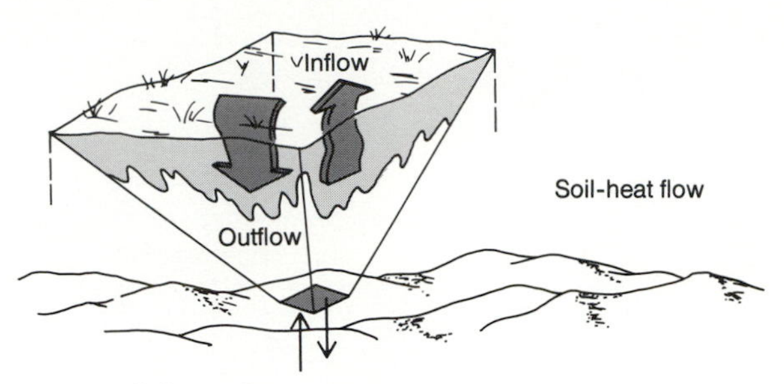

1. *Soil-heat flux:* This entails heat transfer (either downward or upward depending on the net radiation) by conduction between the surface and the soil material (or water) beneath it. For water, the heat transfer from the surface into the underlying water is mainly by convection because water has the capacity for mixing motion.

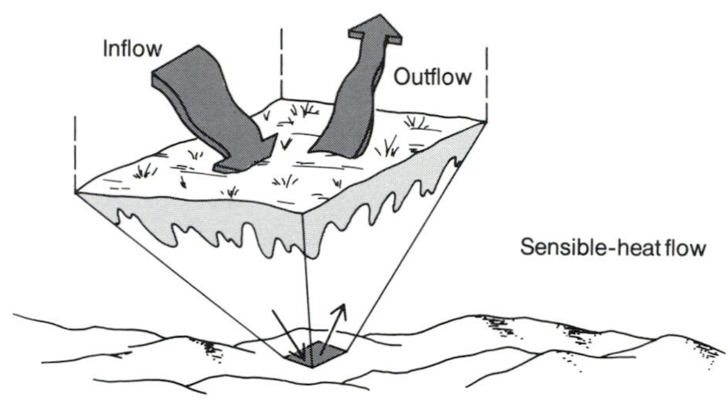

2. *Sensible-heat flux:* This involves heat exchange between the surface (ground or water) and the overlying air by conduction and convection. *Sensible heat* is the sort of heat that is measured with a standard thermometer, and is often referred to as the *heat of dry air*. As with the soil-heat flux, the net radiation is the principle determinant of the direction of sensible-heat flux.

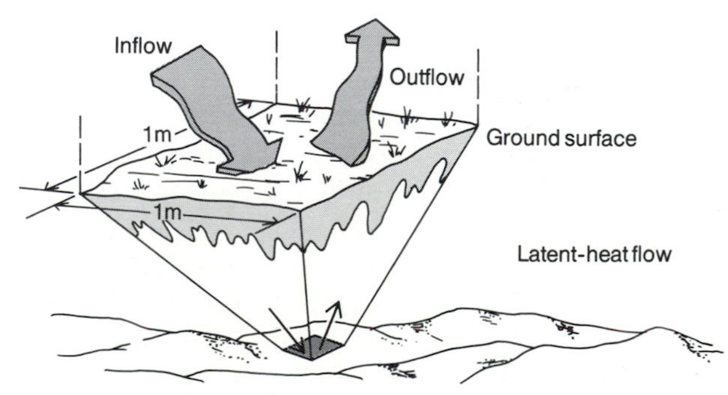

3. *Latent-heat flux:* This is also a heat exchange between the surface and the overlying atmosphere, but moisture is the vehicle of transfer instead of the molecules of dry air. The vapor-

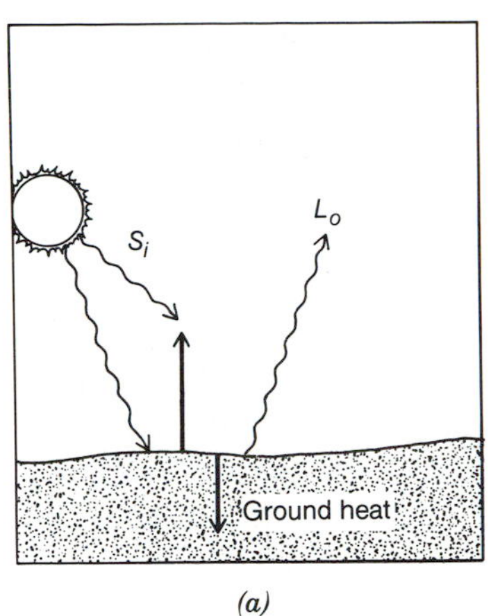

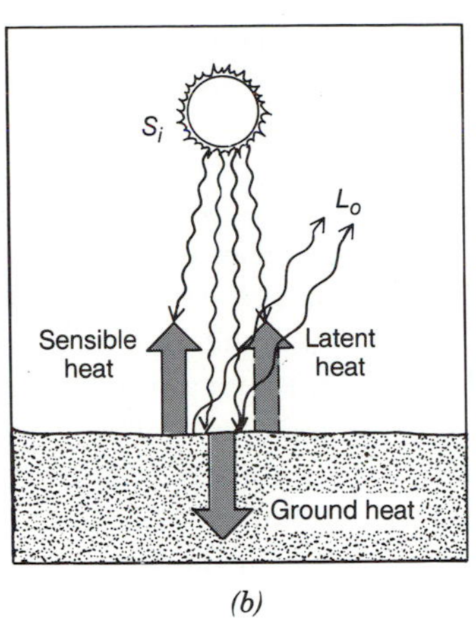

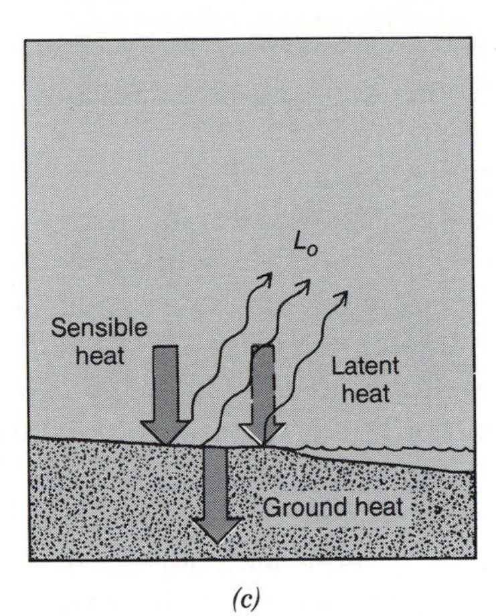

Figure 4.1 Situations (*a*) and (*b*) depict a typical summer daytime condition: Surface heating from incoming radiation is sufficiently strong to produce both upward and downward outflows of heat. Situation (*c*) depicts the nighttime sequel: Longwave radiation outflow produces surface cooling, thereby inducing inflows of sensible heat, latent heat, and soil heat. Latent-heat flow is evidenced by the formation of dew on the ground.

ization of liquid water or ice requires energy in the form of heat to drive the process. Thus, when vaporization takes place heat is taken up and the surface is cooled, just as your skin is cooled when perspiration evaporates. The reverse happens when moisture condenses or freezes on a cool surface; heat is brought to the surface. As a rule, when water leaves the surface for the air, the net radiation is positive (i.e., heat is available), and when water leaves the air for the surface, the net radiation is negative.

More will be said later in this chapter about latent heat, including the amounts of energy involved when water changes states.

Many important questions in physical geography concern processes driven by thermal and radiant energy at the earth's surface: the evaporation of water, the heating of air, the freezing of soil moisture, and the photosynthesis and growth of plants. In order to address these questions scientifically, the individual flows of energy must be measured and computed. Any computation of the energy balance for a soil or water surface must begin by defining (1) the unit size of the measurement area (either one square meter or one square centimeter), (2) all the essential fluxes of heat (soil, sensible, and latent) and radiation (shortwave and longwave), and (3) each flow as positive or negative depending on whether it is bringing energy to the surface or taking it away. A general energy-balance equation must include the net radiation value (NR) plus or minus sensible-heat flux (S), soil-heat flux (G), and latent-heat flux (L). Together, these four factors must equal zero for any moment of time:

$$\mathrm{NR} \pm \mathrm{S} \pm \mathrm{G} \pm \mathrm{L} = 0$$

In all cases *any flow away* from the surface (i.e., upward into the atmosphere or downward into the soil or water) is negative, and *any flow toward* the surface (i.e., downward from the atmosphere and upward from the soil) is positive. It is therefore important to remember that the land (or water) surface is always the plane of reference for describing energy flows.

One of the clearest illustrations of the energy balance is a daytime/nighttime situation for a small parcel of land. Figure 4.1*a* and *b* show a positive net radiation, typical of the hours when the sun is rising in the sky, when energy is building up on the surface and producing flows of heat into the soil and air.

In the afternoon and evening as the receipt of solar radiation declines, the balance of radiation flow shifts from solar flowing to the surface to longwave flowing from the surface. The outflow of longwave radiation produces surface cooling, which results in a vertical distribution of soil and air temperature near the surface that is basically the reverse of the situation of midday. Thus, the heat stored earlier in the day in the soil and air flows back to the surface, where it feeds the flow of outgoing longwave radiation (Fig. 4.1*c*).

These two conditions (*a* and *b* in Fig. 4.1) deserve some thought, for they are the very heart of the energy-balance concept. Do not be surprised if you find situation *b* a little confusing because the cool surface is producing an outflow of longwave radiation. We have to bear in mind that even if, as is the case late at night, the ground surface is the coolest point in the whole scene (and is therefore the recipient of heat flows from soil and air), it nonetheless generates longwave radiation. This is important, for the model is illogical unless we understand that the surface will always produce an outflow of longwave radiation *as long as its temperature is above absolute zero.* There is no surface on earth whose temperature is even close to absolute zero; even the coldest frozen ground radiates longwave radiation. For example, at a temperature of −30°C, it is still +243° Kelvin (see Table 2.1). Therefore, the earth is constantly emitting longwave from its entire surface.

These basic principles of net radiation and heat flow apply throughout the world to land, seas, ice, and even manufactured surfaces. We employ them in dealing with a variety of everyday problems, including maintenance of body climate and control of climate inside buildings, automobiles, greenhouses, and even in open agricultural fields where the concern is often with killing frosts and the evaporation of soil moisture from heated ground.

HEAT BEHAVIOR IN EARTH MATERIALS

To this point we have not said much about the mechanisms and controls of heat movement in soil, air, and water. Although it may seem to be a narrow focus of concern for physical geographers, with their interest in the broad-scale patterns and process of the earth's surface, such detailed attention is absolutely essential if we are to understand how the different climates and conditions of life, especially those close to the ground, originate on the planet.

Thermal Property Concepts

What controls the rate at which heat flows into or through earth materials? There are just two factors to consider: (1) thermal gradient, and (2) thermal conductivity.

Thermal Gradient In order for energy in any form to be transferred under any circumstances, whether it be heat from the stove into your morning eggs or water rolling down a hillslope, there must be an *energy gradient.* This is defined as the difference in the level of energy from one point to another in a substance or over some continuous space. The thermal gradient is a measure of the change in heat energy through a substance. It is defined on the basis of temperature differences; in soil, for example, it is based on the temperature at or near the surface and a temperature at some depth divided by the distance between the two readings. The thermal gradient is not a thermal property of a substance as such, but rather a thermal condition—that is, a state of affairs that varies with circumstances. (For example, the thermal gradient of a roof differs depending, among other things, on whether or not the sun is shining on it and warming it.) The thermal gradient tells us two important things: (1) the direction of heat flow, and (2)

TABLE 4.1 THERMAL PROPERTIES OF SOME COMMON EARTH MATERIALS

Substance	Thermal Conductivity[a]	Volumetric Heat Capacity[b]
Air		
Still (at 10°C)	0.025	0.0012
Turbulent	3,500–35,000	0.0012
Water		
Still (at 4°C)	0.60	4.18
Stirred	350.00 (approx.)	4.18
Ice (at −10°C)	2.24	1.93
Snow (fresh)	0.08	0.21
Sand (quartz)		
Dry	0.25	0.9
15% moisture	2.0	1.7
40% moisture	2.4	2.7
Clay (nonorganic)		
Dry	0.25	1.1
15% moisture	1.3	1.6
40% moisture	1.8	3.0
Organic soil		
Dry	0.02	0.2
15% moisture	0.04	0.5
40% moisture	0.21	2.1
Asphalt	0.8–1.1	1.5
Concrete	0.9–1.3	1.6

[a]Heat flux through a column 1 m^2 in joules per second per linear meter when the temperature gradient is 1°K per meter.
[b]Millions of joules needed to raise 1 m^2 of a substance 1°K.

the relative rate of heat transfer in a particular substance: the steeper the gradient the higher the rate of flux.

Thermal Conductivity By contrast, *thermal conductivity*, is a thermal property specific to every substance. It is convenient to think of this property as the resistance a substance poses to heat transfer through it. To conserve heat in a house you would want walls with high resistance to heat transfer: that would translate into walls with *low thermal* conductivity. Conductivity varies substantially among landscape materials, depending on two factors: (1) composition and (2) capacity for mixing motion. Among still (nonmoving) materials, ice, sand, asphalt, concrete, and water have relatively high conductivities, whereas dry organic soil, air, and snow have low conductivities (Table 4.1). When motion is added to air (creating wind) and to water (creating waves and currents), conductivity increases dramatically because the principal mechanism of heat transfer is changed from conduction to convection.

A second important question about the thermal properties of earth materials is, how great an increase in temperature will be produced when heat is added to a substance?

Volumetric Heat Capacity For this answer we must examine a property called *volumetric heat capacity*, defined by the amount of heat that is needed to raise the temperature of one cubic meter of a substance by 1 degree Kelvin (or 1°C). (Volumetric heat capacity is similar to *specific heat*, which is also widely used in science. Specific heat is a measure of the amount of heat needed to raise one gram of water 1°C.)

A substance with a low volumetric heat capacity, such as air, gains a high temperature with the addition of a moderate amount of heat, whereas a substance with a high volumetric heat capacity, such as water, gains little temperature with the addition of the same amount of heat (Table 4.1). A cubic meter of water needs 4.18 million joules of heat to raise its temperature 1°K; the same volume of air needs only 1200 joules of heat to raise its temperature 1°K. Thus, a given volume of water at any temperature contains far more heat than an equal volume of air at the same temperature. Volumetric heat capacity is important in the development of our thinking about physical geography because it is necessary to understand that the temperatures we measure in the different materials of the landscape represent significantly different levels of heat, that is, energy content.

Heating of Land, Water, and Air

Consider the heating of equal-sized blocks of soil and water. Assume that the starting temperatures of the substances are the same and that equal amounts of

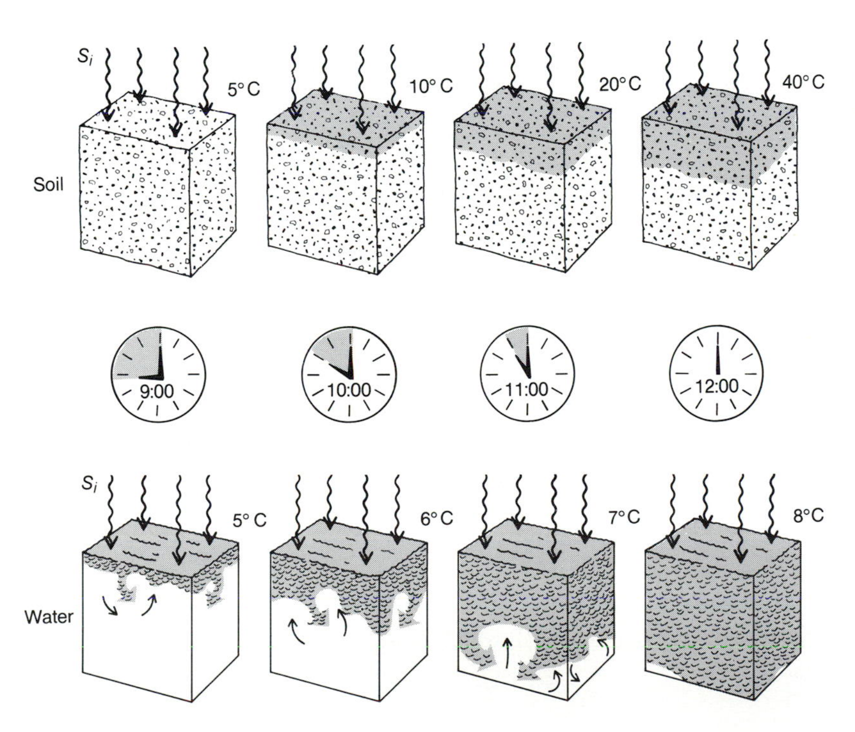

Figure 4.2 The relative rates of heat flow into water and soil. Note that as time passes, heat is diffused through a much larger volume of water than of soil because of the mixing motion of water. Also note that for comparable inputs of radiation (S_i), the temperature of water is not raised as high as that of soil.

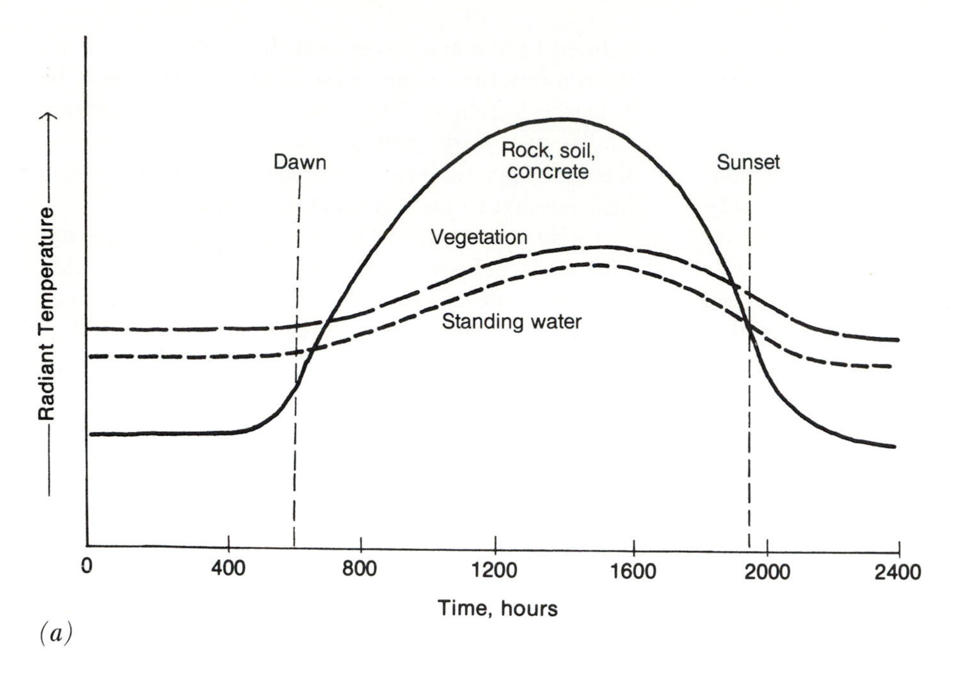

(a)

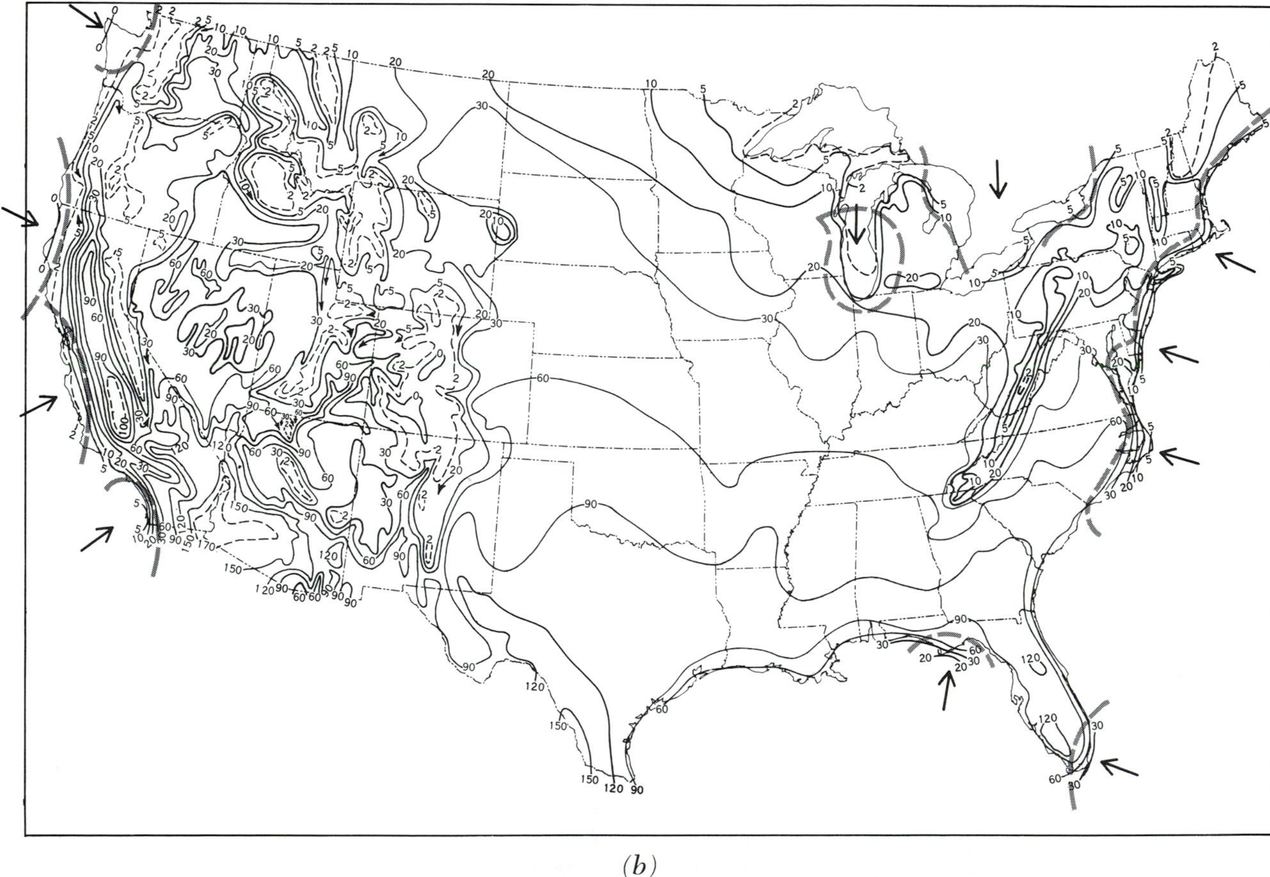

(b)

Figure 4.3 (a) Relative surface temperatures of dry, water, and vegetated surfaces on a summer afternoon as revealed on a thermal infrared image. (b) locations in the United States strongly influenced by the thermal effects of water revealed by the distribution of the annual number of days of 90°F or more in coastal areas.

radiation are absorbed over several hours by the upper surface of each. Because of the difference in heat capacities, the surface temperature of the soil will rise faster and reach a level about four times higher than that of the water (Fig. 4.2). *If* the water is held *motionless*, the conduction of the heat into the underlying mass will be about two times greater in the water than in dry soil. However, motionlessness is not characteristic of the restless seas; hence, we will use the conductivity value for stirred water, which is approximately 1400 times that of dry soil. The result is clear: the temperature of the soil rises spectacularly, but little heat is conducted beyond the surface layer, while the temperature of water rises only slightly and a large quantity of heat is conducted through a large volume of water.

These differences are very important to the heating of the atmosphere. Because of its low volumetric heat capacity, air responds readily to the surface temperature of whatever material it happens to be over. The air over land, therefore, reaches a much higher summer temperature than the air over water does (Fig. 4.3*a*). In addition, the heat in water has been conducted through more mass than the heat in land, and thus it takes much longer for it to be released to the atmosphere. Therefore, we find that although water does not induce high air temperatures, it does tend to induce a steadiness (that is, low variations) in air temperatures from day to night and from winter to summer, which is in sharp contrast to the pattern of temperature (the thermal regime) over land which tends to fluctuate radically on a day/night and seasonal basis. One influence of water on summer high temperatures is illustrated in Figure 4.3*b*.

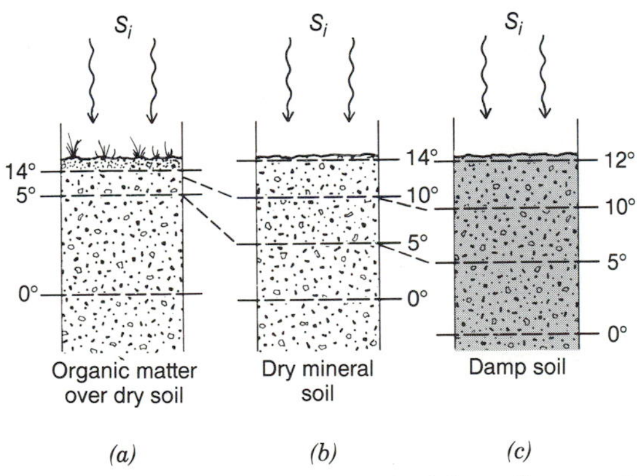

Figure 4.4 Schematic portrayal of the influence of organic matter and soil moisture on the thermal conductivity of soil. Organic matter has high resistance to heat flow (that is, poor conductivity), dry mineral soil has less resistance, and damp soil has the least resistance. The depth of penetration of a specified temperature (as opposed to transmission of a quantity of heat) is termed *thermal diffusivity*.

THERMAL CONDITIONS OF THE SOIL

Heat flows in and out of the soil mainly by conduction from particle to particle, but the rate may vary considerably with particle composition and soil-moisture content. Organic matter reduces the thermal conductivity of the ground; generally, organic soils are about ten times more resistant to heat flow than mineral soils are. In most soils, organic matter forms a thin surface layer, called *topsoil*, which helps to insulate the underlying mineral soils from thermal extremes on the surface (Fig. 4.4*a*).

Water increases the thermal conductivity of soil because when water is added to soil, it takes the place of air between the particles. Soil air has very low conductivity (around 0.025) because in the minute spaces between the soil particles, any mixing motion is extremely limited (Fig. 4.4*b*). Therefore, when water displaces air the thermal conductivity in the spaces between the mineral particles jumps from 0.025 to 0.60, a twenty-four-fold increase. This produces an overall increase in the thermal conductivity of the soil—taking both water and mineral particles into account—as great as ten times for many soils from a dry to a 40 percent moisture condition (Fig. 4.4*c*).

Seasonal and Daily Variations in Soil Heat

The *net annual* soil-heat flow in and out of the soil at any place must, on the average, equal zero; otherwise, the soil would get hotter (or colder) each year. Annual and longer term imbalances do occur, but they are difficult to detect because they usually involve long periods of time, and for most places we do not have many years of records on soil temperatures. Imbalances over shorter periods have been documented, however, and they correlate with daily temperature changes, weekly weather changes, and seasonal changes in the availability of heat at the earth's surface (Fig. 4.5).

The seasonal changes give rise to distinctive gradients, or *temperature profiles*, in the upper two meters or so of the soil. A representative set of soil-temperature profiles for a midlatitude location is given in Figure 4.6. Similar profiles also appear in miniature

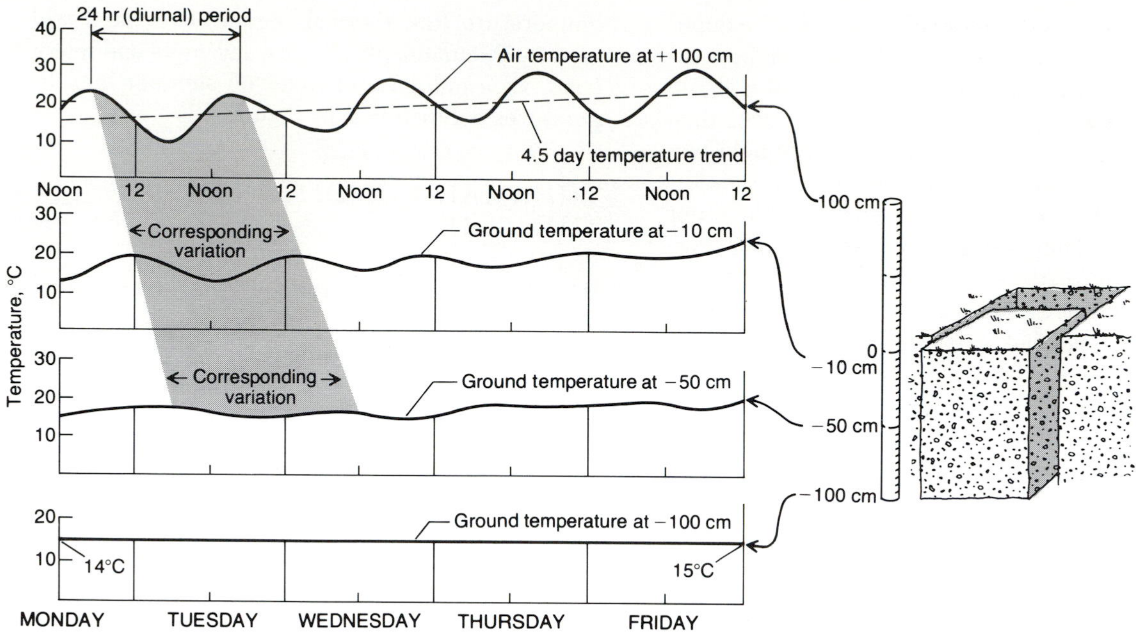

Figure 4.5 Air temperature variation at 100 cm above the ground over 4.5 days (top graph) and the corresponding ground temperatures at depths of 10 cm, 50 cm, and 100 cm (lower graphs). Note that the warming trend in the upper graph also appears in the ground-temperature graphs but is less pronounced at lower levels. Also note that the diurnal variation in air temperature appears at 10 cm and 50 cm, but the variance is smaller and is offset by as much as 12 hours at 50 cm. Data are from a grass-covered site in the Midwest.

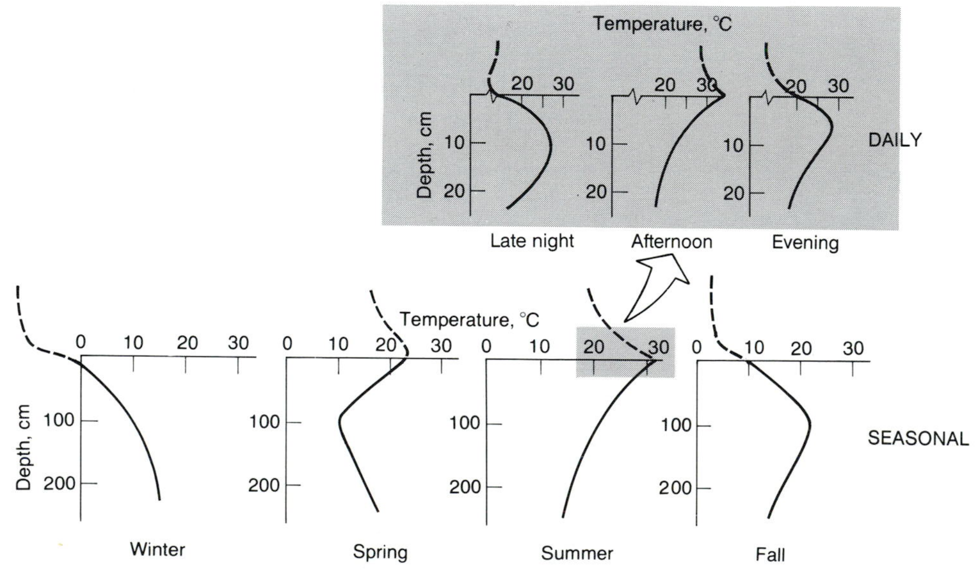

Figure 4.6 Soil-temperature profiles for a typical winter, spring, summer, and fall day. In spring and fall, the heat flux is both upward and downward, indicating that the subsoil is warmer than the layers above and below it. In winter, the flux is upward, and in summer, it is usually downward.

on a *daily* basis in the upper 10 to 20 cm of soil and are superimposed on the seasonal pattern. That is, the soil-temperature profiles near the surface typically flop back and forth from day to night, as the enlargement in the Figure 4.6 inset illustrates. The surface is coolest in early morning and warmest in early afternoon.

Because heat transfer into and from the soil takes time, the coolest and warmest temperatures usually lag behind the maximum and minimum air temperatures. The deeper the level, the longer the temperature lag; at a depth of 3 m, it may be as long as two months. Thus, in midlatitude locations in the Northern Hemisphere the warmest soil temperature at this depth is not often reached until the end of September. Daily temperature fluctuations do not occur much below 20 cm, called the *diurnal damping depth*, and in most places seasonal change cannot be detected much below 3 meters.

Air temperature changes little in the tropics over a year, and for this reason soil heat varies little there from season to season. Average monthly air temperatures in the tropics vary less over the year than they do from day to night. Consequently, the soil temperatures vary more diurnally than they do seasonally.

Permafrost

Perhaps nowhere is the seasonal soil-heat flow more vividly expressed than in the permafrost regions of the arctic and subarctic. Permafrost is ground in such a low heat state that it is continuously frozen, but usually only at depths greater than 1 to 3 meters. The upper zone, called the *active layer*, freezes and thaws seasonally, just as do the upper 10 cm to 20 cm of midlatitude soils. The thickness of the active layer varies with the thermal conductivity of the different soil materials (Fig. 4.7).

As heat flows out of the soil in the fall, frost forms at the surface. As cold weather progresses, the frost expands down through the active layer toward the permafrost in the subsoil. In spring and summer, the trend reverses: the frost recedes back toward the surface as heat flows down from the surface. In fall, the thaw line (marked by a temperature close to 0°C) progresses downward, and the temperature profile steepens, reflecting a growing departure of soil-surface temperatures from that of the permafrost at several degrees or more below freezing.

With the penetration of new frost from the surface in fall, the temperature profile straightens somewhat;

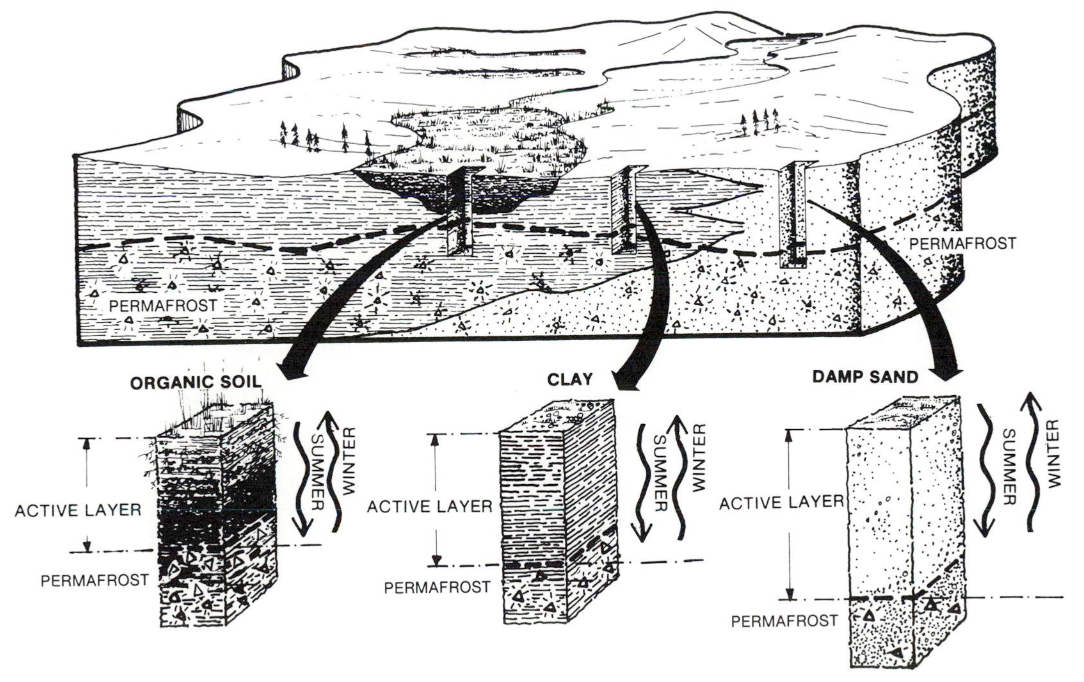

Figure 4.7 The thickness of the active layer in permafrost regions can reflect the influence of soil composition and moisture content on seasonal heat flow. The directions and depths of seasonal heat flows are shown in the soil sections.

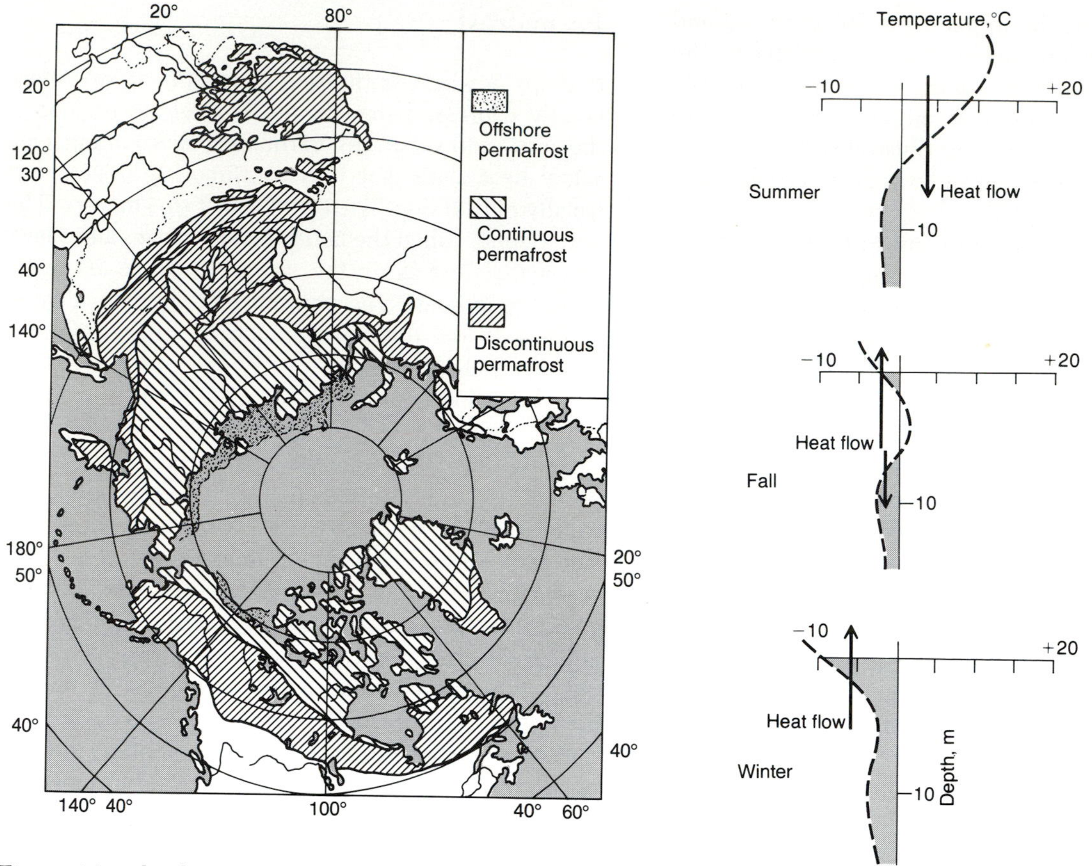

Figure 4.8 The distribution of permafrost in the Northern Hemisphere, with seasonal variations in soil heat in a permafrost environment shown in the diagrams and graphs. Summer heat flow is downward, producing thawing of the active layer. In fall the surface freezes, and the active layer loses heat to the frozen ground above and below it. By winter the active layer is frozen out (except for some pockets), and the surface is so cold that the heat flow, though slight, is upward from the permafrost.

in ensuing months of winter, when surface temperatures fall far below 0°C, the gradient is fully reversed from that of summer (Fig. 4.8). Paradoxically, the permafrost is the primary source of soil heat in winter. Upward gradients as great as 4° to 5°C per meter confirm this fact.

Permafrost is generally found in regions where the mean annual surface temperature is below 0°C, although this can vary with local soil, heat and radiation conditions. Three zones of permafrost are recognized: *continuous* in very cold regions where frozen ground blankets the terrain without interruption; *discontinuous* in less cold (usually subarctic) regions where the coverage is somewhat patchy; and *offshore* where frozen ground extends under the sea, a relic of the last glaciations when sea levels were much lower (Fig. 4.8). Today permafrost occupies about 25 percent of the world's land area, most of which lies in Eurasia, North America, Antarctica, and high mountain areas such as Tibet.

HEAT FLOW IN THE ATMOSPHERE

Energy Transfer as Sensible and Latent Heat

Two forms of heat flow, or transfer, occur between the ground and the overlying atmosphere: sensible-heat transfer and latent-heat transfer. *Sensible-heat* transfer takes place whenever the temperature of a soil or water surface is different from that of the overlying air. The transfer process is carried out in the thin layer of air immediately over the ground by conduction between the lowermost air molecules and the soil or water surface. Above a height of several centimeters, transfer also takes place by convection, and at much greater heights it is overwhelmingly by convection. The direction of transfer may be up or down depending on the thermal gradient.

Latent-heat transfer takes place when water changes phase, that is, when water goes from one physical state to another. Water can exist in three phases: solid (ice), liquid (water), or gas (water vapor). A change of phase always produces an exchange of energy. A change from solid to liquid (*melting* or *thawing*) requires an energy input of 0.334×10^6 joules per kilogram of water, or 80 calories per gram of water. (The scientific notation 10^6 stands for 1,000,000; therefore, $0.334 \times 10^6 = 334{,}000$.) A phase change from liquid to vapor (evaporation) at a temperature of 0°C requires 2.5×10^6 joules per kilogram, or 597 calories per gram. At higher temperatures, the value is lower; for example, at 100°C, it is 2.3×10^6 J/kg (540 cal/g).

Change directly from ice to vapor (*sublimation*) requires 2.834×10^6 J/kg (676 cal/g), the sum of the melting and vaporization values. An example of sublimation is provided by clothes hung outdoors to dry in winter; the clothes first freeze and then dry as the ice vaporizes. The reverse changes—vapor to liquid (*condensation*), liquid to solid (*freezing* or *fusion*), or vapor to solid (also called *sublimation*)—result in the release of equivalent quantities of heat (Table 4.2).

The air lying over water bodies often becomes heavily loaded with water vapor, and when the air cools a little (as at night) the vapor condenses (to form fog) and the latent heat is released into the air. If the air warms later in the day, the fog droplets evaporate and sensible heat is taken from the air to drive the vaporization.

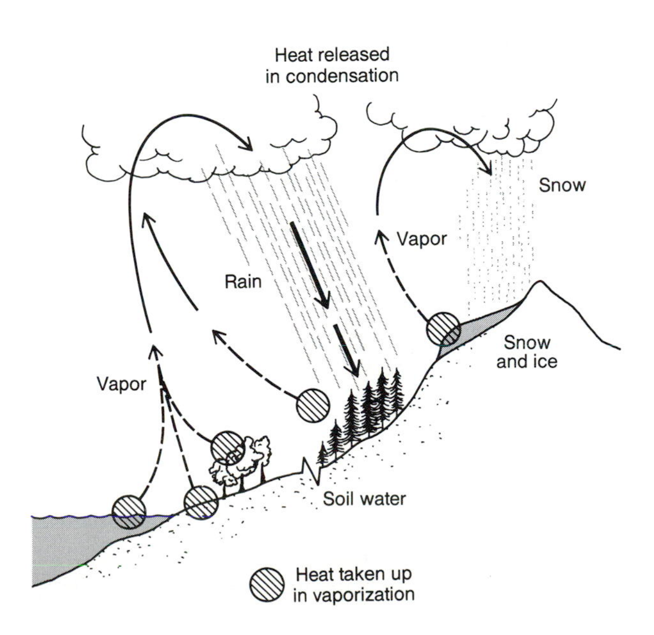

When liquid water or ice on the earth's surface vaporizes and the vapor passes into the atmosphere, energy (heat) is taken off the surface. Conversely, when vapor in the air near the ground condenses on the surface as dew or frost, energy is transferred to the ground. Any latent-heat transfer between the ground and the air above requires an energy gradient, but with latent heat, it is defined by a humidity gradient rather than a thermal gradient.

Humidity is a measure of the vapor content of air. It is measured in a variety of ways, two of which are discussed in Chapter 6. The important facts to note here are that the capacity of air to hold vapor varies directly with its temperature and that this capacity increases rapidly as temperature increases (Fig. 4.9). For example, a temperature increase from 10°C to 15°C increases the holding capacity by only about 2 grams per cubic meter of air (from 8 to 10 grams), whereas a temperature increase from 35°C to 40°C increases holding capacity by about 13 grams per cubic meter of air (from 37 grams to 50 grams). Colder air is therefore generally drier: this explains why in winter when cold air passes over open water and is

TABLE 4.2 LATENT-HEAT TRANSFERS IN PHASE CHANGES OF WATER

Phase Change	Process	Direction of Transfer	Rate
Vapor → Liquid	Condensation	Heat given up ↑	2,500,000 J/kilogram (597 cal/gram)
Liquid → Vapor	Evaporation	Heat taken up ↓	2,500,000 J/kilogram (597 cal/gram)
Vapor → Solid	Sublimation	Heat given up ↑	2,834,000 J/kilogram (676 cal/gram)
Solid → Vapor	Sublimation	Heat taken up ↓	2,834,000 J/kilogram (676 cal/gram)
Solid → Liquid	Melting	Heat taken up ↓	334,000 J/kilogram (80 cal/gram)
Liquid → Solid	Freezing	Heat given up ↑	334,000 J/kilogram (80 cal/gram)

warmed a little, evaporation rates from the water are extremely high.

How effective is latent-heat flux as a heat exchange mechanism? We have all seen photographs of big-city dwellers hosing down sidewalks, streets, and building fronts during a hot afternoon. Just fun for the kids? Not at all; the practice utilizes a sound energy-balance principle. Because latent-heat flux into the atmosphere is powered by surface heat, the more water that can be evaporated, the greater the surface cooling. Considering that the urban landscape is poor in natural sources of moisture, spraying hot surfaces is about the only means of cooling them quickly—unless it rains, of course. Note 4 takes a brief look at health problems related to heat extremes in urban centers.

This is not the case with most plant-covered surfaces, however. Plants draw moisture from the soil and release it to the atmosphere through *transpiration,* that is, vaporization of water through tiny openings (stomata) in the foliage (leaves). Moisture is also released from the soil by direct evaporation. Latent-heat flux from such landscapes increases as summer temperatures rise, with the largest contributions coming from plants—as long as the soil moisture lasts. In eastern United States, latent-heat flux from such surfaces reaches $60{,}000 \times 10^6$ joules ($15{,}000 \times 10^6$ calories or 15 billion calories) per acre on many summer days. It is easy to see that this is a highly efficient mechanism for disposing of surface (ground) heat and sensible heat in the air near the ground.

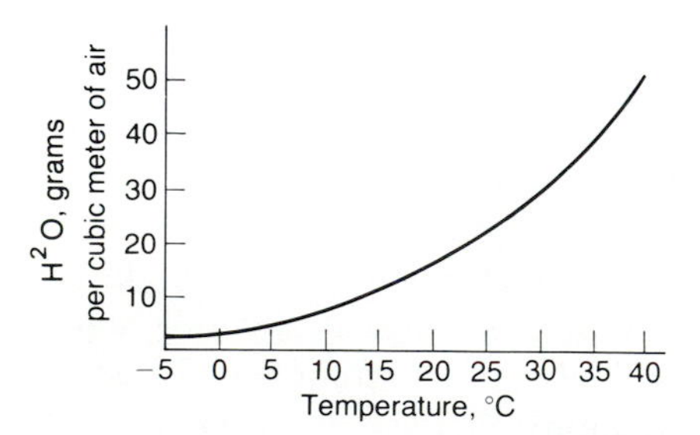

Figure 4.9 The maximum amount of water vapor that can be held at a given temperature. Note that the vapor-holding capacity increases at an increasing rate with temperature.

The Role of Wind in Energy Transfer

Wind increases both sensible and latent-heat flux from the ground to the atmosphere. You sense this phenomenon on windy days in winter when you feel colder at higher wind velocities because heat flow from your skin is increased. We call this the *wind chill.* Wind, however, never reaches your skin or the ground directly because a film of calm air, called the *laminar sublayer*, blankets all surfaces, soil, vegetation, buildings, and creatures alike. (Tiny bugs, by the way, live inside this layer, which is why they are seldom blown away by the wind.)

The laminar sublayer changes in thickness with wind speed and surface roughness. When wind is light, the sublayer is thicker than when it is strong. If the surface is rough, for example, in the case of a hair-covered leaf, the sublayer is thicker because airflow near the surface is slowed down owing to friction.

We are interested in the laminar sublayer because it affects heat flow. Both sensible and latent heat must cross this layer by conduction before entering the zone of convective transfer where things really move. Thus, when the wind is light and the sublayer is thick, heat transfer is exceedingly slow; as a result, heat and vapor build up over the surface. This buildup, in turn, reduces the energy gradients at the interface between ground and air, and transfer declines correspondingly. On a warm, humid day when this buildup occurs on your skin you usually feel very uncomfortable, an air condition described as "close." Figure 4.10 indicates how heat flux from the soil declines as the temperature and humidity gradients grow smaller in the tiny (micro-) layer over the surface.

If stronger wind is brought to the surface, a large

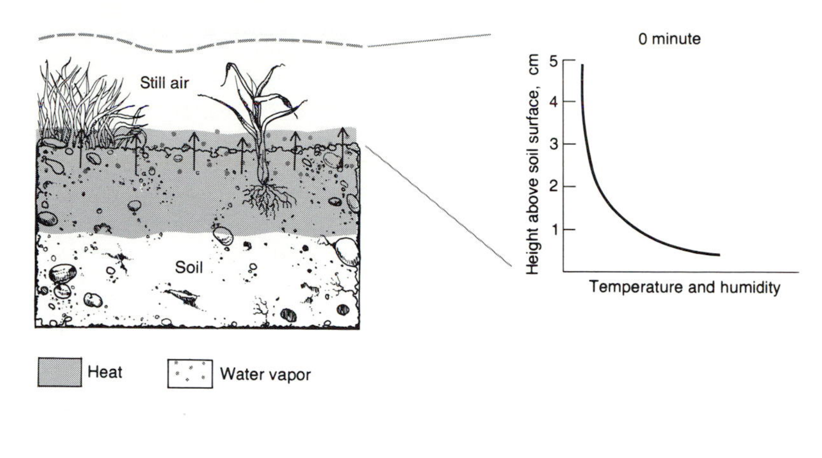

Figure 4.10 The flow of sensible and latent heat into the laminar boundary layer of still air. Initially (0 minute), the energy gradient between the soil surface and air is steep, and the rate of heat flow is relatively high in the first 10 minutes, after which the rate is very slow, owing to the weakness of the gradient over the surface. At 50 minutes, the heat and vapor content of the air has increased only slightly.

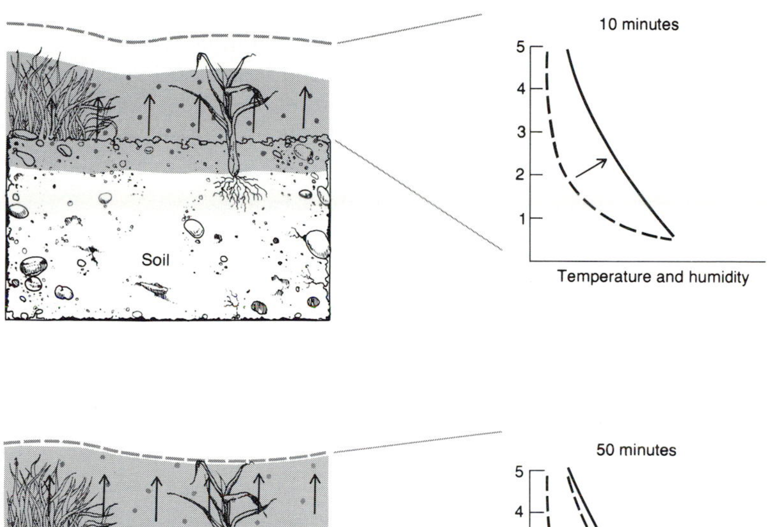

part of the energy-saturated surface layer is swept away and replaced by cooler, drier air. As a result, the energy gradients are reestablished; if the wind continues, the gradients can be maintained, thereby ensuring a relatively high rate of latent- and sensible-heat transfer. The actual transfer rates achieved, of course, depend on the wind speed as well as the temperature and humidity differences between the air and surface.

The action of the wind in the transfer of energy over small distances is referred to as *convection mixing*. It includes not only vertical movement, but also a wide range of complex mixing (swirling and whirling) movements near the surface. Convectional mixing is produced in two ways. The first is by wind movement in response to density (pressure) differences in the atmosphere resulting from variations in the heating of air. This is called *free convection*, or *thermal turbulence*, and is associated with vertical movement of unstable parcels of air both near the ground and at

THE VIEW FROM THE FIELD / NOTE 4

The Thermal Climate and Heat Syndrome in Humans

In the period 1950 through 1967, more than 8000 persons in the United States died from the effects of heat and solar radiation. According to NOAA, these were identified by health and medical authorities as direct casualties. How many deaths in the aged and infirm were brought on by excessive heat or solar radiation—but were not directly attributable to it—is not known because medical records do not usually identify climatic factors as the cause of death in such cases.

Heat syndrome refers to several clinically recognizable disturbances of the human thermoregulatory system. The disorders generally have to do with a reduction or collapse of the body's ability to shed heat by circulatory changes and sweating, or a chemical (salt) imbalance caused by too much sweating. Ranging in severity from the vague malaise of heat asthenia to the extremely lethal heat stroke, heat syndrome disorders have one thing in common: the individual has been subject to overexposure or overexercise for his or her age and physical condition and the thermal environment.

Two climatic conditions are associated with most epidemics of heat syndrome: regional heat waves and thermal microclimates. Heat waves are usually accompanied by increased mortality, especially among the elderly; this correlation can be expected without the complicating factors of high humidity or air pollution. Studies of heat syndrome show that it affects humans of all ages, but, other things being equal, the severity of the disorder tends to increase with age; heat cramps in a seventeen-year-old boy may be heat exhaustion in someone of forty years and heat stroke in a person over sixty years of age.

There is evidence that heat waves are worse in the brick and asphalt canyons of the "inner cities" than in the more open and better vegetated suburbs and rural areas. The July 1966 heat wave in St. Louis is a case in point. Most of the 236 deaths attributed to excessive temperatures occurred in the more heavily builtup areas of the city. Records also show that the death rate soared when the daily high temperature exceeded 100 degrees F (38°C) and that the highest tolls lagged behind temperature peaks by about a day.

For people with heart disease, thermal stress is worse than for others. In a hot, humid environment, impaired evaporation and water loss hamper thermal regulation, while physical exertion and heart failure increase the body's rate of heat production. The ensuing cycle is vicious in the extreme.

In a healthy person, the body acclimates to heat by adjusting perspiration-salt concentrations, among other things. In the midlatitude continental climates, this concentration changes in winter and summer just as it does when one moves from Boston to Panama. The body seeks an equilibrium in which enough water is lost to regulate body temperature without upsetting its chemical balance. Females appear to be better at this than males because females excrete less perspiration and so less salt; therefore, heat syndrome usually strikes fewer females.

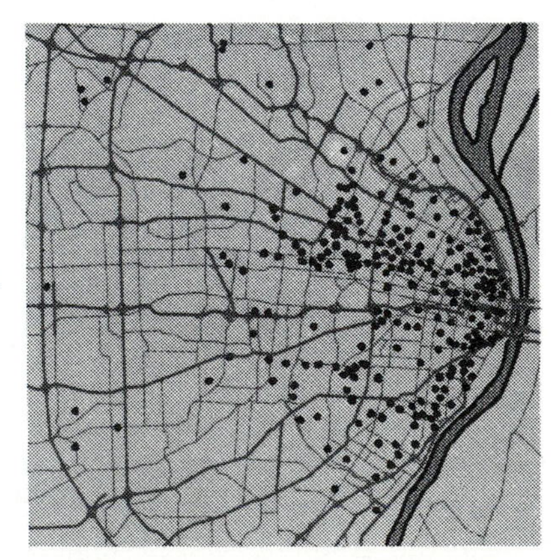

DISTRIBUTION OF HEAT DEATHS ST. LOUIS, MO. JULY 1966

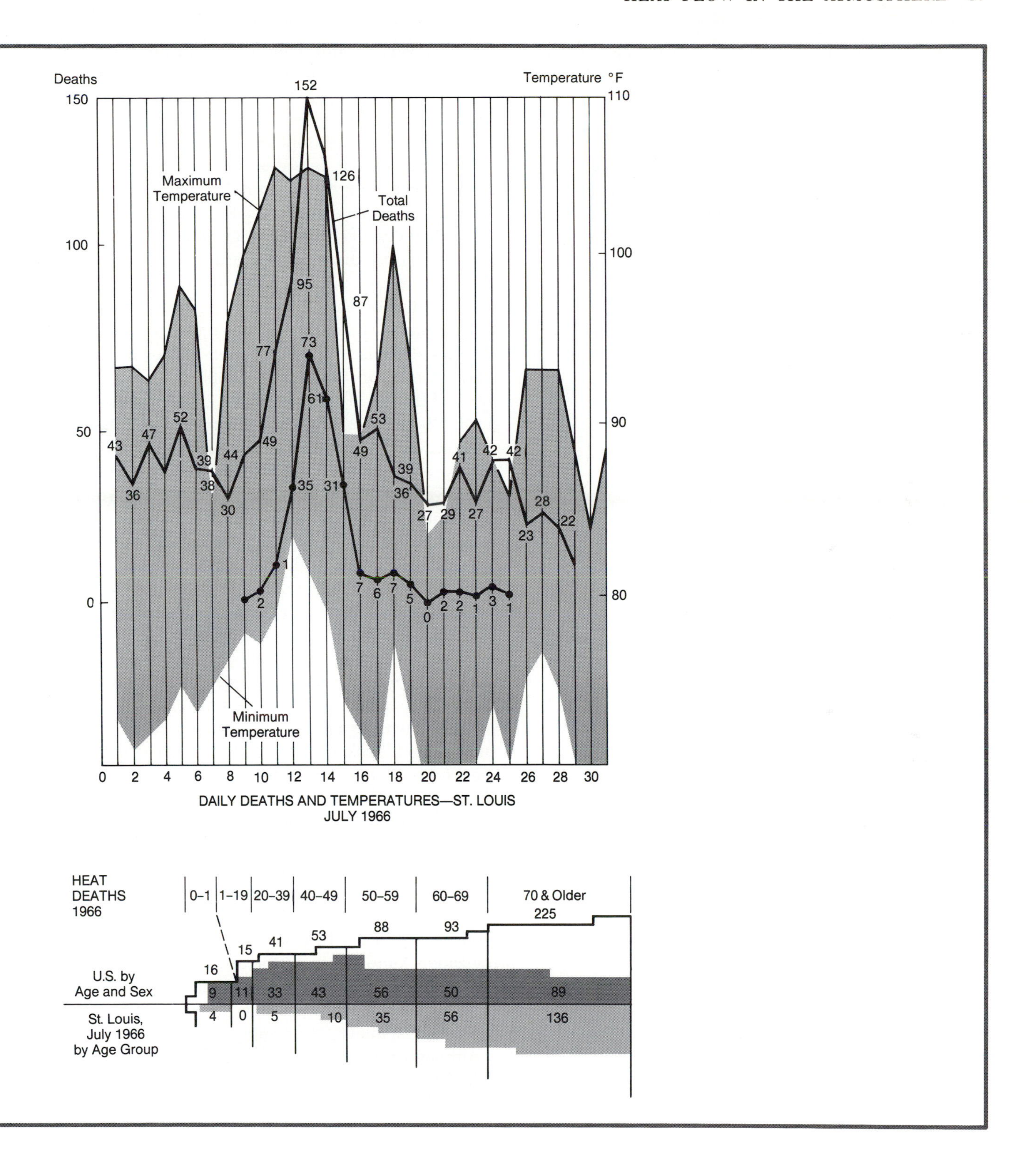

Deaths
Temperature °F
Maximum Temperature
Total Deaths
Minimum Temperature
DAILY DEATHS AND TEMPERATURES—ST. LOUIS JULY 1966
HEAT DEATHS 1966
0–1
1–19
20–39
40–49
50–59
60–69
70 & Older
U.S. by Age and Sex
St. Louis, July 1966 by Age Group

higher elevations (as in a thunderstorm). The second is by movement of wind over rough surfaces in the landscape. This is called *mechanical*, or *forced*, *convection* and is characterized by a swirling flow similar to that of river water moving through rapids. Both types of convection are essential mechanisms in transferring sensible and latent heat between the surface and the atmosphere. Let us examine them in greater detail.

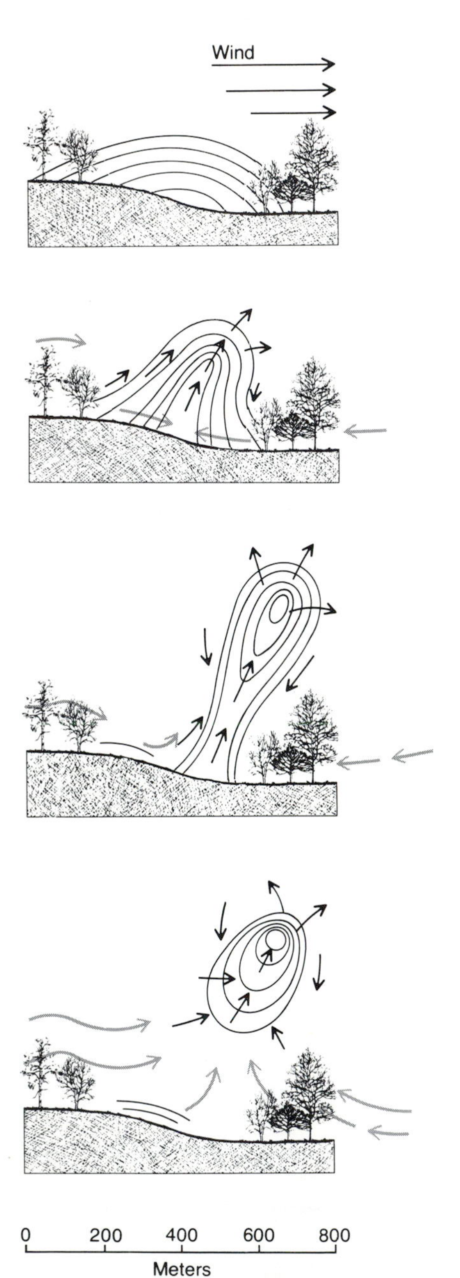

Figure 4.11 The development and ascent of a heated parcel of air. The vertically flowing air, called a thermal, is one of the most prevalent forms of free convection.

Free Convection and Atmospheric Instability Any body immersed in a fluid (liquid or gas) is displaced upward if its density is less than that of the fluid or downward if its density is greater than that of the fluid. It explains the rise of bubbles in beer and the sinking of most bars of soap in water. To apply this principle to the atmosphere, we need only imagine a body of gas containing some pockets with densities markedly lower than that of the body as a whole. Such bubbles of air should be displaced upward.

Parcels of air that are lighter or heavier than the surrounding air—and therefore sink or rise—are said to be *unstable*. In studies of the atmosphere, vertical movements caused by density differences in air are referred to as *instability*. Instability is the cause of much of the world's precipitation, a topic we will take up in Chapter 6. Here we want to examine its connection with wind and heat transfer.

Parcels of light air form over the land as a result of intensive outflows of sensible heat from the ground. As the air is heated, it expands and becomes buoyant, eventually breaking free of the surface. As it ascends, it generates an upward-flowing wind, called a *thermal* (Fig. 4.11). Such instability is a mild form of free convection. A thunderstorm, on the other hand, represents one of the most intensive forms of convection, but it, too, is usually initiated by thermals. Thermals draw in a train of surface air which in turn produces surface winds that sweep heated air off the ground. The air drawn upward to high altitude eventually cools and descends back to earth. Thus, the free convection causes a boiling action in the lower atmosphere, exchanging heat between the surface and air aloft.

Instability may be caused not only by surface heating, but also by condensation. In the case of a rising parcel of air that ordinarily would be stable at some height were it not for the occurrence of condensation, the latent heat released raises the air temperature and induces instability. In essence, the latent heat of vaporization supplies an extra charge of energy that drives the parcel even higher. This is typically the case in most rainstorms. (For more details on this aspect of instability, see Chapter 6.)

Although free convection is probably best exemplified in its most dramatic form, the thunderstorm, it is important from our standpoint to remember that this process is usually initiated at or near the earth's surface, that it draws energy off the surface, and that it occurs frequently at the local scale in the lower 10 to 100 m or so of the atmosphere. At this scale, convection can often be related to local variations in surface heating as in the case of warm and cool hillslopes

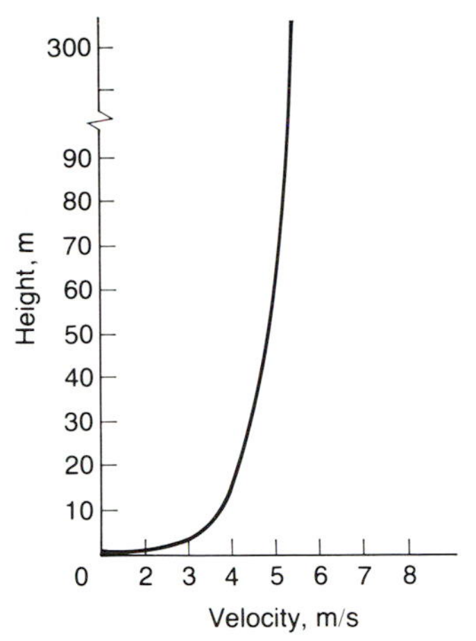

Figure 4.12 The change in wind velocity with height above the ground. Note that the graph line, or velocity profile, changes most rapidly just above the surface. From 20 to 300 m, velocity increases at a lower rate, and in the zone above an elevation of 300 m, it generally holds steady with increasing elevation.

and warm urban surfaces adjacent to cool water surfaces.

Mechanical Convection The second kind of convection—*mechanical convection*—results when turbulence (mixing motion) is created as air moves over a rough surface. To understand this form of convection, we must know a few things about wind as it flows over the landscape. Every wind has a distinctive velocity profile in which velocity increases with height above the surface (Fig. 4.12). The sharpest increase comes in the first 10 m; at a height of 20 m or so, the rate of increase declines, and for the next 280 m it increases only slightly. You can verify this profile by observing the contrasting degrees of motion in the upper and lower parts of a large tree on most days.

At the bottom of the profile, wind velocity falls to zero at some height just above the surface. This height is called the *roughness length,* and it defines the ceiling of calm air layer that lies over the ground. This

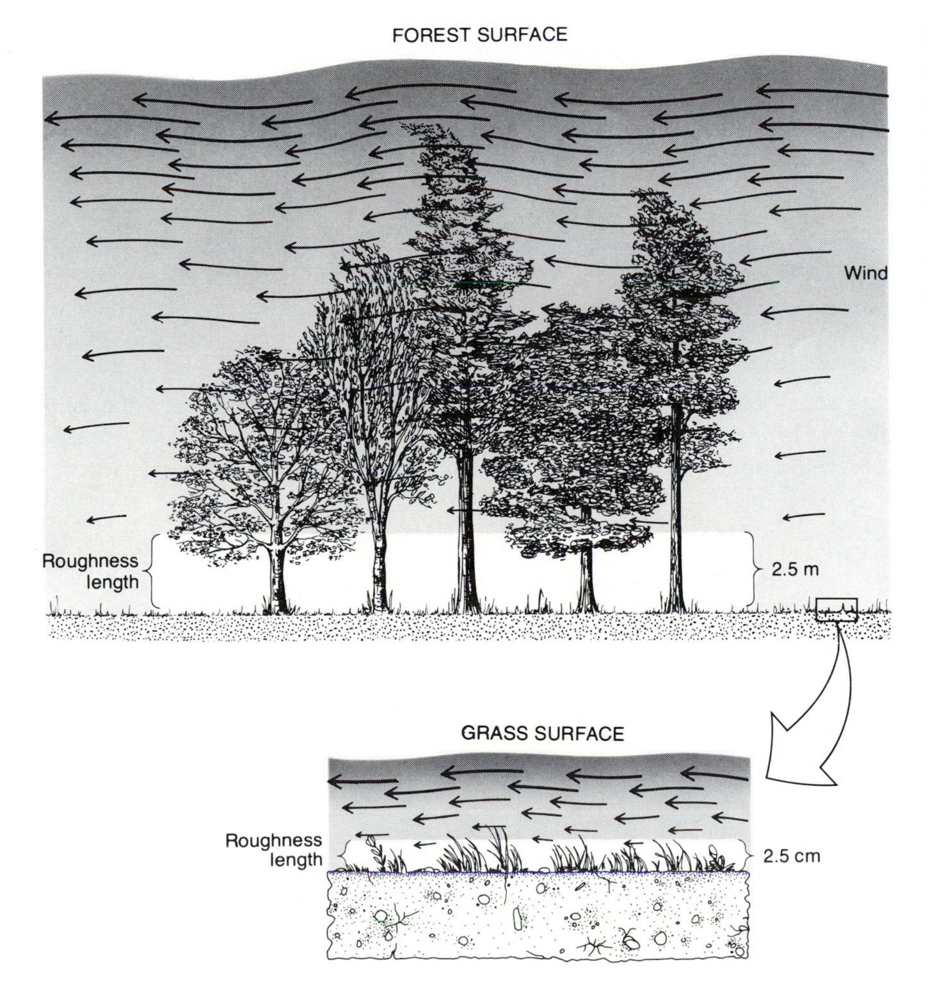

Figure 4.13 Conceptualization of roughness length, or the depth of the zone of calm air which envelops a surface. As surface roughness decreases, so too does roughness length. The ceiling of the calm layer (at 2.5 m and 2.5 cm, respectively) marks the bottom of the zone of wind turbulence.

Figure 4.14 Wind flow over an irregular surface. Note the eddy mixing on the lee side of obstacles such as buildings.

layer varies in thickness with the roughness of the landscape (Fig. 4.13). Over smooth, flat surfaces, such as mud flats or snow, it may be less than a millimeter; over rough surfaces, such as forests,[1] it may be as great as several meters (Table 4.3).

The roughness length affects mechanical convection in two ways: (1) by limiting the proximity of wind to the soil surface, and (2) by influencing the amount of turbulence above the surface. Most large roughness lengths are characterized by the presence of some objects (tall trees or buildings, for example) which protrude much higher into the air than others. This produces an extremely irregular aerodynamic surface resulting in complex air movement and high levels of convective mixing (Fig. 4.14). The mixing motion is characterized by both vertical and horizontal currents that sweep heated air away, or when the ground is cold, bring heated air to the landscape.

The behavior of wind over a rough surface is well illustrated by a windy day in a city. Flow is irregular, or gusty, with eddies (whirls) developing off the corners of buildings. The eddies of the strongest gusts typically violate the roughness length, whipping down to ground level and sweeping heat and vapor away. Where the buildings are tall and closely spaced, however even the strongest gusts are unable to permeate the roughness length. As a result, a layer of surface air develops whose heat and humidity characteristics are decidedly different from those in the air above; this surface air is also slow to change with the daily weather.

[1]Here we are using the word *surface* to mean two different things. In one sense the forest itself—that is, the top of the tree canopies—is the surface of the landscape. The *surface* of the land is the ground at the floor of the forest. With respect to the roughness length, we are talking about zero velocity of wind through the forest at some number of meters above the ground level, not at some distance above the treetops.

Summary

Net radiation, ground (or water) heat flux, sensible-heat flux, and latent-heat flux are the major components of the atmospheric energy balance and are related by the equation:

$$\underbrace{Si\text{-}So + Li\text{-}Lo}_{\text{radiation}} \pm \underbrace{G \pm S \pm L}_{\text{heat}} = 0$$

In a still atmosphere, energy flows essentially up and down. Shortwave radiation from the sun enters the atmosphere, much of it passing through and striking the earth's surface, where it is absorbed or reflected. Absorbed radiation heats the surface, producing temperature gradients downward into the cooler soil or water and upward into the cooler air.

Heat flows in both directions as long as the temperature of this surface is greater than that of the adjacent air and soil (or water). The rate of downward heat flow in soil depends mainly on its water content and particle composition. The rate of heat flow in water depends on the mixing motion of waves and currents.

Heat flows into the air in two forms—sensible and latent. Both sensible and latent heat enter the atmosphere by conduction and move through the at-

TABLE 4.3 REPRESENTATIVE ROUGHNESS LENGTHS FOR DIFFERENT TYPES OF SURFACES

Surface Type	Roughness Length
Smooth mud flats	0.001 cm
Dry lake bed	0.003 cm
Smooth desert	0.03 cm
Grass field	2.5–15 cm
Corn field	0.75–1.25 m
Buildup urban area	1.50–3.00 m
Large forest	2.50–5.5 m

mosphere by convection, created by atmospheric instability and wind. The greater the instability and the higher the wind velocity, the greater the transfer of sensible and latent heat.

Energy is lost from the surface as reflected shortwave radiation and as flows of longwave radiation, sensible heat, latent heat, and soil heat. The soil heat ultimately flows back to the surface when the soil surface cools (as at night or in winter), causing a reversal in the soil-temperature gradient. The net soil-heat flow over a long period of time equals zero over most of the continents. Net sensible- and latent-heat flows, on the other hand, typically do not equal zero even over short time periods (weeks and months) because heat transferred into the atmosphere can be moved to other places, as we will see in Chapter 5.

Concepts and Terms for Review

- scientific model
- energy-balance model
- heat flow
  - soil-heat flux
  - sensible-heat flux
  - latent-heat flux
- negative and positive energy flow
- heating of earth materials
  - thermal gradient
  - thermal conductivity
  - volumetric heat capacity
  - specific heat
- heat in soil
  - role of water
  - temperature profile
  - damping depth
  - permafrost
  - active layer
  - permafrost zones
- heat in the atmosphere
  - phase changes in water
  - role of wind
  - laminar sublayer
  - convectional mixing
  - stability and instability
  - roughness length
- heat syndrome

Sources and References

Black, R.F. (1954) "Permafrost—A Review." *Bulletin of the Geological Society of America* 65: pp. 839–855.

Fitch, J.M., and Branch, D.P. (1960) "Primitive Architecture." *Scientific American* 203: pp. 133–144.

Gates, D.M. (1962) *Energy Exchanges in the Biosphere*. New York: Harper and Row.

Geiger, Rudolph. (1965) *Climate Near the Ground*, 4th ed. 611 pp. Cambridge, Mass.: Harvard University Press.

Landsberg, Helmut E. (1956) "The Climate of Towns." In *Man's Role in Changing the Face of the Earth*, pp. 584–606. Chicago: University of Chicago Press.

Mackay, J.R. (1972) "The World of Underground Ice." *Annals of the Association of American Geographers* 82: pp. 1–22.

Mitchell, J.M., Jr. (1962) "The Thermal Climate of Cities." In *Symposium: Air Over Cities*, Cincinnati, Ohio: U.S. Public Health Service, Taft Sanitary Engineering Center, Technical Report A62-5.

Oke, T.R. (1978) *Boundary Layer Climates*. London: Methuen. New York: Halsted Press.

Thurow, Charles. (1983) *Improving Street Climate through Urban Design*. American Planning Association, Report 370.

Woolum, C.A., and Canfield, N. (1968) "Washington Metropolitan Area Precipitation and Temperature Patterns." U.S. Weather Bureau Tech. Memo.

Chapter 5

The Circulation of the Atmosphere and Oceans

The most prevalent surface material on earth is air; water is second and land, third. As fluids, air and water have the capacity to circulate over broad reaches of the planet. What drives this circulation? Are there distinctive geographic patterns to atmosphere and oceanic circulation? What is the role of circulation systems in the distribution of heat and moisture?

How is the vast surface of the earth with its widely separated land masses linked together by the circulation of air, water, and organisms? Or is it? Is, perhaps, the earth geographically partitioned in such a way that the behavior of the atmosphere, oceans, and life in one region has nothing to do with that in another?

Without some geographic knowledge of the world beyond our local region, we would probably be inclined to take such a provincial view of the earth's processes. Indeed, before the Age of Exploration and the emergence of natural science, this sort of thinking was common. Granted, seafarers as far back as the Phoenicians seem to have had some understanding of the huge scale at which the atmosphere circulates, because they used the systems of prevailing winds to navigate the broad expanses of ocean. But sailors were a small and inbred society, and if they did understand atmospheric circulation, there is little evidence prior to the eighteenth century that the ideas spread to land-based societies. With the eighteenth century, however, certain scholars became interested in the big picture of atmospheric circulation. Drawing on the knowledge of the seamen and combining it with emerging knowledge of the earth's motion, size, and shape, they proposed new schemes to explain atmospheric circulation. Most of their models were incomplete or inaccurate, but they deserve much credit for stretching the framework of thought about atmospheric circulation to a global scale.

Questions about atmospheric circulation logically lead to questions about oceanic circulation. Here too, the seafarers, particularly those of the eighteenth and nineteenth centuries, provided many of the earliest answers. Voyages in those days were slow, and after many ocean crossings, ship masters became very sensitive to the patterns of forces that worked for and against their vessels. These were principally the winds and currents, and in the Atlantic it was common knowledge that a great stream of warm water coursed across the ocean from North America to Northern Europe. Benjamin Franklin (1706–90) appears to have been the first to write about this current, the Gulf Stream, but his interests at the time were in navigation. Within the next 100 years, understanding of the Gulf Stream was advanced tremendously, going beyond mere description of the current to observations about its apparent influence on climate. The geographer M. F. Maury, a pioneer in oceanography, wrote in 1844 that the warm water of the current "spreads itself out for thousands of leagues over the cold waters around, and covers the ocean with that mantle of warmth which tends so much to mitigate, in Europe, the rigors of winter."

We owe a great deal to explorer-scientists like Maury, who, without the aid of aircraft or satellite, expanded our geographic perspective of the planet by documenting how far the winds and currents go and by suggesting some of their influences on rainfall, temperature, vegetation, and so on. The geographic insights gained from these efforts led in turn to questions about the magnitude of the air and water systems, what drives them, and how much they influence land and life. To the twentieth century geographer goes this challenge.

The chapter opens on the role of atmospheric and oceanic circulation in the global redistribution of heat and then goes on to examine the principles of atmospheric circulation and the great systems of winds and pressure cells. Oceanic circulation, which is driven by the wind systems, is taken up last.

GLOBAL IMPLICATIONS OF THE ENERGY BALANCE

In order to maintain an energy balance, the earth must release as much radiant energy as it receives. Recall that shortwave is the principal form of radiation coming into the atmosphere and that longwave is the principal form of radiation leaving it. In Figure 5.1, curve I represents the total radiation received by the earth and its atmosphere, whereas curve II represents the amount of longwave radiation and reflected shortwave radiation leaving the atmosphere. The graph shows that an area of radiation surplus exists at latitudes lower than about 38° in both hemispheres, whereas a radiation deficit exists poleward of these latitudes.

If no energy were exchanged between the zones of energy surplus and deficit, the tropics would grow hotter, ultimately reaching some equilibrium temperature much higher than the temperatures in the area today. The polar areas, on the other hand, would become colder and colder, eventually stabilizing at a much lower temperature. But because of the oceans of air and water on the planet, energy exchange does take place between the equatorial and polar zones. The redistribution of energy by the atmosphere and the oceans is possible because these bodies are continuously moving or circulating over the planet.

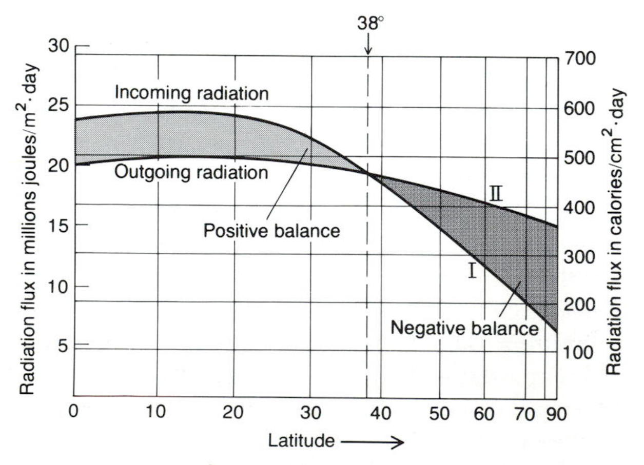

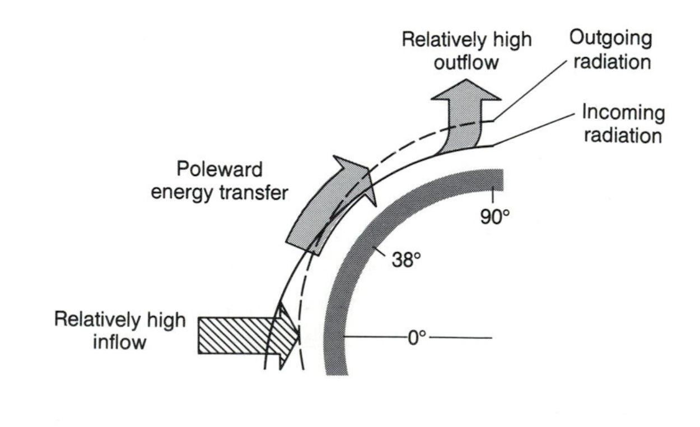

Figure 5.1 Distribution with latitude of incoming and outgoing radiation. Curve I shows the solar radiation absorbed by the earth and atmosphere; curve II, short- and longwave radiation leaving the atmosphere. The balance is positive in low latitudes and negative in high latitudes.

ATMOSPHERIC CIRCULATION: FACTORS AND CONTROLS

The circulation of the atmosphere is influenced by many factors which work in combination to produce the patterns of winds on earth. Among the important influences are the gross differences in heating between low and high latitudes, the rotation of the earth, and the differences in heat and pressure associated with land and water.

Hypothetical Circulation

If the planet did not rotate and were uniformly covered with one material, say, barren rock, the atmosphere would circulate in response to the latitudinal differences in the heat generated from incoming solar radiation. The overall circulation pattern should be very simple compared to that of a rotating planet. Air would be heated most intensively in the equatorial regions, where it would become unstable and would rise. A low-pressure area would thus develop along the equator. In contrast, high pressure would develop over the poles, where heating would be weakest. Between the equator and the poles there would be a pressure gradient, and a gigantic convectional system would form, in which the equatorial low would be fed with air from the polar highs. At the surface, air would flow toward the equator; aloft, it would flow toward the poles. That is, in the Northern Hemisphere the winds at the surface would blow from the north. This circulation is shown in Figure 5.2.

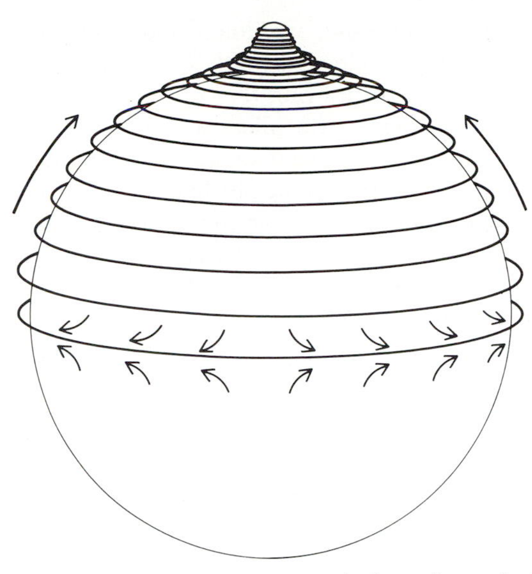

Figure 5.3 Schematic portrayal of one form which air circulation in the upper atmosphere could take about a rotating planet. The velocity of circulation would increase poleward as the radius of orbit decreased. The poleward trend would be caused by the general thermal convection between the equatorial and polar zones illustrated in Figure 5.2.

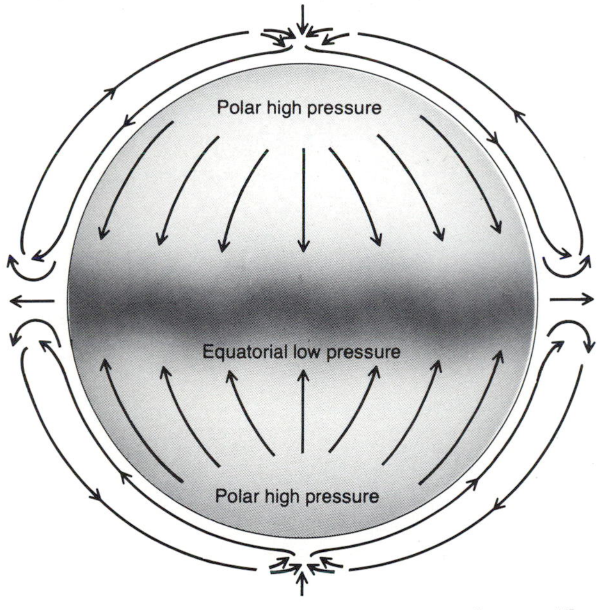

Figure 5.2 Atmospheric circulation pattern that would develop on a nonrotating planet. The equatorial belt would heat intensively and would produce low pressure, which would in turn set into motion a gigantic convection system. Each side of the system would span one hemisphere.

But, of course, the earth does rotate, and the addition of rotating motion to the earth transforms the simple north-south circulation into a pattern that tends toward east-west circulation. Thus, a particle of air driven into the atmosphere above the equatorial zone should circulate about the earth, gradually spiralling toward the pole (Fig. 5.3). As the particle gets nearer the pole, however, an important change takes place: its velocity increases. This is explained by the principle of the conservation of *angular momentum,* or the momentum of a mass moving in a circular path: the velocity at which a particle circulates about the earth will increase as the radius of its orbital path (that is, the distance to the earth's axis) decreases.

When you were a child, you probably applied this principle many times. If, for example, you sat on a swing and "wound yourself" up by twisting the swing ropes and then unwound, you found that you could control your angular velocity by either extending your legs to slow down or drawing them in to speed up. Another example of this principle is the ice skater who folds his or her arms in order to spin more rapidly. In the atmosphere, this principle helps explain the extraordinary wind speeds near the center of a tornado. In all of these examples, the rotating mass remains constant, but since the radius of rotation changes, velocity must change.

As the particle in the atmosphere moves poleward, the earth beneath it is moving more slowly than the particle because one rotation of the earth (24 hours) represents shorter travel distances (or radii of rotation) at higher latitudes. To consider an extreme example, at latitude 45°N, a point on the earth's surface would have a velocity of 328 m/sec (734 mph). Assuming the particle is situated 10 km above the earth, its velocity at this latitude would now be 658 m/sec (1472 mph), 330 m/sec (738 mph) faster than the surface beneath it. Such extreme velocities are not possible in the earth's atmosphere because of factors such as atmospheric density and the earth's size and motion. Thus, the atmospheric circulation pictured in Figure 5.3 is impossible, especially at latitudes higher than the tropics.

What actually happens—and we are speaking of the real earth now—is that the poleward spiralling circulation system breaks down, that is, the fluid motion of the air becomes irregular, and instead of continuing poleward it begins to descend toward earth around latitudes 25° to 30° (Fig. 5.4*a*). This descending air produces a buildup of atmospheric mass over the earth's surface leading to a great body of *subtropical high pressure* in both hemispheres. Like the zone of equatorial low pressure, these zones of high pressure more or less encircle the world.

Fed with air forced aloft in the equatorial low, the subtropical high pressure zones cycle air back to the equatorial low pressure zone as surface winds (Fig. 5.4*b*). This circulation system is known as the Hadley Cells (one in each hemisphere), after George C. Hadley, the British physicist who formulated the first theoretical explanation for intertropical surface circulation in 1735.

Isobaric Pressure and Pressure Cells

At this point, it may be well to review and clarify some of the terminology associated with atmospheric pressure. You will recall that air pressure represents the force exerted on a square meter of earth's surface by the total mass of molecules and particles in the overlying air and is measured in millibars (mb). The air pressure readings we see on isobaric maps represent pressure relative to normal (or mean) pressure at sea level which is 1013.2 mb.

Various terms are used to describe pressure features at global and regional scales. The circumglobal zones of pressure such as the equatorial low and the subtropical high are called *pressure belts.* Within these belts are geographic centers of pressure, called *cells,* and they are classed as *high* or *low* pressure depending on whether their pressure at sea level is above or

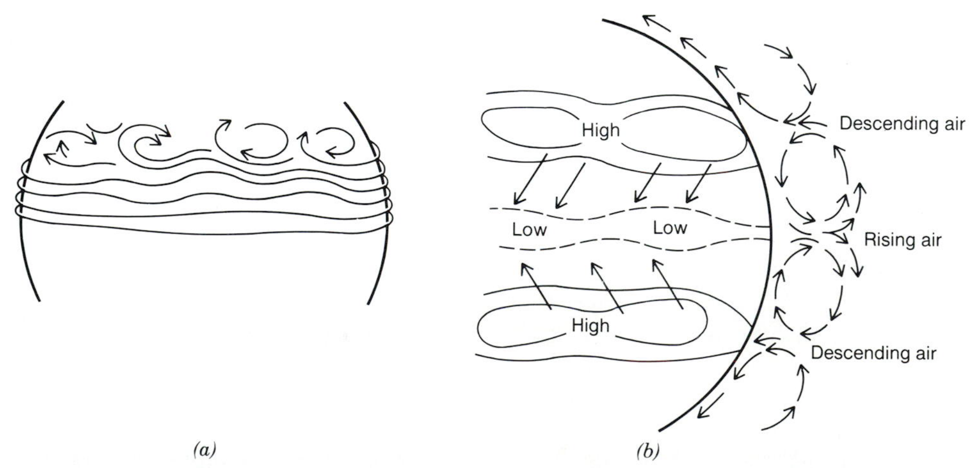

Figure 5.4 (*a*) The breakdown of the upper atmospheric circulation and the formation of the subtropical high-pressure cells. (*b*) The resultant pattern of circulation between the equatorial low and subtropical highs.

below 1013.2 mb. In climatology and meteorology the terms *anticyclone* and *cyclone* are applied to high-pressure cells and low-pressure cells, respectively.

In general, we identify two classes of pressure cells based on their origins: thermal and dynamic. *Thermal cells* are caused mainly by the heating of air over warm land surfaces or the cooling of air over warm surfaces; cells in the equatorial low-pressure belt which develop over land masses are thermal cells. *Dynamic cells* are caused by mechanisms of fluid circulation, such as the breakdown of upper atmospheric flow in the case of the subtropical highs. In addition, pressure cells are classed according to size and duration. The largest cells cover great sections of continents and persist so long that they consistently appear on weekly, monthly, and annual isobaric maps. These cells are called *semipermanent* pressure cells. At the other extreme cells such as a thunderstorm may cover less than a square kilometer and may last only 8 to 12 hours. Between these scales is the *synoptic scale*, which is what we see on the daily weather charts depicting cells of intermediate sizes and lifetimes usually measured in days (see Fig. 1.10). These include midlatitude cyclones, tropical storms, and hurricanes, all of which we will discuss in detail in the next chapter.

Circulation Patterns and the Coriolis Effect

Air pressure is the force that drives the flow of air (wind) over the earth's surface. The direction of this force is determined by the air-pressure gradient, which always runs perpendicular to the isobars of surface pressure. This is shown by the arrow marked PG in Figure 5.5. If we inspect a synoptic weather chart, however, we will find that surface winds *do not* conform to the exact direction of the pressure gradient, for they consistently veer off the line of the pressure gradient, crossing the isobars at an angle (see arrow marked SW in Fig. 5.5).

In the Northern Hemisphere, the winds veer to the right of the pressure gradient, and in the Southern Hemisphere, they veer to the left of the pressure gradient. This shift in wind direction is caused by the *Coriolis effect* (named for the French physicist Georges de Coriolis). The Coriolis effect influences the direction of movement of all airborne objects, including birds, aircraft, and wind as well as ocean currents. It functions independently of the direction of movement of an object, and its magnitude depends only on the angular velocity of earth rotation and the relative speed of the moving object.

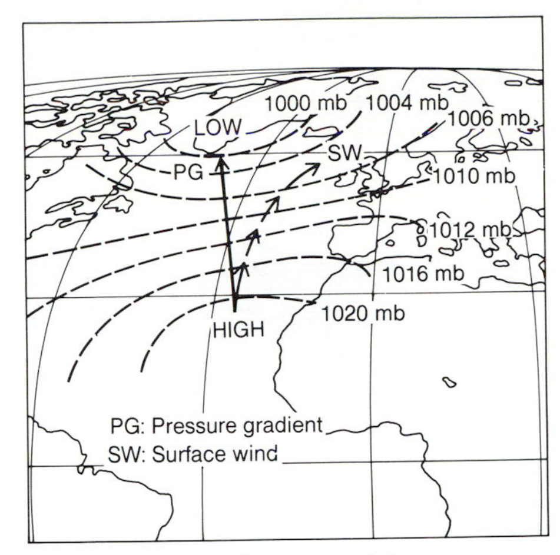

Figure 5.5 The direction of the pressure gradient between a high and low-pressure cell over the Atlantic Ocean and the resultant flow of surface wind. The deviation from the pressure gradient path is due to the Coriolis effect.

When viewed above the North Pole, the earth rotates counterclockwise; when viewed from above the South Pole, the earth rotates clockwise. This explains why the Coriolis effect is to the right in the Northern Hemisphere and to the left in the Southern Hemisphere. But what happens at the equator? Do we find a sudden switch in directions? No, instead we find that there is no Coriolis effect at the equator. From the poles, where it is strongest, the Coriolis effect decreases with decreasing latitude, becoming zero at the equator.

Although the Coriolis effect is an important consideration in explaining airflow patterns, it is very weak overall, even near the poles. The fact that you do not feel a car being pulled to the right when you drive on a straight highway in the midlatitudes suggests this. However, if you were navigating an airplane over a great distance, you would notice the influence of the Coriolis effect because the airplane would veer far off course unless directional corrections were made in flight. The difference in the influence on car and aircraft is explained by the fact that the frictional resistance of the car's tires on the road surface is very great, whereas the frictional resistance of air to changes in direction of the aircraft is very slight. Winds and currents moving through the atmosphere and the oceans also meet with slight resistance from the surrounding fluid to changes in direction. Thus, their courses, like that of an aircraft, are markedly influenced by the Coriolis effect.

In describing the movement of air from a high-pressure cell or into a low-pressure cell, we have to

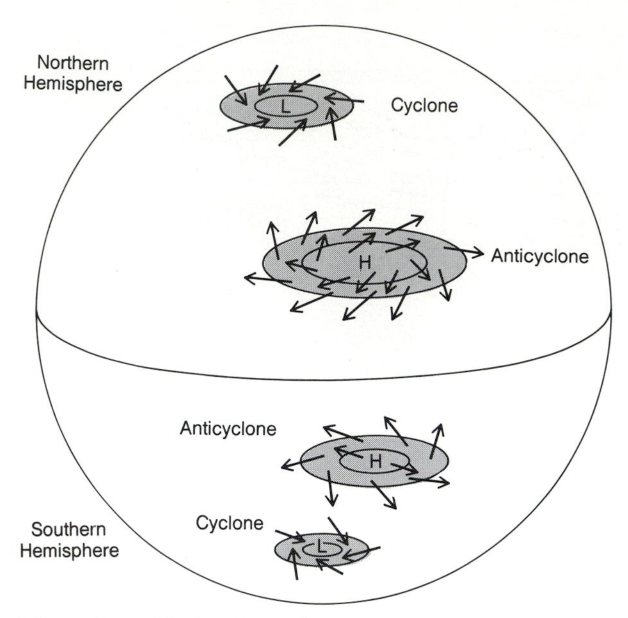

Figure 5.6 Idealized circulation patterns associated with cyclones and anticyclones in the Northern and Southern hemispheres.

consider first the direction of the pressure gradient and second the Coriolis effect. All high-pressure cells (anticyclones) have divergent airflow, but the direction of deviation from the pressure gradient changes with the hemisphere. In the Northern Hemisphere, anticyclones have a clockwise whirl, and in the Southern Hemisphere, a counterclockwise whirl. Low-pressure cells (*cyclones*) are basically the opposite: they have convergent airflow in both hemispheres, counterclockwise motion in the Northern Hemisphere, and clockwise motion in the Southern Hemisphere (Fig. 5.6).

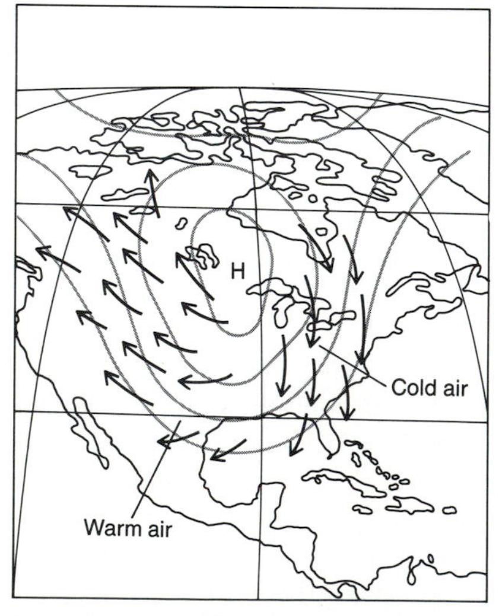

Figure 5.7 Cold and warm airflow associated with a large anticyclone over North America.

This pattern of circulation is an important factor in regional weather patterns. The geographic coverage of both cyclones and anticyclones is often so vast that contrasting types of air are circulated in different sectors of the system. In the case of an anticyclone in the Northern Hemisphere, for example, the general circulation is northerly along the west side and southerly along the east side. Thus, if the cell is situated over the midlatitude zone of a continent—as anticyclones tend to be in the winter over Northcentral Eurasia and North America—the southerly airflow brings warm air northward and the northerly airflow brings cold air southward (Fig. 5.7). In the midlatitudes this circulation helps to break down the zonal differentiation between the belts of tropical and polar air.

Circulation Aloft and the Formation of Midlatitude Cells

The influence of the earth's surface in creating friction on moving air is limited to the lower 300 to 500 m of the atmosphere. Most of the atmosphere lies above this level, of course, and it is important to ask what is the influence of the Coriolis effect aloft and what is the nature of circulation in the upper atmosphere. In the absence of friction, wind is controlled only by the pressure gradient and the Coriolis effect. Air set into motion between high- and low-pressure cells will increase in velocity until the Coriolis effect exactly balances the pressure gradient. Because one force offsets the other, the resultant wind should blow parallel to the isobars. Such winds do occur in nature, and they are called *geostrophic winds* (Fig. 5.8).

Geostrophic winds are the fastest winds on earth, reaching velocities of 150 m/sec (325 mph). They are limited to zones above 25° latitude and to elevations greater than about 500 meters. Nearer the equator, the Coriolis effect is too slight to allow their development. At elevations lower than about 500 meters at all latitudes, friction between air and the earth's surface prevents wind from attaining geostrophic velocities and directions of flow. This is the chief difference between surface winds and geostrophic winds: surface winds stabilize at lower velocities and in directions falling between the isobars (the geostrophic direction) and the pressure gradient. Thus, as you ascend from ground level into the atmosphere, wind direction becomes more parallel to the isobars until it attains true geostrophic flow. The shift in direction with altitude is, of course, dependent on the hemi-

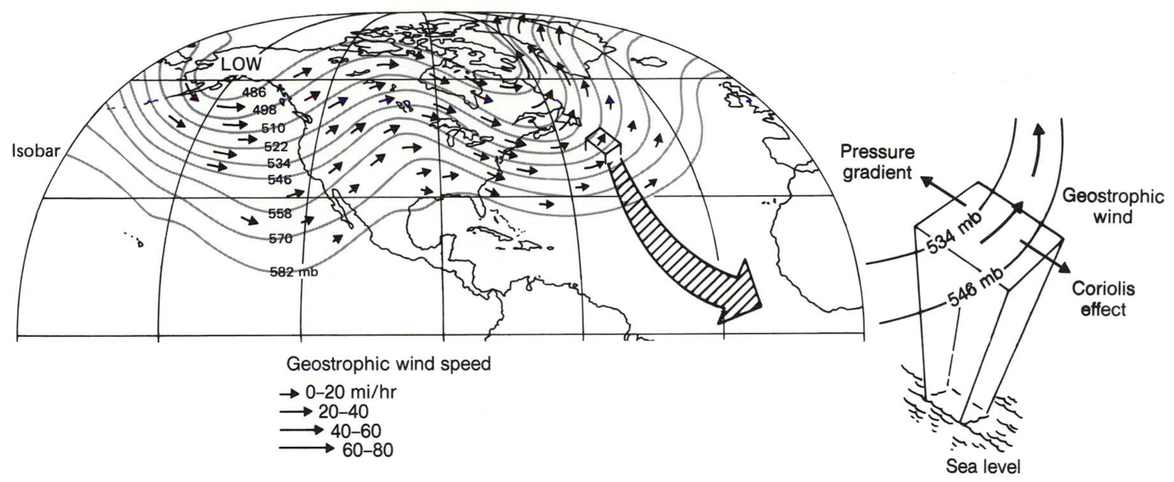

Figure 5.8 A geostrophic wind on December 19, 1979, at an altitude of about 5500 m, where pressure is approximately half that at sea level. Note that the wind travels parallel to the isobars as the Coriolis effect and the pressure gradient offset each other (inset).

sphere. In the Northern Hemisphere, wind direction shifts clockwise upward through the lower atmosphere; in the Southern Hemisphere, the shift is counterclockwise.

In the midlatitudes the geostrophic winds form a great belt of westerly flow such as the one shown in Figure 5.8. This belt develops a huge meander pattern, called Rossby waves (after C. G. Rossby, who first described them quantitatively), which usually loop equatorward over the continents and poleward over the oceans. To the polar side is the zone of cold air; to the equatorial side is the zone of warm air. The leading edge of the cold air, called the *polar front*, is often marked by a stream of concentrated geostrophic flow, called the *jet stream*. Because the flanks (east and west sides) of the meanders trend north and south, there is a strong tendency toward exchanges of air along the polar front between the tropics and the high latitudes. This takes the form of migrating air masses whose movement generally follows the pattern of the upper airflow as it meanders across the continents and oceans.

The mixing of warm and cold air masses along the polar front produces atmospheric disturbances which in turn lead to large-scale cyclonic cells. Although these cells are transitory, moving eastward in the general circulation scheme, they are prominent and frequent enough, especially over the oceans, that semipermanent low pressure is assigned to the upper middle latitude and subarctic zones over water. These cells are called the *subarctic lows*, and in the great scheme of global pressure cells, they are attributed to dynamic factors, that is, to complications (mixing) in the fluid flow between the polar and subtropical/midlatitude zones.

Thermal Effects of Land and Water on the General Circulation

Pressure gradients also result from thermal differences between land masses and water bodies. On a local scale, the distances which air travels are very short, and so we can ignore Coriolis effects. An example of local-scale circulation is the land-sea breeze. In coastal areas, the land is typically colder than the water at night but warmer during the day. During the day, then, a low-pressure area forms over the land, and a sea breeze results, the wind blowing along the pressure gradient from water to land. At higher elevations, where cooling of the rising air over the land has taken place, the pressure gradient may be reversed, in which case the air aloft will flow seaward over the water. At night the situation is reversed (Fig. 5.9).

Much of the difference in the thermal character of land and water is related to the depth to which heat can penetrate each. Recall that heat moves into the land by conduction; as a result, only the upper several meters of the land exchange much heat with the atmosphere on an annual basis. Moreover, land materials generally have low volumetric heat capacities and in turn produce strong thermal gradients into the atmosphere. For these reasons, heat absorbed by the soil in summer is quickly lost in the ensuing cool

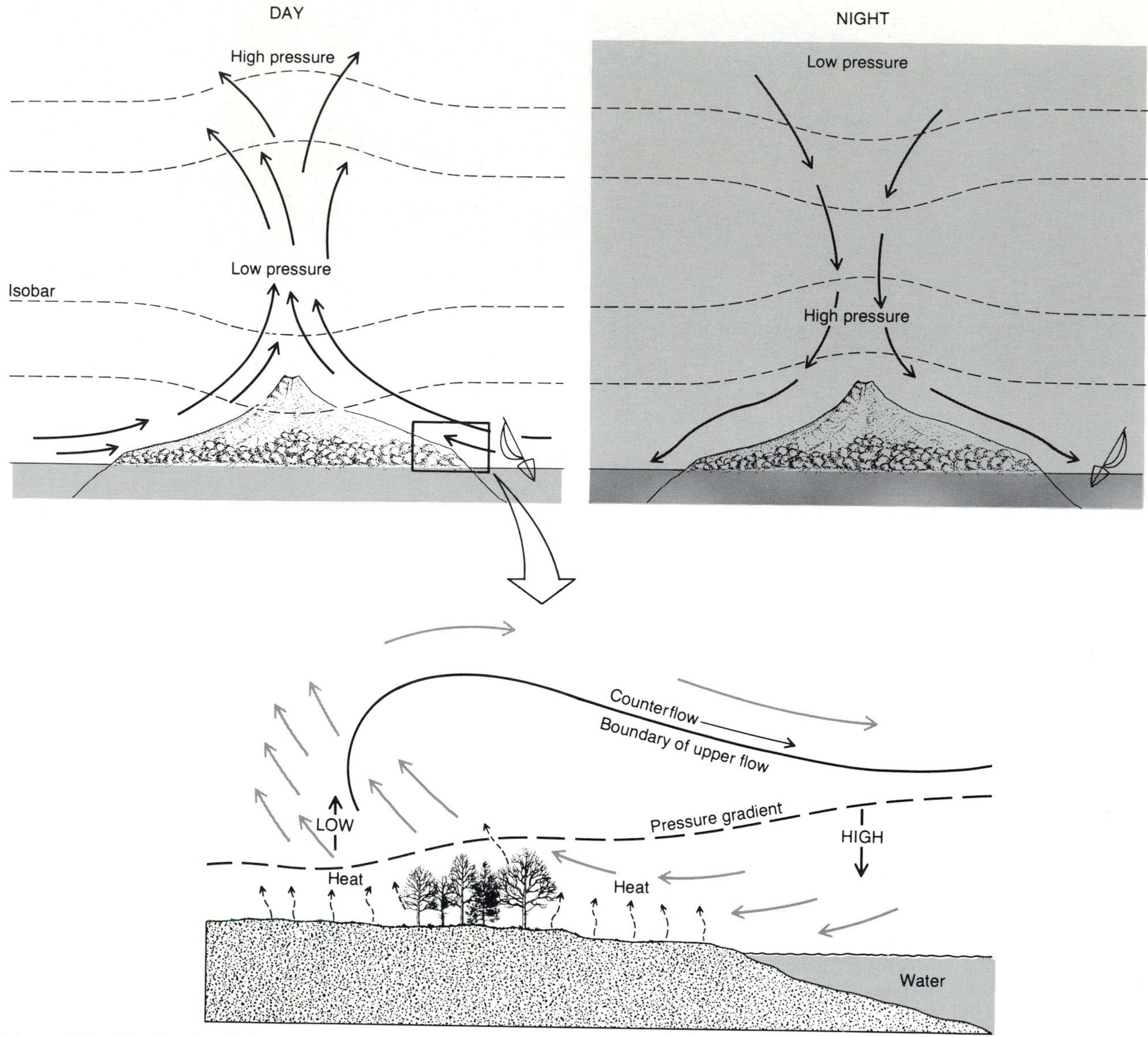

Figure 5.9 Day/night differences in airflow in response to changes in pressure over a land area. Note that because of the vertical component of airflow, the pressure that develops aloft is always opposite that near the ground.

season. In contrast, heat moves into the ocean by convection and is carried to much greater depths than it is on land. In addition, water has a high heat capacity and, because it is usually cooler than the air, fails to develop strong upward thermal gradients into the overlying atmosphere.

On balance, then, the ocean gives up its store of heat much more slowly than the land does. Therefore, in winter land areas are typically much colder than the oceans, whereas in summer they are much warmer. (See the July 20 isotherm over the East Pacific and North America in Fig. 5.12). For example, the North American cities of San Francisco and St. Louis are at approximately the same latitude and would, under the same atmospheric conditions, receive about the same amount of heating from solar radiation. But San Francisco is near the ocean and is downwind from it, whereas St. Louis is inland. The coolest month in St. Louis is January, with a mean temperature of 0°C (32°F), and the warmest is July, with a mean temperature of 26°C (79°F). In contrast, January in San Francisco averages 10°C (50°F), and the warmest month is September, with a mean temperature of 17°C (62°F).

Such differences in heat storage are essential to explaining certain pressure differences in the atmosphere. The large land masses of the world, particularly North America and Eurasia, typically develop very high pressures in the winter and low pressures in the summer. This difference is particularly well developed over Eurasia because of its very large size. The resulting winds also reverse seasonally, and in Asia these are called the *monsoon* (from the Arabic word *mausim*, meaning "seasonal wind"). During the summer, warm, moist winds, called the summer monsoon, blow from the Indian Ocean and Arabian Sea toward the land mass. In the winter, the strong pressure system that develops over Asia causes the winds to blow from the land mass to the ocean, and this is called the winter monsoon (Fig. 5.10).

THE GENERAL CIRCULATION OF THE ATMOSPHERE

The general circulation of the atmosphere refers to the global pattern of prevailing winds and pressure cells. This includes not only those wind systems and cells that appear consistently year round on the weather charts, but also those that appear seasonally year after year.

Equatorial and Tropical Zones

Figure 5.11 is a schematic portrayal of the circulation of the atmosphere at the earth's surface. Note the low-pressure zone along the equator. This is called the *intertropical convergence zone* (ITC), and it results from surface heating caused by intensive incoming solar radiation and from the convergence of tradewinds near the equator. The variable winds in the ITC are often calm and frequently caused grief during the days of ocean sailing, when ships attempting to cross the equator would be becalmed, sometimes for

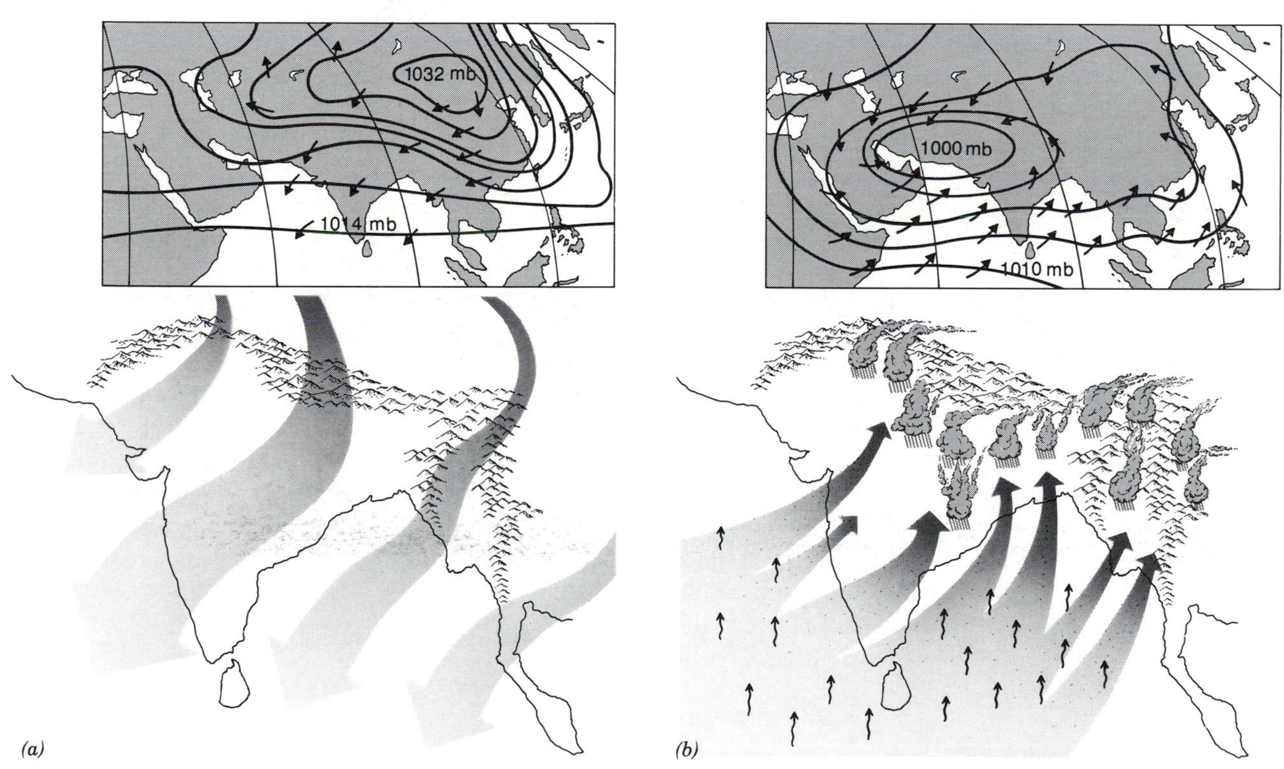

Figure 5.10 Monsoon circulation over South Asia in (*a*) winter and (*b*) summer. The summer monsoon develops in response to low pressure over land (see inset) and gains large amounts of vapor as it passes over tropical oceans. In winter the flow reverses as high pressure forms over Central Asia.

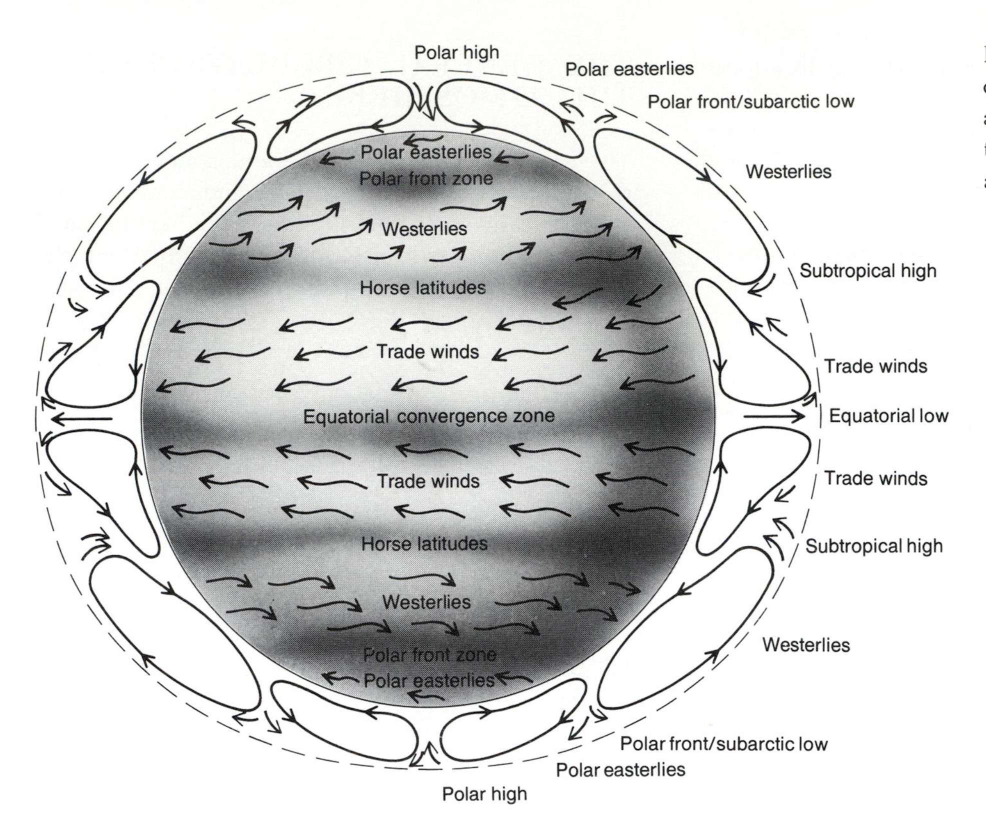

Figure 5.11 Idealized circulation of the atmosphere at the earth's surface, showing the principal areas of pressure and belts of winds.

weeks. This region was also called the *doldrums*, a word that is sometimes used to portray a weary, depressed state of mind.

Poleward of the ITC in both hemispheres are the trade winds. In the Northern Hemisphere, they blow from the northeast to the southwest and are called the *northeast trades*. In the Southern Hemisphere, they are called the *southeast trades*. The Coriolis effect causes the directions to deviate from true north (in the Northern Hemisphere) and true south (in the Southern Hemisphere).

Aloft, at an elevation of about 12 kilometers, the air converging on the ITC after moving up the convectional cells fans out and moves back toward the poles. But it never reaches them because at latitudes 25° to 30° north and south, in the zone called the *horse latitudes*, the air descends, forming large, subtropical high-pressure cells. In this pattern of cyclical circulation, the trade winds at the surface bring air to the equator, and the *antitrades* aloft bring air back to the subtropical high-pressure cells.

Monsoon circulation represents a major departure in the pattern of trade winds. The summer low pressure over Southcentral Asia is so strong that the northeasterly trade winds over the northern Indian Ocean and southwestern Pacific are reversed, blowing as southwesterly winds onto India, Southeast Asia and China. In the winter, with the development of high pressure over Asia, the flow of the northerly trade winds resumes across the northern Indian Ocean and the southwestern Pacific.

In the sixteenth century during the period of Portuguese trade with India, ship captains timed their voyages from Europe to the Indian peninsula so that the vessels crossed the Arabian Sea from Africa to India during the summer monsoon. They then waited in India until the onset of the winter monsoon. The first fleet to sail to India (in 1497–98), commanded by Vasco da Gama, did not do this, but instead crossed to India in May, at the beginning of the summer monsoon; the voyage from Malindi on the east coast of Africa required less than a month. The return voyage at the end of August, with the summer monsoon still in progress, required three months to cover the 4200 km (2700 mi.) distance (Fig. 5.12; color plate 1).

Midlatitude Circulation

From the poleward flanks of the subtropical highs, air flows northeastward and eastward across the midlatitudes. This belt of wind, called the *westerlies*, is found in both hemispheres between 35° and 55° latitude (Fig. 5.12). In the Southern Hemisphere, the westerlies form a continuous belt that lies principally over the oceans. Only the southern portions of South America, Africa, and Australia extend into this zone;

they feel the strong effects of the westerlies in the winter months as cyclonic storms generated along the polar front migrate eastward with the airflow.

In the Northern Hemisphere, the belt of the prevailing westerlies is interrupted by the North American and Eurasian land masses. In summer, a strong southerly airflow develops over these land masses. In North America, a monsoon-like airflow develops as warm, moist air sweeps into the center of the continent from the Gulf of Mexico/Caribbean region. At the same time, the subtropical highs over the Atlantic and Pacific intensify and shift northward, driving the westerlies into the subarctic zone. Over Asia a region of low pressure envelops the bulk of the continent, generating the summer monsoon on its southern flank and a northeasterly airflow on its northern flank. The westerlies effectively disappear in this region while this low pressure cell is strong, but in winter, they return with the development of high pressure over Central Asia (Fig. 5.12).

In the upper atmosphere, the midlatitude airflow is also westerly (Fig. 5.13). The strongest flow is associated with the polar front jet stream along the equatorward limits of the zone of polar air. This boundary, which is frequently associated with geostrophic wind velocities of 100 m/sec (225 mph), is the most dynamic part of the upper atmosphere and, as noted earlier, is the source of great eddies in the form of midlatitude cyclones. In contrast, in the intertropical zone, upper atmospheric winds are light and easterly, showing little tendency to meander in the fashion of the polar front jet stream.

High Latitude Circulation

The polar zones of both hemispheres are dominated by high pressure which generates an easterly airflow above the Arctic and Antarctic circles. As wind systems, the *polar easterlies* are neither strong nor seasonally permanent; rather, they vary with changes in the strengths of the polar highs. In the Northern Hemisphere, they also vary with reversals of pressure over the land masses.

The General Circulation in Overview

At the first level of approximation, the circulation in each hemisphere can be characterized by three great belts of prevailing winds generated from four zones

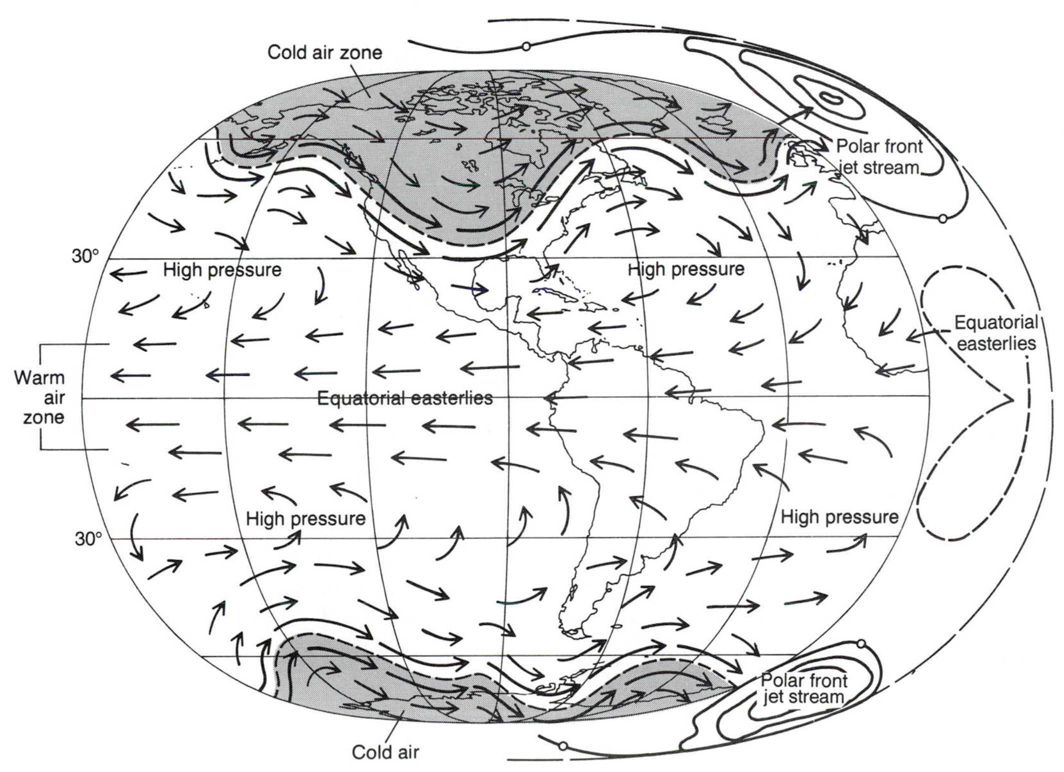

Figure 5.13 Generalized circulation in the upper troposphere, showing a typical configuration of Rossby waves along the edge of the cold air zones. The location of these waves is often marked by the polar front jet stream. The tracks of cyclones generally follow this airflow.

of pressure: the equatorial low, subtropical high, subarctic low, and polar high zones. Within these zones are nodes of pressure that vary from land to water and produce regional variations in the circulation scheme. At the next level of approximation, the seasonal changes in pressure and wind must be taken into consideration. Indeed, the circulation in many regions makes sense only as a two-part seasonal system. The monsoon circulation of South Asia is the most striking example, but many other examples of seasonal contrasts in circulation can also be cited. These include areas that lie alternately under the subtropical highs and in the belt of the westerlies and the bulk of the intertropical zone of latitude which experiences seasonal changes in pressure and wind as the ITC zone migrates north and south.

In addition to shifting latitudinally with the seasons, the pressure cells also enlarge and contract seasonally. The subarctic lows are very strong in winter but weaken almost to nonexistence (and shift poleward) in summer. Even more extreme is the pressure over Asia, which reverses from winter to summer in response to changes in the thermal regime of the continent. The subtropical highs change with the position and strength of the equatorial low. As the ITC zone shifts into the Northern Hemisphere and intensifies, more air is forced aloft into the antitrades and ultimately into the subtropical highs. The highs in turn expand significantly, enveloping areas which, during the winter months, were well within the belt of influence of the midlatitude westerlies. In coastal areas, this change produces a summer dry climate associated with the westerlies and occasional cyclonic storms.

Secondary Circulation

We have emphasized the general circulation of the atmosphere because it is the primary influence on the general character of world climates. There is also a second level of influences, called *secondary circulation features*. These features are best described as discrete disturbances that result in relatively short-lived migratory cells. The most common are the midlatitude cyclones that originate along the polar front and about which we will say more later. Less common, though often greater in magnitude and duration, are tropical disturbances. *Easterly waves* are troughs of low pressure that develop within the belt of trade winds and migrate westward. They develop over the oceans and are fed with moist air that produces rain and a source of heat (latent) to maintain the low pressure. A much larger and more intensive tropical storm is the *hurricane* or *typhoon*, which also develops over water and, during its early stages, also migrates westward in the belt of trade winds. Generally considered the most fearsome of all atmospheric disturbances, hurricanes often move poleward well beyond the tropics and wreak havoc on midlatitude land masses. These storms will be examined in greater detail in Chapter 6.

OCEANIC CIRCULATION

Increasingly, scientists are learning that the oceans exert an important control on global climate—not just in the coastal areas but also over entire zones of latitude including oceans and continents. In sharp contrast to the atmosphere, the seas have a high volumetric heat capacity and thus are able to retain great quantities of heat at modest temperature levels. Furthermore, through wave and current circulation, the oceans are able to redistribute heat to considerable depths and over great areas of the ocean surface. Much of this redistribution is east-west or west-east. But much is also meridianal, resulting in the transfer of energy across the midlatitudes from the tropics to the subarctic, thereby augmenting the overall poleward heat transfer in the atmosphere. In addition, certain coastal areas are profoundly affected by warm currents or cold currents that give rise to unique coastal climates. Because wind is the principal driving force for waves and currents, we begin our discussion of oceanic circulation with a look at water motion and wind. (For a more detailed discussion of waves and currents, turn to Chapter 21.)

Generation of Currents By Wind

Although waves are generated by a variety of forces including earthquakes and volcanic eruptions, the waves that dominate the oceans are wind generated. Wind waves cover a broad range of sizes from ripples a few centimeters high to great storm waves 10 or more meters high. In large waves there are two types of motion. The principal motion is a circular action in the water under the wave form. This produces only a rotation of water particles but no transfer of water across the surface and hence no current type movement. The second motion occurs on the water surface when a wave breaks or its crest is blown over by wind and sends water spilling ahead. The net effect of this motion is a downwind transfer of surface water, or a *current*.

Current direction and velocity are governed by the direction, velocity, and duration of wind. If brisk winds blow consistently in one direction over a broad reach of ocean, a strong current will develop. It follows that we should find major currents corresponding to the earth's prevailing wind systems, and indeed this is the case. If the earth were covered entirely by water, there would be a great belt of west-east currents in each of the zones of the prevailing westerlies, a great belt of east-west currents in the zones of the trade winds, and a belt of east-west currents in the zones of the polar easterlies.

The Influence of the Continents

The continental land masses interrupt the circumglobal circulation described previously. The currents are deflected by the land masses, resulting in large circular current patterns called *gyres*. In an ocean such as the Atlantic, three gyres form in each hemisphere corresponding to the prevailing wind systems (Fig. 5.14). The largest are the subtropical gyres. They form when the westward-flowing *equatorial currents* in the trade wind zones run into the east coasts of the continents, are deflected poleward, and enter the zone of the prevailing westerlies. A strong current called the *west-wind drift* is set up here which carries warm water northeastward (southeastward in the Southern Hemisphere) to the west coasts of the bordering continents. Note 5 examines the relationship between ocean currents and the migration patterns of sockeye salmon and Atlantic eel.

Not all the water pushed against the western rim of the oceans by the equatorial currents is deflected to the western drift; some is also deflected equatorward and redirected into a return current called the *counter-equatorial current*. With this current flowing eastward along the equator, a second set of gyres is set up. Smaller and weaker than the subtropical gyres, these equatorial gyres form a narrow loop along each side of the equator.

Poleward of the subtropical gyres are the subpolar gyres. The currents that flow toward the poles carry warm water to higher latitudes, whereas those that flow toward the equator carry cool water to lower latitudes (see Fig. 22.8*b*).

The actual circulation of the oceans is remarkably similar to that of our hypothetical circulation (Fig. 5.15). The principal exception is found in the area north of Antarctica, where the ocean basin is not bounded by continents, and the westwind drift circles the entire earth. This is extremely important because without continents to deflect currents northward or southward, not much energy can be exchanged between tropical waters and cooler waters at higher latitudes. (Also see color plate 3).

The Gulf Stream is an example of a warm current that flows into the cold North Atlantic. The temperature of the Gulf Stream is usually only a few degrees warmer than that of the water of the Mid- and North Atlantic. But the huge volumes of water that this current carries northward contain enough additional heat to increase by more than twofold the sensible- and latent-heat flux into the atmosphere (Fig. 5.16). As a result, the atmosphere over the North Atlantic and the climate of Northwest Europe are measurably warmer and wetter than they would be otherwise. The opposite tends to be true where cold currents, such as the Humboldt Current along the Pacific Coast

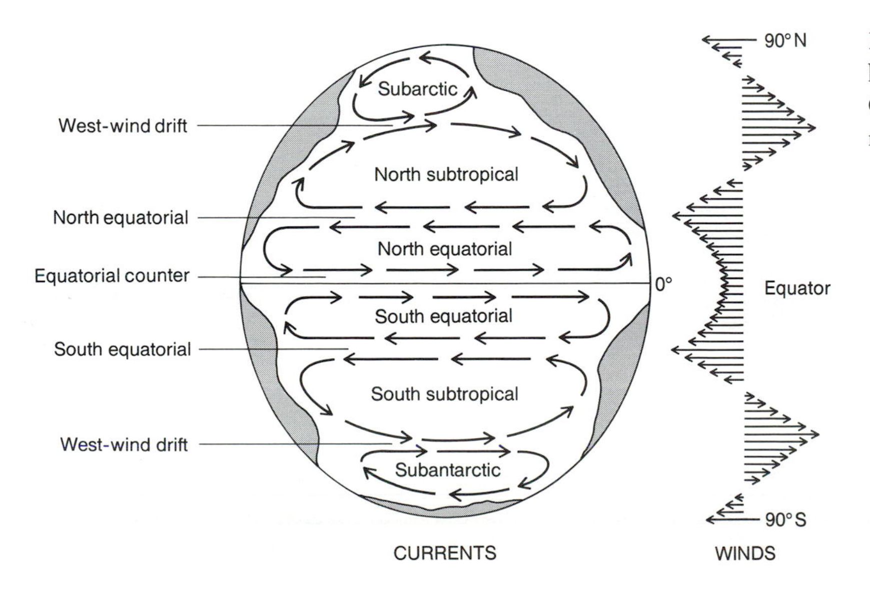

Figure 5.14 Surface currents for a hypothetical ocean surrounded by land. Corresponding winds are shown on the right.

THE VIEW FROM THE FIELD / NOTE 5

Marine Migration and Ocean Circulation

We began this chapter with some remarks about the development of early thought on ocean circulation. Navigators, however, are not the only travelers to take advantage of ocean currents, for many marine animals, including numerous fish species, mammals, and eels, also use them to aid their migrations. The life of most salmon are adapted to ocean currents. Salmon are *anadromous* fish, meaning that they spend most of their life at sea where they feed and grow and then return to the fresh water of their birth to reproduce and die.

The sockeye salmon is one of six salmon species in the North Pacific. Sockeyes are born in streams and lakes along the southern coast of Bristol Bay, Alaska. Young sockeyes remain in fresh water for about two years as they develop to a stage known as a *smolt*. The smolt migrate to the sea where they enter a period of heavy feeding and growth that takes them on a journey driven by ocean currents. The Oyashio Current takes them from Bristol Bay into the Pacific Ocean where they ride eastward with the North Pacific Drift. Off the coast of British Columbia the North Pacific Current is deflected by the land mass into a northward and southward arm. The northern arm, the Alaska Current, takes the salmon, now grown, along the Alaska Panhandle, then westward with the Aleutian Current to the southern margin of Bristol Bay and the streams of their birth where they spawn and die.

The Atlantic eel's migratory pattern is the reverse of that of the salmon. This eel spawns at sea and then migrates into fresh water lakes and streams where it grows to maturity. Although the details of the eel's migration pattern are cloudy, it appears that the Atlantic eel is born far out to sea in a region of the Atlantic called the Sargasso. The early life form is a minute organism, a larva, which rides the Gulf Stream and the North Atlantic Current from the Sargasso toward the coast of Western Europe. Somewhere off the European coast the larvae separate into two populations, one going to northwestern Europe and the other to southern Europe via the Mediterranean Sea. When they appear in the streams and lakes of these regions, they are designated as two different species. Here they mature and then complete their lives by returning to the Sargasso to reproduce.

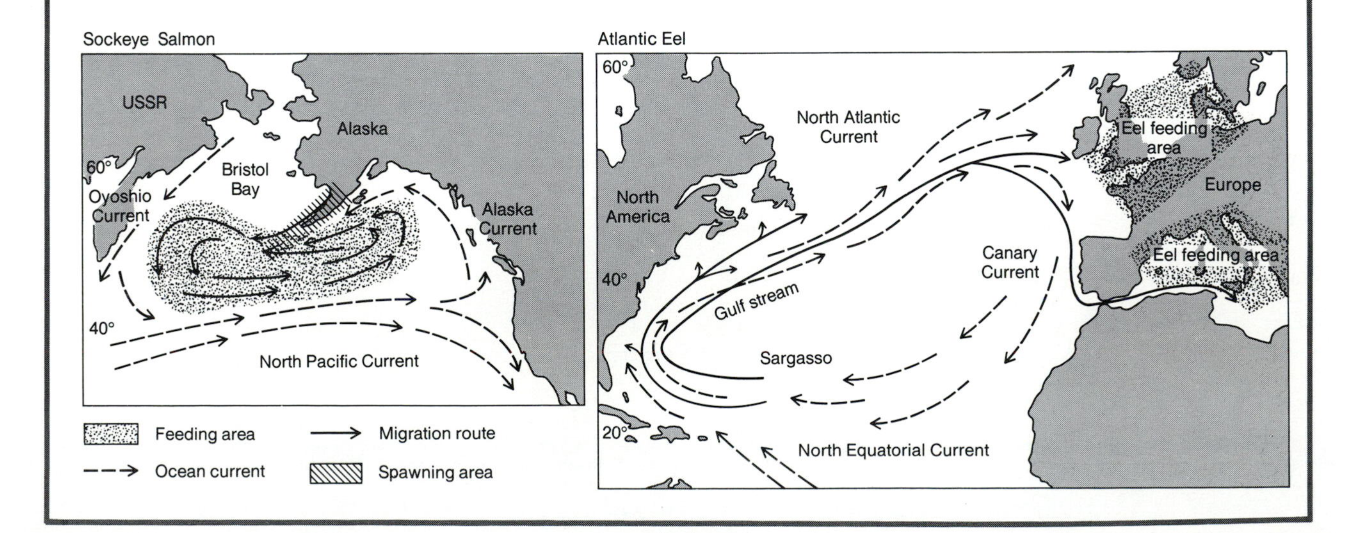

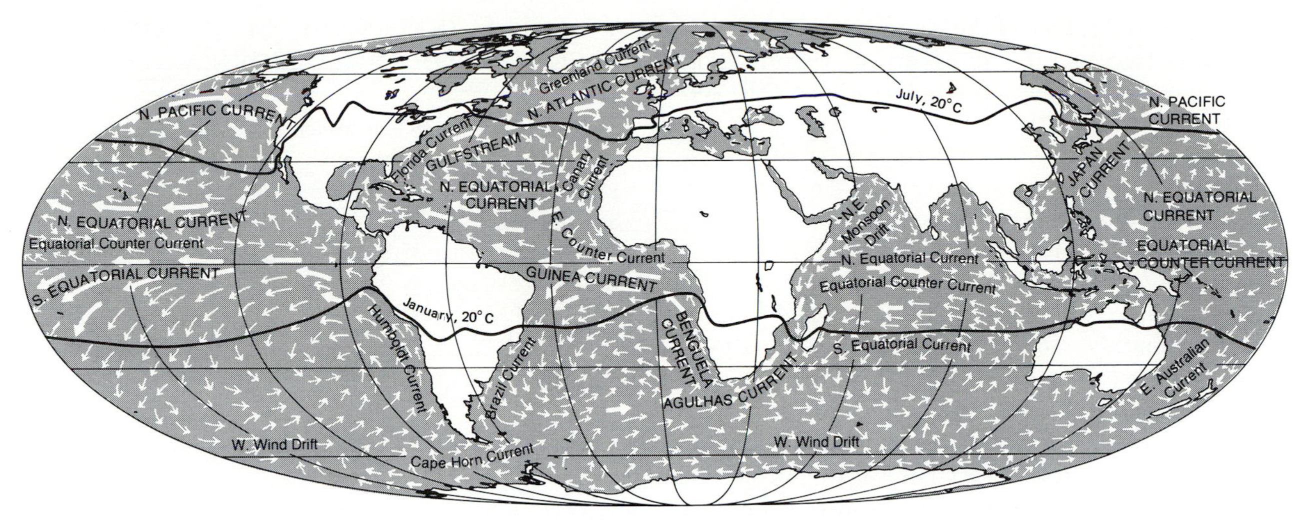

Figure 5.15 The actual circulation of the oceans. Major currents are shown with heavy arrows.

of South America or the California Current along the California Coast, move against a warm coastline (see Fig. 2.5). The coastal climate is drier and cooler because the current cools the air over it while giving up little moisture to evaporation. The details of ocean evaporation are taken up in Chapter 7.

One of the recent discoveries about the Gulf Stream, made possible by satellite monitoring of the current's behavior, is that great sections of the current break off to form whirls or rings in the ocean. These rings originate as meander loops in the Gulf Stream and rotate for months after they become detached from the current. Rings are released into waters on both the north and south sides of the current. Those on the north stand out as warm rings; those on the south, as cold rings; and each is identifiable in the satellite imagery not only by its thermal pattern, but also by the ecological conditions represented by phytoplanckton populations associated with it (Fig. 5.17; color plate 2).

The Influence of the Oceans on the Atmosphere

The oceans have a strong influence on the atmosphere, on this point there is little disagreement among scientists. But on the nature of the relationship and the mechanisms involved, there are many unan-

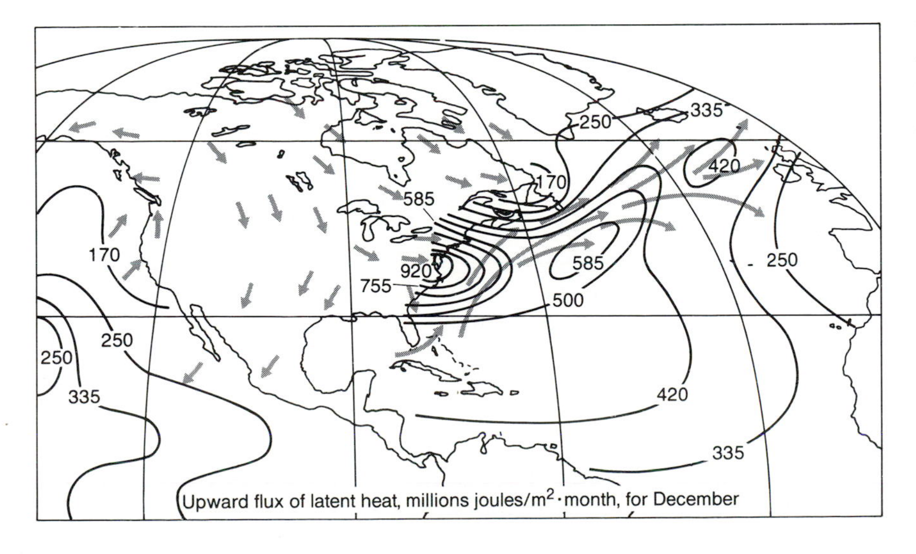

Figure 5.16 Latent-heat flux from the ocean into the atmosphere for the North Atlantic and Pacific Southwest of North America in December. The Gulf Stream and the prevailing surface winds are shown by the arrows. The large energy flux off the East Coast is related to the warm water brought northward by the Gulf Stream and the cold, dry condition of the air flowing off North America.

swered questions. One of the "hot" issues in this connection centers on the discovery of a correlation between ocean temperatures and weather trends and events in the midlatitudes.

We have long known that ocean temperatures oscillate over periods of several years in a manner broadly similar to that of atmospheric temperatures. In the past decade or so, we have been able to monitor temperature changes in the ocean surface with the help of satellite sensors and to build detailed maps of weekly, monthly, and seasonal thermal trends. A trend that is receiving particular attention from North American scientists is the growth and decline of a body of warm water in the equatorial zone of the eastern Pacific Ocean. Known as El Niño ("The Infant" in Spanish—an illusion to the Christ Child because it typically begins around Christmas), this mass of water develops (expands) every 5 years or so from an area off the coast of Peru. It is apparently initiated by changes in atmospheric pressure in the eastern Pacific which result in a decline in the easterly trade winds: when the trade winds fall, the eastward-flowing equatorial countercurrent rises (because of reduced resistance from wind), pumping thousands of cubic kilometers of warm water into the eastern Pacific. As El Niño builds up, warm water spreads outward into the Pacific as well as along the tropical west coasts of the Americas, where it displaces the cooler waters of the California and Humboldt currents. In the past, El Niños have been recorded in 1891, 1925, 1941, 1957, 1965, 1972, and 1982.

One of the strongest El Niños began in 1982. The warming trend started earlier in the year than in any episode of the past 30 years and quickly built up a great mass of warm water stretching from British Columbia to Chile. Air temperatures over this great area increased by 1 to 2°C. Upper atmospheric circulation over the Pacific and North America also changed as the jet stream assumed a winter position somewhat farther south than in past winters. Some scientists reason that these changes were behind much of the nasty winter weather of 1982–83 in the American West (Table 5.1). That winter saw huge storms ravage the California coast and exceptionally heavy precipitation and runoff in Utah, Colorado, and Nevada (Fig. 5.18).

Were the events associated with El Niño in 1982–83 unique to that particular El Niño occurrence? Apparently not, for an examination of past climatic records indicates that weather conditions oscillate on a fairly regular basis in response to El Niño events. One of the important manifestations of this weather variation, called the Southern Oscillation, is a dislocation of precipitation patterns resulting in dry conditions in some areas of substantial average precipitation and wet, stormy conditions in some areas of normally light precipitation.

TABLE 5.1 INFLUENCE OF EL NIÑO ON PRECIPITATION IN CALIFORNIA

Location	30-Year Mean inches (cm)	July 1982–June 1983 inches (cm)	Percent Increase
Eureka	98 (38.5)	151 (59.4)	54%
San Francisco	49 (17.3)	97 (38.2)	97%
Los Angeles	35.5 (14.0)	79.5 (31.3)	124%
San Diego	24 (9.3)	46.5 (18.3)	96%
Sacramento	45.5 (17.9)	95 (37.5)	109%
Bakersfield	14.5 (5.7)	25 (9.9)	74%

Source: Data from U.S. Weather Service, NOAA.

Summary

The transfer of energy by the atmosphere and oceans is critical to the earth's energy balance. Without a poleward transfer of energy from the low latitudes, the tropics (or intertropics) would be warmer and the high latitudes colder than they are. The general circulation of the atmosphere is a response to latitudinal differences in solar heating of the earth's surface, the distribution of land and water, the mechanics of the atmosphere's fluid flow, and the Coriolis effect. The Coriolis effect causes winds to deviate from a direct path along the pressure gradient. This explains both the great curved paths that winds take in moving from high- to low-pressure zones and the presence of the prevailing easterly and westerly winds between the major zones of pressure.

Airflow aloft tends to counterbalance flow near the surface; in the tropics this takes the form of Hadley cells. Oceanic circulation generally follows that of the prevailing winds, except that the currents are deflected where they meet land masses. The result is the formation of gyres in the ocean basins, with warm currents on one side and cold currents on the other. The influence of the oceans on the atmosphere is significant and one manifestation of this is El Niño and the Southern Oscillation.

Figure 5.18 Photograph showing some of the effects of the exceptional storms of the 1983 winter.

Concepts and Terms for Review

concept of global circulation
zonal redistribution of energy
Hadley circulation
pressure cells
  thermal and dynamic
  cyclone and anticyclone
wind direction
  pressure gradient
  Coriolis effect
upper atmospheric wind
  geostrophic wind
  Rossby waves
  jet stream
intertropical convergence zone
monsoon
trade winds
subtropical high pressure
westerlies
subarctic low pressure
polar easterlies
secondary circulation
oceanic circulation
  gyres
  equatorial current
  west-wind drift
ocean-atmosphere system
  El Niño
  Southern Oscillation

Sources and References

Aleem, A.A. (1967) "Concepts of Currents, Tides and Winds Among Medieval Arab Geographers in the Indian Ocean." *Deep-Sea Research,* Vol. 14, pp. 459–463.

Armi, L. (1978) "Mixing in the Deep Ocean—the Importance of Boundaries." *Oceanus,* Vol. 21 (Fall), pp. 14–19.

Bryan, K. (1978) "The Ocean Heat Balance." *Oceanus,* Vol. 21 (Fall), pp. 18–26.

Carpenter, W.B. (1870) "The Gulf Stream." *Nature,* Vol. 2, pp. 334–335.

Dietrich, G. (1963) *General Oceanography.* New York: Wiley/Interscience.

Dutton, J.A. (1976) *The Ceaseless Wind: An Introduction to the Theory of Atmospheric Motion.* New York: McGraw-Hill.

Emig, M. (1967) "Heat Transport by Ocean Currents." *Journal of Geophysical Research,* 72:10.

McDonald, J.E. (1952) "The Coriolis Effect." *Scientific American,* May, pp. 72–76.

McGowan, J.A. (1984) "The California El Nino, 1983." *Oceanus,* 27:2, pp. 48–51.

Palmén, E., and Newton, C.W. (1969) *Atmospheric Circulation Systems.* New York: Academic Press.

Rasmusson, E.M. (1985) "El Nino and Variations in Climate." *American Scientist,* 73:2.

Stewart, R.W. (1967) "The Atmosphere and the Ocean." *Scientific American,* September, pp. 27–36.

Chapter 6

Atmospheric Moisture, Precipitation, and Weather

On the average, the atmosphere precipitates nearly 100 cm (40 in.) of moisture over the earth each year. Not surprisingly, precipitation processes constitute one of the most important elements of weather and climate. What causes the condensation leading to precipitation? What weather processes are associated with the occurrence of different types of precipitation and how are they distributed geographically?

About midnight Joe awoke, and called the boys. There was a brooding oppressiveness in the air that seemed to bode something. The boys huddled themselves together and sought the friendly companionship of the fire, though the dull dead heat of the breathless atmosphere was stifling. They sat still, intent and waiting. The solemn hush continued. Beyond the light of the fire everything was swallowed up in the blackness of darkness. Presently there came a quivering glow that vaguely revealed the foliage for a moment and then vanished. By and by another came, a little stronger. Then another. Then a faint moan came sighing through the branches of the forest and the boys felt a fleeting breath upon their cheeks, and shuddered with the fancy that the Spirit of the Night had gone by. There was a pause. Now a weird flash turned night into day and showed every little grassblade, separate and distinct that grew about their feet. And it showed three white, startled faces, too. A deep peal of thunder went rolling and tumbling down the heavens and lost itself in sullen rumblings in the distance. A sweep of chilly air passed by, rustling all the leaves and snowing the flaky ashes broadcast about the fire. Another fierce glare lit up the forest, and an instant crash followed that seemed to rend the treetops right over the boys' heads. They clung together in terror, in the thick gloom that followed. A few big raindrops fell pattering upon the leaves.

"Quick, boys! go for the tent!" exclaimed Tom.

They sprang away, stumbling over roots and among vines in the dark, no two plunging in the same direction. A furious blast roared through the trees, making everything sing as it went. One blinding flash after another came, and peal on peal of deafening thunder.

And now a drenching rain poured down and the rising hurricane drove it in sheets along the ground. . . . It was a wild night for homeless young heads to be out in.

But at last the battle was done, and the forces retired with weaker and weaker threatenings and grumblings, and peace resumed her sway. . .

This was Mark Twain's description, in *The Adventures of Tom Sawyer*, of a thunderstorm—the most intensive source of rainfall on earth. It makes for exciting reading, especially when set in the context of the three runaway boys living on an island in the Mississippi River. But it is also of scientific interest because events such as these coupled with the general character of precipitation over the seasons and years are important considerations whenever we examine the climate of a place on earth.

How much does it rain? Does it snow? When does it rain? Is there a rainy season? Is rainfall characterized by drizzles or torrential downpours? These questions can usually be answered by examining the climatic data recorded at a weather station. But the reasons for the particular characteristics of the precipitation of a place are more difficult to discover. In this chapter we take up the question of why it rains and snows, examining the atmospheric processes that cause condensation of water vapor and the types of precipitation.

WATER VAPOR IN THE ATMOSPHERE

Since precipitation comes from the atmosphere, it must originate with the change of water vapor into particles of liquid or solid water. The clouds we see in the atmosphere in fact represent an important phase in that process because they are made up of tiny liquid droplets or ice crystals formed from the condensation of water vapor. Clouds may or may not produce precipitation. Precipitation occurs when cloud droplets become too heavy for the atmosphere to support and they fall to the earth.

To undertand how and why clouds and precipitation form, it is necessary to examine water vapor. First, recall that water vapor is the most variable gas in the lower atmosphere, ranging from 0 to as much as 4 percent. Second, it is important to understand that the maximum amount of water vapor that can be held in air is dependent on air temperature. The warmer air is, the greater the amount of water vapor it can hold; conversely, the colder air is, the smaller the amount of water vapor it can hold. This helps explain why—considering the range of temperatures that exist in the atmosphere—the water vapor content of the atmosphere is so variable. When a body or parcel of air is holding the maximum amount of water vapor for its temperature, it is said to be *saturated*.

Measures of Humidity

The water vapor content of air is referred to as *humidity*. Generally, when we say that air is humid we mean that its vapor is high. But what this means in terms of the actual moisture content is uncertain unless we specify what measure of humidity we are using. Humidity can be expressed in a variety of ways, and we want to familiarize ourselves with three expressions: absolute humidity, specific humidity, and relative humidity.

Absolute humidity is a measure of the weight (in grams) of the water vapor in a parcel of air with a volume of one cubic meter. Absolute humidity is a useful means of expressing the water content of a large body of air, such as the great air masses that migrate across the midlatitudes, where we are interested in how much moisture is available for precipitation. On the other hand, it has a distinct disadvantage in weather analysis because it is based on a volumetric

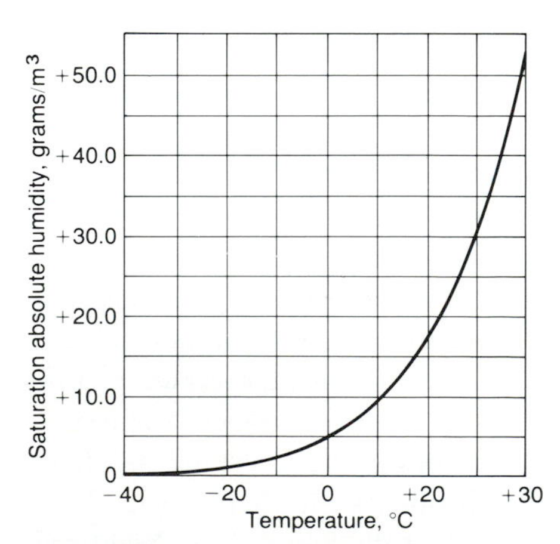

Figure 6.1 The relationship between temperature and saturation absolute humidity.

unit and the volume of air changes with air pressure. As a cubic meter of air expands upon being elevated into the atmosphere, its absolute humidity declines, even though no vapor has been gained or lost. The maximum amount of vapor that can be held by a cubic meter of air is called *saturation absolute humidity*, and it increases rapidly with air temperature (Fig. 6.1).

Specific humidity is a measure of the weight of water vapor (in grams) to the weight of the air holding it (in kilograms, including the weight of the water vapor). This expression is widely used in weather analysis because the moisture content of air can be evaluated regardless of changes in its volume. For example, a parcel of air can undergo expansion as it rises within a thunderstorm, but if no vapor is lost or gained, its specific humidity remains the same because the total number of molecules of air—that is, the weight of the gases—has not changed.

Relative humidity, the most commonly used measure of water vapor content of air, expresses vapor content relative to the amount of vapor that can be held in a parcel at saturation. Relative humidity is always expressed as a percentage; the higher the number, the nearer the air is to saturation. Relative humidity can be changed by varying either the temperature of air or its vapor content. If vapor content is held constant, relative humidity goes up when air temperature is lowered and down when air temperature is raised. Although air at 100 percent relative humidity is described as saturated, under certain conditions humidity can rise above 100 percent. Such air is called *supersaturated* air, and it usually involves dramatic cooling of air which is resistant to condensation (usually very clean air, for reasons that will become apparent in the next section). The temperature at which air is saturated is called the *dew point* of the air.

How to Measure Humidity

The humidity of the air can be measured in a number of ways. Two of the most common devices for making such measurements are the psychrometer and the hair hygrometer. A *psychrometer* is a simple instrument for measuring humidity based on the principle of cooling by evaporation. Evaporation of water requires that the latent heat of vaporization be given up; therefore, an evaporating surface will be cooler than a dry surface if other environmental conditions are equal. But the rate of evaporation and thus the amount of cooling depend on the humidity of the receiving air. If air is saturated, no evaporation can take place; if it is dry, say, 20 percent relative humidity, lots of evaporation is still possible; therefore, the loss of surface temperature is controlled by atmospheric humidity.

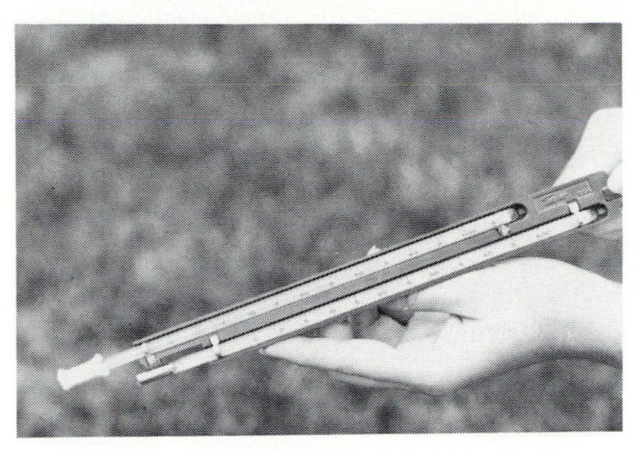

Figure 6.2 Sling psychrometer. The wetbulb thermometer is covered with the gauze sock.

The psychrometer is made up of two thermometers—one with a water-saturated gauze sock attached to the bulb and one without. If we circulate air over the thermometers, the drybulb thermometer will record the actual air temperature, whereas the wetbulb thermometer will record a depressed temperature caused by the evaporation of water from it. Humidity is computed on the basis of the air (drybulb) temperature and the *difference* between the dry- and wetbulb temperatures (Fig. 6.2).

A *hair hygrometer* is even simpler than a psychrometer; it is based on the expansion and contraction of hair with humidity changes. Like many other organic materials, human hair expands when wet and contracts when dry. A typical expansion value for a human hair is 2.5 percent from 0 to 100 percent relative humidity. Humidity can be measured by connecting a bundle of hairs to a pen arm which is attached to a dial. We commonly see hair hygrometers mounted on a desk or wall in homes; however, they are not very accurate as a whole and should be used as a humidity indicator rather than a scientific instrument.

AIR MOVEMENT AND MOISTURE CONDENSATION

Condensation is the physical process by which water changes from the vapor to the liquid phase. Condensation occurs when air is cooled below its dew point. The simplest way air can be cooled is for it to pass over a colder surface. This is what happens when you open a freezer; warm air rushes over the cold surfaces and cools instantly, briefly forming a light cloud of condensation near the door. In the atmosphere, this process of warm air cooling upon contact with a cool surface is known as *advection*, and it is very common in coastal areas.

On the west coast of North America, coastal fog often originates when warm, moist air from the Pacific Ocean blows over cold currents that run along the shore. The fog, which is called *advection fog*, is driven over the land by the westerly winds, but within a kilometer or so inland, the air is warmed and the fog dissipates (Fig. 6.3). Given the reverse arrangement of warm water and cold land, condensation increases inland. During the winter in the Great Lakes region, for example, air blowing from the relatively warm lake surfaces (at about 1 degree C) over the cold land surface (at about −5 degrees C) causes heavy snowfalls on the east sides of the lakes.

But the cooling processes that induce most precipitation involve rising air and usually form some kind of *convection*. Air may rise spontaneously because it is warmer and lighter than surrounding air, or it may be forced aloft as an air mass moves over a mountain range or a colder, denser air mass. In all such cases, the rising air undergoes pressure and temperature changes that may lead to cloud formation and precipitation.

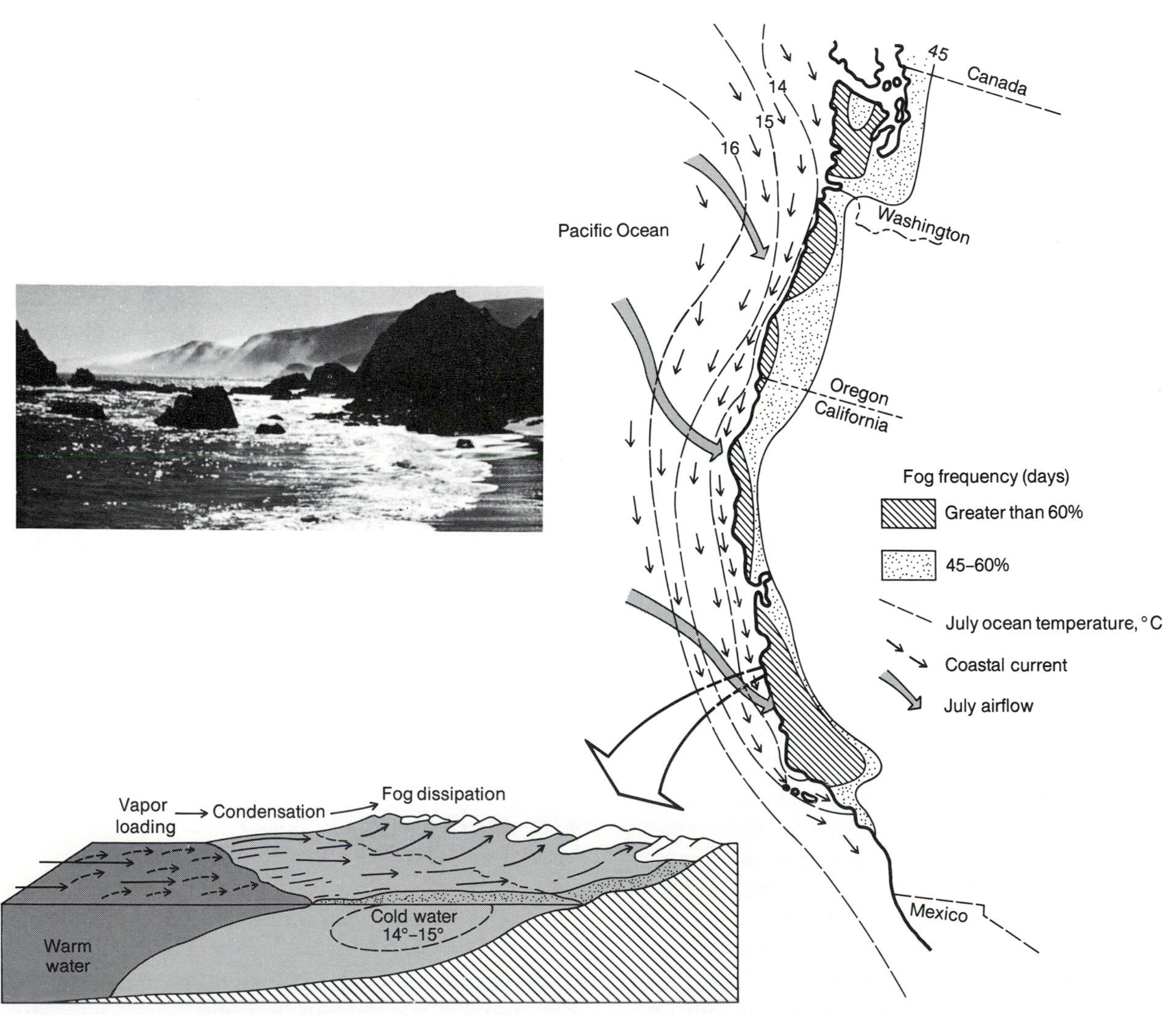

Figure 6.3 The distribution of coastal fog produced by advective cooling of onshore winds along the middle part of the west coast of North America.

The spontaneous ascent of air in the atmosphere is termed *instability*. We examined it briefly in Chapter 4 in connection with the generation of wind by free convection. We now want to examine it in connection with condensation and precipitation.

Atmospheric Instability

The concept of atmospheric instability can be illustrated using a model called the *parcel method*. This model takes a large bubble of air, called a parcel, and measures its behavior when displaced upward into the atmosphere. The model assumes that no mixing takes place between the parcel and the surrounding air.

When the parcel is displaced, just two results are possible: (1) The parcel accelerates upward, indicating that atmospheric conditions are unstable; or (2) the parcel returns to its original elevation, indicating that atmospheric conditions are stable. Two conditions determine which of these will take place; (1) the rate of temperature change within the ascending air parcel, called the *adiabatic lapse rate*, and (2) the rate of change with elevation in the air through which the parcel is moving, called the *ambient atmospheric lapse rate*. The reason we use temperature instead of density in problems of atmospheric instability is that air temperature changes directly with changes in density, and because temperature is so easy to measure, temperature is taken as the measure of density. When the parcel has a higher temperature than the surrounding air, it is less dense (that is, lighter) than the surrounding air, and it rises; likewise, when the parcel is cooler than the surrounding air, it is heavier (denser) and sinks. But remember that ambient air pressure (density) changes with altitude; therefore, the process also involves temperature changes *within* the parcel as it rises or falls in the atmosphere.

When air rises, it expands because of the decreased pressure at higher altitude. This expansion involves work and since work is driven by energy, we must account for an energy source whenever it takes place. Heat is the source of energy in expanding air; therefore, when expansion occurs heat is removed from the air, and its temperature falls. Likewise, when air is forced to descend, it is compressed; the work done is converted into heat, and the temperature of the air rises.

The rate of cooling in rising air, called the *dry adiabatic lapse rate*, is −0.98 degrees C/100 m, or approximately −1 degree C per 100 m. We use the minus sign to indicate that the air cools as its height increases. For example, a parcel of air at 20 degrees

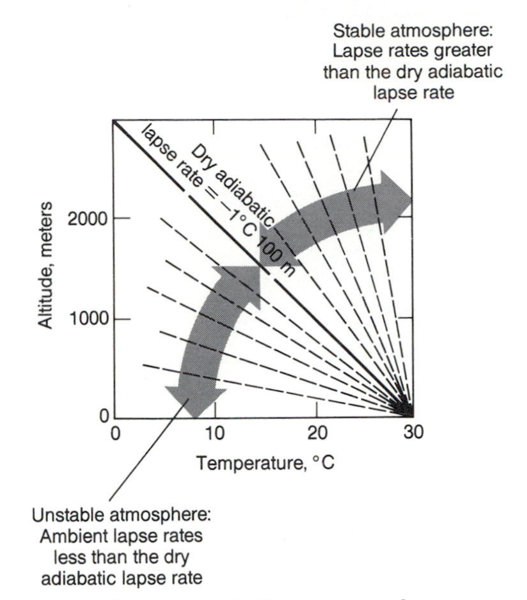

Figure 6.4 A graph illustrating ambient atmospheric lapse rates conducive to atmospheric instability (less than the dry adiabatic rate) and atmospheric stability (greater than the dry adiabatic rate).

C at sea level will cool to 13 degrees C if raised to 700 m. If the air is brought back down again, it will warm to 20 degrees C. Note that these temperature changes are a result solely of expansion and compression, *not* of interchange with air of a different temperature.

The dry adiabatic lapse rate is the *same* for all air parcels, as long as no condensation takes place within them. (If condensation takes place the rate changes, a condition we will discuss later.) On the other hand, the ambient atmospheric lapse rate *is* highly variable from time to time and place to place, and therefore turns out to be the major control on atmospheric stability (Fig. 6.4). To understand this fact, imagine a situation in which that ambient lapse rate is *greater* than −1 degree C per 100 m, say, −1.5 degrees C per 100 m, so that as a parcel rises and cools at the dry adiabatic rate, the air around it is always cooler as it goes up. After 1000 m of altitude, the parcel will have cooled 10 degrees C while the surrounding air will have cooled 15 degrees C; being warmer than the air around it—and therefore, lighter—the parcel will continue to rise (Fig. 6.5*a*).

Now let us take the parcel up another 1000 m. But in this case the ambient lapse rate is *lower* than the dry adiabatic rate, say, only −0.5 degrees C per 100 m; therefore, as the parcel goes up, it will cool twice

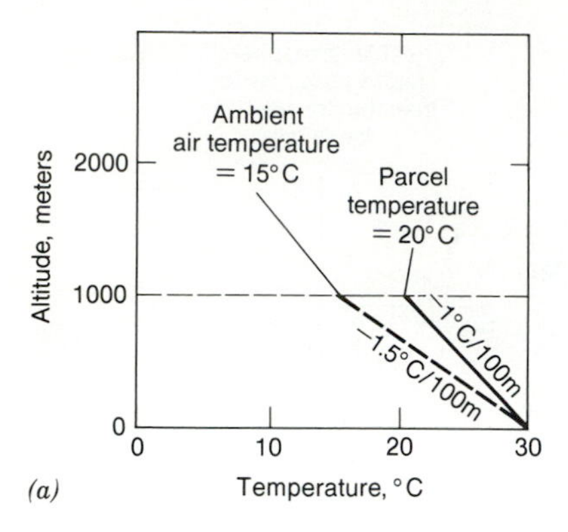

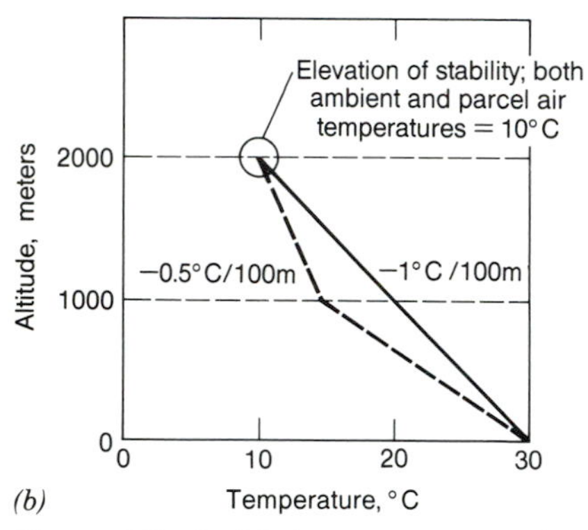

Figure 6.5 In (*a*) the temperature of a parcel of air (20°C) relative to the ambient air temperature (15°C) after an ascent of 1000 m; (*b*) the same parcel after an additional 1000 m in which the ambient atmospheric lapse rate was −0.5 per 100 m. At 2000 m, the parcel and surrounding air are at the same temperatures and stable.

as fast as the temperature declines in the surrounding air. At the 2000 m mark, the parcel will have cooled by a total of 20 degrees C (10 degrees in the first 1000 m plus 10 degrees in the second 1000 m), and the temperature of the surrounding air will have cooled by a total of 20 degrees C (15 degrees C in the first 1000 m plus 5 degrees C in the second 1000 m) (Fig. 6.5*b*). Inasmuch as both the air parcel and the air around it are at the same temperature, they are also of equal densities, and the condition is *stable*.

This illustration of instability and stability in a 2000 m layer of atmosphere is helpful in understanding a form of extreme stability called a *temperature inversion*. In a temperature inversion the ambient atmospheric lapse rate reverses, that is, goes from cool to warm with altitude rather than from warm to cool (Fig. 6.6). With cool (denser) air below and warm (lighter) air above, no vertical mixing is possible. A parcel displaced upward enters a warmer zone and being much colder (denser) returns to its original position. Temperature inversions are caused by a number of processes—cold air sliding under a warmer body of air, cooling of ground-level air by loss of radiation, for example—and we will discuss some individual examples of them later in this chapter.

If a rising parcel of air containing moisture cools to its dew point, it will be saturated. Any further increase in elevation will cause more cooling, and condensation will set in, an event that will be marked by cloud formation. When the water vapor condenses, heat will be given off, causing warming of the air parcel.

The effect of this process on the rising parcel is to make it more unstable by reducing the rate of adiabatic cooling. The saturated air will still cool as it rises, but at a lower rate, called the *wet* (or saturated) *adiabatic lapse rate*. The wet adiabatic lapse rate is variable, depending on the amount of condensation taking place. Table 6.1 gives the range of values for

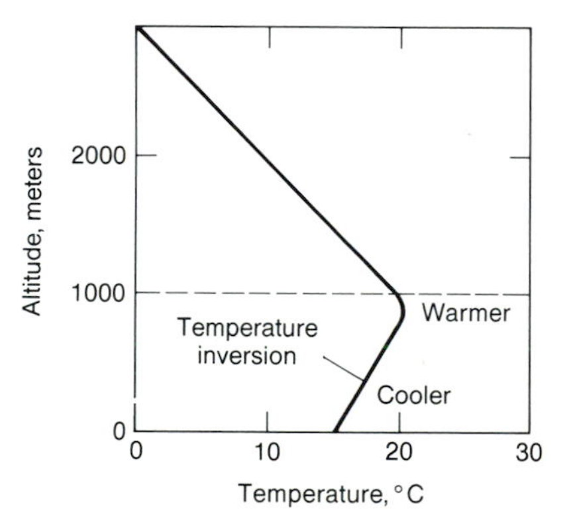

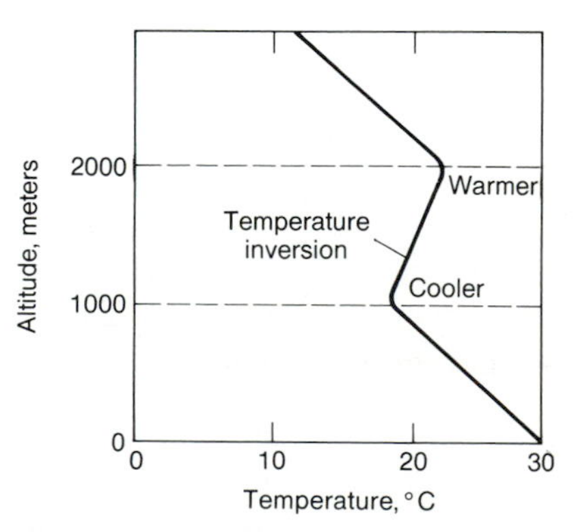

Figure 6.6 Ambient atmospheric temperature profiles illustrating inversions; one from 0 to 1000 m and the other from 1000 m to 2000 m. Also see Figure 6.11.

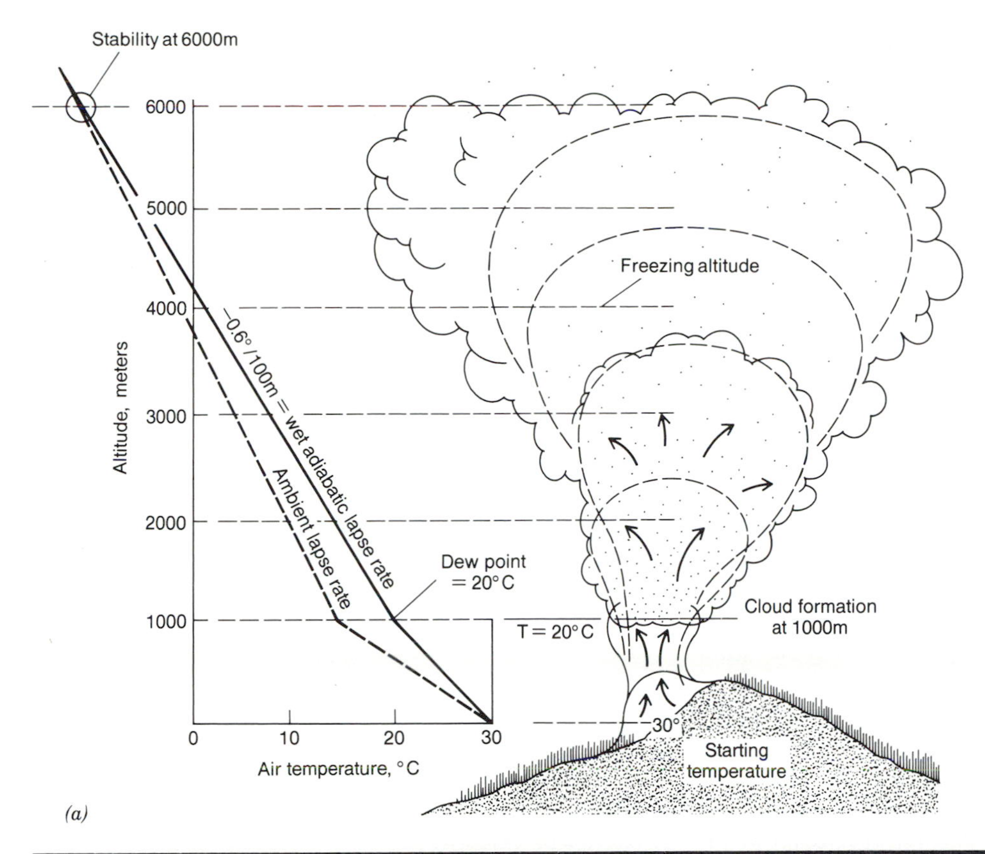

(a)

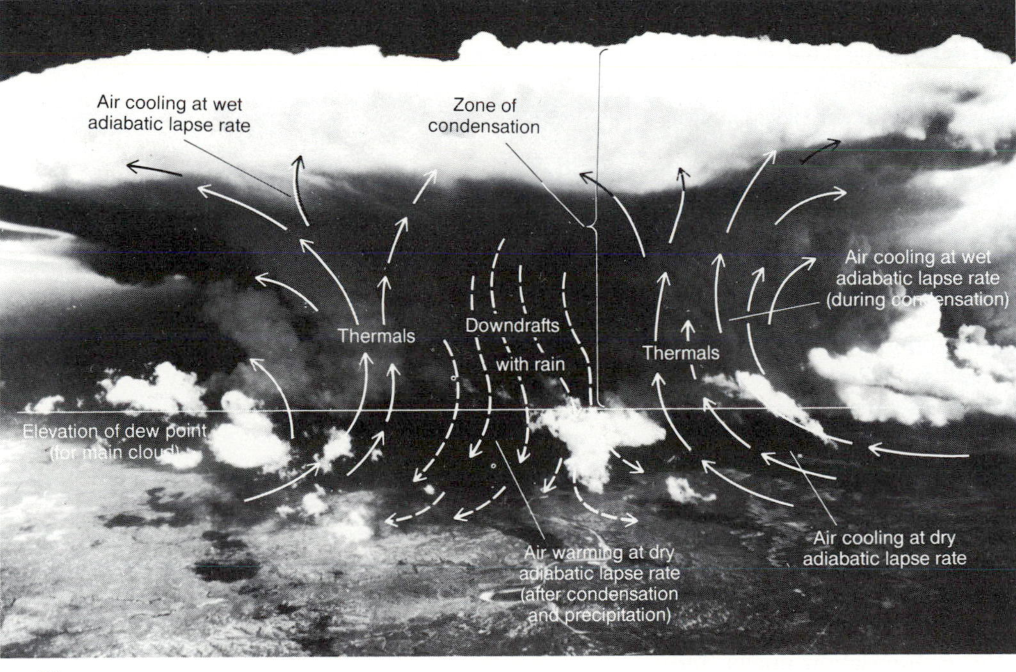

(b)

Figure 6.7 (*a*) Dry and saturated adiabatic lapse rates associated with a rising parcel of air starting at a ground temperature of 30°C. Ambient air temperature is the temperature of the air around the rising parcel, represented by the broken line. In (*b*) the wet and dry adiabatic lapse rate associated with upward and downward moving air in a convection cell.

TABLE 6.1 SATURATED ADIABATIC LAPSE RATE (°C/100 m)

	Pressure (mb)			
Temperature (°C)	1000	850	700	500
40	−0.30	−0.29	−0.27	
20	−0.43	−0.40	−0.37	−0.32
0	−0.65	−0.61	−0.57	−0.51
−20	−0.86	−0.84	−0.81	−0.76
−40	−0.95	−0.95	−0.94	−0.93

Note: At very cold temperatures, the saturated adiabatic lapse rate is almost equal to the dry adiabatic lapse rate of −0.98°C/100 m because very cold air, as noted earlier, can hold very little water vapor. At higher temperatures, the difference is much greater because more water is condensing.

the adiabatic lapse rate; an average value is −0.6 degrees C/100 m. Note that the adiabatic lapse rate applies only to air that is saturated. For purposes of calculating temperature change with elevation change, air at 80 percent relative humidity is considered "dry" until it reaches its dew point (Fig. 6.7*a*).

Because most of the water vapor that condenses will fall from the air as precipitation, the saturated adiabatic lapse rate, unlike the dry, is generally not reversible. Therefore, air that is lifted beyond its elevation of saturation and releases its water vapor will be *warmer* if it is returned to its original elevation. This will happen because the air is dry (nonsaturated) on its descent and therefore warms at the dry adiabatic rate (+1 degree per 100 m), which is nearly a half degree greater per 100 m than its wet cooling rate on the way up (Fig. 6.7*b*).

Precipitation Processes and Forms

In the atmosphere, condensation takes place on *condensation nuclei,* very small particles in the atmosphere which are composed of dust or salt. When condensation initially occurs, the droplets or ice particles are very small and are kept aloft by the motion of the air molecules. For example, in a fog (a ground-level cloud) the water droplets remain suspended in the air. In order for the droplets to fall, they must grow by coalescing with one another.

Once minute droplets have formed, the droplets themselves begin to act as condensation nuclei. Ordinarily, such droplets would tend to repel one another, but in the presence of an electric field in the atmosphere, they attract one another and grow larger, eventually falling through the atmosphere. Some of them may evaporate in the atmosphere between the clouds and the ground.

Most precipitation falls in the *form* of *rain,* meaning that it arrives at the ground in liquid form. It may have condensed as water, or it may have condensed into a frozen state and melted while falling. If the dew point at the condensation altitude is below freezing, the water vapor in the air condenses into *snow* crystals and may fall to the ground in this state. Small amounts of liquid water may condense onto the snow crystals as it falls, and in this case the snow is said to be *rimed.* When a snow crystal collides with a raindrop, a small frozen ball, or *graupel,* is formed. *Sleet* is rain that freezes as it falls. *Hail* forms under conditions of strong atmospheric turbulence; a small snow crystal drops into the zone where temperatures are below freezing, collects additional water, is lifted into the frozen zone again, descends once more, is swept aloft yet again, and so on over and over. In this fashion hailstones can grow to large sizes, several centimeters in diameter.

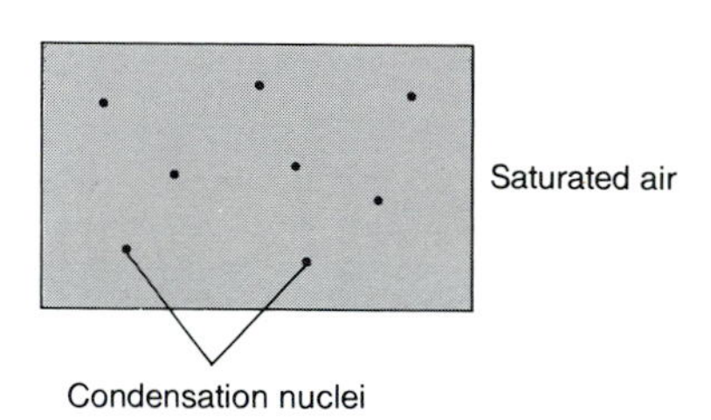

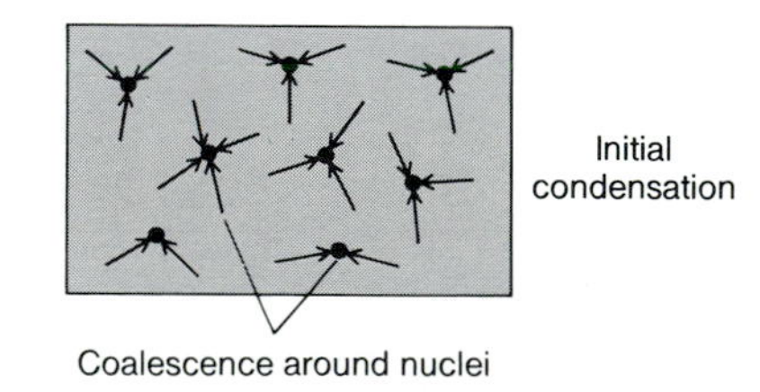

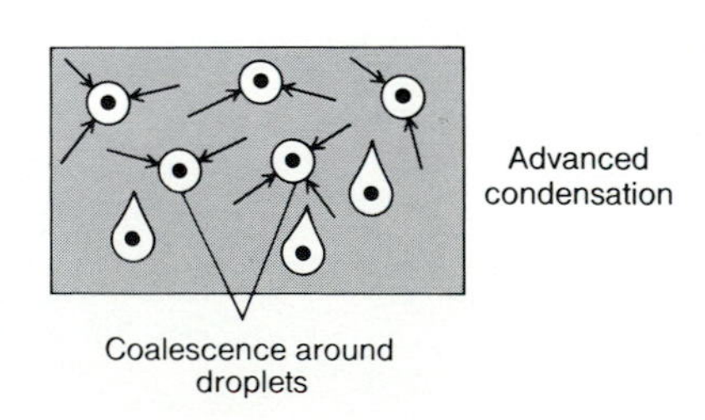

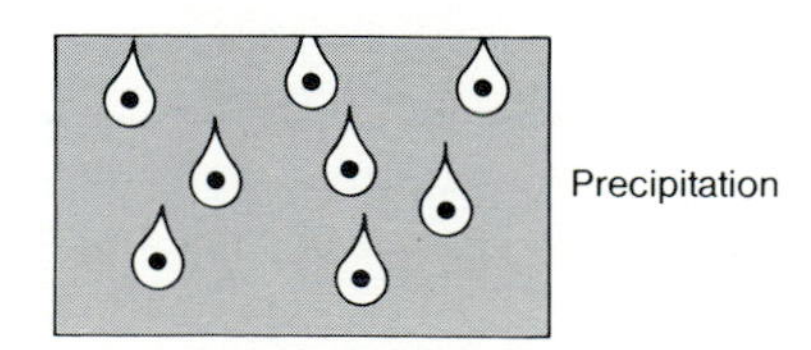

PRECIPITATION TYPES AND ASSOCIATED WEATHER

Because virtually all precipitation results when air cools by rising through the atmosphere, we can classify precipitation according to what causes the rise. Four principal causes are identifiable, and they produce four *types* of precipitation:

- *Orographic*-caused by airflow over high terrain, usually a mountain range along which air is forced upwards.
- *Cyclonic/frontal*-caused by the meeting of air masses of different densities, in which warm air is pushed upward by colder air.
- *Convectional*-caused by the free rise of unstable surface air, usually due to the intensive heating of air near the ground.
- *Convergent*-caused by barometric or topographic sinks, where air is drawn into large low-pressure areas and then rises.

Orographic Precipitation

Orographic precipitation is the easiest type to describe. It results when moisture-laden air is forced to pass over a mountain range, inducing cooling and condensation (Fig. 6.8). In areas where mountains lie in the paths of moist, prevailing winds, annual precipitation rates can be quite high. Virtually all areas with rainfall above 500 cm per year are in such situations. Examples of areas of heavy orographic rainfall are the mountains of Hawaii, the Himalayan mountain front, and the west coasts of North and South America. An extreme example often cited by climatologists is the town of Cherrapunji in the Assam Hills of northeastern India, where rainfall between October 1, 1860, and September 30, 1861, amounted to 26.5 meters (87 feet)! And 8.9 m (350 inches) fell in July alone! In Note 6 geographers Bryce Decker and John Street comment on the remarkable rainfall gradients associated with orographic precipitation in the Hawaiian Islands.

Most cases of extremely heavy orographic rainfall result from a combination of the forced rise of air as

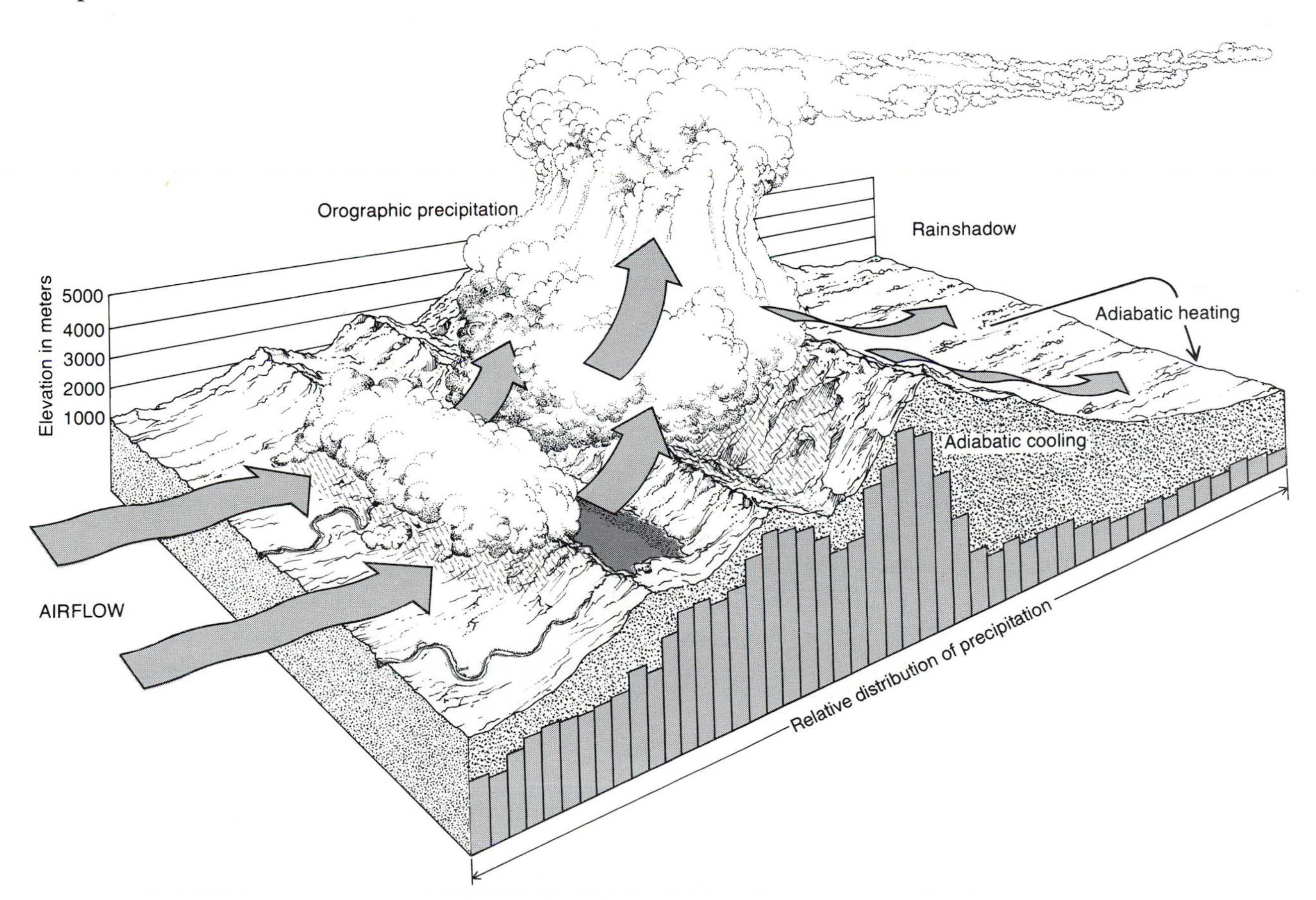

Figure 6.8 Orographic precipitation results when moist air is forced over a mountain range. Relative humidity drops when the air descends the leeward slope and heats adiabatically.

THE VIEW FROM THE FIELD / NOTE 6

The Amazing Precipitation Gradients of Hawaii

Waikiki Beach, one of America's favorite sun spots, lies only about 6 miles (10 km) from a mountain ridge that is shrouded in clouds most of the year. Between Waikiki and this ridge lies the City of Honolulu, and near the foot of the ridge itself is the campus of the University of Hawaii. From the summit of the ridge to Waikiki the annual average rainfall decreases 125 inches (310 cm), equivalent to one inch (2.5 cm) of change per city block! Incredibly, from the campus side of Honolulu to Waikiki the climate changes almost as much as it does from southern to central Mexico, a distance of more than 500 miles.

Even more extreme is the precipitation gradient on the Island of Waialeale, northwest of Honolulu. On the windward slope of Waialaele, annual average rainfall changes at a rate of 118 inches per mile (186 cm per km) over a distance of 2.5 miles. On Mount Waialaele, elevation 5148 ft. (1569 m), the average annual rainfall is 486 inches (1234 cm)—one of the highest rates in the world. Nineteen miles away, in the tropical desert at Barking Sands, annual precipitation averages less than 20 inches (50 cm). (See accompanying graphs.)

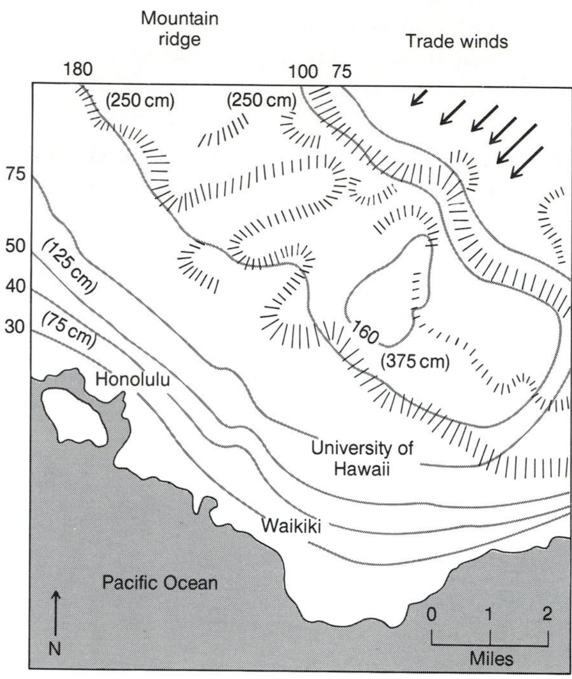

Rainfall isohyets: inches of rainfall per year

These contrasts in rainfall make the Hawaiian Islands one of the most varied geographic settings in the world. The reasons behind this diversity are related to (1) the geographic situation of the Hawaiian Island chain, lying in the path of the easterly trade winds, and (2) the mountainous terrain of the islands. When the trade winds cross the islands, a strong orographic effect is created as this otherwise stable air is cooled in its flow up the mountain slopes. The trade winds, which blow from the northeast, prevail 50 to 80 percent of the time in the winter months and 80 to 95 percent of the time in summer.

Without the influence of the islands, the trade winds would produce little precipitation, about 25 inches (63 cm) per year. The stability of trade-wind air in general is associated with a high-level temperature inversion, often referred to as a trade-wind inversion. This inversion is produced by adiabatic heating of high-altitude air descending in the eastern part of a subtropical high-pressure cell that lies over the Hawaiian region. Near Hawaii the inversion persists at elevations between 5000 and 7000 feet (1500 to 2100 m). It is an effective lid to instability, so that only very exceptionally do thunderstorms develop over the sea.

Above the trade-wind inversion, the air is bone-dry: relative humidities are as low as 5 percent. The broad volcanic summits that project above it—such as Mauna Loa and Mauna Kea, both above 13,000 feet (4000 m) elevation—are cool deserts. What little precipitation they do get—less than 15 inches (40 cm) annually—occurs mostly when storm systems pass through the region and disrupt the normal trade-wind regime.

Below the inversion, however, the trade-wind air is warm and humid. As it sweeps up the slopes of the Hawaiian mountains, it yields shower after shower from the cap of cloud that results as adiabatic cooling transforms its abundant vapor to rain and mist. Rainforest districts are confined to the lower slopes of the great volcanoes on the islands of Maui and Hawaii, where maximum average annual rainfall of about 300 inches (75 cm) occurs around 3000 feet (900 m) above sea level. On other islands, most elevations lie below the inversion, so that the

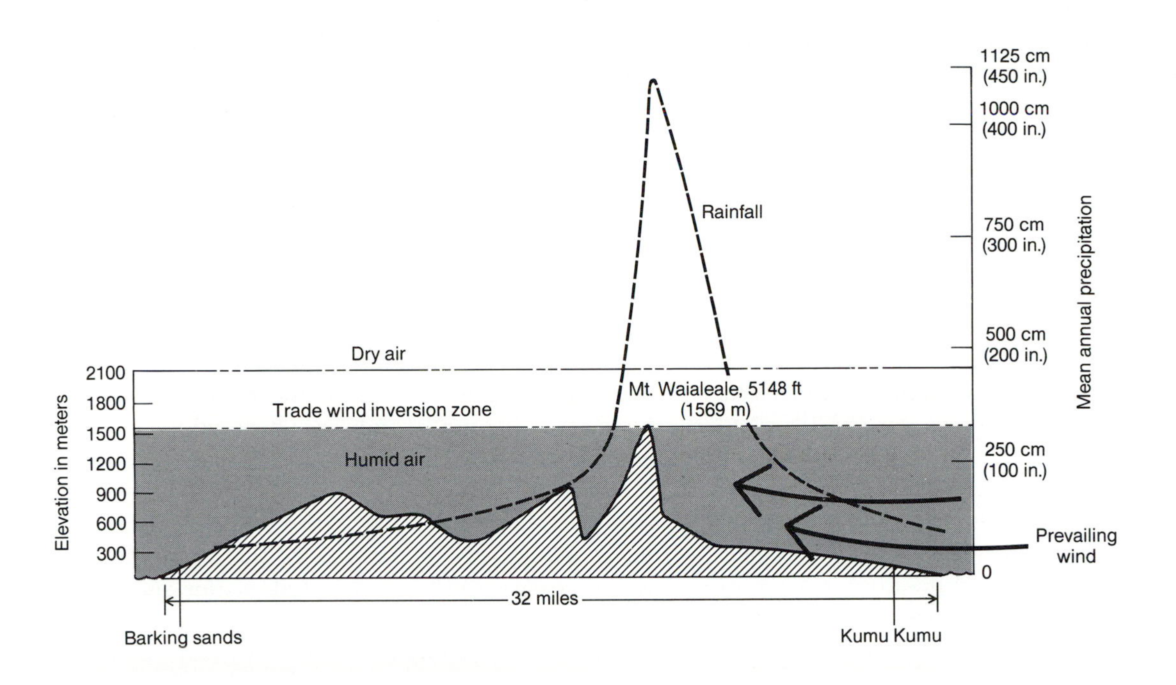

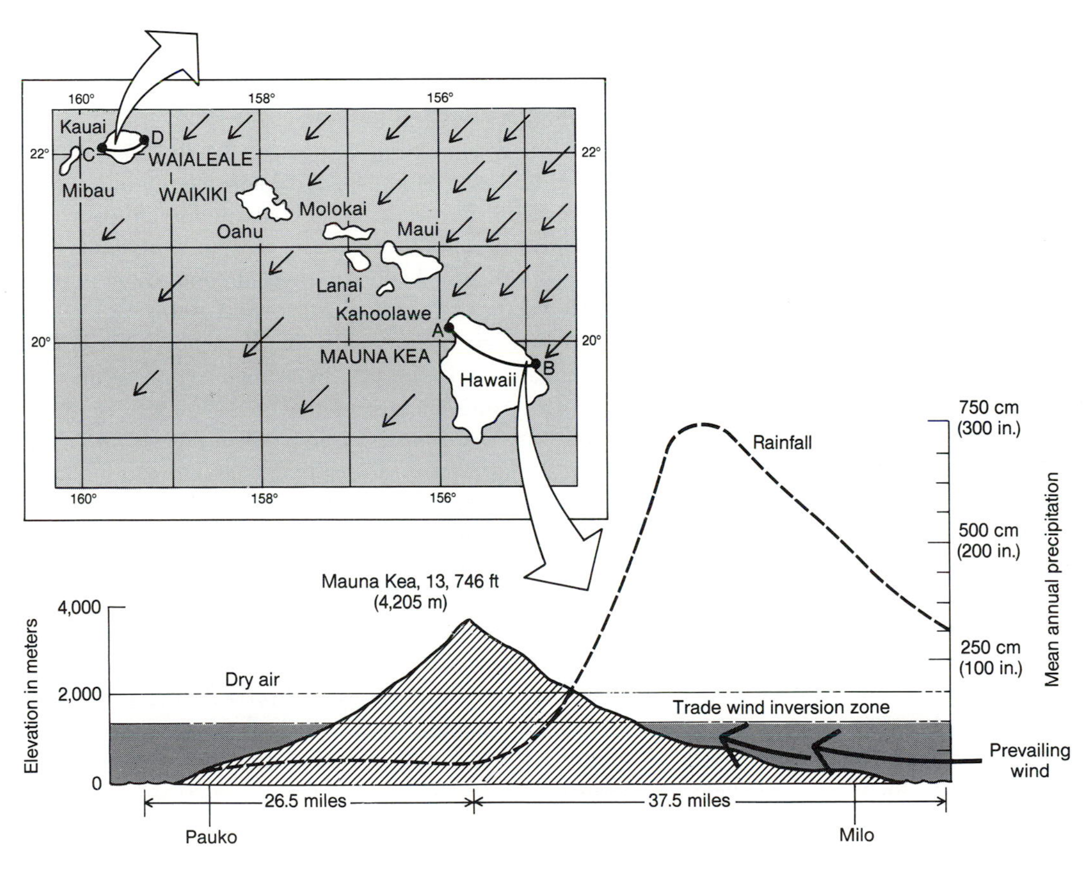

trade winds can blow over the summits, and the heaviest rainfall nearly coincides with the highest elevations. Leeward shores lie in rainshadows that are distinctly drier and sunnier than the windward coasts, and their beaches are favorite resort-hotel sites.

Bryce Decker and John Street
University of Hawaii

it passes over the mountain barrier and thermal convection caused by the release of heat in condensation. When massive amounts of latent heat are released, thunderstorms may form, building up thousands of meters over the mountain range. Orographic precipitation in the midlatitudes is often associated with an additional condition: air mass movement, which takes place as large bodies of maritime air are swept against coastal mountain ranges by westerly airflow.

Figure 6.8 illustrates the sequence of orographic precipitation. The leeward side of the mountain range—that is, the side away from the prevailing wind—is typically quite dry and is termed the *rainshadow*. When the air descends the leeward slopes, it warms adiabatically and can become very hot and dry, resulting in rapid desiccation (drying) of the land. When such winds occur during the winter, they may produce a 20° to 25°C rise in temperature in a matter of hours. The effects can be dramatic: rapid melting of snow, avalanches, and sprouting of spring flowers. In areas where these winds are common, they are significant events in the lives of people living in their path. For example, they appear to affect behavior: the incidences of mental disorders ("transient situational disorders," in the language of the psychologist) and of violent crimes are higher than in neighboring areas that are not affected by these winds. In Germany, Switzerland, and Austria such a wind is termed a *Föhn*, the German word for hairdryer; east of the Rocky Mountains it is the *chinook*.

Cyclonic/Frontal Precipitation

Cyclonic/frontal precipitation results from the cooling that takes place as air is driven upward along a line of contact with an air mass of lower density. This line or zone of contact is called a *front*. In order for fronts to form, air masses of contrasting thermal composition must be juxtaposed. The juxtaposition of such air masses usually takes place when one type of air mass migrates into the region of another. The zone on earth where this is most likely to take place is the midlatitudes because (1) this zone lies in contact with sources of very cold (polar) and very warm (tropical) air; and (2) the meandering pattern of upper air circulation in the midlatitudes facilitates air-mass migration.

Air Mass Origins and Movements The atmospheric machinery that drives the movement of air masses across the midlatitudes appears to be tied to the Rossby waves (see Fig. 5.14). These great meanders of geostropic winds mark the lower (equatorward) edge of the huge bodies of cold air that cap both poles (see Fig. 5.13). As the pattern of the Rossby waves shifts, so shift the locations of cold and warm air masses. Sudden changes aloft can trigger dramatic air-mass movements, and unquestionably the most dramatic are the outbreaks of polar and arctic air that can rush across the continental midsections at velocities of 100 km/hr or more. The most extreme of these can cause violent weather and a drop in temperature of 25 to 35°C in a matter of several hours.

Not surprisingly, much of the efforts of weather forecasts in the midlatitudes are devoted to analyzing and tracking air masses of different origins. This always involves classifying air masses based on the temperature and moisture properties that the air acquired in its source region. Air masses that originate over water are wet and are denoted *m* (for maritime), whereas continental air masses are dry and are denoted *c* (for continental). The temperature characteristic is specified as arctic (A), polar (P), tropical (T), or equatorial (E) in order, from coldest to warmest. The system is illogical, as polar really ought to be colder than arctic, but the convention has been established, so we will have to use it.

The predominant air masses in North America are mP, cP, and mT. Most of the cold air masses are polar, but occasionally arctic air masses penetrate into the northeastern United States. The maritime polar air masses in North America originate mainly over the North Pacific, whereas the principal source of maritime tropic air masses is the Gulf of Mexico and the Caribbean Sea (Fig. 6.9). There are a few cT air masses in North America, owing to the small size of their source area—Mexico and Central America. The source area that makes up Eurasia, on the other hand, is large enough to produce substantial cT air masses over that continent.

When air masses move from their source areas, there is a strong tendency for them to migrate into the midlatitude zone, where they meet with other air masses and are driven eastward by the westerly winds. This movement appears to be directed by airflow aloft, principally the jet stream, which also initiates formation of disturbances that become midlatitude cyclones.

Cyclogenesis and Frontal Weather Activity Most midlatitude cyclones, or large, low-pressure cells, appear to begin at points along the polar front where the pressure of air aloft is relatively low. This causes surface air to ascend, creating low pressure at ground

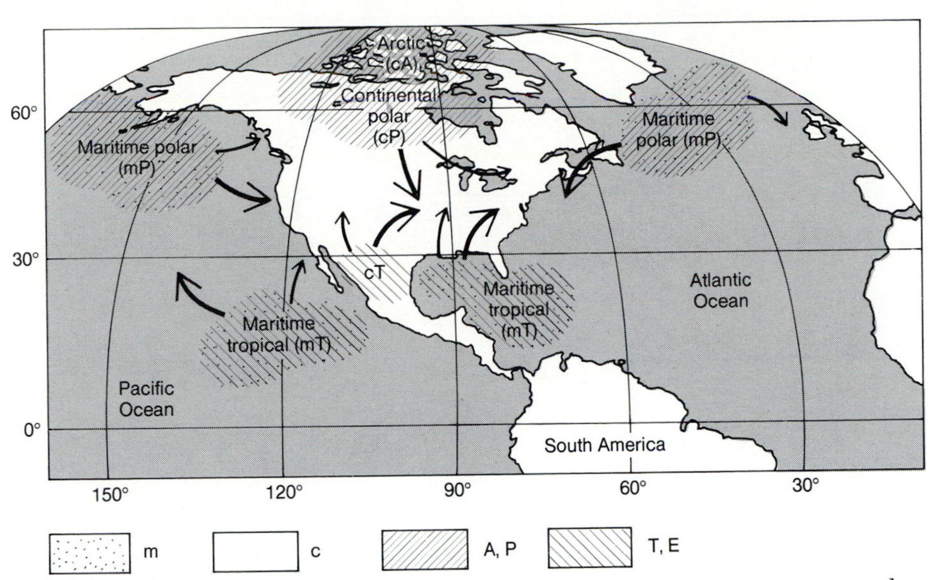

Figure 6.9 Principal air-mass source regions and paths of air-mass movement in and around North America.

level which triggers the convergence of air (wind). Aloft, the ascending air is diffused by the jet stream in a process termed *upper air divergence*. This mechanism is critical to the maintenance of the young pressure cell because it prevents the buildup of pressure within the system.

On the surface, air is drawn into the cell from all directions, but the air has only two sources: cold or cool air from the polar side and warm air from the tropical side. The leading edges of these two types of air form a broad, S-shaped wave in the polar front, the limbs of which become the cold and warm fronts we see on the daily weather maps. In the Northern Hemisphere, the cold front usually forms on the west or southwest and the warm front, on the east or northeast (Fig. 6.10*a*).

Three facts about the budding cyclone are critical: First, because of the density differences in the two types of air, the warm front and the cold front are going to be structured differently; second, the inflowing air is influenced by the Coriolis effect; therefore, as the cell develops, it is set into a whirling motion; and third, with the convergence of air on the body of low pressure, an upward flow develops in the interior of the cell.

In a cold front, the leading edge of the cold air mass drives under the warm air, forming a relatively steep contact slope over which the warm air is forced. In a warm front, the contact angle is reversed and very gentle as the warm air slides over the cold air. Because cold fronts are steeper and move faster than warm fronts, the rate of uplift along them is usually greater. Consequently, precipitation and turbulence can be very intensive along cold fronts. Strong cold fronts advancing on warm, moist air often produce a *squall line*, which is characterized by thunderstorms, gusty winds, hail, and heavy rainfall (Fig. 6.10*b*).

Cloud development along strong cold fronts typically reaches elevations of 10,000 to 15,000 m and is characterized by great billowing clouds, called *cumulonimbus* (or *cumulus* if the cloud is not producing precipitation; in cloud terminology the suffix *nimbus* or the prefix *nimbo* indicates precipitation). Because of their great heights, the tops of the cumulonimbus clouds may reach into the zone of fast upper airflow where they are drawn downwind ahead of the front and often ahead of the entire cyclone. These high-altitude clouds, called *cirrus* clouds, have a wispy appearance, and their presence in the sky is often taken as a precursor of bad weather. Cirrus clouds belong to the high-altitude family of clouds. The three other families of clouds and descriptions of the basic cloud types in each are given in Table 6.2.

Weather conditions along warm fronts contrast sharply with those along cold fronts. Warm-front cloud formation is mainly horizontal, leading to a broad zone of stratus clouds (nimbostratus with precipitation) at the base of the warm air. Turbulence is usually neg-

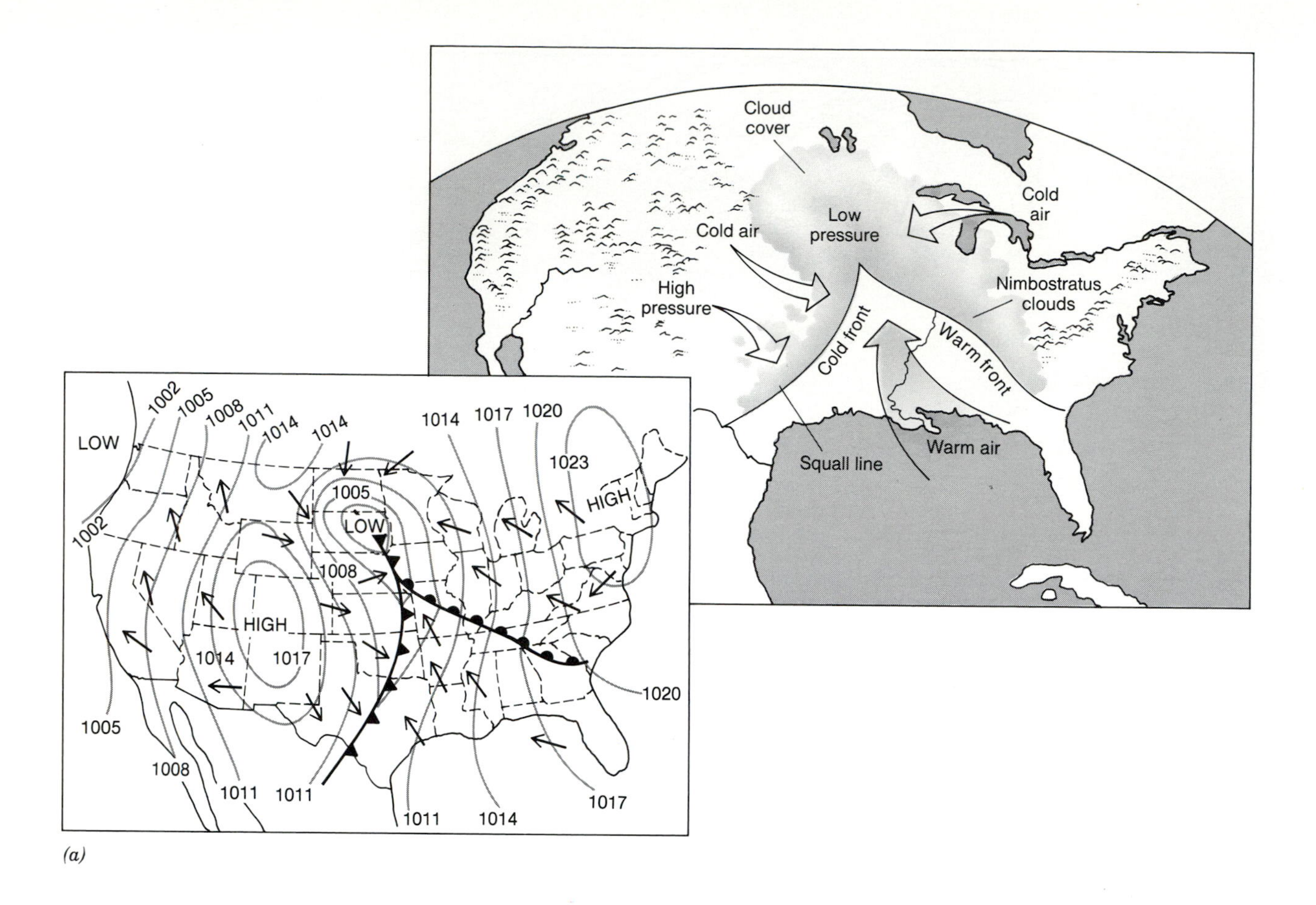

(a)

TABLE 6.2 CLOUD TYPES AND CHARACTERISTICS

Family	Types	Description
High clouds (altitude of cloud base above 6000 m)	Cirrus Cirrostratus Cirrocumulus	Generally wispy with hairlike character; appearing in bands or patches and sometimes covering the sky with light film
Middle clouds (altitude of cloud base 2000 m–5500 m)	Altostratus	Grayish or bluish cloud layer with uniform appearance; thin enough to show sun vaguely
	Altocumulus	White or gray patches, sheets, or layers in mounds or rolls often arranged in regular patterns
Low clouds (altitude of cloud base under 2000 m)	Nimbostratus	Gray layer emitting precipitation; dense enough to blot out sun
	Stratocumulus	Gray or whitish patches, layers, or sheets in rounded masses and rolls often arranged in regular patterns
	Stratus	Gray layer with fairly uniform base; not dense enough to blot out sun
Vertical clouds	Cumulus	Detached, round-shaped forms rising in bulges, domes, and towers; brilliant white in sunshine; often associated with fair weather
	Cumulonimbus	Associated with thunderstorms; huge towers, heavy and dense; top may be flattened, anvil-shaped, or plumed; precipitation, thunder, and lightning.

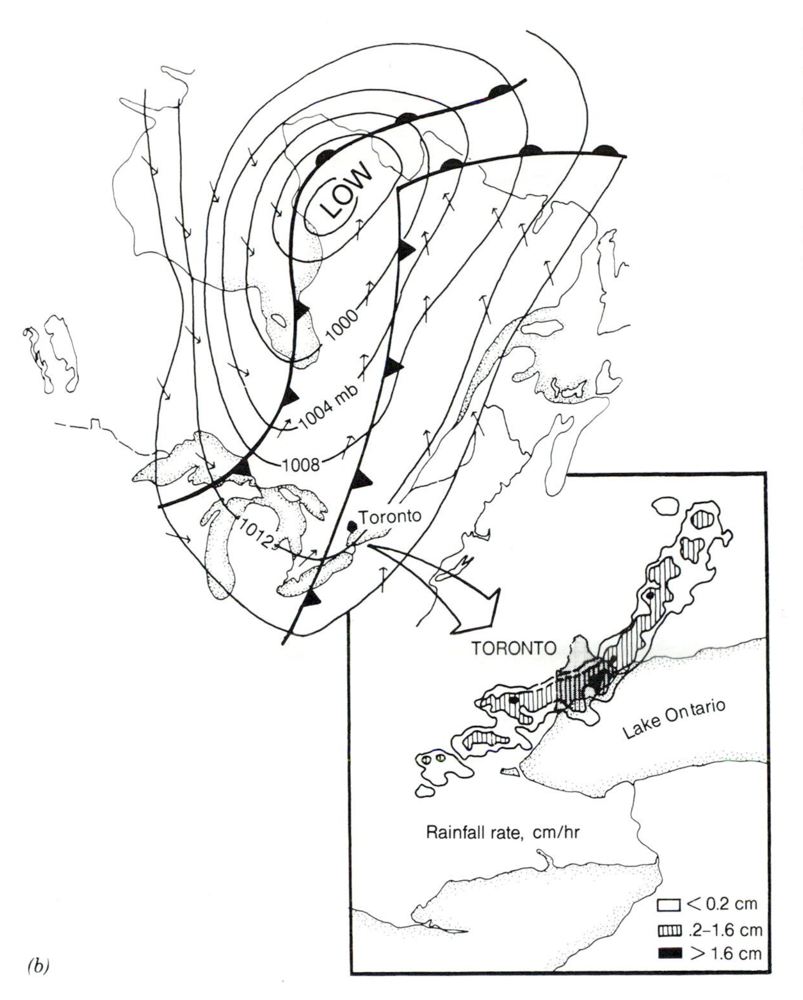

Figure 6.10 The configuration of cold and warm fronts and the general circulation associated with a midlatitude cyclone; (*a*) turbulence is greatest above the cold front, but cloud development is more extensive along the warm front; (*b*) a cold front and the precipitation produced along its squall line near Toronto, August 28, 1976.

ligible and precipitation prolonged. If the cold air under the front is below freezing, rain may freeze on contact with surface objects, resulting in the formation of glaze which damages vegetation and makes all modes of travel hazardous (Fig. 6.10*a*).

With the warm air aloft and cold air below, warm fronts form pronounced temperature inversions that prevent vertical mixing of the surface air. If the frontal system is slow-moving, this air may become stalled, or stagnated, over an area for several days. In areas of heavy air pollution emissions, such events can reduce the flushing capacity of the lower layer of air to negligible levels. When this happens, pollutants will build up, reaching levels dangerous to human health (Fig. 6.11*a*). Some of the worst air pollution disasters in the midlatitudes have been caused by cyclonic inversions related to warm fronts and low-level cloud masses created by combined warm and cold fronts. In 1952 an inversion over London led to the formation of smog (fog combined with pollutants such as sulfur dioxide and airborne particulates) so thick that 1600 persons died of respiratory difficulties in just a few days. Serious air pollution episodes are also related to inversions of other origins—in particular, adiabatic warming associated with anticyclones and airflow over mountains, and advection (Fig. 6.11).

Tornadoes Under conditions of extreme turbulence along cold fronts—especially when atmospheric lapse rates are very steep—*tornadoes* may form. These are the most intensive storms on earth, but they are so small geographically that they do not show up on synoptic weather charts. Moreover, their distribution along cold fronts is highly random, making it nearly impossible to forecast where they will strike; however, we do know that they are always associated with heavy

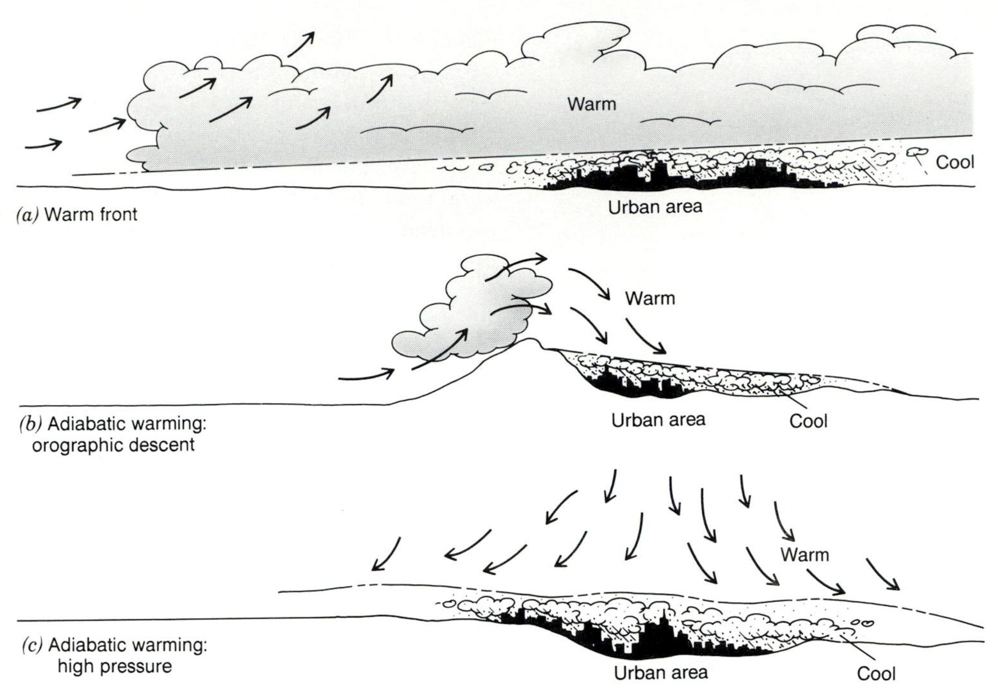

Figure 6.11 Three types of inversions associated with serious air pollution episodes in urbanized areas: (*a*) warm front, (*b*) orographic, and (*c*) anticyclonic.

thunderstorms. The uncertainty about the occurrence of tornadoes helps explain why tornado watches are so important for public safety during periods of threatening weather.

Tornadoes occur with greatest frequency in the plains and prairie states of the United States (Fig. 6.12). Their extraordinary destructive capacity is related to super high wind speeds and extremely low pressure which causes buildings to explode under the force of their own internal air pressure. It appears that rough terrain may reduce the destructive capacity of tornadoes by disrupting the funnel cloud's contact with the landscape.

Cyclonic Development and Movement Midlatitude cyclones are characterized by three principal motions. The first is that of airflow converging on the center of the cell and then rising into the troposphere. The rate of this flow is dependent mainly on the pressure gradient, which steepens as the cell develops. The geographic pattern of airflow around the cell is influenced strongly by the Coriolis effect; some air may travel from one side to the other by the time it takes its ascent. Generally, winds tend to be northerly in the northern or northwestern half of the storm and southerly in the southern or southeastern half. Therefore,

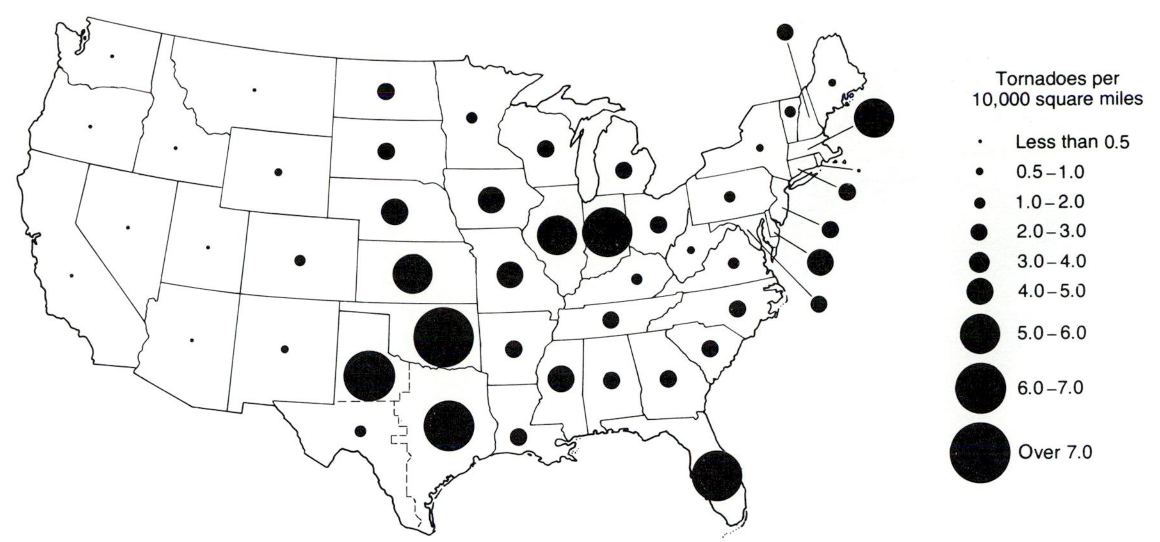

Figure 6.12 Incidence of tornadoes per 10,000 square miles of area, 1953–76.

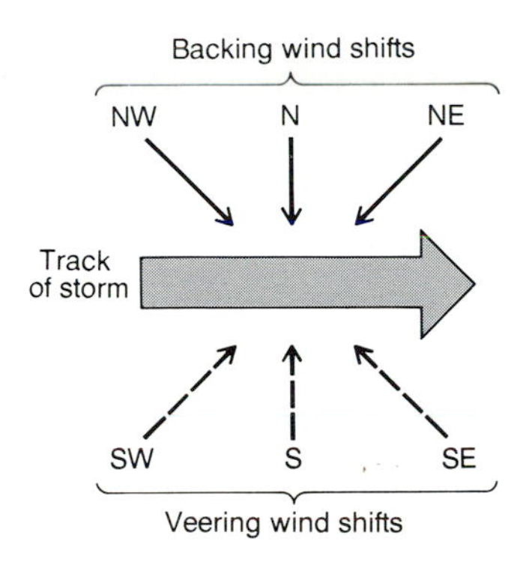

as the storm crosses an area, the types of wind shifts experienced will be different depending on which side of the cell you are located. In a storm moving eastward, two patterns of wind shifts are experienced: *backing* wind shifts (NE to N to NW) in the northern half of the cell, and *veering* wind shifts (SE to S to SW) in the southern half.

The other two motions of the cyclone are the movement of the fronts themselves and the migration of the entire storm cell along the polar front. We need to understand these motions in order to understand the sequence of weather changes associated with the storm. In the early stage of development, called the *initial wave,* the cold front and warm front are positioned a great distance apart (Fig. 6.13). Between them lies a sector of warm air where southerly winds are drawing high-energy mT air into the cell. As the storm develops, the cold front advances rapidly on the warm air sector (usually from the west or northwest), closing the gap between the fronts. When the gap (warm sector) between the fronts approaches an angle of 90 degrees or so, the cyclone has reached the *mature* stage of development.

At this point, weather conditions contrast sharply among the various sectors of the storm. The warm air sector is marked by a southerly flow of warm, humid air. Along the warm front this air slides over the cold, forming a broad belt of stratus clouds and low-intensity rain that dominates the eastern/southeastern sector of the cyclone. The western/southwestern sector, on the other hand, shows rapidly changing conditions behind the cold front as the frontal turbulence gives way to cool, clear cP or cA air from the northwest.

Within a matter of days after formation of the mature wave, the faster moving cold front usually overtakes the warm front. This marks the beginning of the *occluded* stage and the end of the warm air sector (Fig. 6.13). When the fronts reach full occlusion, the remaining warm air has been driven above the cold air, forming a great body of stratus, nimbostratus, and stratocumulus clouds.

The movement of the cyclone as a pressure cell appears to be controlled by upper atmospheric circulation along the polar front. Therefore, the movement almost always has an eastward component, but its tracks usually show marked southerly or northerly trends as well (Fig. 6.14). Over North America a common track for winter cyclones begins in the south-central part of the United States, where the cyclone is fed with maritime tropical air from the Gulf of Mexico, and trends northeastward, leaving the land over New England or the Canadian Maritime Provinces as an occluded wave.

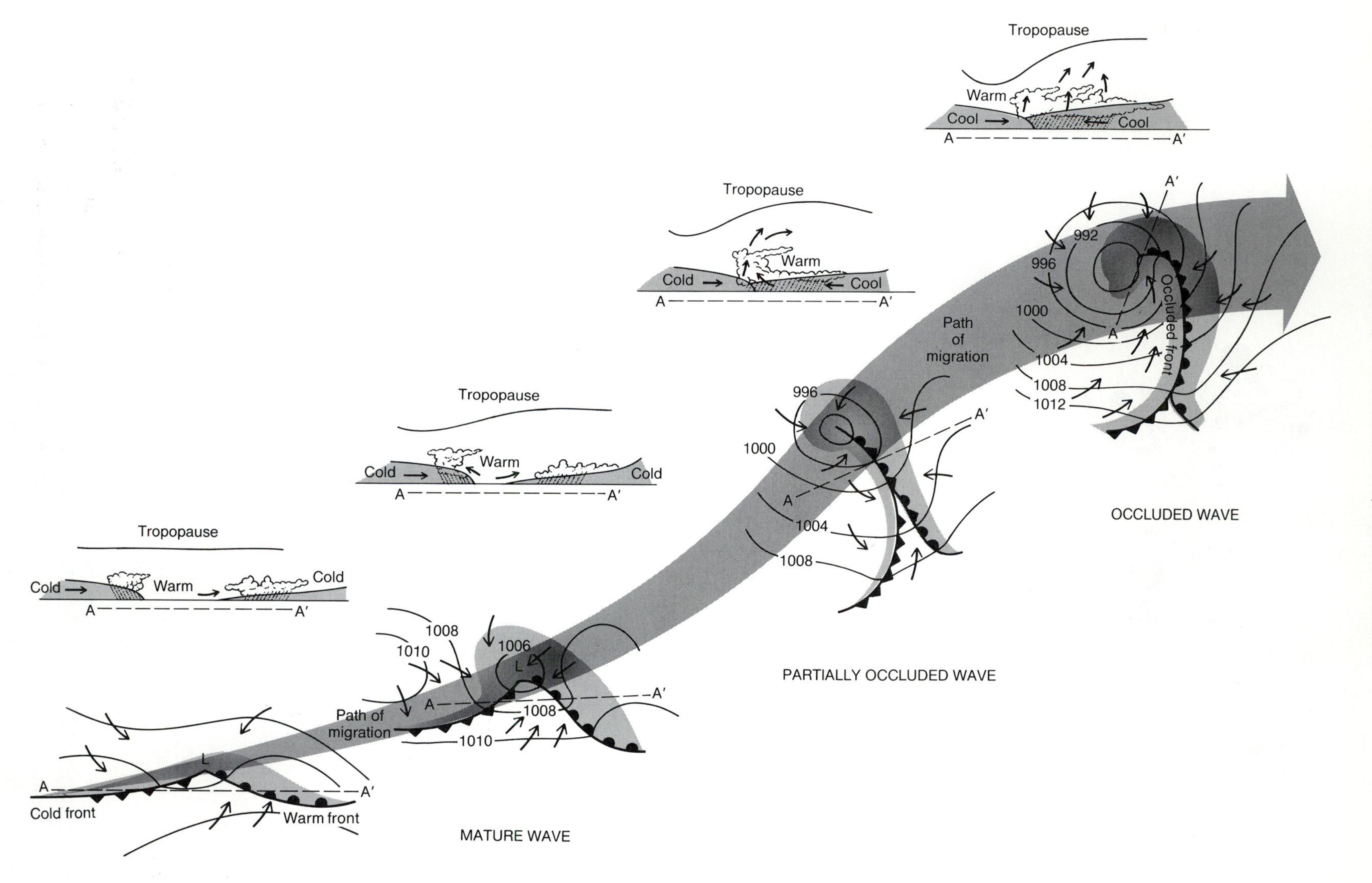

Figure 6.13 The stages of development of a midlatitude cyclone. The cyclone begins as a wave along the contact—polar front—between tropical and polar air. After the occluded stage, the cyclone dissipates.

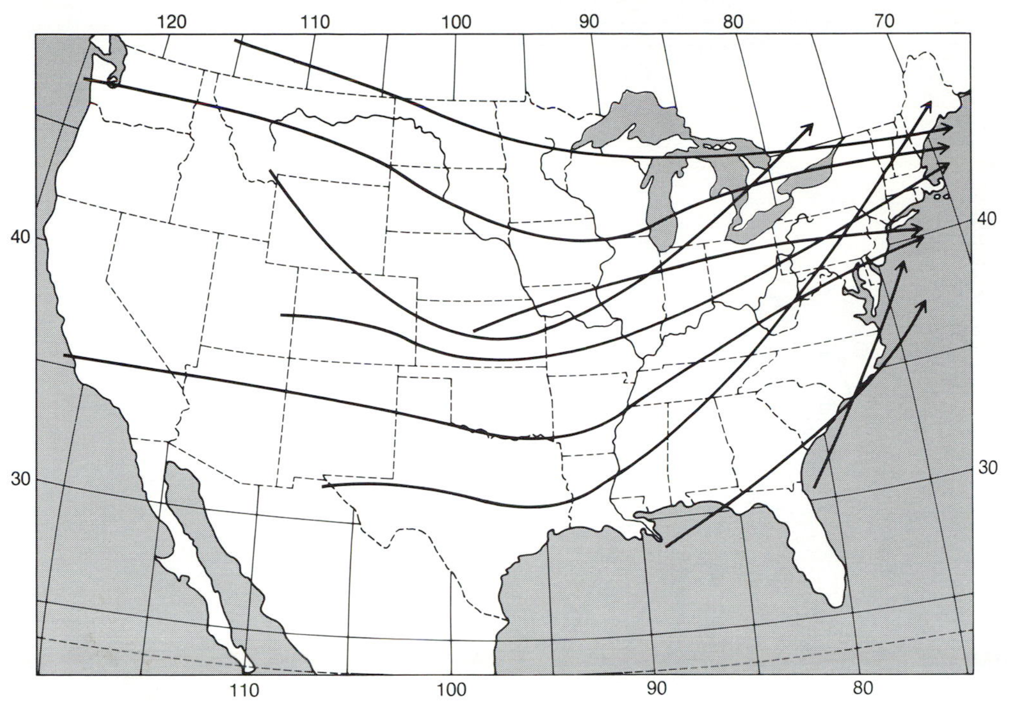

Figure 6.14 Typical paths of cyclones appearing in various regions of the United States.

Finally, it is important to add that the cyclonic tracks shift north and south with summer and winter, and this is accompanied by a change in the frequency and magnitude of cyclonic storms. Invariably, cyclones exhibit the greatest strength and frequency of occurrence in winter when it is not uncommon to find the polar front populated with families of storms, each in a different stage of development (Fig. 6.15).

Convectional Precipitation

Convectional precipitation is caused by cooling caused by the spontaneous ascent of unstable surface air. It is almost always associated with extreme surface heating and is marked by the formation of individual convective columns extending thousands of meters into the air. The convective columns are formed by vertical currents of warm air called *thermals*. As the thermals rise, they are cooled adiabatically, which in turn triggers condensation and precipitation (see Fig. 6.7).

Although our focus in this chapter is on precipitation, we should note in passing that convective columns may develop without precipitation. *Dry thermal convections* (which would be invisible were it not for the dust they often stir up) are common in arid lands; and *cumulus convections*, which produce a cloud but no precipitation, are common over dry landscapes where surface heating is modest. In humid regions convection is fed with an ample supply of moisture, which leads to massive levels of condensation and formation of a large cloud, precipitation, and accelerated instability. Above the level of cloud formation instability, you will recall, is caused by the release of latent heat with condensation (marked by the changed lapse rate in the example shown in Figure 6.7*a*). This enables the storm to reach a much greater magnitude than would be possible in convection driven only by the sensible heat of surface air.

Thunderstorms A thunderstorm is a convectional storm that emits thunder. Thunder (acoustical energy) is caused by the explosive expansion of a narrow band of air that is suddenly heated by a lightning discharge. The power source for lightning is electrical energy that builds up in the cloud from the intense friction between air molecules and precipitation droplets. As this electrical field takes shape, positive charges and negative charges develop in different parts of the cloud. The lightning discharge itself is an electrical arc between clouds, or between clouds and the ground.

Thunderstorms are usually considered the largest

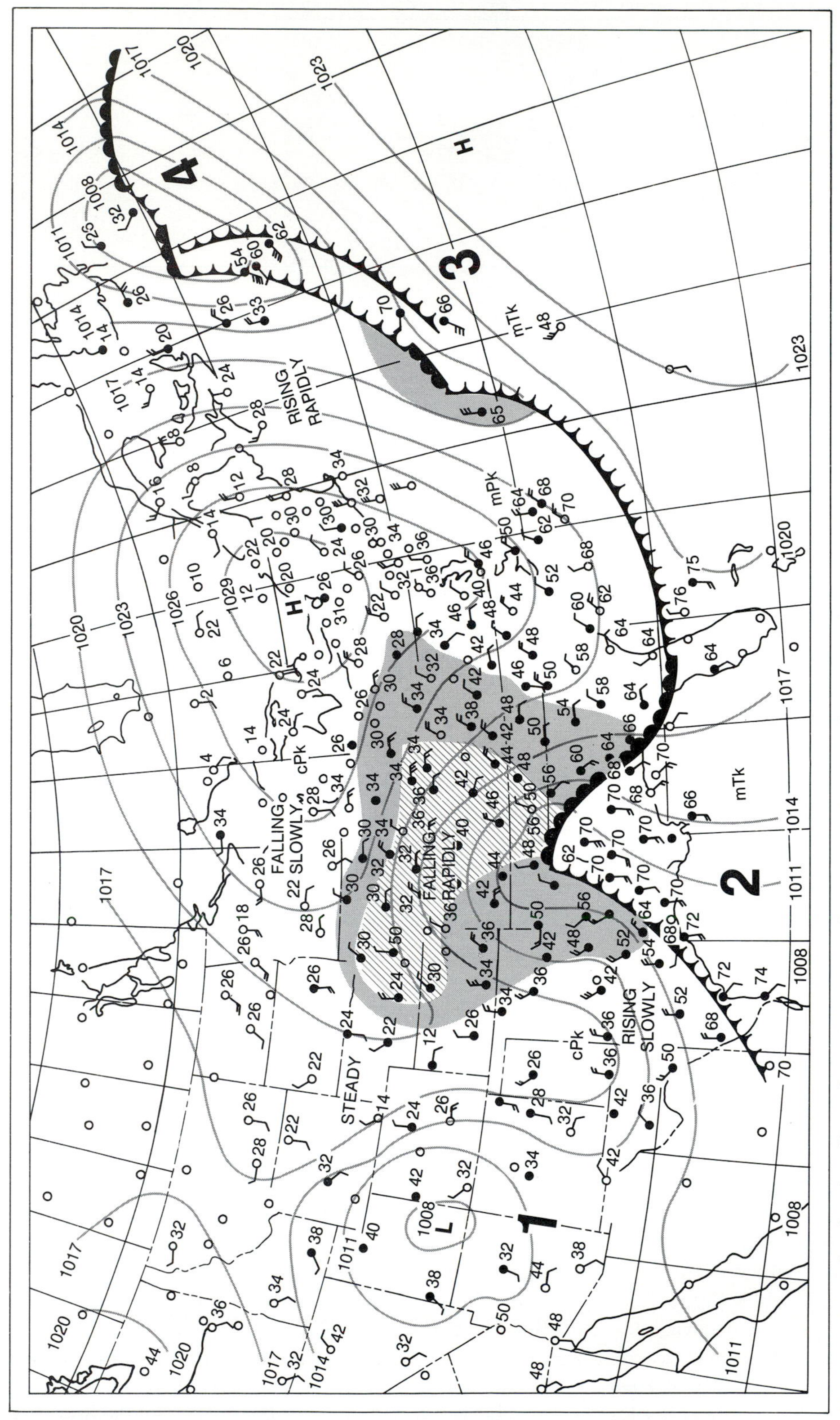

Figure 6.15 A family of midlatitude cyclones extending across the United States and the Eastern Atlantic.

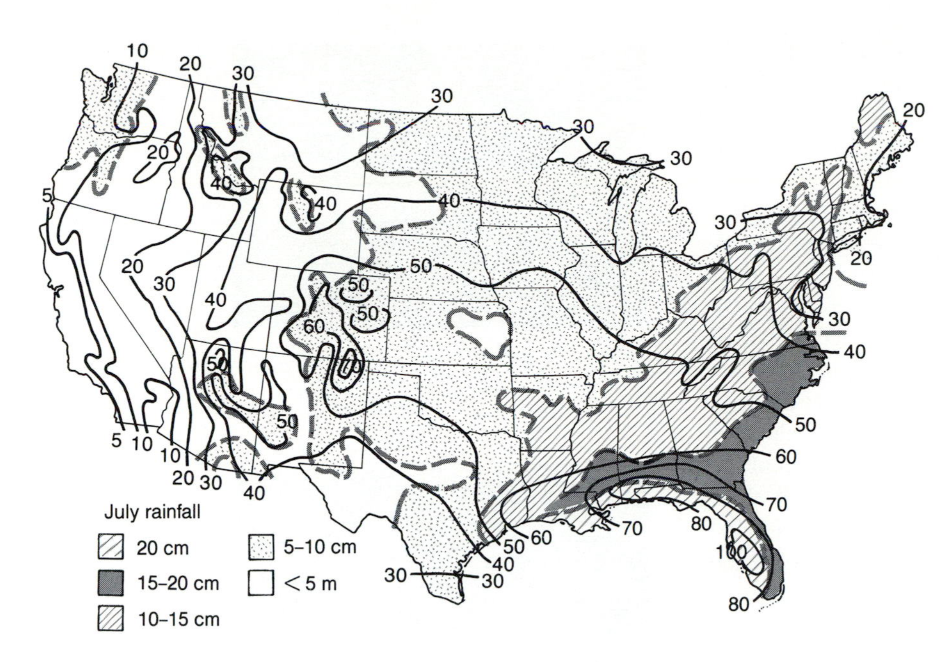

Figure 6.16 Mean annual number of days with thunderstorms and mean July rainfall. The incidence of thunderstorms in the Gulf Coast is two to three times higher than in the northern states and as much as fifteen to twenty times higher than at comparable latitudes on the West Coast.

convectional storms, commonly reaching 10,000 m into the atmosphere. Although our discussion here is directed at thunderstorms caused by free convection, let us not forget that they are also initiated by forced convection along cold fronts and orographic zones. The distribution of thunderstorms in the United States correlates well with the intensity of surface heating and the availability of moisture (Fig. 6.16).

Generally, thunderstorms generated by free convection can be described as having three stages of development: cumulus, mature, and dissipating (Fig. 6.17). The *cumulus stage* is characterized by strong

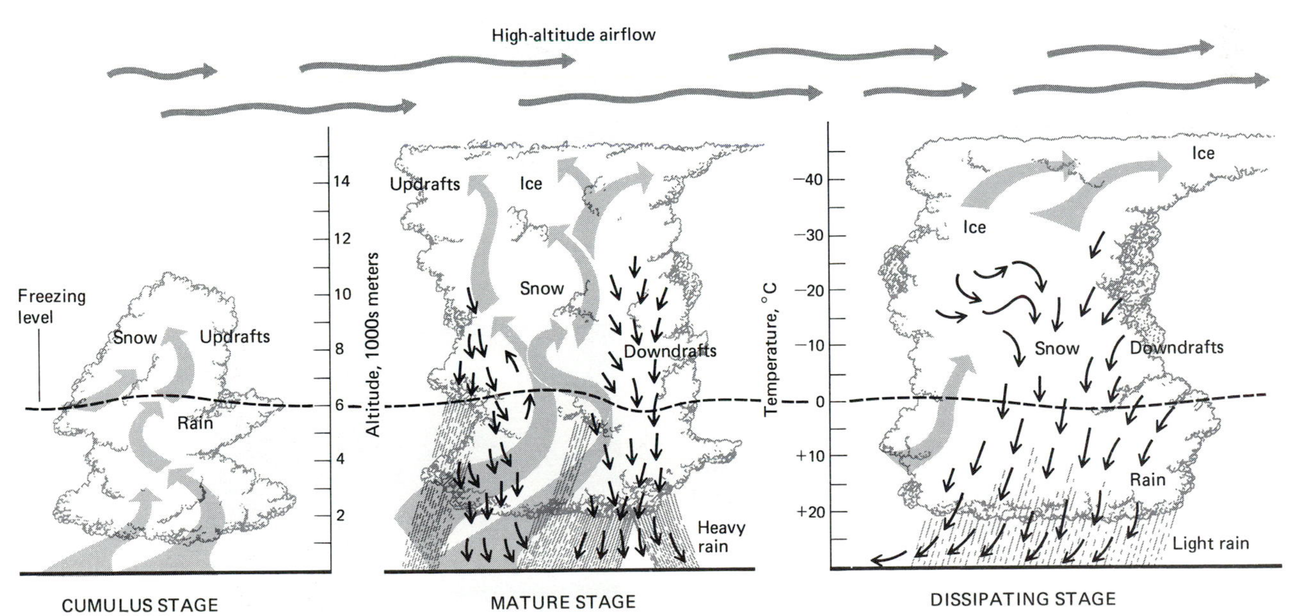

Figure 6.17 The formation of a thunderstorm can be described in three stages: cumulus, mature, and dissipating. The first is characterized by updrafts, the second by updrafts, downdrafts, and heavy rain, and the third by downdrafts and light rain.

vertical development and cloud formation. Precipitation droplets form quickly, but their fall is delayed by powerful updrafts within the cloud. In the *mature stage* the storm has reached full height, and the cloud billows out to form a huge head, often several kilometers in diameter. Internal circulation is complex, with downdrafts as well as updrafts occurring simultaneously in different parts of the storm cell. Lightning and thunder are usually abundant in this stage. Rainfall is generally heavy and may be accompanied by hail in the opening moments of precipitation. The *dissipating stage* marks the waning hours of the storm. Rainfall is light, and the lower part of the cloud is dominated by downdrafts. By this time, the top of the cloud has often been blown downwind by winds aloft, and the entire storm cell often has drifted somewhat from its place of origin. The entire life cycle of the thunderstorm is usually only 6 to 8 hours.

Megastorms Because most thunderstorms are initiated by surface heating, they typically form in the hottest part of the day, reach their zenith in mid- or late afternoon, and decline with the cooler temperatures of dusk and evening. The two exceptions to this life cycle are (1) those thunderstorms associated with frontal passages (squall lines), which can occur anytime during the day or night, and (2) those thunderstorms associated with *megastorms.*

Megastorms are great swarms of thunderstorms that cover areas as large as the State of Illinois or Iowa. They begin as individual storms or small groups of storms, which merge together, developing a total magnitude that exceeds the combined magnitudes of the original individual storms or storm groups. Once formed, the megastorm functions as a single storm system, with heating and ascent of air on a regional scale. The lifetime of the megastorms ranges from 12 to 18 hours, about twice that of the average midlatitude thunderstorm but much less than that of a midlatitude cyclone. Because megastorms have such short lifespans and large dimensions, meteorologists were unaware of their existence before the advent of weather satellites in the 1960s (Fig. 6.18). The storm shown in Figure 6.17 passed directly over northeastern Missouri and may well have been the sort of thunderstorm described by Mark Twain in the opening paragraphs of this chapter. From our standpoint, these megastorms are significant because they are capable of producing huge quantities of rain over a large area.

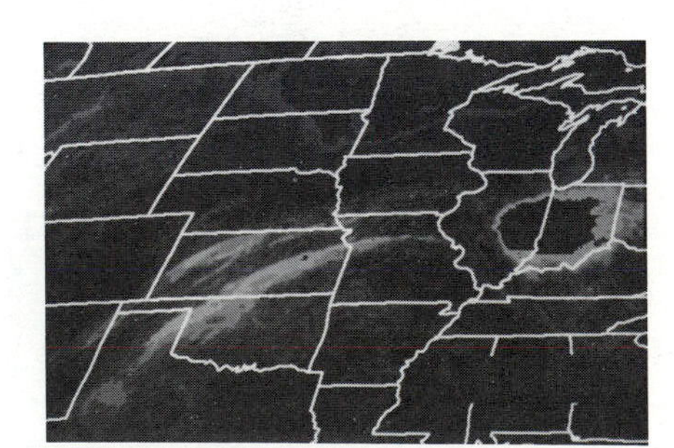

Figure 6.18 Satellite images of a megastorm complex, June 7, 1982.

Convergent Precipitation

Convergent precipitation results when air moves into a low-pressure trough or enclosed topographic area from which it can escape only by flowing upward. The ITC (intertropical convergence zone), which is fed by the trade winds, is such a trough of low pressure. Although individual storm cells may be convectional in origin, the net upward flow in the ITC zone is fundamentally convergent, especially over the oceans. In certain areas along the ITC, a very broad cell of general low pressure may develop because of special circumstances in circulation and surface heating. Thus, weak lows may produce especially heavy convergent rainfall but, typical of most low pressure cells near the equator, do not develop into cyclonic systems.

Various types of disturbances and storms of tropical origins, however, do produce convergent precipitation. Three of these are especially noteworthy. The simplest is a trough of low pressure, called an *easterly wave*, that forms in the belt of easterly trade winds. Easterly waves always form over the oceans at latitudes above 5°N and S and are characterized by a line of showers and thunderstorms. The wave moves slowly (300 to 500 km per day) westward as moist surface air, fed by the trade winds, converges on the trailing (eastern) side of the trough.

Tropical depressions and *storms* are another source of convergent precipitation. These disturbances appear to be caused by several factors, one of which is a weak cold front associated with an incursion of polar air into the tropics. These polar outbreaks and the weak storms associated with them may reach equatorward to 15° latitude. Understandably, their occurrence is limited to regions, such as North America, which produce strong arctic air masses that can penetrate into the tropical zone, such as the Caribbean.

Hurricanes (or typhoons) are the behemoths of tropical storms. These tropical cyclones are the strongest, most feared, and most destructive storms on earth. They are distinguished from other tropical storms mainly by the intensity of wind; for a storm to qualify as a hurricane, winds must exceed 65 knots (75 mph), according to international agreement. The most prominent characteristics of hurricanes are as follows:

- They form only in certain seasons and certain regions of the tropics.
- They form over oceans with high surface temperatures.
- They do not develop in connection with fronts, nor do they form fronts once developed.
- They do not occur with any regularity and can develop over any tropical ocean.
- They are many times more intense than midlatitude cyclones.

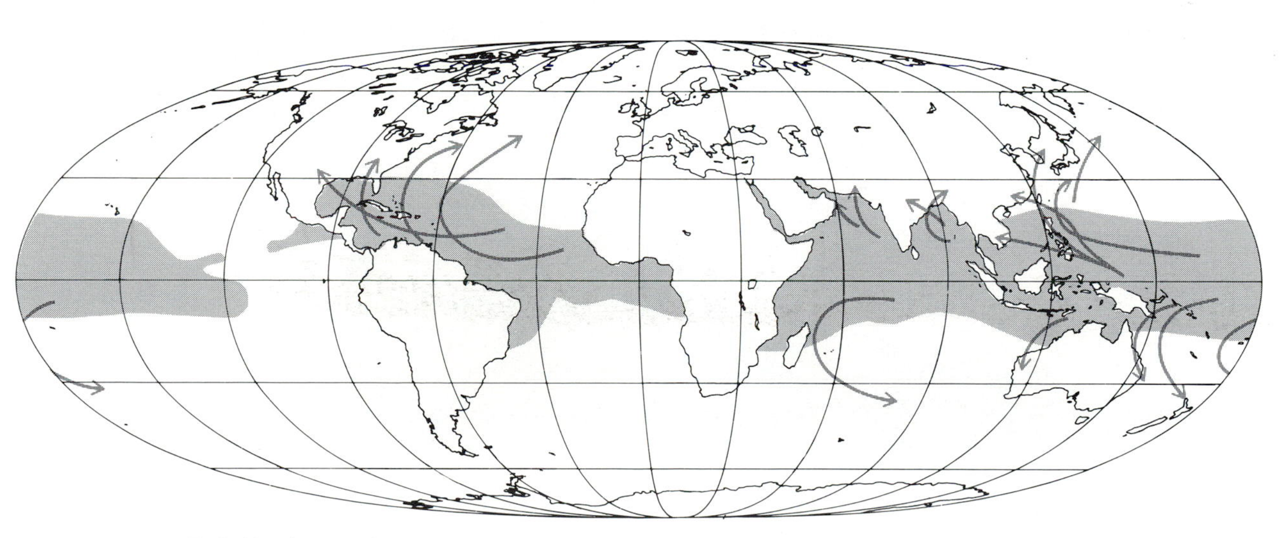

Figure 6.19 Hurricane Allan, August 8, 1980, two days before landfall on coastal Texas. Maximum wind speed at this time was 150 mph. The eye of the hurricane is clearly visible.

Hurricanes begin as tropical storms and intensify as they gain energy from the latent heat of condensation supplied by moist equatorial and tropical air. As with midlatitude cyclones, hurricanes are characterized by a great spiralling circulation scheme that draws in air from all directions. In the equatorial zone, the Coriolis effect is not great enough to produce this circulation, and so most hurricanes originate poleward of 10° latitude. Once formed, they move eastward in the belt of easterly trade winds, veering increasingly poleward as they go. This takes many hurricanes into the subtropics and beyond, where they may become caught up in the westerly airflows of the midlatitudes and (in the Northern Hemisphere) loop back to the northeast or east. Those that pass onto land masses decline quickly (usually in a matter of days) because they are no longer fueled by energy-rich maritime air.

The detailed structure of hurricanes varies greatly from storm to storm, but the one feature they all possess is an internal circulation pattern characterized by the ascent of massive amounts of air around an interior "eye" (Fig. 6.19). The eye is the cloud-free vortex of the storm where air descends in a narrow chimney. The stability of this air is attributed to adiabatic heating, which contrasts sharply with the adiabatic cooling that takes place in the great body of rising air around it where pressure typically falls to 965 mb.

Hurricanes are especially significant geographically because of the destruction they wreak on coastal areas (Fig. 6.20). The force of hurricane winds regularly destroys forests and buildings and related facilities, and the waves generated by the storms can cause massive coastal erosion. In addition, the pressure of the storm winds on the ocean surface causes a rise in

Figure 6.20 A section of Florida coastline during a hurricane. Notice the high wind, waves, and elevated water surface.

sea level, called a *storm surge*, which creates coastal flooding and promotes wave erosion.

Summary

Water vapor makes up less than 4 percent of the gas in the troposphere, is highly variable geographically, and is subject to rapid phase changes associated with energy exchanges between the earth's surface and the atmosphere. The total amount of vapor that can be held in air is dependent on air temperature; as temperature increases, the total vapor that can be held at saturation also increases. Absolute humidity, and specific humidity are conventional scientific expressions for the vapor content of air, although Relative humidity is the most commonly used expression in media weather reporting.

Precipitation occurs when air is cooled below its dew point, condensation sets in, and moisture falls to the surface. Cooling can be produced by advection or convection. Convection involves atmospheric instability in which rising air cools at the dry and wet adiabatic lapse rates.

Most precipitation is related to free or forced rise of air and falls in the form of rain. The types of precipitation are classified according to the cause of cooling. Orographic precipitation yields the heaviest annual averages of precipitation, exceeding 500 cm in many locations; cyclonic/frontal precipitation is prevalent in the midlatitudes where warm air and cold air masses meet; convectional precipitation develops with greatest intensity and frequency in areas of intensive surface heating and is widely known for the thunderstorms it produces; and convergent precipitation results from the upflow of air when opposing winds meet as in the ITC zone. Hurricanes are tropical cyclones of convergent origins, and they are the most powerful storms on earth.

Concepts and Terms for Review

- measures of humidity
  - absolute humidity
  - specific humidity
  - relative humidity
- psychrometer
- condensation
- dew point
- atmospheric instability
  - adiabatic lapse rate, dry and wet
  - ambient atmospheric lapse rate
  - temperature inversion
- precipitation forms and types
- orographic precipitation
- cyclonic precipitation
  - air masses
  - fronts
  - tornadoes
  - cyclonic stages
- convectional precipitation
  - thermals
  - thunderstorm
  - megastorm
- convergent precipitation
  - easterly wave
  - tropical depression
  - hurricane

Sources and References

Barry, Roger, G. (1981) *Mountain Weather and Climate.* New York: Methuen.

Battan, L.J. (1983) *Weather in Your Life.* San Francisco: Freeman.

Byers, H.R., and Braham, R.R. (1949) *The Thunderstorm.* Washington, D.C.: U.S. Weather Bureau.

Eagleton, J.R., et al. (1975) *Thunderstorms, Tornadoes, and Building Damage.* New York: Heath.

Fendell, F.E. (1974) "Tropical Cyclones." *Advances in Geophysics,* Vol. 17, pp. 1–100.

Fritsch, M.J. (1983) "Megastorms." *Natural History,* 92:7 (July).

Godske, C.L., et. al. (1975) *Dynamic Meteorology and Weather Forecasting.* Boston: American Meteorology Society.

Ludlam, F.H. (1980) *Clouds and Storms.* University Park, Pa.: Pennsylvania State University Press.

Mather, John R. (1974) *Climatology Fundamentals and Applications.* New York: McGraw-Hill. 412 pages.

Neiburger, Morris, Edinger, James G., and Bonner, William D. (1973) *Understanding Our Atmospheric Environment.* San Francisco: Freeman.

Winstanley, D. (1973) "Rainfall Patterns and General Circulation." *Science,* Vol. 295, pp. 190–194.

Chapter 7

The Hydrologic System Over Land and Water

The earth possesses a great reservoir of moisture in its oceans. The land depends on this reservoir as its sole source of water. How is the water transferred from ocean to land? How is the moisture distributed over the land, and how does the land finally give up its moisture?

In the science fiction classic, *Dune*, by Frank Herbert, the planet Arrakis is portrayed as so severely arid that humans must wear condensation suits to recycle water emitted by the body. An important measure of a person's physical essence is total body fluid, and upon death this moisture is extracted from the body rather than committed to the land. Rainfall is nonexistent, and nowhere on the planet is there open water.

Frank Herbert drew on our knowledge of earth's deserts to build a geographic framework in which to stage his extraordinary tale. Most readers detect this, of course, and cannot help to pause and wonder about earth's relative dryness. The earth *can* be very dry: in a few desert locations, precipitation has not been recorded for decades. The land in these places may occasionally be moistened by dew, but it is quickly lost as the land heats with the morning sun. If you set a vat of water out in the sun, it would evaporate more than 10 feet of water per year. Overall, nearly 20 percent of the earth's land is classed as dry.

These facts notwithstanding, earth actually falls at the opposite end of the planetary water spectrum from fictitious Arrakis. In a word, earth is a moist planet, with, for example, about 100 times more water than its sister planet Venus. Water is the dominant surface material on earth, covering nearly 75 percent (in liquid and solid forms) of its 510 million square kilometer area. Average annual precipitation worldwide, including the driest and wettest places, is 39 inches.

Water is fundamental to an understanding of the earth's physical geography, especially its weather and climate, not only because of the vast geographic coverage of the oceans and their ability to store and transfer heat, but also because of the rapid rate of water turnover between the atmosphere and the surface. The energy that drives this exchange is heat derived from the solar radiation that is absorbed by the surface layer of the continents and oceans. The water vapor that is driven into the atmosphere represents a tremendous flow of latent heat which, upon condensation, is converted into sensible heat and released into the atmosphere. From a geographic standpoint, it is important to realize that in the brief time between evaporation and condensation, the vapor may have traveled thousands of kilometers over the earth and delivered great amounts of energy to other regions.

This chapter examines the basic hydrologic system of earth and its role in shaping the world's climates. Not only is it important to understand the flows of water from ocean to atmosphere, atmosphere to land, land to atmosphere, and land to ocean; it is also important to realize that these flows are complex and uneven over time and geographic space. The controls on these flows are tied mainly to the temperature and circulation of the atmosphere and oceans. In the end our real focus must be on the water balance of the continents, so that we may understand the nature of arid and humid lands.

THE HYDROLOGIC CYCLE

Precipitation processes—the thunderstorms, hurricanes, and so on that deliver water to the earth's surface—are supplied with moisture by an elaborate circulation system called the *hydrologic cycle* (Fig. 7.1). This water system is made up of five basic components or phases which are listed in Table 7.1 according to the percentage of the earth's total water supply held by each. These phases are linked together principally by flows of vapor and liquid water (rainfall, runoff, and ocean currents) and to a lesser extent by flows of ice (snowfall and glaciers).

The oceans, ice caps, and glaciers together are enormous reservoirs that contain a total of 99.5 percent of the world's water. The amount in the atmosphere is surprisingly small. If *all* of the water held in the atmosphere at a given moment condensed out and rained evenly all over the earth, it would result in only 2.5 cm (1 inch) of precipitation. Contrast this figure with the worldwide average for precipitation of about 100 cm (39 inches) per year. This means that there is a complete exchange of atmospheric moisture about forty times every year, once every nine days! Thus, the atmospheric segment of the hydrologic cycle is highly dynamic. As we will see, however, this dynamism includes a great deal of geographic variation.

TABLE 7.1 PROPORTIONS OF WATER IN THE MAJOR PHASES OF THE HYDROLOGIC CYCLE

Phase	Proportion
Oceans	97.6
Ice caps and glaciers	1.9
Ground- and soil water	0.5
Rivers, lakes	0.02
Atmosphere	0.0001

To begin with, precipitation is unevenly distributed between land and water: nearly 80 percent of the annual global precipitation simply falls back on the oceans. Furthermore, much of the water vapor that is blown over the continents is not precipitated there but is blown back to the sea. For that falling on the continents, three subcycles are common: (1) rain falling through a dry atmosphere often vaporizes before reaching the ground; (2) rain reaching the ground and forming runoff in deserts rarely reaches the sea before being lost to evaporation; and (3) in some areas a large share of the water stored as groundwater is today pumped out for irrigation and lost to evaporation (Fig. 7.2).

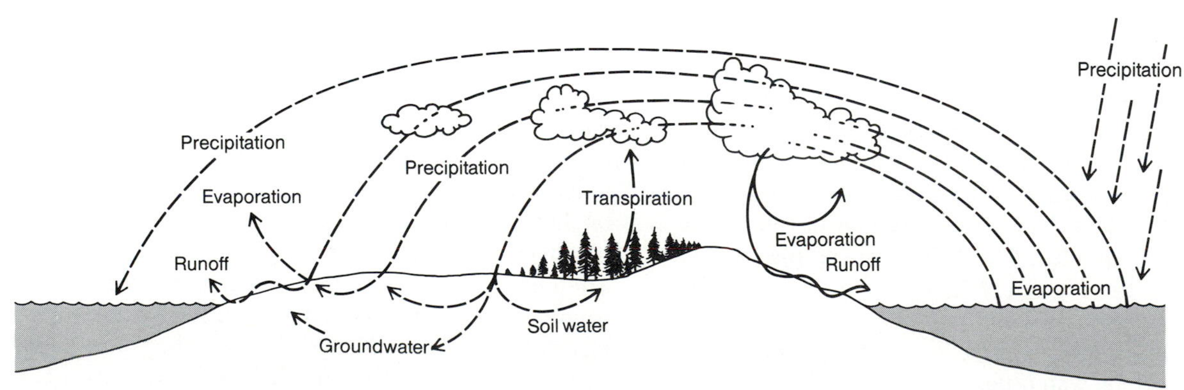

Figure 7.1 The hydrologic cycle.

Rain falling over the dry grasslands of Outer Mongolia. Most of the rain evaporates before reaching the ground.

In geologically active environments, relatively small amounts of water become locked in rock formations and may not return to the oceans for millions of years. In addition, huge amounts of water are held in glaciers for hundreds and thousands of years before being released to the atmosphere or the oceans. During the last Ice Age, continental glaciers formed a hydrological bottleneck, holding enough water on land to lower global sea level by more than 75 meters (Fig. 7.2).

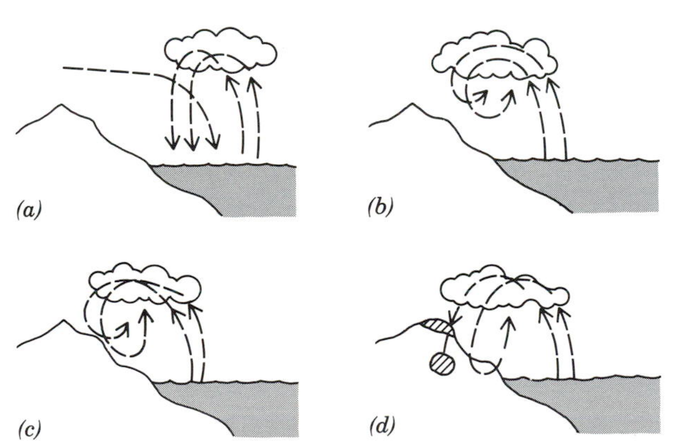

Figure 7.2 Subcycles of the hydrologic cycle: (*a*) sea-atmosphere-sea; (*b*) sea-atmosphere-atmosphere; (*c*) sea-atmosphere-land-atmosphere; and (*d*) sea-atmosphere-land.

The Oceanic Phase: Input to the Atmosphere

The oceans supply 88 percent of all atmospheric moisture. Virtually all of this vapor comes from sea surfaces between 60 degrees north and 60 degrees south lat-

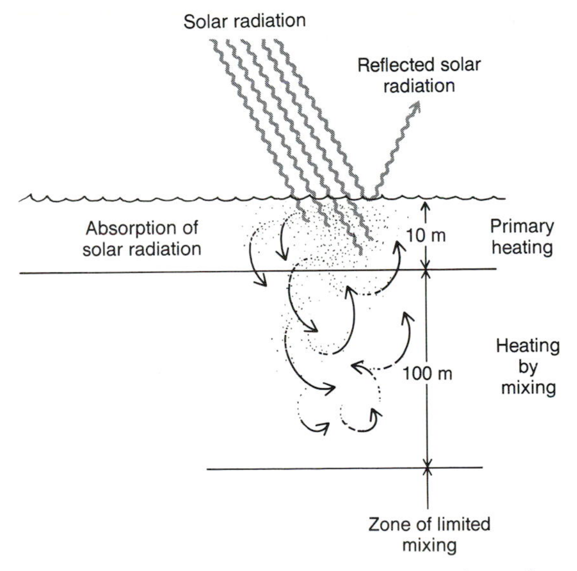

Figure 7.3 Solar heating of the oceans is limited principally to the upper 10 m of water; this water is mixed to a depth of 100 m.

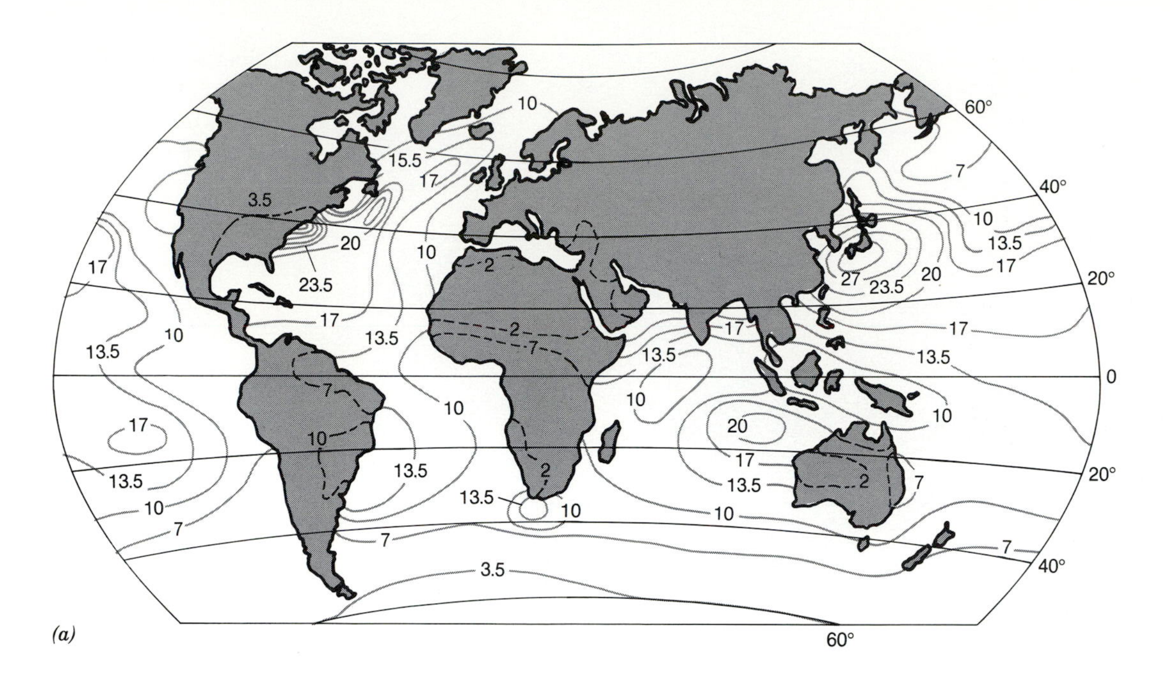

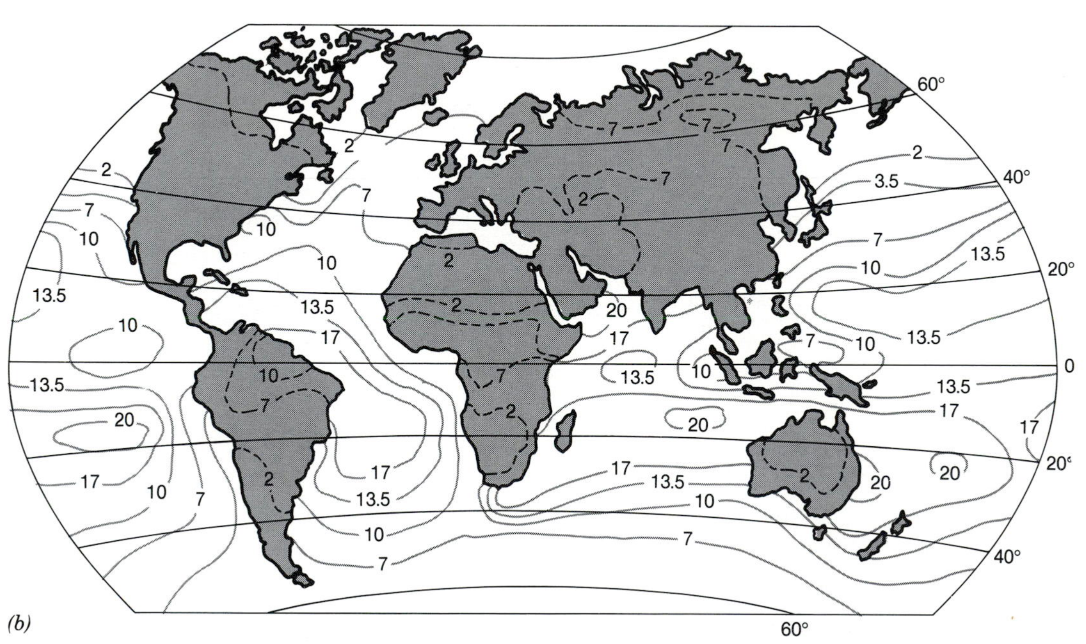

Figure 7.4 (*a*) Evaporation rates for December in centimeters. The highest rates are associated with the Gulf Stream and Kuroshio currents where cold, dry continental air is blown over warm ocean water. (*b*) Evaporation rates for June in centimeters. The Southern Hemisphere does not develop the extreme rates that are associated with the NorthernHemisphere in winter because of the absence of large midlatitude land masses to supply cold, dry air.

itude. Within this broad zone, vapor production varies greatly, depending on water temperature, air temperature, humidity, and wind speed.

Water temperature is an extremely important factor in controlling vapor production because the energy that drives the evporation process comes largely from the water surface. Ocean water is heated principally by solar radiation (Fig. 7.3). The absorption of shortwave radiation is very high in the low latitudes because solar radiation there is great and because the albedo of the water surface is low owing to the steep sun angles throughout the year. Conversely, shortwave absorption is low in the high latitudes because of low incident radiation and high albedos (owing largely to the low sun angles). Ninety percent or more of the radiation penetrating the water surface is absorbed in the upper 10 m. In still water, the surface layer may become very warm, but the oceans are rarely still, and heat is mixed more or less evenly throughout the upper 100 m of water.

The highest ocean surface temperatures (25 to 30°C) are found mainly in the intertropical zone. Although evaporation rates in this zone are substantial, averaging about 13 cm per month, they are not the highest among the world's oceans. The highest evaporation rates for large tracts of ocean occur off the eastern coasts of Asia and North America in the winter months (Fig. 7.4*a*). During January, evaporation rates in these areas average above 20 cm, meaning that water losses from the surface total a depth greater than 20 cm. The explanation for this lies in the juxtaposition of warm waters, driven northward by the Gulf Stream and the Kuroshio currents, and cold, dry air from the land, driven over the water by the westerlies. The energy gradient from water to air is very steep here, inducing vapor flows as great as 10 kg/m^2 per day in some places. In summer, the thermal gradient between water and air reverses with the flow of warm air off the land, and the evaporation rates fall substantially (Fig. 7.4*b*).

The tendency for winter evaporation rates to be high holds true throughout the world, even in the tropics. Again the explanation is found in the fundamental contrasts in the thermal regimes of air and water. Water has a much higher heat capacity than air and is slow to change with the seasons. This sets up a great winter heat reservoir that feeds the warm currents, and where cool, dry air is blown over this water, the uptake of water vapor is great. Conversely, evaporation is low where warm air falls into contact with cool water. Even where the air is extraordinarily dry, as along the arid west coast of South America and southern Africa, the presence of cold currents limits winter evaporation to less than 10 cm/mo. Evaporation rates are also low in the high latitudes, although for a different reason: ice cover. See color plate 16.

The Delivery Phase: Precipitation

The vapor-holding capacity of the atmosphere increases with air temperature, and where moist air is cooled, precipitation can be expected. The details on the mechanisms and processes of precipitation were discussed in Chapter 6, but let us repeat the fact that the principal cause of cooling is the ascent of air, either by free (thermal) convection or forced mixing. On a global scale, annual precipitation varies with (1) the general circulation scheme as it influences vertical cooling of air, and (2) the availability of precipitable vapor, that is, water in the troposphere that is subject to the formation of precipitation.

Worldwide, there are significant zonal variations in the water content of the atmosphere (Fig. 7.5). More than 60 percent of the precipitable water vapor in the atmosphere is found in the low latitudes, between 30 degrees north and south latitude. Within this broad belt, which covers a little over 50 percent of the earth's surface, the atmosphere is deeper (the troposphere reaches an altitude of 16 to 18 km) and therefore greater in volume than at higher latitudes. Moreover, this belt has a greater vapor-holding capacity because of its high temperatures. (See Fig. 1.12.)

Poleward of this belt, in the zone between 30 and 60 degrees latitude, the vapor content declines to a total of 29.2 percent (Northern and Southern hemispheres combined). A large part of this decline is explained simply by the lower volume of atmosphere at these latitudes, but the distribution of land and water also plays a part. This is evident when we compare this zone in the two hemispheres. Note that the Southern Hemisphere contains 2.6 percent more precipitable water. There is also a significant difference in the two hemispheres in the zone between 60 and 90 degrees latitude; however, in this case it is the Northern Hemisphere that is higher because of the relative abundance of ocean surface at the high latitude mainly in the form of the Arctic Ocean and North Atlantic.

Figure 7.6 shows the general profile of mean annual precipitation from pole to pole for the continents and oceans. The gross shape of the profile approximates that for precipitable water vapor in Figure 7.5, but within this scheme are some important variations corresponding to the major pressure cells. First, the belt dominated by the ITC zone stands out with high val-

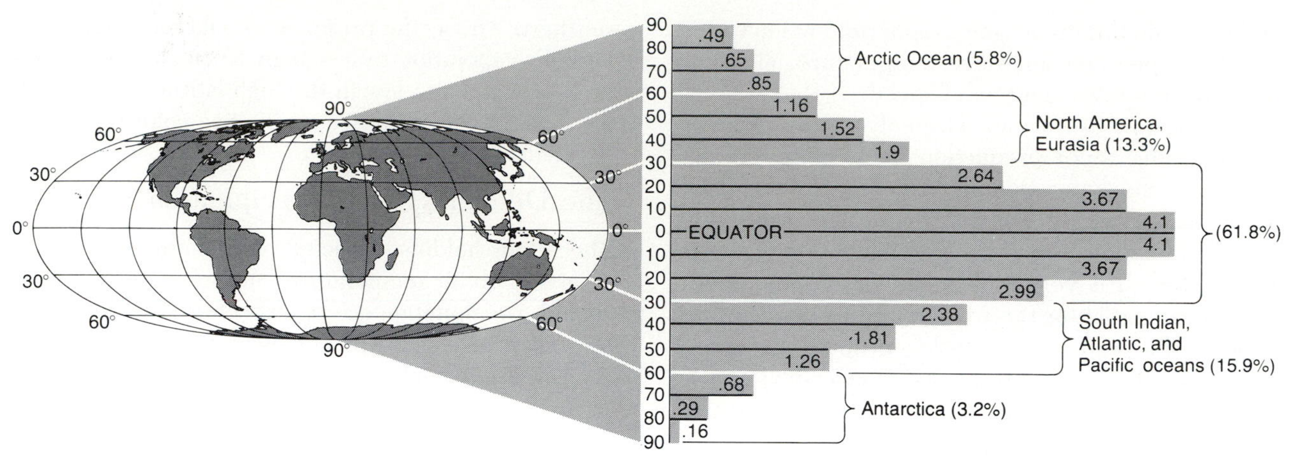

Figure 7.5 Precipitable water vapor in the atmosphere by latitude. Note the differences between the Northern and Southern hemispheres in the 30° to 60° and the 60° to 90° latitude ranges.

ues over both land and water. The fact that the oceans are even higher than land reflects the strength of the convergent mechanism in generating rainfall in this zone.

Second, low precipitation marks the tropical highs, located around 30° latitude. The stable, cloudless weather in this zone also contributes to extremely high solar radiation values and intensive surface heating; coupled with the low precipitation rates, the world's greatest deserts are the result.

Third, precipitation rises in the zone of subarctic low pressure (upper midlatitudes and subarctic) except over land in the Northern Hemisphere where the size of the land masses has a limiting effect on the development of the low-pressure cells. Finally, precipitation declines to less than 20 cm in both polar zones, though it is substantially greater in the Arctic owing to the large area of ocean coverage. The global distribution of average annual precipitation for the continents is shown in Figure 7.7 and color plate 4.

The annual precipitation received by various regions of the world varies not only geographically, but also over time. Indeed, in many regions the pattern of delivery from year to year is so irregular that figures representing annual, seasonal, or monthly mean quantities are not very meaningful. Precipitation variabil-

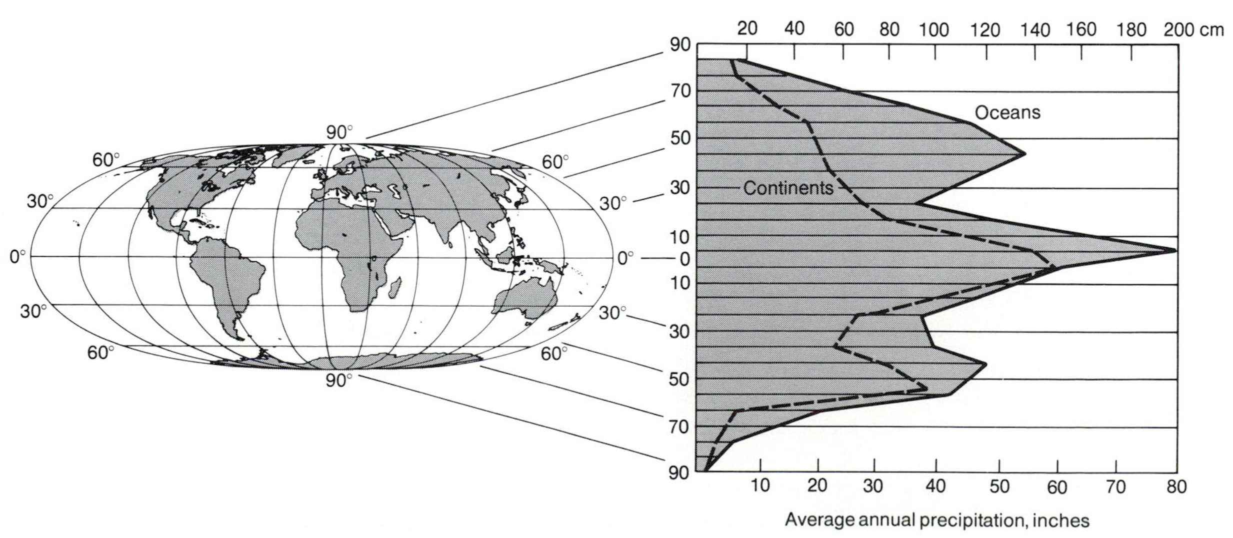

Figure 7.6 General distribution of precipitation for the oceans and continents from pole to pole.

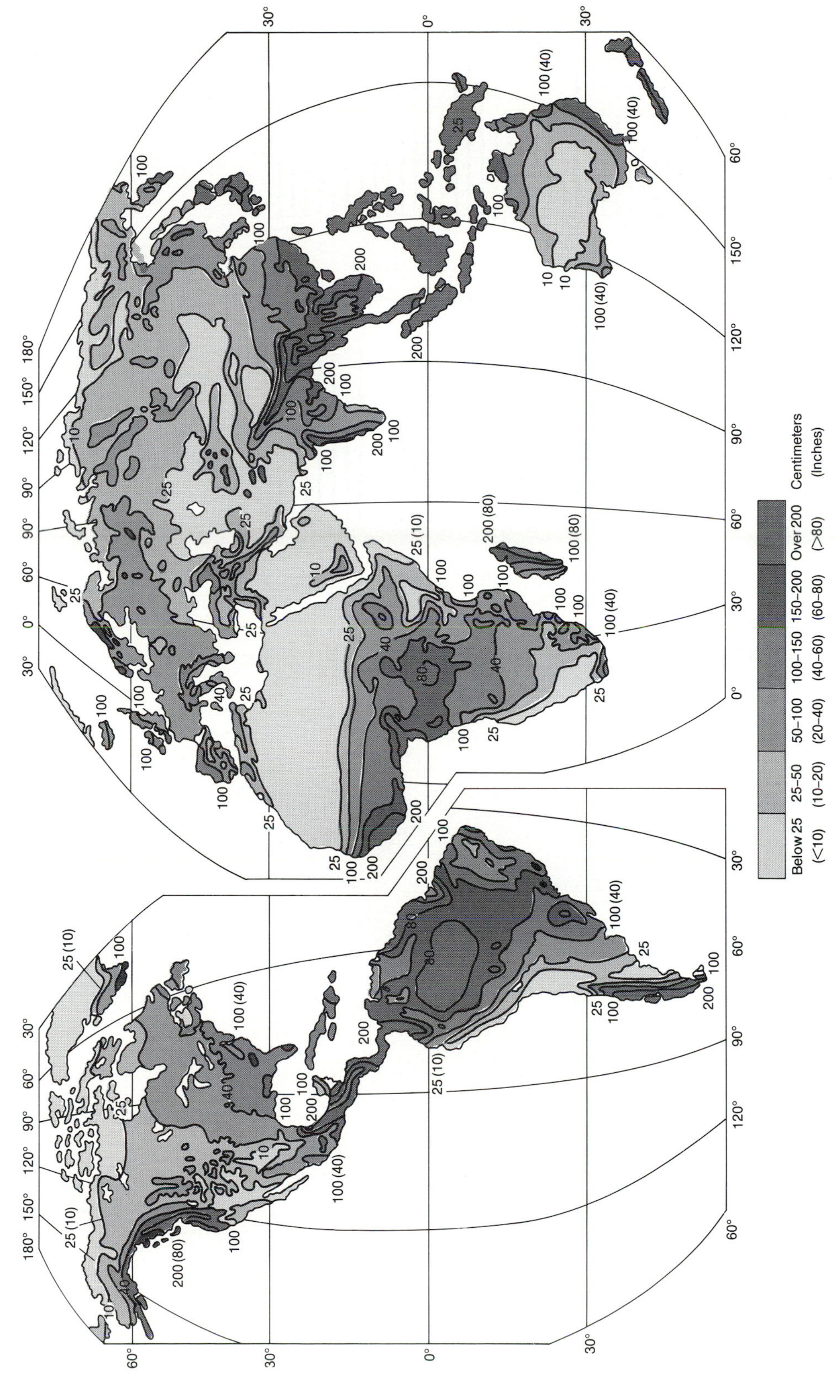

Figure 7.7 Average annual precipitation for the world's land areas, excepting Antarctica.

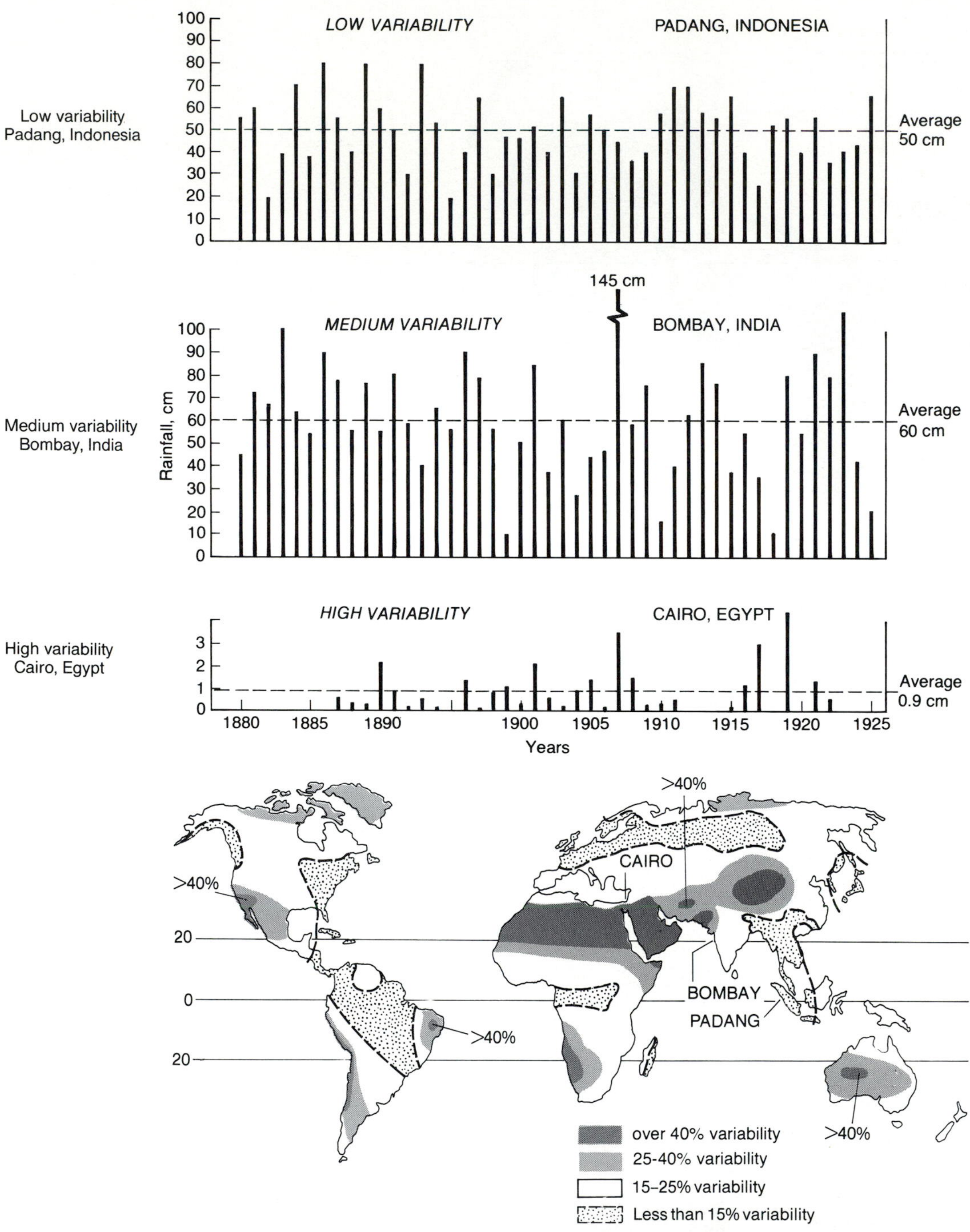

Figure 7.8 The graphs show the total annual rainfall of Padang, Indonesia; Bombay, India; and Cairo, Egypt, for the period 1880–1925. Rainfall at Cairo is highly variable, which is the norm for deserts, as the world map suggests. If precipitation is highly variable, most other parts of the hydrologic cycle must also be highly variable.

ity—or dependability—is a measure of the relative departure of the precipitation amount in individual years from the mean annual precipitation (based on many years of records). In general, there is an inverse relationship between variability and mean annual precipitation. In deserts, for example, variability is high—more than 40 percent in the harsh deserts such as the Sahara, Arabian, and Mongolian, where mean annual precipitation is less than 10 cm (Fig. 7.8). In moderately dry and semiarid environments, as illustrated by Bombay, India, in Figure 7.8, variability falls in the 25 to 40 percent range and precipitation averages 30 to 60 cm per year. The polar regions, which generally average less than 50 cm of precipitation yearly, also fall in this range. The lowest variability is found in areas averaging more than 75 cm of precipitation per year. In the humid midlatitudes and the wet tropics, annual variability is less than 15 percent. (Fig. 7.8).

Figure 7.9 A moisture-limited environment in a humid climate. This volcanic ash, located in a wet coastal climate, has such a high infiltration capacity that it retains virtually none of the moisture precipitated on it.

WATER BALANCE OF THE CONTINENTS

The continents receive much more water from the atmosphere than they contribute to it. The continents, which constitute about 30 percent of the earth's surface area, contribute only 12 percent of the vapor to the atmosphere. From the atmosphere, however, the continents receive 21 percent of the earth's precipitation. This amounts to a total of 108,000 km^3 of water delivered to the land each year. About 62,000 km^3 of this precipitation is given back to the atmosphere as water vapor, the remaining 46,000 km^3 going to the sea as runoff.

Moisture-Limited and Moisture-Abundant Environments

Generally, the continents are composed of two types of hydrologic environments: moisture limited and moisture abundant. *Moisture-limited* environments are those in which free surface water is scarce or available only occasionally. These are best exemplified by the arid zones where precipitation is light and evaporation rates, or the potential for them, is very high.

On a smaller scale, moisture-limited environments are formed under two other circumstances: (1) where runoff rates are so high that surface water is removed as rapidly as it falls (this is common in urbanized areas where hard-surface materials cover 80 to 90 percent of the ground and in mountainous areas where bedrock is the primary surface material) and (2) where precipitation is lost so rapidly through infiltration into surface deposits that water cannot be retained on the surface or within the surface layer (this is common in loose, sandy soils such as sand dune deposits, and in rock rubble deposits on mountain slopes) (Fig. 7.9). Both high runoff and high infiltration environments are, like true deserts, exceedingly poor sources of water vapor.

There are essentially two types of *moisture-abundant* environments: *open water* such as lakes, streams, and reservoirs; and *moisture-holding soils.* Moisture-holding soils are able to retain large amounts of water among and within the soil particles. Clay particles and organic matter improve the moisture-holding capacity of soil, whereas sand, pebbles, and larger particles tend to reduce it. (The details of soil moisture and its dynamics are presented in Chapter 13.)

The rate at which moisture is given up in vaporization from open water is highest where solar heating of the surface layer is great, humidity is low, and airflow at ground level is substantial. Measurements made from shallow pans and from lakes and reservoirs show that in the United States mean annual loss rates are three times higher in the Southwest than they are in the northeastern part of the country (Figs. 7.10*a*

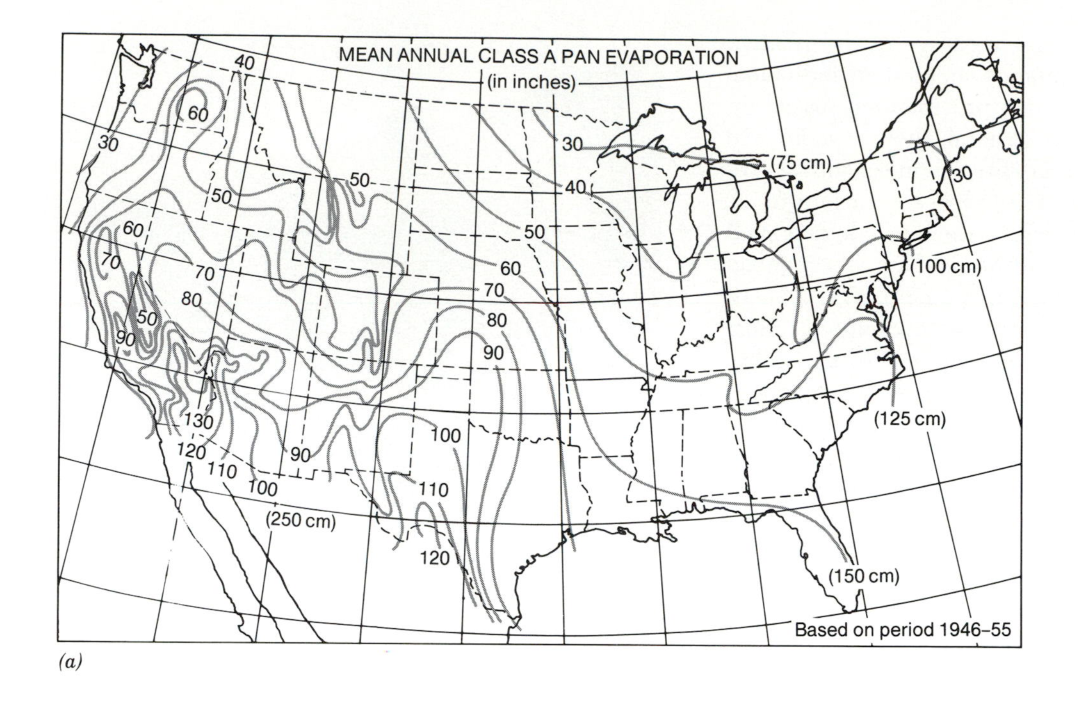

(a)

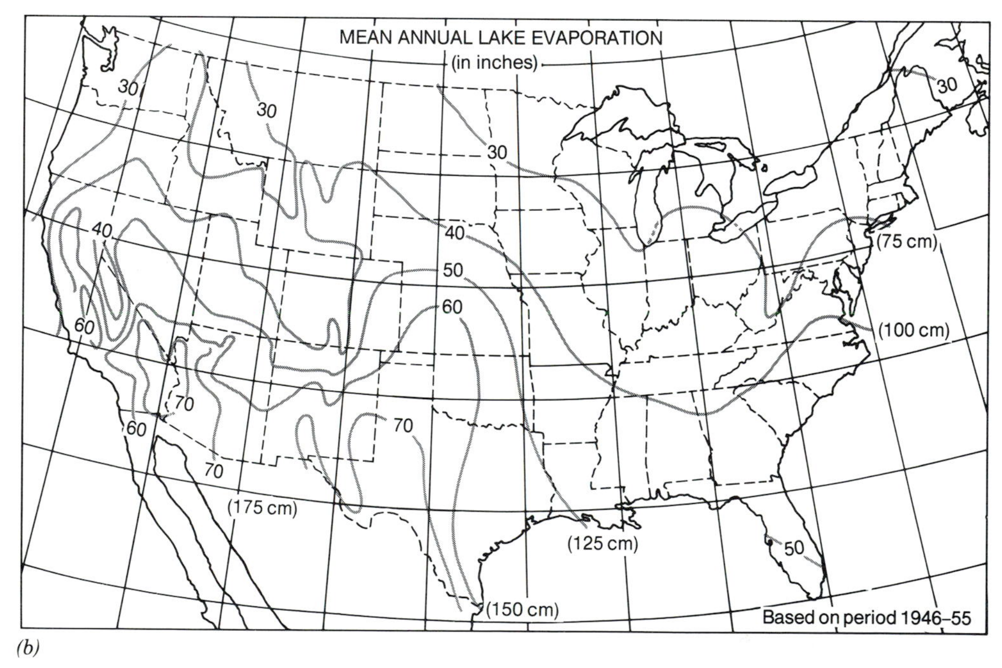

(b)

Figure 7.10 (a) Pan and (b) lake evaporation rates for the coterminous United States.

and 7.10b). Within the Southwest are three areas of exceedingly high evaporation: West Texas, the Lower Colorado River Valley, and the San Joaquin Valley of California. Pan evaporation in these areas reaches 229 cm (90 in.) or more and lake evaporation 165 cm (65 in.) or more in the average year. How does this compare with precipitation rates (Fig. 7.11)? The annual average precipitation for these areas is less than 50 cm (20 in.); thus, the moisture balance for open water surfaces such as farm ponds and reservoirs is strongly negative because three to four times more water is lost from each m^2 or acre than is gained. By compar-

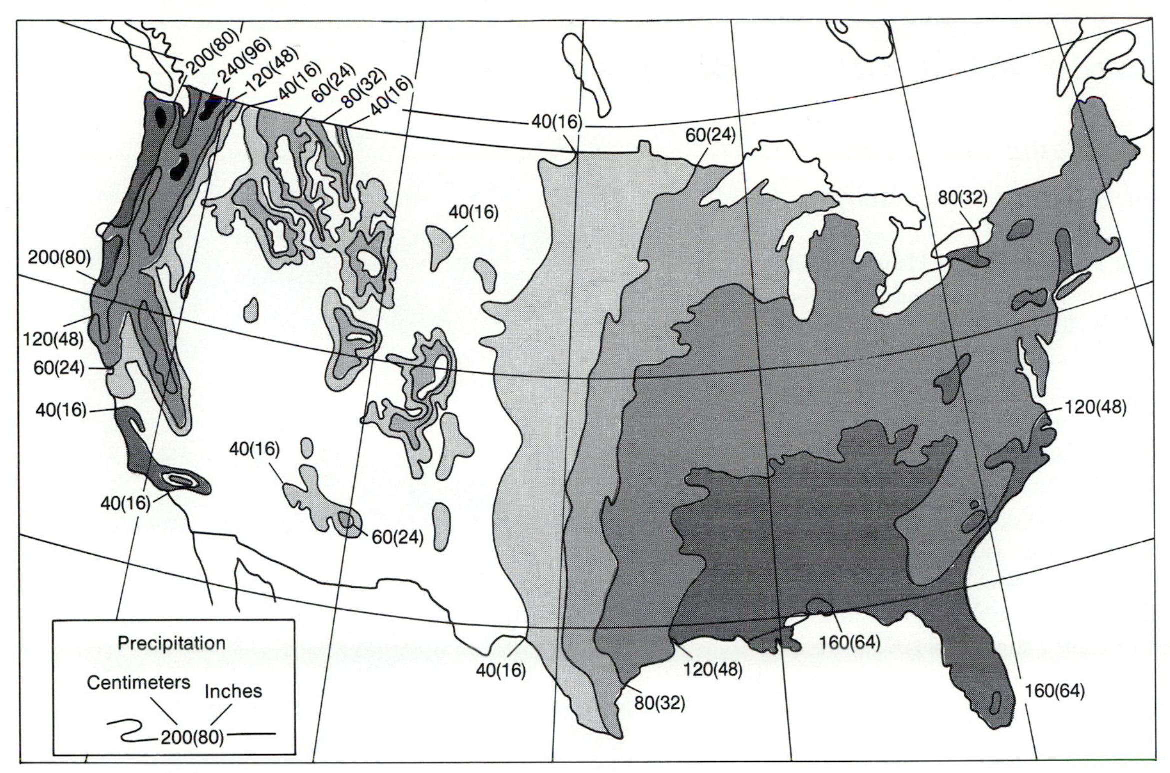

Figure 7.11 Mean annual precipitation for the United States in centimeters and inches.

ison, in the upper Midwest and the Northeast, evaporation rates from water bodies are approximately equal to mean annual precipitation (Fig. 7.10*b*). In southern Canada, north of the Great Lakes, they are substantially less than precipitation, which helps account for the abundance of fresh water in that region. In Note 7 Walter A. Schroeder of the University of Missouri offers an interesting glimpse at the changing character of the hydrographic landscape in America and questions its influence on the hydrologic system.

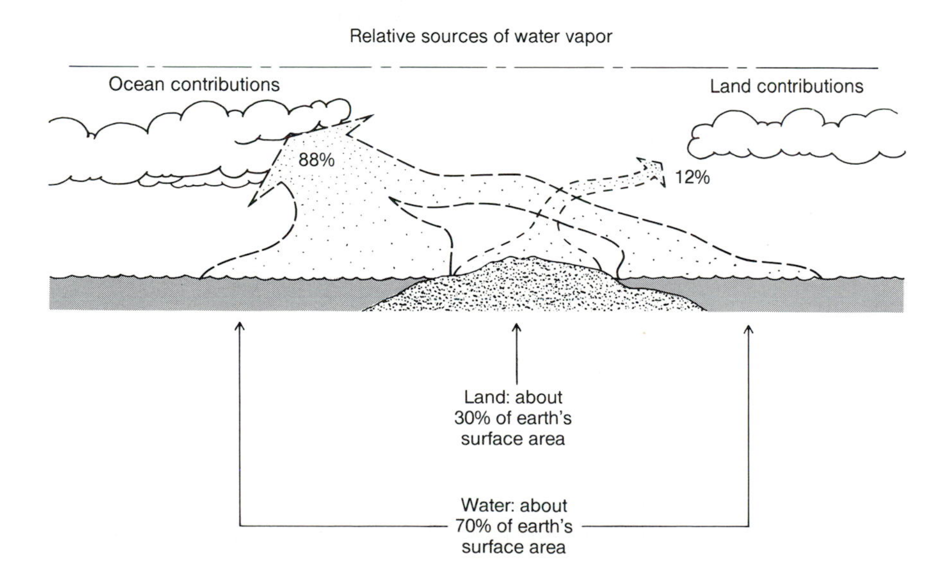

THE VIEW FROM THE FIELD / NOTE 7

The Changing Face of the Hydrographic Landscape

No one flying over the Central Plains can fail to be impressed by the sight of thousands upon thousands of ponds and small lakes dotting the landscape. In late afternoon, when the angle of sunlight is low, these myriad water surfaces glint like countless diamonds strewn among fields, pastures, and woodlands. How many are there? Missouri alone has an estimated 300,000 small ponds, or an average of four per square mile. (This does not include 3331 larger, known ponds and lakes with dams over 6 feet high and more than 50 acre-feet of water.) Kansas, Oklahoma, and Texas probably have even more ponds and lakes. A detailed study of a 126-square mile tract in the hilly terrain of Boone County in central Missouri enumerated 1561 ponds, or an average of 12 per square mile, one every 60 acres. The size of the ponds averages half an acre, about half the playing area of a football field.

It was not always so. These ponds are manufactured features, and that diamond-studded landscape seen from the air is really an artificial hydrographic landscape. Earlier, in a pre-pond landscape, ubiquitous windmills pumped groundwater for livestock, or it came, less dependably, from local creeks. In the 1930s, as part of a national concern over the deterioration of soil and land resources, farm ponds began to be promoted by county extension agents, by the U.S. Soil Conservation Service, and by fish and wildlife agencies. More recently, an affluent American Society has begun to prize waterside locations for rural residences and, in general, considers a water-feature landscape to be a more aesthetic and pleasing one.

The reasons for these ponds and small lakes, therefore, are quite varied. Some are for irrigation, livestock water, or other functional agricultural purposes. Some are for fishing and swimming and for enhancing wildlife and waterfowl habitat. Some may have been built expressly for soil or water conserva-

The Concept of Potential Moisture Balance

Open water surfaces make up a small percentage (perhaps 2 percent) of the land areas worldwide. The majority of the area of the continents is covered by a mantle of porous soil that can absorb and retain surface water. The loss of water from these materials takes place by direct evaporation or by transpiration from plants; the rate of moisture loss from the soil—for reasons that will become apparent later—is much lower than that from open water. Figure 7.12 shows the calculated or potential rates of *evapotranspiration* (evaporation + transpiration) assuming an inexhaus-

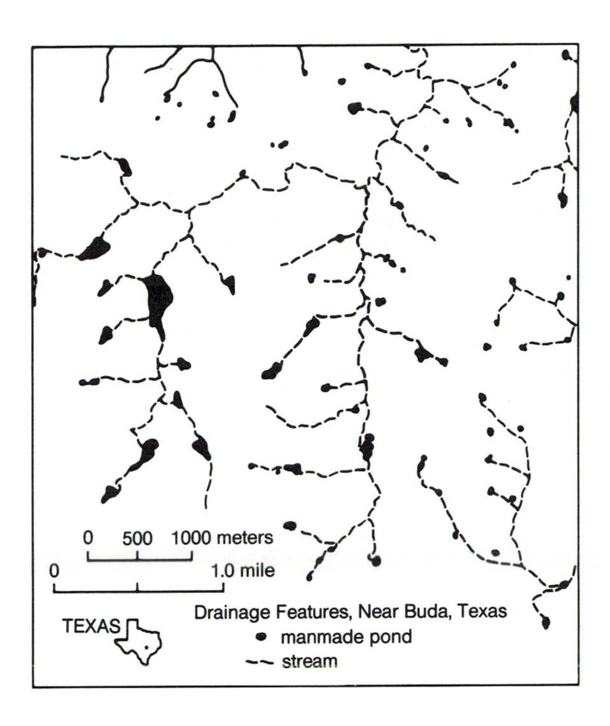

tion. Most ponds serve several primary purposes, or have major secondary purposes, such as a water supply for fire protection in rural areas. Larger ponds and lakes serve boating and other water-related recreational activities.

Surprisingly, the effects of this widespread change in surface hydrology are relatively unknown, unlike the counterpart effects of the urbanization of the landscape. Some major questions need to be addressed.

What is the net effect of these ponds on runoff? If a 2-square mile drainage basin now has twenty ponds along its drainageways, how have its discharge (flow) characteristics changed? Does more or less surface water now leave the basin? Is the stream now subject to fewer and lower peak discharges? Have periods of low or zero discharge become longer, or have they become shorter because baseflow (groundwater) has increased? Is streamflow more variable or less variable?

How have the ponds affected sediment yield of the stream? If one assumes the ponds collect sediment, is streamflow downvalley now less sediment-laden, or does erosion increase downvalley from the ponds at streams restore their sediment loads? Do these small ponds have any effect on soil conservation, or do they only collect eroded soil temporarily in its inexorable movement downstream? When projecting into the future, as planners do, what is the lifespan of these small ponds before they "fill up" with sediment and lose their water and soil retention capabilities?

How do these ponds affect surface water infiltration to groundwater? What water loss is there to evaporation and to transpiration by plants around their edges? How does this evapotranspiration loss compare with the amount of water sent downstream or lost to the water table?

How do all of these effects relate to size and depth of the pond? Do the small ponds affect hydrologic systems differently from larger ponds and lakes?

These are fundamental questions of geographic hydrology that point the way toward new research problems, the answers to which will help us understand the nature and dimensions of another aspect of human influence on hydrologic processes and systems.

Walter A. Schroeder
University of Missouri, Columbia

tible supply of soil moisture in a plant-covered landscape.

Where the potential evapotranspiration rates exceed mean annual precipitation, the soil-moisture balance has a negative potential, meaning that the projected outflow of moisture (measured by the energy available to produce vaporization) exceeds mean annual precipitation. It is necessary to use the word *potential* because a truly negative balance can be sustained for a short time only, specifically the time it takes the soil to give up the water it is holding near the surface. Inasmuch as the vaporization process results in the conversion of sensible heat to latent, when available soil moisture has been depleted, the ratio of

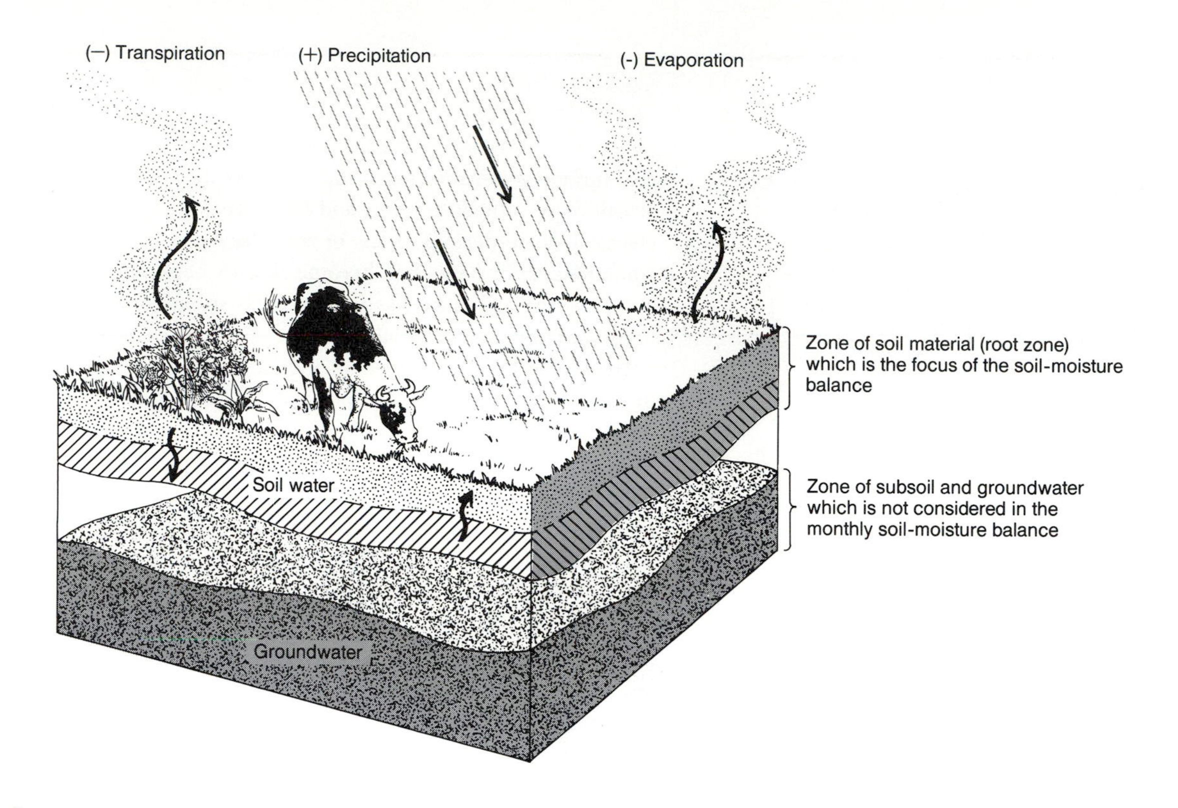

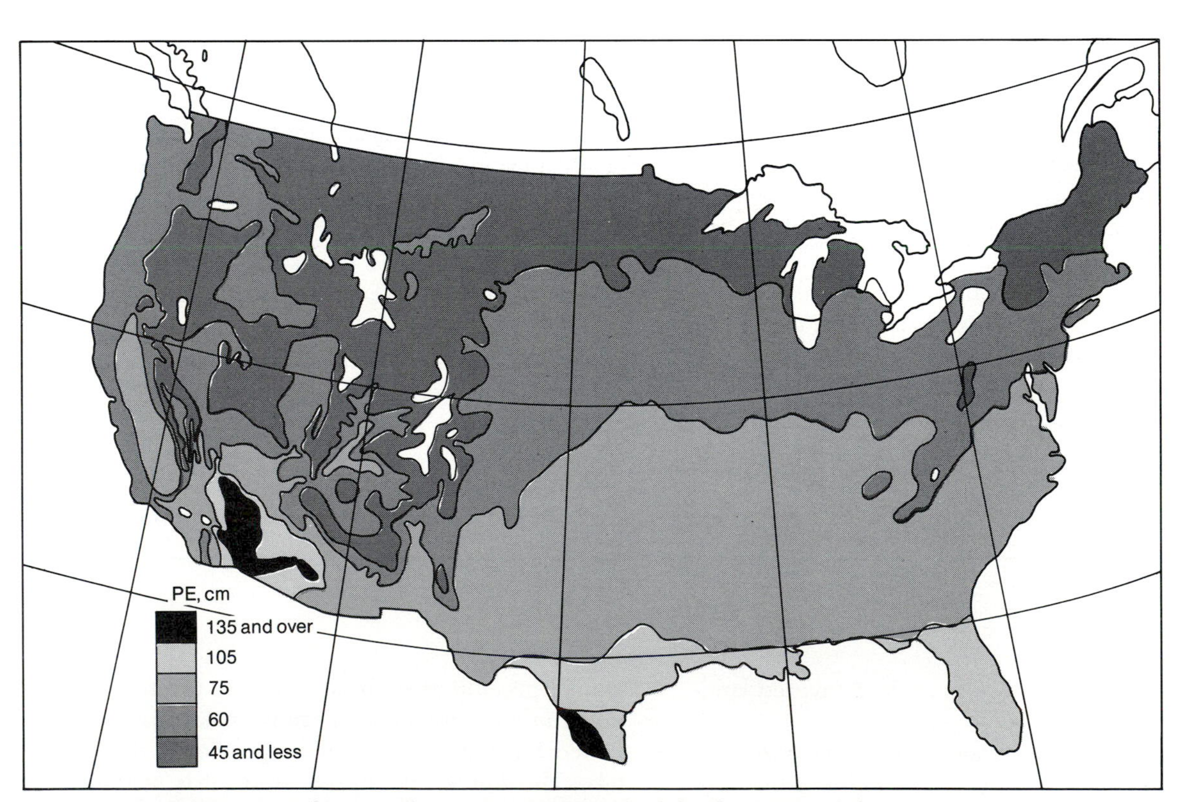

Figure 7.12 Average annual potential evapotranspiration (PE) for the continental United States. These values represent the amount of moisture that would be lost from the soil given an inexhaustible supply of soil water.

sensible- to latent-heat flux rises. This ratio, called the *Bowen ratio*, is a good indicator of moisture-abundant and moisture-limited conditions for areas of similar net radiation (Fig. 7.13).

The loss of soil moisture is generally greater where living plants cover the ground than where it is barren because plant roots can draw on moisture from a large volume of soil below the surface layer. This is very important because in most soils, direct evaporation of soil moisture is limited to the upper half meter or so of the soil. This results in the formation of a dry layer that functions as a barrier to moisture loss from greater depths. Any moisture transfer across this barrier is limited to two mechanisms: (1) movement through the vascular systems of plants (tissues of roots, stems, and foliage), and (2) movement by capillary action, which is the transfer of water by molecular cohesion among damp soil particles. In most soils, capillary action is less effective than vegetation in bringing up moisture, especially where the dry layer is greater than 0.5 to

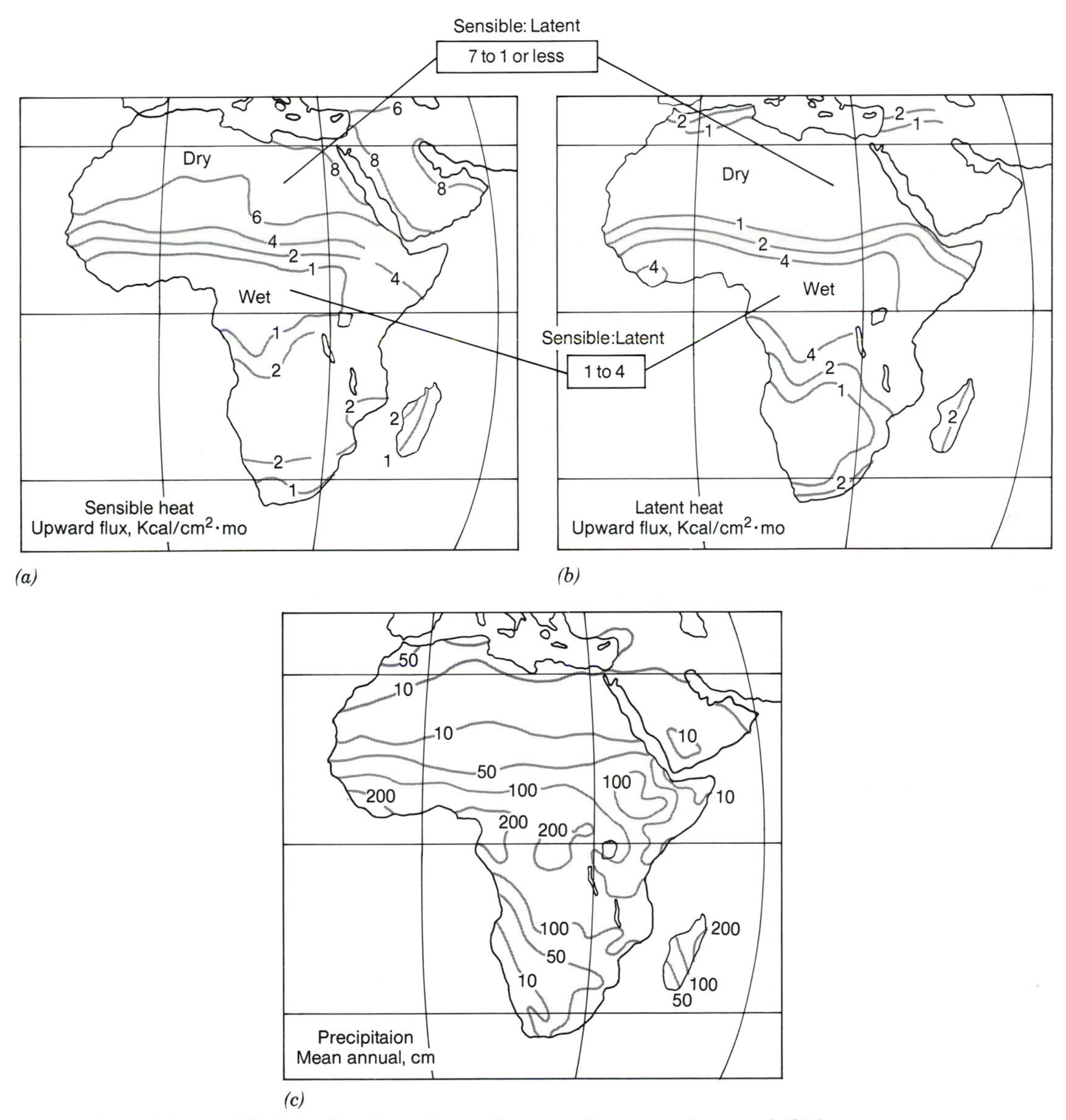

Figure 7.13 (*a*) Sensible-heat flux from the surface into the atmosphere and (*b*) latent-heat flux into the atmosphere for the month of June. Westcentral Africa is a region of forest cover and heavy precipitation (map *c*) where the Bowen ratio contrasts sharply with that of North Africa where precipitation averages 10 cm or less annually.

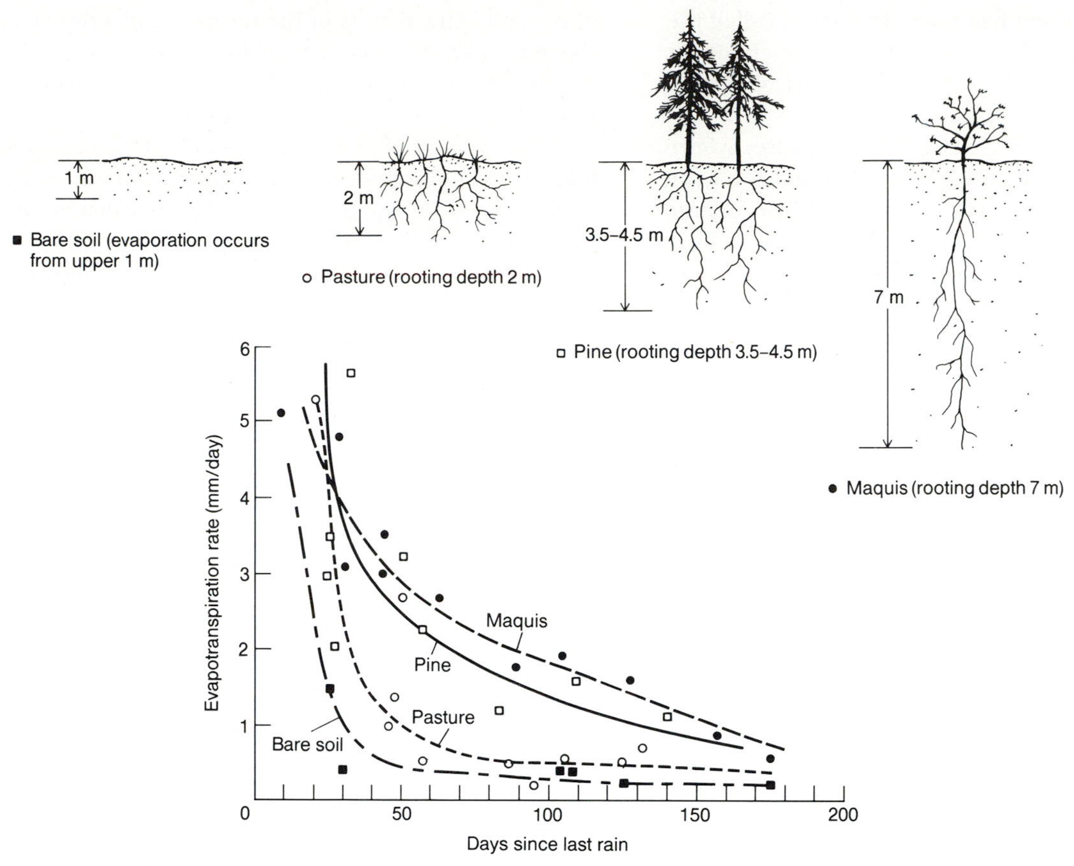

Figure 7.14 Decline in actual evapotranspiration from four different covers as the soil dries out after the last rainfall. The influence of rooting depth is apparent in the higher rates for pine and maquis.

1.0 m deep. This is supported by the results of field experiments, such as those shown in Figure 7.14, in which evapotranspiration rates are given for bare soil and for three types of vegetation with different rooting depths. Notice that the rates for the days following the last rainfall are highest for maquis which has a 7 m rooting depth, intermediate for pine which has a 3.5 to 4.5 m rooting depth, and lowest for bare soil where water loss was limited to 1 m.

THE SOIL-MOISTURE BALANCE

Like radiation and heat, soil moisture can be examined in the context of a simple balance problem based on the losses and gains of soil moisture. At the heart of the water balance are moisture flows between: (1) atmosphere and soil; (2) soil and plants; and (3) plants and atmosphere. Precipitation constitutes the sole inflow of water to the soil and evapotranspiration the sole outflow of water from the soil. The *soil-moisture balance* for some period of time, say, a month, is equal to precipitation minus *evapotranspiration* (evaporation plus transpiration) plus or minus any change in soil moisture.

The concept of the soil-water balance has grown largely from the work of American climatologist Charles W. Thornthwaite (1899–1963). The Thornthwaite method for computing the soil-water balance employs the energy-balance principles described in earlier chapters. It assumes a vegetated landscape and a reserve of soil moisture in the root zone (upper soil) and is based mainly on air temperature and solar radiation data.

Potential and Actual Evapotranspiration

An important feature of the soil-moisture balance model is the distinction between *potential* evapotranspiration (PE) and *actual* evapotranspiration (AE). Actual evapotranspiration is the true quantity of moisture lost from the soil; potential evapotranspiration, as noted previously, is the quantity that could be lost given an inexhaustible water supply. To create a moisture balance favorable to agriculture, for example, the gap between the actual and potential evapotranspiration, called the *moisture deficit*, must be made up through irrigation.

By knowing the values of evapotranspiration, precipitation, and moisture content of the soil, we can compute fairly accurately the water balance of most soils in the midlatitudes. Figured on a monthly basis, the water balance often reveals seasonal soil-water surpluses, soil-water deficits, potential runoff rates, and changes in soil moisture available to plants, each of which is directly or indirectly an important factor in the formation and maintenance of local and regional landscapes.

More than any other problem of the "balance" type, the soil-moisture balance resembles an accounting problem. It begins with a reserve on account in the form of soil water. This account, however, is unique because it is limited to a maximum amount, fixed by the moisture-holding capacity of the soil. Into and out of this account flows water (see the lower diagrams in Fig. 7.15). Summer is a time of big outflows and the reserve is drawn down, perhaps nearly exhausted, but (unlike a financial account) never overdrawn. During fall and winter the reserve is rebuilt, and in spring the fully recharged reserve rejects additional inflows, which as surplus water are converted to runoff.

A graphic model of the annual water balance for a typical location in the humid midsection of North America is shown in Figure 7.16. Note in the center (summer) part of the graph that the *actual evapo-*

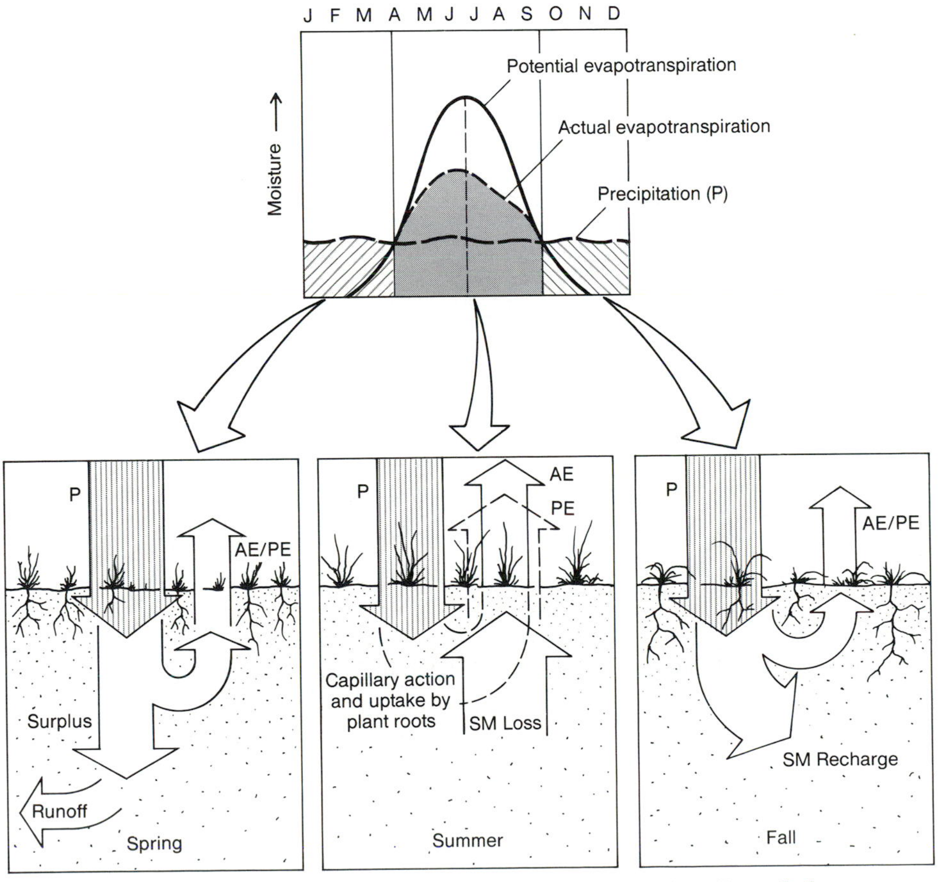

Figure 7.15 The basic soil-moisture balance model for a humid midlatitude location. The three principal phases of the year are represented by the diagrams below.

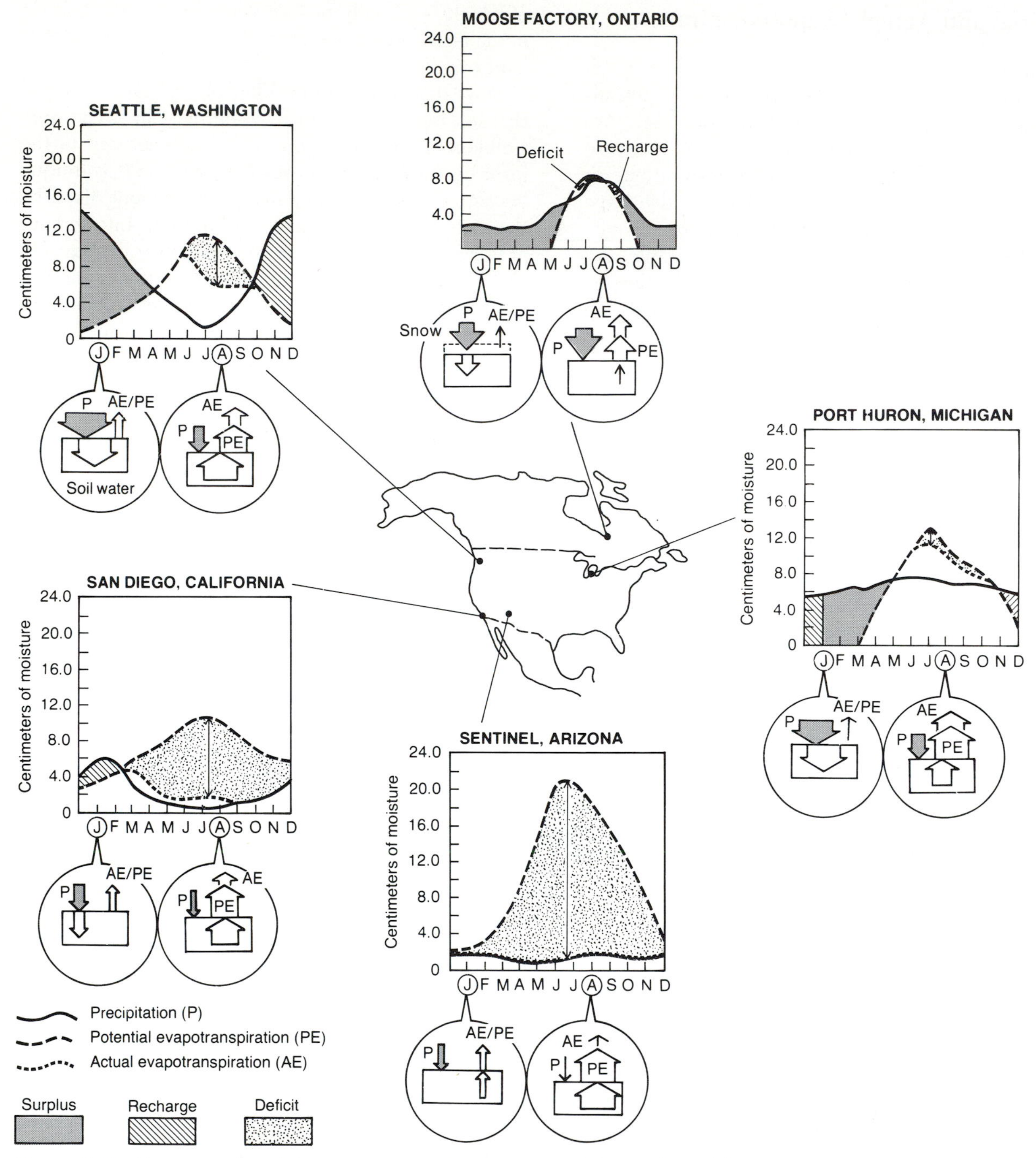

Figure 7.16 Five examples of the average annual soil-moisture balances from different climatic zones of North America.

transpiration is fed by soil water brought up by plant roots and capillary action in the soil itself. But the supply of soil water is limited, and as it is drawn down, the remainding water is more and more difficult for plants to take up or to evaporate directly. As a result, the upflow is usually too slow to meet the demands of plants and atmosphere, and the moisture deficit grows, represented by the area between the PE and AE curves. Note that AE can never exceed PE, as PE is the maximum possible evapotranspiration. *Recharge*, the replacement by precipitation of that soil moisture lost during a deficit period, is possible only until the soil-moisture reserve is refilled. After recharge is completed (and evapotranspiration has taken its share), any water remaining either trickles through the soil or accumulates as snow (storage water) and eventually becomes runoff or groundwater.

Examples of Soil-Moisture Balance from the United States and Canada

Sentinel, Arizona (see Fig. 7.16), exemplifies an area of severe soil-moisture deficit. Note that there is almost no recharge and the deficit period is a year long. This means that plants must either have low water requirements or be artificially watered through irrigation. Sentinel has an annual PE of 119 cm and annual precipitation (P) of 12 cm. This gives you an idea of the amount of irrigation water needed to maintain a lawn in this desert area.

San Diego, California (see Fig. 7.16), provides an example of a water balance in a highly seasonal climate. Precipitation is received almost entirely in the winter months, but the greatest demand (PE) is in the summer. This results in a large summer deficit that is only partially made up by recharge in the winter months, leaving a net annual deficit of 53 cm. This markedly seasonal water balance is characteristic of the Mediterranean climate found along the west coast of California.

Port Huron, Michigan (see Fig. 7.16), is also a seasonal station in that temperature varies widely from summer to winter. Here, however, the precipitation is much more evenly distributed throughout the year than in Sentinel or San Diego. When the temperature falls below freezing, PE is negligible, contrasting with a sharp peak in the summer. The even distribution of precipitation throughout the year provides enough recharge to minimize the deficit in midsummer. By midwinter, the deficit is made up; that is, recharge is completed, and the remaining precipitation is surplus water, much of which is stored in the form of snow until the spring melt. Port Huron has an annual PE of 61 cm and receives 74 cm of precipitation.

Seattle, Washington (see Fig. 7.16), is an example of a site with an adequate annual precipitation (84 cm), which is quite seasonal in its distribution. The precipitation minimum in the summer months causes a strong deficit for that season, but heavy precipitation in the winter is more than enough to offset the deficit. Note that the maximum moisture deficit in midsummer occurs a few weeks after the precipitation minimum and the PE peak. This illustrates the time lag in the flow of soil-moisture from the soil.

Moose Factory, Ontario, located at the southern tip of Hudson Bay, receives a modest amount of precipitation (annual mean of 53 cm [21 in.]), but summers are short and cool, therefore, PE is low. In addition, precipitation rates in summer are more than twice those of the winter months. As a result, Moose Factory has a very small moisture deficit, which appears to be made up by recharge in a month or so in the fall. In the winter, surplus moisture is stored on the surface as snow which is released to the soil in spring.

Summary

Water is an important control on climate not only because of the heat storage capacity of the oceans, but also because of the heat transfer affected by the hydrologic cycle. The hydrologic cycle is a complex system that produces a complete exchange of atmospheric vapor on the average of forty times per year. The oceans supply 88 percent of water to the atmosphere, and the highest rates of evaporation are found in midlatitude areas where cold, winter air sweeps over warm water.

On the land two general types of moisture environments are recognized: moisture limited and moisture abundant. The latter is comprised of open water surfaces and moisture-holding soils. Evapotranspiration rates are influenced principally by surface heating, the availability of moisture on or near the surface and the abundance of vegetation. Where potential evapotranspiration is high and moisture is limited, Bowen ratios are large. Soil is an important reservoir of moisture that fluctuates with season climate related to input from precipitation and losses from evapotranspiration.

Concepts and Terms for Review

- hydrologic cycle
  - sources of moisture
  - surface-atmosphere exchange
  - subcycles
- oceanic phase
  - global evaporation rates
  - high evaporation zones
- precipitation
  - precipitable water vapor
  - precipitation rates
  - variability
- moisture-limited environments
- moisture-abundant environments
- open water evaporation
- lakes, reservoirs, ponds
- hydrographic landscape
- evapotranspiration
  - potential
  - actual
- soil-moisture balance
  - moisture deficit
  - recharge
  - climate relations

Sources and References

Bowen, I.S. (1928) "The Ratio of Heat Losses by Conduction and by Evaporation From Any Water Surface." *Physics Review,* 27: pp. 179–87.

Brooks, C.E.P., and Hunt, T.M. (1930) "The Zonal Distribution of Rainfall over the Earth." *Mem. Royal Meterological Society,* Vol. 3:28, pp. 139–185.

Budyko, M.I., et al. (1962) "The Heat Balance of the Surface of the Earth." *Soviet Geograph. Rev. Transl.* 3, 5: pp. 3–16.

Ferguson, H.L. et al. (1970) "Mean Evaporation over Canada." *Water Resources Research,* Vol. 7, pp. 1618–1633.

Hasse, L. (1963) "On the Cooling of the Sea Surface by Evaporation and Heat Exchange." *Tellus,* Vol. 15, pp. 363–366.

Penman, H.L. (1961) "Weather, Plant and Soil Factors in Hydrology." *Weather,* Vol. 16, pp. 207–219.

Perry, A.H., and Walker, J.M. *The Ocean-Atmosphere System.* New York: Longman.

Sopper, W.E., and LuU, H.W., (1967) *Forest Hydrology.* Pergamon Press: Oxford.

Starr, V.P., and Peixoto, J.P. (1958) "On the Global Balance of Water Vapor and the Hydrology of Deserts." *Tellus* 10, 2: pp. 188–194.

Thornthwaite, C.W., and Mather, J.R. (1955) "The Water Balance." *Publication No. 8.,* Laboratory of Climatology, Centerton, New Jersey.

Chapter 8

Climates of the World

We ascribe a great deal of importance to differences in climate from place to place. Climate not only helps shape our sense of place, but it also has a profound influence on vegetation, soil, and land use. But how do we actually measure and characterize climates? What are the preferred climate classification schemes used by geographers? What are the major climate zones and what is the general character of land and life in each of these zones?

"Arizona's generally warm and sunny, but every now and again there's a devil of a thunderstorm." "Toronto's nice for most of the year, but the winters can get mean; lots of snow in January and February in some years." "Spring is lovely in Kansas, but there's always the threat of tornadoes." "Panama's eternally hot, the humidity's high, and it rains nearly every day.."

These are descriptions of four different climates as perceived by persons who live in them. They are informative remarks, but for three reasons they are not very meaningful portrayals of climates. First, their emphasis is more on comfort and aesthetics than on atmospheric conditions. Second, they are inconsistent and imprecise in their use of the terms *hot, cold, rainy,* and so on. And third, they are subjective in that no numerical reference is given for these terms.

Climate is the word we use to describe the representative conditions of the atmosphere at a place on earth. Climate is more than an average of the weather over a period of time because extreme and infrequent conditions—which are not evident in averages—are important traits of climate. In geography, climate refers to the conditions of the lower atmosphere that most directly affect the landscape and the organisms in it.

In this chapter we describe the general climates of the world and how they are classified by geographers. We will build on the principles of atmospheric circulation, oceanic circulation, the hydrologic cycle, and weather processes to explain why certain climates form where they do on the continents. We will also try to draw out some of the apparent influences of climatic conditions on land and life, emphasizing the relationship of climate to water supplies, vegetation, and land use.

CLASSIFYING CLIMATES

For the purposes of the geographer, description and classification of an area's climate should be based on (1) the recorded amounts of rain, heat (temperature), humidity, and so on, as well as (2) some measure of the atmospheric processes themselves. We would want to know, for example, not only how much rain falls, but also when it falls and what causes it. Classification schemes in turn have two important attributes: (1) they must be readily understandable—that is, not so complex that they are unwieldy; and (2) they must be built on a known body of data, that is, recorded observations. And on the second attribute the notion of an ideal scheme must be set aside, for we lack suitable data on atmospheric processes for most of the world.

The international data base on weather in the past 50 years or more has been largely limited to readings on temperature, precipitation amounts and forms, wind speed and direction, atmospheric pressure, and humidity. Only in the past decade or two have countries begun to collect data on atmospheric processes, such as the types of storms producing precipitation. Much of this effort has been made possible by weather satellites, but as yet the results have not been built into the climate classification schemes.

From the vast pool of available data, which do we use to classify climate at the global scale? Precipitation and temperature have been the traditional choices. What quantitative values or levels of these parameters are most meaningful, assuming that we have daily data from a worldwide network of weather stations to work with?

In the case of temperature, only one temperature comes to mind immediately—namely, 0 degrees C—by which one can distinguish climates that experience frost from those that do not. But how should frost, or freezing temperatures, be defined? Should it be defined in terms of monthly mean temperature, daily minimum, or daily mean temperature?

The answers are not easy to come by, and geographers have argued over the years about what definitions are most meaningful. So far, there is no single classification scheme or set of values that serves all purposes well; some are better for studying agriculture, others for gauging human comfort, and still others for studying natural vegetation. Moreover, there is a problem in mapping climates.

Climatic Boundaries

Of the major geographic features of the earth's surface, climate is probably the most difficult to delineate in discrete geographic regions. In contrast to the border between land and sea, for example, which falls within a narrow transition between high tide and low tide, climatic borders are generally broader and more complex because most climatic features change gradually over long distances. Where to place a climatic boundary line is, within certain limits, a matter of one's objectives and the data base being used. This often boils down to a problem of selecting the appropriate criteria for climatic classification, such as average annual precipitation and seasonal temperature ranges based on monthly mean temperatures; then establishing the quantitative values by which the different classes are defined; and, finally, translating the results into a map showing the spatial extent of each climate type.

There are three types of climatic boundaries. First, in the broad interiors of those continents for which climatic data are abundant, such as in North America and Europe, boundaries are often established according to statistical trends over great distances. The critical issue is where to place the boundary along a set of gradients. To make matters more difficult, in virtually all such cases the values shift from year to year and decade to decade; thus, it is necessary to select one period of record to serve as the mapping base. In the United States, the 30-year period from 1950 to 1980 is often used as the data base for climate mapping.

In sharp contrast to the boundaries representing broad transitions are boundaries based on precipitous gradients in terrain. The rainfall gradients associated with the mountains of Hawaii are a case in point (see Note 6). This type of boundary is usually found along the margins of land masses where maritime and continental climates are differentiated by mountain belts.

A third type of climate boundary is one established on the basis of some indicator feature of the landscape. Vegetation is the feature most commonly used because certain types of vegetation are thought to be climate-sensitive; therefore, their presence or absence is taken as a clue to one or another climate type.

CLIMATE ON A HYPOTHETICAL EARTH

When faced with a complex problem, such as the geographic arrangement of climates on a complex planet, it is sometimes helpful to set up a model of the situation before diving into reality. The model gives us a simplified representation of reality that enables us to sort through the various influencing factors, or variables, in order to see more clearly how certain factors influence the problem. Here we will begin our

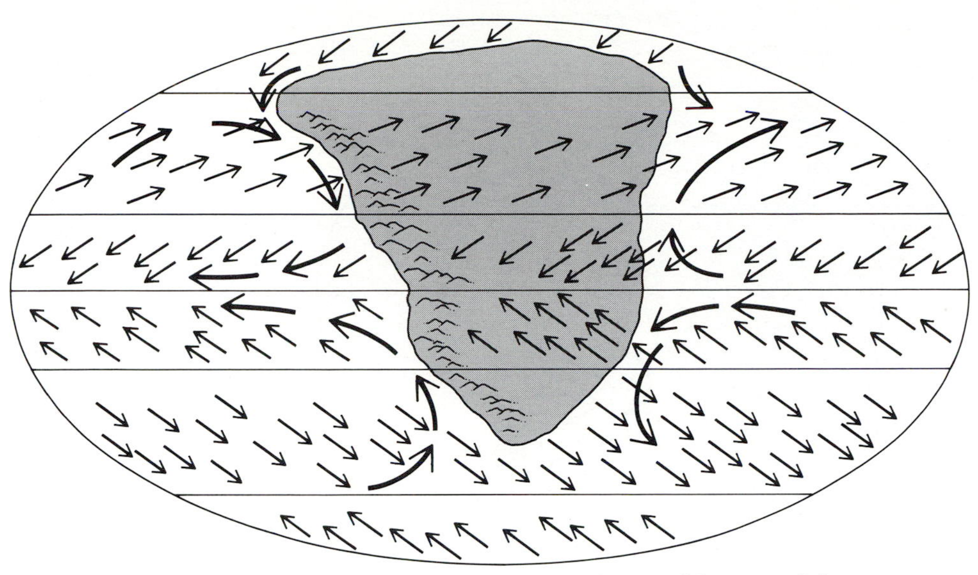

Figure 8.1 Hypothetical continent in the approximate shape and location of the Americas and Eurasia/Africa combined. The arrows represent the prevailing winds and the major ocean currents.

examination of world climates with a hypothetical model of the world.

Suppose we set up a world with one great continent whose location and gross features approximate those of the Americas and Eurasia/Africa combined (Fig. 8.1). In the Northern Hemisphere, it reaches beyond the Arctic Circle; in the Southern Hemisphere, it reaches only into the midlatitudes, to about 50°S latitude. The principal topographic feature of this supercontinent is a great mountain chain running along its west coast.

The Thermal Regime

Let us first consider the most basic question: the thermal controls. There are three levels or scales of thermal influence: (1) zonal (zones of latitude), (2) continental-maritime, and (3) regional (mainly subdivisions of the continents).

At the *zonal scale*, we must recognize the principal zones of solar heating and the seasonal variations in solar radiation with latitude. We know that the intertropics, between 23.5°N and 23.5°S latitude, has high sun angles most of the year and so season of abbreviated daylight. Therefore, this zone should have the highest and least variable temperatures on earth—barring, of course, special regional circumstances. The thermal climate in this broad belt should, in fact, be essentially seasonless.

The midlatitudes should experience more seasonal conditions because of the great differences there in sun angle and length of day between winter and summer. Generally, we should expect to find warm summers and cool to cold winters. The seasonal sun angles are ample evidence for this assertion. At 45°N latitude, for instance, the summer high sun is 68.5° above the horizon, approximately the same sun angle as that at the equator on the same date. (The sun angle at the equator on June 20–22 is 66.5°, 2 degrees lower than at 45°N latitude!) Conversely, the low sun angle at 45° latitude is 21.5°, which is comparable to the sun angle at the poles during the summer solstice.

In the high latitudes (above 66.5° latitude), summer sun angles are low (23.5° at 90° latitude), the days are long, and the summer season is cool and short. Winter is the dominant season, and at altitudes above 75° the thermal climate is essentially seasonless: winter prevails virtually the year round.

At the *continental-maritime scale* in our hypothetical model of the world, however, the basic zonal pattern described previously is altered by the uneven distribution of land and water and by the pattern of large-scale atmospheric circulation. In the midlatitudes over the large land masses, the seasons are made more extreme by the continental regime found there; summers in the interior are hotter, and winters are colder than average for that latitude. Over the oceans, by contrast, the solar seasons are modified by mari-

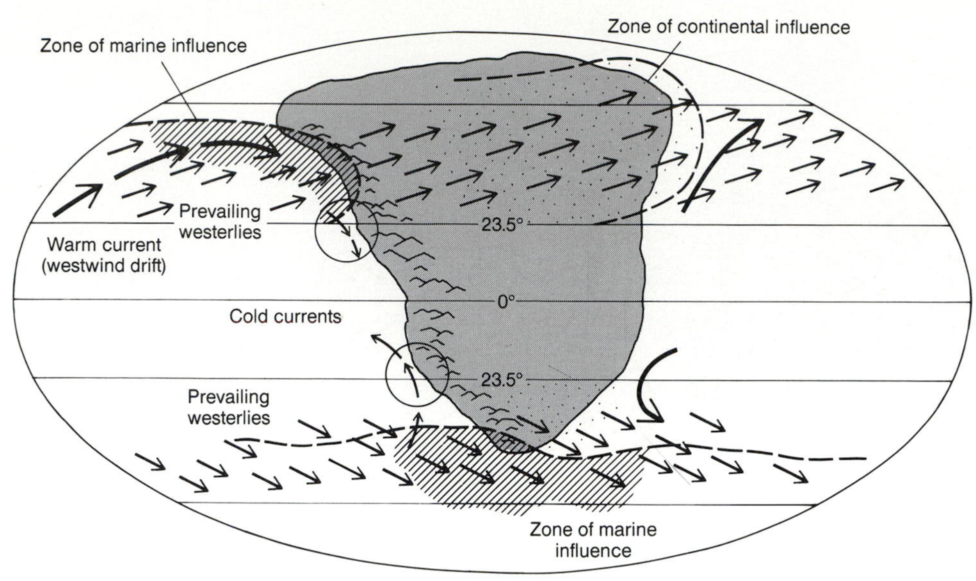

Figure 8.2 Influence of the westerlies and the west-wind drift in creating a west coast marine climate in the midlatitudes and subarctic. On the subtropical west coast, cold currents cool an otherwise hot desert climate.

time regime, resulting in small annual temperature ranges. In the zones of the prevailing westerlies, these maritime conditions encroach onto the continental west coast, displacing the continental regime eastward. Marine climates thus dominate this section of the midlatitude west coast, whereas a continental regime dominates the rest of the land mass, including the midlatitude east coast (Fig. 8.2).

At the third level of consideration—the *regional level*—we must examine the influences of the ocean currents on the pattern of thermal climates. The major currents formed by the great gyres distort the zonal patterns of air temperature by several degrees or more. The west-wind drift carries warm water northeastward, thereby stretching the zone of marine influence poleward along the west coast and into the high latitudes (Fig. 8.2). In the zone of subtropical high pressure, cold currents run along the west coast which greatly modify an otherwise hot desert climate in the coastal zone. Because the landward airflow at this latitude is weak, however, this effect is limited to the coastal fringe.

The Influence of Air Masses

Let us now consider the distribution of air masses and their areas of influence in our hypothetical world model. The intertropical zone is dominated by two sources of air: equatorial air associated with the trough of equatorial low pressure and tropical air associated with the easterly trade winds. Some air masses from these sources, especially maritime tropical air under the influence of the easterlies, migrate onto the southeastern part of the continent in both hemispheres. Many of these air masses are carried into the continental midlatitudes where they mix with cold air along the polar front to form cyclonic storms (Fig. 8.3).

In the Northern Hemisphere, the cold air masses of the midlatitudes are derived from the arctic and polar source regions and are both continental and maritime in makeup. In the Southern Hemisphere, these air masses are exclusively maritime for the simple reason that there is only water at these latitudes in the Southern Hemisphere. In both hemispheres, the movement of the maritime polar air masses is strongly influenced by the westerly flow of air which brings them onto the west coast of the continent (Fig. 8.3).

On the western side of the continent near the tropics of Cancer and Capricorn, the large cells of subtropical high pressure extend well into the land mass. These are a source of stable, dry air (continental tropical). In both hemispheres, the west coast mountains favor a poleward expansion of these sources (on the leeward side of the mountains) because they block the entry of moist air into the continent's midsection. In addition, the great size of the land mass in the North-

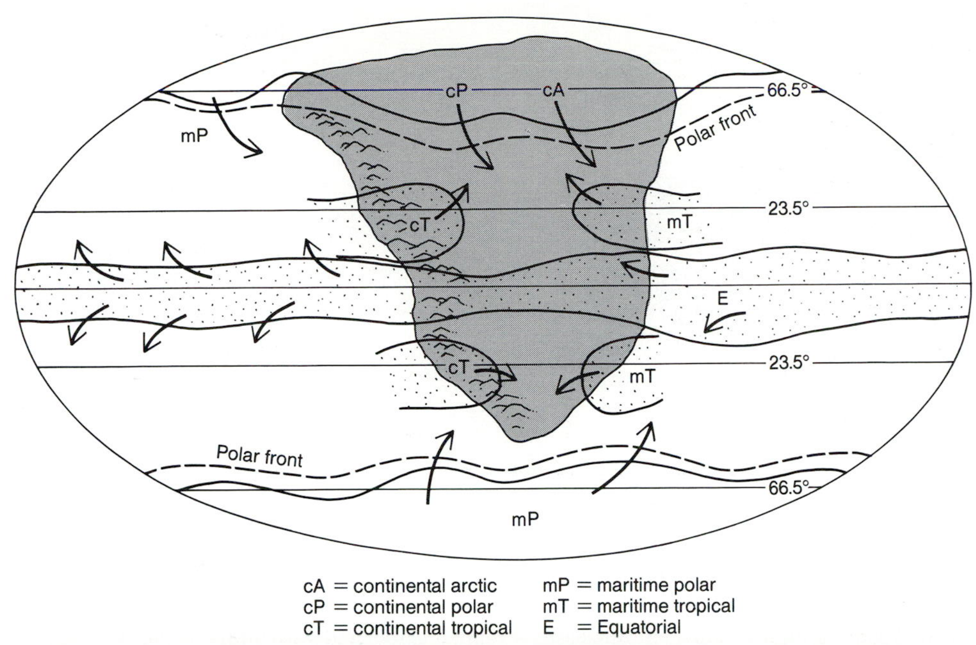

Figure 8.3 Distribution of air masses and their tracks of movement on the hypothetical continent.

ern Hemisphere also encourages the development of a strong body of continental tropical air, just as Central Asia does in the real world.

In both hemispheres, in the zone poleward of the subtropical highs, the atmosphere over the oceans is increasingly dominated by maritime air masses. Air masses from this region are moved eastward by the westerlies, resulting in heavy precipitation and modest temperature regimes along the west coast.

In the Northern Hemisphere above 60° latitude, polar and arctic air masses are dominant throughout the year. In winter, these air masses intensify greatly over the continent and expand into the midlatitudes. The leading edge of this air, the polar front, fluctuates north and south over the winter depending on the pattern of jet stream circulation and development of cyclonic storms.

Over the oceans the polar front does not extend much beyond the subarctic latitudes (55° to 66.5°) in both hemispheres. Cyclones over the oceans are formed from polar maritime and tropical maritime. Fed with ample moisture, they usually produce heavy precipitation.

Precipitation Patterns

With an understanding of the basic thermal conditions and the air-mass types and movements in our hypothetical world, we should be able to piece together a picture of the precipitation patterns in it. (See Fig. 8.4 for the area corresponding to each letter in the following discussion.) Generally speaking;

a. Near the equator heavy rainfall should be produced in all seasons from convergent and convectional mechanisms associated with the Intertropical Convergence Zone (ITC) (Fig. 8.4).
b. Poleward of this equatorial wet zone, but still in the intertropics, rainfall should be seasonal because the ITC is overhead only during high sun; during the low-sun months, the subtropical high-pressure cells slide into this zone producing a dry season (Fig. 8.4).
c. The poleward shift of the ITC should be greater in the Northern Hemisphere because of its greater land mass. This could sometimes result in a monsoon condition: strong airflow onto the continent and heavy rainfall.
d. On the tropical east coasts, the easterly trade winds should bring a steady supply of rainfall in all seasons.
e. Under the subtropical high-pressure cells, we would expect the atmosphere to be clear and very dry. In addition, solar heating of the landscape should be intensive, creating a high po-

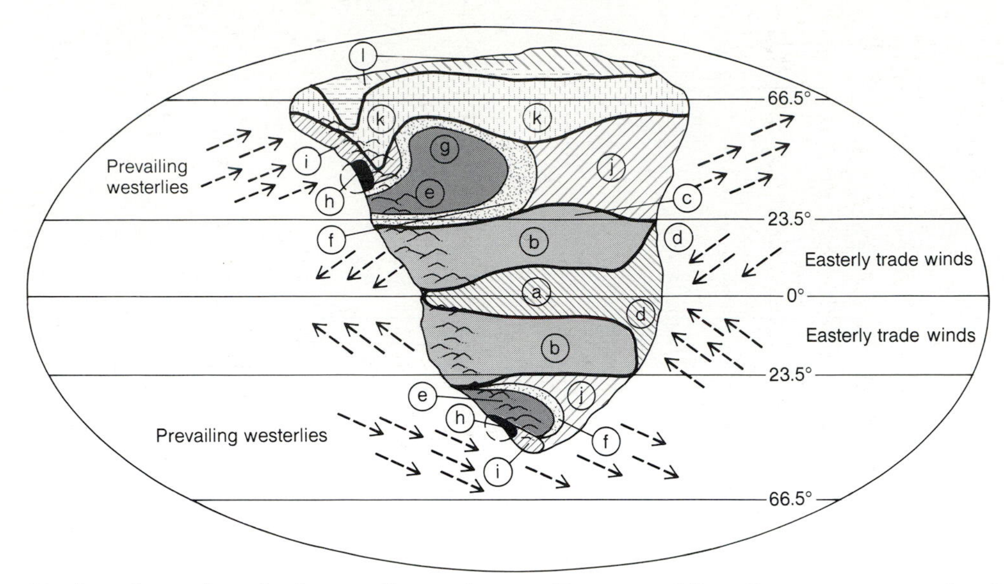

Figure 8.4 Principal moisture/precipitation zones of the hypothetical continent. The letters represent locations of precipitation zones described in text. a) Equatorial belt of heavy rainfall in all seasons associated with the ITC zone; b) seasonal rainfall associated with the shifts in the ITC zone; c) area representing greater northward shift of ITC zone in response to large Northern hemisphere land mass; d) coastal zone of heavy rainfall in all seasons associated with Easterly Trade Winds; e) clear, dry zone associated with subtropical high pressure cells; f) semiarid zone on margin of dry zone; transitional to moister climates; g) arid and semiarid zone of middle latitudes; h) seasonal rainfall associated with shifts of the subtropical highs and moist westerlies; i) coastal marine zone influenced by moisture westerlies; j) continental moist zone of cyclonic convectional precipitation; k) cold zone along Arctic Circle, only in Northern Hemisphere; and l) polar zone above the Arctic Circle.

tential evapotranspiration. The net result should be a severely arid climate.

f. On the fringe of the arid zone, we should find a sizable transition zone on all sides to the moist climates. This zone would be characterized as semiarid (Fig. 8.4).

g. In the Northern Hemisphere, the belt of arid and semiarid climate should extend northward into the interior of the continent. This would be caused by the lee effect of the west coast mountains which stand in the way of westerly moving moist air and by the great distance between the continental interior and moisture source areas in general.

h. On the poleward edge of the subtropical highs is a section of the coast where the precipitation should be strongly seasonal: dry in the summer under the influence of the subtropical highs and moist in the winter under the influence of the westerlies.

i. Poleward of the above area, the westerlies are dominant in all seasons, which should result in a wet or marine-type climate (Fig. 8.4).

j. In the eastern half of the continent poleward of the tropics is a broad zone where cyclonic precipitation should be prevalent. The southern part of this zone should also experience appreciable thunderstorm activity in summer related to surface heating and a supply of moisture from maritime tropical air masses.

k. Along the Arctic Circle and stretching across the Northern Hemisphere should be a cold zone where precipitation is generally light (on account of low moisture-holding capacity of the cold air) and mostly in the form of snow. The comparable zone is beyond land's end in the Southern Hemisphere.

l. Above the Arctic Circle, is a polar climate with very light precipitation in the form of snow. This climate should extend farthest equatorward along the axis of the mountain range and in the interior of the land mass where the continental thermal regime is most pronounced (Fig. 8.4).

With some background on what to expect in the way of climatic conditions in and around the land masses of the world, we can now examine the climate classification schemes used in science and the types of climates found in the real world.

SCHEMES FOR CLASSIFYING CLIMATES

Most descriptions of the world's climates draw on two basic classification schemes: the Köppen-Geiger system and the Thornthwaite system. The Köppen-Geiger system was developed early in this century when many leading geographers were interested in synthesizing data and observations on land, life, and climate and building global-scale schemes for classifying climate, vegetation, and soil. The name of this scheme is derived from its founders, the German climatologist Vladimir Köppen and his student Rudolph Geiger, also a climatologist. It is based mainly on temperature and moisture, with an emphasis on the seasonal character of each. The Köppen-Geiger system has some shortcomings, particularly in its definitions of dry climates and in its placement of certain climate borders based on vegetation. However, it does provide a very useful framework for classifying climates at the global scale mainly because it gives us an idea of what the seasons (or representative years in seasonless climates) are like in different parts of the world.

The Thornthwaite system was developed in 1948 by C. W. Thornthwaite, the American climatologist whose work we mentioned in Chapter 7 in connection with the soil-moisture balance. This system is built directly on the moisture balance concept, which is in turn based on moisture and potential evapotranspiration. Geographers have found that the Thornthwaite system is best suited to the classification of the dry climates, and in this respect it makes up for one of the weaknesses in the Köppen-Geiger system.

The Köppen-Geiger System

The Köppen-Geiger system uses three basic levels of definition. Each level is coded with a set of letters, upper case for the most general level and lower case for the more specific levels. At the first level the earth is divided into five great zones covering the globe from the equator to the poles:

- A = tropical and equatorial rainy climates
- B = dry climates: the potential exists for evapotranspiration to exceed precipitation
- C = temperate moist climates: long, warm summers, cool winters
- D = cold, snowy climates with forest: short summers, cold winters
- E = polar climates: long, cold winters, brief or no summers (no forests)

At the second level, each of these zones is further divided according to temperature and rainfall, seasonal variations, and certain relations to vegetation.

TABLE 8.1 THE TWELVE PRINCIPAL CLIMATIC TYPES IN THE KÖPPEN-GEIGER CLASSIFICATION

Main Zones	Climate Types (Symbol)
A. Tropical rainy climates	Tropical rainforest (Af)
	Savanna (Aw)
B. Dry climates	Steppe (BS)
	Desert (BW)
C. Temperate rainy climates	Warm, with dry winter (Cw)
	Warm, moist in all seasons (Cf)
	Warm, with dry summer (Cs)
D. Cold snow-forest climates	Snow forest, moist in all seasons (Df)
	Snow forest, dry winter (Dw)
E. Polar (snow) climates	Tundra (ET)
	Perpetual snow and ice (EF)
	High mountain, nonpolar (ETH)

Climates with adequate rainfall in all seasons are designated *f*; climates with a winter dry season are designated *w*; and those with a summer dry season are designated *s*. These apply only to the A, C, and D climates, the humid climates capable of growing "high-trunked" trees (Table 8.1).

The dry climates (B) are divided into arid (desert) and semiarid (steppe) and are given a second capital letter designation of *W* and S, respectively. The distinction between steppe and desert is based on the relative dryness (severity of drought) and is reflected in the vegetative cover, steppe being grassland and desert being lands with very little or no vegetation.

The polar climates (E) are also divided into two varieties: EF, the most severe, which is characterized by perpetual ice and snow and no month with an average temperature above 0°C; and ET, which is characterized by tundra landscapes and at least one summer month without snow cover when the average monthly temperature reaches beween 0°C and 10°C. A variant of the ET climate, denoted ETH, is found in high mountain areas outside polar regions (Table 8.1).

The third level in the Köppen-Geiger system is used to indicate seasonal temperature conditions in the C and D climates and the overall temperature regime of the B climates:

- a = with hot summer (average temperature of warmest month exceeds 22°C)
- b = with warm summer (average temperature of warmest month less than 22°C, but at least 4 months average over 10°C)
- c = with cool, short summer (average temperature of warmest month less than 22°C, less than 4

months average over 10°C, and the coldest month averages above −38°C)

d = with very cold winter (D climates only; same as above, except that coldest month averages below −38°C)

h = hot desert or steppe (average annual temperature exceeds 18°C)

k = cool desert or steppe (average annual temperature less than 18°C and average temperature of warmest month exceeds 18°C)

Another letter (n) is also given to desert climates in coastal areas where fog is frequent.

The Thornthwaite System

The Thornthwaite system relies on two criteria for climate classification: *potential evapotranspiration* and the *moisture index*. Potential evapotranspiration is a measure of energy available in the landscape based mainly on heat (air temperature) and light. The moisture index represents the difference between precipitation and evapotranspiration. It is scaled from positive (+) 100 for the very humid climates to negative (−) 100 for the very dry climates. The boundary between the humid and dry climates, or realms, is marked by a moisture index of 0.

The Thornthwaite system provides a refined classification of climate based on moisture (compared to the Köppen-Geiger system). The Köppen-Geiger system uses two basic classes of dry climates (BW and BS), whereas the Thornthwaite system adds a third class, the *subhumid*. This climate is transitional between the semiarid and the humid climates where precipitation and potential evapotranspiration are about equal (moisture index of 0). To help understand Thornthwaite's motivation for devising a moisture-based classification, we must recognize that much of his work was carried out during the Great Depression of the 1930s. This was a time of extensive drought in the Great Plains, the West, and the Midwest, and one of its most disastrous consequences was the infamous Dust Bowl. It was apparent to geographers such as Thornthwaite that better understanding of drought and geographic variation in the boundary separating humid and semiarid climates would be necessary to better plan and manage agriculture in the Great Plains (see Fig. 8.12).

Thornthwaite's classification of world climates uses five major moisture categories (dry to wet: arid, semiarid, subhumid, humid, perihumid) and two cold thermal categories (subpolar and polar). These are

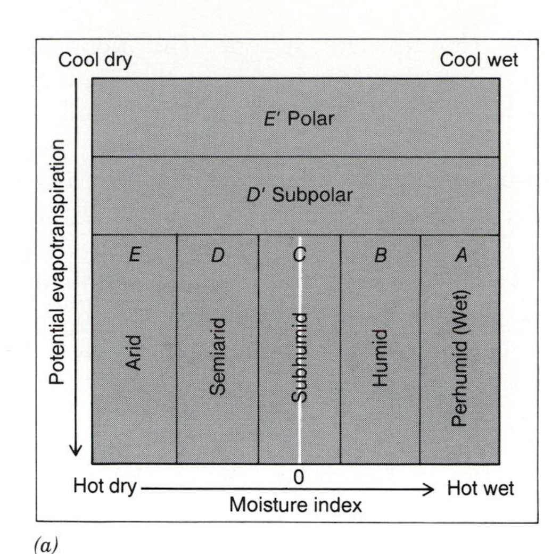

(*a*)

Figure 8.5 (*a*) The seven major categories of the Thornthwaite climate classification; (*b*) the moisture regions (zones) of the coterminous United States according to the Thornthwaite classification.

shown diagrammatically in Figure 8.5*a*. These in turn were subdivided according to energy (potential evapotranspiration) and moisture into a total of sixty-five subclasses which were related to the global patterns of vegetation, soils, precipitation, and runoff. At this scale the system has proven to be too complex for general use; however, it is widely used in textbooks at the regional scale as is illustrated for the coterminous United States in Figure 8.5*b*.

WORLD CLIMATES

In this section we describe fifteen major climates of the world. The descriptions draw on both the Köppen-Geiger and Thornthwaite systems. In addition, the principal mechanisms controlling precipitation and the general character of the landscape of each climate type are discussed. As you read these descriptions, it will be helpful to refer frequently to the world climate map in Figure 8.6 (color plate 5) for the distribution of the climates described, and to the annual precipitation and moisture graphs in Figure 8.7 for stations (locations) representative of these climates.

Tropical Climates

These climates form a belt centering on the equator and in places extending poleward to the tropics of Cancer and Capricorn. They are warm in all seasons

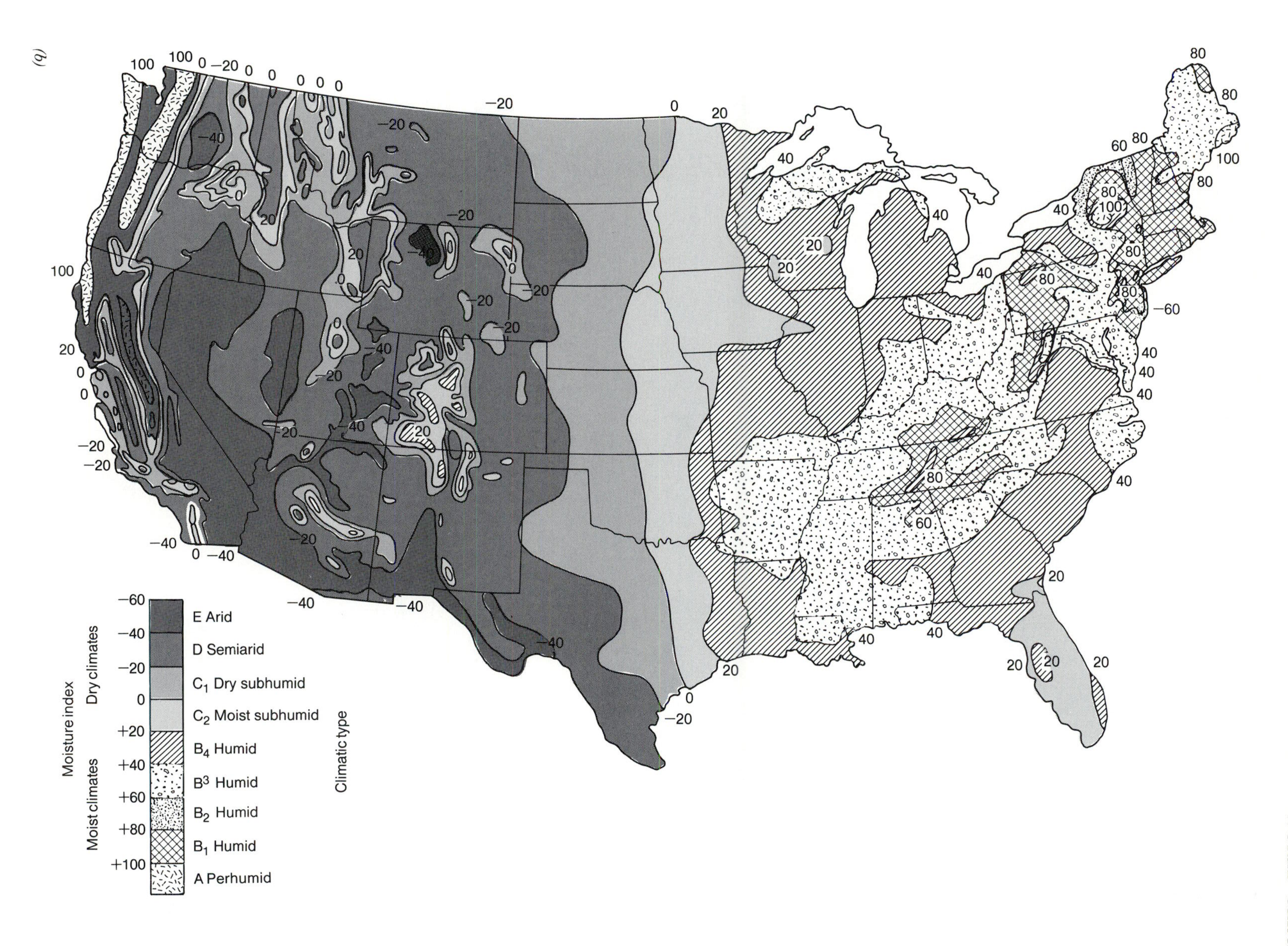
Moisture index
Dry climates
Moist climates
−60
−40
−20
0
+20
+40
+60
+80
+100
Climatic type
E Arid
D Semiarid
C_1 Dry subhumid
C_2 Moist subhumid
B_4 Humid
B^3 Humid
B_2 Humid
B_1 Humid
A Perhumid

(b)

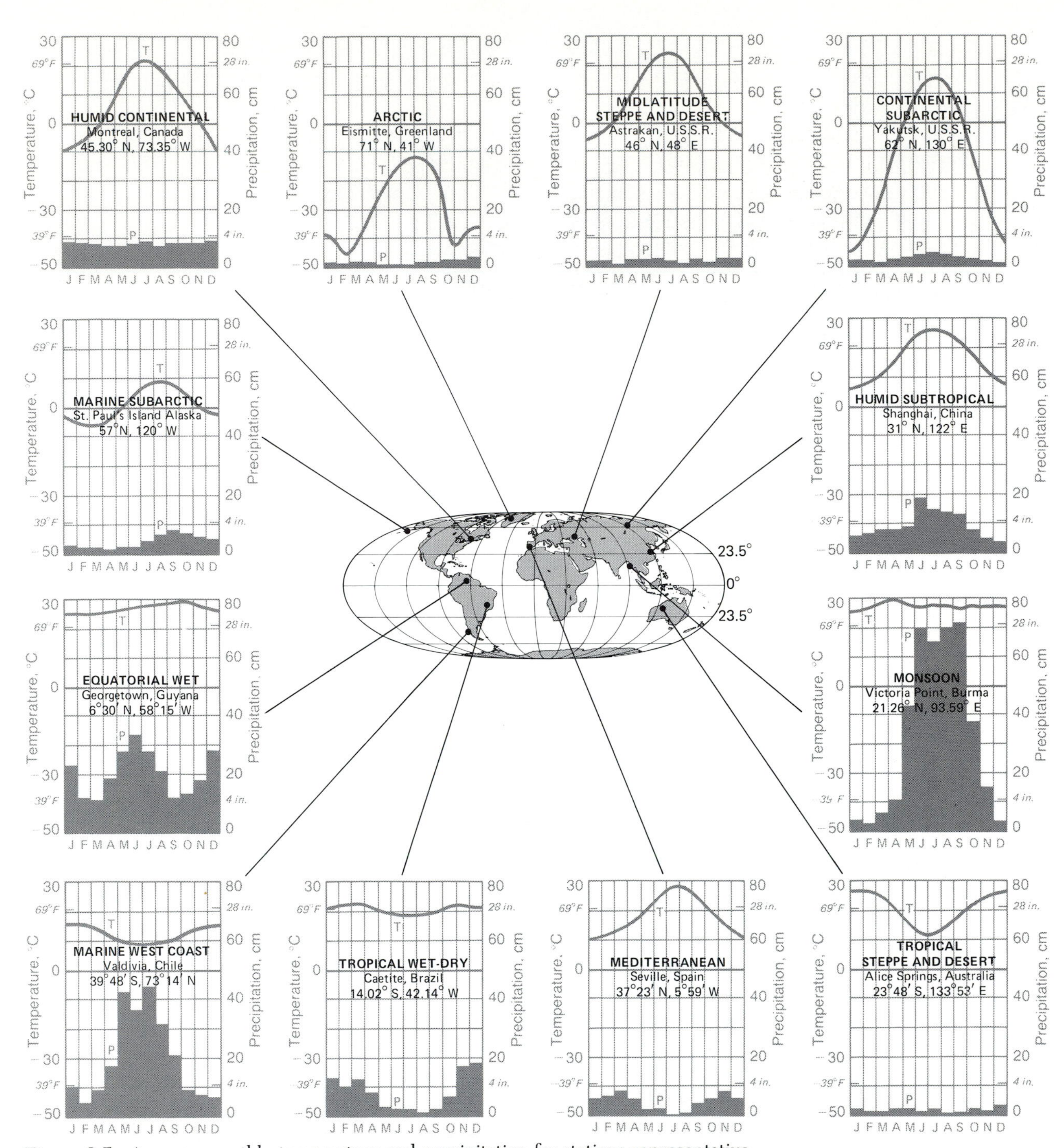

Figure 8.7 Average monthly temperature and precipitation for stations representative of twelve major climates.

and are dominated by equatorial (E) air associated with the belt of equatorial low pressure and tropical (T) air associated with the trade winds.

Equatorial Wet (Af) Along the equator and as much as 10 to 15 degrees north and south of it, the climate is rainy, with high humidity and mean monthly temperatures around 30°C throughout the year. The ITC zone is overhead or nearby in all seasons. As a result, every month has substantial precipitation, and clouds are prominent for a part of most days in the year (see Fig. 5.13).

Rainfall is generally due to convergent and convectional mechanisms, and over land it is associated with two distinct rhythmical patterns. The first is a daily rhythm characterized by rain showers in the afternoon of most days. The second is seasonal, characterized by heavy precipitation when the ITC is overhead and somewhat lower amounts when it is not. Over water, precipitation is predominantly convergent because surface heating is not intensive enough to initiate convection.

Near the equator the seasonal pattern results in two rainfall peaks during the year as the ITC shifts from one hemisphere to the other with the migration of the sun. The graph in Figure 8.7 for Georgetown, Guyana (latitude 6°30′N) shows this pattern. Notice, however, that the peaks of rainfall do not coincide with the equinoxes when the sun is highest at the equator but rather lag behind by a few months. This delay is caused by the time it takes for surface heat to build up and for the ITC to slide into place.

The principal land areas dominated by the tropical wet climate are found in South America, Africa, and Indonesia. There are, in addition, three narrow bands of this climate that extend poleward to about 25° latitude. Located on the coasts of Madagascar, southeastern Brazil, and Middle America, these bands are produced by the easterly trade winds, which bring a steady supply of moisture to these coastlines. In months around the equinoxes, tropical storms may migrate onto these coasts, and some of the hurricanes are highly destructive. The landscape of the equatorial-wet climate is predominantly heavy forest, called *rainforest*, characterized by abundant plant species and rapid growth rates.

Tropical Wet-dry (Aw) North and south of the equatorial-wet climate regions are areas that are in the ITC zone during the high-sun season, but under the subtropical high-pressure cell during the low-sun season. Thus, the areas have a wet season in the "summer" and a dry season during "winter." Temperatures are warm throughout the year, however (Fig. 8.7).

Understandably, the soil moisture available to plants is depleted in the first 1 to 2 months of the dry season, and vegetation declines until the landscape is brown. Streams also decline, and by midseason water holes (mainly deep stream pools and low spots where groundwater seeps out) are the only remaining surface water. With the return of summer rains, the soil-moisture reservoir is recharged, vegetation is revived, and streamflow is renewed.

The duration of the dry season is variable from year to year because of variations in the extent of the poleward migration of the ITC zone. Hence, areas on the poleward margins (the desert sides) of the wet-dry climate are likely to experience variations in moisture that may have serious consequences for land and life. Both crop and pastoral farming in these zones are heavily dependent on summer rains, and when rains fall short of expected amounts, as they often do, crop failures and food shortages result. If drought conditions continue for many seasons, many of those inhabitants who do not migrate starve—as has happened more than once in the Sahel zone south of the Sahara in the past decade.

Much of the tropical wet-dry climate is occupied by savanna landscapes, which in Africa and Australia are characterized by scattered trees among large expanses of grass. In the summer as the rains sweep across the landscape, the area is vibrant with life, but with the coming of the dry season it grows brown and relatively lifeless. In Africa the great herds of grazing animals, such as the wildebeest, migrate with these changes.

Monsoon (Am) The monsoon climate, which is limited to South and Southeast Asia, can be thought of as an extreme version of the tropical wet-dry climate. The climate is named for the seasonal winds that dominate this region: the wet monsoon and the dry monsoon. Summer rainfall here is heavy, generally greater than that of the tropical monsoon landscape, and this is usually evident in a heavier tree cover. Precipitation is caused by convective and orographic mechanisms that are fed by massive amounts of moist air from the Indian Ocean and southwestern Pacific Ocean (see Fig. 5.10). During the peak months (June through September) of the wet monsoon, rainfall in the foothills of the Himalayas typically reaches 50 to 100 cm per month (Figs. 8.6 and 8.7).

In winter, with the development of strong high

pressure over Central Asia, airflow reverses. Now the air is dry and stable, producing a clear winter atmosphere. The duration of the dry monsoon season varies from year to year, and in years when it extends well into the spring, the resultant drought can be very damaging to agriculture. Marginal cropland in India is especially susceptible to this condition. Most of these croplands are located on the dry side of the monsoon zone in central and southwestern India. Conversely, excessive precipitation in the wet season can be equally damaging to agriculture because it causes flooding and soil erosion.

Although the monsoon climate is limited to South and Southeast Asia, monsoon-type circulation also occurs in North America; however, it does not produce the distinctive wet-dry seasons because the seasonal airflow is weaker owing to the smaller size of the land mass and lower magnitudes of the seasonal pressure cells. Nevertheless, the southerly airflow in summer from the Gulf of Mexico onto the Coastal Plains and into the Great Plains greatly increases precipitation between June and September (Fig. 8.8).

Tropical Steppe and Desert (BSh, BWh) Poleward of the tropical wet-dry climate regions are broad regions dominated throughout the year by the stable subtropical high-pressure cells. The airflow near the centers of these cells is downward, resulting in substantial adiabatic heating by the time the air reaches the ground. This is important because this air starts out dry (by virtue of the fact that it originated some 5 to 8 km aloft) and gets even drier as its relative humidity falls with rising temperature. This condition, coupled with the fact that air masses with precipitable

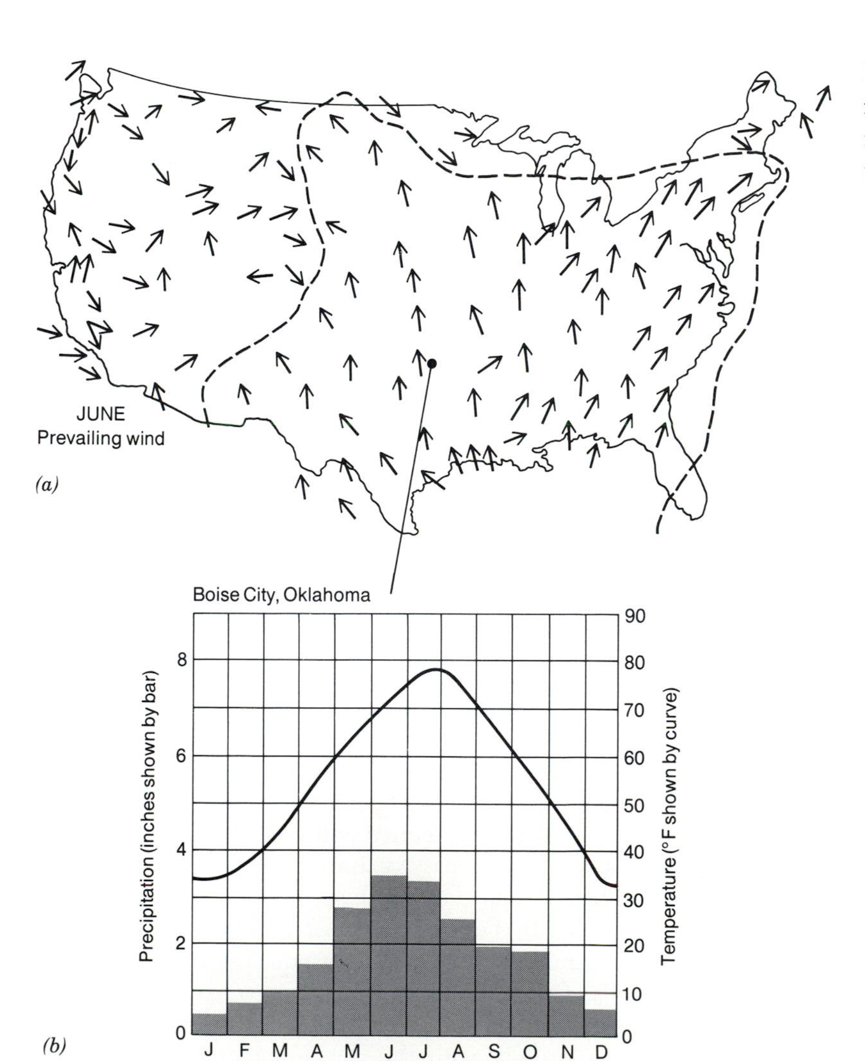

Figure 8.8 (*a*) Prevailing June airflow for middle North America and (*b*) the monthly pattern of precipitation and temperature for Boise City, Oklahoma.

moisture rarely penetrate these regions, gives rise to the driest landscapes on earth. In addition, cloud cover is scarce (in many areas, less than 30 days per year), with the result that intensive solar heating occurs at ground level. Therefore, any available surface moisture is rapidly driven into the atmosphere, giving rise to high-potential evapotranspiration rates and extraordinary soil-moisture deficits (color plate 16).

The world's major deserts, such as the Sahara (Africa), Rub 'al Khali (Arabian Peninsula), Sonoran (North America), and Great Australian, are classical examples of deserts produced by the subtropical highs. Their life-threatening conditions for humans are legendary: solar radiation on exposed skin can be lethal if prolonged; heat exhaustion from loss of body fluids and salts threatens those who risk physical exertion in the hottest hours of the day, which typically exceed 40°C. The temperature falls drastically at night because the dry air has such a low capacity to retain longwave radiation.

The transition zones between the deserts and the tropical (A) and the midlatitude (C) climates are the semiarid steppes. Toward the equator the steppes are characterized by a gradual change from perennially dry on one side to tropical wet-dry on the other. On the poleward border the steppe grades into the Mediterranean climate, which is also a wet-dry climate.

Along certain coastlines of the tropical steppe and deserts, the climate may be cool and foggy. This variant climate (BWh) is associated with cold ocean currents, which not only set up conditions for ground-level condensation and the occurrence of frequent fog, but also add significantly to the stability of air near the ground. As a result, precipitation in this climate may be even less than it is in the BWh; in places in the Atacama Desert of southern Peru, which is influenced by the cold Humboldt current, precipitation has *never* been recorded (see Fig. 5.15). Despite the paucity of rainfall, the combination of cool marine air and cool desert nights produces condensation on desert surfaces. This moisture is often the only source of water for plants and animals in these environments.

Patterns of traditional life in all the arid and semiarid climates have many similarities throughout the world. Overall population is very sparse. Settlements (villages, towns, and cities) are limited to those special locations, called *oases*, where a permanent supply of water is available. These exotic places are found where the desert is fed by a river such as the Nile or by springs at the foot of a mountain range. Elsewhere over the broad reaches of the desert the traditional mode of land use is nomadic herding, characterized by small groups (tribes or large family units) and their grazing animals shifting about the desert in search of pasturage. Unfortunately, this way of life is rapidly being eliminated as national borders and central governments in Africa, the Middle East, and Asia become more and more restrictive in their policies toward nonsedentary populations.

Midlatitude Climates

Midlatitude climates lie in the zone of prevailing westerly winds. Variations in the geographic and seasonal character of these climates depend on the interplay of air masses, from maritime and tropical sources, along the polar front. The polar front weakens and shifts poleward in the summer with the decline of arctic and polar air masses. This is accompanied by a growth of and poleward shift in the subtropical highs and a decline in the magnitude and frequency of midlatitude cyclones. Offsetting the decline in frontal/cyclonic precipitation is an increase in convectional precipitation over much of the midlatitude land masses, except for those areas dominated by the subtropical highs.

Mediterranean (Csa and Csb) This climate is distinctive for its long sunny summers and short moist winters. This combination is caused by seasonal shifts in the subtropical high-pressure cells: during the summer, the cells expand poleward to cover the southwestern coastal areas of the continents; in winter, the cells shift equatorward, and these areas fall under the influence of the westerlies and the midlatitude cyclones (Fig. 8.9). Thus, you may think of the Mediterranean lands as belonging to the humid midlatitude climates in winter and the dry (B) climates in summer. Moreover, these areas tend to be thermally modified by the adjacent ocean, though summer temperatures are typically warm, averaging about 30°C in the warmest month.

The soil-moisture balance is one of the most distinctive in the Mediterranean climate because of the seasonal extremes that occur there. In winter, evapotranspiration rates are low and the soil-moisture recharge is high. Summers are essentially the opposite; under the clear, sunny skies evapotranspiration rates are very high, and most of the available soil moisture is depleted by early summer or by midsummer. The landscape at this time turns brown, trees may drop leaves and go into a drought dormancy state, and streamflows decline or disappear.

THE VIEW FROM THE FIELD / NOTE 8

Adaptation to the Thermal Climate: Traditional Housing

The thermal objective in the design and construction of human dwellings is to reduce the variation in the daily and seasonal supply of natural heat. This entails raising winter temperatures or lowering summer temperatures, or both, depending on the particular climate and season. From the standpoint of the thermal climate, buildings constructed of earthen materials such as sod, stone, or clay brick can be considered as extensions of the soil environment. Such buildings possess thermal characteristics similar to those of soil, and in certain climates they thereby have a clear advantage over buildings constructed of wood or metal in terms of both energy economy and living comfort.

Recent history is rich in examples of the mismatch between building design and materials on one hand and climate on the other. One of the most notable examples was the adoption of the wood frame house by the Eskimos (or Inuit) on the northern coast of Alaska. In addition to calling for lumber in a treeless environment, these houses proved to be a very poor second, thermally, to the Eskimo's smaller, traditional houses built of rock, sod, and animal skins.

Though no longer extensively used by the Indian in the American Southwest, the adobe, or mud masonry, house is still the principal house type of desert people throughout the world. The thick walls and closed design (see diagram) of the adobe house may seem inappropriate for the desert climate. However, on closer examination we find that it is an especially good adaptation to the daily thermal regime of the desert because of the low thermal conductivity of the material used for the roof and walls. This material usually consists of a mixture of clay, sand, and some organic matter, such as straw or sticks, all of which have relatively low thermal conductivities when dried. Thus, if the roof and walls are thick enough, relatively little of each day's heat will penetrate all the way through them and reach the interior of the structure. Therefore, the temperature of inside air remains relatively low, especially if circulation with hot outside air can be minimized. The thermograph illustrates just how effective the adobe house can be in this regard; the midafternoon difference in temperature between the rooftop and inside air is 33 degrees C (60 degrees F).

Most of the heat absorbed by the surface of the roof penetrates only 5 to 7 cm. Then, after sundown, when the roof's outer surface cools down and the temperature gradient in the roof material reverses, the heat flows back to the surface, where it is released into the outside air. Some heat, however, flows through the roof, which accounts for the late-afternoon and early-evening rise in indoor tem-

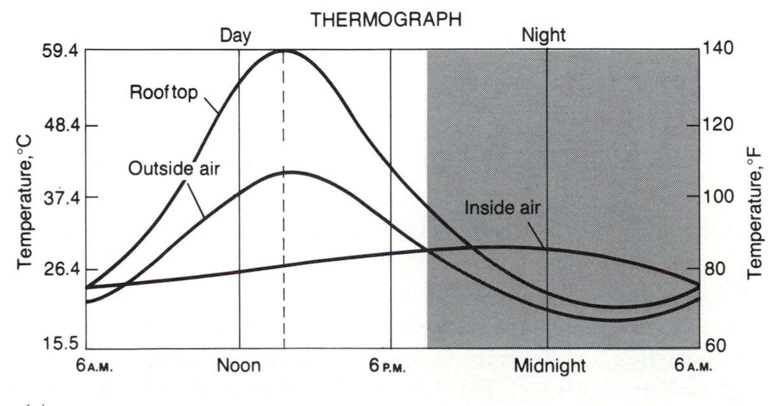

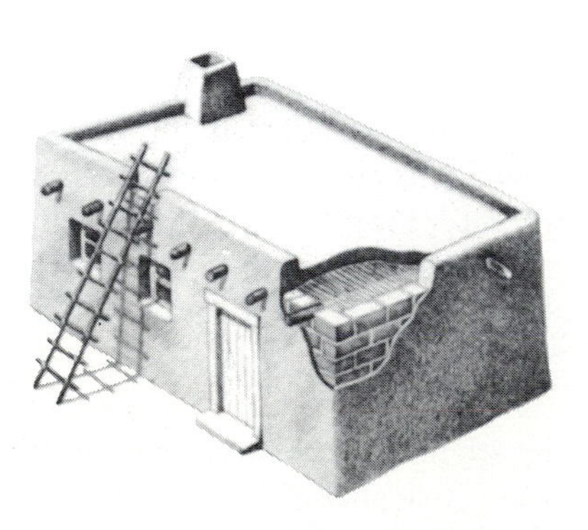

(a)

perature. On balance, in its environment the house is the coolest place during the day and the warmest place at night.

Today architecturally modern residential facilities, such as multistory residential buildings, have been adopted by many governments and developers in traditional countries in arid regions of the Middle East. In order to build these facilities with their complex infrastructures (streets, sewers, water lines, electric lines, gas lines, and so on), many of the thermal (microclimatic) advantages of traditional architecture have to be compromised. For example, modern structures are much larger, constructed of different materials (steel and glass, for instance) and provide fewer opportunities for the daily mix of indoor and outdoor activities which is a very important tradition in Middle Eastern life. As a result, from the standpoint of human comfort and energy costs (for air conditioning), the modern alternatives are proving to be a poor second to traditional housing.

Cairo, Egypt.

City of the Dead, near Cairo.

The Mediterranean climate is found on all the continents except Antarctica, but its total coverage is not very great. In North and South America it is limited to relatively small areas on the west coast between the marine west coast climate on the poleward side and the dry climates on the equatorward side. Only in Eurasia does the Mediterranean climate extend eastward far beyond the continental west coast. The Mediterranean Basin (sea) accounts for this extension, of course, lying as it does between the westerlies and the subtropical high. The Mediterranean climate here borders the sea and extends beyond eastward to the southern end of the Caspian Sea.

Marine west coast (Cfb and Cfc) Poleward of the Mediterranean climate regions are the marine west coast areas. These areas lie in the path of cyclonic storms that originate over the ocean and drift onto the continents with the westerlies. Where the coasts are mountainous, as in North and South America, the orographic mechanism induces heavy precipitation, often exceeding 500 cm per year. In the summer, precipitation declines as the cyclonic storms weaken and decrease in frequency.

Seasonal temperatures in the marine west coast climates are very modest compared to locations of the same latitude in the continental interiors. The annual temperature range based on mean monthly temperatures is typically 10 to 15°C (14 to 21°F), less than half that of continental counterparts. Summers are cooler, and the growing season is much longer than at inland locations. Coupled with the favorable moisture balance, these conditions are conducive to the formation of heavily forested landscapes. In fact, in the areas of heaviest rainfall, the forests are of such stature that they are termed rainforests.

The total land area classified as marine west coast is small, roughly comparable to that of its neighbor, the Mediterranean climate. This climate is not found in Asia, Africa, and, of course, Antarctica. The areas occupied by the marine west coast in North America, South America, Australia, and New Zealand are narrow, coastal belts, 200 to 400 km wide, which abut against high mountain ranges (Fig. 8.6). Europe is the exception where the marine west coast climate extends from the British Isles and France 2000 km or more inland into East Germany. The reason for this is related to the orientation of mountain ranges; in Europe the large mountain ranges (the Alps and the Pyrenees) trend east-west, thereby allowing the marine conditions to penetrate inland from the Atlantic. In Norway, on the other hand, the marine west coast

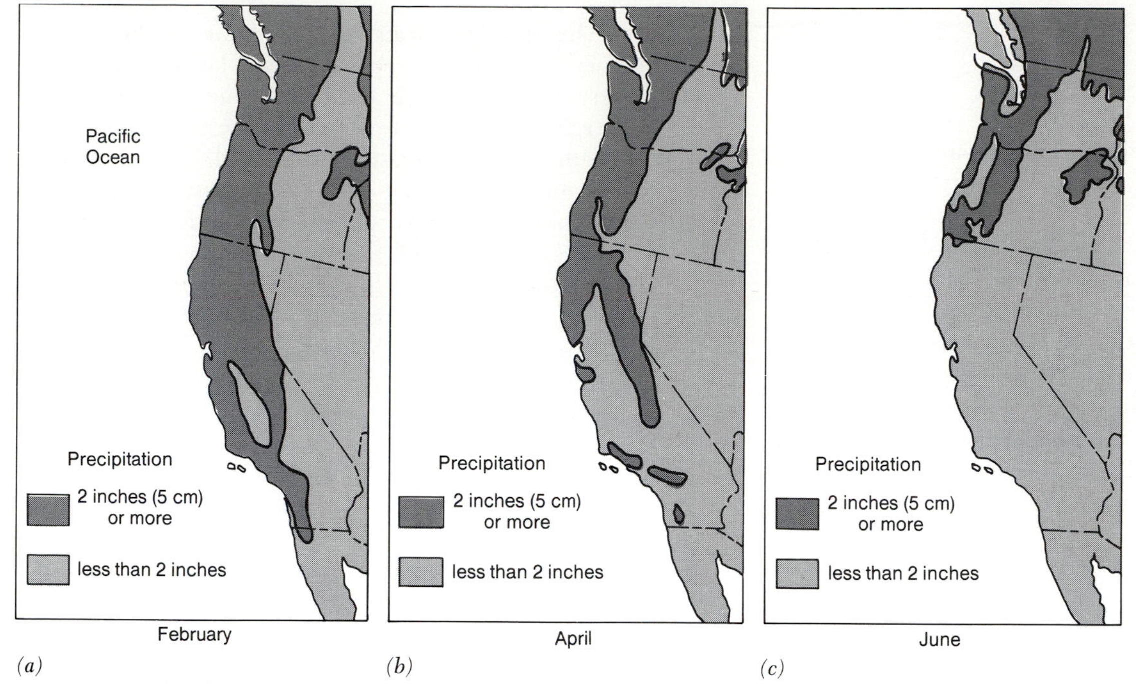

Figure 8.9 The changing pattern of precipitation on the North American West Coast, February through June. This is the principal feature of the Mediterranean climate of California.

situation is more like that of the Americas because the climate is limited by north-south trending mountains to a belt along the coast.

Humid subtropical (Cfa) In the subtropics on the southeastern sides of large continents, cyclonic and convectional storms are the principal sources of precipitation. Midlatitude cyclones are predominant in winter, but in spring and fall tropical cyclones occasionally cross the coastal areas. In North America and Asia, frequent incursions of polar air masses occur each winter (see Fig. 6.9).

As the polar front weakens in summer, the frontal precipitation declines; however, this is more than offset by convectional precipitation which is very pervasive throughout the warm months. In coastal areas, especially in China, an airflow of the monsoon type also augments precipitation (Fig. 8.10). In contrast to Mediterranean areas on the west coasts, these areas lack the thermal influence of the oceans; therefore, seasonal temperatures tend to be more extreme. The largest areas of humid subtropical climate are found in China, Argentina and the American South.

A variation of the humid subtropical climate, characterized by a dry winter season, is found in five or six small areas of the world. These climates (denoted

Figure 8.10 One of the landscapes of the humid subtropical climate of south China. The curious-shaped hills are caused by weathering and erosion of limestone under the warm, humid climate conditions.

Figure 8.11 A typical street scene during a blizzard in eastern United States. On the average, such storms occur several times each winter in the humid continental climate of North America, Europe, and Asia.

Cwa) are found in or near regions generally dominated by tropical wet-dry or tropical monsoon climates, but are too cool to be given the tropical designator.

Humid Continental (Dfa and Dfb) Poleward of the humid subtropical climate, winters are more severe, with temperatures averaging below freezing in several months. The polar front lies in this zone most of the year, and along it cyclonic storms are generated. Although winters are generally cold, they are subject to extremely frigid "cold snaps" with incursion of arctic air masses and warm spells ("thaws") associated with incursions of tropical air masses. These incursions are associated with major shifts in the position of the midlatitude jet stream and the polar front (Fig. 8.11).

Midlatitude cyclones are the principal cause of precipitation in the humid continental climate, although convection contributes appreciable rainfall in summer. Winter precipitation is mainly in the form of snow, and snow cover is present for at least 1 to 2 months in most years; however, monthly precipitation rates are greater in summr than in winter. Violent weather events in the form of intensive thunderstorms and tornadoes are not uncommon in this climatic zone, especially in the southern reaches of it. Because there are no sizable land masses in the midlatitudes of the Southern Hemisphere, humid continental conditions are not found there.

Midlatitude Steppe and Desert (BWk and BSk) In North America, Eurasia, and the southern part of South America, precipitation declines toward the interior of the continents, eventually giving way to steppe and desert. In addition, these areas are shielded by mountain ranges from the maritime air driven landward by the westerlies, thereby creating a rainshadow effect. Midlatitude steppe and desert differs from its tropical counterparts in that it tends to be less severely arid, especially in winter, and more seasonal in temperature, with winter temperatures often falling below freezing.

Some of the most damaging droughts in history have taken place in the midlatitude steppes. During average or better-than-average years these lands are extraordinarily alluring to grain farmers and grazers. But the good times are deceptive in dry regions because they are inevitably followed by drought, as the farmers of the American Great Plains found in the 1930s. In this instance, the effects of the drought were compounded by a national economic depression and a farm economy that had pushed itself well beyond the geographic limits of profitable grain farming. The costs of

Figure 8.12 A scene from the American Dust Bowl of the 1930s. Grassland was reduced to barren, eroded ground as a result of failed croplands and drought.

disasters such as the American Dust Bowl often include the permanent loss of native grass covers and severe soil erosion, and, in other regions, famine and death (Fig. 8.12).

SUBARCTIC AND ARCTIC CLIMATES

Continental Subarctic (Dfc and Dfd) In the northern interiors of North America and Eurasia, winters are fairly dry and very cold. The dryness is due to the stability of the high-pressure cells that develop over the land in response to the winter cold. Indeed, in northeastern Siberia, the winters are so dry that some areas are given a Dw designation. The subarctic zones are the source areas of continental polar air masses, which move southward and eastward into the humid continental zones and beyond. Summers may be mild, with long hours of sunlight, but there are only two to three months without frost.

At depths of a meter or more in the ground, permafrost can be found throughout much of the continental subarctic zone, but its distribution is discontinuous (see Fig. 4.8). Light and heat are sufficient, however, to support extensive boreal forests in both Eurasia and North America. These forests are floristically simple, usually made up of only four to six tree species (e.g., spruce, balsam fir, tamarack, and white birch), and growth rates are very slow. In the northern belt of the subarctic zone, the trees decline in size and the forest cover thins out as the boreal forest gives way to the tundra landscape.

Marine Subarctic (Dfc) Along the shifting polar front, particularly on west coasts, the climate remains cold and wet throughout the year. Because of the thermal influence of the sea, however, temperatures in all but one or two months are higher than they are in the interiors and hover around freezing throughout much of the winter. These coastal regions nevertheless have some of the cloudiest, windiest weather on earth. Indeed, in the language of the Aleuts, the native inhabitants of the Aleutian Islands, there is no way to say "It's a nice day." The phrase used, when literally translated, corresponds more closely to "The wind isn't blowing so hard today." The marine subarctic climate is associated mainly with tundra vegetation because summer does not produce even a month of temperatures above the critical 10°C level. Many tree species are usually found in this landscape, but they are always characterized by dwarf forms.

Tundra and Ice/snow (ET and EF) Poleward of the subarctic climates is the tundra climate, named for the landscape comprised of diverse herbs, shrub-sized woody plants, and wet soil conditions. On the average, only one month has an average temperature above 0°C (but less than 10°). Most snow melts away in summer, and virtually the entire landscape is underlain by permafrost.

Farther poleward is the earth's coldest climate overall. It is characterized by permanent snowfields and ice caps; average air temperatures do not exceed 0°C in any month. Despite the abundance of snow and ice, little snow is precipitated from the dry arctic air; the annual snowfall in liquid water is less than 25 cm. Greenland, Antarctica, and North America and Eurasia above 70°N latitude are the principal areas of arctic climate.

Summary

Climate is the representative condition of the atmosphere at a place on earth. An understanding of global climatic patterns requires knowledge of thermal conditions related to latitude and continental and maritime environments, air mass-movement, and atmospheric and oceanic circulation. Climate can be classified according to many factors; in geography, we are concerned primarily with temperature and moisture conditions, how they vary with the seasons, and what atmospheric processes and controls regulate them. Two classification schemes are widely used today: the Köppen-Geiger system and the Thornthwaite system. The first provides a useful framework for describing world climates, whereas the second is best suited to describing the dry, semiarid, and subhumid climates.

Broadly speaking, the climatic extremes of the earth are characterized by perpetually warm conditions in the tropical and equatorial climates and perpetually cold conditions in the polar climates. Between these zones, the climates can be divided into those that are wet and those that are dry. The dry climates are either arid or semiarid and may be warm or cool depending on latitude. The wet climates are distinctly seasonal, characterized either by warm/cool (or cold seasons, such as the humid continental climate, or wet/dry seasons, such as the Mediterranean climate.

Concepts and Terms for Review

- problems in climatic classification
- climatic boundaries
- climate on a hypothetical continent
  - thermal regime
  - air masses
  - precipitation
  - patterns
- Köppen-Geiger classification
- Thornthwaite classification
- tropical climates
  - equatorial wet
  - tropical wet-dry
  - monsoon
  - tropical steppe and desert
- midlatitude climates
  - Mediterranean
  - marine west coast
  - humid subtropical
  - midlatitude steppe and desert
- subarctic and arctic climates
  - continental subarctic
  - marine subarctic
  - tundra and ice/snow
- climate and architecture

Sources and References

Ayoade, J.O. (1983) *Introduction to Climatology for the Tropics*. New York: John Wiley & Sons.

Chang, J.H. (1968) *Climate and Agriculture: An Ecological Survey*. Aldine: Chicago.

Dregne, E., ed. (1970) *Arid Lands in Transition* (*Publication 90*). American Association for the Advancement of Science, Washington, D.C.

Dregne, H.E. (1978) "Desertification: Man's Abuse of the Land." *Journal of Soil and Water Conservation* 33, 1: 11–14.

Huntington, E. (1945) *Mainspring of Civilization*. New York: John Wiley & Sons.

Kendrew, W.G. (1962) *The Climates of the Continents*. London: Oxford University Press.

Köppen, W., and Geiger, R. (1954) *Klima der Erde* (map). Darmstadt, Germany: Justus Perthes (and Chicago: Nystrom).

Landsberg, Helmut E., ed. (1969–present) *World Survey of Climatology*. Amsterdam: Elsevier Science, 15 volumes.

Lockwood, John G. (1984) *World Climatic Systems*. Baltimore: Edward Arnold.

Oliver, John E., and Hidore, John J. (1984) *Climatology: An Introduction*. Columbus, Ohio: Charles E. Merrill.

Rosenberg, Norman J., ed. (1978) *North American Droughts*. Boulder: Westview Publishing Co.

Thornthwaite, C.W. (1943) "Problems in the Classification of Climates." *Geographical Review* 33 2, 233–255.

Thornthwaite, C.W. (1948) "An Approach Toward a Rational Classification of Climate." *Geographic Review*, Vol. 38.

Trewartha, Glenn T. (1981) *The Earth's Problem Climates*, 2nd ed. Madison: University of Wisconsin Press.

Trewartha, G.T. (1968) *An Introduction to Climate*. New York: McGraw-Hill.

Chapter 9

Climate Change: Then and Now

Weather certainly changes; is it not reasonable to suppose that climate also changes? What evidence is there of past climatic change and have such changes been great enough to alter land and life? What are the causes of alterations in climate and how frequently do such alterations take place? How does man fit into this picture as change agent and victim?

Prediction is perhaps the noblest achievement of science, and it is also one of its most difficult. In the time of Sir Isaac Newton, scientists believed that nature operated according to strict laws and that, if they could discover these laws, they could base absolute and universal predictions on them. In the past century, however, the concept of scientific prediction changed to one of probability, meaning that for each cause there is only a certain degree of chance that the predicted effect will occur. Even gravity is expressed as a probability because detailed measurements show that it is not constant over the earth's surface, nor is it constant through time.

Gravity is a reasonably simple natural phenomenon. Many other phenomena, such as changes in weather and climate, appear to be far more intricate and puzzling. Meteorologists measure and analyze the day-to-day changes in the atmosphere and have achieved a basic understanding of what causes the different patterns of weather. Even with this knowledge, however, they have been only moderately successful in making accurate short-term (over hours or days) weather predictions; weather forecasting in the United States has only about a 60 percent accuracy record—which is to say that it is only about 10 percent better than the accuracy record one would compile over time by flipping a coin: "Heads, it will rain tomorrow, tails, it won't." Long-term weather forecasting (over periods of weeks or months) is much less accurate, and when we try to forecast climatic trends (over centuries, millennia, and longer periods), our track record is very poor indeed.

Can the past provide some clues about climatic change? Although weather records do not go back much more than 100 years or so, there are other sources of information. Certain organisms can be used as yardsticks of past climates because they are known to be heat- and moisture-sensitive. The presence of their fossilized ancestors in old rocks can sometimes be taken as indicators of the climatic conditions of former environments. In addition, archaeological data about ancient societies, as well as historical information from the past several thousand years of civilization, can provide important clues about past climates.

But what can knowledge of past climates tell us about future climates? Can it be used to make predictions? Strictly speaking, the answer is no. But knowledge of patterns of past change can tell us something about the types and rates of changes that could be expected in the future (without knowing when or why they would take place, of course). Furthermore, it tells us something about the nature of climate in our own times and whether we are living in a period that is climatically unique. After all, we do think we are a rather special organism; is it not reasonable to infer that the climates that have fostered our origins and development are also unique?

The chapter opens with a brief description of climatic change in geologic, prehistoric, and historic times, and then goes on to explore the implications and causes of climatic change. The second half of the chapter deals with the influence of urbanization on climate, highlighting the differences between urban and rural climatic conditions.

CLIMATIC CHANGE IN PERSPECTIVE

Climatic change has received much scientific attention in recent years. Beyond the fact that it is an interesting and challenging topic it is important because such change is now taking place and is of sufficient magnitude to affect the distributions of large populations of plants and animals, including humans. We know that at times during the earth's history the global distribution of climates was very different from what it is today and that some of the new distributions of climate were caused by movements of great sections of the earth's crust as a part of a still ongoing process which we call *plate tectonics*. (See Chapter 16 for more details on this process.) This process produced displacements of thousands of kilometers in North America, South America, India, Australia, and Antarctica over a period of hundreds of millions of years, in some cases moving continents across several zones of latitudes and apparently moving from one or more climatic zones to another. The geographic arrangement of oceans and land masses also changed with plate movement; for example, the Atlantic Ocean itself developed over the past 200 million years as North and South America drifted away from Europe and Africa, and this development created the maritime climatic regime of Western Europe. Many other factors have also contributed to climatic change, often over much shorter periods of time. We know that such changes will continue and that they will influence the capacity of the planet to support life.

The Time Frames of Change

How do present climatic conditions compare with those at various times in the past? Today the average earth temperature is colder than it has been over most of the last billion years. During the period of about 70 to 200 million years ago, climate on most continents appears to have been wetter and warmer than at the present. Dinosaurs inhabited much of the earth, and the lush plant growth that is now the source of our coal and oil was widespread. But the extent to which these climatic conditions can be attributed either to the geographic position of the continents in the tropical latitudes at that time as a result of plate tectonics (see Note 16) or to conditions peculiar to the atmosphere itself is difficult to say. In any case, about 65 to 70 million years ago, the global climate cooled, perhaps suddenly, and the dinosaurs died off.

About 2 million years ago the climate cooled again, and the present Ice Age began. This period has been marked by several dramatic expansions and contractions of the world's ice volume. Over the past million years, global temperatures (equilibrium surface temperature) have fluctuated by about 4°C (Fig. 9.1*a*). After the last glacial peak, about 16,000 years ago, climate grew warmer until 6000 years before the present (4000 B.C.) (Fig. 9.1*c*). Since then climate has generally cooled, but the cooling trend has been interrupted by oscillations of about 2°C every 2000 years. Scientists generally agree that the earth is still in the last ice age, probably in a warm period between major glaciations, called an *interglacial period* (Fig. 9.2).

Our knowledge of climate change for the past 3000 years is a little more detailed because of the information provided by historical materials and archaeological finds. This evidence indicates that the deserts of North Africa and Southwest Asia were less harsh in the several centuries before Christ than they are today. Populations of nomads lived in parts of the deserts where they cannot survive today. Caravans crossed sections of desert that could not be crossed by similar caravans in the nineteenth and twentieth

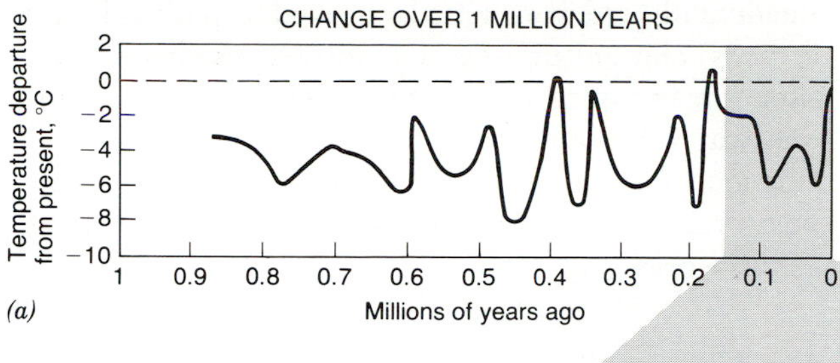

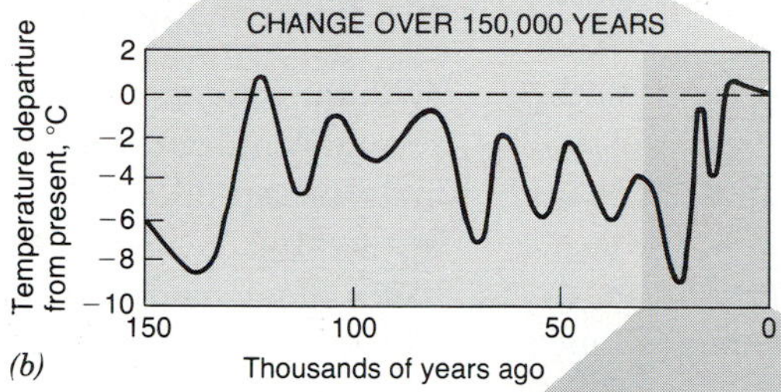

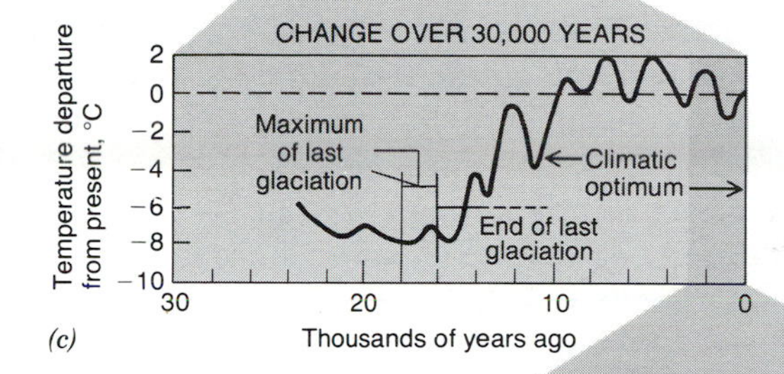

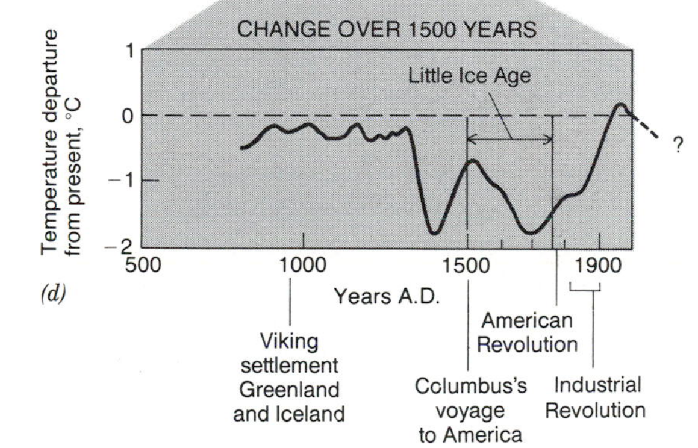

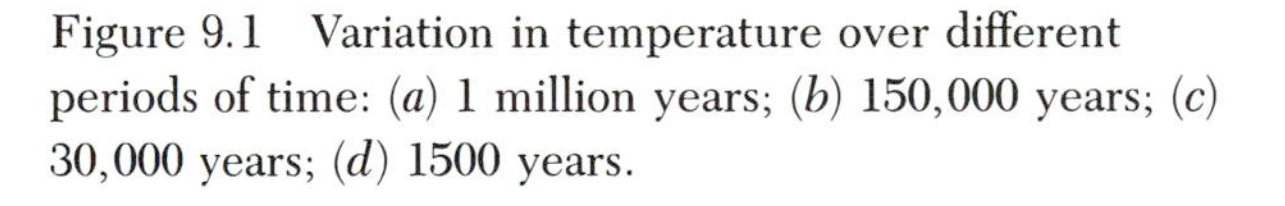

Figure 9.1 Variation in temperature over different periods of time: (*a*) 1 million years; (*b*) 150,000 years; (*c*) 30,000 years; (*d*) 1500 years.

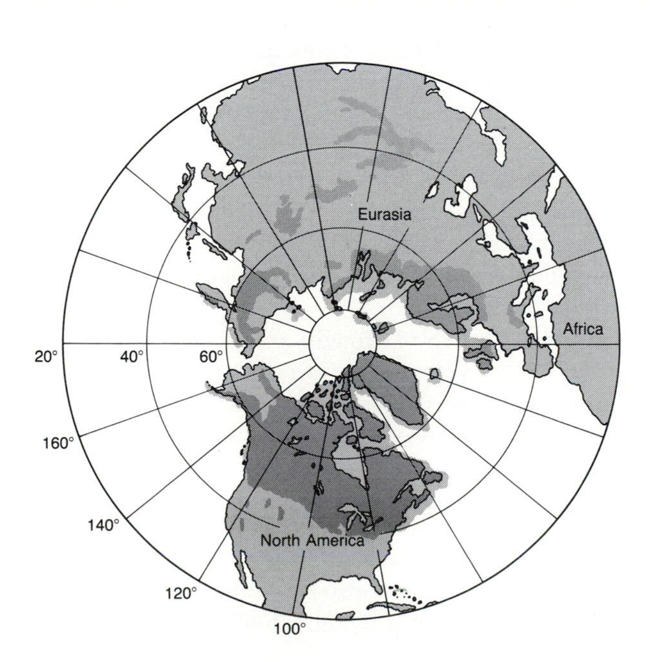

Figure 9.2 Extent of glacial ice in the Northern Hemisphere at the peak of the last glaciation.

centuries. Surface water was somewhat more plentiful; not only were there more oases, but many groundwater aquifers grew, and some of these reserves—such as those in the Nubian sandstone of Egypt—are still being used today.

After Christ, several temperature trends are clearly identifiable. A major warming trend culminated in the period A.D. 800–1200 (Fig. 9.1*d*). During this period the Vikings colonized Iceland, Greenland, and North America, and grapes grew in England (Fig. 9.3). The vines are still there, but they do not bear much fruit. Historical records indicate that, in Europe, storminess increased and winters grew colder after 1200 A.D. or so.

In the 1300s and 1400s the Viking settlements in North America and Greenland were abandoned, and a severe outbreak of the Black Death (plague) in 1348–50 reduced the world population by almost half. Glaciers advanced; famines were frequent; and the period from the fourteenth to the eighteenth centuries is sometimes called the "Little Ice Age." Around 1750 a warming trend began, apparently lasting until the 1940s, when temperatures began to get cooler, although perhaps not in the Southern Hemisphere. The manifestations of climate change in the landscape are often complex. In Note 9, geographers Tim F. Ball and Geoffrey A. J. Scott examine the northern limits of forest in central Canada in relation to recent climate change.

In addition to the rather sudden and dramatic historical events mentioned previously, many gradual changes in human populations have been associated with climatic fluctuations in the past 3000 years or so. Although some of these changes (both sudden and gradual) were related to climatic variations only coincidentally and were actually caused by political (such as wars) and cultural events (such as medical inventions), many were undoubtedly related to climate in one way or another. We can only guess about such relationships, however, because as noted above climatic records go back only 100 years or so for much of the world.

Figure 9.3 A modern farm in Iceland. A decline in temperature of the magnitude that occurred in the Little Ice Age would hamper productivity or even drive the farmer out of business.

Modern Implications of Climatic Change

With the burgeoning of human populations in the past several centuries, human survival has grown increasingly susceptible to fluctuations in climate. One reason for this change is that large numbers of people have been forced to seek out a living in lands that are climatically marginal for agriculture and settlement. Climatically marginal lands generally fall into two classes: (1) moisture marginal; and (2) thermal marginal.

In thermally marginal lands, such as shown in Figure 9.3, the growing season is just long enough to sustain a crop in cold years, which occur once or twice a decade. If the mean temperature drops by 2 degrees C, it might well shorten the growing season 30 days or so, causing certain crops to fail. Because the farming economy is already marginal, a few crop failures could depress it below the survival threshold and farmlands would fall out of production. It is essential to realize that much of the agricultural area of Canada and the Soviet Union is thermally marginal.

Moisture-marginal lands include the desert fringes, steppe, and dry savanna regions where precipitation is light and variable and the balance between a stable and modestly productive landscape and an unstable, desolate one is very tenuous. The stable landscape represents a delicate balance among soil, its moisture-holding capacity, vegetation, the forces of the atmosphere such as wind, solar heating, evaporation, and precipitation, and land use. Soil is especially critical because it helps buffer the landscape against variations in precipitation associated with climatic oscillations (see Fig. 8.12).

Extended periods of drought are inevitable in such climates. To survive a major drought without serious degradation of the moisture-marginal landscape usually requires land use adjustments such as reduced grazing, population redistribution, and changes in crops. In many parts of Africa and Asia such adjustments are, for many reasons, not possible, and in order to maintain life over the short run, farmers are forced to attempt to squeeze more out of a declining landscape. The result is overgrazing and expansion of croplands, which in turn lead to loss of groundcover vegetation, increased soil erosion, and an overall decline in the landscape's carrying capacity. The downward spiral of land and life are usually irreversible, and the moisture-marginal landscape under the combined stress of drought and land use pressure is transformed into a truly arid landscape, a process termed *desertification.*

The Search for a Cause of Climatic Change

In searching for a cause of climatic change, a major difficulty is that we do not understand very much about the long-term feedback mechanisms between the atmosphere and the ocean. Although we might be able to explain what effects should be expected from certain causes, such as the buildup of the El Niño warm-water body, predicting the magnitude of these effects on the atmosphere is very difficult; hence, evaluating the net result of conflicting effects is seldom possible.

In this presentation of some possible mechanisms for climatic change, we will focus our attention on the last million years. During this period some significant changes have occurred, characterized by the advance and retreat of great glaciers, yet the locations of the continents have not changed appreciably. We know, of course, that a shift in the position of a continent owing to plate tectonics can cause its climate to change. But we also know that climatic changes do not necessarily depend on such shifts, especially in the last million years, because in this short time a continent could have shifted only 20 km to 60 km.

In the last million years, the continents have been favorably positioned for widespread glaciation. The basic requirement for a major glaciation (an ice age) is that more snow falls than melts over a large land area. A suitable geographic arrangement of the continents for glaciation is one that has a great deal of land at high and middle latitudes, preferably some of it at high altitude above the Arctic Circle. This by no means ensures glaciation because, if winters are too cold, little snow falls. (Recall that cold air can hold little moisture, and, as a result, continental subarctic and arctic climates are quite low in total precipitation.) The conditions most favorable to glaciation are relatively warm winters and relatively cool summers. For these conditions to exist, open ocean water must be present at high latitudes because the water provides a source of moisture in winter and helps to depress temperatures in summer. In cold winters, by contrast, sea ice builds up, limiting the flux of moisture to the air, which in turn limits snowfall and nourishment of glaciers.

But what forces drive climatic change? A first place to look for mechanisms of change is in the amount of radiation which the earth receives from the sun. There is no evidence that the total output of radiation has varied significantly; however, regular changes in the earth's orbit would cause changes in the seasonal and latitudinal distribution of solar radiation. There are three orbital variations to consider:

1. Variations in the shape of the earth's orbit—called *eccentricity*—from one that is more elliptical to one that is nearly circular. This change takes place within a period of 90,000 to 100,000 years. At present, the orbit is nearly circular and is becoming more so, making seasonal differences less pronounced.
2. Variation in the angle between the earth's axis and a line perpendicular to the plane of the orbit—called *obliquity*. This angle varies from 21.8 degrees to 24.4 degrees, with a period of about 40,000 years. Today the angle is 23 degrees 27″ and is decreasing. As the angle decreases, seasonal differences become smaller.
3. Variations in the time of the year of the equinox—called *precession*—which has a period of about 21,000 to 23,000 years. This means that perihelion, the time of the year when the earth is closest to the sun, which is currently January 5, varies with this period. At present, the sun is closer to the earth during the Northern Hemisphere winter, hence making the winters relatively warmer and the summers relatively cooler there. In the Southern Hemisphere this position makes the seasons more different.

Figure 9.4 shows the variation in total ice volume over the last million years. There is a clearly evident periodicity between maximum of about 100,000 years and a very short time (only about 10,000 years) between maxima and minima. In addition, temperatures and the corresponding periodicities in the rate of deep sea sediment deposition shows a strong correlation of peaks at 100,000, 41,000, and 23,000 years.

For variations on scales shorter than 21,000 years, the evidence for astronomical influences is not so compelling. Variations in solar output with the 11-year sunspot cycle have not been convincingly related to any widespread changes in climate. There is no evidence of any *regular* variations in climate on a time

THE VIEW FROM THE FIELD / NOTE 9

Climate Change and the Geographic Limits of Forest in Central Canada

In the mid-1700s Canadian explorer Samuel Hearne made several expeditions from Churchill, Manitoba, on the southwest side of Hudson Bay to Coppermine River, some 1400 km (900 mi.) away on the Arctic Coast. The route he used followed the border between the zone of the boreal forest and the tundra. Hearne prepared a map in 1772 showing his route which he labeled the "wood's edge."

Hearne's map came at a unique time in climatic history, near the end of the Little Ice Age. The Little Ice Age was a period between the late 1400s and 1750 when world temperatures cooled markedly. In central Canada this meant that summers were shorter, winters longer and colder, and the polar front probably occupied a position south of where we find it today. Evidence indicates that the settlement of York Factory, Manitoba, located 180 km (120 mi.) south of Churchill, had a colder climate in the first half of the 1700s, similar to that of Churchill's. After 1760, York Factory had a warmer climate similar to that of the boreal forest zone with more summer rainfalls, more thunderstorms, and more southerly winds.

The northern limits of tree growth, that is, the border between the boreal forest and tundra, is understood by modern scientists to be primarily temperature-controlled. Therefore, in comparing Hearne's map with the modern boreal-tundra border, we would expect Hearne's "wood's edge" line to be located distinctly south of today's border. However, we find that such a comparison is not so easy to make because what constitutes the modern boreal-tundra border is not clear. Indeed, what is described as the border or tree-line in books and maps is actually a broad transition zone, 100 km in width, in which the dominant vegetation is a boreal-tundra mix. The southern edge of this transition zone is defined as the tree-line, whereas the northern edge is defined as the poleward limit of tree forms. The term *tree form* pertains to the life forms (size and shape) of common tree species such as spruce, tamarack, and balsam fir. Beyond the tree-form border, tree species can exist, but they take on other life forms, such as ground shrubs or krummholtz. (See accompanying illustrations.)

When we compare the modern tree-form border with Hearne's "wood's edge" of the 1700s, we find that in the southern half of this zone the current border is farther north and in the northern half it is farther south. In other words, change in the south is what we would expect based on the evidence—the trees have advanced poleward since the Little Ice Age.

Such shifts in the boreal-tundra border appear to have taken place several times since the last episode of continental glaciation in Canada, which ended about 10,000 years ago. Studies of buried pollen, wood, and other parts of trees show that in the 2000 to 3000 years after glaciation the border shifted every two or three centuries, and there were periods when temperatures were even warmer than today and the tree-line was farther north than today's. Some scientists suggest that the tree-line was also much farther north 3500 years ago, arguing that the present outliers of the boreal forest (large islands of trees within the zone of tundra) are relics of that old tree-line.

This suggestion raises some interesting questions concerning the magnitude and duration of climate change necessary to produce major shifts in vegetation. We know that conditions at the tree-line are too harsh in central Canada for trees to reproduce sexually; therefore, the only way they can spread is by vegetative regeneration, which is a less efficient means of migration for most tree species than sexual reproduction is. Therefore, it appears that climate would have to improve significantly for trees to make a major advance like those that have been documented.

But the picture is complicated by the complex nature of the interplay of plants and environment and the variety of environmental factors involved. For example, trees help create their own microclimate by reducing wind at ground level and trapping snow, both of which reduce soil-heat loss and frost penetration into the ground. Presumably, this would

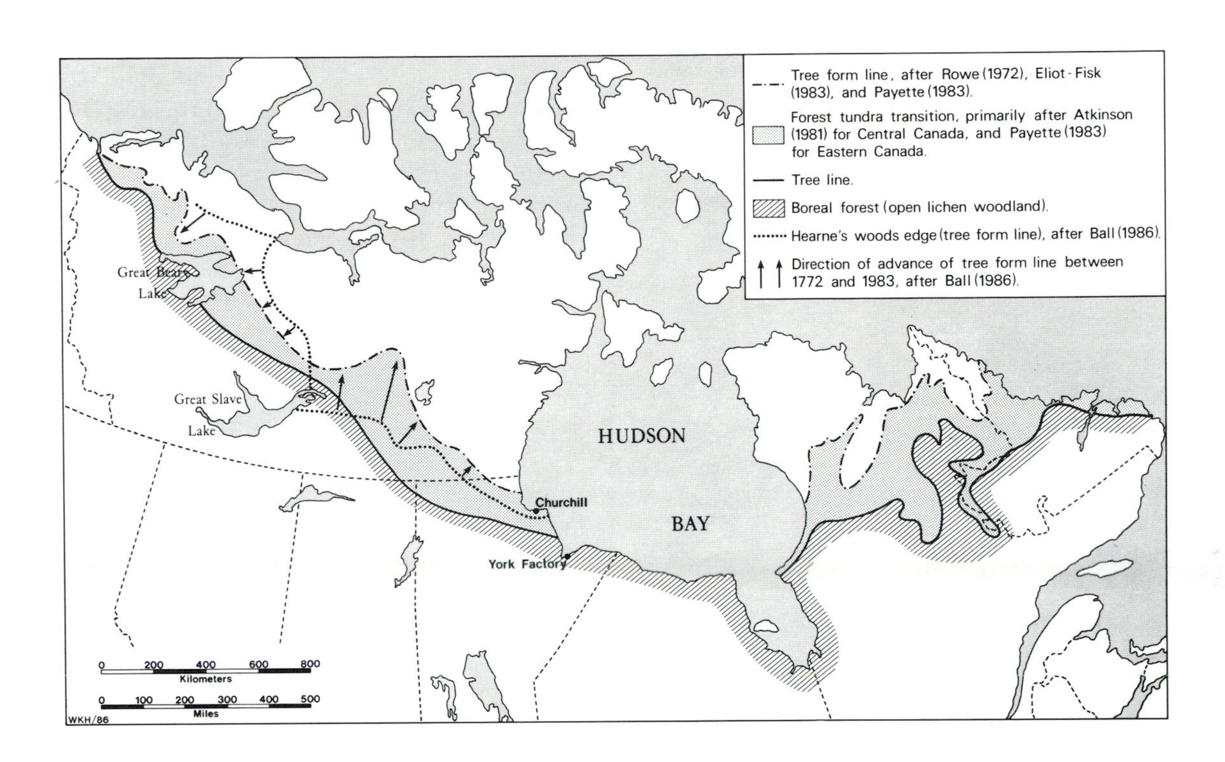

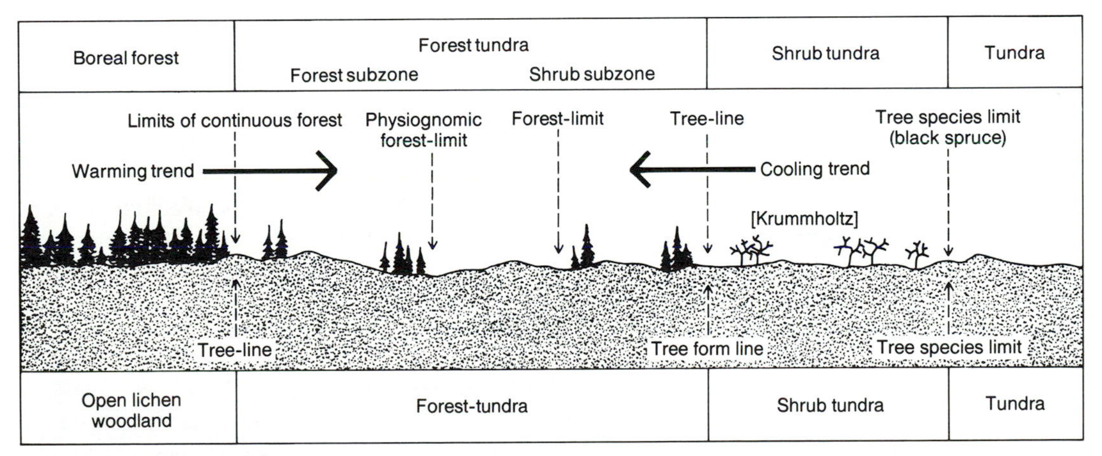

(*a*) The Canadian boreal forest-tundra transition.
(*b*) Vegetation types and associated features of the boreal forest-tundra transition.

help buffer trees against the lower tempertures of colder periods. The records of the Hudson's Bay Company document the extent of the woodcutting activities, particularly in the area of Churchill. Wood was used for construction, firewood, and in lime kilns for the production of mortar by the latter part of the eighteenth century. A limited wood supply and a slow rate of tree regrowth meant that wood had to be transported 80 miles (128 km) up the Churchill River from the forested area to the south.

Tim F. Ball and Geoffrey A.J. Scott
University of Winnipeg

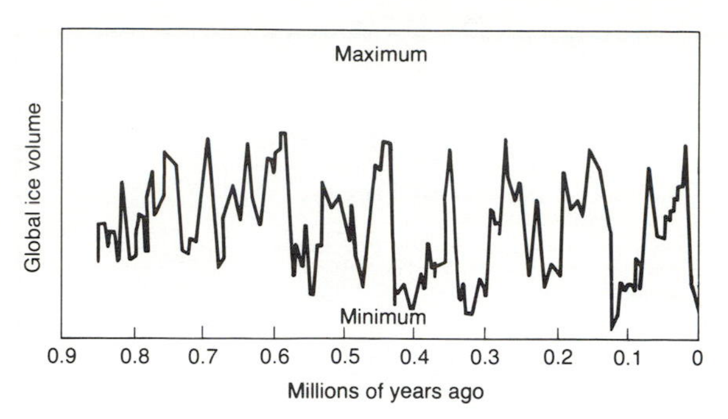

Figure 9.4 Estimate of the global ice volume over the last million years.

scale of hundreds or tens of years. For these reasons, we must seek to discover the forces that drive the climate in the earth-ocean-atmosphere system. There are many plausible mechanisms, only a few of which can be mentioned here.

- The climatic system has variation, and the *feedback* between the effects is poorly understood. Here feedback means that the consequences of certain effects will either amplify a variation (positive feedback) or dampen it (negative feedback). For example, increased volcanic activity will put more dust in the atmosphere, which will reduce solar radiation; reduced solar radiation, in turn, leads to lower temperatures and more ice coverage on land and sea, which increases the albedo of the affected regions and thereby lowers net solar radiation. Spring comes later, cyclone tracks are closer to the equator, and ice-covered seas and a lowered sea level reduce the poleward transfer of heat, causing still lower temperatures in the higher latitudes. These effects are what we often call *positive feedback*, in that the original cooling trend caused effects that led to further cooling. However, there are checks and balances that would tend to prevent the development of another ice age. Cool temperatures and increased cloudiness would reduce the outflow of longwave radiation to space, and the increased storminess in the midlatitudes would lead to increased heat exchange with the tropics.
- Changes in atmospheric gases and particulates have climatic effects. Recently, a great deal of attention has been given to reductions in solar radiation caused by sudden, massive increases in atmospheric debris. Two natural sources of debris are identified: volcanic eruptions and explosions from asteroid or comet collisions with earth. (An interesting concept in favor of the second source proposes that in the past 225 million years the earth has been bombarded by asteroids every 26 million years, resulting in increased backscattering of solar radiation leading to lower atmospheric temperatures and widespread plant and animal extinctions.) Increased solid and liquid particles (called *aerosols*) from volcanic, industrial, or agricultural sources can reduce solar radiation, causing cooling in the lower atmosphere but warming higher up in the stratosphere. Increased carbon dioxide, on the other hand, reduces outgoing longwave radiation in the atmosphere, thereby warming the lower atmosphere and the earth's surface. Fluorocarbons (freons) added to the atmosphere in the past few decades from refrigerators and spray cans have been widely hypothesized to have destructive effects on the ozone layer, which protects life by absorbing a large portion of the incoming ultraviolet radiation.
- Human settlement and land use can have a pronounced influence on ground-level climate because when land is developed, the vegetative cover is altered and new materials with different heat properties are introduced to the landscape. For example, eradication of forests for agriculture can affect the heat balance of a region by increasing solar heating on the soil surface and decreasing latent heat flux into the atmosphere; thus, ground-level air temperatures are higher. Urbanization, as we will see in the next section, can have an even stronger influence.

A volcanic plume: dust decreases the atmosphere's capacity to transmit solar radiation.

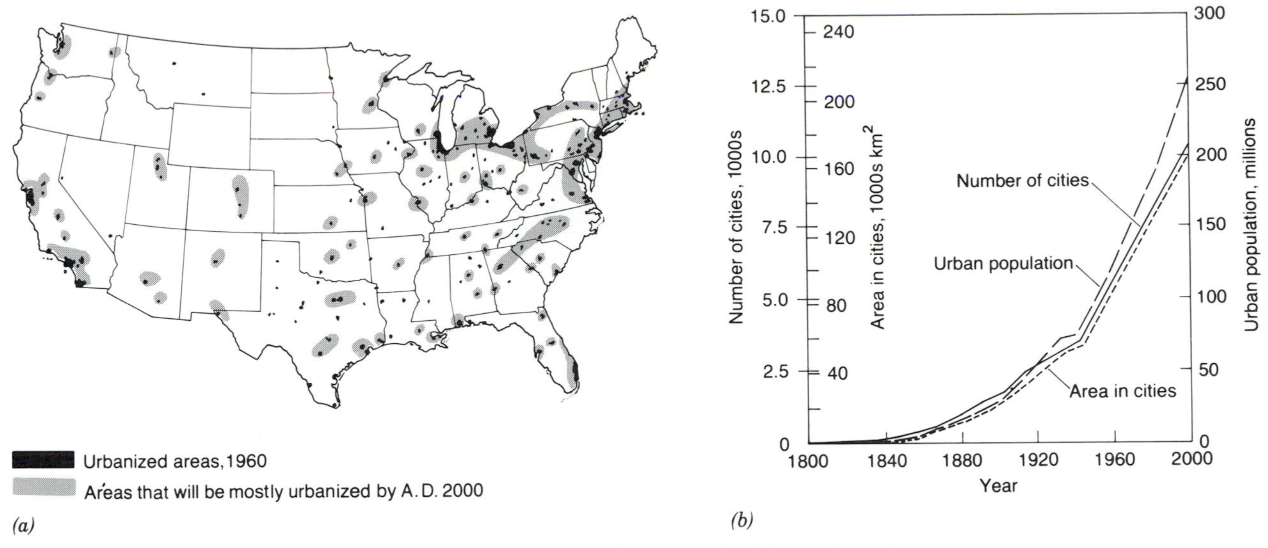

Figure 9.5 (*a*) Urbanized areas in the coterminous United States. The black areas show the 1960 pattern; the gray areas, the projected pattern of urbanization in the year 2000. (*b*) Trends in the number of cities, urban population, and the area of cities between 1800 and 2000.

URBANIZATION AND CLIMATIC CHANGE

Cities are the most intensive form of land use yet devised, and it is not surprising that some of the most dramatic alterations of local climate, or microclimate, have traditionally been associated with them. In the past several decades urbanized areas have undergone a geographical explosion and now cover such large areas that it is no longer appropriate to call them "localized environments" (Fig. 9.5). The greater metropolitan areas of cities such as New York, London, and Los Angeles occupy areas of 1000 square kilometers or more, for example.

Comparing the climate of cities to that of the countryside of the regions in which they are located, we find that solar radiation is generally less intensive in cities. However, temperature may be as much as 5 to 10°C higher in cities than in the countryside. Wind at ground level is less strong in cities, and fog is characteristically much more common in urban settings than elsewhere. As for other components of climate, such as humidity, snowfall, and rainfall, it is not altogether clear whether urban areas are consistently responsible for inducing change.

Variation in climate associated with cities can be readily appreciated when examined in the context of the energy balance. We can identify several changes in the forms and relative quantities of radiation and heat which are brought about by urbanization.

Alterations in the Flow of Radiation

Large quantities of gases and particulates are expelled into the atmosphere from cities. If the rate of expulsion exceeds the rate of removal by airflow through the city, these materials build up in the air. Thus, for a given volume of air, the mass balance of pollutants is a product of the flushing capacity of the atmosphere as well as the rate of production of airborne materials by human activities. Most of the time airflow in cities is insufficient to remove the contaminated air. As a result, the air not only smells and tastes poorer than country air, but is visibly less transparent. On sunny days the atmosphere appears discolored and hazy, and the sun itself often has a certain dullness about it. Much of the haziness and discoloration is due to airborne particulates whose concentrations are usually some ten times greater over cities than over rural lands. Studies show that in the lower 1000 m of atmosphere above cities, the intensity of solar radiation may be reduced by more than 50 percent. A large part of this loss is attributable to backscattering (upward) of shortwave radiation from the dome of particulates that blankets the city (Fig. 9.6).

In addition to retarding the entry of incoming short-

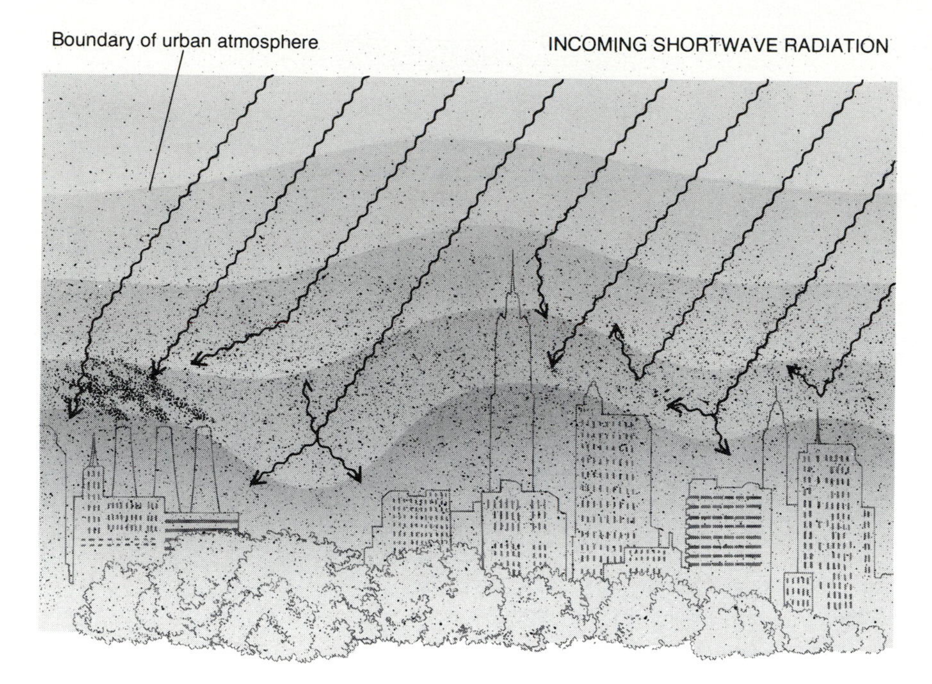

Figure 9.6 Backscattering of incoming shortwave radiation by the urban dust dome.

wave radiation, the dust dome and associated gaseous pollutants decrease the rate of release of longwave radiation from the city. This, of course, is a greenhouse effect, and though it exists in an undisturbed (i.e., natural) atmosphere, it is greatly increased over cities. There it involves mainly atmospheric absorption by carbon dioxide, ozone, and aerosols. As a result, heat is retained in the city atmosphere longer than it is in a cleaner atmosphere, thereby inducing higher air temperatures.

The Heat Island

The process of urbanization is most vividly expressed in changing patterns of land use and land cover. This change may take a variety of forms; closely spaced high rise buildings in the inner city, sprawling suburban neighborhoods, industrial parks, massive highway systems, and shopping malls, but each has important energy-balance and climatic implications. Locally, albedos may be reduced with the construction of dark roofs and streets, but overall, urban surfaces have a combined albedo comparable to that of rural areas. On the other hand, the change to urban land uses may greatly influence the latent-heat flux because hard-surface materials such as asphalt, concrete, and brick nearly eliminate the natural moisture flow from the soil to the air. As a result, the ratio of sensible-heat flux to latent-heat flux (the Bowen ratio) is increased significantly. Because a greater proportion of energy is converted to sensible heat, the ground-level air temperatures in cities are usually higher than those in the neighboring countryside.

Another important land-cover factor in the energy balance of cities is the thermal properties of hard-surface materials. Because of the low volumetric heat capacities of asphalt, concrete, and roof materials, the temperatures of these materials tend to rise more rapidly and to reach a higher level with the absorption of radiation than do the plants and soil they replaced. Measurements by Rudolph Geiger, a pioneer in the study of microclimate, showed that in early summer the temperature of asphalt would often be double that of the grass. Given both the high Bowen ratio and the low volumetric heat capacities of concrete, asphalt, and brick, it is understandable why the air above hard surfaces tends to heat rapidly (Fig. 9.7*a*). One can verify the remarkable difference in the heating of hard surfaces and vegetated soil surfaces by touching, say, an asphalt parking lot and an adjacent lawn during a sunny afternoon in any city. Often just walking on them is sufficient to reveal the difference. Little wonder that foot-patrol cops and mail carriers have traditionally complained about their hot, swollen feet! From the standpoint of urban planning, this heat differential is a strong argument for incorporating parks, greenbelts, and water features into neighborhoods (Fig. 9.7*b*).

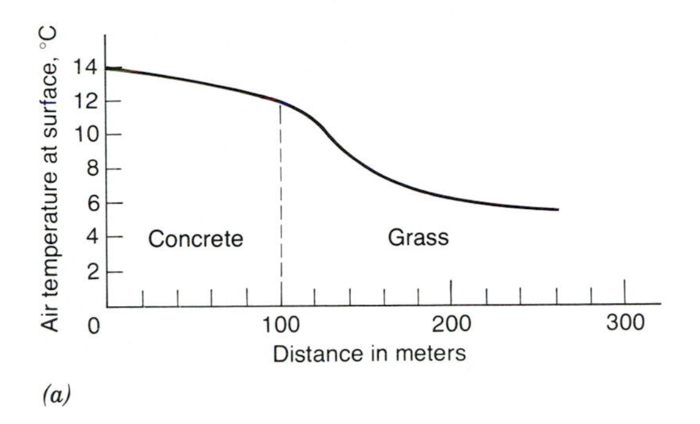

(a)

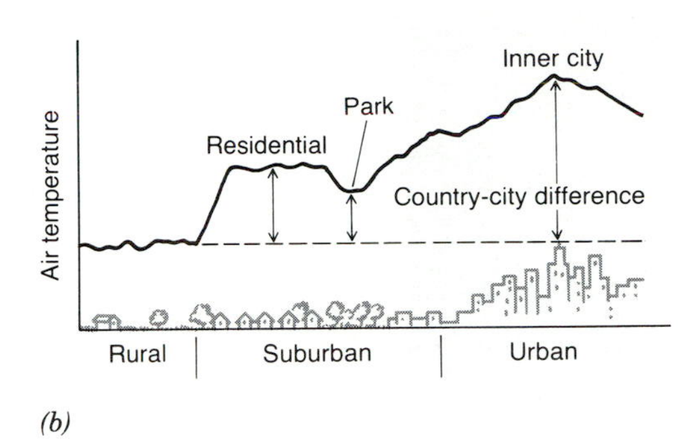

(b)

The high emittance of heat from hard surfaces raises the overall air temperature of cities, creating a warm spot, or *heat island*, in the land (Fig. 9.8). According to geographer T. R. Oke of the University of British Columbia, heat islands usually drop off sharply where the city gives way to the rural landscape, forming a "cliff" in the temperature profile (Fig. 9.7*b*). The magnitude of the heat island—defined as the difference between urban and rural temperatures—increases with city size, based on population.

For any city, the magnitude of the heat island fluctuates from day to night. Figure 9.9 compares an inner-city location with a suburban location in Vienna, Austria, over a 24-hour period in winter and summer. The magnitude of the heat island is represented by the difference between the urban (solid) and suburban (broken) lines. In this case—which is typical for many midlatitude cities—it is greatest in evening and at night because the urban surface materials generate more sensible heat than materials in the suburban landscape do. If this graph were extended to include a rural location, the difference would be even greater.

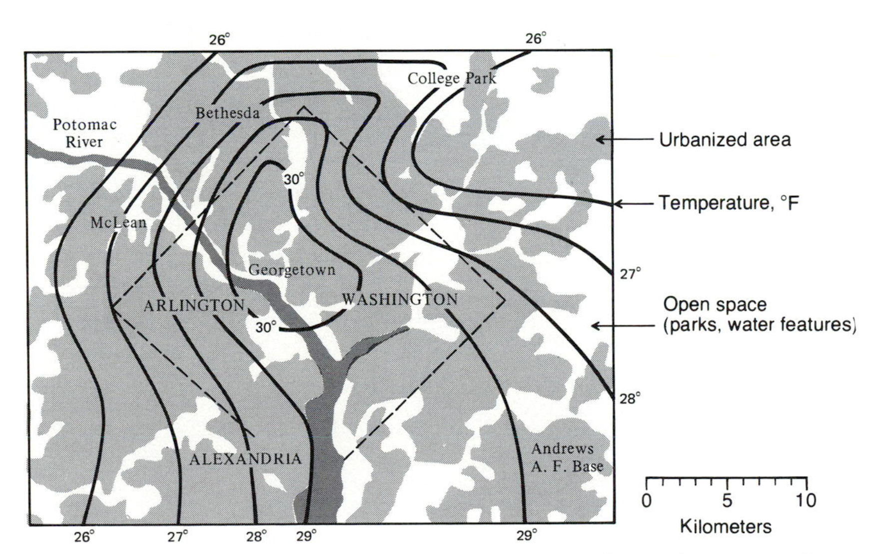

Figure 9.8 Mean minimum winter temperature over the Washington, D.C metropolitan area, based on data for the period 1946–65. This is one definition of the heat island; others would include daily high summer temperature and mean maximum and mean minimum summer temperatures.

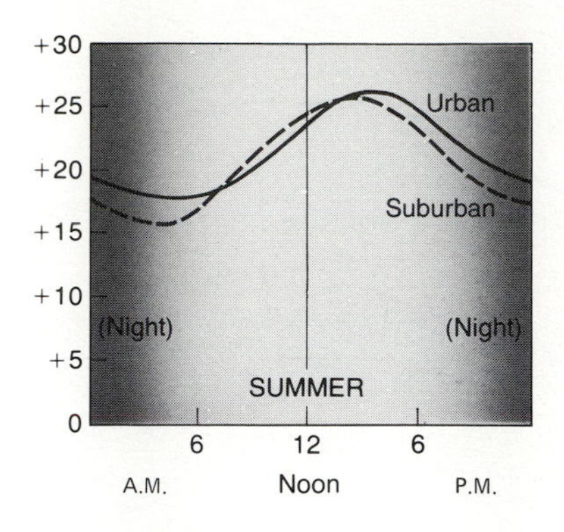

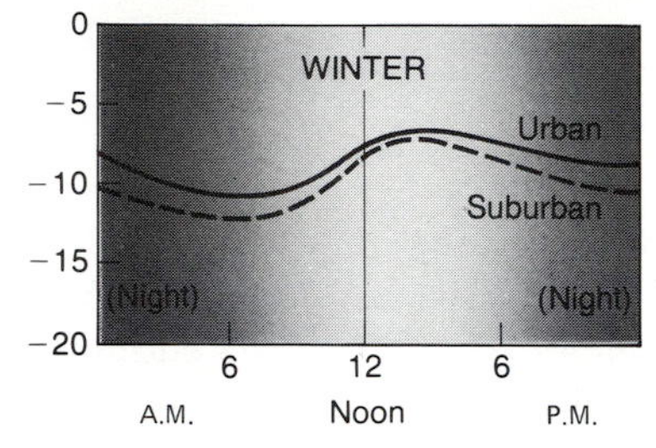

Figure 9.9 Day/night variations in the daily temperatures of an urban location (solid line) and a suburban location (broken line), Vienna, Austria.

The Not-So-Windy City

Wind is a critical factor in the energy balance and the climate of cities because it is the principal disperser of dirty, heated air. Cities influence wind in two ways. First, and most important, they increase the roughness of landscape, thereby raising the roughness length. As wind moves from a rural area toward the inner city, its velocity profile is displaced upward when it encounters taller, more closely spaced buildings (Fig. 9.10). This is a response to the stronger frictional, or drag, resistance posed by the irregular terrain formed by the mass of buildings and other large urban structures. Average wind speed may be 20 to 30 percent lower and extreme gusts, 10 to 20 percent lower over cities than over the nearby countryside. Consequently, the mixing and removal of the surface air is less efficient in cities than in rural landscapes. In other words, surface air tends to linger over urban areas, allowing more time for it to be altered by surface heat and human activities (Fig. 9.10).

Convection related to the urban heat island is the second influence on wind. During periods of calm atmosphere, a gentle inflow of surface air from the suburbs to the inner city may develop. It is especially likely to occur if the calm is associated with a strong heat island over the city. The heat island produces a weak pocket of unstable low-pressure air, which generates internal upflow and lateral inflow along the surface. Some studies conducted over New York City have recorded not only a marked upward flow over heavily builtup Manhattan Island, but also a downward flow over the nearby Hudson and East rivers. This circulation pattern is consistent with the temperature and pressure regimes that would be expected for surfaces with energy characteristics as different as these two locations.

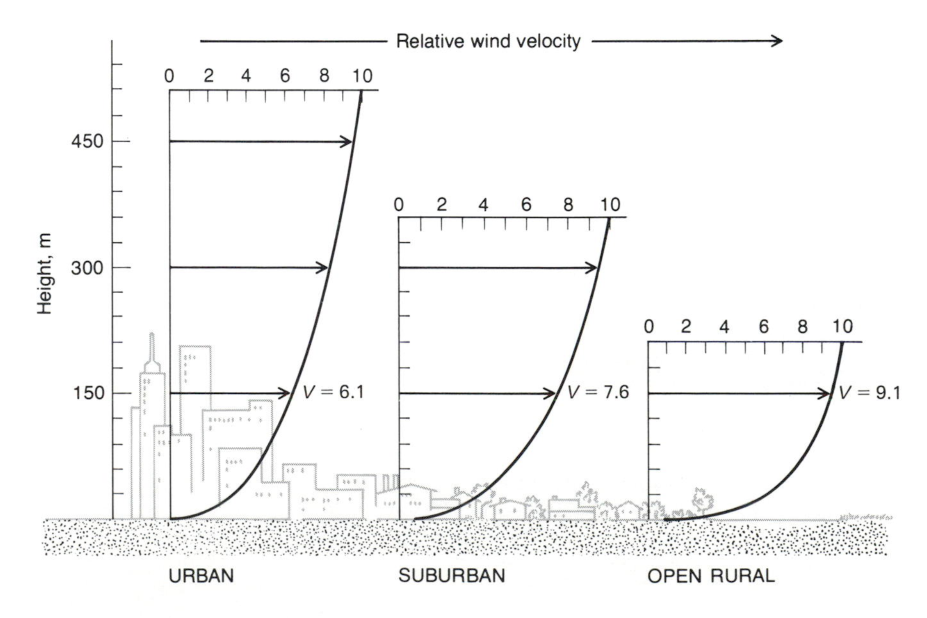

Figure 9.10 Wind velocity profiles over urban and rural surfaces. Note that the rate of velocity increase over the city is much lower than it is over the suburban and rural surfaces.

Precipitation

Does the urban heat island and related conditions influence precipitation? In many areas it appears that convectional rainfall is greater in and around the urban region, but the magnitude of the urban influence is often masked by the broader regional weather and climatic patterns. A 1976 study in the Detroit metropolitan region, however, suggests that precipitation in some urbanized areas is significantly greater than in the outlying suburban and rural areas (Fig. 9.11).

With respect to cloud cover and fog, there is no doubt that both are more frequent in most urban areas, especially large ones. Although this frequency may be related to instability associated with heat island circulation, it is probably related more directly to the abundance of minute airborne particles from air pollution, which serve as condensation nuclei for moisture droplets. The concentration of condensation nuclei over medium-sized cities is typically five to seven times greater than that over rural areas.

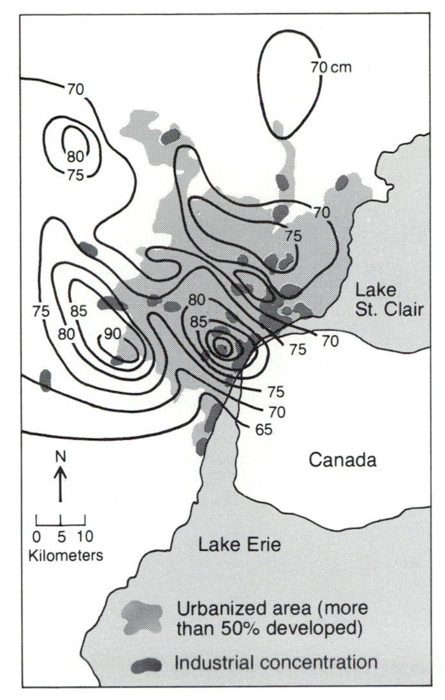

Figure 9.11 Average annual precipitation (cm) in the Detroit metropolitan region, based on 13 years of records from fifty stations.

Heat from Artificial Sources

Finally, we must consider the influence on the urban energy balance of heat generated by artificial means (mainly, internal combustion and power generation) from automobiles, factories, homes, and institutional establishments. The heat from these sources escapes into the air around them, and if wind is absent or light, this air can be heated rather quickly. The greatest influence from artificial sources is in northern cities during winter. In Manhattan in New York City, heat from combustion in midwinter is typically 2.5 times (250 percent) greater than that of the solar energy reaching ground level. In summer, on the other hand, this factor drops to about one-sixth (17 percent) of solar energy at ground level.

THE URBAN CLIMATE

Modern urbanization has a pronounced influence on most components of the energy balance and related atmospheric conditions. Generally, the larger the city and the denser its development, the greater the magnitude of change. Many factors contribute to this change, and much remains to be learned about them. Table 9.1 provides a qualitative summary of changes in radiation flux, heat flux, and airflow and the related cause(s) in the urban environment.

TABLE 9.1 SUMMARY OF ENERGY-BALANCE CHANGES IN THE COMPONENTS OF THE URBAN CLIMATE

Component	Change	Cause
Incoming shortwave radiation	Decrease	Dust dome backscattering
Outgoing longwave radiation	Decrease	Dust dome greenhouse effect
Sensible heat	Increase	Low thermal capacities of hard-surface materials; heat from artificial sources
Latent heat	Decrease	Hard surfaces over soil; sparse plant cover[a]
Mechanical turbulence (ground level)	Decrease	Large buildings and other structures
Convectional turbulence	Increase	Heat island

[a]Cities in arid climates may be an exception with regard to plant cover. Here shrubs, shade trees, and lawns are maintained, with the aid of irrigation, in densities greater than those present under natural conditions.

Trends in Urban Climate

Urbanization has produced significant climatic change but the change is often localized geographically and definitely limited by the size of the city. For large urban areas, the magnitude of the change is sufficiently strong to warrant the special designation "urban climate" or "climate of cities." Table 9.2 presents a summary of the climatic changes associated with urbanization. The data were compiled from studies conducted mainly in North America and Europe. The figure given for each climatic element represents the general variation in that element between the city and its neighboring countryside.

Modern urbanization is advancing and changing at dramatic rates, and much remains to be learned about its influence on the energy balance and climate. Virtually all findings to date are based on research in midlatitude European, North American, and Japanese cities, which are characterized by industrialization, large buildings, an abundance of hard surfaces and automobiles, and a high rate of heat output from artificial sources. In contrast, cities in nonindustrialized Asia and Africa are less built up, have fewer automobiles and factories, generate less artificial heat, have less area covered by hard-surface materials, and are generally smaller in size. Whether the influence of these cities on climate is correspondingly less than that of Western cities is not known.

TABLE 9.2 CLIMATIC CHANGES IDENTIFIED WITH URBANIZATION

Element	Comparison with Rural Environs
Temperature	
Annual mean	0.55–0.83°C higher
Winter minima	1.1–1.7°C higher
Relative Humidity	
Annual mean	6% lower
Winter	2% lower
Summer	8% lower
Dust Particles	10 times more
Cloudiness	
Clouds	5–10% more
Fog, winter	100% more
Fog, summer	30% more
Radiation	
Total on horizontal surface	15–20% less
Ultraviolet, winter	30% less
Ultraviolet, summer	5% less
Wind Speed	
Annual mean	20–30% lower
Extreme gusts	10–20% lower
Calms	5–20% more
Precipitation	
Amounts	5–10% more
Days with <0.5 cm	10% more

Source: Based mainly on Helmut E. Landsberg, "City Air—Better or Worse," in *Symposium: Air Over Cities*, Cincinnati, Ohio, U.S. Public Health Service, Taft Sanitary Engineering Center Technical Report A62-5, 1962.

Acid Rain: A Regional-Scale Problem

Cities also influence climate at a regional scale. As was noted earlier, airborne particulates from air pollution can increase fog formation and precipitation. Where a large urban dust dome is blown downwind (forming a plume), precipitation may be increased over the receiving area. A serious corollary of this phenomenon is acid rain.

Acid rain is precipitation (both rain and snow) whose pH has been significantly lowered because of air pollution. Unaltered rainwater as a pH of around 5.6 (a pH of 7 is neutral; greater than 7 is basic and less than 7 is acidic), whereas acid rainwater has a pH of between 1.5 and 5.6. (See Chapter 13 for a definition of pH.) The increased acidity can be caused by many pollutants, but the most common sources are sulfur and nitrogen oxides. After being discharged into the atmosphere, these pollutants can be converted into sulfuric acid and nitric acid through the process of oxidation. Oxidation apparently takes place after contact with a condensation particle or precipitation droplet.

Because of the westerly airflow across the center of North America, acid moisture is transported from the industrial Midwest and precipitated in New England and southeastern Canada. After many years, the pH of the water in lakes and ponds in these regions has been gradually lowered from 5.6–6.0 to 4.5 or less. At pHs of less than 4.5, most fish may be eliminated, and thus, many lakes have become ecologically sterile (Fig. 9.12).

Global and Hemispheric-Scale Changes

What is the influence of urbanization and related land uses on the climate over an entire hemisphere or over the whole earth? For many years the answer to this

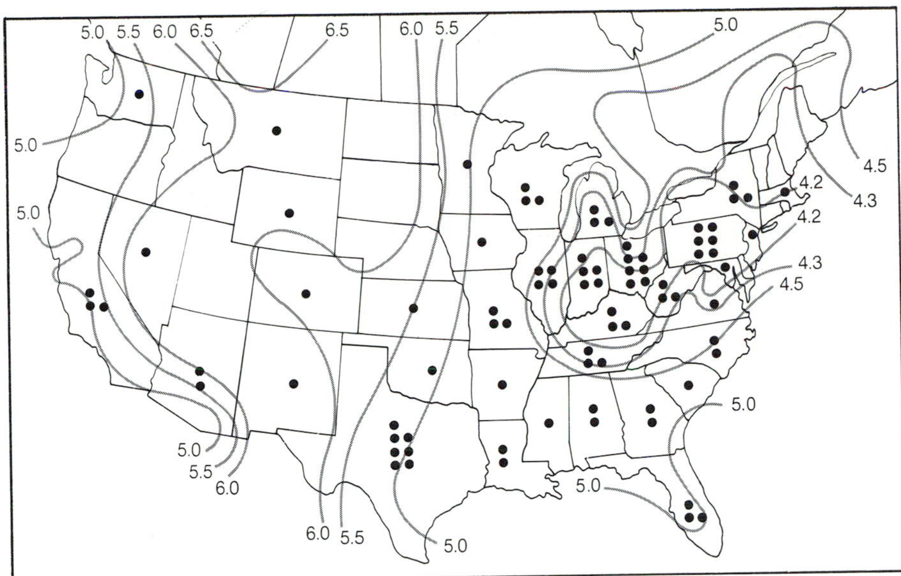

• 500 metric kilotons of sulfur and nitrogen oxide emission per year

Figure 9.12 The pattern of acid rain in eastern North America and the states with heaviest sulfur and nitrogen oxide emissions.

question was largely speculative, but today it is centered around two opposing lines of thought: *cooling* of global climate related to increased particulates in the atmosphere and *warming* of global climate related to increased carbon dioxide in the atmosphere. The cooling idea is based on the observation that airborne particulates have increased in this century for the atmosphere in general. This is supported by measurements in various nonurban areas, such as the summit of Mauna Loa in Hawaii, and some climatologists have shown evidence of temperature trends which indicate a corresponding decline in world temperature after 1940 or so (see Fig. 9.1*d*). Like the dust dome over a city, these particulates increase backscattering of incoming shortwave radiation, reducing the total intensity of incoming radiation in the lower atmosphere and at the earth's surface. We also know that dust from large volcanic eruptions can have the same effect; records show quite clearly that several years of unseasonably cool temperatures occurred in many parts of the world following major eruptions such as Krakatoa in Indonesia in 1883 (see Chapter 16 for a description of that explosion).

The warming idea is based on abundant data showing an increase in atmospheric carbon dioxide over the past 100 years. This gas has increased 11 to 12 percent (of its original 0.03 percent of the atmosphere) since about 1850, and were it not for the increase in particulates, it seems almost certain that an increase of this magnitude would have produced a rise in global temperature. It is likely that the cooling effect of particulates has been offsetting the heating effect of carbon dioxide. But world population is growing at an astonishing rate, and with it cities, industry, and agriculture are also growing and producing more and more carbon dioxide while large areas of forest, which consume carbon dioxide, are being eradicated. As of this date, most experts argue that the balance in global climatic change favors carbon dioxide and greenhouse warming. Just how strong the warming trend is or will be is uncertain. Scientists are certain, however, that it will continue for some time and that by the end of this century, it could influence changes in precipitation, the soil-moisture balance, and the magnitude and frequency of drought great enough to seriously alter the life-support capacity of the planet.

Summary

Climate change is common to earth, a fact scientists have understood for many years. We also understand that climatic change is not limited to the earth's past but is taking place now and is part of the drama of human affairs and survival itself. The cause of climate changes that occur over intervals of millions of years is very difficult to ascertain. One idea recently advanced is based on atmospheric cooling associated with asteroid collisions with earth that occur in intervals of 26 million years.

Changes with periods of 20,000 to 100,000 years

are probably related to astronomical factors, such as eccentricity, obliquity, and precession, whereas shorter term changes appear to be complexly related through earth-atmosphere-ocean feedback to many factors including volcanic activity, land-cover changes, urbanization, and air pollution. In the industrialized world, massive urban areas are causing substantial climatic changes, which for midlatitude cities, are characterized, among other things, by heat islands, reduced solar radiation, reduced wind speeds, and increased fog and cloud cover. The influence of urban areas on regional atmospheric conditions is illustrated by acid rain, and at a global scale, urbanization is a major source of carbon dioxide and airborne particulates which cause, respectively, atmospheric warming and cooling.

Concepts and Terms for Review

- prediction in science
- evidence of past climates
- role of plate tectonics
- climate change
  - age of dinosaurs
  - present Ice Age
  - interglacial period
  - Little Ice Age
  - boreal forest border
- implications of climate change
  - human survival
  - marginal lands
  - desertification
- causes of change
  - atmospheric-ocean feedback
  - eccentricity
  - obliquity
  - precession
  - air pollution
  - land use
- urban climate
  - radiation flow
  - heat island
  - windiness
  - precipitation
  - artificial heat
- regional acid rain
- global scale change
  - carbon dioxide trend
  - particulate trend

Sources and References

Budyko, M.I. (1982) *The Earth's Climate: Past and Future.* Orlando, Florida: Academic Press.

Hays, J.D., et al. (1976) "Variations in the Earth's Orbit: Pacemaker of the Ice Ages." *Science*, 194, pp. 1121–1132.

Kellog, W.W., and Mead, M. (1977) *The Atmosphere: Endangered and Endangering.* Wash., D.C.: U.S. Department of Health, Education and Welfare.

Lamb, H.H. (1972) *Climate: Present, Past and Future.* London: Methuen.

Namais, J. (1978) "Long-Range Weather and Climate Prediction." In *Geophysical Predictions*, Washington, D.C., National Academy of Sciences.

National Academy of Sciences. (1975) *Understanding Climatic Change.* Washington, D.C.: U.S. Government Printing Office.

National Research Council. (1985) *The Effects on the Atmosphere of a Major Nuclear Exchange.* Washington, D.C.: National Academy Press.

Oke, T.R. (1973) "City Size and the Urban Heat Island." *Atmospheric Environment*, Vol. 7, pp. 769–779. New York: Pergamon Press.

Peterson, J.T. (1969) *The Climate of Cities: A Survey of Recent Literature.* Raleigh, N. C.: U.S. Department of Health, Education and Welfare.

Rotberg, Robert I., and Rabb, Theodore K. (1981) *Climate and History.* Princeton: Princeton University Press.

Wendorf, Fred, et al. (1985) "Prehistoric Settlements in the Nubian Desert." *American Scientist*, 73:2.

Chapter 10

Basic Plants and Vegetation of the World

Of the total mass of living matter on earth all but a small fraction is vegetation. Vegetation is comprised of hundreds of thousands, perhaps millions, of plant species, each with its own forms, origins, and roles in the environment. How do we begin to understand the types of plants that inhabit earth and how they combine to form different types of vegetation? And what is the geographic relationship among vegetation, climate, and human activity?

Plant geography is the term traditionally applied in physical geography to the study of the plant cover. Although the Greek Theophrastus was probably the first plant geographer, plant geography did not emerge as a formal science until the 1800s. The main activity of early plant geographers was to identify, name, and map the flora (plants) in various regions of the world. As members of expeditions sponsored by national governments, or as independent adventurer-explorers, they traveled to remote lands collecting specimens and recording their observations on the conditions of land and life. The German Alexander Von Humboldt (1769–1859) and the Englishman Charles Darwin (1809–82), for instance, studied the vegetation of South America; another Englishman, J. D. Hooker (1818–1911), explored and recorded the flora in the islands of the South Pacific; and the American Asa Gray (1810–88) documented thousands of plants in North America. Although these earlier plant geographers were concerned mainly with the distribution of species (a part of the field called *floristic plant geography*), their contributions were much broader. In particular, they helped lay the groundwork for the development of world vegetation and climatic maps as well as many important ecological concepts.

In the late 1800s plant geographers became increasingly interested in knowing *why* plants grow where they do and how they are related to the environment. This gave rise to plant ecology, the study of interrelationships between plants and their environment. Today plant ecology is studied principally in the biological sciences where the emphasis is on the interrelations of plants and the biological environment, that is, with animals, microorganisms, and other plants. Plant ecology (or *ecological plant geography*) in physical geography, on the other hand, is oriented more toward the study of vegetation (the total assem-

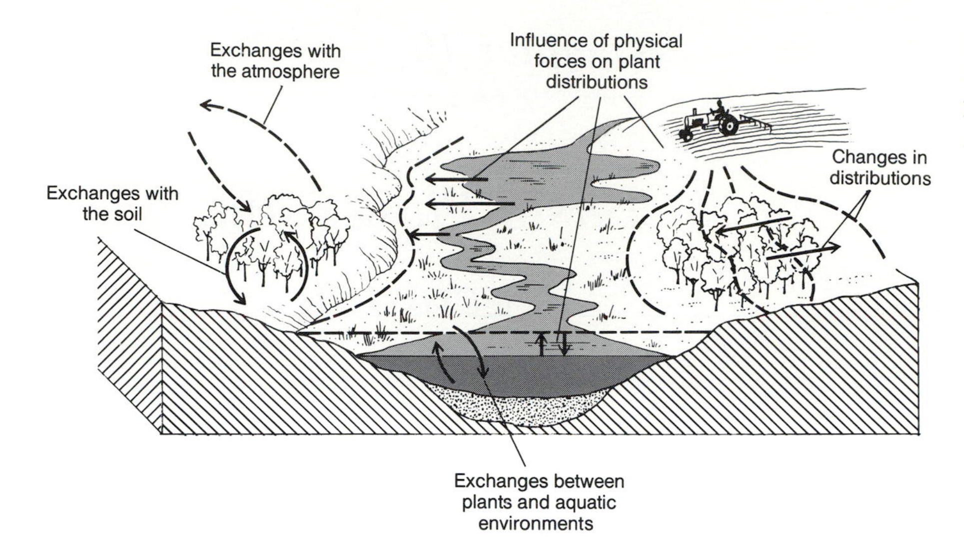

Figure 10.1 Schematic illustration of some of the principal topics studied in plant geography.

blage of plants in a landscape) and its relationship with the physical environment (especially climate, soils, water, and land use activities) as the basis for understanding vegetation distributions at global, continental, and regional scales.

In this chapter we first examine some of the schemes used to classify plants and then describe the basic types of vegetation that are prominent in different landscapes and climatic zones of the world. In the following two chapters (11 and 12) we will focus on the processes and controls of plant growth, the concept and functions of an ecosystem, the forms of plant adaptation, and the concepts used to explain changing plant distributions (Fig. 10.1).

THE HIGHER PLANTS

Various schemes are available to help us organize our description of the complex body of life (plants, animals, and microorganisms) that inhabits the land and sea. Those systems devised by biologists are based mainly on their understanding of the evolutionary relationships among them: that is, which groups of plants evolved from which. According to one scheme, the plant kingdom is divided into three major groups: (1) fungi and algae, (2) mosses and liverworts, and (3) vascular plants. In general, this sequence of plants ranges from simple to complex, small to large, and on the evolutionary ladder from those that arose early and are considered primitive (algae) to those that came later and are considered advanced (vascular).

The Vascular Plants

Plant geography is concerned mainly with the *vascular plants*, which are generally considered to be "higher plants" because of their biological complexity, size, and evolutionary history. Plants in this group have bodies differentiated into stems, leaves, roots, and reproductive organs (Fig. 10.2). The roots both anchor the plant in the soil and absorb water and nutrients. The stem holds the leaves and the reproductive organs aloft. Leaves are the plant's interface with the atmosphere and the medium through which it exchanges gases, releases water, and receives light for the manufacture of food substances. The reproductive organs provide the primary means by which the plant regenerates its biological line.

All parts of all plants are composed of minute, boxlike cells. The key feature of the vascular plants is the arrangement of certain cells into vascular, or conducting tissues, known as *xylem* (woody, water-conducting tissue) and *phloem* (specialized, nutrient-conducting tissue). These tissues form a sort of pipe system in the plant through which water, nutrients, and manufactured foods are conducted. In the stems of woody plants, the xylem is surrounded by a cylinder of phloem and associated cells known as bark (Fig. 10.2*b*).

Groups of Vascular Plants

Botanists classify vascular plants into three main groups: (1) *angiosperms*, (2) *gymnosperms*, and (3)

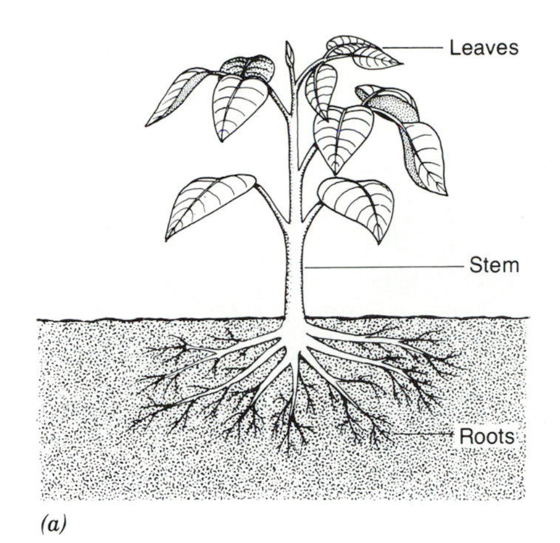

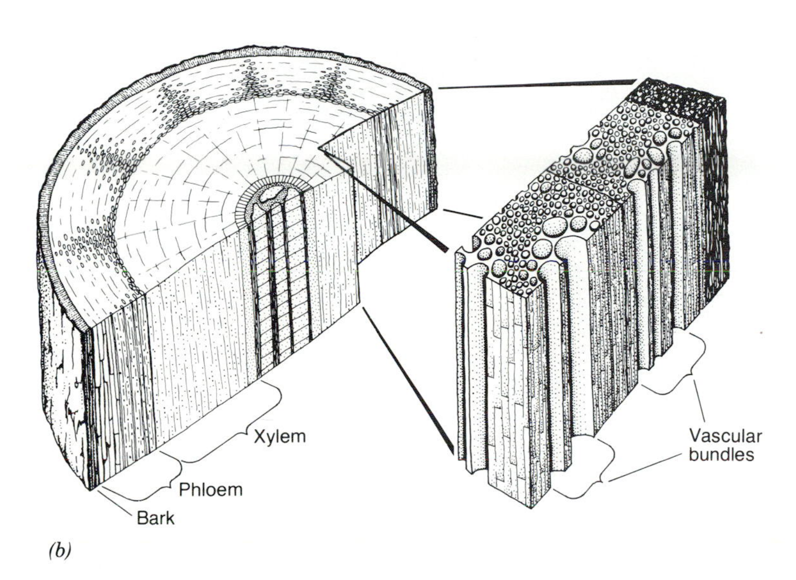

Figure 10.2 A vascular plant: (*a*) major anatomical features; (*b*) major components in the stem.

pteridophytes (Fig. 10.3). *Angiosperms* are the flowering plants. They produce seeds and are divided into two major groups on the basis of whether the seed sprouts one leaf, called a cotyledon, or two. The largest group is the *dicotyledons*, which encompasses the broadleaf trees, most shrubs, and familiar small plants such as peas, buttercups, and roses. The *monocotyledons* are mainly nonwoody plants with blade-shaped leaves; familiar examples include grasses, lilies, orchids, and palms.

Gymnosperms are plants that bear naked seeds (that is, seeds not enclosed by a mature ovary, or fruit, as they are in angiosperms), although we usually think of them as the cone-bearing plants. They are dominated by one group, the conifers, comprised of woody needleleaf plants such as pine, redwood, fir, and cypress. The *pteridophytes*, small, nonwoody plants, also are dominated by one group, the ferns. Unlike the angiosperms and gymnosperms, the pteridophytes regenerate by broadcasting *spores* rather than seeds.

WAYS OF CLASSIFYING PLANTS

Three types of plant classification schemes are used in plant geography: floristic, life-form, and ecological.

Floristic Classification

The universal system of botanical names is based on a classification scheme called the *floristic* system. Under this system, the plant kingdom is made up of

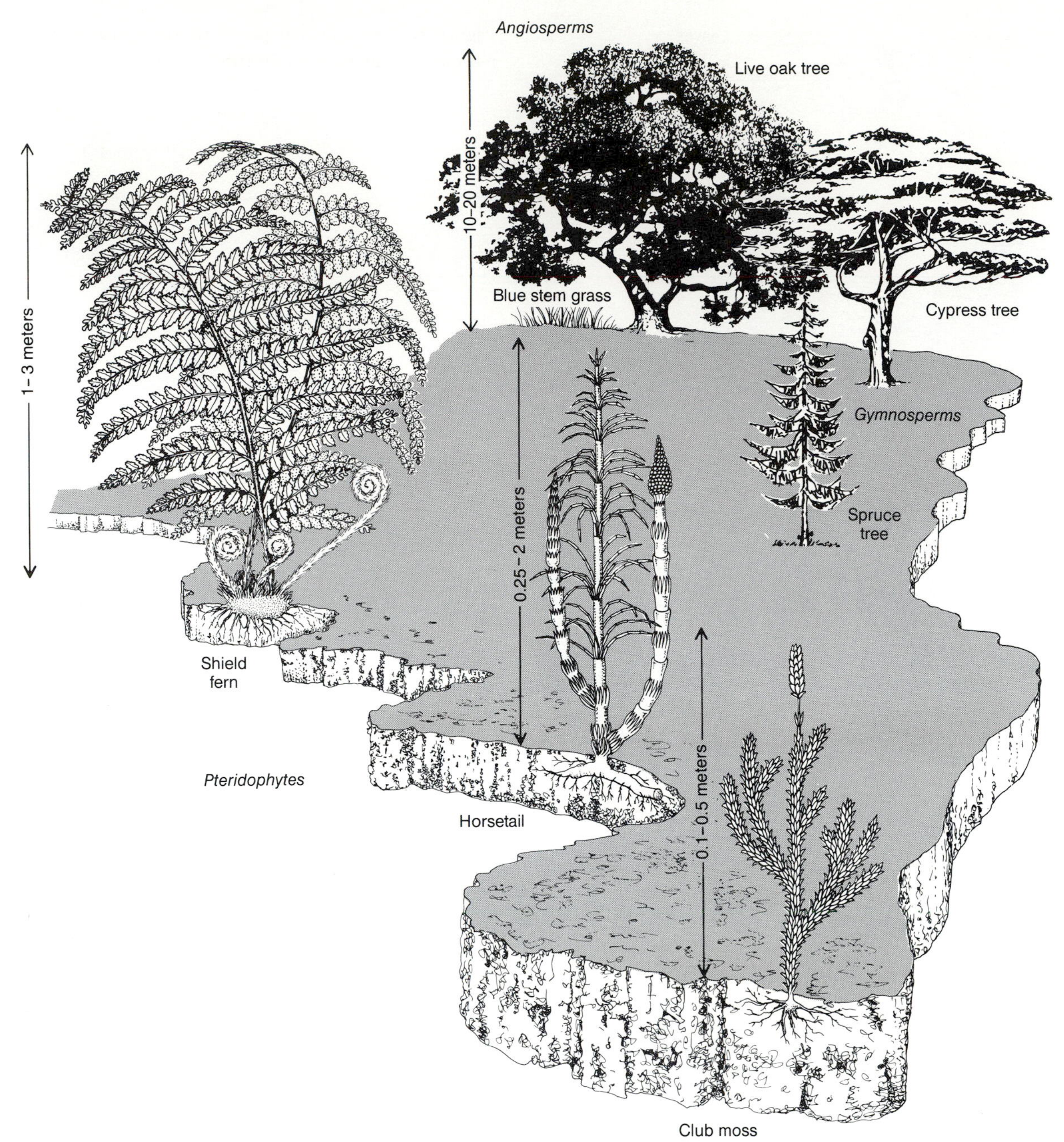

Figure 10.3 Examples of angiosperms, gymnosperms, and pteridophytes. Ferns comprise the majority of the pteridophytes. Shrub and tree forms are found only among the angiosperms and gymnosperms.

TABLE 10.1 FLORISTIC CLASSIFICATION, BEGINNING AT THE CLASS LEVEL

Taxon	Size[a]	Example
Class	Largest	*Gymnospermae* (cone-bearing plants)
Order	↓	*Coniferales* (evergreen, needleleaf)
Family		*Pinaceae* (pine family)
Genus		*Pinus* (true pines)
Species	Smallest	*Pinus strobus* (white pine)

[a]Based on number of members.

divisions, each of which is subdivided into classes, and then into smaller and smaller classification units, or *taxons* (Table 10.1). Each taxon has a Latin name or a name with a Latin suffix (usually *-is*, *-us*, *-ae*, *-aceae*, *-i*, or *-a*). The more well-known plants also have common names, but these names often vary from country to country or from region to region within a country.

The *species* taxon represents the smallest classification unit in general use today. A species (pronounced spee-shees or spee-sees for both singular and plural) is comprised of individual plants that are able to freely interbreed among themselves but are unable to breed with members of other groups. Although members of a species are not identical, they usually look so much alike that we would have trouble telling them apart. Similarity of traits, or characters, especially in the reproductive organs and leaves, is the basis for classification at any level in the floristic system. The rationale here is that a high degree of physical similarity is indicative of a common heritage, that is, a genetic affiliation during the evolution of the plants. The importance of evolution studies and the need for an international plant-names system make the floristic scheme essential to the modern biological sciences. For the purposes of plant geography, however, other classification schemes are equally or more useful.

Life-Form Classification

Life-form schemes are based on either the form of individual plants or the overall form, or structure, of the vegetative cover. The most familiar life-form scheme classes plants primarily according to individual form and size. *Trees* are large woody plants with a main stem, or trunk, that supports branches from which the foliage grows. Most tree species are angiosperms, although the very largest trees and some of the most extensive forests on earth are coniferous (Fig. 10.4).

Shrubs are also woody, branching plants, but are much shorter than trees and thus tend to be dominated more by branches than by stems. Most shrubs are dicotyledonous angiosperms, although a few familiar conifers (e.g., yew and juniper) are shrubs. Under stressful environmental conditions, certain trees can take on a shrub form, a phenomenon called *dwarfism*. Dwarfism is especially pronounced in mountainous, arctic, and arid environments: in the North American arctic, birch and spruce are common dwarf trees. Many ferns and monocotyledons (e.g., banana) have shrub forms and even tree heights, but because they are nonwoody, they are classed as herbs (Fig. 10.4).

The *lianas* are another life-form common to woody plants. Lianas are climbers (vines) that use trees for support; they are chiefly dicotyledons and are found in many different families. In the midlatitude forests, grape and poison ivy are common lianas.

Herbs are generally the smallest of the vascular plants. Having light superstructures, they are supported by stems composed of nonwoody tissue. Herbs are limited to the pteridophytes and angiosperms and can roughly be arranged into three groups: (1) ferns, (2) blade-leaf monocotyledons, and (3) *forbs*, or broadleaf herbs (Fig. 10.4). Although most herbs are ground plants, some live as parasites on trees, and one group, called *epiphytes*, simply uses trees for support without rooting in soil. Epiphytes rest on tree branches and, as in the case of Spanish moss, dangle their roots into the air, where they are able to gain their moisture through condensation and rainfall interception.

If we draw back from individual plants and allow the various forms in the plant cover to merge into a whole, we can identify differences in the general structure of the vegetative mass. These differences can be extraordinary, even over short distances, and we know the various formations as forests, grassland, tundra, or steppe, for example. In this respect, this scheme is little more than an extension of certain descriptive words from various folk cultures.

Classification According to Environment and Ecology

Because our concern is with the plant environment as well as with the plants themselves, we cannot overlook systems that classify plants according to the hab-

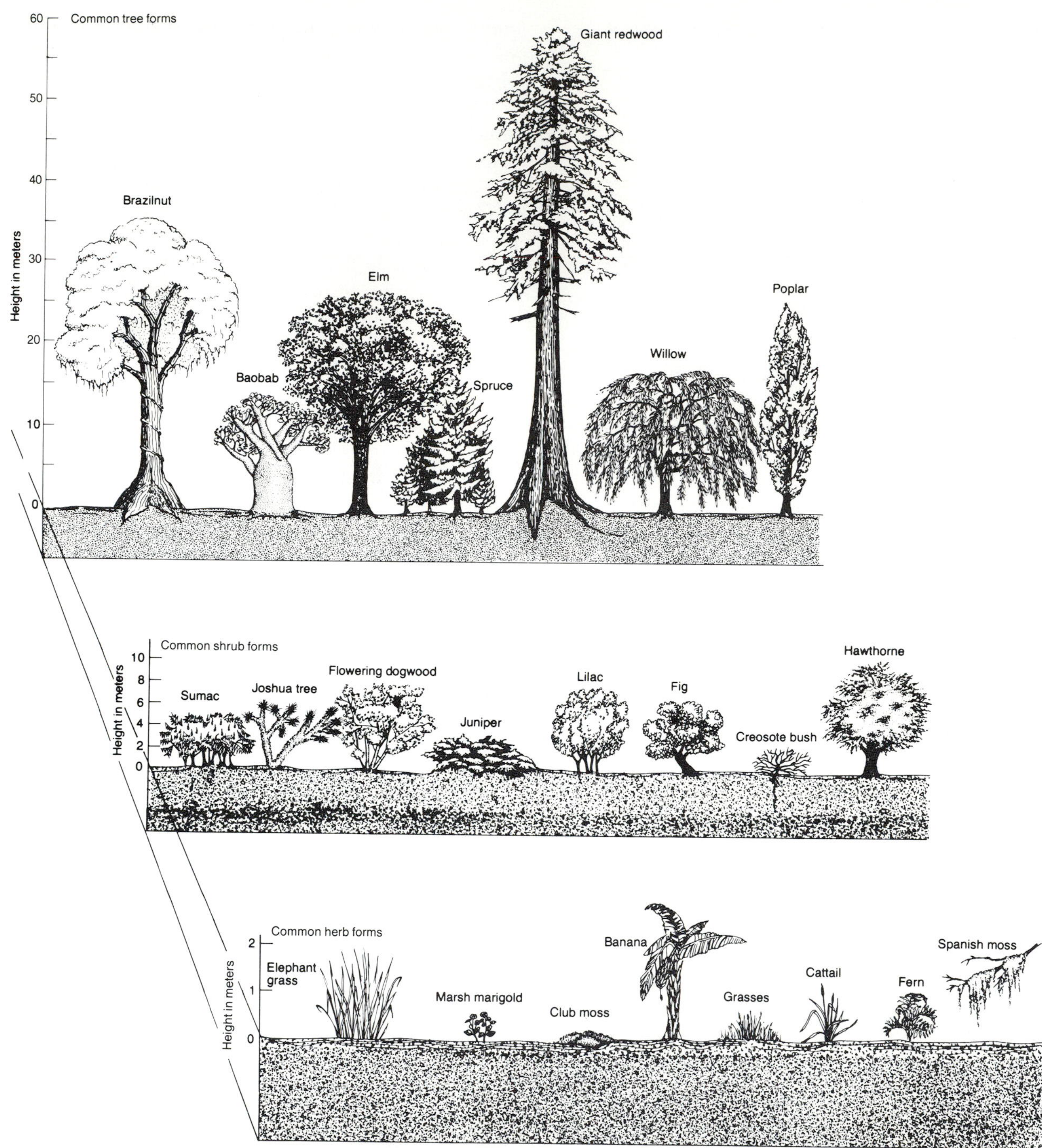

Figure 10.4 Common examples of tree, shrub, and herb forms.

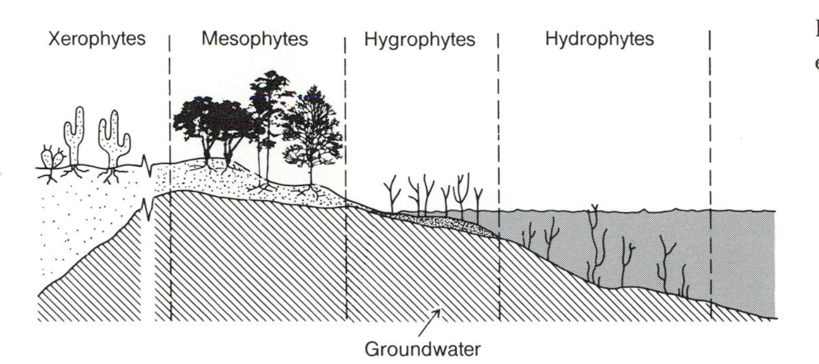

Four classes of plants based on their moisture environment.

itats in which they live. Geographers and ecologists use the word *habitat* to describe a particular kind of surrounding or local environment, such as a stream valley or a lake shore. Habitats are usually characterized by a certain combination of soil, drainage, and climatic conditions associated with a certain topographic setting in the terrain. Many different plants (and animals) with similar environmental needs and limitations usually occupy the same habitat. Ecological classification schemes group plants on the basis of common habitat.

Although no comprehensive ecological system is presently complete, two simple systems based on temperature and moisture have been in use for many years. One of these, the *moisture-based system,* uses four main classes for virtually all plants. Those that live in water, such as water lily, are called *hydrophytes;* those that live in saturated soil, such as cattail, are called *hygrophytes.* In contrast are the drought-tolerant plants, called *xerophytes,* which are best exemplified by cactus. Plants that occupy intermediate moisture environments are termed *mesophytes.* In the midlatitudes, mesophytes grow in well-drained sites that are neither saturated nor severely dry for more than several weeks a year. Maple, beech, elm, and hickory trees are good examples of mesophytes.

MAJOR VEGETATION TYPES OF THE CONTINENTS

Despite its uneven distribution over the planet as a whole, the vegetative cover is truly massive, comprising hundreds of thousands of plant species that exist in billions of tons or kilograms of biomass. No animal life, from microscopic protozoans to mammals including ourselves, can survive without this mass of greenery; therefore, as dependent organisms, we should know where the plants grow on the planet, and as students of the land, we should know what forms and roles they take on in the different landscapes.

Biomes and Vegetation Formations

All plants and animals belong to ecosystems, and all ecosystems are grouped into three major types: (1) salt water; (2) fresh water; and (3) terrestrial. Each major type of ecosystem is subdivided into large biogeographical units called biomes which cover vast zones of the earth. *Biomes* are characterized by a particular combination of vegetation and animals whose geographic distribution is related to global and regional patterns of climate and soil. Our concern is mainly with the terrestrial biomes, namely, forest, savanna, grassland, desert, and tundra (Fig. 10.5). Each biome contains many ecosystems, but we will save our examination of ecosystems until the next chapter.

The forest biome is characterized by a cover of trees that forms a continuous canopy over the ground and is found in humid climates with an average temperature of at least 10 to 15 degrees C in two to three of the warmest months of the year. Although herbs and other plant forms are found in every forest, trees are the dominant vegetation. In the savanna biome, a mixture of forest and grassland, patches of trees and shrubs are scattered among tall grasses (Fig. 10.5). The savanna biome is widely associated with the wet-dry tropical climate. A blanket of grasses dominates the grassland biome, and only infrequently is it interrupted, usually by trees and shrubs along stream courses. Most grasslands are found in semiarid climatic zones. The desert biome, which is found in tropical and midlatitude arid zones, is characterized by a highly discontinuous plant cover comprised mainly of low herbs and shrubs. The tundra biome is found in cold (arctic, subarctic, and high mountain)

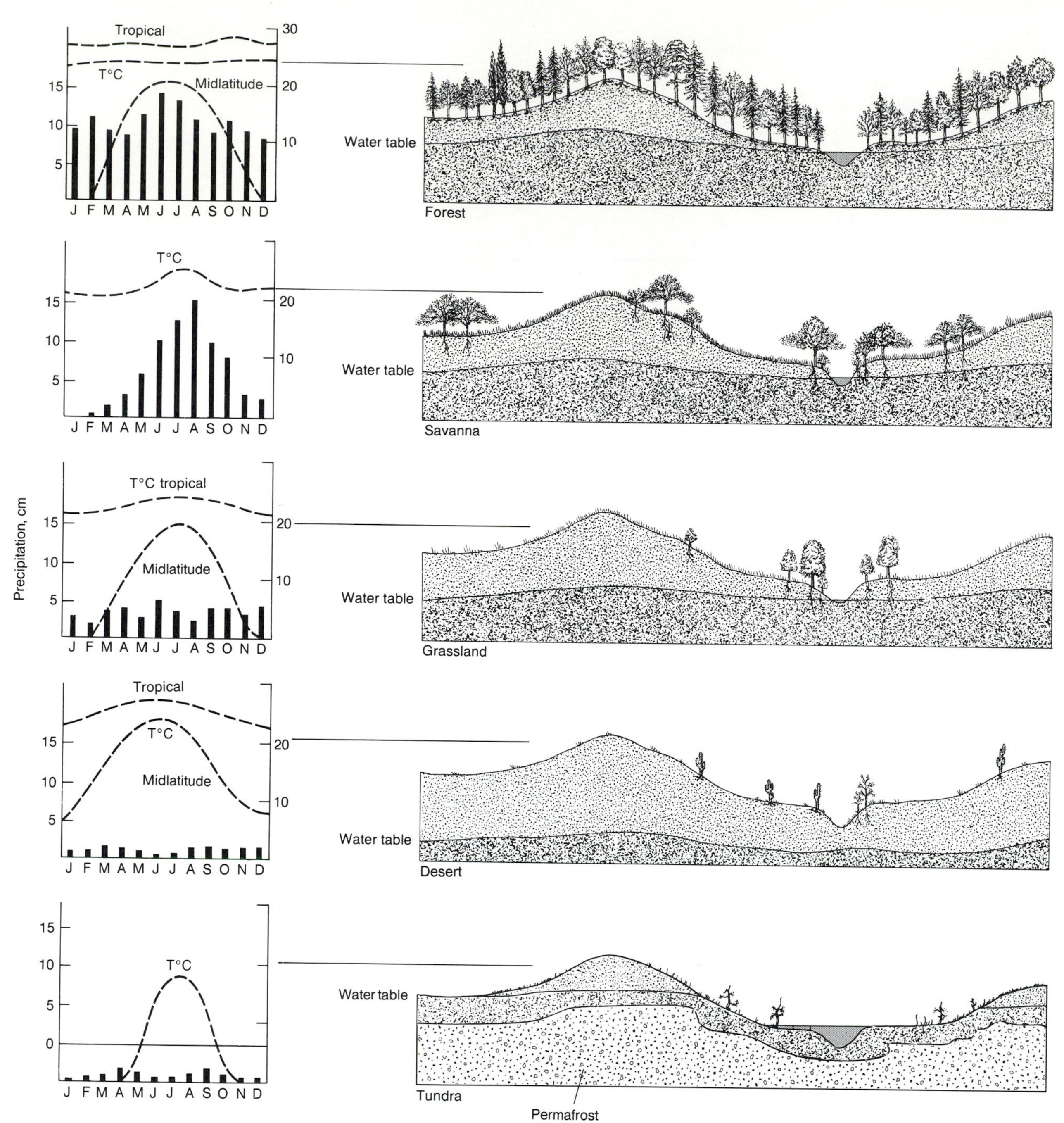

Figure 10.5 The five terrestrial biomes and the representative patterns of monthly temperature and precipitation associated with each.

climates where vegetation is limited to herbs and shrubs (Fig. 10.6, color plate 6).

Because the overwhelming majority of the biomass of terrestrial ecosystems is made up of plants, the physical character of the biomes is dominated by vegetation. As geographers, one of the first things we want to know about vegetation is its structure, that is, the shape, size, and form of the plant assemblage that covers the land. The structure of the vegetation in each biome can be classified according to *formation classes*, such as open forest and closed forest. Some of the important terms used in describing the structural characteristics of vegetation are given in Table 10.2.

Our descriptions of the vegetation of the terrestrial biomes draw on both structural and floristic (plant type) criteria. We will call each kind of vegetation a *vegetation type*, and besides defining its basic composition and forms, where appropriate, we will also identify its relationship to climate and other landscape features such as soils, water supplies, animals, and human activity. Let us begin by introducing some generalizations about vegetation at the global scale, especially with respect to the extremes in global environments:

- Vegetation is smaller in stature with slower growth rates in the more rigorous environments (cold and/or dry climates) than in the less rigorous environments (wet and warm climates).
- Vegetation is less dense with a lower biomass (total weight of living matter) in the more rigorous environments.
- Species diversity (the number of different species per square kilometer) is lower in the more rigorous environments; conversely, the populations of individual species, especially in forests, are greater in the more rigorous environments than in the less rigorous ones.
- Flora of the cold environments (arctic, subarctic, and high mountain) are newer (having been destroyed in large part only 10,000 to 20,000 years ago by glaciation) than tropical and equatorial flora, which accounts in part for the low species diversity in the vegetation of the cold zones.
- Human impact on vegetation has been great throughout most of the world, but particularly in the midlatitude zones and in the wet-dry tropics (including the monsoon regions), and is currently increasing rapidly in the wet tropics, equatorial, and subarctic zones.
- As natural (or relatively undisturbed) landscapes are altered by human land use, the rate of extinction is increasing and the earth's species diversity is declining.

Forest Biome

Equatorial and Tropical Rainforest The forests of the equatorial and tropical zones are characterized by a very dense vegetation dominated by large trees 30 to 50 meters high. The trees, such as the Brazilnut (*Bertholletia excelsea*) of South America, form massive crowns that intertwine to form an almost continuous canopy. Although the canopy may appear to be comprised of equally tall trees, several different stories can be discerned in most rainforests.

Because of the great density of the forest canopy, sunlight cannot penetrate in sufficient quantity to nurture a heavy groundcover of shrubs and herbs. Thus, the forest floor is relatively open (except for the tree trunks themselves), and the rainforest is classed as an *open forest* formation. This forest structure gives rise to three life zones in the rainforest: an *upper zone* at the top of the canopy which is well lighted, windy, and occupied principally by birds; a *middle zone* within the canopy which is climatically less rigorous than the one above and is occupied by one of the richest assortments of plants and animals on earth; and a *lower zone*, beneath the canopy, where it is humid, shaded, and inhabited by relatively light plant and animal populations (Fig. 10.7).

Most of the rainforest trees and associated plants are *evergreen*, that is, they do not have a dormant season during which foliate is lost. Rather, foliage is shed and replaced during all months in a continuous cycle, so there is usually no period in the year when ground shade is interrupted. The trees provide habitats for a rich variety of smaller plants, including epiphytes, thick woody lianas vines, and strangler vines (see Fig. 10.7) and an amazing variety of birds and arboreal mammals. The floristic composition of rainforest vegetation is the most diversified on earth. Many species, commonly on the order of thousands per square kilometer of land, but each with small populations of individual plants, distinguish this plant cover.

Vast areas of equatorial and tropical rainforest are found in the Amazon Basin of South America, in West Central Africa, and in Southeast Asia, but all are shrinking rapidly because of lumbering, mining, and land use development (Fig. 10.6, color plate 6). The

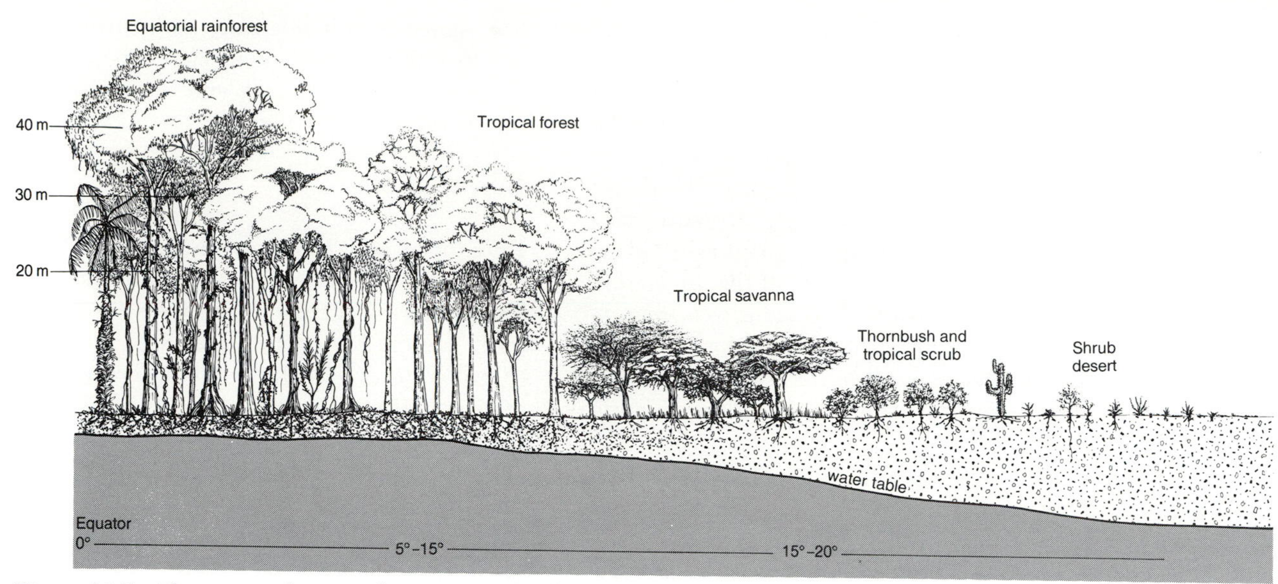

Figure 10.7 The principal types of vegetation: forest—tropical rainforest, temperate forest, sclerophyll forest, boreal forest; savanna—thornbush savanna, tropical savanna; grassland, prairie, steppe; desert—shrub desert, dry desert.

TABLE 10.2 TERMS USED IN DESCRIBING THE STRUCTURAL CHARACTERISTICS OF VEGETATION

Dominant species:	A plant species that occupies the greatest amount of space in an area as measured by the extent of its foliage or root system.
Dominant stratum:	A number of plants whose combined foliage makes up the principal layer (story) in a vegetative formation.
Story:	A layer or level of tree crowns in a forest.
Canopy:	The roof of foliage formed by the crowns of trees in a forest.
Cover density:	The percentage of areal coverage by vegetation of all types in an area; *groundcover density* (coverage by ground plants such as grasses) and *canopy density* (coverage by a forest canopy) are also used to describe density.
Stand:	A floristically uniform growth of vegetation, often of similar size and age, that dominates an area.
Open forest:	A forest that is relatively free of ground vegetation, so that standing on the forest floor you can see a long way into it.
Closed forest:	A forest that is filled in with vegetation from ground-level upward so that it is impossible to see more than a few meters into it.

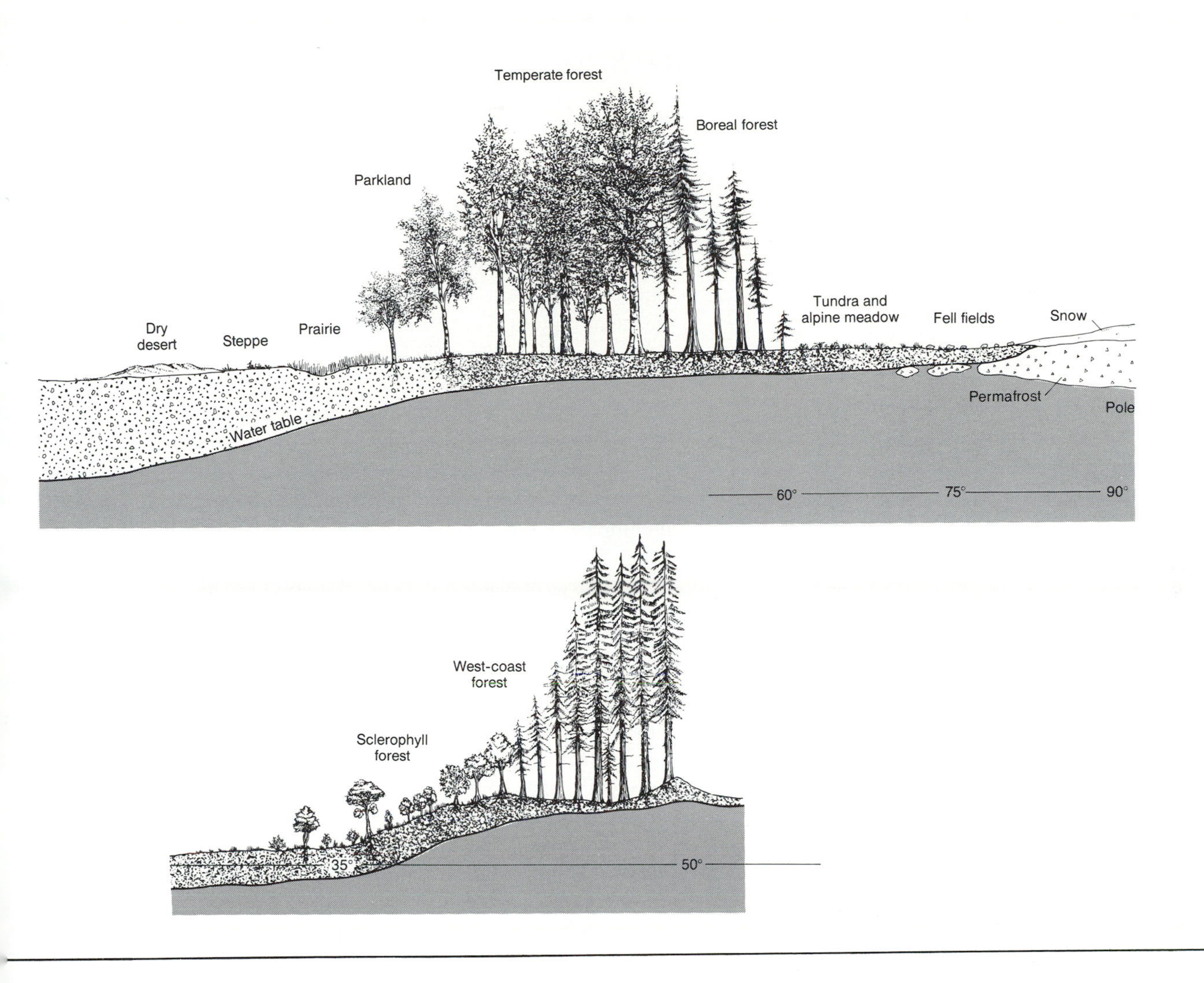
Temperate forest
Boreal forest
Parkland
Dry desert
Steppe
Prairie
Tundra and alpine meadow
Fell fields
Snow
Water table
Permafrost
Pole
60°
75°
90°
West-coast forest
Sclerophyll forest
35°
50°

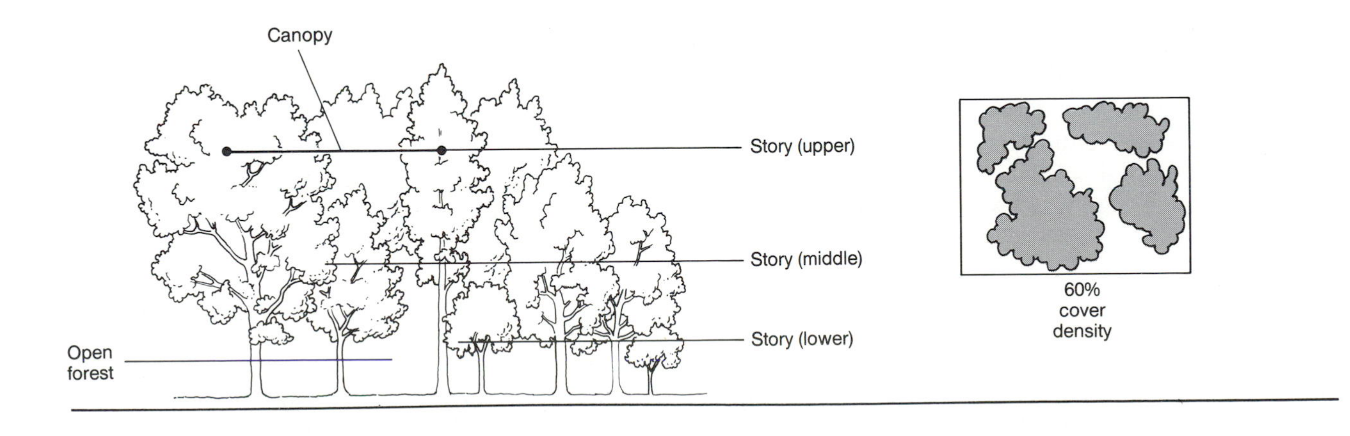
Canopy
Story (upper)
Story (middle)
Story (lower)
Open forest
60% cover density

THE VIEW FROM THE FIELD / NOTE 10

Biogeographical Diversity on a Shrinking Planet

Landscape change as a direct result of human activity is usually harsh and far reaching, especially in the past several centuries. Landforms, water features, and soils are seriously altered when land is lumbered, farmed, burned, mined, or developed. Compared to the biota, however, the magnitude of change in these features might be considered modest. The reason is that landscape change by humans leads not only to wholesale displacement of organisms but to extinction in many species. As natural landscapes are disturbed or destroyed, extinction advances and the earth's biological diversity—that is, the richness of its species mix—declines.

Measuring the magnitude of the loss, however, is proving to be difficult because we are uncertain about the earth's true biogeographical diversity. Since the scientific system of species identification was introduced in 1753 about 1.7 million species of plants, animals, and microorganisms have been formally catalogued. But most experts agree that this count grossly underestimates the true species diversity of the earth. In the past few decades, the search for unrecorded species in remote habitats, such as the deep ocean floors, and complex habitats, such as the tropical rainforests, has resulted in evidence of thousands more species. Some scientists now suggest that earth's species count is greater than 10 million, and perhaps more than 30 million. The great majority of these are tropical invertebrates such as insects, mites, crustaceans, and worms.

Episodes of mass extinction have occurred many times in earth's history. Two notable episodes are the great reptiles about 65 million years ago and the large mammals at the end of the last glaciation, 10,000 to 15,000 years ago. Many ideas have been offered to explain these extinctions, and one currently favored for the large mammals is overkill by humans. Animals such as the wooly mammoth, the giant bison, and the mastodon fell victim to early hunting parties or to habitat change caused by humans. The loss of herbivores led in turn to the loss of dependent predators; all in all, dozens of prominent mammals disappeared.

The present episode of extinction is the most massive of all—400 times the extinction rate of recent geologic time and increasing rapidly according to biologist E. O. Wilson. The loss is greatest in the tropical rainforests. In prehistoric times these species-rich habitats covered 5 million square miles; today they cover 3.5 million square miles and are being destroyed at rates estimated between 24,000 and 60,000 square miles annually. The effect on species diversity follows this biogeographical rule: When a habitat is reduced to 10 percent (that is, by 90 percent) of its original area, the number of species will ultimately decline to 50 percent. Unlike landforms or soil formation, species evolve by irreversible processes. Once lost, they are lost forever.

climate in these areas is characterized by high temperatures, which from day to day vary little from the 20 to 25 degrees C annual range, and by abundant precipitation throughout the year. Light is abundant, although the wet tropics as a whole receive somewhat less light than do the subtropical deserts because they are overlain by a substantial cloud cover during most days of the year. Soils are heavily leached and low in organic content, but this generally does not deter vegetation because nutrient recycling between decaying plant remains and living vegetation, a topic taken up in Chapter 13. All in all, the wet tropics are ideally suited to plant growth, and this suitability is manifested in three principal ways: (1) high biomasses, (2) high rates of productivity, and (3) great variety of plant species.

Tropical Forest Where the rainfall regime of the wet tropics is broken by a dry period or a season of low rainfall, a less luxuriant forest may be found. Such *tropical forests* are typified by the *monsoon forests* of India and Southeast Asia. Trees tend to be smaller than those of the rainforest, and many exhibit a marked seasonal rhythm by dropping their leaves in one season, although they are evergreen, given adequate moisture. Others, however, are deciduous. Auxiliary vegetation in the form of epiphytes and vines is less abundant here than in the tropical rainforest.

In both the tropical forests and the rainforests, any opening in the canopy nurtures the development of a lower and denser *closed forest*. The absence of a continuous canopy allows light to penetrate to the ground, where it fosters thick growth of herbs, shrubs, and small trees. This formation is the *tropical jungle*, and, although Edgar Rice Burroughs may lead us to believe otherwise, the jungle is certainly not conducive to vine swinging. Although jungle formations occur naturally along openings created by stream channels and shorelines, they are most commonly related to lands cleared by humans as a part of lumbering or an agricultural practice called shifting agriculture, about which we will say more in a later chapter. It is currently estimated that the tropical and equatorial rainforests are being destroyed at a rate as high as 60,000 square miles (156,000 km^2) per year—an area about equal to the State of Georgia. Note 10 examines another aspect of vegetation—biological diversity.

Temperate forest partially cleared for agriculture.

Temperate Forest The original plant cover of the humid midlatitudes was usually continuous forest, but with a significantly lower biomass than that of the wet tropics. Three large regions of temperate forest were once found in the Northern Hemisphere: the eastern United States, Europe, and eastern China (see Figs. 10.6 and 10.7). In the forest tracts that remain today, the trees are generally 15 to 25 meters high and are associated with less abundant life in the form of vines and epiphytes than is the case in the tropics. Mature temperate forests are usually open but may support a light cover of ground plants: ferns, club mosses, and lovely flowering plants such as violets, geraniums, and spring beauty.

The trees of temperate forests may be evergreen or deciduous, needleleaf or broadleaf, and all undergo marked dormancy during the winter, when insufficient light and heat are available for growth. The most extensive tracts of the temperate forests are located in the Northern Hemisphere, where representative tree types include various kinds of oaks (*Quercus*), maples (*Acer*), and pines (*Pinus*), as well as ash (*Fraxinus*), beech (*Fagus*), walnut (*Juglans*), elm (*Ulmus*), and hickory (*Carya*). Although the needleleaf trees tend to be more abundant on the northern side of this forest zone, where they are often mixed with stands of broadleaf trees, they may also occur in appreciable numbers on the subtropical side. Ground plants, vines, and epiphytes are more abundant in the subtropics, and in general the biomass is significantly greater there than it is farther north.

Temperate forests are often comprised of many stands of different ages and tree types that together give the forest a patchwork character. The stands may consist of a single species, such as white pine, or of two or three species, such as oaks and hickory. The origins of these stands appear to be tied both to differences in site conditions (soil, drainage, slope, and microclimate) and to past disturbance from fires, windstorms, land use, and disease.

For the most part, the temperate forests of Asia, Europe, and North America have been destroyed or drastically altered over the past several millennia by land use, lumbering, and related activities. For example, China's forests were all but totally obliterated by A.D. 1000, and records show that England was practically treeless by 1700 or so. These and many other midlatitude countries have instituted reforestation and forest management programs. Managed forests, which are of natural origins as opposed to planted, are often distinguishable from unmanaged (natural) ones because the former are often closed, a result of selective cutting practices that encourage multiple-storied woods, and usually consist of several species. Planted forests show unnatural uniformity of tree size and age, and are easily detected by their open structure, uniform spacing between trees, and regular planimetric (map) patterns.

Boreal Forest Poleward of about 45 to 50 degrees north latitude, the mixed forests of the midlatitudes give way to more homogeneous subarctic forests, where only three or four tree species may dominate extensive forest stands. Conifers such as spruce (*Abies*) and tamarack are the principal trees, but a few hardy broadleafs, such as birch (*Betula*) and tag alder (*Ulnus*), are also common in the great boreal forests of Canada and the Soviet Union. Boreal forests are also known widely by their Russian name, *taiga*. (See Figure 10.6, color plate 6.)

On their northern boundaries, the boreal forests grade into the treeless tundra. Virtually all borders between major vegetational formations are broad transition zones, and the transition zone here is one in which the characterictic tundra cover is interspersed with patches of trees, many of which exhibit dwarfism owing to the stressful arctic environment (see Note 9). The same type of transition also appears on midlatitude mountains around an elevation of 3000 meters, where the forest grades into the grassy Alpine meadows.

West Coast Forest One notable exception to the humid midlatitude forests described previously is the West Coast forest, which is often composed of very large trees such as the redwood (*Sequoia*), fir (*Pseudotsuga*), and pine (*Pinus*) of the American Northwest. Here high humidity, relatively abundant precipitation, and a moderate temperature regime that precludes a long dormant season combine to create an environment that has nourished the largest trees on earth, some of which exceed a height of 100 meters. In locations where the West Coast climate yields especially heavy rainfall, say, more than 300 cm per year, the forests may be extraordinarily dense and harbor dense growths of ground mosses, epiphytes, and related plants. Such forests are often called *midlatitude rainforests*, and they are limited principally to the marine west coast climatic zone of North America.

Sclerophyll Forest Equatorward from the West Coast forests the climate grows drier, and the trees become much shorter and more widely spaced. This is the sclerophyll forest, or hardwood evergreen forest, found in areas of Mediterranean climate in both hemispheres. The sclerophyll forests' canopy gener-

Boreal forest.

West Coast forest.

ally covers only 25 to 60 percent of the terrain; in many places it is more like a tropical savanna than a forest (see Figs. 10.6 and 10.7).

The sparseness of the sclerophyll forest is attributable to two factors. The first is the severe moisture stress of the warm, rainless Mediterranean summer, which has limited the sclerophyll forest to a light cover of drought-resistant trees. These include many types of oaks, notably live oaks (*Quercus spp.*), cork oak (*Quercus suber*), and white oak (*Quercus lobata*), several species of pines, and numerous shrub-sized trees, such as wild lilac (*Ceanothus sp.*) and olive (*Olea spp.*). The second factor is the long tenure of human settlement in Mediterranean regions, especially in the Old World. Where more luxuriant forests may once have stood, fires, agriculture, grazing, and wood harvesting long ago destroyed them, and the cover we see today is probably a very poor reflection of the past. This appears to be especially so in the case of the scrub vegetation of Mediterranean Europe and southern California. In Europe it is called *maquis* (or *macchia* or *garique*) and consists of dense shrubby thickets; in California it consists of dwarf forests of oaks mixed with shrubs and is called *chaparral.*

Savanna Biome

Between the forests and the open grasslands is a biochore of grass and forbs, mixed with a light cover of trees and shrubs. This is the *savanna*, and although strictly speaking the word *savanna* refers to a tropical vegetation, this biome also includes a similar formation in the midlatitudes called *parkland.* In general, the abundance of savanna grasses results from the harsh drought conditions of the winter season, which limit the establishment of a forest cover. But there are likely several other contributing factors as well, in particular fires set by humans and forest clearing for agriculture.

Tropical Savanna We have come to stereotype this formation by the tall-grass savanna of Africa (see Figs. 10.6 and 10.7). Here extensive areas of grass such as elephant grass, which may reach heights of 3 m, are interrupted by scattered individual trees or groves of trees. The trees are umbrella-shaped and, like the grasses, tend to flourish in the wet season.

The grasses provide excellent pasture for grazing animals, such as wildebeest and zebra, which are found in great herds across the African savanna. In contrast, during the dry season the grasses turn brown, and the trees, which are drought-tolerant, may lose part or all of their foliage. The herds of grazing animals retreat to wooded areas, waterholes, and diverse rocky knolls to wait out the dry season. This is also the time when herders may burn large tracts of the savanna in an effort to improve grazing quality and to enlarge the grasslands at the expense of nearby forests.

Thornbush and Tropical Scrub In many regions with a wet-dry tropical climate, short, thorny trees and shrubs are found in place of the classical savanna trees and grasses. This vegetation may form a nearly continuous cover, thereby eliminating most grasses, or it may be broken, allowing grasses and other herbs to fill the intervening space (see Figs. 10.6 and 10.7). The most impoverished thornbush formations are those where only barren soil is present between the woody plants. Generally, the thornbush and scrub savannas are thought to be responses to longer and more intensive dry seasons, but there are undoubtedly many other contributing factors in various regions, including fire and cultivation. Many different regional names are given to this formation; for example, in northeastern Brazil it is called the *caatinqa,* and in South Africa it is called the *dornveld.*

Parkland Parkland can be described as prairie that is broken by patches and ribbons of broadleaf trees. It was apparently fairly common in Kentucky and Illinois when these areas were settled in the early 1800s. In both states, most of the forest groves were subsequently destroyed as the land was converted to agriculture. A parkland formation, which the English call *deerpark*, is found in southern England, northwest France, and other regions of Western Europe where pasture land is partitioned by hedgerows and interspersed with small groves of deciduous trees and shrubs. In a general sense, the abandoned farmfields in the eastern half of the United States also qualify as parkland, since they too are a mixture of grasses, forbs, trees, and shrubs. An extensive belt of parkland is also formed in Canada where the grasslands of the northern Great Plains give way to the boreal forest. See Note 14 for additional information on this area.

Grassland Biome

The great regions of grasslands in the world are located in the midlatitudes between the forests and the deserts. Summer drought is strong enough to prohibit the growth of trees and shrubs, but not severe enough to prohibit the growth of abundant grasses and forbs.

Prairie Prairie is found on the humid side of the grassland biome where the annual moisture balance is just about even. Prairie consists mainly of tall grasses that grow over a lighter cover of smaller forbs. Trees and shrubs are not absent altogether, but are limited to depressions such as stream valleys and floodplains, where water is more plentiful. In North America, grasses such as big blue stem (*Andropogon gerardi*) and little blue stem (*Andropogon scoparius*) were the predominant plants in the famous prairie of Illinois, Iowa, and adjacent states. Virtually all of this original cover, however, has been replaced by agriculture. This is also the case with the other prairies of the world, for example, the *Pampa* of Argentina and the *Puszta* of Hungary. (See Figure 10.6, color plate 6.)

Steppe Farther toward the desert, the moisture balance grows poorer and the grass cover shorter and thinner. This formation is sometimes termed short-grass prairie, but it is more widely known by the Russian word *steppe*. In North America it consists of grasses such as buffalo grass (*Buchloe dactyloides*) and black gramma grass (*Bouteloua eriopoda*), which tend to grow in bunches that are often separated by barren soil. The most extensive areas of short-grass prairie are found in the Soviet Union and the Great Plains of North America.

Desert Biome

Desert vegetation is one of the lightest plant covers on earth and can generally be divided into that having

Short-grass prairie (steppe) on the American High Plains.

virtually no plants and that having a conspicuous cover of shrubs and herbs (see Figs. 10.6 and 10.7).

Dry Desert In the harshest deserts, such as the Atacama of northern Chile, where in places measurable precipitation has never been recorded, the landscape is virtually barren of plants. Only in select microenvironments are a few plants able to survive, but they are very small and isolated. Moreover, their survival is dependent on either special physiological traits, such as the capacity for water storage, or on exceptionally deep root systems to draw on deep sources of water. But even with the special adaptations, plant growth and reproduction in the desert are exceptionally slow.

In addition to severe drought, erosional processes can also restrict plant growth in dry deserts. Typically, desert surfaces are very active, owing to the absence of a strong plant cover to hold the soil in place. Running water is the most powerful erosion agent and during infrequent periods of runoff can move massive amounts of material, including plants. Wind is also an effective erosional agent in deserts. Active sand dunes, such as those of the Empty Quarter of Saudi Arabia, shift so rapidly that they preclude the establishment of plants altogether. In other areas severe wind stress and the absence of soil over the bedrock may be important limiting factors.

Shrub Desert At the other extreme are deserts such as those in the American West, which the respected plant geographer Nicholas Polunin calls *near-deserts*. These are characterized by diverse plant forms ranging in some places from sahuaro cactus (*Carnegia gigantea*), which reaches heights of 10 to 15 m, and various shrubs at heights of 1 to 2 m, to tiny forbs barely 3 or 4 cm above the ground. Together they may cover 10 to 20 percent of the ground, and in special locales, for example, along dry river beds, where moisture is more plentiful, the coverage may be substantially greater. Although xerophytic vegetation (plants specially adapted to dry conditions) represents one of the most interesting collections of plants in the world, it is not highly diversified floristically. Compared with forest formations at the same latitude, there are fewer types of plants in most deserts. Of these, cacti are probably the most famous, but they are overwhelmingly limited to the New World. (Cacti were once exclusive to North America, but prickly pear cactus are now reported in parts of North Africa,

American shrub desert, Arizona.

a result of human introduction.) Other plant groups prevalent in deserts are the *Euphorbiales*, which often resemble cacti in their thorny, fleshy bodies, and the lily family (*Liliceae*), which includes many desert palms and small succulents (plants with water storage capability).

Tundra Biome

The tundra biome is found beyond the thermal limits of tree growth in the high latitude and high mountain zones. Like the desert biome, the vegetation is generally small in stature, irregularly distributed geographically, and limited to very low growth rates. Most of the tundra biome is underlain by permafrost.

Tundra and Alpine Meadow Beyond the boreal forests of the subarctic and above the tree-line in mountainous areas, there may occur a grassy, treeless, prairielike formation called, respectively, tundra and alpine meadow. Here low temperature rather than scant moisture prohibits establishment of a forest cover. Winters are fiercely cold, and summers are short (1–2 months) and cool. Short grasses, such as cottongrass (*Eriophorum*) and arctic meadow grass (*Poa arctica*), and forbs are abundant. In sites protected from the harshest weather and ground conditions, shrubs and dwarf trees can sometimes be found. Because of the geographic diversity and the dynamic character of arctic and alpine environments, grassy tundra and alpine meadow have localized distributions in many areas, particularly in mountain regions.

Tundra vegetation.

Fell Fields Some polar and high mountain regions contain areas with a light cover of lichens, mosses, and flowering plants that resemble desert formations. These areas are fell fields—sites of active weathering and erosion where there is a heavy concentration of rock fragments. They may be places where frost action or deposition of rock debris is especially pronounced. Coupled with the harsh polar or high mountain climate, these sites limit plants to scattered patches, which together constitute less than 50 percent coverage. Antarctica and Greenland, as well as portions of insular Canada, Alaska, and Siberia, are occupied by fell-type landscapes.

Changing Geography of Vegetation

The distribution of vegetation appears to be controlled by many factors, chief of which are climate and people. Under natural conditions, the availability of heat and moisture probably has the greatest influence on the global patterns of vegetation. Changes in climate as well as other factors such as fire and disease, however, help to produce complex geographic patterns that make scientific correlations with heat and moisture exceedingly difficult.

Added to this is the influence of humans over the past 10,000 to 20,000 years or more. We know a good deal about the effect of human actions on vegetation in the twentieth century, but we know very little about the extent and nature of such actions in earlier periods, especially where historical records and archaeological evidence are scarce. Some scientists reason, for example, that the savanna of Africa is due largely to forest clearing and burning by earlier hunters, herders, and farmers. In the case of burning, how much was natural and how much was caused by people is mainly speculation. In North America, on the other hand, forest clearing has taken place mostly within the past 300 years. Therefore, it is possible to measure the gross changes in this biochore and to identify the causal factors. The extent of cultivated land in the world today gives a good indication of how much humans have changed the vegetative cover (Fig. 10.8).

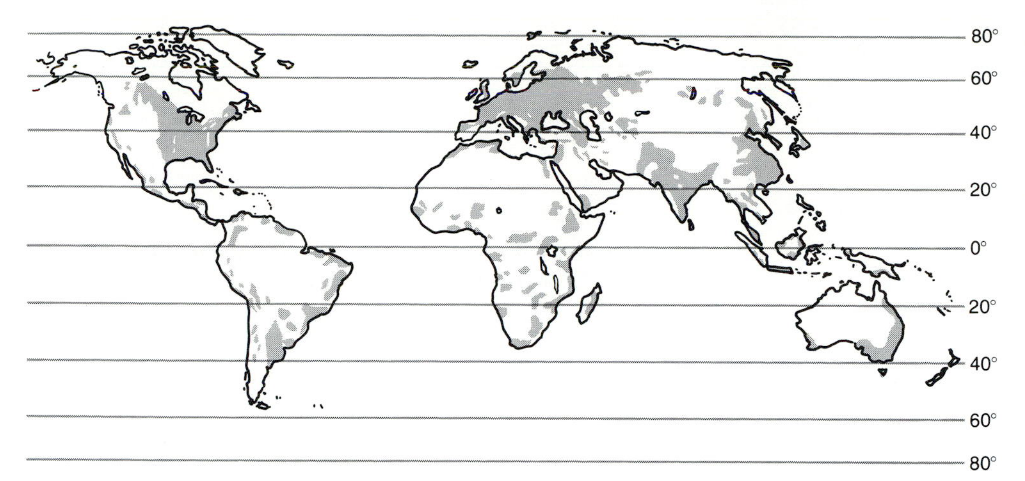

Figure 10.8 In agricultural areas, cultivated ground constitutes more than 50 percent of the land cover.

Summary

Plant geography is concerned with the composition, structure, and distribution of the higher plants, or vascular plants, and with how these interrelate with other components of the landscape such as soil and climate. There are three main groups of vascular plants: the angiosperms, which are the dominant plants in most landscapes; the gymnosperms, which still cover extensive areas in subarctic and mountain environments; and the pteridophytes, which are mainly ground plants such as ferns. Plants may be classified according to floristic, life-form, or ecological schemes, their genetic affiliation, their life-forms, or their ecological habitats. The floristic classification is the most widely used scheme in the plant sciences.

Terrestrial vegetation is grouped into five biomes: forest, savanna, grassland, desert, and tundra. The global patterns of these biomes are mainly a response to climate, but human activity has undoubtedly been important in shaping the tropical savanna and the forests and grasslands of the midlatitudes. Forest cover the world over has been greatly reduced by humans and with it has come the eradication of many species of plants, animals and microorganisms. But extensive tracts of forest still remain in the equatorial and tropical zones, the humid midlatitudes, and the subarctic. Savanna vegetation is the most extensive in the wet-dry tropics, and the largest tracts of grasslands are located in the semiarid zones of the midlatitudes. Tundra is mainly limited to the Northern Hemisphere poleward of the boreal forests. The lightest plant covers are found in the deserts and polar regions, which together occupy about 25 percent of the world's land area.

Concepts and Terms for Review

- plant geography
- vascular plants
  - angiosperms
  - gymnosperms
  - pteridophytes
- classification systems
  - floristic
  - life-form
  - ecological
- biome
- vegetation types
- formation classes
- global vegetation
- species diversity
- stature and growth rates
- human influence
- forest biome
  - equatorial and tropical rainforest
  - tropical forest
  - boreal forest
  - West Coast forest
  - sclerophyll forest
- savanna biome
  - tropical savanna
  - thornbush and tropical scrub
  - parkland
- grassland biome
  - prairie
  - steppe
- desert biome
  - dry desert
  - shrub desert
- tundra biome
  - tundra and alpine meadow
  - fell field
- biogeographical diversity
  - landscape change
  - extinction

Sources and References

Anderson, Edgar. (1956) "Man As a Maker of New Plants and New Communities." In *Man's Role in Changing the Face of the Earth.* W.L. Thomas, Jr., ed. Chicago: University of Chicago Press.

Bourliere, F., ed. (1983) *Tropical Savannas.* Amsterdam: Elsevier Science.

Canfield, Catherine. (1984) *In the Rainforest.* Alfred A. Knopf: New York.

Conniff, Richard. (1986) "Inventorying Life in a Biotic Frontier Before it Disappears." *Smithsonian* 17:6, pp. 80–90.

Kosztarab, M. (1984) "A Biological Survey of the United States." *Science*, 223:4635.

Kuchler, A.W. (1967) *Vegetation Mapping.* New York: Ronald Press.

Mather, J.R., and Yoshioka, G.A. (1968) "The Role of Climate in the Distribution of Vegetation." *Annals of the Association of American Geographers* 58, 1: pp. 29–41.

Nitecki, M.H., ed. (1984) *Extinctions.* Chicago: University of Chicago Press.

Ovington, J.D., ed. (1983) *Temperate Broadleaved Evergreen Forests.* Amsterdam: Elsevier Science.

Polunin, N. (1967) *Introduction to Plant Geography.* London: Longmans.

Raup, Hugh, M. (1942) "Trends in the Development of Geographic Botany." *Annals of the Association of American Geographers* 32, 4: pp. 319–354.

United Nations FAO. (1982) "Tropical Forest Resources." *Forestry Paper 30*, Rome.

Wilson, E.O. (1985) "The Biological Diversity Crisis: A Challenge to Science." *Issue in Science and Technology*, 2: 1.

Chapter 11

Plant Processes in the Ecosystem

Through photosynthesis plants have evolved a means of combining certain forms of energy and materials at the earth's surface to produce energy in the form of organic matter. The rate of production, however, varies sharply with the supply of these essential ingredients. What is the geographic character of the production processes over land and water? How does this system of organic energy relate to the larger system of energy flow to the earth's surface?

Now that we have looked at some basic types of plants and vegetation, we can begin to examine the processes of plant growth and the production of organic matter. Why should physical geographers be concerned with the biological processes of plant growth? There are several good reasons. First, plants represent an important architectural element of the landscape which influences, for example, the flows of solar radiation, wind, and rain. If we are to understand climate and hydrology near the ground, we need to know about how plants grow; why some are large and some are small; and why they shed and grow their leaves. Second, through these processes plants produce organic materials and introduce these materials to the landscape, thereby influencing the chemical composition of soil and the thermal and hydrologic properties of the ground. Third, plants do work as they grow—by moving mass, for example, as they lift water from the soil and discharge it into the atmosphere, thereby increasing the rate of soil-moisture loss over what it would be without vegetation. Fourth, plants are dependent on the physical environment for their basic growth resources, but unless we understand the processes of growth and development we cannot begin to discover the mechanisms that explain the nature of the plant/environment relations and thus the basic controls on plant distributions. Discovering explanations for the distribution of phenomena on the earth's surface is, after all, much of what geography is about.

To understand the role of plants in the landscape, we must consider two sets of factors: (1) the processes by which plants develop, maintain themselves, and grow; and (2) the forces and processes of the environment that influence these plant processes. The environmental factors fall into two categories: (a) those that provide the resources essential to plant life and growth; and (b) those that are not essential to plant life but that limit plant growth and survival. The factors in the second category are called *disturbances;* they are processes such as fire and drought, and we will examine them in the next chapter. In this chapter we will examine the elementary physiological processes of the higher plants and the environmental factors that influence them. We are interested in understanding how the plant is related to the environment via the processes that supply it with water, nutrients, and energy for development and growth.

Photosynthesis, *respiration*, and *transpiration* are the essential physiological processes of green plants. *Photosynthesis* is the process by which green plants convert light energy into chemical energy in the form of plant materials. *Respiration* is the set of biochemical processes that oxidize foods (sugars and other organic compounds) and thereby release the energy that maintains the *metabolism* of the plant. *Transpiration* is the process by which the plant releases water.

TRANSPIRATION AND MOISTURE FLOW

In living plants, water flows from the roots through the main body of the plant and into the inner tissue of plant foliage, called *mesophyll. Transpiration* takes place as the water is released as vapor from the *stomata,* or pores, of leaves (Fig. 11.1). Transpiration is the principal mechanism in maintaining the internal water balance of the plant as well as an important regulator of leaf temperature. Because this process involves the transformation of liquid water into vapor, heat is taken from the leaf in transpiration.

Water uptake by the plant begins when moisture is drawn from the soil into the filamentous plant roots. Within the vascular tissues of the plant, the water moves upward to the foliage. The processes that move the water are exceedingly efficient, especially in large trees, where water is moved to heights of 75 meters or more above the ground. Once in the foliage, a source of energy is needed to drive the vaporization process and transfer the moisture to the atmosphere. This energy is provided by heat in the leaf and in the air immediately around the leaf.

Influences of the Atmosphere

Solar radiation influences transpiration in two ways. First, it initiates photosynthesis, which results in the opening of the stomata; second, it heats the leaf sur-

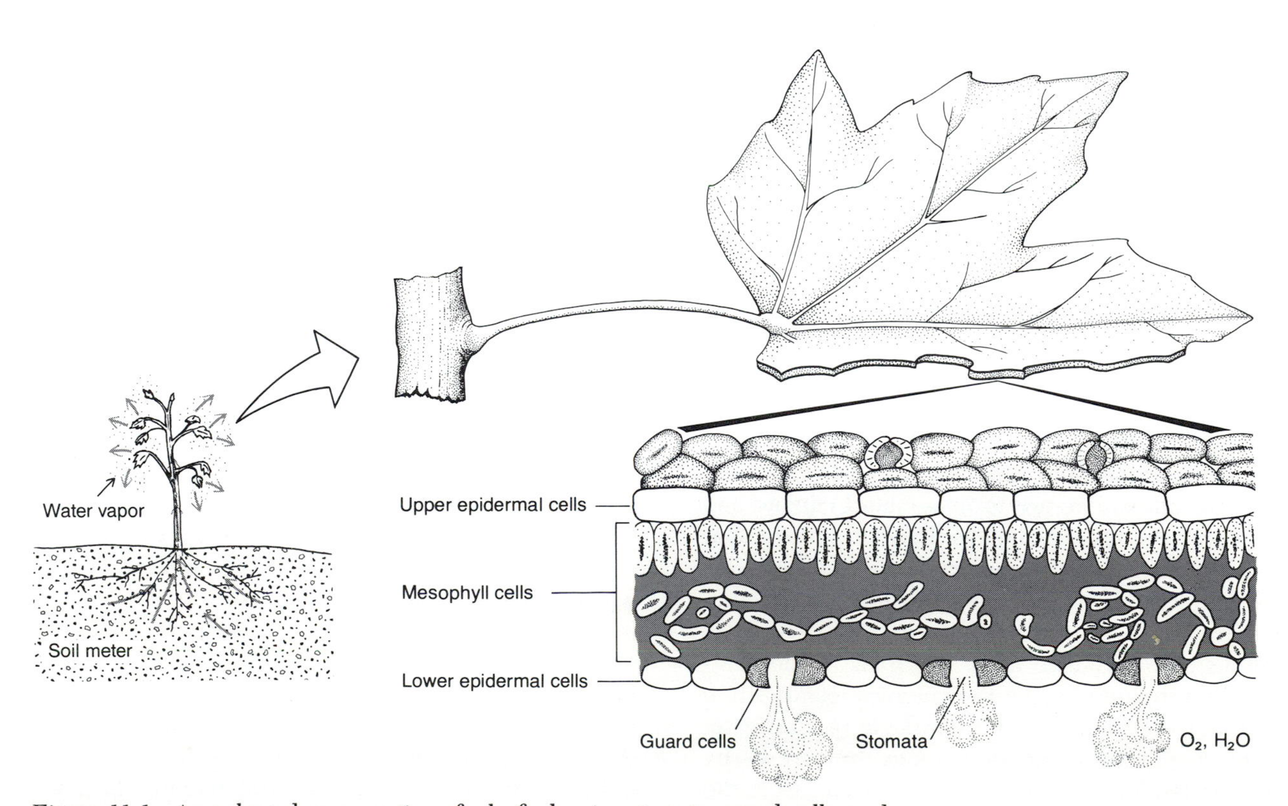

Figure 11.1 An enlarged cross-section of a leaf, showing stomata, guard cells, and related features.

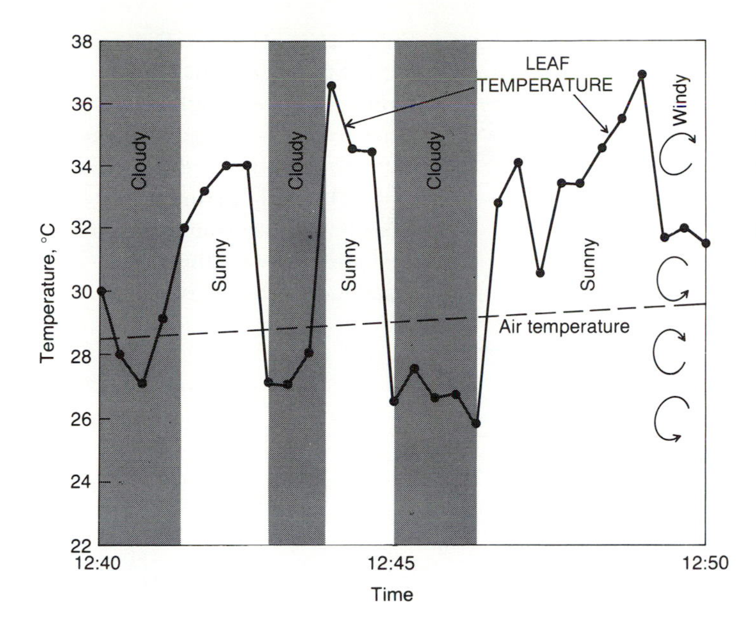

Figure 11.2 Temperature variations of the leaf surface of a poplar tree during sunny, cloudy, and windy conditions in Colorado. Note that the temperature fluctuates far above and below air temperature. In the absence of direct-beam sunlight, leaf temperatures lower than air temperature are possible because heat is being released from the leaf through transpiration.

face, inducing further stomatal enlargement and vaporization of leaf water. Conversely, as transpiration increases, latent-heat flow increases, lowering the temperature of the leaf and inducing some reduction in the size of the stomatal openings, followed, in turn, by a decline in transpiration. Sensible-heat loss resulting from wind generally produces the same effects. Figure 11.2 shows the changes in leaf temperature associated with sunny, cloudy, and windy conditions during a 10-minute midday period on a summer day.

In addition to leaf temperature, transpiration is influenced by (1) air temperature, (2) the speed of wind moving over the leaf, and (3) atmospheric humidity. As with evaporation from the soil, it is the combination of low humidity, high temperature, and fast wind that produces high transpiration rates in plants.

Influences of the Soil

Although the atmosphere supplies the land with water, it is the soil that supplies the plants with water. The soil receives and stores precipitation, giving it up to plants in limited amounts over periods of time much longer than the duration of rainfalls. Most land plants depend on a type of soil water, called *capillary water*, which is held among the soil particles. The amount of capillary moisture held in the soil depends not only on how much water is supplied by precipitation, but also on the soil *field capacity*. Field capacity is the maximum amount of capillary water that can be held by a soil, and it is controlled mainly by soil texture and organic content. The depth to which plants draw on soil moisture varies with the structure of the root systems, some of which extend down 10 m or more. The great majority of plants are shallow-rooted, however, drawing most of their water from the upper 20 to 100 cm of the soil (see Fig. 7.14).

Capillary moisture moves through the soil by molecular cohesion from spots of high-moisture content to spots of low-moisture content. During the growing season, when the upper soil dries out, the movement is usually upward, toward the *soil root zone*. However, the rate of movement is seldom fast enough to meet the water needs of the plants, especially in the middle and late summer when plant demands are high. The resultant differential (represented by the capillary lag in Figure 11.3) between the upflow of capillary water to the plant roots and the water demands of the plants is called the *soil-moisture deficit*.

Two conditions can produce large soil-moisture deficits: (1) low soil capillary, that is, low capacity to move water via capillary processes, owing to coarse soil texture; and/or (2) a small soil-moisture reserve in the soil root zone (Fig. 11.3). Most plants can tolerate weak moisture deficits without loss of vigor. However, strong deficits can produce *moisture tension* within the plant. As a result, respiration declines and the leaf mesophyll loses water. When this happens, water pressure within the leaf declines and the leaf begins to contract, or lose *turgor*, which gives it a puckered, droopy appearance.

There is, of course, a considerable range of tolerance among different species to drought and moisture deficits. For most plants—with the exception of the

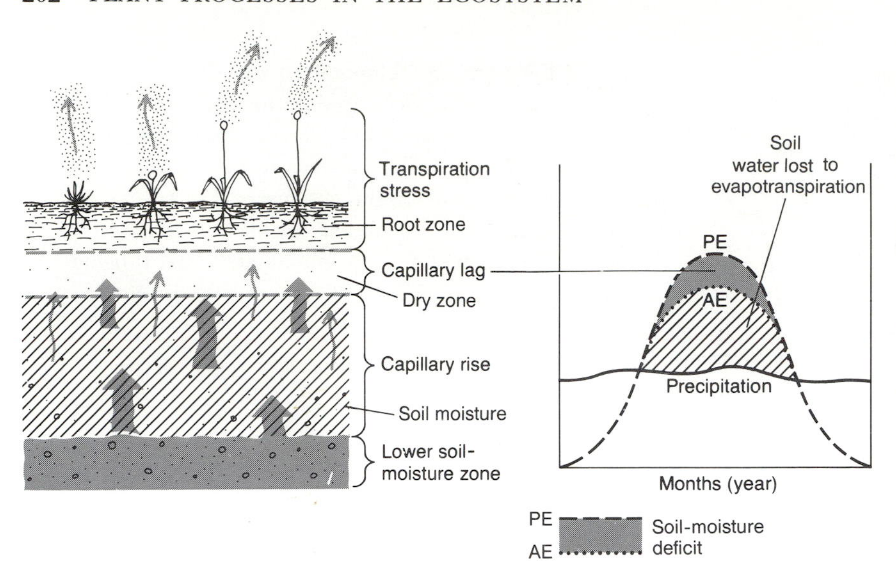

Figure 11.3 Moisture flow from the soil through plants. Under conditions of low moisture, a gap (capillary lag) forms between the upflow of moisture and the plant roots. This is shown in the moisture-balance graphs as the difference between the actual evapotranspiration curve (AE) and the potential evapotranspiration curve (PE).

xerophytes which are specially adapted for drought—if soil moisture falls below a critical level, the plant wilts. If this condition is prolonged, the plant's cells will be damaged and it may ultimately die. The particular level of soil-moisture tension at which wilting occurs, called the *wilting point*, is higher for the xerophytes, whose physiologies are adjusted to extended periods of drought. (See Chapter 13 for a detailed discussion of soil moisture, drought, and vegetation.)

RESPIRATION AND PHOTOSYNTHESIS

We have already mentioned one important aspect of heat in plant physiology: its relationship to transpiration. Even more fundamental, however, is the internal temperature of the plant organs, for heat is a primary control on the essential biochemical processes of photosynthesis and respiration.

The Role of Heat

Although many plants can survive freezing temperatures, in most plants very little photosynthesis and respiration occurs below 10°C. At temperatures between freezing and 10°C, biochemical activity is negligible regardless of the availability of light, moisture, and carbon dioxide. But the rates of most biochemical reactions double with each 10-degree increment until some optimum temperature is reached, generally around 30°C. At this temperature the rate of biochemical activity in the plant is at a maximum (provided, of course, that all other necessary forms and amounts of energy are available). Above 30°C photosynthesis and respiration decline, and at some high temperature—around 40°C for many plants—the organism enters a stage of severe heat stress and suffers physiological damage (Fig. 11.4).

The influence of temperature on vegetation (as opposed to individual plants) takes many different forms. Among them is the influence on size and growth rates of vegetation. In the maritime arctic and subarctic regions, for example, coastal vegetation is often stunted even though winter temperatures never get extremely low. The reason for this dwarfism has to do not with cold-season temperatures but with warm-season temperatures. Because of the maritime influence, summer temperatures in these latitudes are held down, never far exceeding the critical 10°C level. Inland, beyond the influence of oceans, summer temperatures reach higher levels (15 to 20°C) and induce much larger and faster growing vegetation despite the short growing season (usually 30 to 60 days) and extremely cold winters (Fig. 11.5).

Photosynthesis and the Principle of Limiting Factors

Photosynthesis is one of the truly astonishing natural processes, for it not only converts radiant energy into chemical energy, but also fixes that energy in the form of organic compounds that do not spontaneously break down and can thus be stored until needed by the plant. This is the essence of photosynthesis, which enables plants to survive and even grow during periods of little or no light.

Photosynthesis takes place only in the green parts of plants in the presence of light, water, carbon dioxide, and heat. Basically, carbon, oxygen, and hydrogen—the elements that make up carbon dioxide and water (two atmospheric gases)—are combined with

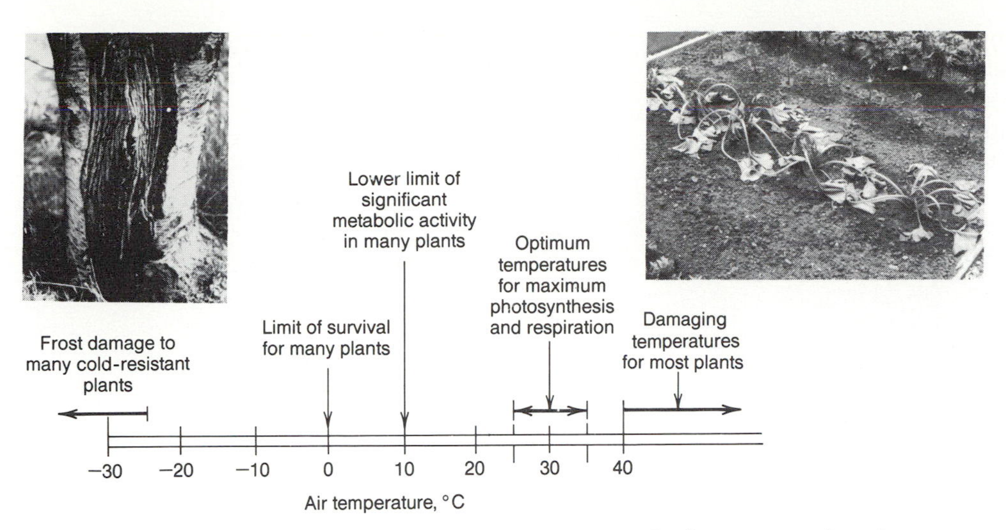

Figure 11.4 Some critical temperatures for plants in general. Photo: Severe heat loss can also cause damage to plants, such as the frost splitting of this tree trunk. Photo: Excessive heat can cause heat stress, which may damage or kill plants.

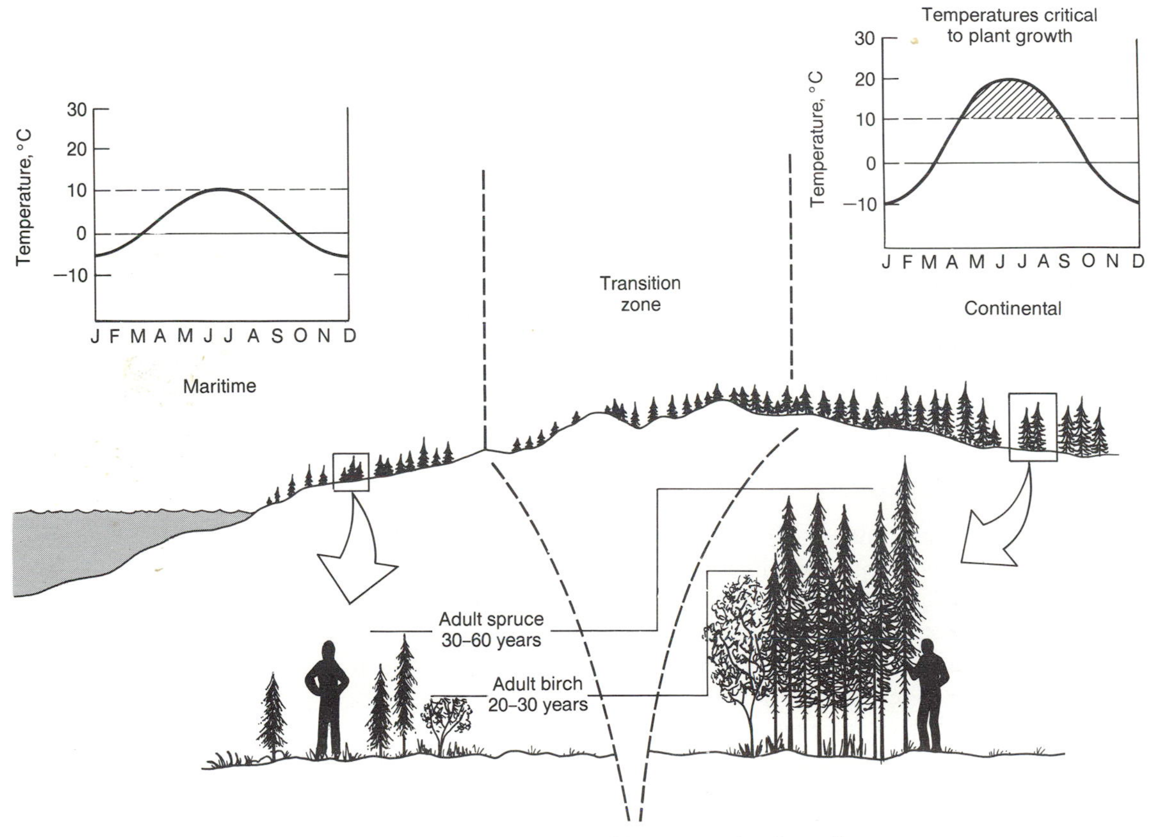

Figure 11.5 The contrasts in the size of trees of comparable ages under the influence of the maritime and continental thermal regimes in the subarctic and arctic zones: in maritime areas, the trees are severely stunted.

light energy, which is absorbed by the chlorophyll in the photoactive cells (called chloroplasts) of plant leaves. Molecular processes rearrange these three elements to produce plant materials (organic compounds) in the form of carbohydrates (sugars and starch) and at the same time release oxygen. The oxygen and water vapor are released into the atmosphere through the stomata (Fig. 11.1). Photosynthesis can be simplified into the following formula:

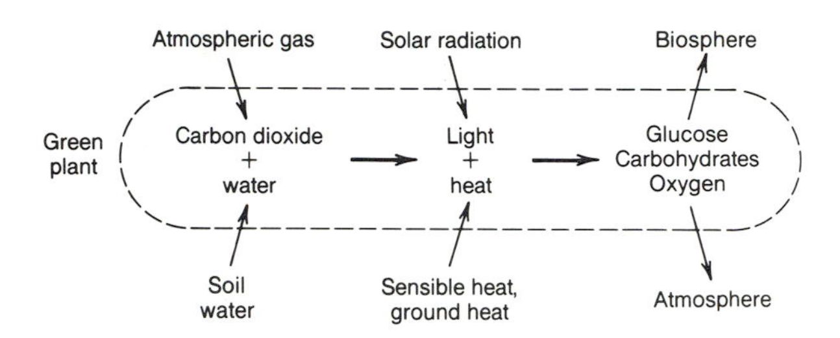

Although light is the preeminent form of energy necessary in photosynthesis, scientific analysis shows that heat, carbon dioxide, and water acting *together* with solar radiation regulate this process. This fact was discovered by plant physiologists in the early part of this century, and it formed the basis for the *principle of limiting factors:* when a process is influenced by several factors, the highest intensity or rate it can attain is controlled by the factor that is in shortest supply. Thus, as we mentioned earlier, without enough heat to maintain a temperature above 10°C, photosynthesis will be minimal no matter how much light, water, and carbon dioxide are available. The same holds true for carbon dioxide. Where it is reduced to levels below the atmospheric average of 0.03 percent, photosynthesis declines markedly. And, of course, the same is true for water. With this principle in mind, let us examine photosynthesis further.

Photosynthesis and Solar Radiation

Photosynthesis begins with the absorption of light particles, called *photons*, by the chloroplasts, which initiates photochemical activity, the first phase of photosynthesis. The rate of photochemical activity depends on, among other things, the wavelength of the radiation, the intensity of radiation, and the duration of the daily light period called the *photoperiod* (Fig. 11.6).

The rate of photosynthesis usually increases as the intensity of incoming shortwave radiation increases, and vice versa. As solar radiation varies with cloud cover, for example, so does photosynthesis (Fig. 11.7).

1 Longwave incoming (from sky)
2 Shortwave incoming (from sky)
3 Shortwave outgoing (reflected)
4 Longwave outgoing (reradiated)
5 Shortwave incoming (reflected from ground)
6 Longwave outgoing (from leaf to ground)
7 Longwave incoming (to leaf from ground)

Figure 11.6 The radiation balance of a leaf includes exchange with both atmosphere above and the ground below.

However, for any plant to achieve a peak rate of photosynthesis, an optimum intensity of light is necessary. Although we normally associate a reduction in the intensity of light or the length of the photoperiod with a decline in photosynthesis, light intensities surpassing the optimum range of a plant can also inhibit pho-

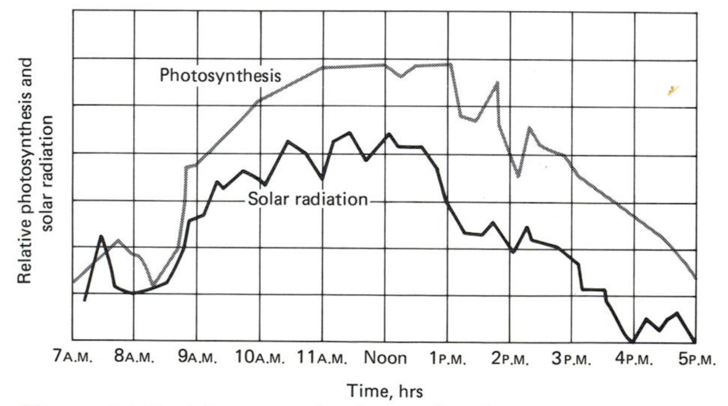

Figure 11.7 The typical relationship between incoming shortwave radiation (S_i) and the rate of photosynthesis (measured by the rate of CO_2 taken in by the plant per hour). The S_i variations are due to variations in cloud cover.

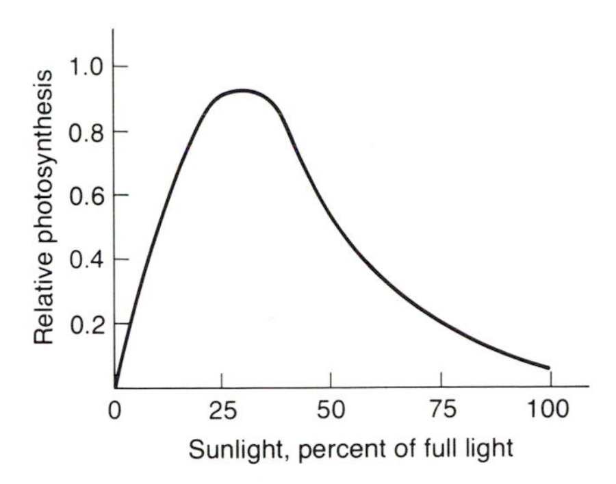

Figure 11.8 The relationship between light intensity and relative photosynthesis in marine phytoplankton. Note that the intensity of photosynthesis is greatest at 25 percent of full light.

tosynthesis. For example, Figure 11.8 shows the relationship between light intensity and relative photosynthesis in marine phytoplankton. Note that photosynthesis for these tiny plants is greatest at an intensity of about 25 percent of full sunlight and decreases with higher and lower light intensities.

PLANT GROWTH AND PRODUCTION

Individual plants are small energy systems. They absorb energy from the environment and convert a portion of it to chemical (organic) forms that can be stored in the landscape within the organic molecules of living or dead plant and animal tissue. These organic molecules can, in turn, be taken up by other plants (and animals) and broken down by them to release the energy and nutrients (such as nitrogen and phosphorus). Thus, individual plants of the same and different species are linked together, along with microorganisms, animals, and components of the physical environment, to form larger energy systems, or *ecosystems.* The output of organic matter by plants as a result of photosynthesis is termed *productivity* or *primary productivity,* and it is the source of energy that drives the ecosystems.

Net Photosynthesis

Plants grow when photosynthesis produces more glucose and starch than are needed to maintain the biochemical processes of respiration. In other words, in order to create an energy balance favorable for growth, the plant must manufacture materials faster than they can be consumed in respiration. Unlike photosynthesis, which is limited by the photoperiod, respiration is unaffected by light conditions and thus continues throughout the diurnal period. Therefore, the *rate* of photosynthesis must greatly exceed the *rate* of respiration. By comparing the rate of *gross* or *total photosynthesis* (plant materials produced) and the rate of respiration (plant materials consumed), we derive *net photosynthesis,* which is simply an expression of a plant's productivity:

Net photosynthesis
= Total photosynthesis − Respiration

If net photosynthesis is positive, total photosynthesis exceeds respiration, and plant material is available for plant growth. If net photosynthesis is zero, respiration is consuming all plant materials but still maintaining the plant's metabolism. If net photosynthesis is negative, too little plant material is available for respiration; respiration declines, and the plant loses weight. For many plants, these three states of net photosynthesis usually occur in early summer, mid to late summer, and fall, respectively. (In winter, photosynthesis ceases for many plants, and so net photosynthesis is negative—but only very slightly so because respiration too has virtually ceased. Plants survive this condition because their very low rates of respiration are fueled by the food energy they stored during periods of positive net photosynthesis.)

Plant Production

Although it is important to understand the productivity of an individual plant, in physical geography our chief concern is with the productivity of large complexes of plants, or vegetation, in different environments. Production, or *net primary production* as it is also termed, is measured in terms of the grams or kilograms of organic matter added to a ground area (or area of water in the case of aquatic environments) of one square meter per day or year. In general, where light, heat, moisture, and carbon dioxide are present in large and dependable quantities, plant life is most abundant and productive. From the maps of the global distributions of solar radiation (Fig. 3.14) and average rainfall (Fig. 7.7), and our general knowledge of the thermal character of the major climatic zones (Fig. 8.6, color plate 5), we can put together an overall picture of the global distribution pattern of the climatic factors that limit vegetative productivity (Fig. 11.9).

The equatorial zone and the tropics are the least limited in terms of heat, moisture, and light. Moreover, the supply of each of these resources in this vast

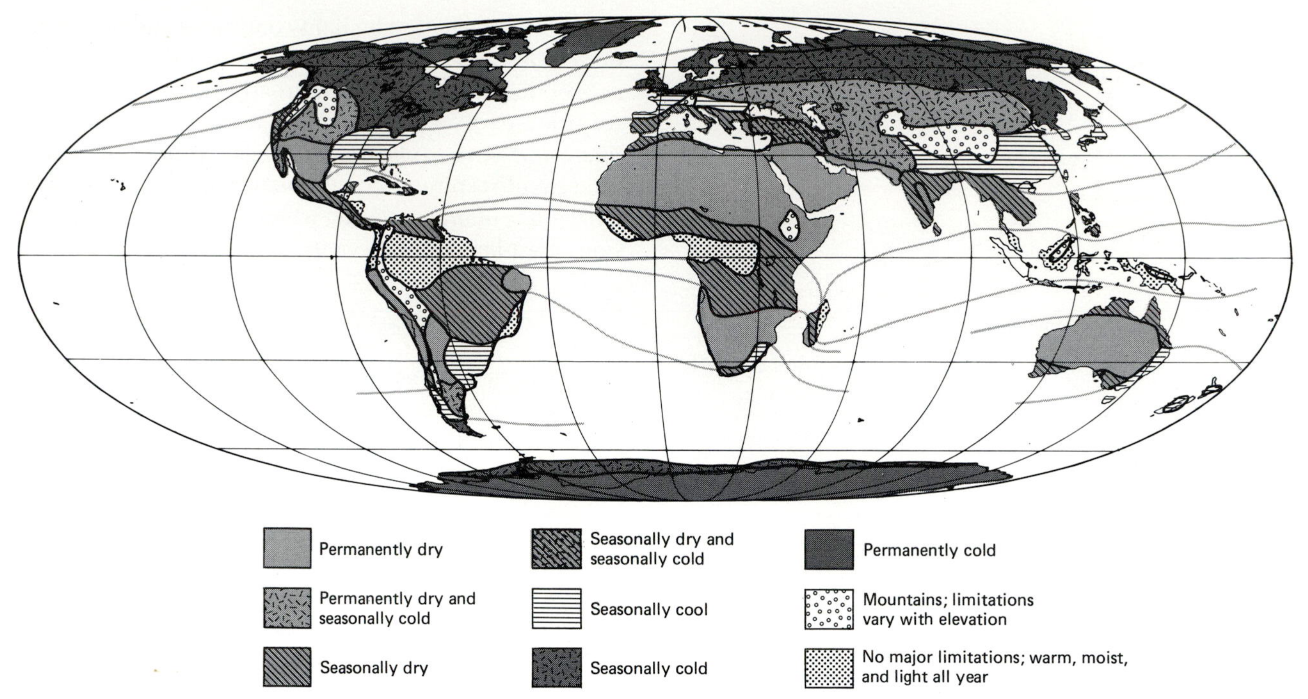

Figure 11.9 The principal climatic limiting factors that govern plant productivity worldwide.

TABLE 11.1 VEGETATIVE PRODUCTIVITY IN THE MAJOR BIOCLIMATIC ZONES

Vegetation	Average Year Productivity, g/m²	Range of Productivity, g/m² · yr	Geographic Coverage, Millions of Km²	Total Productivity per Year, Millions of Tons	Percentage of World Productivity
Tropical rainforest	2200	1000–3500	17.0	37.4	(32%)
Tropical forest	1600	1000–2500	7.5	12.0	(10%)
Temperate evergreen forest	1300	600–2500	5.0	6.5	(5.5%)
Temperate deciduous forest	1200	600–2500	7.0	8.4	(7%)
Boreal forest	800	400–2000	12.0	9.6	(8%)
Woodland and shrubland	700	250–1200	8.5	6.0	(5%)
Savanna	900	200–2000	15.0	13.5	(11.5%)
Temperate grassland	600	200–1500	9.0	5.4	(4.5%)
Tundra and alpine	140	10–400	8.0	1.1	(1%)
Desert and semiarid scrub	90	10–250	18.0	1.6	(1.5%)
Extreme desert, glacial and fell	3	0–10	24.0	0.07	(.05%)
Cultivated land	650	100–4000	14.0	9.1	(8%)
Swamp and marsh	3000	800–6000	2.0	6.0	(5%)
Lake and stream	400	100–1500	2.0	0.8	(.7%)
Total	782		149	117.5	

Source: From R.H. Whittaker and G.E. Likens, "The Biosphere and Man," in *Primary Production of the Biosphere,* 1975. Used by permission of Springer-Verlag, New York.

region is quite constant; for example, the annual variability in rainfall is generally less than 15 percent from the average, as compared with more than 40 percent for the major deserts. With such ample and dependable supplies of energy, it is understandable that vegetative productivity should be higher in the intertropics than in any other large region on earth. And measurements bear this out: the equatorial and tropical rainforests produce, on the average, 2200 grams (almost 5 pounds) of organic matter per square meter per year—a rate 100 or more times greater than that for many deserts (Table 11.1).

Poleward from the tropics, one or more of the four resources declines in availability, dependability, or both. The deserts, for example, are often amply supplied with light and heat, but they are severely deficient in moisture, and this deficiency limits photosynthesis, with the result that growth rates are very low (Table 11.1). Survival itself is possible only for plants, such as xerophytes, which are especially adapted to aridity.

Another severely limiting climate is the polar climate, where low temperatures induce appreciable levels of biochemical activity for plant growth. Water, too, is a limiting factor here: water, though abundant in this environment, is in the frozen state most of the time and is not available to plants. Light is plentiful during the summer but severely limited during the winter. Except for a few lower plants, such as mosses, lichens, and liverworts, much of the land area in this climatic zone is devoid of a plant cover, and productivity amounts to only several grams per year on the average (Table 11.1).

Although the average productivity values given in Table 11.1 illustrate the general relationship of productivity to regional climate, it is also important to recognize the variation in productivity *within* the individual bioclimatic regions. This is especially significant from a geographic standpoint because it tells us something about the spatial diversity of different landscapes. Notice, for example, that productivity for desert and semiarid scrub ranges from 10 to 250 grams per m^2 per year, whereas the ratio between the high and low values (1000 to 3500 grams per m^2 per year) for tropical rainforest is much smaller. The explanation for this difference is that the harsh environments tend to produce highly diverse landscapes in the way of soils, surface hydrology, and microclimate, which gives rise to wonderfully diverse plant distributions and patterns of productivity (Fig. 11.10). In contrast,

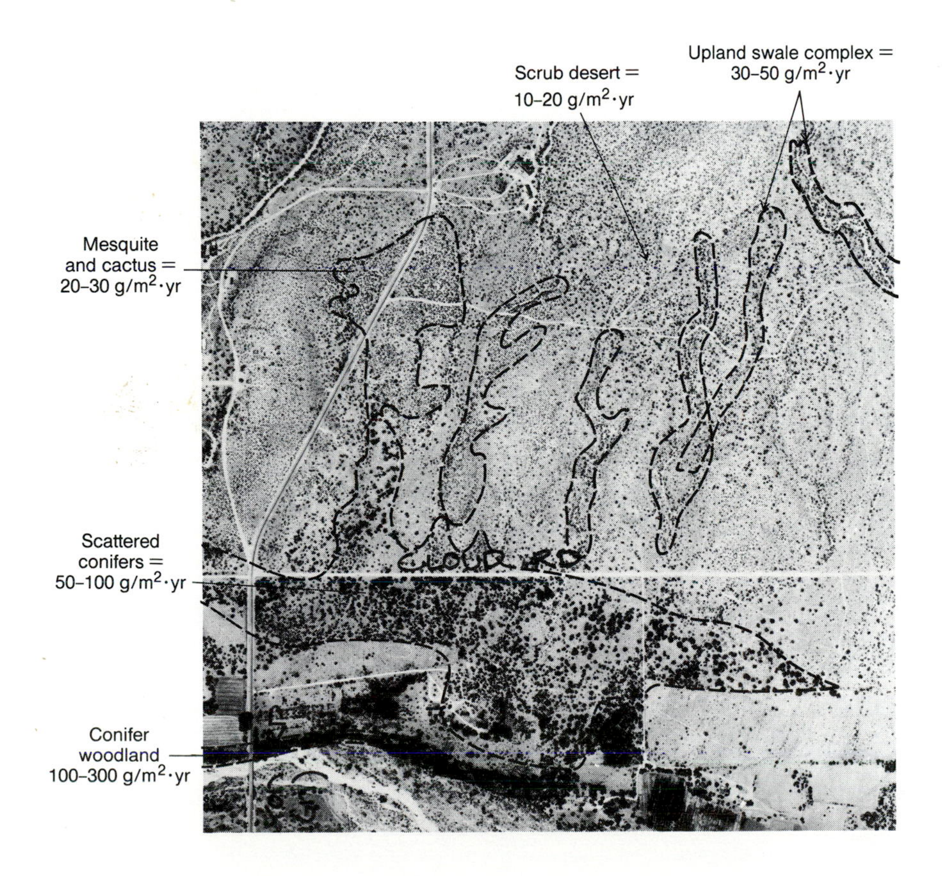

Figure 11.10 Variations in annual plant productivity over a small tract of desert landscape near Tucson, Arizona.

THE VIEW FROM THE FIELD / NOTE 11

Eutrophication: A Case of Unwanted Productivity

Among the many problems caused by water pollution, nutrient loading is one of the most serious and widespread in North America. Nutrients are dissolved minerals that nurture growth in plants, including aquatic plants such as algae, as well as bacteria. Among the many nutrients found in natural waters, nitrogen and phosphorus are usually recognized as the most critical ones because when both are present in large quantities they can induce accelerated rates of biological activity. That is, they can induce much faster rates of growth and denser populations in plant communities.

Massive growth of aquatic plants in a lake or reservoir has several important results. First, the increase in photosynthesis causes a change in the balance of dissolved oxygen and carbon dioxide. Second, the number of plants and microorganisms increases. Third, the total production of organic matter increases. These changes lead to further alterations in the aquatic environment, most of which are decidedly undesirable from a human use standpoint. For example, the rate of in-filling of the lake basin by dead organic matter increases; water clarity decreases; fish species change to rougher types such as carp; there is a rise in unpleasant odors; and the cost of water treatment by municipalities and industry generally increases.

Together, the processes of nutrient loading, accelerated biological activity, and the buildup of organic deposits are known as *eutrophication.* Often described as the process of aging a water body, eutrophication is a natural biochemical process that works hand-in-hand with sediment deposition from soil erosion to in fill water bodies. Driven by natural forces alone, an inland lake in the midlatitudes may be consumed by eutrophication within several thousand years (though the rate varies widely with the size, depth, and bioclimatic conditions of the lake). In practically every instance, however, land development accelerates the rate by adding a surcharge of nutrients and sediment to the lake. This surcharge is especially pronounced in agricultural areas with sediment from soil erosion and nutrients from fertilizers, and in urban areas with sediment and nutrients washed off streets, lawns, and industrial lands by stormwater. So pronounced is this increase that scientists refer to two eutrophication rates for water bodies in developed areas: natural and cultural.

When introduced to the landscape in the dissolved form (from either natural or cultural sources), phosphorus and nitrogen show different responses to runoff processes. Nitrogen, which is generally more abundant, tends to be highly mobile, moving with the flow of the soil water and groundwater to receiving water bodies. If introduced to a field as fertilizer, for example, most of it may pass through the soil in the time it takes rainwater to percolate through the soil column, as little as weeks in humid climates. In contrast, phosphorus tends to be retained in the soil, leaching into the groundwater very slowly. As a result, under natural conditions most waters tend to be phosphorus-limited. When a surcharge of phosphorus is directly introduced to a water body, it is likely to trigger accelerated productivity. Accordingly, in water management programs aimed at limiting eutrophication, phosphorus control is often the primary goal.

---

the equatorial and tropical wet environments are far less differentiated geographically, reflecting a more uniform distribution of essential resources.

In the intermediate (temperate) climates, moisture is adequate, and temperature and light range seasonally from high to low. Although the plant cover in these midlatitude regions may be dense, growth rates average about half those in the wet tropics. The rates are much greater than those of the desert and polar climates, however (Table 11.1).

If we extend our examination of productivity to include coastal and marine environments, the contrasts are even more striking. The most productive earth environments are the tropical coastlines—specifically,

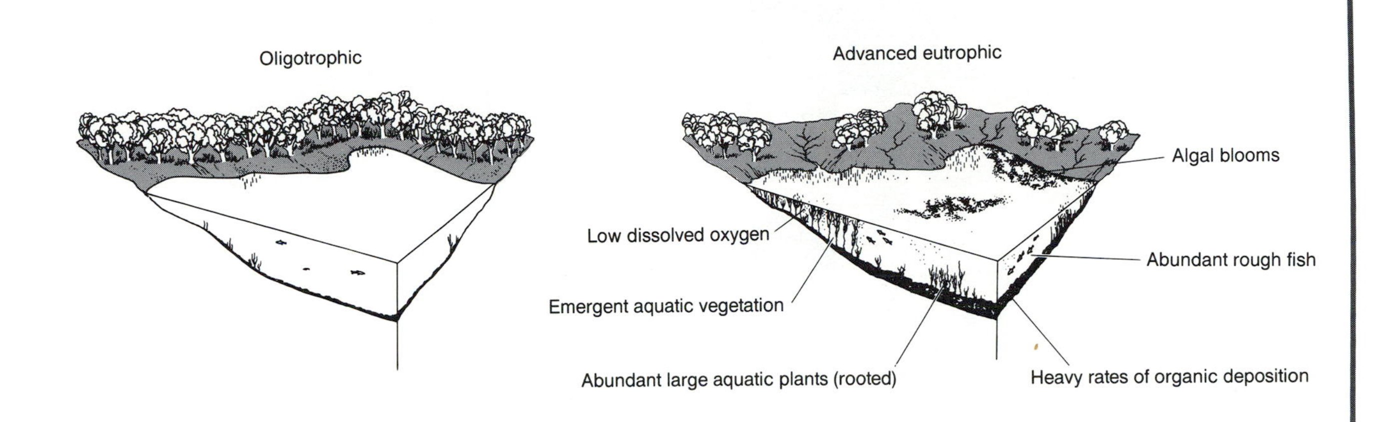

LEVELS OF EUTROPHICATION BASED ON DISSOLVED PHOSPHORUS

Level	*Total Phosphorus, mg/l*	*Water Characteristics*
Oligotrophic (pre-eutrophic)	less than 0.025	no algal blooms or nuisance weeds; clear water; abundant dissolved oxygen
Early eutrophic	.025–.045	↓
Middle eutrophic	.045–.065	
Eutrophic	.065–.085	
Advanced eutrophic	greater than .085	algal blooms and nuisance aquatic weeds throughout growing season; poor light penetration; limited dissolved oxygen

Representative mean annual values of phosphorus in phosphorus-limited water bodies.

tidal marshes, estuaries, and coral reefs (Fig. 11.11). Here yearly productivity is as high as 6000 grams per square meter. Only a few kilometers offshore, however, the annual rate drops, and beyond the continental shelf, over the deep ocean, the rate is less than 250 $g/m^2 \cdot$ year.

Certain pollutants can cause the productivity of aquatic ecosystems to increase tremendously. Measurements from a South Dakota pond that received untreated sewage showed that its productivity during some summer days reached an annual rate of 10,000 grams per square meter. Sewage water from households is highly enriched with nitrogen and phosphorus, nutrients essential to high rates of productivity in most ecosystems (see Note 11).

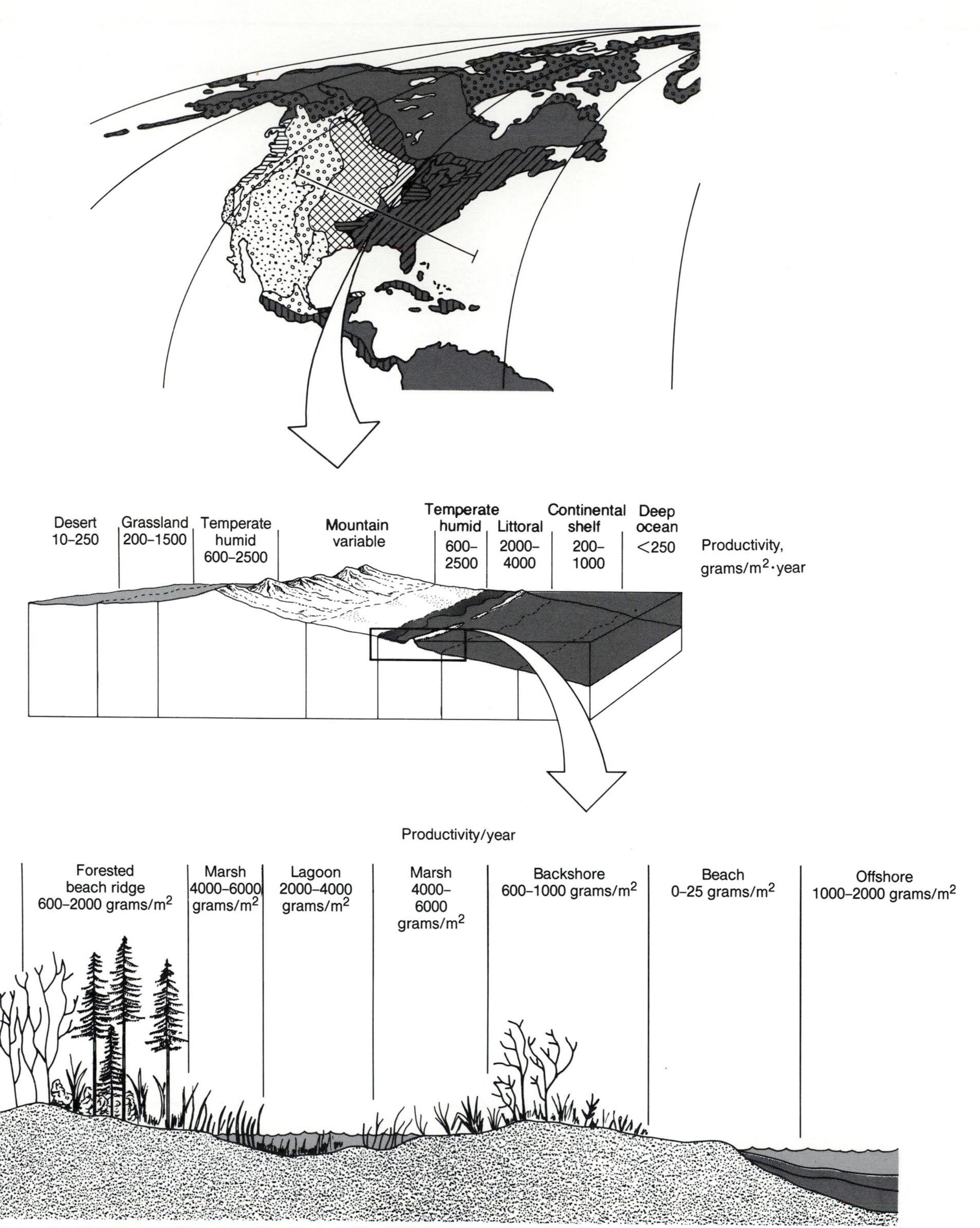

Figure 11.11 Zones of productivity with a detailed portrayal of vegetation and productivity in the coastal zone.

ENERGY FLOW IN PLANT ECOSYSTEMS

As we noted in our examination of the energy balance of the atmosphere, the vegetative cover has an important influence on heat flow and radiation near the ground. This effect is so profound in areas of dense forests that the major climate boundary between atmosphere and land is not at the soil surface, but rather the forest canopy. Here most of the incoming radiation and rainfall is intercepted and the brunt of the wind received (Fig. 11.12). As a result, a microclimate is created between the forest floor and the canopy where the energy flow patterns are distinctively different from those in the climate above the canopy.

Now let us turn to another aspect of energy flow in the landscape: energy flow through the plant cover. On a summer day in the midlatitudes, the hourly receipt of solar radiation by the plant cover amounts to 2 million J/m^2 or so. About 20 percent of this total is reflected by a broadleaf forest, and the remainder, 1.6 million J/m^2, is absorbed mainly by the foliage. From this point, the absorbed energy can be traced along two different paths: heat and photosynthesis. The heat path, which represents the bulk of the energy, provides energy for transpiration, heats up the wood and foliage, and ultimately leaves the plant in outflows of latent heat, sensible heat, and longwave radiation. These outflows together account for nearly 79 percent of the absorbed energy (Fig. 11.13).

The remaining 1 percent (20,000 J/m^2) is converted by photosynthesis into plant materials. This percentage is a measure of the *efficiency* of a plant cover, that is, the level of effectiveness in converting solar radiation into plant materials. Some plant covers may have efficiencies as high as 6 to 8 percent, but in general, the efficiency of vegetation the world over is around 1 percent. Ultimately, a substantial share of this energy is spent in respiration, leaving less than 1 percent to plant growth.

The Energy Balance of Ecosystems

The fraction of energy utilized in respiration is dissipated by internal biochemical processes and ultimately released from the plant in the form of heat. The fraction utilized in plant growth is incorporated as chemical energy for living plant matter—leaves, stems, roots, and reproductive organs. As the plants complete their growth cycles, most of this matter is deposited on the soil or consumed by animals.

Most of the fraction that is consumed by animals is released from them as heat from respiration; however, a small proportion of it goes into the growth of animal tissue. Through a series of prey-predator relations, much of the energy represented by this tissue is

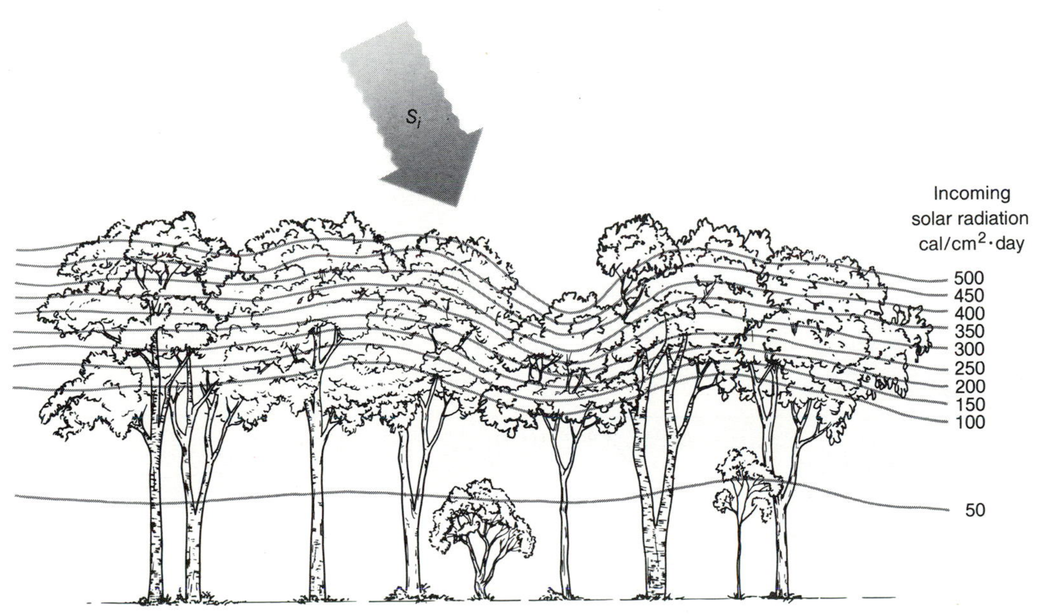

Figure 11.12 Attenuation of solar radiation associated with a temperate deciduous forest with full canopy in summer.

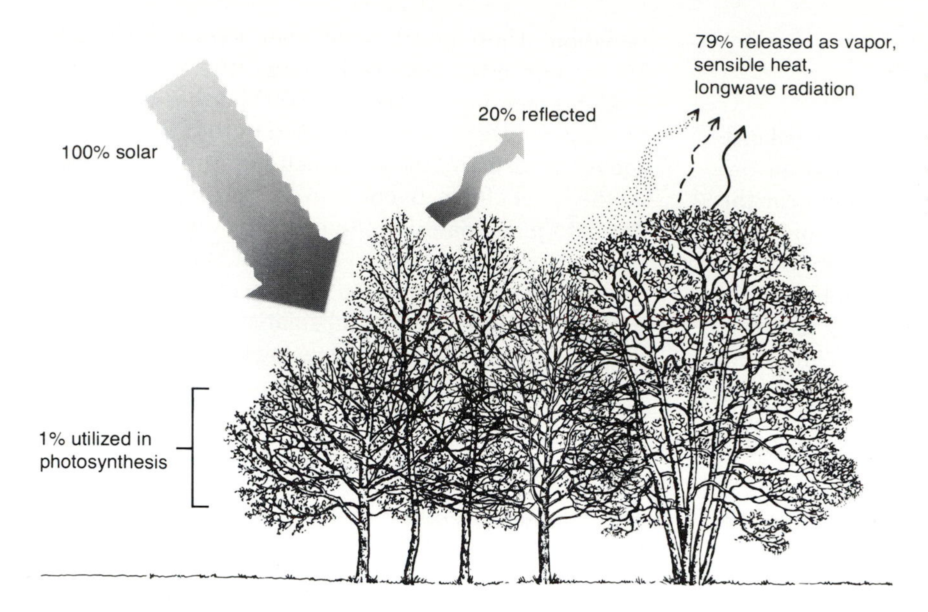

Figure 11.13 The fate of solar energy in a temperate forest. Note that only 1 percent is utilized in photosynthesis and that only a fraction of this produces plant growth.

passed from one set of animals to the next. These linkages are known as *food chains;* in most ecosystems, food chains are made up of three or four levels called *trophic levels.*

Trophic levels represent the energy conversion phases in an ecosystem. Because the total amount of energy is smaller at each succeeding level, the energy flow in an ecosystem may be characterized as a pyramid. Above the production level—that is, the level of the primary producers, the green plants—the organisms in each level are called *consumers: primary consumers, secondary consumers,* and *tertiary consumers.* The primary consumers are microorganisms such as bacteria which feed on organic litter, and herbivores, including animals such as earthworms and grazing and browsing animals such as goats and elephants. The secondary consumers are animals (carnivores) such as weasels and cheetahs which feed on

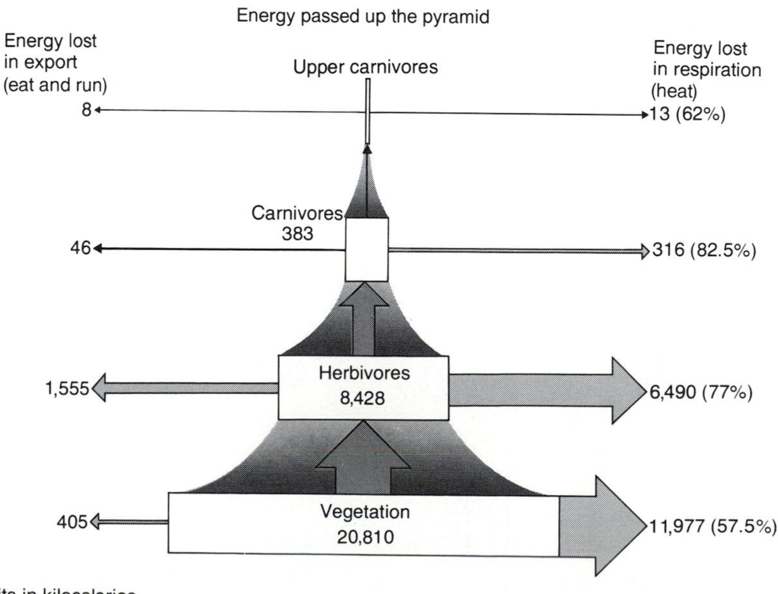

Figure 11.14 The energy pyramid for a fresh water ecosystem at Silver Springs, Florida. Note that, at each level, most of the energy is lost in respiration as heat.

herbivores; and the tertiary consumers, or upper carnivores, are animals such as eagles and sharks which feed on animals at both the primary and secondary levels.

The Energy Pyramid

Because it is very difficult to isolate and measure the actual flow of organic energy from one group of organisms to another, we have relatively few complete sets of data on the energy balance of a specific *ecosystem.* To make such a determination, the total organic mass, called the *biomass,* must be measured for each level in the ecological pyramid, and an energy equivalent must be assigned to each unit of organic matter. The energy equivalent turns out to be 4 kilocalories (17,000 joules) per dry gram of organic matter for plants and 5 kilocalories (21,000 joules) for animals. Ecologist Howard T. Odum provided a good example of the energy pyramid of a fresh water ecosystem at Silver Springs, Florida (Fig. 11.14). His data demonstrate how rapidly energy attenuates from one level to the next. The respiration of soil microorganisms and herbivores (secondary consumers) alone accounts for the dissipation of almost 90 percent of the original mass of energy represented by vegetation.

In wet environments such as bogs, swamps, and lakes, thermal conditions and the balance of gases are not always conducive to high rates of consumer activity. Consumption and decomposition of each year's production may be incomplete, and a fraction of organic energy is retained or, more appropriately stated, detained on the earth's surface. Ironically, we have come to depend on ancient preserved deposits of this fragile reserve, in the form of peat, coal, and petroleum, as our chief energy resource. It is estimated that the earth's total reserve of fossil energy is equivalent to a total of only 3600 years of production of plant material by photosynthesis.

Summary

Photosynthesis, respiration, and transpiration are the essential physiological maintenance processes of green plants. (Other essential processes are growth and reproduction.) Certain amounts of water, light, carbon dioxide, and heat are required by these processes. Water is drawn from the soil and is limited not only by the amount supplied by precipitation but also by the soil's capacity to store and transfer moisture. Light varies mainly with latitude and season, as does heat; but heat is also influenced by the thermal differences of land and water. Carbon dioxide is the least variable geographically. According to the principle of limiting factors, the one resource in least supply controls the rate of photosynthesis. Because there is substantial geographic variation in the supply of moisture, light, and heat over the earth's surface, the earth's potential to support plant life is also variable geographically.

Plant growth can take place only when total photosynthesis exceeds respiration. Production by land plants is greatest in the wet tropics and smallest in dry and cold regions. The average efficiency of plants in utilizing available solar energy, however, is only about 1 percent the world over. In all environments, plants form the foundation for complex chains of organisms called ecosystems, which are linked together by the flow of organic energy from one group, or level, to the next.

Concepts and Terms for Review

- essential physiological processes
- disturbances
- transpiration
  - stomata and mesophyll
  - air temperature
  - soil moisture
  - moisture tension
- respiration
  - heat
  - dwarfism
- photosynthesis
  - essential resources
  - sugars, starch, and oxygen
  - photoperiod
- principle of limiting factors
- net photosynthesis
- primary production
- productivity by bioclimatic zone
- spatial diversity
- ecosystems
  - energy flow
  - efficiency
  - food chains
  - trophic level
  - energy pyramid
- eutrophication

Sources and References

Anderson, E. (1948) "Hybridization of Habitat." *Evolution* 2: pp. 1–9.

Billings, W.D. (1970) *Plants, Man and the Ecosystem,* 2d ed. Belmont, California: Wadsworth.

Deshmukh, Ian. (1986) *Ecology and Tropical Biology.* Palo Alto, California: Blackwell Scientific Publications.

Graham, L.E. (1985) "The Origin of the Life Cycle of Land Plants." *American Scientist,* 73:2, pp. 178–186.

Grime, J.P. (1979) *Plant Strategies and Vegetation Processes.* New York: John Wiley and Sons.

Myers, Norman. (1983) *A Wealth of Wild Species: Storehouse for Human Welfare.* Boulder, Colorado: Westview Press.

National Cooperative Soil Survey. (1970) *Soil Taxonomy.* Washington, D.C.: Government Printing Office.

Odum, H.T. (1955) "Trophic Structure and Productivity of Silver Springs, Florida." *Ecological Monographs* 27: pp. 55–112.

Oldfield, M.L. (1984) *The Value of Conserving Genetic Resources.* Washington, D.C.: U.S. Department of Interior.

Raup, H.M. (1975) "Species Versatility." *Journal of the Arnold Arboretum 56:* pp. 126–163.

Ridley, H.N. (1930) *The Dispersal of Plants Throughout the World.* Ashford, England: L. Reeve.

Treshow, M. (1970) *Environment and Plant Response.* New York: McGraw-Hill.

West, N.E., ed. (1982) *Temperate Deserts and Semi-Deserts.* Vol. 5 of *Ecosystems of the World.* Amsterdam: Elsevier Science.

Chapter 12

Patterns and Processes of Vegetation Distributions

The environment is an uncertain place for plants over much of the earth. Uncertainty is improved through adaptation, but the environment rarely holds steady long enough for plants to obtain just the right distribution. As a result, the geographic arrangement of plants and vegetation is in a constant state of change. What drives these changes in different geographic settings? Are there distinctive patterns to the change processes and do they lend themselves to ready descriptions of vegetation dynamics?

Looking down on the land from an aircraft or in aerial photography, one cannot help but be impressed by the fantastic array of patterns in the earth's vegetative cover. These patterns appear as different colors, textures, tones, and forms that overlap in various ways to form a complex plaid and paisley cloak on the land. In reality, the patterns represent variations—at different scales—in vegetative *structure, cover density, floristic composition,* and often the *age and condition* of different stands of vegetation.

Although the geographic patterns in the vegetative cover may often appear to be impossibly complex, their explanation cannot be dismissed as the willy nilly result of random processes in the landscape. Indeed, one of the major tasks of plant geography is to explain why plants and vegetation are distributed as they are based on scientific knowledge of the interrelations of plants and environment.

The search for explanations takes geographers not only into the realm of natural phenomena, such as the influence of climatic factors on plants and vegetation, but also into the realm of human phenomena, such as the influence of land use on the distribution of plants and vegetation. Although vegetation—as opposed to individual plants—is the traditional center of attention, the distributions of species are also of concern to geographers because certain species are good indicators of how larger groups of plants are distributed. Furthermore, vegetation is a composite of the populations of many species, and knowledge about the distributions of species contributes toward an understanding of vegetation in general.

In this chapter we examine the factors that influence the distribution of plants and vegetation. This involves examining the life cycle of plants, the nature of the forces that the physical environment exerts on the plant cover, and the ways in which certain plants adapt to these conditions. Finally, we will describe three concepts or models that plant geographers use to help explain the patterns of vegetation we see in different parts of the landscape.

PLANT DISTRIBUTION

The geographic coverage of a particular species, genus, or family is called its *range.* Plant ranges exhibit all sorts of shapes and sizes, from those that occupy many separate regions at different locations around the globe to those that are restricted to a single area of only several hundred square kilometers. The ranges of plants change as environmental conditions change and as the plant itself changes through the process of evolution. Ranges may grow, shrink, subdivide, change shape, or change location. Some of the most apparent changes in ranges have taken place in the past several thousand years with the spread of human populations and land use over the earth. Our species has purposely favored certain plants, particularly the agricultural plants, which we have spread far beyond their original ranges; others, the weeds, we have spread inadvertently with the spread of certain land uses; still others, including many forest species, we have decimated, sharply reducing their geographic coverage.

Dispersal and Migration

Plants become established through a two-stage process: dispersal and migration. *Dispersal* is the distribution of a disseminule from a parent plant to a new spot. *Disseminules* are any part of a plant from which another plant can be established; they include seeds, fruits, spores, or vegetative parts such as roots or stems. *Migration* involves the successful growth and establishment of a plant in a new place.

When we compare the number of disseminules produced by a plant with the number of plants that become established, we find that the established plants constitute a very small percentage of the former. In overcoming these unfavorable odds, plants have evolved a capacity to produce massive quantities of disseminules. Most end up in unfavorable environments: goat stomachs, the oceans, on city streets, and so on. For those few that land in favorable spots and start to grow, very few survive to become adults capable of producing disseminules themselves.

This suggests that for many plants the environment is not an easy place; indeed, it is an inhospitable place. Moreover, the environment that affects plants is almost constantly changing, making one year's relationship to the environment not quite right in another year.

TOLERANCE, STRESS AND DISTURBANCE

The survival of a plant depends not only on proper supplies of the basic requirements for photosynthesis, respiration, and growth, but also on its tolerance to a wide variety of potentially damaging or killing influences from the environment. These two sets of influences are termed stress and disturbance. A *stress* is a deficiency in a factor that affects photosynthesis—light, heat, water, carbon dioxide, and certain minerals. A *disturbance* is a force such as fire, extreme wind, massive snow and ice accumulations, flooding, disease, and human activities that can adversely affect a plant's establishment, growth, and survival. *Tolerance* refers to the range of stress or disturbance that the plant is able to withstand without a decline in photosynthetic activity, physical damage, or loss of certain functional capacities.

Stresses and disturbances act on plants individually and collectively, causing damage, loss of reproductive capacity, or death itself. Tolerance is variable from phase to phase in the life cycle of plants, as well as from group to group within a single plant species. Over time, the level of disturbance or stress produced by the environment, measured, for instance, by changes in the amount of soil moisture or the intensity of flooding, is also variable.

The Life Cycle

The life cycle of plants evolves many phases, and for each phase the combination of growth requirements and tolerances may be different—often greatly different. In some phases, for example, a plant can withstand severe cold or drought that would kill it in any other phase of its life. Figure 12.1 shows the general phases in the life cycle of flowering annual plants.

The life cycle begins when a mature plant resumes its growth after a period of dormancy. First comes the introduction of leaves. This is followed by the development of sexual organs, which in turn produce pollen that is dispersed by wind and insects. The flowers impregnated by the pollen then form embryos, which mature into seeds. Upon their release from the plant,

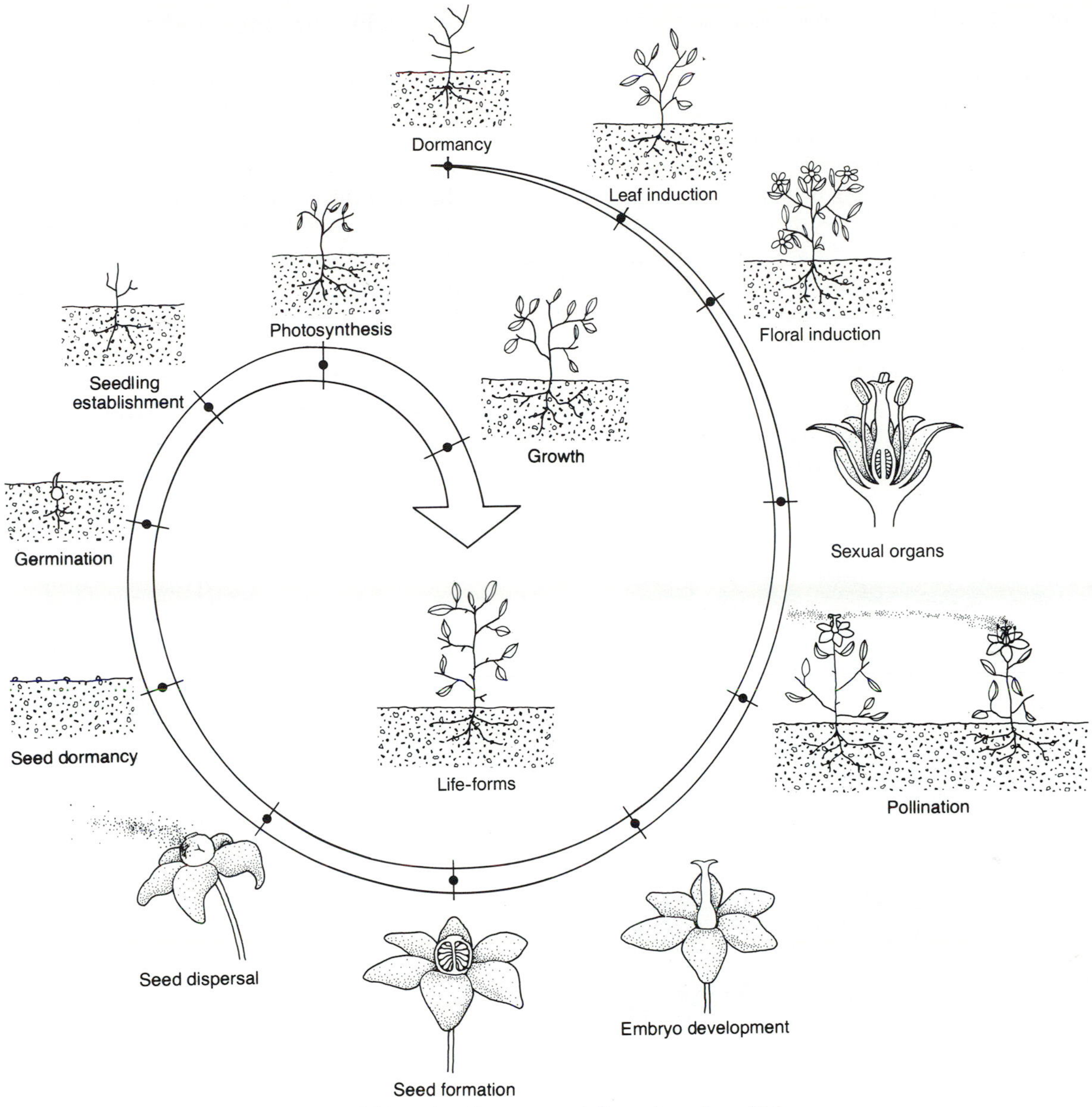

Figure 12.1 The reproductive (or life) cycle of a perennial flowering plant. Tolerances and requirements may be different at various phases in the cycle. Incompletion of any phase generally eliminates the plant from a locale.

the seeds are dispersed and eventually deposited somewhere. Here they usually lie dormant for some time, usually over the winter season or dry season. When conditions become favorable, some of the seeds germinate and grow to seedlings.

Interruption of any phase in the life cycle stops the reproduction process and hence can eliminate the plant from the locale. Of all the phases, seed germination is identified by plant physiologists as one of the main "bottlenecks" in the cycle of many plants. To induce germination in most plants, specific combinations and intensities of light, heat, moisture, soil chemicals, and aeration as well as a certain level of ground stability are required. Because these factors can be highly variable over small areas, germination can be very irregular. Slight variations in topography, for example, can greatly influence germination potentials because sun angle varies with slope, and surface heating, snow melt, and evaporation in turn vary with sun angle.

The story becomes all the more complex when we recognize that different members of a particular species vary in their tolerance levels and requirements. These variations, taken together, are a measure of a population's *ecological amplitude* or *habitat versatility*. In tree farming, for example, seeds are often transported to a plantation several hundred kilometers from their birthplace. Although the new location is well within the species' range, the trees often do not do as well as those of the same generation that were left at home. The U.S. Forest Service recognizes the importance of habitat versatility in tree species and as a rule will not move seed to plantations more than 100 miles or so from its place of origin.

The Nature of Disturbance and Stress

The forces exerted on plants by the environment can be geophysical, biological (including human), and geochemical. These forces influence plants in different ways, but the patterns of influence over time tend to be similar because the intensity or magnitude of the forces is almost constantly changing. The extreme magnitudes (high or low) of an occurrence, or *event*, happen infrequently, whereas the low-magnitude events happen often. For example, picture the flow of a river: small rises and falls take place every few days, but a big flood flow or the complete loss of flow takes place only once in a great while, say, every 100 years or so.

Change in the level of disturbance of stress, how-

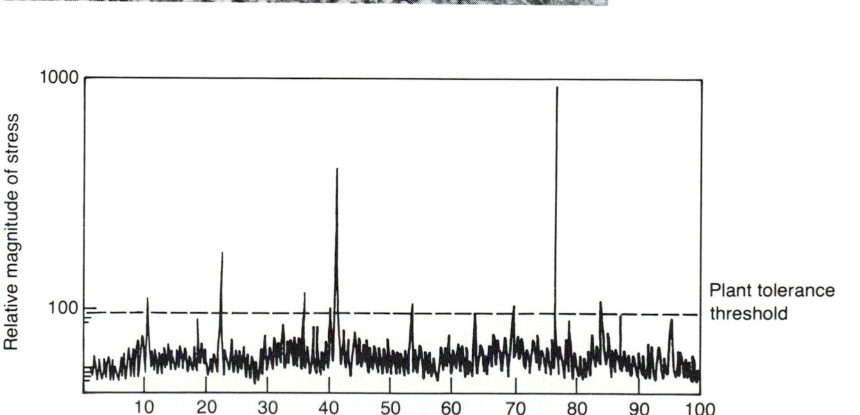

Figure 12.2 The photographs in (*a*) show the Pisgah Tract of the Harvard Forest before (left) and after (right) a large hurricane passed through New England in 1938. Judging from the age of trees that were blown down, a windstorm of this magnitude may happen in New England once every century or two. The graph in (*b*) shows the distribution of the magnitudes and frequencies associated with processes such as wind, river flow, and fire. The broken line represents a hypothetical threshold of plant tolerance.

ever, is not the only important measure of the effectiveness of disturbance or stress. Also important is the fact that the force exerted on a plant by any event, such as a windstorm, increases rapidly with the magnitude of the event. For example, the force of a wind, measured in terms of the stress exerted on a tree, increases as the cube of wind velocity. Thus, wind that occurs at an average frequency of once every 10 weeks may not be just ten times stronger than one wind that occurs on an average of once per week, but perhaps twenty to fifty times stronger. Floods, storms, and fires tend to follow the same pattern. In short, the plant's physical environment is not more or less static, varying little from the average on a seasonal, annual, and longer term basis; rather, it is highly dynamic, varying greatly from the average on a seasonal, annual, and longer term basis. The probability of great variations in process intensities increases with time; thus, the chances of a truly powerful episode of stress or disturbance are greater over a century than over a year, for an example see Figure 12.2*a*.

This describes the magnitude-and-frequency principle. Simply put, this principle states that the level of disturbance or stress produced by the environment is continually undergoing fluxes of intensity. Most of these fluxes fall below the tolerance thresholds of the various phases in the plant's life cycle, but every so often some exceed a tolerance threshold (Fig. 12.2*b*). When this occurs, the population may be reduced in some way, for example, by a lowering of population density or a dieback and loss of range. Although the magnitude-and-frequency principle is applicable to any environment, it is most apparent in certain high-energy environments, especially shorelands, sand dunes, and mountain slopes.

ADAPTATION TO THE ENVIRONMENT

Where plants are faced with a high intensity of stress or disturbance, they may improve their chances of survival through adaptation. Two types of adaptation are possible: *acquired* and *genetic*. Acquired adaptation occurs when an individual plant undergoes morphological or physiological change during its own lifetime in response to some stress or disturbance. For example, dwarfism in trees and shrubs can be induced by the practice of Bonsai (binding of roots). This results in extremely stunted individuals; however, if they could reproduce, the dwarf trait would not be transmitted to their offspring. Dwarfism among trees in the marine arctic environments is also an example of acquired adaptation; if their offspring are transplanted to zones of warmer summers, they acquire normal size for their species. This is in contrast to genetic adaptation, whereby new traits that appear in members' population become part of their genetic code and *are* passed on to succeeding generations (see Note 12).

Adaptation Strategies

When we examine general means of adaptation that occur in plants, we find that there are three principal modes or *primary strategies:* competition, stress toleration, and disturbance toleration. The *competition strategy* involves the development of life-forms and mechanisms that enable plants to exploit the resources for photosynthesis more successfully than other plants. This strategy is well developed among plants that inhabit environments which produce relatively low levels of disturbance and stress (such as the wet tropics as contrasted to deserts) and is common among plants that live in dense communities of many species. This strategy often takes the form of morphological adjustments such as plant height and extent of root system. In forest communities, for example, one means which certain tree species have developed to improve their exploitation of solar radiation is increasing height over competing tree species.

The *stress toleration strategy* is characterized by the development of endurance in conditions of limited productivity. These conditions are associated primarily with cold environments (such as polar lands), those where light is limited (such as polar lands and heavily shaded forest floors), and those with little soil moisture (such as desert lands). Adaptation that employs the stress toleration strategy often takes the form of special physiological adjustments. For example, in both cold and dry environments plants are able to survive if their growth rates are very slow and their periods of dormancy long. Stress-tolerant plants are often evergreen and, therefore, capable of taking up nutrients and photosynthesizing during brief periods of favorable conditions, such as the short subarctic summers.

The *disturbance toleration strategy* involves the development of adaptative mechanisms to endure high levels of disturbance. Generally, the environments that induce disturbance toleration are highly active settings, such as seashores, or plowed fields in the humid tropics or temperate midlatitudes that would

THE VIEW FROM THE FIELD / NOTE 12

Speciation and Geographic Change

Speciation is the splitting away of a new line of organisms from a parent population. Our knowledge of processes of speciation has grown tremendously since Charles Darwin brought the subject to world attention in the nineteenth century. Today, this is one of the most actively researched and controversial topics in natural science.

At the base of the concept of speciation is the relationship of organisms to their environment. In order to live and reproduce, an organism must have the capability to feed, defend itself, and so on within its habitat. For the match between the organism and its surroundings to continue, a *population* of organisms must possess the capacity to adjust over time, or *adapt*, to a changing environment. This is necessary because over long periods no environment is without major change. This we know from analysis of all kinds of data about the earth, especially geologic data on past environments. As the environment changes, the population's characteristics must also change; otherwise, these organisms will face stresses and disturbances beyond their tolerance limits and will trend toward extinction.

Biologists are in wide agreement that speciation occurs by a mode known as *allopatry*. Allopatric means "occurring in another place," and, according to the allopatric theory, a new species begins as a small group on the periphery of the parental range. The group becomes geographically isolated from the parent population and thus functions biologically as an independent population. Genetic variations (that is, mutations) that arise in the group (and are favorable to survival) can spread quickly through the small population—which is not as likely in large groups such as the parent population.

Isolation can occur when a barrier such as a mountain range or a desert forms across the geographic range of a species, thereby breaking the population into two separate groups. In time, the peripheral group may evolve along such different lines that eventually its members are no longer genetically compatible with the parent population. At this point—when interbreeding ceases—a new species has formed.

The accompanying diagram shows the distribution of two varieties of a California shrub called *Potentilla gladulosa*. This mountain plant may be in the process of evolving into different species, as its environment is subdivided with the formation of the High Sierras and the arid San Joaquin Valley.

otherwise be highly productive. The processes of disturbance fall into two categories: (1) fires, grazing, mowing, erosion, and other processes that kill plants or plant organs and remove plant material (biomass) from the habitat, and (2) frost, disease, drought, and other processes that kill plants or plant organs but do not remove biomass from the habitat. Disturbance-tolerant plants, which are often termed *ruderals*, are characterized by special structures or functions that enable them to overcome disturbance that eliminates most other plants. These include organs such as specially adapted buds that enable a plant to reestablish itself after being knocked down by a flood and buried by sediments and seeds, many of which survive the passage through a grazing animals' alimentary canal.

Forms of Adaptation

Plants have adapted in countless ways to stresses and disturbances. One of the most common types of adaptation involves an adjustment in the timing of certain phases in the life cycle to the rhythm of the seasons. These sorts of adjustments are termed *phenological* adaptations, and *annualism* is one of the most prevalent expressions of it among herbs in the midlatitudes. Plants that have developed this habit are called *annuals*, and their life cycle is characterized by a seed-dormancy phase that coincides with the high-stress season, normally winter. All other phases of the life cycle are completed during the spring and summer, leaving the dormant seed—the toughest phase

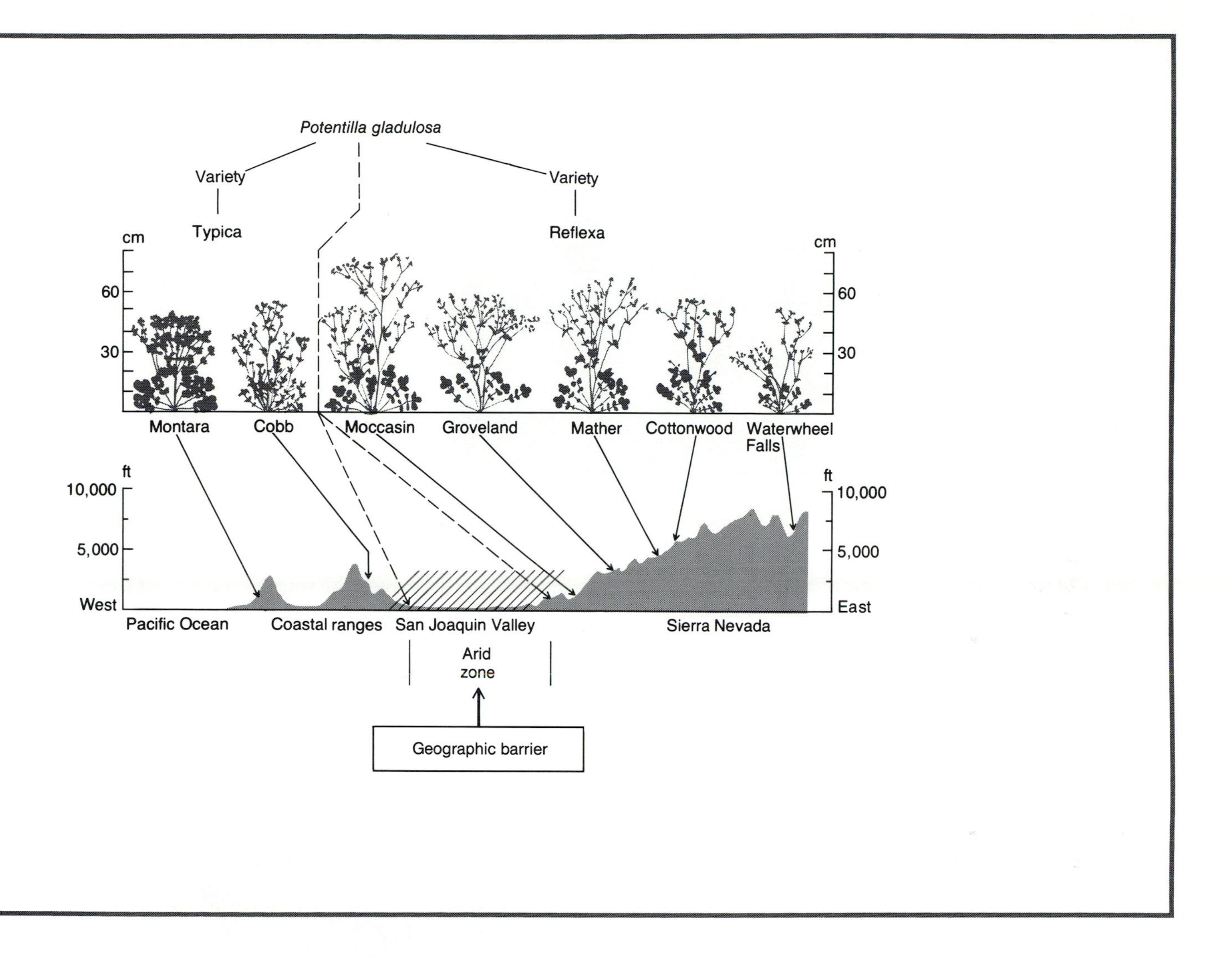

in the life cycle—to carry the plant through the winter.

Another type of phenological adaptation involves the adjustment of the plant life cycle to a particular event in the environment. This sort of adaptation is characteristic of desert herbs called *ephemerals.* These plants have evolved the capacity to maintain seed dormancy for extended dry periods, easily several years in duration, and then to suddenly germinate, mature, and flower in a matter of days or weeks when soil-moisture conditions are favorable. In the desert such brief episodes produce brilliant, short-lived floral displays.

Among the *perennial* desert plants—that is, those that live for several years—adaptation to persistent drought, called the *xeric habit,* is the most pervasive type of adaptation. The *xerophytes* have developed two main forms of adaptation to drought. The first is the *succulent* habit, well typified by the cacti. Three physiological traits are pronounced in the succulents: (1) In order to reduce vapor loss, leaves are small and few in number, with small and widely spaced stomata. (2) Bodies are thick and fleshy for storage of water. (3) Root systems are shallow and fairly extensive in order to maximize water uptake when rainfall wets the upper soil. Thanks to these adaptations, the succulent is able to withstand prolonged periods of drought that would kill most other plants.

The second type of adaptation in desert perennials is characterized by the formation of deep roots that

Various types of cacti, shrubs (including mesquite) and ephemerals in the Sonora Desert of Arizona.

enable plants to draw on water at depths far beyond the reach of most plants. Mesquite, a shrub found in the American Southwest, is a good example: Its roots penetrate to depths of 25 m or more. This is truly exceptional because even the roots of the largest trees, such as redwoods and certain pines and firs, are lim-

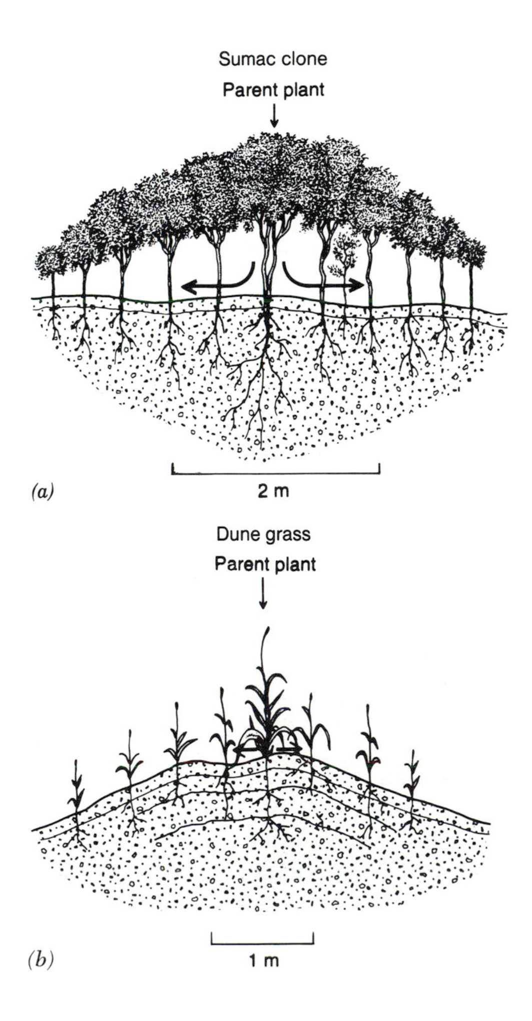

Figure 12.3 Vegetative regeneration: (*a*) Sumac clone, the stems of which generally decrease in age outward from the present stem. (*b*) Dune grass (marram), which spreads not only laterally, but also vertically in response to burial by wind-deposited sand. The photograph shows willow and sand cherry clones interspersed with dune grass.

ited to depths of less than 10 m, with the bulk of the root mass within 1 to 2 m of the surface (see Fig. 7.14).

In environments frequented by high-magnitude disturbances, it is difficult for plants to complete the reproductive phases of the life cycle. This is especially so in settings such as sand dunes, shorelines, and certain mountain slopes where the soil surface is so changeable that establishment from seed is impossible for most plants. In order to survive in these environments, some plants have evolved the means to regenerate by nonsexual processes. This *vegetative regeneration* involves the production of new plants from a living part of the parent plant such as a runner or lateral root. Many such organs may radiate from a parent plant, and from them hundreds of stems can be produced. Together, the parent and its stems may form a single plant, called a *clone*, which has the aspect of many individual plants. Among the most successful vegetative reproducers are dune grass, willow, and sumac (Fig. 12.3). Sumac is particularly common along fence lines and in abandoned farmfields in eastern North America.

GEOGRAPHIC INFLUENCES ON PLANT DISTRIBUTIONS

The landscape is a patchwork of vegetation types. The patches are all different sizes, but each represents an individual group of plants. Along their perimeters, the patches tend to merge into one another; ecologists use the term *ecotone* to describe the transition zones between two groups, or zones, of plants. Along lawns, plowed fields, or shorelines, ecotones are relatively sharp, more like borders, marking the edge of an area dominated by powerful forces that limit the area to only certain plants. In most places, however, ecotones are broad and gradual and tend to be rather geographically irregular in terms of actual changeover from one group to the other (Fig. 12.4). In this section we examine a few of the most important factors that control the location of a group of plants. Some of these factors are easy to ascertain, especially where human activity is involved, for example, lumbering, agriculture, urbanization, and war. The effects of some of the other factors are more difficult to determine—especially climate, erosional processes, topography, and soils.

Climatic Controls

Centuries of experience gained by European vineyardists and farmers provided early plant scientists with ample evidence that plant growth and survival were related to climate. Such knowledge led nineteenth-century plant geographers to search for the critical relationships between broad patterns of vegetation and the major climatic zones. Some of the scientists, believing that they were able to identify certain correlations, went so far as to recommend that the vegetation in some regions be used to delimit climatic borders. Although later studies cast doubt on the reliability of some of these correlations, certain borders that were identified still appear on modern climatic maps (see Chapter 8).

The relationship of vegetation to climate is exceedingly complex, and so many variables are involved that attempts at correlations of the two have generally proved inconclusive, particularly where vegetation of

(a)

(b)

(c)

Figure 12.4 Three examples of ecotones: (*a*) wetland-forest; (*b*) arctic fell-tundra; (*c*) alpine forest-alpine meadow.

complex composition is involved. Three factors are particularly important in this regard: (1) Humans have drastically altered the distribution of vegetation in the past 10,000 years; thus, in some regions vegetation borders have little to do with climate directly. (2) The earth's climatic patterns have undergone major shifts in the past 10,000 years, and some zones of vegetation are still in the process of adjusting their distributions to these shifts; therefore, their borders may not correspond well with their potential climatic ranges. (3) Plant migrations have been highly uneven in the past 100,000 years or so, with rapid diffusion in certain directions and none where barriers such as oceans or mountain ranges stand in the way; therefore, not all vegetation has had equal opportunity to respond to climatic patterns.

So what can we say about climate and the distribution of vegetation? Short of making gross generalizations about worldwide patterns, two statements are possible. First, the strongest relationships appear between individual plants and a key feature of climate such as the moisture balance. An example of moisture balance is provided in Figure 12.5, which shows the east-west ranges of eight plants as a function of soil-moisture conditions. Second, strong relationships also appear where climate is sharply differentiated geographically, as in mountainous regions. From the windward to leeward side of a coastal mountain range, for instance, vegetation may change from heavy forest to grassland or desert in response to a sudden decline in total precipitation and an increase in evaporation. A similar change is also apparent from the sunny to the shaded slopes of mountains in semiarid regions, where the difference is attributed to change in soil moisture (Fig. 12.6).

Topography and Mountain Vegetation

Probably the most celebrated relationship of vegetation to mountain climates is found in the vertical zonation of vegetation on high mountains. Temperature declines with altitude, resulting not only in less heat higher up, but lower evapotranspiration rates as well. Coupled with stresses such as heavy winds, avalanches, and landslides, these factors produce different belts of vegetation at various elevations (Fig. 12.7*a*).

In the American West, the lower slopes of the Rocky Mountains are often dry and covered with grasses and shrubs, whereas the cooler middle slopes, having a better moisture balance, are usually inhabited by conifers and hardy broadleaf trees. At the upper limit of the forest, called the *tree-line,* the trees are smaller and give way to alpine meadows of herbs. Here the limiting factor is heat. If the mountain is high enough, the alpine meadows will grade into a colder zone of snow and ice that is virtually devoid of plants. See Note 9 for details on the tree-line associated with the boreal forest-tundra border in Central Canada.

Another interesting observation about mountain vegetation is that the pattern of vertical zonation often approximates the zonal pattern of vegetation over great expanses of latitude. In this respect, the mountain gives us a microcosm of much larger vegetation patterns. The number of zones on any mountain depends on the height of the mountain and its latitude (Fig. 12.7*b*). High mountains (above 5000 m) in the tropics may exhibit all the major zones of vegetation of an entire hemisphere from tropical rainforest at the base to arctic fell near the summit. Mount Kilimanjaro, located less than 5 degrees south of the equator in Tanzania, is capped with snow but supports coffee plantations and tropical forests on the lower southern slopes. In the high latitudes, however, the base of a mountain lies in the fell or tundra zone, and higher up the vegetation only grows sparser; consequently, the entire mountain is covered by only one or two vegetation types.

Geomorphic Influences

As geomorphic processes reshape the land by erosion and deposition, they may also influence the plant cover. Running water, wind, and moving glacial ice exert force against the land, and are capable not only of eroding soil, but also of disturbing and destroying the plant cover.

Considerations of Scale Geomorphic influences are usually most pronounced at the local scale in places such as shorelines, sand dunes, mountain slopes, and river valleys, where erosional processes are very powerful. At the regional scale, the most evident influence of a geomorphic agent on vegetation is found in areas of continental glaciation. In North America, the great sheets of Pleistocene ice destroyed vast areas of flora as they spread over the land. When the ice retreated, a new flora, the *boreal forest,* became established in the glacial region. The present borders of this forest cover are thought by some geographers to coincide with the area occupied by the ice 6000 to 8000 years ago. They suggest that the newness of the boreal forest is reflected in its low diversity of species. Beyond this example, however, we know very little about the re-

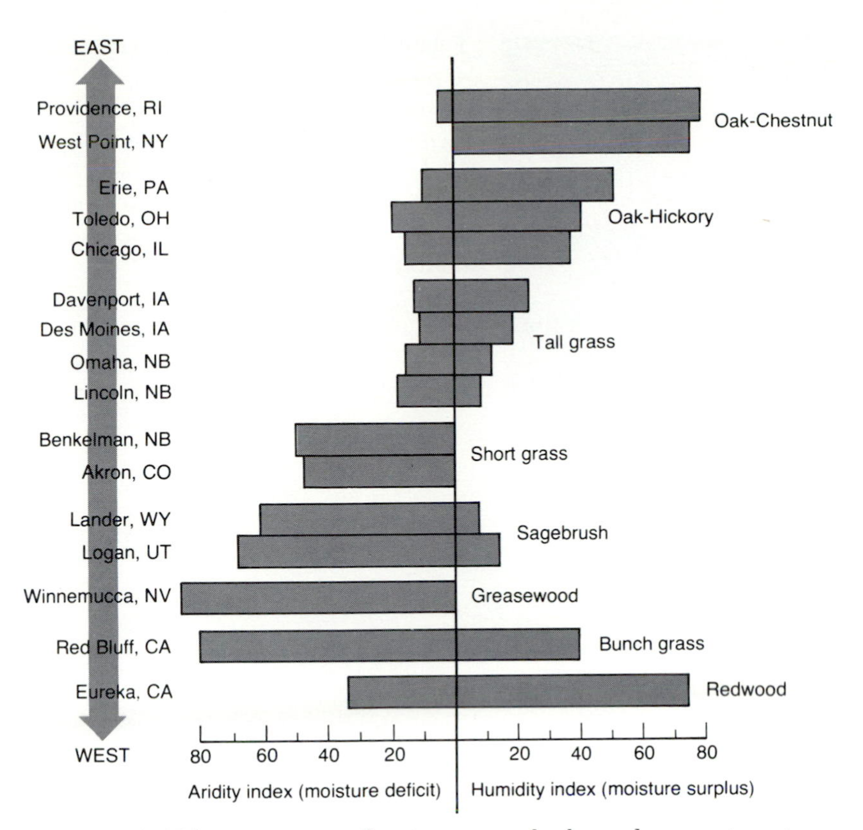

Figure 12.5 Tolerance ranges of various trees, shrubs, and grasses to water-balance deficit and surplus along an east-west United States transect at latitude 41°N.

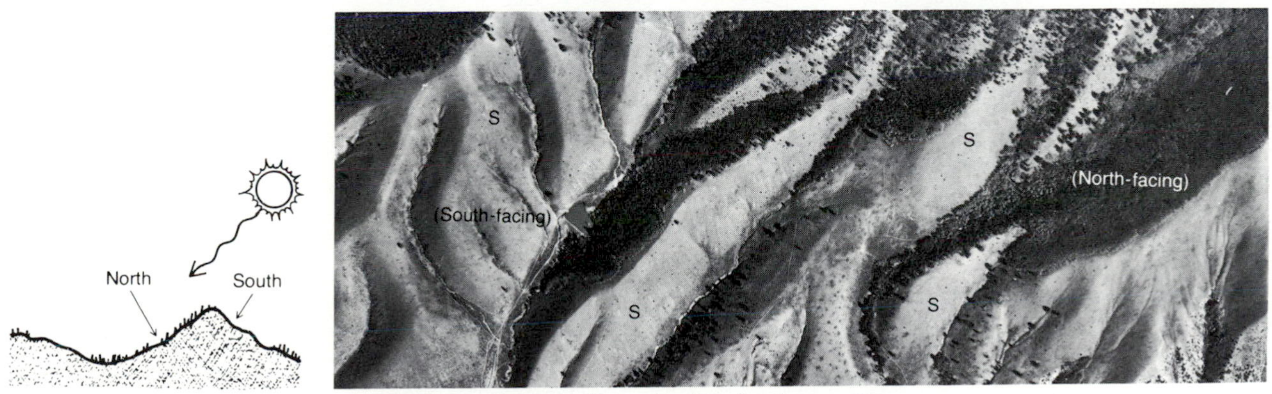

Figure 12.6 Aerial view of the distribution of forest and grassland in the Sangre De Cristo Mountains of southern Colorado. Because the sun angles are more direct on the southfacing slopes, heating is more intensive, and the moisture balance is much poorer than on northfacing slopes. As a result, southfacing slopes are limited to grass covers, whereas northfacing slopes can support forests, as shown in the diagram.

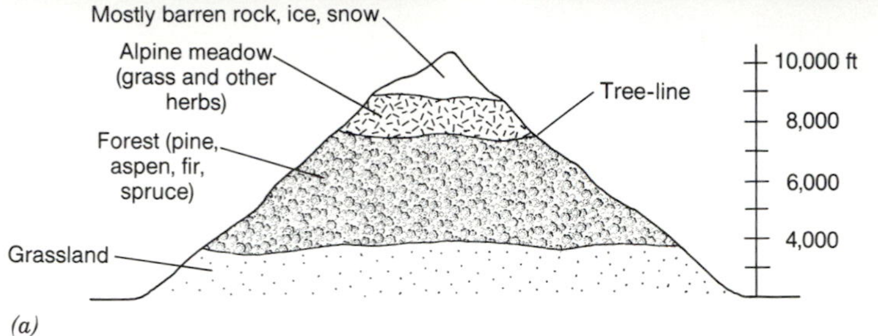

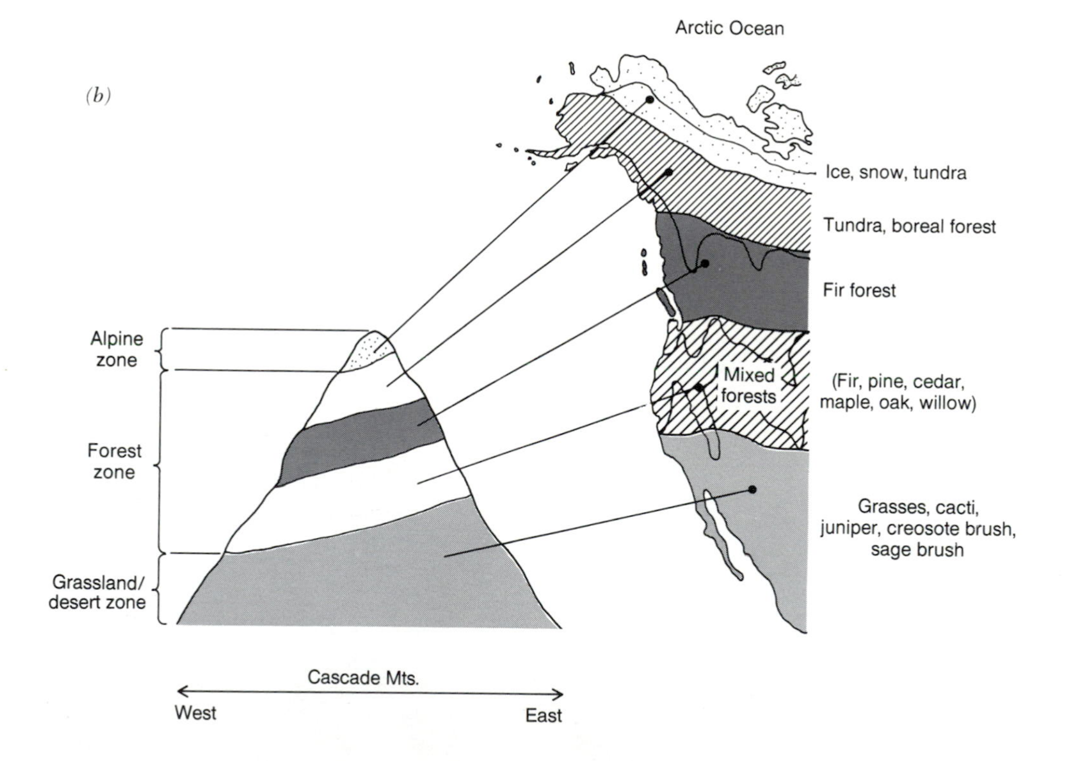

Figure 12.7 (*a*) The vertical zonation of vegetation in the northern Rocky Mountains. Three major zones are identifiable: alpine, forest, and grassland. In (*b*) the relationship of altitudinal vegetation zones to the latitudinal zones of vegetation in the Pacific Coast mountains of North America.

lations of vegetation to geomorphic factors at a regional scale.

Little is known about vegetation and geomorphic systems at a global scale. We do know that the location of the continents in the world has changed substantially in the past 50 to 200 million years, which is described by a theory called *plate tectonics* (see Chapter 16). Plate tectonics may provide some important information about changes in the locations and climates of the earth's land masses, which will help to explain some puzzling similarities and differences in widely separated flora.

Since the relationship of plants and geomorphic phenomena is best illustrated at the local scale, let us look at examples of two local settings: river valleys and sand dunes.

Rivers River channels tend to shift back and forth on the floors of their valleys. In the process, vegetation is destroyed or damaged by bank erosion, powerful flood flows, and heavy sediment deposits. But where fresh deposits, such as sand bars, form in the river channel, new plant habitats are created. Because of their nearness to the river, however, such sites are disturbed by flooding, erosion, and further deposition. As a result, it is difficult for plants to establish themselves on these sites and to maintain themselves once they are established (Fig. 12.8). Therefore, only

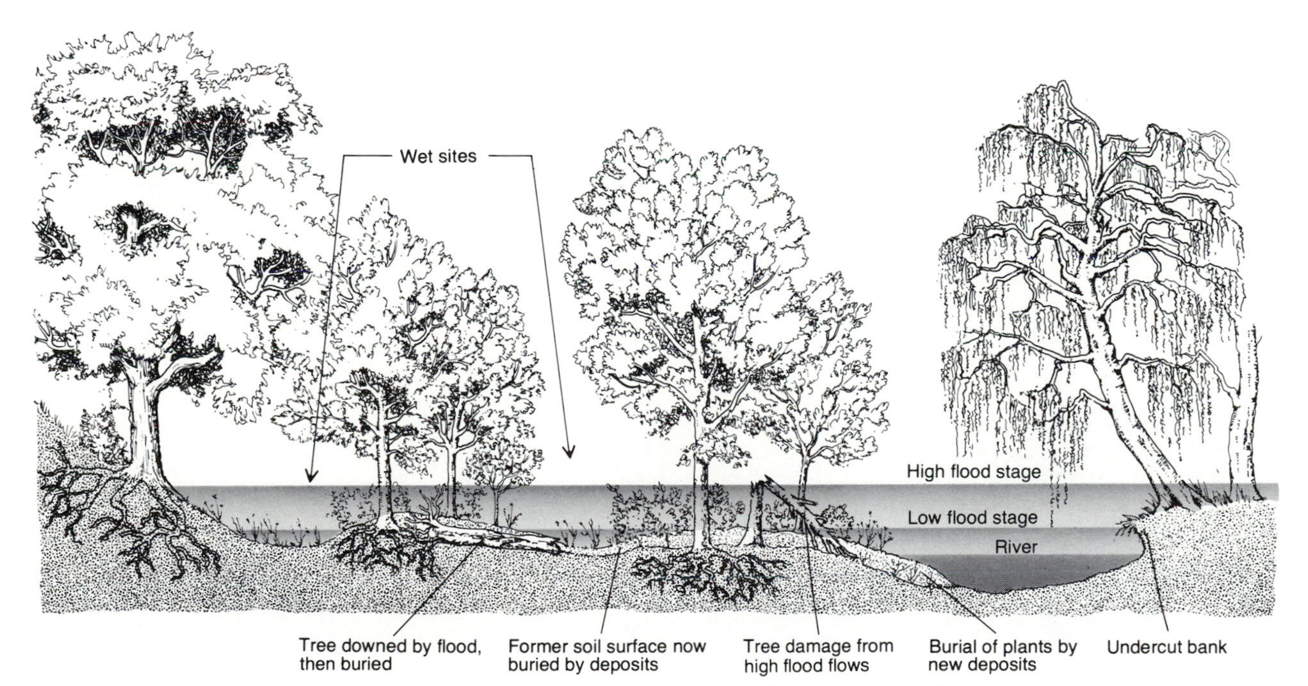

Figure 12.8 Some of the important relations between vegetation and the processes and features of river valleys. The most stressful sites are nearest the river, where deposition and erosion are commonplace. Farther away from the river, plants are influenced by high floods and the soil drainage related to old river deposits. The aerial photograph shows the spatial relationship between vegetation and a river in flood stage.

plants with high tolerance to these particular processes can grow on such sites. Willow, cottonwood, and sycamore are among the most successful trees. The pattern of these plants on the valley floor often reflects the distribution of channel deposits or the impacts of strong flood flows (see photograph in Fig. 12.8).

Sand Dunes Sand dunes are highly dynamic environments, characterized by rapid sand erosion and deposition by wind. One of the chief factors governing the survival of plants in the dune environment is their tolerance to being covered by sand and to root exposure by erosion. Figure 12.9 gives the approximate ranges of tolerance to erosion and deposition for six plants found in and around coastal dune fields in the midlatitudes. Dune grass (or marram), willow, and sand cherry have the highest tolerance to burial. Studies show that dune grass prefers being covered with 10 to 20 cm of sand per year for proper growth and development. At the other extreme is white birch, which seems able to tolerate only slight burial of its roots. This is one of the reasons why this lovely tree has limited success as a yard plant.

Because dune plants have different tolerances to burial and erosion, they do not all compete for the same space in the dune environment. Where deposition rates vary down the lee slope of a dune, for example, the plants tend to pattern themselves correspondingly. Dune grass usually grows highest on the slope, followed by sand cherry, willow, mosses, and various trees (Fig. 12.9). As the pattern of deposition changes from year to year, the distribution of these plants tends to change as well.

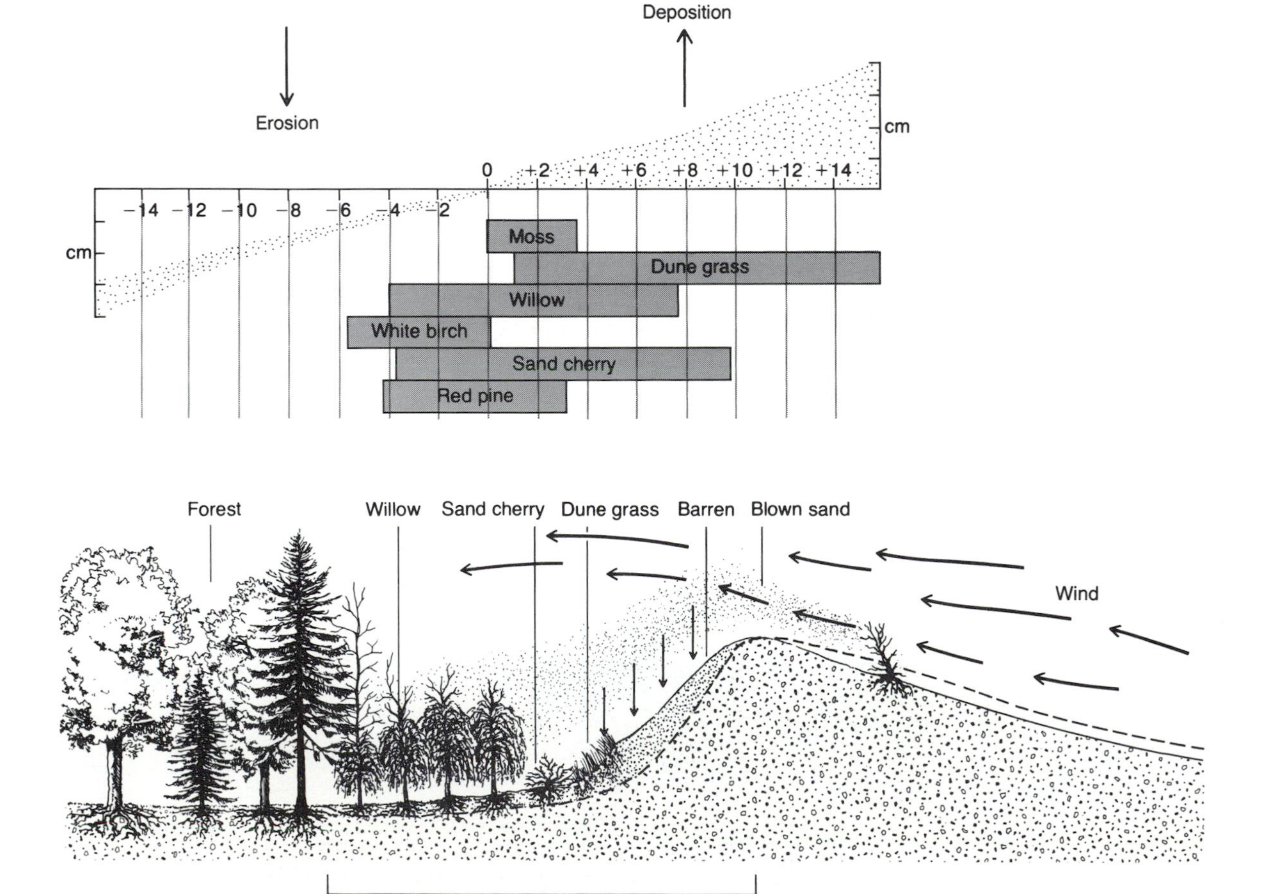

Figure 12.9 Tolerance ranges of six plants to sand erosion and deposition and the corresponding patterning of these plants on the depositional slope of an active sand dune in the Great Lakes region.

Fire

Fire ravages vegetation in all natural environments except the coldest and wettest ones. Most fires in vegetation are caused by natural means, usually lightning strikes, of which there are an estimated 3 billion on earth each year. The effect of fire on vegetation depends on many factors, including the type of fire, the makeup of the flora, and the phase of the life cycle that individual plants are in at the time of the fire.

Three types of vegetation fires are recognized: surface, crown, and ground. *Surface fires* are relatively cool and short-lived, generally causing little damage to mature trees. *Crown fires* move in the upper parts of trees; they are hotter and more damaging than surface fires. Under the right conditions, they may burn the foliage from a tree and kill it. *Ground fires* are all-consuming, burning not only trees and ground plants but topsoil as well. In areas of deep organic soils such as peat bogs, fires in the organic mass may burn for years, destroying the bog vegetation (shrubs, bladeleaf plants, and forbs). Records show that destroyed peat bogs often revert to swamp vegetation comprised of trees (Fig. 12.10).

Figure 12.10 A forest ravaged by fire.

It is widely recognized today that fires play an important role in the maintenance of grasslands and forests. They help destroy diseases and insects, release nutrients to the soil, open the cones of trees such as jack pine, and promote floral diversity in vegetation. Thus, fires are both vegetation-maintaining and vegetation-destroying. In either case, we have many examples of the influence of fire on vegetation patterns. The original forests of northern and northeastern United States, characterized as a mosaic of floristically uniform stands of trees, appear to have been burned out intermittently. The prairies encountered in Kentucky, Indiana, and Illinois in the 1800s also appear to have been a result of fires that were probably set by the Indians.

MODELS OF CHANGE IN THE PLANT COVER

The question of how change in the distribution of plants takes place has been a lively topic among geographers and ecologists for most of this century. The problem is difficult because of the diverse and complex relationships between the various plant species and the environment. Today it is a fair certainty that no one scientific model or concept of vegetation change is universally applicable. In this section we examine three models that have been devised to describe and explain the nature of change in vegetation. The first, the community-succession model, is the most popular of the three. The other two—the individualistic model and the disturbance model—provide interesting alternatives to the first, although they are less widely used by scientists.

The Community-Succession Model

We know that plants are able to change various aspects of the environments they inhabit, especially the climate near the ground, the nutrient and moisture content of the soil, and the erodibility of the soil. This capability of plants to change the environment they inhabit is the key feature in the community-succession concept.

A description of the community-succession concept begins with an area of land or water that has been denuded—that is, made barren, by some process, say, glaciation, fire, or human intervention. Into this environment migrate certain hardy plants, such as mosses and grasses, which are referred to as *pioneers*.

These plants effect changes in the environment by, for example, holding the soil against erosion and adding nutrients to the soil, thereby making it suitable for other plants. Once this has taken place, the pioneers are replaced, or succeeded, by a second group of plants, which in turn renders further change in the environment. Each group to inhabit this environment is called a *community* because it is composed of plants that live together in an ecologically interdependent fashion.

One community of plants succeeds another until eventually one of them achieves stability (that is, is able to inhabit the environment on a relatively permanent basis). This group of plants is called the *climax community*, and its presence indicates that a state of equilibrium has been reached in the plant-soil-atmosphere system. In other words, the climax community represents a steady state between the plant cover and the physical environment. Should some outside force change the climate or soil, this state is interrupted, and a different climax community will probably evolve. Should the climax community be destroyed altogether, succession will begin anew and continue until a climax community is reestablished.

Many examples of plant succession have been described by modern scientists. One of the most frequently cited examples involves the filling of ponds, lakes, bays, and estuaries by plant succession. Initially, only aquatic plants, such as water lilies and algae, and semi-aquatic plants, such as reeds and rushes, inhabit the water body. As these plants grow and die, their remains are deposited on the bottom, and after many years a thick, organic layer may develop (see Note 11). In the shallow water near the shore, where plant productivity is greatest, the organic layer eventually builds up to the water surface, where it provides new habitats for certain mosses, grasses, and shrubs. These plants both stabilize the new soil and add organic matter to it. In addition, streams contribute sediment to the organic mass. Little by little, the community of plants encroaches on the open water, adding much of each year's production to the growing organic mass (Fig. 12.11*a*). When the water body is completely filled, the aquatic and semi-aquatic plants no longer have a place to live, and eventually the types of trees that grew only near the original shore overgrow the entire environment, forming the climax community (Fig. 12.11*b*).

The Individualistic Model

In the 1930s the American plant taxonomist and ecologist Henry A. Gleason developed a model counter to the community-succession concept. The main thesis of his model, called the *individualistic concept*, is that patterning and change in the plant cover can be explained on the basis of the probability of recurrence of individual plants rather than on the basis of the relationship of an entire community to the environment. Gleason argued that when a plant dies and

(a)

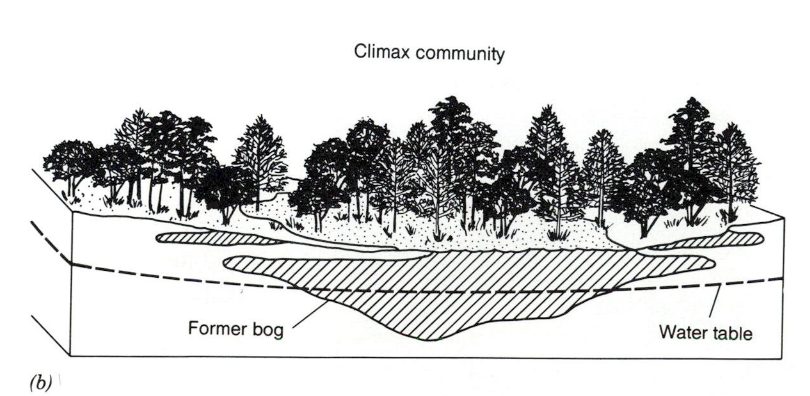

(b)

Figure 12.11 The community-succession concept: (*a*) Seventy years of succession. This pond formed on the mass of debris produced by the Gros Ventre landslide near Jackson Hole, Wyoming, in 1906. Early accounts indicate that the original plant cover was completely obliterated by the slide. (*b*) Schematic example of a bog that has been completely filled and overgrown with a climax forest community.

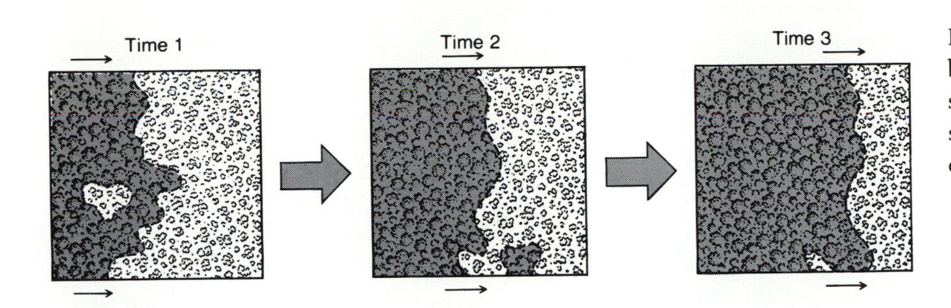

Figure 12.12 The shift in the boundary between two plant species, one of which is a more successful migrator than the other one.

space is made available for a new plant, it is a matter of probability as to which species will fill the space. Those species that are (1) nearest to the site, (2) present in greatest numbers, and (3) most efficient in dispersal have the greatest probability of inhabiting the vacant space.

Two points about the individualistic concept are pertinent here: (1) Changes in the plant cover that are commonly called *succession* are attributable to the advantage certain species have over others in establishing themselves at new sites (Fig. 12.12). This advantage is often due to the fact that some plants are more efficient dispersers than are others. This is clearly the case for many sand dune plants, for example, because they have the advantage of vegetative reproduction. (2) Climax communities are stable because probabilities favor recurrence of existing species, not because they are better suited to the particular environment than are other groups of species. For example, when a pine tree dies in the middle of a pine grove, the chances are very high that it will be replaced by another pine.

The Disturbance Model

According to the disturbance model, the distribution of a plant population fluctuates in response to the magnitude and frequency of the forces that operate in and around its range. We know that every plant has a relatively fixed tolerance to stresses and disturbances—heat, light, soil moisture, disease, fire, human actions, and geophysical processes such as landslides and flooding. When the level of stress or disturbance exceeds the tolerance threshold of a plant species, the members living in the affected area are either eradicated or so severely damaged that they can no longer reproduce. As a result, dieback might occur or the plant loses its edge over a competing species, giving up part of its range.

But events powerful enough to cause significant setback in a plant population usually occur so infrequently that, in the intervening years, the plant may begin to reinhabit the area that was lost. According to the disturbance model, while this is taking place, many smaller, but yet substantial, events are likely to occur which may delay or set back the plant's recovery in the area. In short, if we could compile a time-compressed film of the plant's distribution, it would appear to be in a state of continuous flux, expanding and contracting in response to the rises and falls in the magnitude of certain forces in the plant population's environment.

Let us look at lake or bog vegetation from the disturbance concept perspective. The water in bogs is derived mainly from groundwater, and the water level corresponds to the level of the water table in the surrounding land. We know that the height of the water table fluctuates with changes in the amount of precipitation; therefore, it follows, that the water level in the bog must also fluctuate. A rise of, say, a half meter in water level floods bog plants that are not hygrophytic; dieback occurs; if the high water level is maintained for several years, the area of bog vegetation may enlarge (Fig. 12.13*a*). Conversely, a comparable drop in water level eliminates many plants whose roots are adjusted to a limited range of water levels. More important, a drop in water level can expose the mass of dead organic material to the air, thereby accelerating its decomposition, and in turn reducing the total organic mass of the bog (Fig. 12.13*b*).

We can thus envision a bog fluctuating in area with major water-level changes. During periods when water level is near average, the bog may fill with organic debris and shrink in area (Fig. 12.13*c*); during highs in the water level, the bog may lose certain plants and open up; during lows, it may lose organic debris, decline in total mass, and shrink in area. In some cases two of these trends may occur in the same time period.

This interpretation of bog formation is markedly different from the one customarily used in the com-

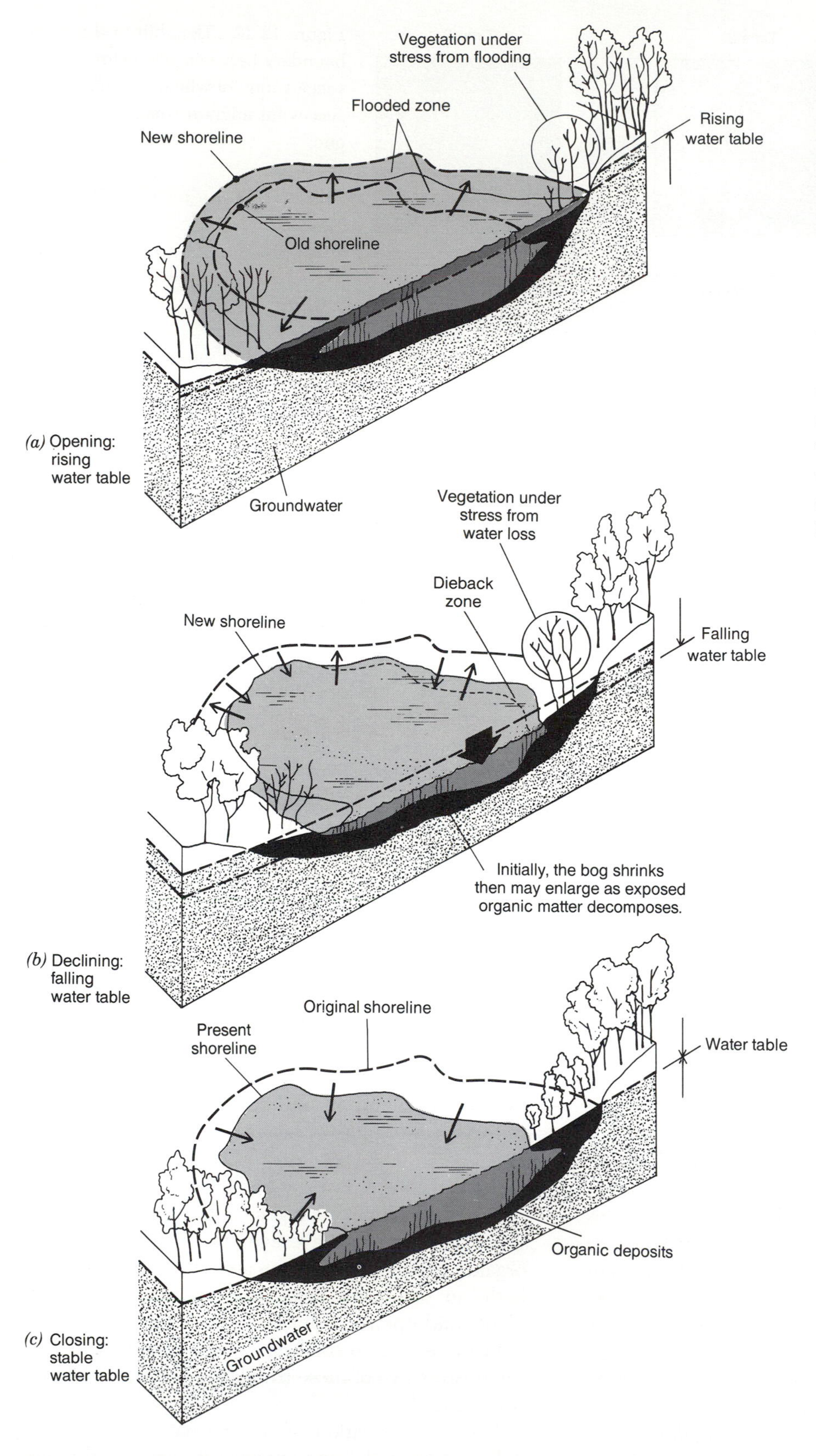

Figure 12.13 Changes in bog vegetation with stable water level and with variations in water level. Water level typically changes with variations in ground water or when beavers dam outlets. If the water table holds steady, the bog will grow inward and close out the pond (*a*). However, the opposite may occur when the water level rises or falls one or more meters from the average (*b*, *c*). Examples of such changes can often be seen where beavers have dammed the outlets to ponds and where established beaver dams have burst.

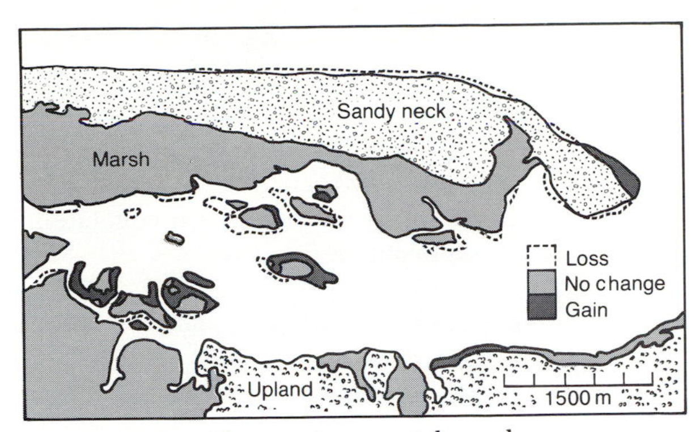

Figure 12.14 Changes in a coastal marsh near Barnstable, Massachusetts, between 1859 and 1957. Note that losses and gains are about equal.

munity-succession concept. The standard succession interpretation is based on a plant-controlled change, whereas the disturbance interpretation is based on environmentally controlled change. In areas where the environment remains fairly stable for hundreds or even thousands of years, succession is undoubtedly the predominant mode of change. The thousands of filled bogs across the Midwest and southern Canada, for example, attest to this. But the data gathered by scientists show that in many environments there is a great variation in the energy flow that drives the essential processes in and around the plant communities. Groundwater, streamflow, soil moisture, and the intensities of atmospheric storms, for example, fluctuate so much that they are capable of drastically altering the plant cover. Therefore, it can be argued that the notion of plant succession, and its assumption that a stable physical environment is the norm rather than the exception, seems to be too simplistic a way to interpret vegetation change in environments as dynamic as some lakes, ponds, coastlines, mountain slopes, river valleys, and sand dune fields.

Plant succession in reverse on an active sand dune.

Summary

Plant distributions are the result of a two-stage process: dispersal and migration. Tolerance is the range of stress or disturbance a plant can withstand without damage and loss of reproductive capacity. Tolerance varies with the phases of the life cycle. In addition, the magnitude and frequency of stress and disturbance produced by the environment are highly variable, especially in dynamic environments such as coasts, river valleys, and sand dunes. Over time (many generations), through the process of evolution, plant species may improve their chances of survival through adaptation. Notable examples of adaptation include the succulent habit, the annual habit, and vegetative regeneration.

The influence of the environment on plant distribution is complex. Climatic influences are manifested in the regional and global patterns of vegetation as well as in vertical zonation in mountain ranges. The effects of geomorphic processes are most evident at the local scale. The effects of fire are variable, but have had a pronounced influence on North American forests and prairies. The community-succession model, the most popular concept of change in the plant cover, is based on the observation that plant communities are capable of changing the environment, thereby giving rise to other plant communities until eventually a stable state is attained between one community and the environment. The individualistic model argues that change is dependent on the probability of recurrence of individual plants. The disturbance model recognizes the variable nature of environment, arguing that plant distributions fluctuate in response to the magnitude and frequency of environmental forces.

Concepts and Terms for Review

range
dispersal
migration
stress
disturbance
tolerance
life cycle phases
ecological amplitude
environmental variance
magnitude and frequency principle
adaptation
- acquired
- genetic

speciation: allopatry
adaptation strategies
- competition
- stress toleration
- disturbance toleration

forms of adaptation
- phenological
- xeric habit
- vegetative regeneration

ecotone
plant distributions
- climatic controls
- topography and mountain vegetation
- geomorphic influences
- fire influences

models of vegetation changes
- community-succession
- individualistic
- disturbance

Sources and References

Arno, S.R., and Habeck, J.R. (1972) "Ecology of Alpine Larch in the Pacific Northwest." *Ecology.*

Bishop, D.M., and Stevens, M.E. (1964) "Landslides on Logged Areas in Southeast Alaska." *U. S. Forest Service Research Paper* NOR-1.

Clements, F.E. (1936) "Nature and Structure of the Climax." *Journal of Ecology* 24, 1: pp. 253–284.

Dansereau, P. (1957) *Biogeography: An Ecological Perspective.* New York: Ronald Press.

Dyrness, C.T. (1967) "Mass-Soil Movements in the H. J. Andrews Experimental Forest." *U. S. Forest Service Research Paper* PNW-42.

Gleason, H.A. (1939) "The Individualistic Concept of Plant Association." *American Midland Naturalist* 12, 1: pp. 92–108.

Grime, J.P. (1979) *Plant Strategies and Vegetation Processes.* Chichester, England: John Wiley and Sons.

Hack, J.T., and Goodlett, J.C. (1960) "Geomorphology and Forest Ecology of a Mountain Region in the Central Appalachians." *U. S. Geological Survey Professional Paper* 347.

Henry, J.D., and Swan, J.M.A. (1974) "Reconstructing Forest History from Live and Dead Plant Material—An Approach to the Study of Forest Succession in Southwest New Hampshire." *Ecology* 55: pp. 772–783.

MacArthur, R.H., and Wilson, E.O. (1967) *Theory of Island Biogeography.* Princeton, N. J.: Princeton University Press.

Raup, H.M. (1964) "Some Problems in Ecological Theory and Their Relation to Conservation." *Journal of Ecology*, British Ecological Society Jubilee Symposium 52 (supplement) 1928.

Watts, D. (1971) *Principles of Biogeography.* New York: McGraw-Hill.

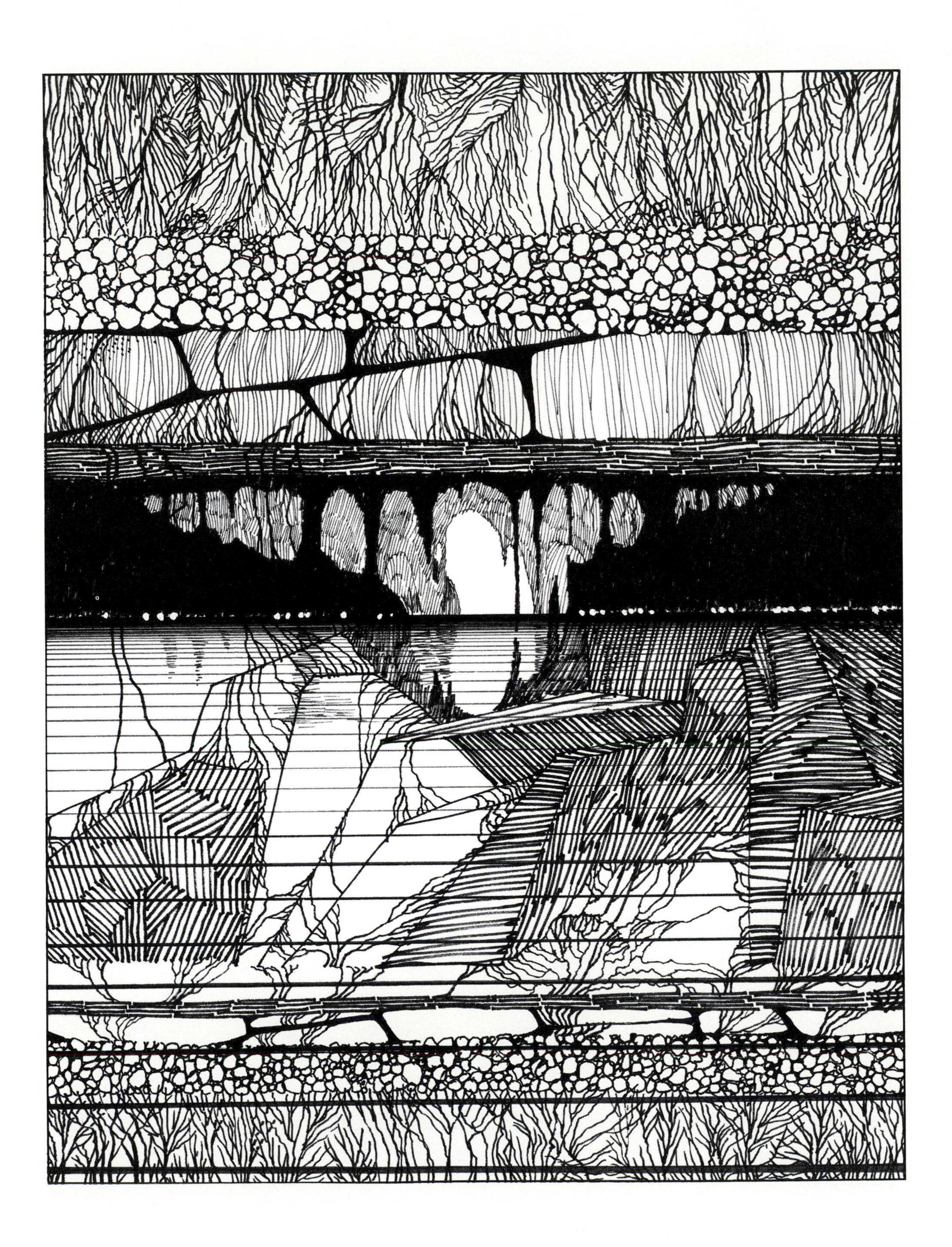

Chapter 13

The Materials and Processes of Soil

The topic of this chapter is soil—that uneven blanket of material on the land in which plants grow and gain sustenance and on which humanity resides. What are the sources of the materials that make up soils? What are the essential traits of soil? What controls soil water content and fertility, which are so important to plant growth? And what are the essential soil-forming processes?

On the face of it, a definition of soil seems to be a reasonably simple matter. As with many natural phenomena, however, we find that it is not so easy; it is especially difficult to come up with a definition that satisfies the different scientists who study soil. The reason is not that scientists are so cantankerous that they cannot agree on anything, but rather that their scientific perspectives differ.

Differences in scientific perspective grow out of the different kinds of questions scientists ask about some aspect of nature such as soil. To an agronomist, for example, questions about soil fertility are meaningful, whereas to a civil engineer they are not. Thus, the definitions arrived at by the various fields that study soil—geography, agronomy, civil engineering, and geology—are likely to be as varied as the questions they ask.

Agronomists would limit soil to the *solum*, the near-surface material capable of supporting plant life. The American soil scientist C. C. Nikiforoff suggests that "agronomy inherited this old concept of soil from the tillers of the land, for whom the soil is just the 'dirt' supporting their crops." The soil, thought of this way, contains gaseous, liquid, and solid portions in both inorganic and organic forms, and the agronomist is interested in what combinations of these materials are best suited for crops.

Civil engineers, by contrast, are concerned about the performance of soil when it is used as footing for buildings or highways. Unlike agronomists, they do not limit their definition of soil to the root zone. Rather, all assemblages of loose material that overlie the bedrock are recognized as soil. Geographers have long been concerned with the relationship among soil, vegetation, climate, drainage and land use and have, in the main, adopted more the agronomists' perspective of soil. Although our definition of soil must be broad enough to include most assemblages of ground materials, the uppermost several meters of soil are usually the locus of attention in physical geography. In this chapter we first examine the materials that make up soil: mineral particles, water, and organic matter. We want to learn about the origins of these materials, the processes that influence their distribution, and their role in soil formation. In the second half of the chapter, we examine the chemical and biological processes that work in the upper soil to shape the solum.

GENERAL CHARACTERISTICS OF SOIL

Before we examine the processes involved in soil formation, let us briefly examine some of the physical characteristics and origins of soil materials.

A System of Particles

The collective mass of particles—deposits and weathered material from various sources which we will discuss shortly—that rests on the bedrock is traditionally called the *soil mantle*. Physically, it forms a great blanket of uneven thickness and composition that covers most of the land masses. In the broad, flattish interiors of the continents, the soil mantle is usually 25 to 50 m thick and in places, 100 m or more thick. In areas of rugged terrain, the soil mantle is very irregular with many thin spots only a few meters thick and holes where bedrock is exposed. On steep mountain slopes, for instance, particles are removed as fast as they appear on the surface. Downslope, these particles build up in the valley bottoms, where they may form extraordinarily deep deposits (Fig. 13.1).

Viewed as a whole, the soil mantle can be seen as a great system of particles that is slowly moving from the elevated interiors of the continents toward the lowlands and the sea. The surface layer of particles is moving at the fastest rates, being nudged along primarily by runoff and wind. Beneath this layer, chemical processes associated with groundwater remove minerals from the soil mantle and deliver them to streams. Ultimately, the particles accumulate in thick deposits on the margins of the continents, and the chemicals are added to the sea. While this is taking place, weathering and erosional processes on the land are producing more particles and chemical residues.

It is appropriate at this point to introduce briefly the question of the influence of land use activity on this system. By clearing land of vegetation and exposing the soil surface—as in land development or agriculture—erosion rates of the upper soil have been

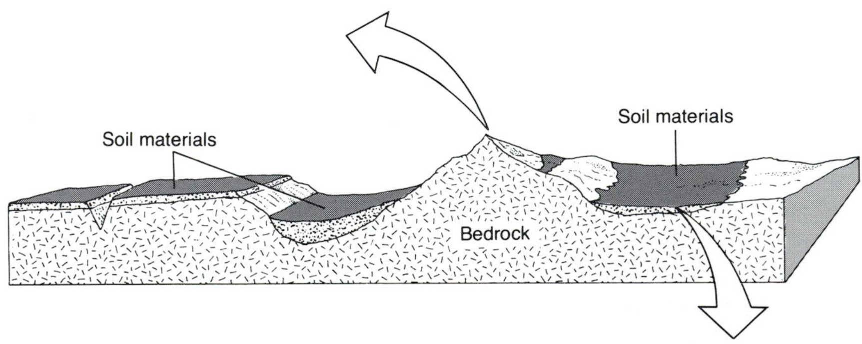

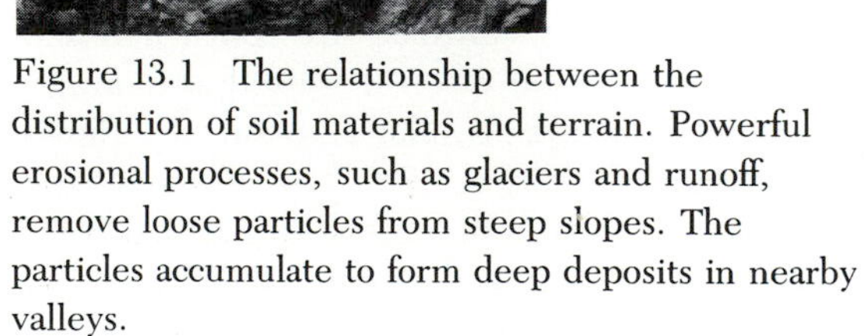
Figure 13.1 The relationship between the distribution of soil materials and terrain. Powerful erosional processes, such as glaciers and runoff, remove loose particles from steep slopes. The particles accumulate to form deep deposits in nearby valleys.

increased significantly throughout the world. In the big picture this means that the rates at which particles are being delivered to the lowlands and the sea are much greater now than in the past when there were fewer humans and less land use activity. Estimates of the balance between the production of new soil particles by nature and the loss of topsoil to erosion made by the World Watch Institute in 1984 place the net annual loss of soil from cropland at 22.7 billion tons worldwide. Clearly, the system of soil particles is seriously out of balance, which poses a serious threat to our future food production capabilities. We will say more about soil erosion later.

Origins of Soil Particles

More than half the total volume of most soils is made up of mineral particles of different sizes and composition. These particles are lodged against one another so that they form a relatively stable skeleton capable of supporting the overlying soil mass as well as surface objects such as forests, water bodies, and buildings. What is the source of these particles?

If we could trace soil particles to their origins, we would find that most came from bedrock, the solid rock underlying all land. When bedrock is exposed to conditions near the earth's surface, it undergoes *weathering*. Weathering produces physical and chemical decomposition of rock, resulting in its breakdown into smaller and smaller fragments (see Fig. 19.5). In general, weathering is dependent on heat and moisture; therefore, it varies with climate. In moist climates, for example, the more soluble minerals, such as salts, are dissolved and carried away in runoff, leaving behind the less soluble minerals, such as quartz. (See Chapter 19 for a more detailed discussion of weathering.)

The remains of these undissolved minerals form residues of particles—fragments of various sizes and shapes—over the bedrock surface. As the bedrock is lowered by weathering, the layer of particles may grow thicker. Soil material formed in this manner is termed *residual parent material.* In order for residual material to form, however, particles freed by weathering must not be carried off by erosional processes such as running water and wind, but very few areas of the world are not affected by erosion (Fig. 13.2).

In most places, weathered particles are indeed eroded and transported away from their place of origin. As sediments in wind, water, or glaciers, these particles are eventually deposited somewhere on land or in the oceans. Those deposited on land are called *surface deposits* and examination of the soil mantle in most places will usually reveal several different types of deposits representing different episodes of deposition associated with different processes. Soil material formed in this manner is termed *transported parent material*, and it is the principal material of the soil (Fig. 13.2).

Looking at large geographic regions, we find that the coverage of certain surface deposits is extremely vast. Figure 13.3 shows three examples for the United States and southern Canada. *Loess,* which is wind-deposited silt, covers large tracts of the prairie states, the Great Plains, and the Lower Mississippi Valley. *Alluvium,* which is river-deposited sediment, extends

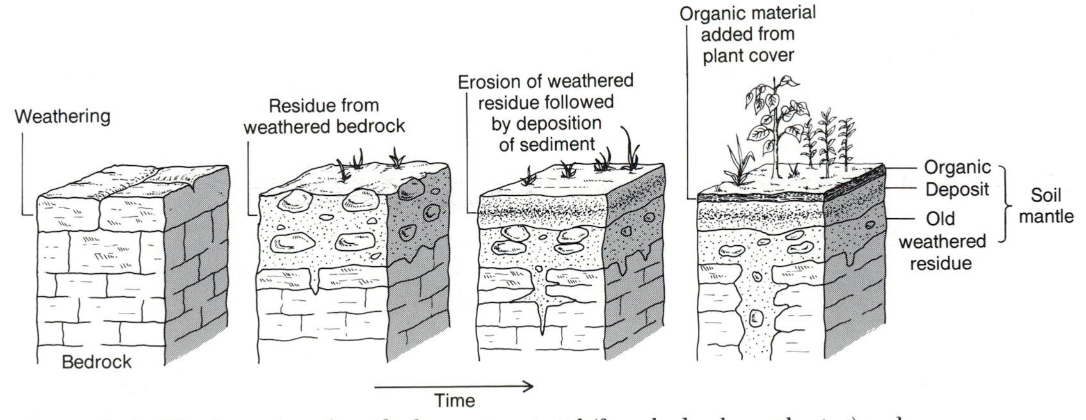

Figure 13.2 The formation of residual parent material (from bedrock weathering) and transported parent material (from surface deposits) which together make up the soil mantle.

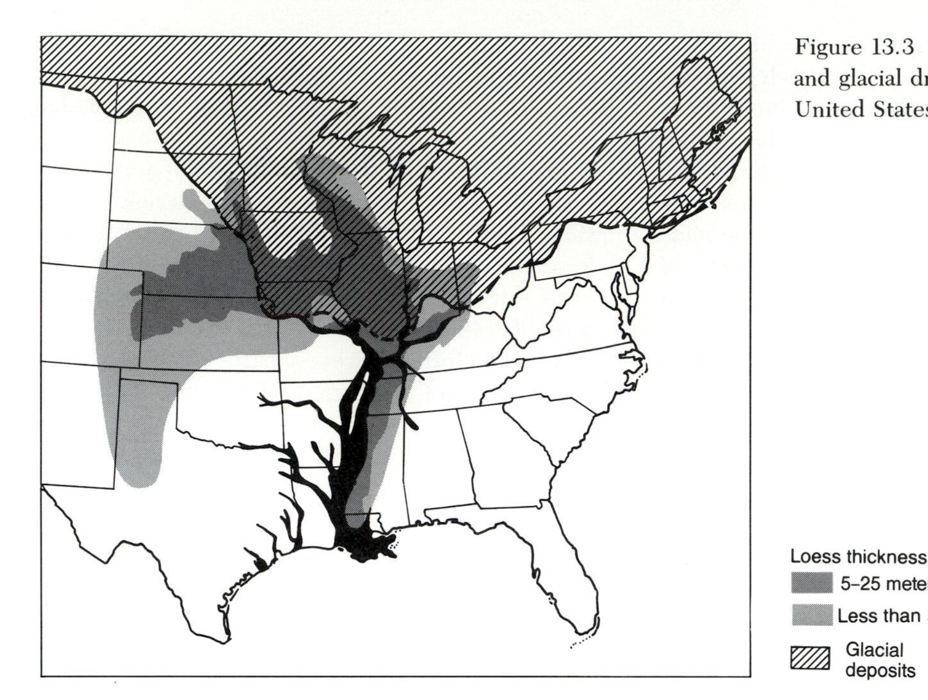

Figure 13.3 Major areas of loess, alluvium, and glacial drift in southern Canada and the United States east of the Rocky Mountains.

along the floor of every major river valley, and in the Mississippi River Valley south of Illinois, it forms a great belt stretching to the Gulf of Mexico. *Glacial drift,* the material deposited by glaciers or by the meltwater draining from a glacier, covers a huge region from the Missouri and Ohio rivers on the south to beyond Hudson Bay on the north.

SOIL PROPERTIES

In order to describe and analyze soils, we must first identify a set of meaningful characteristics that can be measured in the field or laboratory. These characteristics are called *soil properties.* The most important of these are composition and texture. *Composition* is generally classed into mineral and organic, mineral being inorganic substances such as particles of quartz and calcium, and organic being largely dead plant tissue in various states of decomposition. Other soil properties include texture, soil structure, and moisture.

Texture

Soil particles vary greatly in size. Some are so small that they can be seen only with the aid of a microscope; others are so large that they can be moved only with the aid of heavy equipment. Average diameter is the criterion used to classify individual soil particles, and three main size classes of soil particles are universally recognized: sand, silt, and clay. Agronomists, engineers, and earth scientists use somewhat different-size scales in classifying soil particles, two of which are shown in Figure 13.4*a*.

Sand, silt, and clay each play an important role in soil formation. For instance, sand enhances soil drainage, and silt and clay facilitate the movement and retention of capillary water in the soil. Small clay particles carry electrical charges that attract ions of dissolved minerals such as potassium and calcium. Attached to the clay particles, the ions are not readily washed away; in this way, clay helps to maintain soil fertility.

The composite of particle sizes in a representative soil sample, say, several handfuls, is termed *texture.* Soil texture is defined in terms of the percentage of weight of particles in various size classes. Since it is unlikely that all the particles in a soil will fall within one textural class, texture is usually expressed by a set of terms. At the center of this weight range is the class *loam,* which is a mixture of sand, silt, and clay. According to the U.S. Soil Conservation Service, the loam class is composed of 40 percent sand, 40 percent silt, and 20 percent clay. If there is a slightly heavier concentration of sand, say, 50 percent, with 10 percent clay and 40 percent silt, the soil is called sandy loam. In agronomy, the textural names and related percentages are given in the form of a triangular graph (Fig. 13.4*b*). If you know the percentage by weight of the particle sizes in a sample, you can use this graph to determine the appropriate soil name.

Because most soils are formed in surface deposits, a correlation can often be found between certain landforms and soil texture. In basin and range terrain, for example, mountain streams deposit their sediment loads in the form of alluvial fans along the foot of the mountain ranges (Fig. 13.5). Coarse sediments are laid down in the upper part of the fan and the finer ones farther downslope. The finest sediments (usually clays) may be carried into shallow lakes on the basin floor, where they settle out. In addition, the abundance of sediment around the alluvial fan often leads to wind erosion and sand dune formation. As we step back and look at the soils in such environments, not only are there measurable textural differences among

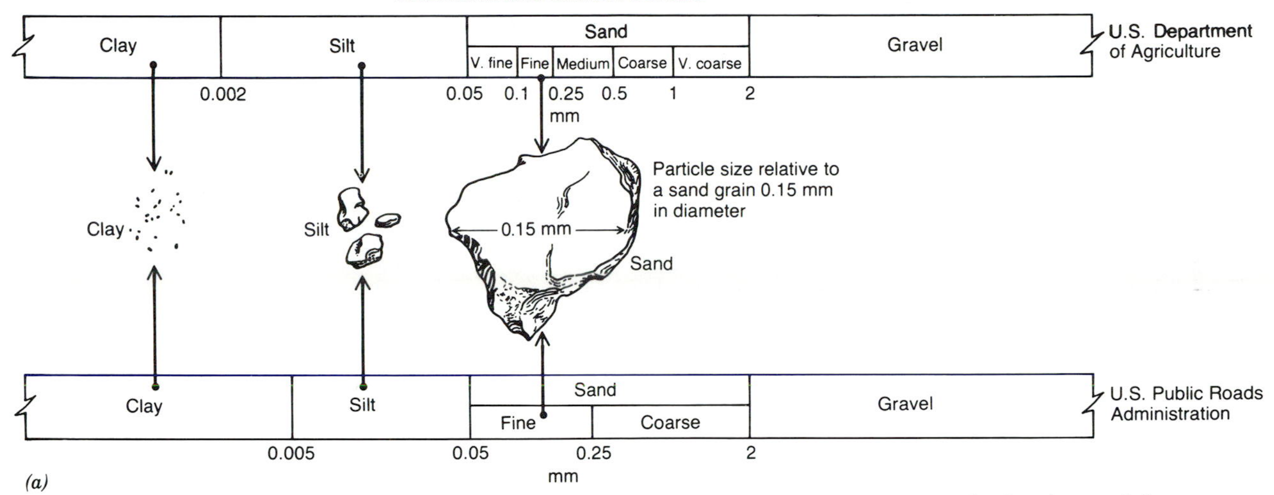

Figure 13.4*a* Standard scales used for classifying soil particles in North America.

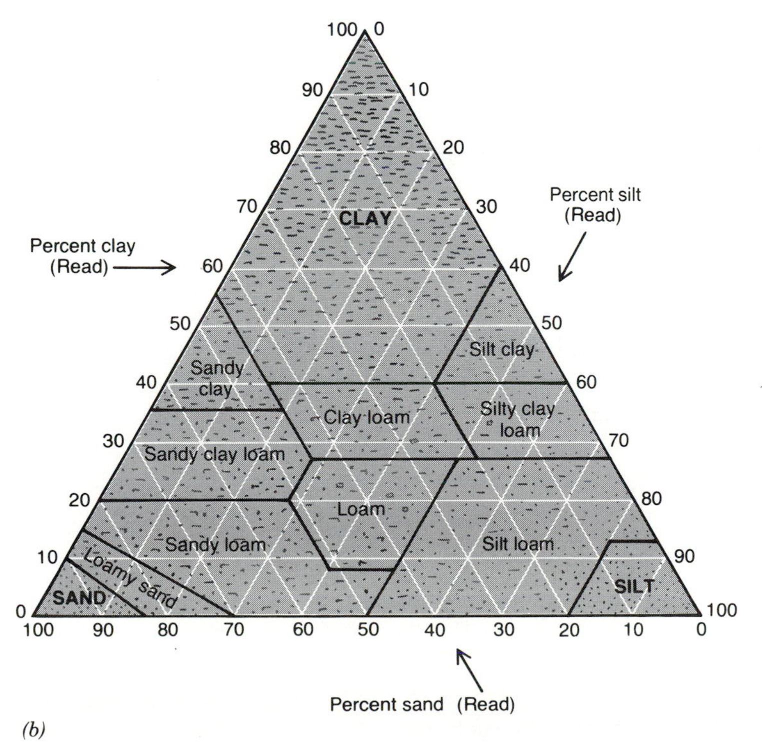

Figure 13.4*b* Soil-texture triangle showing the percentages of sand, silt, and clay comprising standard soil types as defined by the U.S. Department of Agriculture.

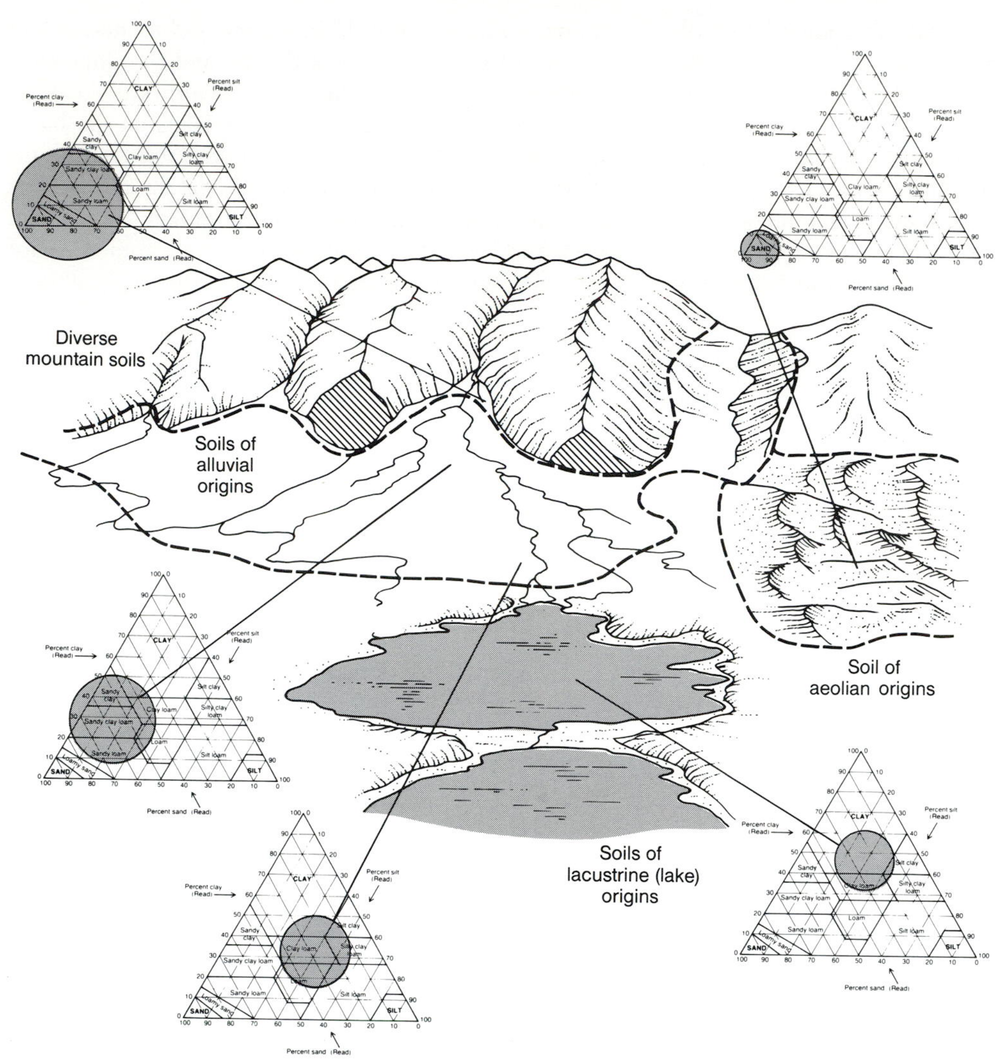

Figure 13.5 The relationship of soils to landforms. Both the texture and the degree of sorting vary with the landform.

the individual landforms, but also textural gradients within certain landforms (such as in the alluvial fan) as well as differences in sorting (homogeneity of particle size). A sand dune, being formed of sands winnowed from alluvial deposits, will be quite homogeneous of sorting compared with an alluvial fan. This is illustrated by the size of the area circled in the textural triangles of Figure 13.5.

Soil Structure

Rarely do we find a soil that is composed solely of free particles. Rather, particles are usually grouped into aggregates, especially if the soil contains clay and humus. Soil scientists refer to these aggregates as the soil *structure*. Individual structures may range from tiny granules smaller than the size of your little fingernail to fist-size blocks. The four main soil structures usually recognized are given in Figure 13.6. Although the processes that produce the different types of structures are not well understood, it appears that clay, humus, and soluble salt content are important influences on structure.

Organic Matter

In addition to the mineral particles, most soils also contain particles of organic matter. These particles are the partially decomposed remains of the plants that reside in and on the soil. Although certain soil environments, such as swamps and marshes, often contain massive accumulations of organic matter, many me-

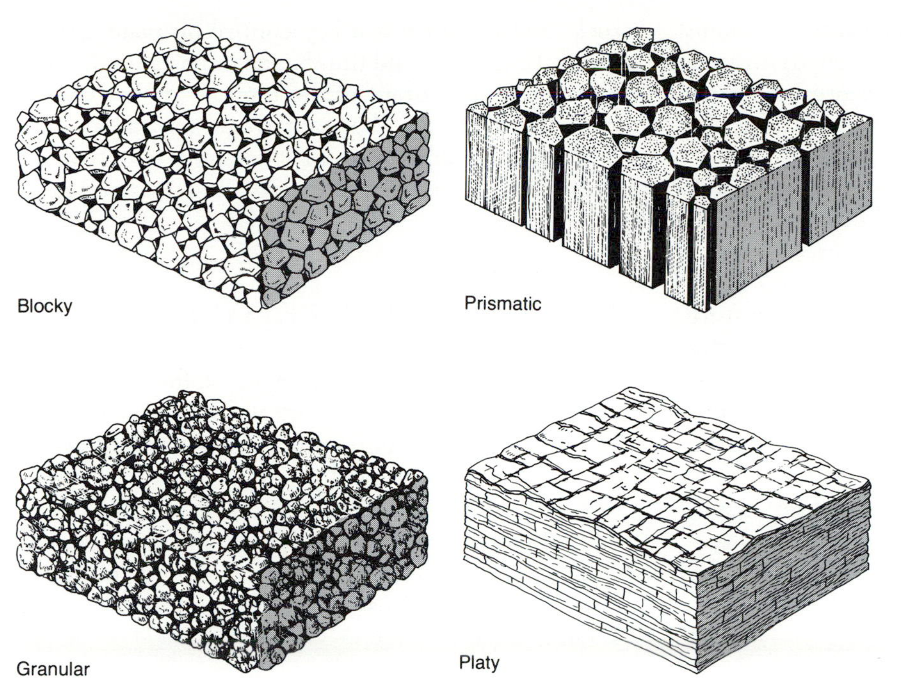

Figure 13.6 The four main types of soil structure. Although the formation of specific structures cannot be easily explained, it appears that the process of particle aggregating is related to the clay, humus, and soluble salts.

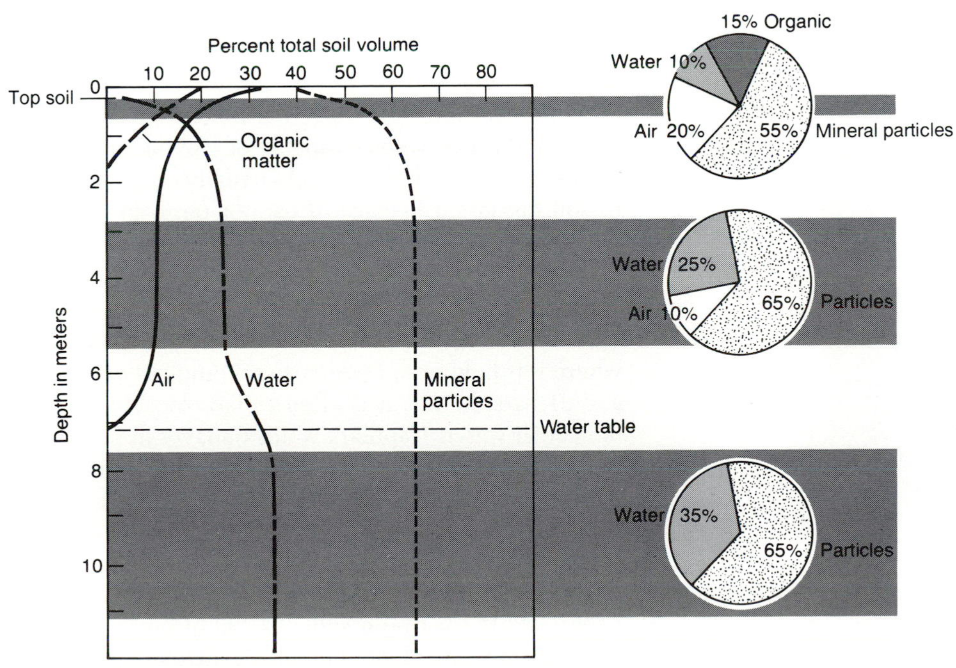

Figure 13.7 The relative proportions of organic, mineral, air, and water with depth into soil. The percentages are generally representative of soils in forested or grass-covered landscapes.

ters deep, the upper meter or so of most soils contains only about 3 to 5 percent organic matter by weight. The uppermost part of the soil, where organic matter represents 20 percent or more of the mass, is generally called *topsoil* (Fig. 13.7).

Organic matter has many important roles in the soil. It is a source of plant nutrients and other chemicals that are released as the tissue decomposes. Organic matter has a high moisture-absorption capacity and is a very effective medium in the transfer of moisture within the soil. Clay-sized particles of organic matter are effective in the adsorption and retention of dissolved chemicals, thereby helping to maintain soil fertility. And as we will learn later in this chapter, soil organic matter is part of a very active organic system and, under most conditions at the soil surface, is subject to relatively rapid decomposition and replacement.

Soil Water

Water is central to virtually every facet of soil. Erosion and deposition by runoff are important determinants of soil thickness and composition. Within the soil, water movement and associated chemical processes are responsible for the dissolution and precipitation of minerals, and, at depth, the weathering of the bedrock. Soil moisture is a key control on plant growth and soil organisms and thus has an important influence on the humus content of the topsoil. These generalizations appear to be basically sound for arid and humid climates alike, and so it is important that we examine the origin, movement, and dissipation of water in the soil.

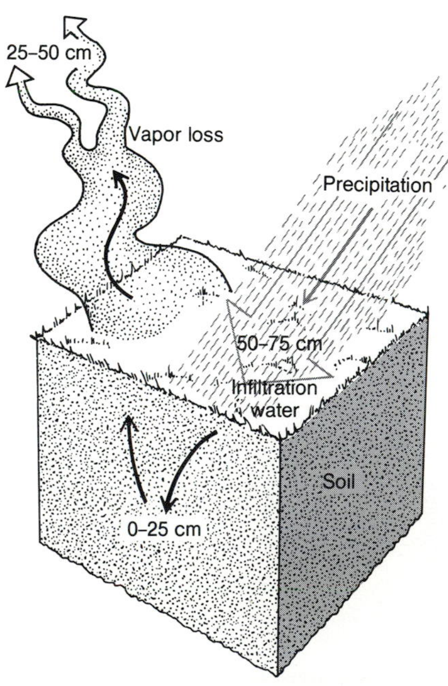

Figure 13.8 The approximate amounts of water entering (infiltration) and leaving (vapor) the soil over a year in a humid midlatitude location.

THE SOIL-MOISTURE SYSTEM

Water is deposited by precipitation on the soil surface, where it may run off or penetrate the soil mass (Fig. 13.8). The rate of penetration is called the *infiltration capacity* and is expressed in millimeters or centimeters of surface water lost to the soil per hour. The infiltration capacity is controlled by many factors, including soil texture, plant cover, existing soil moisture, and ground frost. (This subject is covered in detail in Chapter 17.)

Forms and Measures of Soil Moisture

As water moves into the soil, it joins existing water that is attached to the surfaces of soil particles in a thin, nearly imperceptible film. This water is called *molecular water* because, unlike the infiltration water, which is controlled by gravitational force, it is controlled by molecular forces. There are two varieties of molecular water: *hygroscopic* and *capillary*. Hygroscopic water is represented by a thin film of water molecules attached directly to the surface of every soil particle. The force holding this film is so great that no natural processes are capable of moving or making use of this water. Thus, it is of virtually no importance to soil formation because it cannot perform physical and chemical work.

Capillary water, on the other hand, is fully capable of movement and is thus highly important in soil formation. This water resides on the hygroscopic film, where it is held under pressure ranging between 0.33 and 31 bars; hence, it is often called "loosely bound" molecular water. Capillary water behaves in much the same way as kerosene does in a wick lantern or water does in a paper towel. That is, it tends to move towards spots of relatively lower quantities of moisture, just as kerosene moves up the wick of the lantern or water, tea, or coffee spreads through a paper towel that has been used to soak up a spill. In soil of uniform-sized particles, the rate of movement is controlled primarily by the steepness of the moisture gradients. However, as we will see, in heterogeneous soils particle size can be equally effective in controlling the rate of movement.

As water is added to a soil and the capillary film around the soil particles grows in thickness, the binding pressure declines until the molecular force is too weak to hold the water around the particles. This is extremely important to vegetation because the amount of capillary water that can be taken up by plants depends in large part on the cohesive pressure under which the water is held. Initially, capillary water is easily drawn off by plant roots, but as the film gets thinner, the remaining fraction becomes increasingly difficult to remove because it is held under greater binding force. When the residual capillary water is held too tightly for plants to draw off amounts sufficient for normal respiration, which begins around 15 bars pressure for most plants, the **wilting point** has been reached.

For any soil, there is a maximum quantity of capillary water that it can hold. When this limit is reached, the soil is said to be at *field capacity*. Inasmuch as capillary water is held on the surfaces of soil particles, the greater the number of particles (and thus surfaces) in a given volume of soil, the higher the field capacity must be. For example, clay can hold nearly six times more water at field capacity than sand can (Table 17.3). Water added to a soil beyond its field capacity is *not* taken up as capillary water but remains in the liquid form, draining through the soil, and is called **gravity water.** As a further aid in this discussion, you may find it helpful to review the illustrations concerning soil water in Chapter 17.

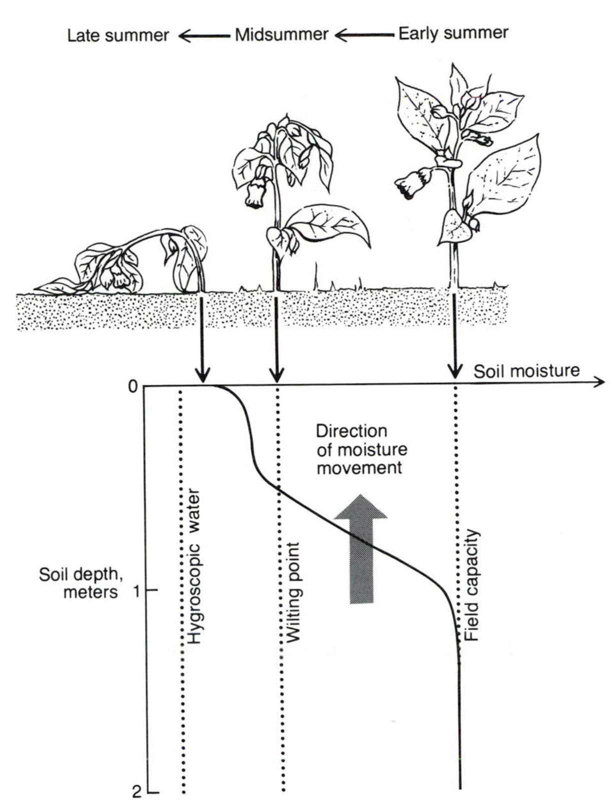

Figure 13.9 Soil-moisture gradient in summer. Soil moisture typically decreases toward the surface from a depth of about 1.5 m. Since evaporation and moisture uptake by plant roots are most concentrated in the upper 25 cm or so of the soil, capillary-water supply is typically lowest there. By late summer, usable moisture is beyond the reach of many plant roots.

Inflow and Outflow of Moisture

A remarkable quantity of water moves into and out of most soils in the course of a year. In the humid regions of North America, for instance, it is not uncommon for the soil to receive 50 to 100 cm of infiltration water annually and to release 25 cm to 50 cm of water vapor to the atmosphere (Fig. 13.8). Inflows and outflows of soil water are identifiable throughout most of the world where precipitation and heat tend to vary sharply from season to season. As the water moves up and down in the soil, it transports chemicals and particles, thereby altering the internal makeup of the soil. Let us trace the movement of water into and out of the soil and define the important controls on soil moisture.

Soil moisture is lost through evaporation and plant transpiration. Loss is greatest when air and soil temperatures are high and wind is strong and dry. In addition, losses are much greater when the soil is plant covered because plants draw moisture rapidly from the upper soil. When the soil reaches an advanced state of drought, it is not completely devoid of moisture; in fact, much of the more tightly bound molecules of the capillary film may remain in the soil. No longer able to take up water from the soil, however, the plants may wilt and eventually die (Fig. 13.9).

Because moisture is lost from the upper soil, soil-moisture gradients are usually toward the surface, which results in an upward transfer of water during dry periods. The rate of moisture transfer or conduction along the gradient is controlled principally by soil texture. Medium textures (silt, fine sand, and loam) have greatest capillary transfer capacity, or *capillarity*, because the particles are small enough to develop strong networks of capillary films, yet large enough to allow ready movement of the moisture film between particles.

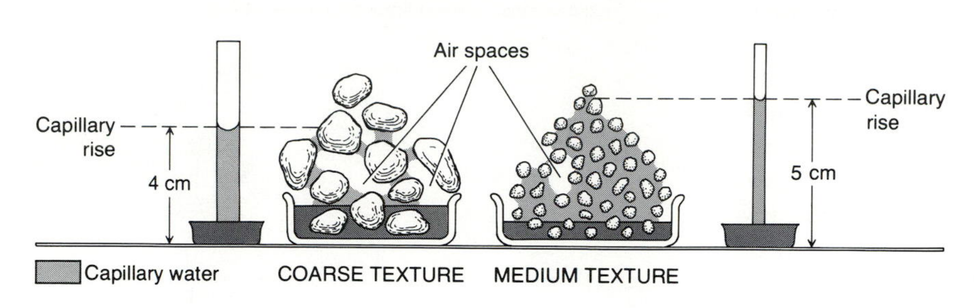

Figure 13.10 Differences in capillary water related to soil texture. The rise is much higher in medium textures than in coarse textures. This can also be illustrated with a simple experiment using large-diameter and small-diameter glass tubes standing in a pan of water.

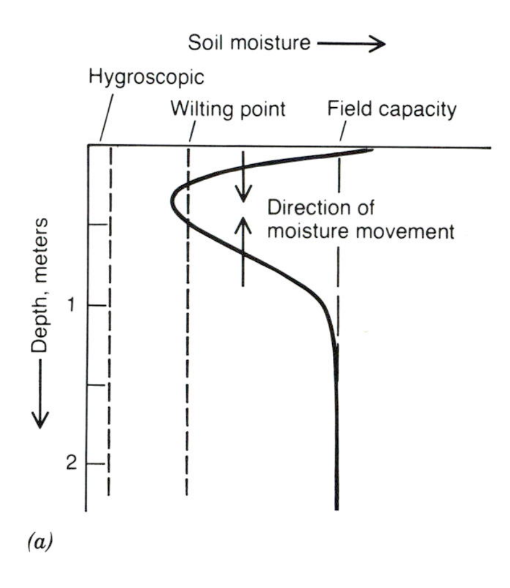

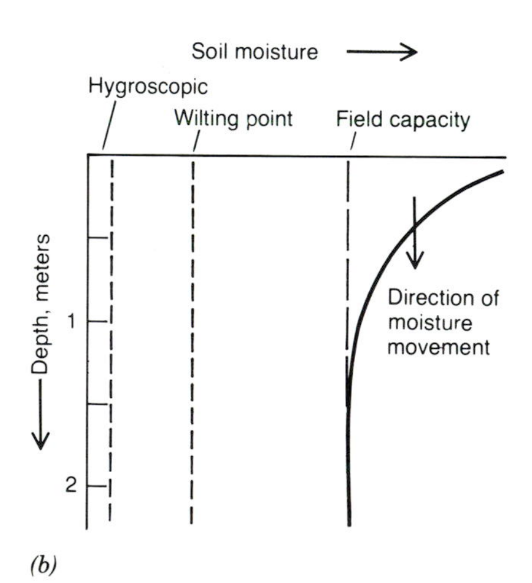

Figure 13.11 Soil-moisture profiles: (*a*) Immediately after a drought has been broken by a substantial rainfall. Infiltration water has recharged the upper 25 cm of soil, and with the upflow of moisture from the lower reservoir, the desiccated layer is being recharged from two directions. (*b*) After the soil has been recharged in excess of field capacity and is incapable of retaining additional water, the surplus is transmitted downward as gravity water.

Soil capillarity is a critical determinant of a soil's susceptibility to drying and drought. The capillarities of both the fine and the coarse textures are relatively low, though for different reasons. In coarse sand or pebbles, for example, the wide spacing between contact points places high stress on the capillary film, making it difficult for moisture to move (Fig. 13.10). In clays, on the other hand, capillarity is often limited by the smallness of the interparticle spaces, especially if the soil is tightly packed. In short, the low capillarities of the fine- and coarse-textured soils mean that moisture differentials are equalized less rapidly than they are in medium-textured soils. Therefore, under similar atmospheric and vegetative conditions, drought can be more severe in soils of the extreme textures.

If moisture from precipitation or another source is added, the uppermost segment of the moisture profile is reversed. And as gravity water is diffused downward, it is quickly transformed into capillary water. Rarely, however, can a single rainstorm recharge the capillary reservoir; thus, the soil is left stratified, with an intermediate dry zone sandwiched between two moist zones. Double gradients are thereby formed (see Fig. 13.16*a*).

Should rainy conditions, coupled with lowered rates of soil-moisture loss, prevail for an extended period of time, as is typically the case during the winter months in the northwestern United States and northwest Europe, the amount of water in the upper soil may greatly exceed field capacity. As a result, enough gravity water can be produced to recharge the capillary reservoir as well as to generate through flow to the water table. Once the soil is at field capacity throughout, the only gradient that exists is the one formed by gravity water (Fig. 13.11).

Soil-Moisture Regimes

If we extend these seasonal observations of moisture flux to different climate zones, three general regimes

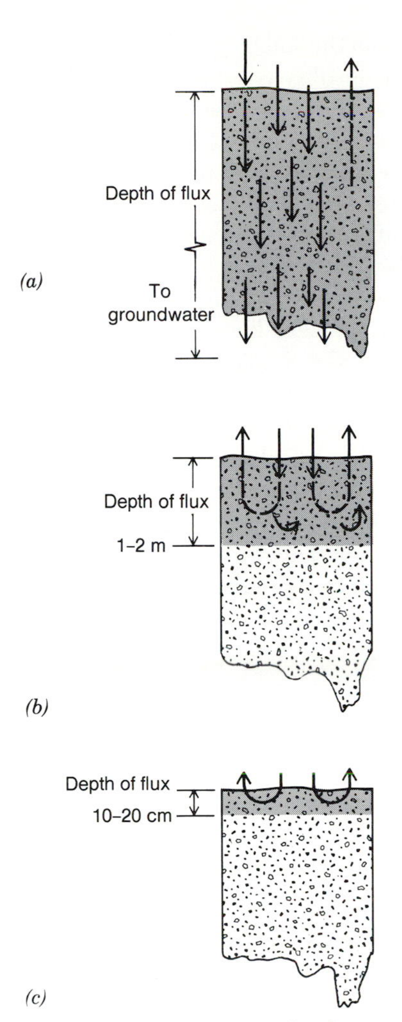

Figure 13.12 Regimes of soil-moisture flux: (*a*) positive moisture balance; (*b*) equilibrium moisture balance; and (*c*) negative moisture balance.

or models of soil-moisture flux emerge (Fig. 13.12). Each is based on the soil-moisture balance concept discussed in Chapter 7. These models are critical to understanding soil development because the chemical and physical changes that take place in the soil layer are driven principally by water. An understanding of soil-forming processes thus rests on an understanding of the directions and relative rates of soil-moisture movement.

In areas of positive soil-moisture balances, the predominant direction of moisture flux is downward. The strength of the flux depends not only on the amount of moisture available, but also on the infiltration capacity and permeability of the soil. Upward movement of soil moisture *will* occur during several or more months each year, but this loss does not offset inflows of surplus water (Fig. 13.12*a*). Thus, the net transport of dissolved minerals and sediment particles follows the net moisture flow.

Zones intermediate between the humid and the arid (that is, semiarid and subhumid zones) are often characterized by a soil-moisture balance that is more or less in equilibrium. The soil-moisture system usually has a strong seasonal component, with massive losses in the summer that are offset by substantial recharge during the fall and winter. The moisture surplus does not develop a through flow of moisture (Fig. 13.12*b*). This soil-moisture regime is sufficient to support a substantial grass cover, but not adequate to promote high rates of decomposition. Therefore, organic matter and its chemical residues tend to be retained in the upper 1 to 2 m of soil.

In arid regions, the moisture balance has a strong negative potential. Precipitation added to the surface penetrates only several centimeters before it is absorbed by the soil, taken up by plants, or vaporized. Minerals dissolved by this water, therefore, are not moved to great depths but remain at or near the surface (Fig. 13.12*c*).

BIOCHEMICAL PROCESSES IN THE SOLUM

The soil is a complex chemical environment comprised of a host of minerals, in different forms, and from

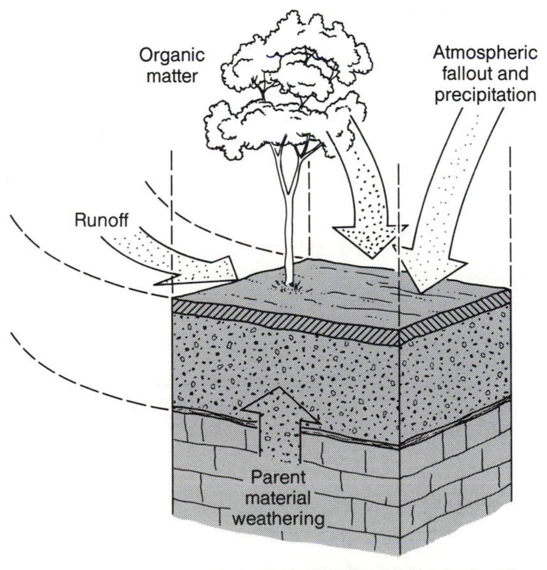

Figure 13.13 Sources of minerals in the soil. The atmosphere, for example, contributes nitrogen, silicon dioxide, carbon dioxide, and phosphorus.

different sources. The soil column functions as a chemical system with chemical inflows, outflows, and related storage and work. The soil receives minerals from the atmosphere (both in precipitation and fallout), vegetation and animals (both on and within the soil), weathered parent material, and runoff (both surface and subsurface flows). The minerals may appear in the soil as particles or as ions and in both organic and inorganic forms (Fig. 13.13). Minerals are lost from the soil in groundwater, runoff, erosion, plant uptake, and, of course, mining by humans.

Chemical Conditions and Activities

Ions—electrically charged atoms or groups of atoms—are the principal agent of the soil's chemistry. An ion's electrical charge may be either positive or negative. *Cations* are positively charged ions; *anions* are negatively charged ions. The minerals that form cations, such as magnesium, calcium, and potassium, are called *bases*; they are important for soil fertility and hence for plant growth.

Also important to soil chemistry are clay-sized particles known as *colloids*. These minute particles carry small electrical charges on their surfaces, usually positive on one side and negative on the other. When ions are released into soil water, they are attracted to these charges, giving rise to swarms of ions on each colloid. This process is called *adsorption*, and it operates according to the principle of electromagnetic attraction in which unlike charges attract and like charges repel. Because the negative charge on colloids tends to be appreciably stronger than the positive, colloids adsorb mainly cations. Thus, when cation-rich colloids are washed from the soil by percolating water, bases are lost and soil fertility is usually reduced.

Ions reside in three different locations in the soil: on the surface of colloids, within the colloids (between the lattices of the clay particle), and in the soil water. Those in the soil water are called *free ions*, and they are the only ones that can be taken up by plants and adsorbed by colloids. When adsorption takes place, the free ions exchange places with the ions already on the colloid surface, a process known as *cation exchange*. (The ions within the colloid are nonexchangeable.) The capacity of a soil to affect this process is termed the *cation exchange capacity*. It is controlled by many factors including the abundance of colloids, the types of colloids, and the moisture and temperature conditions of the soil.

In general, clayey soils have much higher cation exchange capacity than coarser grained soils simply because they have more colloids and thus more adsorption sites per unit volume of soil. Other things being equal, the coarse-textured soils therefore tend to be poor in bases, whereas clayey soils tend to be rich in bases. Soils rich in colloids derived from decomposed organic material have the highest base retention capacity because these colloids have the highest cation exchange capacity of all the clay minerals.

When cations are scarce, colloids tend to be relatively heavy in hydrogen ions. These ions are supplied by the atmosphere, plants, and soil water. Most are supplied by water molecules that give up hydrogen ions through a process called *dissociation*. In this process the two hydrogen atoms of the water molecule break away, forming one hydrogen ion (H^+) and one

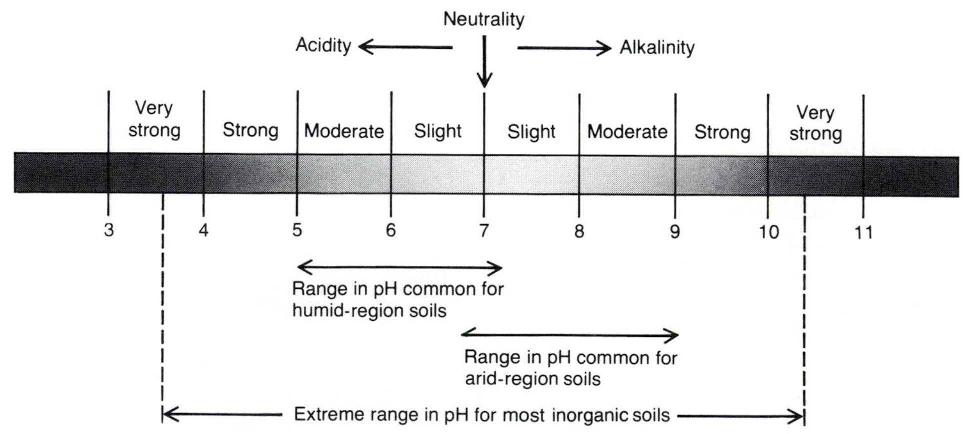

Figure 13.14 A standard soil pH chart. Most pH values fall between 5 and 9. Soils in humid regions generally range from 5 to 7; soils in arid regions, from 7 to 9.

TABLE 13.1 pH RANGES FOR SOME COMMON GARDEN PLANTS

Plant	pH
Sweet clover	7–8
Asparagus	7–8
Cauliflower	6.5–8
Peas	5.5–6.5
Carrots	5.5–6.5
Corn	5.5–6.0
Tomatoes	5.5–6.0
Potatoes	4.5–5.5
Blueberries	3–4

hydroxyl ion (OH^-). Hydrogen ions are usually very abundant in soil, and if space is available, they collect on the adsorption sites of the colloids. If cations are introduced, however, the hydrogen ions lose their place to the cations and reenter the soil solution.

The relative abundance of hydrogen ions in the soil is used as a measure of the soil pH factor. Soil pH is a numerical scale, ranging from 0 to 14, which is based on the ratio of hydrogen ions in the soil solution to hydroxyl ions (Fig. 13.14). At pH 7 the concentrations of hydroxyl ions and hydrogen ions are equal, and the pH is said to be *neutral*. Pure water is thus neutral. If hydrogen ions are abundant, the pH is low (below 7) and the soil is classed as *acidic*. If hydroxyl ions are relatively abundant, the pH is high (above 7) and the soil is classed as *alkaline*, or *basic*.

In agriculture, the soil pH value determines whether or not a soil will need lime in order to be suitable for most crops. The lower the pH, the greater the amount of lime that must be added to "sweeten" the soil sufficiently to return high yields on crops such as corn (Table 13.1). Lime, which is finely powdered limestone, is a sedimentary rock composed chiefly of the mineral calcium carbonate ($CaCO_3$). When lime is applied to soil, this mineral is dissolved by soil moisture into calcium cations (Ca^+) and carbonate anions (CO_3^-). Because soil colloids adsorb mainly cations, more calcium cations are retained in the soil than carbonate anions; the soil's net ion balance becomes positive (or more strongly positive); and the soil is thereby made more basic, or "sweeter."

Eluviation, Leaching, and Illuviation

The removal of materials in solution from a layer of soil is known as *leaching*. This process is characterized by the relocation, or "washing out," of ions to other levels in the soil. If the washing out involves colloids

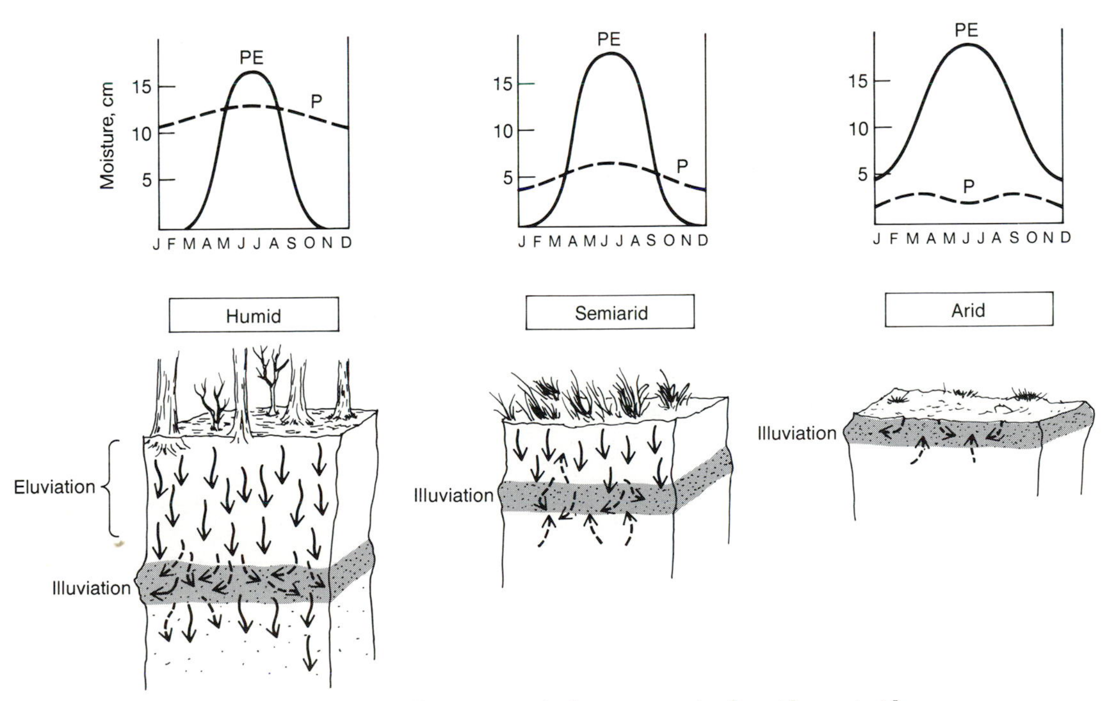

Figure 13.15 Formation of zones of illuviation and eluviation under humid, semiarid, and arid climates.

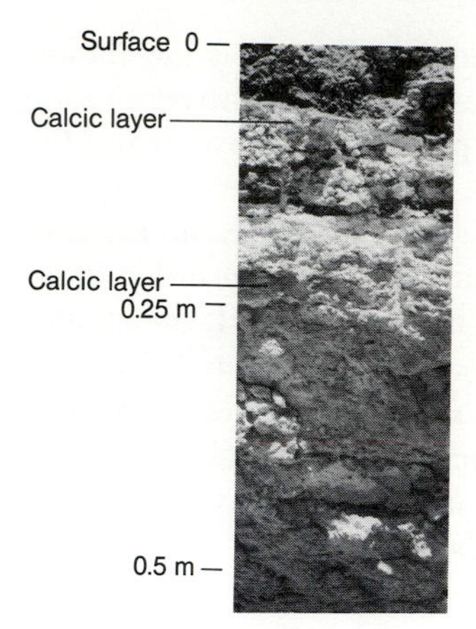

Caliche (calcic) layers in a New Mexico soil.

as well as ions, the term *eluviation* may be used to describe it. The layer or zone in the soil which loses the materials is called the *zone of eluviation* (Fig. 13.15).

The downward transfer of ions and colloids often ceases at some depth, generally anywhere from a few centimeters to 5 m down. Here the materials may be deposited, or *illuviated*, forming a *zone of illuviation* (Fig. 13.15). In time, this zone can collect such a mass of colloids and minerals that the interparticle spaces become clogged, cementing the parent particles together. In heavily leached soils, the zone of illuviation takes the form of a concretelike layer called *hardpan* or *duripan*. Many a farmer has cursed this layer because it can make digging holes for fence posts a very nasty job. In very wet climates, such as the wet tropics, minerals not illuviated in the lower soil are carried by groundwater to streams and rivers.

Where the moisture balance is negative, as in the arid southwestern United States, mineral-rich water does not pass through the soil but penetrates only the first 10 to 30 cm below the surface (Fig. 13.5). The water evaporates, leaving its ion load, commonly calcium carbonate, in the upper soil. In some places wind adds deposits of calcic dust to the surface. When rain falls on the dust, some of it is dissolved and washed into the upper soil where it is deposited as the water evaporates. With each succeeding rainfall, more calcium is deposited, and eventually a hard crust, traditionally known by the Spanish word *caliche*, develops. Caliche is also known to form in other ways; it is sometimes formed at depths of a meter or so in soil, and is also known as the *petrocalcic* or *calcic* layer.

The Soil's Organic System

Soil is influenced by many types of biological activities, especially those related to the plant cover. Plant

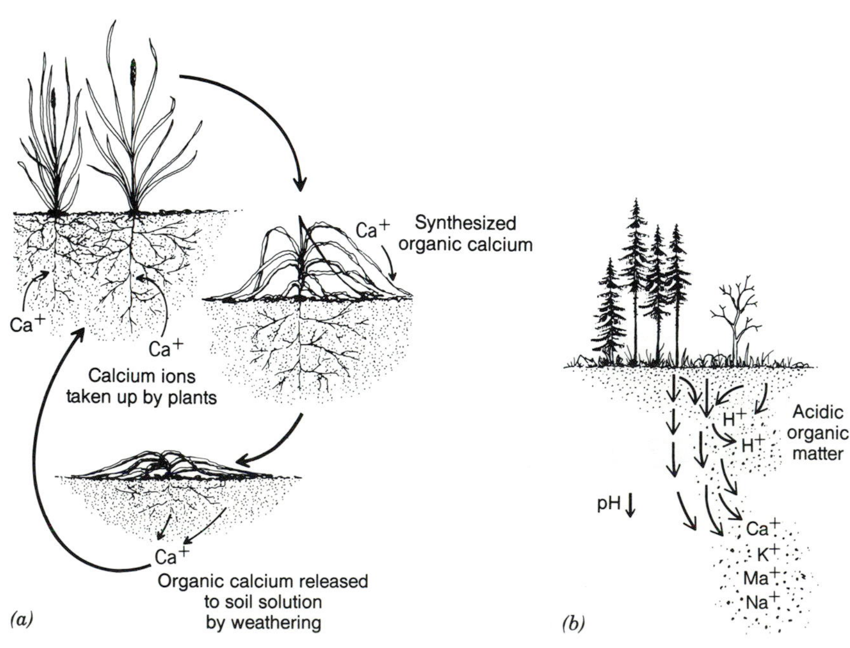

Figure 13.16 (*a*) The nutrient cycle for calcium (Ca) between plants and soil. A change to plants that are less efficient in holding nutrients can change the soil chemistry. (*b*) Leached-out bases are replaced by hydrogen ions, making the soil acidic.

growth is affected not only by heat and moisture, but also by the chemical makeup of the soil. For example, grasses need calcium and magnesium bases (alkalines) in order to grow well. If these minerals are present in the soil, they are taken up with the moisture absorbed by the plant roots and are stored in the plant tissue. When the plant dies, the minerals are returned to the soil, as Figure 13.16*a* illustrates. In this way plants help to maintain the fertility of soil by bringing up the mineral bases and redepositing them at the soil surface in the form of dead leaves, stems, and so on. Indeed, this sort of recycling is critical to the maintenance of the lush rainforest vegetation of tropical and equatorial areas.

For plants with low base requirements, the organic debris in the topsoil yields organic acids and has a low pH. Organic acids are especially strong under coniferous vegetation. These acids accelerate the weathering of soil particles and bedrock and tend to replace the leached calcium, potassium, magnesium, and sodium ions which are so important to growth in most other plants. Thus, in cold, damp climates, for instance, where conifers are abundant and the humus is inherently weak in bases, the soil tends to be strongly acidic (Fig. 13.16*b*). In contrast, deciduous trees growing under the same conditions may give rise to less acidic soils, with base-saturation levels two to three times higher in the upper soil.

Topsoil contains diverse populations of small plants and animals that consume the organic matter produced by vegetation. The chief consumers of organic matter are tiny plants called *soil microflora*, primarily microscopic-size bacteria, algae, and fungi, which are responsible for 60 to 80 percent of the soil's biological activity. This activity results in the production of humus. *Humus* is an altered form of organic matter, reduced both chemically and physically from its original form and composition. Being older, it is usually found below the newer organic matter on the soil surface (Fig. 13.17).

Helping the microflora to consume the organic matter are many small animals, principally insects and worms. These creatures are called *primary consumers*, such as earthworms, and they eat not only leaves, twigs, and pieces of bark that fall to the ground, but the microflora as well. Their droppings and dead bodies also contribute to the formation of humus. *Secondary consumers* are predators, such as spiders and centipedes, that feed on primary consumers. They, in turn, are prey for another, smaller group of pre-

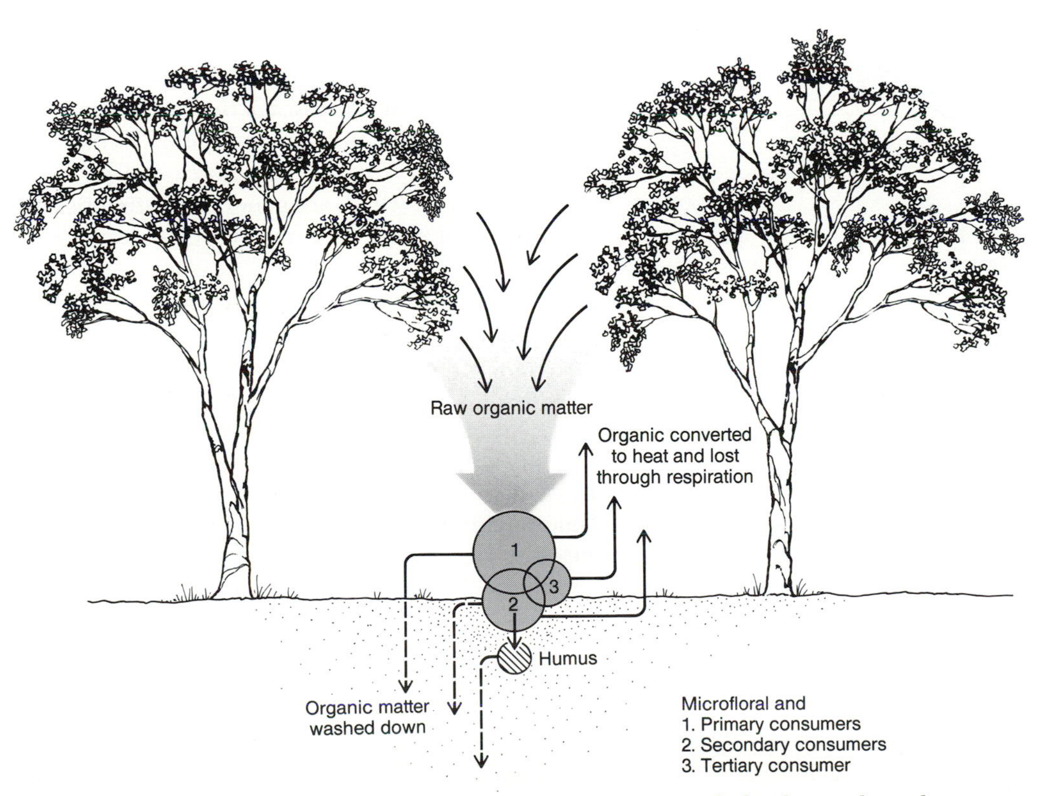

Figure 13.17 The progressive decomposition of organic matter with depth into the soil.

dators, *tertiary consumers*. Humus is produced by each consumer; however, the fraction of organic energy returned to the soil for each calorie of organic matter consumed diminishes rapidly from consumer to consumer because most of the energy is dissipated by the animals in bodily heat and locomotion (Fig. 13.17). Together, these interrelated groups of organisms and the vegetal material they live on constitute the *soil ecosystem*.

The amount of organic matter in any soil is the product of the rate of plant productivity (total growth) less the rate of destruction of organic matter by organisms, leaching, and erosion. Microflora and other soil organisms are very sensitive to heat and moisture conditions. In a cold climate, such as the arctic tundra, microfloral action is slow, and the rate of organic decomposition usually does not exceed the rate of plant growth; therefore, organic matter may build up (to a point) over time.

In arid regions, the lack of moisture in the upper soil prohibits microfloral activity. But since plant productivity is exceedingly low in deserts, not much organic matter accumulates anyway. Under semiarid conditions, however, moisture is often adequate to foster substantial productivity by grasses and other herbs, but inadequate to foster comparable rates of destruction by microflora; hence, considerable organic buildup may occur and soils are very fertile. Figure 13.18 shows the relationship between organic production and temperature under humid conditions. In tropical and equatorial areas, where conditions are hot and moist, plant productivity is high, but microfloral activity is even higher; thus, it is impossible for an organic layer to build up. (Exceptions are found in swamps, where standing water limits microfloral activity, but not productivity, resulting in heavy accumulation of organic material.)

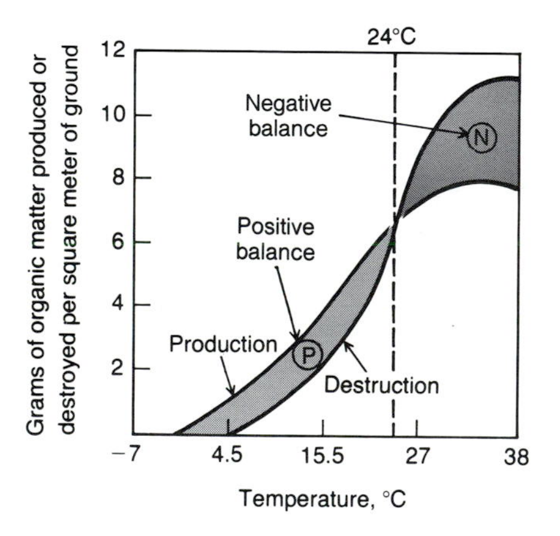

Figure 13.18 Rates of production and destruction of organic matter as a function of temperature under humid climatic conditions. At a temperature of less than 24°C, the destruction of organic matter exceeds the production of organic material by plants, and so it accumulates as humus. But above 24°C—the temperature regime characteristic of tropical and equatorial areas—the rate of destruction is potentially much greater than the rate of plant production. Therefore, all organic matter is consumed, leaving the soil with virtually no humus.

Microflora can also increase soil fertility. In the presence of certain plants, some soil bacteria can convert gaseous nitrogen from the atmosphere into a form that can be utilized by plants. This process, called *nitrogen fixation*, is especially pronounced in plants of the legume (pea) family and is one of the most important means by which plants can increase the nutrient level of the soil well beyond its original level. As for larger organisms, such as insects and small animals, their role is largely one of reworking and mixing the soil through burrowing. This activity tends to increase aeration of the upper soil and in turn often promotes bacterial activity and moisture infiltration.

Soil Horizons

Working in combination, the various chemical, biological, and physical processes of the soil environment are capable of producing horizons within the upper 5 m or so of soils. *Horizons* are horizontal or nearly horizontal layers, which are generally distinguishable on the basis of color, texture, and chemical composition.

In humid environments where the upper soil is stable, well drained, and plant-covered, we can usually expect to see at least three horizons: an organic rich upper layer; a zone of eluviation under the organic layer; and a zone of illuviation beneath the zone of eluviation. In arid regions, only one horizon may be apparent: a calcic layer at or near the surface. In unstable environments, no horizons may be found because the parent material may be newly deposited or continually shifting (as in sand dunes) and, therefore, not altered sufficiently to produce internal differentiation.

The sequence of horizons in a section of soil is called a *soil profile*, and it is the traditional framework used for describing most soils. As with so many other natural phenomena, the boundaries between soil hori-

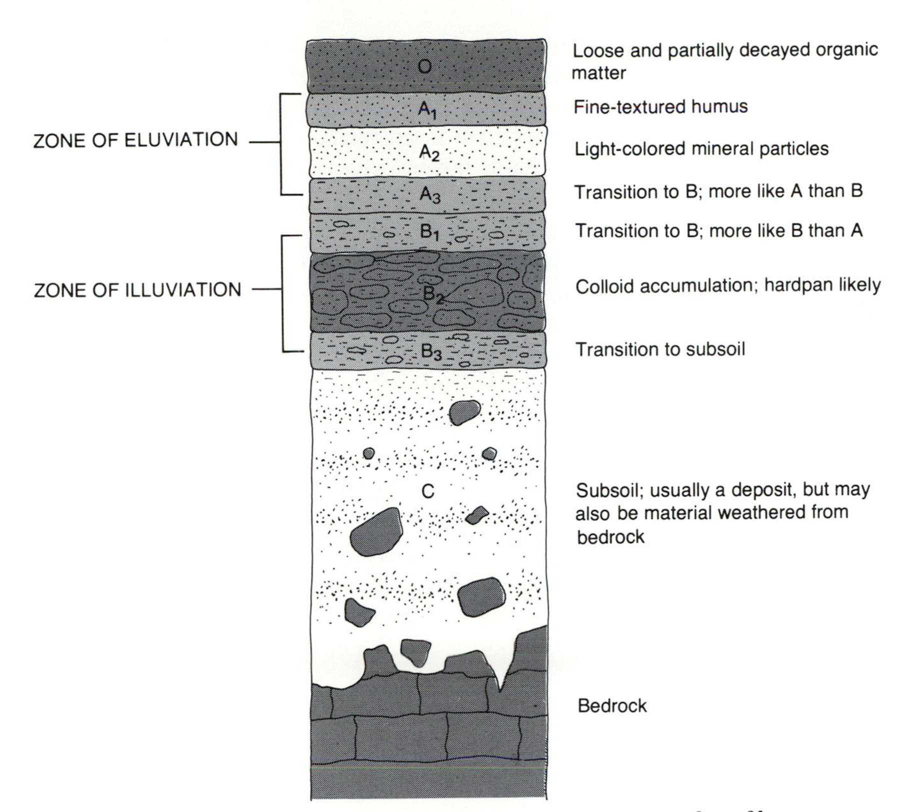

Figure 13.19 Horizons and subhorizons in a representative soil profile.

zons are typically indistinct, as one horizon usually grades almost imperceptibly one into another. These transitional zones are classed as *subhorizons,* several of which are usually identified in each of the upper two horizons (Fig. 13.19).

Soil horizons and subhorizons are often coded for ease of reference and classification. Four principal horizons, denoted O, A, B, and C, are widely recognized. The O horizons are the organic layers of the upper soil. The A horizons are characterized by mixed organic and mineral matter, and under the humid climates, they are also the zone of eluviation. The B horizons are the zones of illuviation; in humid climates they lie under the A horizons, but in arid climates they lie at or very close to the surface. The C horizons are usually characterized as the zones of weathered parent material and transitional to the relatively unaltered nonsolum at depth. The subhorizons are traditionally coded by adding a number to the letter; for example, B1 (transitional between A and B, but more like B), B2 (center of B; most representative of B), and B3 (transitional to C but more like B than C).

Since we know that horizons are the work of soil-forming processes, we can learn about the conditions under which soils have formed by the presence of certain horizons, called *diagnostic horizons,* in the soil profile. According to concepts developed by soil scientists in the United States, two groups of diagnostic horizons are significant: surface horizons (called epipedons) and subsurface horizons. The surface horizons, which would fall into the O horizon category, are distinguished on the basis of organic content (thickness), color, mineral composition, and general texture and consistency. The subsurface horizons, which would fall mainly into the B horizon category, are distinguished according to clay content, oxide content, the presence of certain salts, and hardness owing to chemical cementing.

Summary

In general, the soil mantle is a complex assemblage of loose materials of varying thicknesses that lies over the bedrock. The materials are called parent material, and they are divided into two classes: residual and transported. Most parent material is transported, and it is composed of mineral particles in deposits laid down by water, glaciers, and wind. These particles are mainly sand, silt, and clay; they form the skeletal framework for most soils and are the particle sizes used for defining soil texture. Water is central to the chemical, biological, and physical processes of the soil. The capacity of soil to hold and transfer capillary water is closely related to texture and organic content.

The upper part of the soil mantle is a mixture of mineral particles, organic matter, water, and air, and is called the *solum.* The surface of the solum is influenced strongly by vegetation, and the organic matter in the upper soil is rich in interrelated populations of organisms, the soil ecosystem. Within the solum, the flux of moisture is associated with chemical activity that results in the transfer of chemicals in and out of the soil. This in turn leads to the formation of soil horizons of different physical and chemical compositions, the products of the soil-forming environment. The composite set of horizons, called the soil profile, is used to classify the soil.

Concepts and Terms for Review

- solum
- soil mantle
- weathering
- parent material
  - residual
  - transported
- soil properties
  - composition
  - texture
  - structure
- topsoil and organic matter
- soil moisture
  - capillary
  - field capacity
  - capillarity
- soil-moisture flux
- soil-moisture regime
- soil chemistry
  - cations and anions
  - colloids
  - cation exchange
  - pH
- eluviation, leaching, and illuviation
- humus
- mineral recycling
- soil ecosystem
  - production
  - consumption
- nitrogen fixation
- horizon and soil profile
- diagnostic horizon

Sources and References

Bidwell, O., and Hole, F.D. (1965) "Man as a Factor of Soil Formation." *Soil Science* 99: pp. 65–72.

Birkeland, P.W. (1974) *Pedology, Weathering, and Geomorphological Research.* New York: Oxford University Press.

Brady, N.C. (1974) *The Nature and Properties of Soils,* 8th ed. New York: Macmillan.

Briggs, D. (1977) *Soils: Sources and Methods in Geography.* Boston: Butterworth.

Foth, Henry D. (1978) *Fundamentals of Soil Science.* New York: Wiley.

Hunt, C. (1972) *Geology of Soils.* San Francisco: Freeman.

Nikioroff, C.C. (1959) "Reappraisal of Soil." *Science* 129, 3343: pp. 186–196.

Reeves, C.C., Jr. (1970) "Origin, Classification and Geologic History of Caliche on the Southern High Plains, Texas and Eastern New Mexico." *Journal of Geology* 78: pp. 352–362.

P6
P6
CREEK
M1
P1
P2
Colorado
River
281
P3
P2
Comanche
Lonesome Dove
Chalk Limestone Prairie
CLAYSTONE SANDSTONE SUITE
Claystone/Shale Uplands
Muddy Sand Prairie

Chapter 14

Soil Formation, Classification, and Problems

As is the case with climate and vegetation, it is necessary to classify soils in order to reduce nature's complexity to an understandable and manageable level. What are the objectives of the major soil classification schemes and what are the principal categories or classes employed by each? Which types of soils are used most by humans for various activities? And likewise, which soils are most abused by humans?

Few ideas in natural sciences have enjoyed as much attention and following as the theory of biological evolution. When Charles Darwin published *The Origin of Species* in 1859, it initiated a great wave of controversy that reverberated through academic circles. Among the natural sciences the impact was clearly greatest in the biological fields: botany, ecology, zoology, and paleontology.

But strains of evolutionary thought also permeated other fields of natural science in the late nineteenth and early twentieth centuries and invoked changes among them as well. In the earth sciences it became fashionable—in a manner of speaking—to emphasize the evolutionary character of the earth's surface by describing natural change in developmental stages.

It was in this atmosphere of thought that the Russian geologist V. V. Dokuchaiev introduced natural science to the formal study of soil in 1883. He proposed that soil should be viewed as a natural body, which develops, or evolves, over time in response to its bioclimatic environment. Thus emerged the concept that soil is essentially *organismal.*

Although the organismal model of soil is no longer granted much validity—because, among other things, soil is not an organism—the idea that soil forms in response to the forces and processes of its environment remains at the heart of modern soil science. In fact, we view the soil-forming environment as a complex of several geographic systems—climatic, ecological, geochemical, and geological—that interact in the upper several meters of the soil to form the solum.

The *solum* is part of the soil on which most life depends. It is shaped by both external processes, particularly erosion and deposition by runoff and wind, and internal processes, principally biochemical processes. Our objective in this chapter is to examine how the soil-forming processes relate to climate, vegetation, animals, geologic factors, and site conditions. We want to define several characterisitc sets of conditions and processes that produce certain soil features and soil types. In the second half of the chapter, we describe some soil classification systems and a number of the important soil problems of the world.

SOIL FORMATION AND THE ENVIRONMENT

When we search for a relationship between the distribution of soils and conditions of the soil-forming environment, we are immediately faced with the problem of geographic scale. At a very broad scale, the general pattern of certain soils correlates well with climatic conditions. At increasingly finer scales, the influence of other factors, such as parent materials, drainage, and vegetation, becomes more apparent. Recognizing this principle, let us describe the relationship between soils and environment by starting at the broad scale and working our way down to the local scale.

General Soil-Climate Relations

If we ignore the polar and mountainous regions, two general classes of soils can be identified in the world. In the traditional terminology these are called the *pedocals* and *pedalfers*. The *al* and the *fer* in pedalfer refer to the concentrations of aluminum and iron (*fer*rous) that are found in the zone of illuviation; the *cal* in pedocal refers to the calcium retained in the soil as a result of weak eluviation stemming from the shallow penetration of water from the surface. The distributions of these two classes correspond to broad areas of negative and positive moisture balances. Pedalfers are generally found where precipitation exceeds 60 cm a year, and the moisture balance is sufficient to produce pronounced leaching. Pedocals form under conditions of negative moisture balance, mainly areas that receive less than 60 cm of precipitation annually. In the United States a north-south line along the 98 degree meridian can be used as the approximate boundary between these soil types, with pedocals found in the drier west and pedalfers in the more humid east (Fig. 14.1).

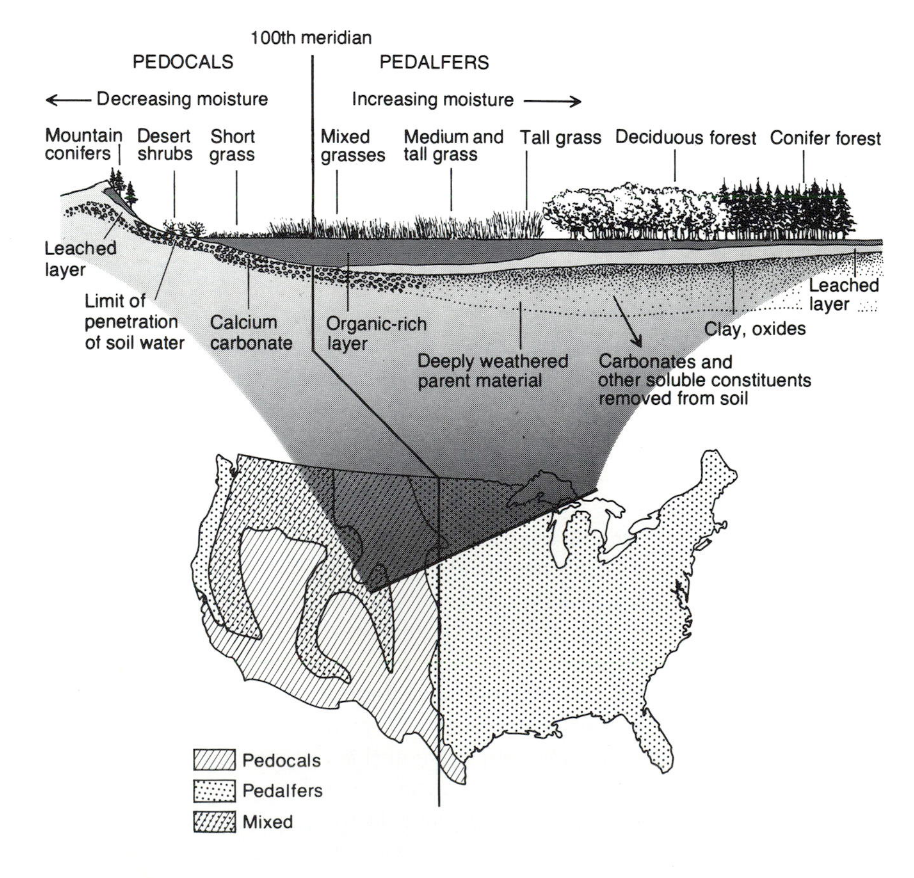

Figure 14.1 The approximate distribution of pedocal and pedalfer soils. This classification is highly generalized, and within the pedocal region are large areas of pedalfer-type soils. This is particularly true for the Pacific Coast, where the annual moisture balance is decidedly positive.

Soil-Forming Regimes

Within this broad climatic framework, we can identify many combinations of environmental factors that produce a distinctive set of soil-forming processes. These factors fall into two main categories: (1) those that are tied mainly to *bioclimatic forces* and (2) those that are tied mainly to *other forces* such as drainage and human action. Several characteristic sets of soil-forming processes, known as *soil-forming regimes*, can be defined for each category. Soils produced by the bioclimatic regimes are found in areas that would be described as well drained and where moisture in the soil column moves relatively freely in response to capillary and gravity forces. In addition, these areas have not been subject to long-term disturbance by land use activities or geophysical forces such as flooding, major soil movements, or erosion.

Bioclimatic Regimes There are three major bioclimatic regimes: podzolization, laterization, and calcification. The *podzolization* (or spodosolization) regime is common to areas with a positive moisture balance in middle to high latitudes or at high elevations in mountainous areas. Conditions in these zones are cold enough to inhibit intensive microflora action, yet warm enough to support substantial forests. Coniferous forests, which require few bases for growth, are abundant and cannot sustain a soil pH much above 5 or 6. Bases, colloids, and oxides are eluviated from the upper soil, leaving a horizon of a sandy, gray character just below the organic layer. This is the *albic* horizon, and it is a distinctive characteristic of a *podzol* or *spodosol* soil. Also characteristic of spodosols are the colloids and iron oxides that accumulate in the B horizon forming a dark layer termed a *spodic* horizon (Fig. 14.2*a*).

Laterization (or oxisolization) occurs under conditions of heavy rainfall and warm temperatures such as those of tropical and equatorial wet climates. Vegetative productivity is high, but intensive microflora action in such regimes rapidly consumes most of the humus so that the topsoil is always thin. Weathering of parent material is also intensive, and most weatherable materials are removed, leaving the soil with a strong *oxic* horizon that is rich in kaolinitic clays with low ion exchange capacity. Iron oxides (sesquioxides) accumulate in the B horizon area forming a *plinthite* horizon, which may become rocklike and massive, and is traditionally termed *laterite*. Some soil scientists argue that the iron oxides are relatively insoluble because strong organic acids are not abundant under this regime. Silica is leached out, and no distinct horizons form other than the *plinthite* horizon and the thin layer of organic matter in the O horizon (see Fig. 14.2*b*).

Calcification occurs in arid and semiarid regions, where the moisture balance ranges from slightly to strongly negative. As there is little leaching by percolating precipitation, calcium, magnesium, and other bases remain in the soil. In semiarid zones grasses may be abundant, and they draw on these bases. Because microfloral activity is limited by low moisture, over time large amounts of organic matter build up, forming a heavy organic layer called a *mollic* horizon (Fig. 14.2*c*). Colloids are not leached out, and they, too, remain in the soil and may accumulate in heavy concentrations with calcium carbonate and other salts. In semiarid settings, calcium concentrations often take the form of pebblelike nodules near the base of the organic layer. In arid settings, salts usually accumulate near the surface in a *calcic* or *petrocalcic* horizon. Under the latter conditions, organic accumulation is negligible, of course—because of the severe moisture stress on plants—and owing to the dominance of salts, the arid extreme of this regime is often termed *salinization* (see lower illustration, Fig. 14.2*c*).

Other Regimes In virtually any bioclimatic region, certain drainage and topographic circumstances can lead to areas that are constantly wet, and so the soil becomes waterlogged. Such places are most prevalent in low-lying terrain, but they can even be found in arid lands where groundwater seeps into small, isolated areas. In most of these water-dominated environments, conditions encourage high rates of plant productivity and low rates of organic destruction, leading to a buildup of organic matter and the formation of a soggy organic soil. Although chemical, thermal, and physical conditions may vary from region to region, water saturation is the dominant soil-forming force; thus, this regime might appropriately be called the *hydromorphic regime*. Let us examine one version of this regime.

In bogs and related wetlands in the cool to cold climates, the hydromorphic regime has traditionally been referred to as *gleization*. The organic layer may reach 10 m or more in thickness, and because of the makeup of the contributing flora and the chemistry of the groundwater, the soil is typically acidic. In time, the organic matter may metamorphose into peat, which is often considered a primitive form of coal and is used as fuel in northern countries such as Ireland. Although decomposition of the organic mass is very

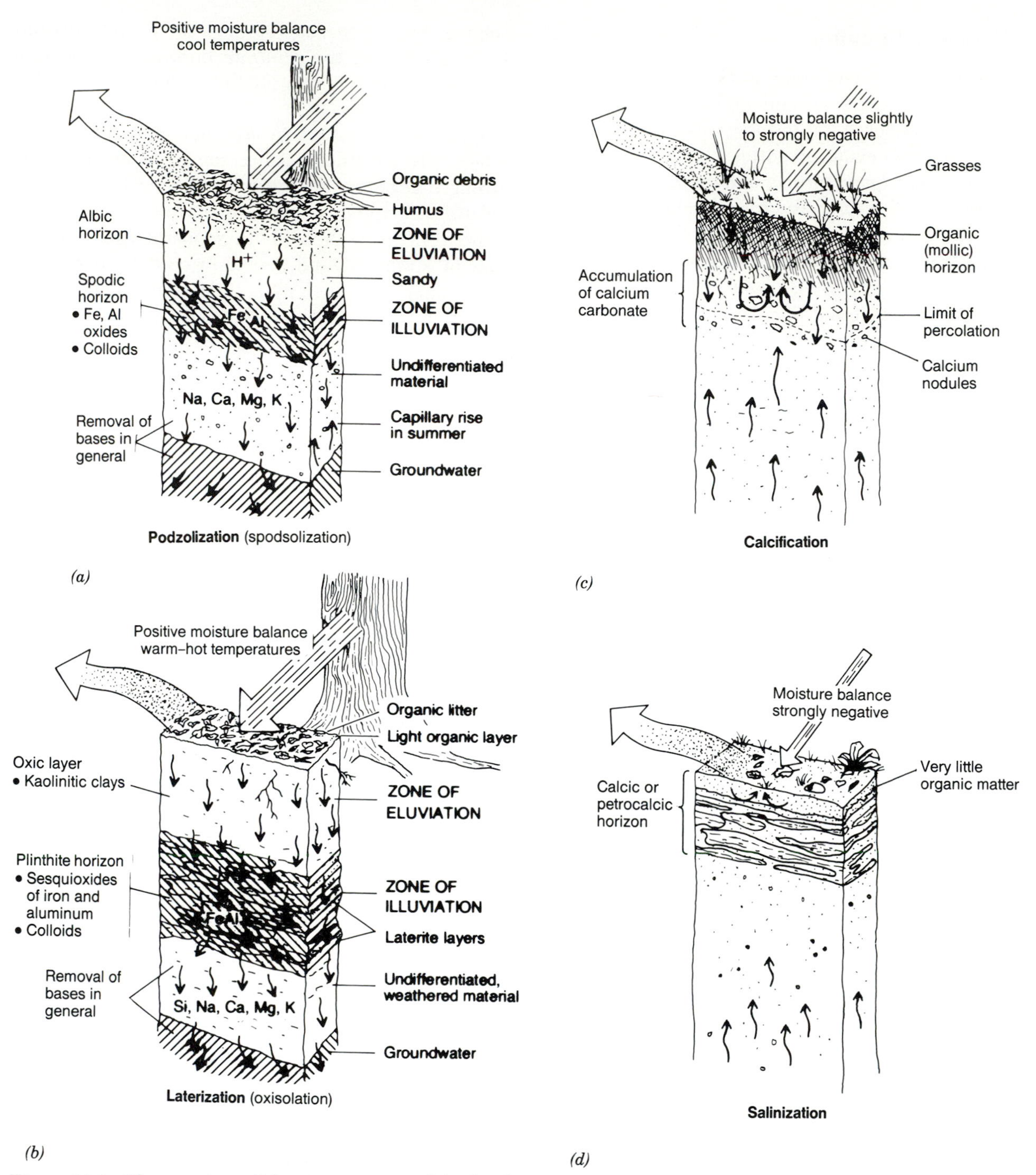

Figure 14.2 Three major soil-forming regimes of the bioclimatic variety: podzolization (or spodosolization), laterization (or oxisolization), and calcification, and salinization.

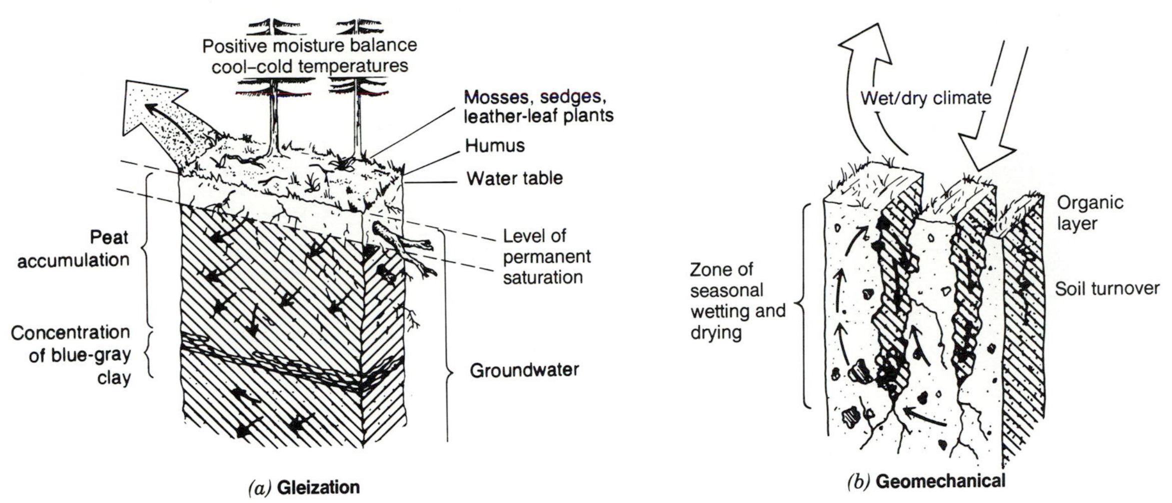

Figure 14.3 Soil-forming regimes related to hydrologic factors (gleization) and parent material (geomechanical) factors.

slow because of the cool temperatures and saturated conditions, eventually a small amount of sticky blue-gray clay forms from internal weathering. This clay is concentrated in a layer called a *gley* horizon in the lower part of the organic mass. The blue-gray color is derived from nonoxidized iron (called reduced iron) derived from groundwater and organic matter in this oxygen-poor environment (Fig. 14.3*a*).

Another type of soil-forming regime, which we will call the *geomechanical regime,* is characterized by mechanical mixing of the upper 1 to 2 m of the soil. One version of this regime is related to parent material containing clays with a high capacity for shrinking and swelling with drying and wetting. This is a common property of clays belonging to the *montmorillonite* type, which are widespread in midlatitude regions. In the dry season the montmorillonite shrinks, forming large cracks in the upper 1 to 2 m of soil. From the walls of the cracks pieces of soil fall off and lodge farther down where the cracks narrow. In the wet season the clay swells, the cracks close, and the pieces are reincorporated into the soil. After many seasons the upper meter or so of the soil is thus mechanically recycled with the material beneath it. This sort of mixing action destroys or greatly weakens the horizons that may otherwise have formed in response to bioclimatic conditions (Fig. 14.3*b*). In permafrost regions, frost action (freeze-thaw) effects a similar mechanical reworking of the upper 1 to 2 m of soil. Whether frost- or moisture-driven, the process is basically the same, and the mixing action exerts a dominating influence on soil formation (see Note 22).

Forces and Factors at the Local Scale

As we stated earlier, soil is the product of combinations of forces; some that act on the surface, some that act within the solum, and others that act below the solum. These forces are generally recognized as *soil-forming factors:* climate, vegetation, and related biological phenomena, parent material, topography and drainage, and land use. These factors represent the different sources and forms of energy which drive the processes of soil formation, and in different settings different combinations of these factors exert a dominating influence in soil formation (Table 14.1). (Traditionally, time is also named as a soil-forming factor, but in reality time is only a measure of the rate of soil formation and not a factor in the same sense as climate, parent material, and so on.)

If you were to examine the soils on a nearby farm or neighborhood park, you are apt to discover considerable variation in soil over small areas. To explain why the soil at a particular location has certain features requires an understanding of the soil-forming processes, not only as we might observe them over many months or any average year, but over much longer periods of time because many of the powerful proc-

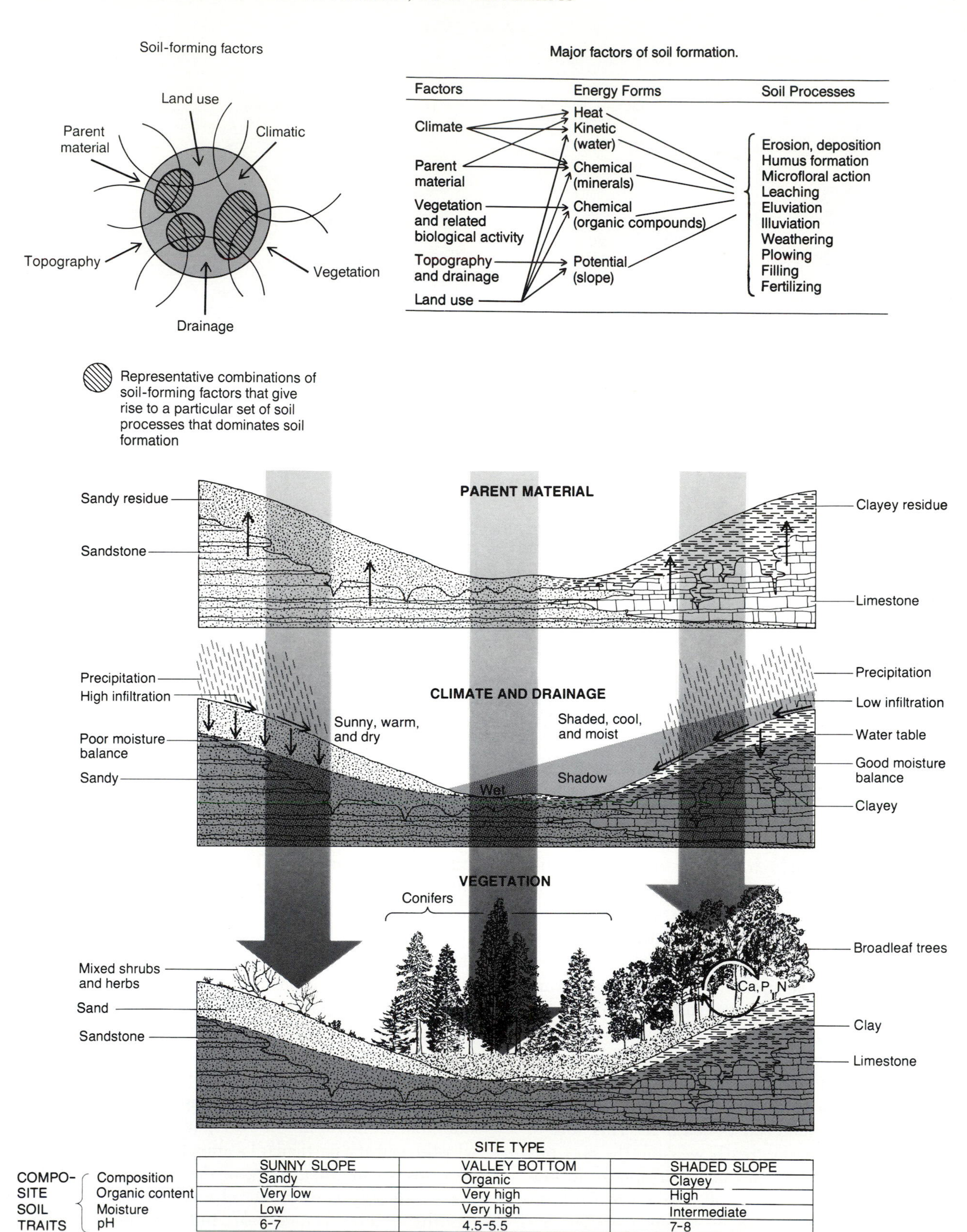

COMPOSITE SOIL TRAITS	SUNNY SLOPE	VALLEY BOTTOM	SHADED SLOPE
Composition	Sandy	Organic	Clayey
Organic content	Very low	Very high	High
Moisture	Low	Very high	Intermediate
pH	6-7	4.5-5.5	7-8

Figure 14.4 An example of the combination of factors associated with different site conditions that produce diversity in soils at the local scale.

esses that shape soils, such as the drought and wind erosion of the American Dust Bowl of the 1930s, occur as infrequent events.

Because we cannot be around to observe and measure all that happens in the soil environment, we often have to deduce that forces and processes are doing the work by the features they produce in the soil. This sort of reasoning begins by defining the spatial correlation between observable soil features and the geographic setting in which they are found (Fig. 14.4). Soil scientists often take diagnostic horizons as the key soil features, and when diagnostic horizons are found that do not correlate with existing geographic conditions, they must assume that the environment has changed and the soil features have been inherited from some earlier conditions or events. Soils exhibiting principally relict features (features that are remnants from an earlier time) are called *paleosols* (old soils) in the traditional terminology.

Whatever the origins of a soil's dominant traits, whether modern or ancient, these traits are the criteria used to define and map the basic soil units. Each environment portrayed in Figure 14.4 gives rise to one or more soil units, called a *pedon*. Pedons are the smallest geographic units of soil defined by U.S. soil scientists. Adjacent pedons of similar characteristics together form *soil bodies*, or *polypedons*. Polypedons vary in area, depending on the diversity of the environment, but they usually cover several acres. Polypedons are in turn combined to form *soil series*; in

TABLE 14.1 SUBORDERS AND GREAT GROUPS OF THE ZONAL ORDER OF SOILS, TRADITIONAL USDA SYSTEM

Suborders	Great Soil Groups	General Traits
• Soils of cold regions	• Tundra soils	• Soils of the active layer in permafrost regions; poorly drained, often water-logged with heavy organic accumulation in 0 horizons
• Light-colored soils of arid regions	• Desert soils • Red desert soils • Sierozem • Brown soils • Reddish brown soils	• Soil-forming regime is salinization or calcification; salt accumulation usually heavy in the upper soil and may form a caliche (or petrocalcic) layer; organic content very low; color usually light, but red and brown are common because of iron and magnesium oxide staining of particles
• Dark-colored soils of semiarid, subhumid, and humid grasslands	• Chestnut soils • Reddish chestnut soils • Chernozem soils • Prairie soils • Reddish prairie soils	• Soils of the calcification regime; substantial organic content in 0 and A horizons (mollic horizon); B horizon rich in calcium carbonate; color ranges from light brown (chestnut) to dark brown (chernozem and prairie)
• Soils of the forest-grassland transition	• Degraded chernozem soils • Noncalcic brown or shantung brown soils	• Soils in midlatitude prairie-forest transition where climate is subhumid; strong organic layer, calcium carbonate weak or absent in B horizon
• Light-colored podzolized soils of forested regions	• Podzol soils • Brown podzol soils • Gray-brown podzolic soils	• Soils of the humid midlatitudes characterized by a strong to moderate organic layer, leached and light-colored A horizon, and a B horizon with distinct concentrations of iron and aluminum oxides
• Lateritic soils of forested subtropical and tropical regions	• Yellow podzolic soils • Red podzolic soils (and terra rossa) • Yellowish-brown lateritic soils • Reddish-brown lateritic soils • Laterite soils	• Heavily weathered (leached) soils of the humid tropics; light organic layer over a deep zone of clays rich in iron and aluminum oxides; heavy accumulations of iron oxides form laterite layers, and aluminum oxides may give rise to bauxite reserves in some areas

Source: U.S. Department of Agriculture, U.S. Government Printing Office, Washington, D.C. (1938).

the United States the U.S. Department of Agriculture has defined a total of 1050 soil series.

SOIL CLASSIFICATION

Many soil classification systems are in use today, but in general they fall into two categories: (1) those developed by agronomists and natural scientists; and (2) those developed by civil engineers. The systems in the first category focus on the solum and the relationship of soil to surface conditions. They are usually designed to provide information on the soil's suitability for supporting plant life and its geographic distribution, but not much about the soil material beneath the solum. The engineering classification systems, on the other hand, are designed to provide information on soil suitability for building sites and construction materials, but next to nothing about soil relationships to the landscape and its distribution over broad areas.

The classification systems used by geographers are mainly those developed by agronomists and soil scientists. These systems, however, tend to vary with different countries. In the United States two systems, both associated with the U.S. Department of Agriculture (USDA), have been developed in this century: the USDA traditional system and the USDA comprehensive system. The comprehensive system was designed to serve as an international system (like the Universal System of Biological Names), but most countries outside the United States have not adopted it. Canada, the United Kingdom, and Australia, for example, each have their own systems, which are specially designed according to soil types and features that are important geographically and economically in those countries.

The Traditional USDA Classification Scheme

In the traditional USDA soil classification scheme, three major levels of classification are used: (1) orders, (2) suborders, and (3) great groups. There are three orders of soil: zonal, intrazonal, and azonal. Zonal soils, the most common order and generally the one accorded the greatest importance (because they are the most prevalent geographically), are distinguishable on the basis of their well-developed horizons. The suborders and great groups of the zonal soils are listed in Table 14.1. The global distribution of zonal soils is presented in Color Plate 7; note the correlations apparent at this scale with the patterns of climate and vegetation (Color Plates 5 and 6).

In any geographic setting certain conditions are necessary for the formation of zonal soils: (1) Erosion and deposition on the surface *must be negligible.* (2) Moisture flux *cannot be impaired* by bedrock, groundwater, ice, or some other substance. (3) The above conditions must persist long enough to allow internal differentiation to take place. Because these conditions are prerequisite to the formation of zonal soils, we can fairly accurately identify those sites in the landscape where they should be found. Basically, the sites must be relatively flat, somewhat elevated, well drained, and geomorphically stable, similar to that illustrated in Figure 14.5*a*.

The formation of zonal soils would be highly unlikely in a river floodplain, along a river valley's walls, in tributary valleys, and on steeper upland slopes. Rather, these sites give rise to *azonal soils,* which are characterized by the absence of horizons (Fig. 14.5*b*). This difference is explained by the differences in the soil-forming processes found in the different sites: external processes, such as erosion and deposition by runoff, rather than internal processes, such as eluviation and illuviation, are dominant in these sites. In the floodplain, for instance, erosion and deposition by floods and channel water occur frequently enough to mask the effects of the horizon-producing processes in the solum. Therefore, the alluvial (river-deposited) soils are heterogeneous in texture and are often arranged in layers created by the original deposits.

Azonal soils are classified according to their particular physical characteristics. Highly mixed, unsorted deposits from landslides, mud flows, and soil creep (a very slow downhill of soil material) are called *colluvial soils;* talus and other stony deposits that lack organic material are called *regosols;* and deposits composed of huge chunks of bedrock are called *lithosols.*

Intrazonal soils form in poorly drained areas. Because the ground in these areas is saturated, organic weathering proceeds slowly, and vegetal debris forms thick deposits; thus, the soil regime may be that of gleization or something similar to it. Although the up-and-down flux of moisture is limited or absent, crude horizons tend to form as a result of an internal differentiation or organic material and clay. Intrazonal soils are found in some low areas and isolated pockets within larger areas of zonal soils (Fig. 14.5*c*).

The USDA Comprehensive Soil Classification

After working with the traditional soil classification scheme for many years, U.S. soil scientists found it unsatisfactory in a number of ways. In particular, they found that it placed too much emphasis on the genetic

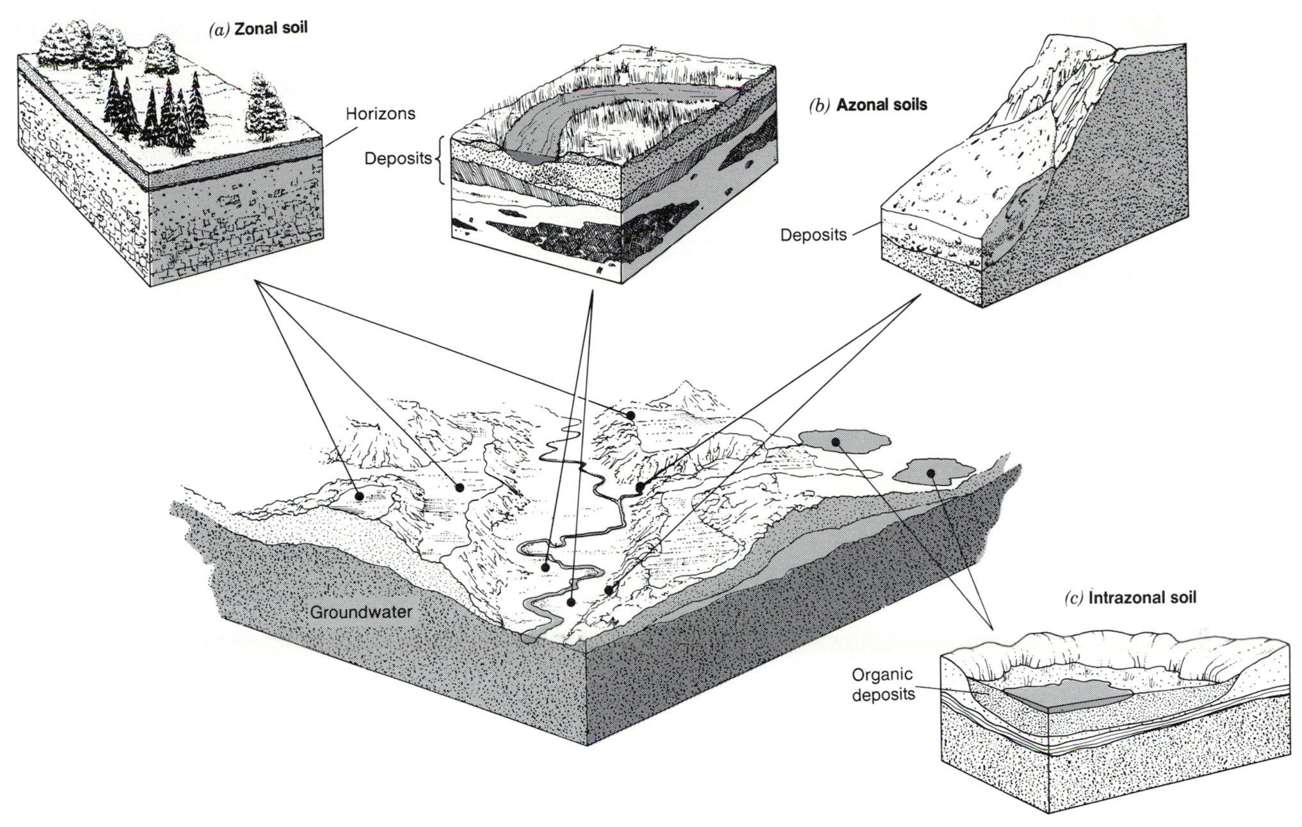

Figure 14.5 (*a*) The formation of zonal soils is limited to relatively flat upland sites. In highly diversified terrain, zonal soils may occupy only a small fraction of the total area. (*b*) Azonal soils are associated with unstable sites, such as steep slopes and floodplains and (*c*) intrazonal soils with poorly drained sites.

(origin and development) aspects of soils and thereby created much uncertainty in soil classification and mapping. In the 1950s USDA soil scientists, with the help of scientists in colleges and universities, devised a new system, the Comprehensive Soil Classification System.

One of the important traits of this system is that soils are classified as you find them in the field, which means that recognition is based on existing, measurable traits and not on what a soil may seem to be, may have been in the past, or could become in the future. Diagnostic horizons are used as the criteria for classification (see Appendix D). These twenty horizons (six surface and fourteen subsurface) provide a much broader range of legitimate soil-forming forces—such as long-continued farming—than was possible under the traditional scheme. Finally, it is worth noting that in the new system the traditional soil names have been discarded for new (but often less interesting) names, each ending in the Latin suffix *sol* (soil) at the order level (see Appendix D).

The USDA comprehensive scheme provides six levels of classification, beginning at the broadest scale with *orders*, which are subdivided into suborders, great groups, subgroups, families, series, and on to pedons. In total, the system establishes about 7000 categories of soil for the United States. Table 14.2 lists the ten orders of the new scheme and identifies the equivalent class in the traditional scheme. The map in Figure 14.6 shows the general distribution of orders in the Western Hemisphere and Color Plate 8 shows this distribution in greater detail. The map in Figure 14.7 shows the distribution of suborders in the coterminous United States and southern Canada. (Appendix D gives the general traits and suborders for each order.)

We will now briefly describe the ten orders of the comprehensive classification system:

Entisols These are mainly soils of recent origins, principally the azonal soils of the traditional USDA scheme. Horizons are very weakly developed or non-

TABLE 14.2 SOIL ORDERS IN THE COMPREHENSIVE USDA SYSTEM AND THEIR EQUIVALENTS IN THE TRADITIONAL USDA SYSTEM

Order System	Meaning of Name	Equivalent In the Traditional System
Entisol	Recent soil	Azonal soils and some gley soils
Inceptisol	Young soil	Some brown forest soils and gley soils
Aridisol	Arid soil	Mainly red and gray desert soils
Mollisol	Soft soil	Mainly chernozem, prairie soils, and chestnut and brown soils
Spodosol	Ashy soil	Podzols
Alfisol	Pedalfer soil	Gray-brown podzolic, prairie soils, weak chernozems, and some intrazonal soils
Ultisol	Ultimate soil	Red-yellow podzolic, reddish-brown lateritic, and some intrazonal soils
Oxisol	Oxide soil	Latosols
Vertisol	Inverted soil	No equivalent
Histosol	Organic (tissue) soil	Intrazonal bog soils

existent. Entisols are most commonly associated with geomorphically active, or recently active, environments such as river floodplains, sand dunes, and mountain slopes. Entisols can be found in any bioclimatic region; understandably, in the lower United States the largest areas of entisols are found in the mountain states, in particular Colorado, Utah, and New Mexico (Fig. 14.7). The agricultural potential of entisols varies with the quality of the deposits in which they form: in the sandy deposits of the deserts and coastlines, they are notoriously poor, whereas in many of the great river deltas, where the deposits are rich in organics and moist, they are highly productive.

Inceptisols These are soils in which horizons are just beginning to form. They are found in a wide range of bioclimatic regions but are limited to humid climates with favorable soil-moisture balances. Although one or more horizons are present, visible evidence of them is typically weak. Inceptisols are associated principally with young geomorphic surfaces such as those in the midlatitudes and the subarctic that were deglaciated 8,000 to 10,000 years ago. In North America, they are widespread in the tundra and in the large river lowlands such as the Mississippi Embayment (Color Plate 8). In the tropics, where many landscapes have been

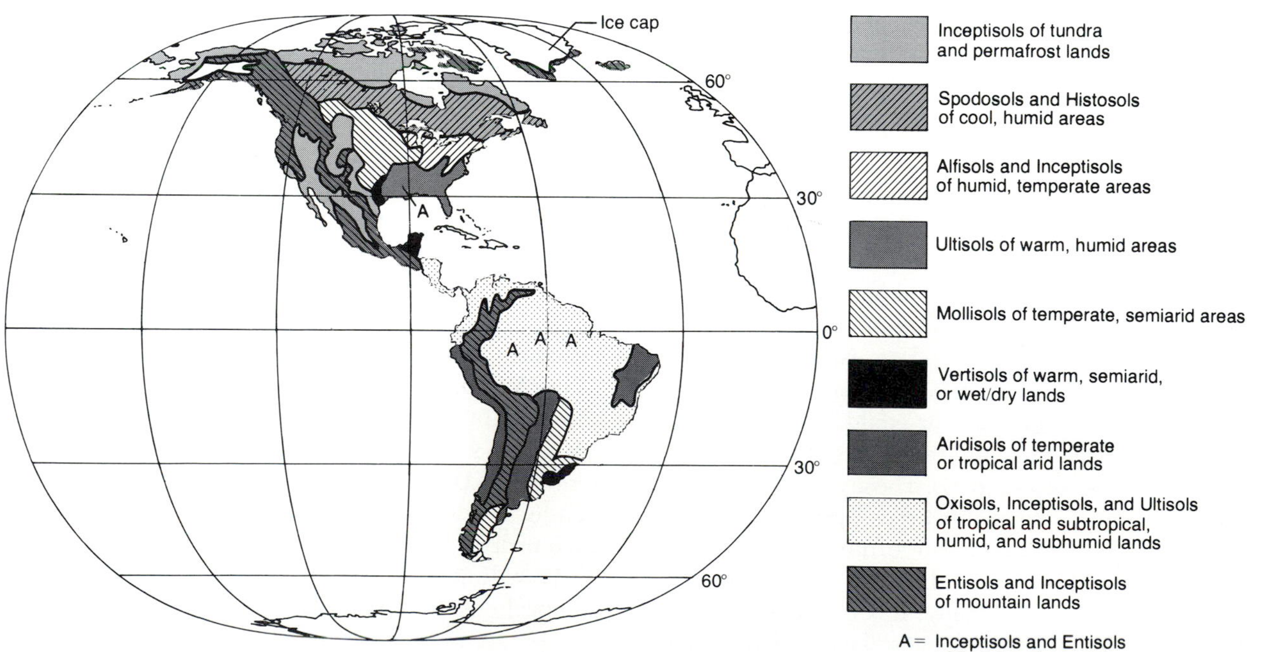

Figure 14.6 Soils of the Western Hemisphere, according to the comprehensive USDA classification. (Also see color plate 8.)

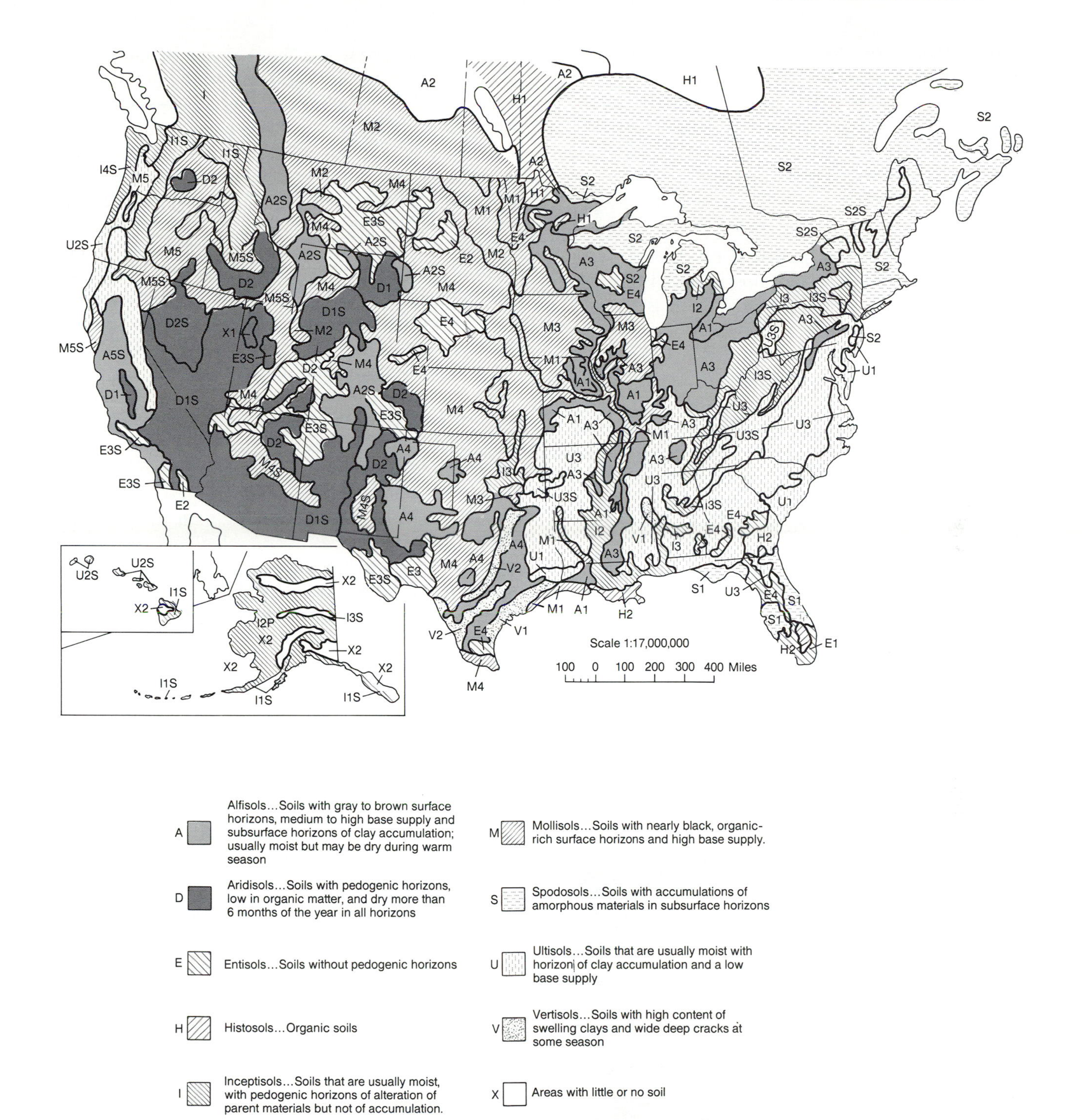

Figure 14.7 Distribution of soil orders in the United States and southern Canada according to the U.S. Comprehensive Soil Classification System.

without large-scale geometric disturbances for millions of years, inceptisols are often found on volcanic ash deposits. A favorable moisture balance generally gives the inceptisols promising agricultural potential, especially where they have formed in rich parent material such as alluvium and volcanic ashes in temperate and warmer regions.

Aridisols As the name implies, the aridisols are soils of dry environments. The moisture balance is strongly negative—or at least the potential for it is strongly negative—and thus there is both little productivity by vegetation and little leaching within the soil column. The soil-forming regime is either calcification or salinization. Aridisols have essentially no organic layer and often have a salt-enriched horizon (either petrocalcic or calcic) near the surface or at depth. Textures in the aridisols vary with the deposits in which they are found and with geomorphic processes, such as wind and runoff, that are active on the surface. The agricultural potential of the aridisols is generally good, but only with the aid of irrigation. Even irrigated aridisols can prove to be a problem, however, as we will discuss later in this chapter. In North America, the aridisols are the dominant soils in the American Southwest and northern Mexico (Fig. 14.7).

Mollisols These soils are generally associated with grasslands and the calcification soil-forming regime. Geographically, they are found on the humid side of the aridisols in the semiarid and subhumid climatic zones (see Color Plate 5). These climates are characterized by a summer moisture deficit strong enough to induce a substantial upflow of capillary water in the soil column. Moisture is sufficient, however, to support grasses that recycle bases in the upper soil and are responsible, together with a slow rate of organic decomposition, for the formation of a strong mollic horizon. This horizon is a dark brown to black epipedon, generally 25 cm (10 in.) or more in thickness, that is very rich in exchangeable bases. Below the mollic horizon, the B and C horizons are usually rich in calcium carbonate, and in the B horizon, this mineral often appears in nodules. The mollisols are among the very best grain-farming soils in the world. In the traditional classification they included the famous chernozem, prairie brown, and chestnut soils of the Great Plains and Prairies of the United States and Canada. In addition to inherently high fertility, the mollisols are friable (easily crumbled) and retain soil moisture well. On the other hand, they are easily eroded by runoff and wind, and they are subject to drought by virtue of their climatic location.

Spodosols These soils are distinguished by a pronounced zone of illuviation called the *spodic horizon*. This is an accumulation of various forms of iron oxides and aluminum oxides together with organic and mineral colloids in a dark (and often hard) B horizon.

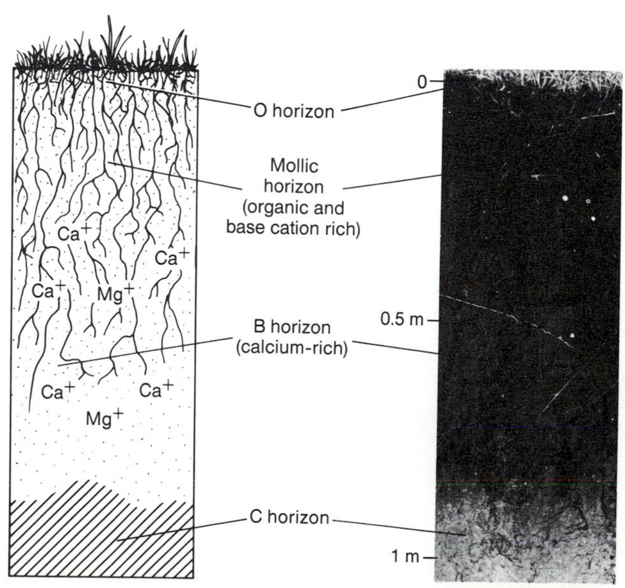

Mollisol from North Dakota.

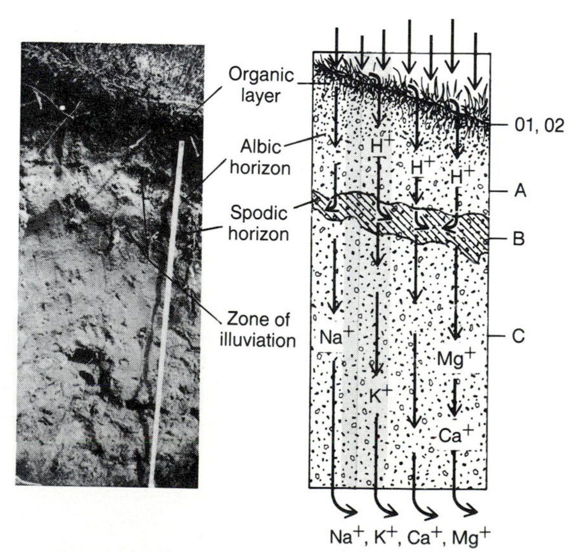

Spodosol soil from Massachusetts.

Immediately above the spodic horizon, many spodosols have a pronounced whitish layer, called the *albic horizon*, from which minerals and fine particles have been intensively eluviated. Spodosols generally form under forest covers in moist, cool climates, but their development is most pronounced under coniferous forests in sandy parent material. The spodosols' fertility is very poor owing to their low base-retention capacity and acidic chemistry. Thus, their agricultural potential is modest and limited principally to crops such as potatoes which are well suited to low soil pH and short growing seasons. Spodosols are widespread in southern and eastern Canada, the Great Lakes states, the northern Appalachians, and the northeastern United States (Fig. 14.7).

Alfisols These soils can be thought of as moist versions of the mollisols. They form under forest covers in humid midlatitude climates; eluviation is moderate, and base retention is fairly high. In the B horizon, clays accumulate, forming a whitish layer known as an *argillic horizon*. Alfisols have fairly good agricultural potential owing to their favorable moisture balance and good fertility. Indeed, these soils in China and Europe have been successfully farmed for thousands of years. In the United States the alfisols are abundant on the older glacial deposits in the Midwest, in the loess deposits in and near the Mississippi Embayment, in the inner Coastal Plain of Texas, and in several large areas within the Rocky Mountain complex.

Ultisols These soils are described as being in an advanced state of development, thus, the prefix "ulti" for ultimate. They are found in warm, moist climates, such as the humid subtropical, where the landscape has been free of major geomorphic disturbances for long periods. Eluviation is pronounced, resulting in a poor supply of bases which grows ever poorer with depth. Light-colored clays form an agrillic horizon (illuvial layer) beneath the A horizon. The C horizon may be deep and underlain weathered bedrock (regolith). Ultisols are the dominant soils in the American South; in China they are found in a similar geographic location (the southeastern part of the country) and are also associated with the humid subtropical climate. Important areas of ultisols are also found in the wet tropical and equatorial climates of South America (Color Plate 8). The ultisols pose serious problems for agriculture because they are quickly exhausted by crops such as corn, tobacco, and cotton; therefore, in underdeveloped countries they induce the practice of shifting agriculture (about which we will say more later), and in developed countries they require much fertilizer (Fig. 14.7).

Oxisols These are soils of warm, humid regions which have undergone intense weathering under the oxisolization (laterization) regime. They form mainly in tropical and equatorial landscapes that have been stable for long periods of time, at least a million years. The organic layer is weak litter and decomposes rapidly in these regions. The underlying soil is rich in kaolinite (a clay) and sesquioxides of iron and aluminum, the residues of weathered parent material. Oxisols do not usually show distinct horizons, though a vague zone of concentrated clays and sesquioxides, called the oxic horizon, is usually found within 2 m of the surface. At greater depths in some oxisols a hard, iron-rich layer, called a *plinthite horizon* (*laterite* in traditional terminology), may be present. Like the ultisols, these soils are rapidly exhausted by crops. Plant nutrients and the cation exchange capacity are low so that, despite a warm, moist climate, agriculture is not productive unless fertilizers are used intensively. Farmers who cannot employ fertilizer in areas of oxisols have traditionally resorted to shifting agriculture.

Vertisols These soils are associated with parent material dominated by the clay montmorillonite. This clay has a high shrink/swell capacity, and under climates with strongly seasonal moisture balances it is mechanically mixed as soil cracks open, pieces of soil fall into the cracks, and the cracks close again. This produces a slow turnover that mixes organic material to a depth of 0.5 m to 1.0 m or so. Vertisols may be streaked vertically, and the sides of cracks and blocks (chunks) may show grooves or scratchlike features called *slickenslides*. In addition, these soils may develop a rough microrelief (mini-topography) characterized by little knobs and hollows called *gilgae*. Vertisols are commonly found in landscapes of grass and savanna vegetation in subtropical and tropical climates with marked seasonal soil-moisture deficits. In the United States they are most common in Texas. Vertisols generally have low to moderate agricultural potential because the clay makes them difficult to plow when wet and they do not give up water to plants very readily. They are, however, quite fertile, being high in exchangeable bases such as calcium and magnesium (Fig. 14.7).

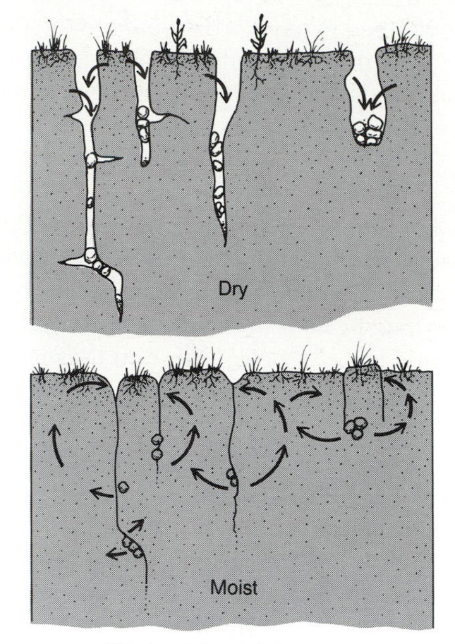

The dry and moist places that characterize the vertisols. Mixing of the surface layer takes place when soil pieces fall into desication cracks.

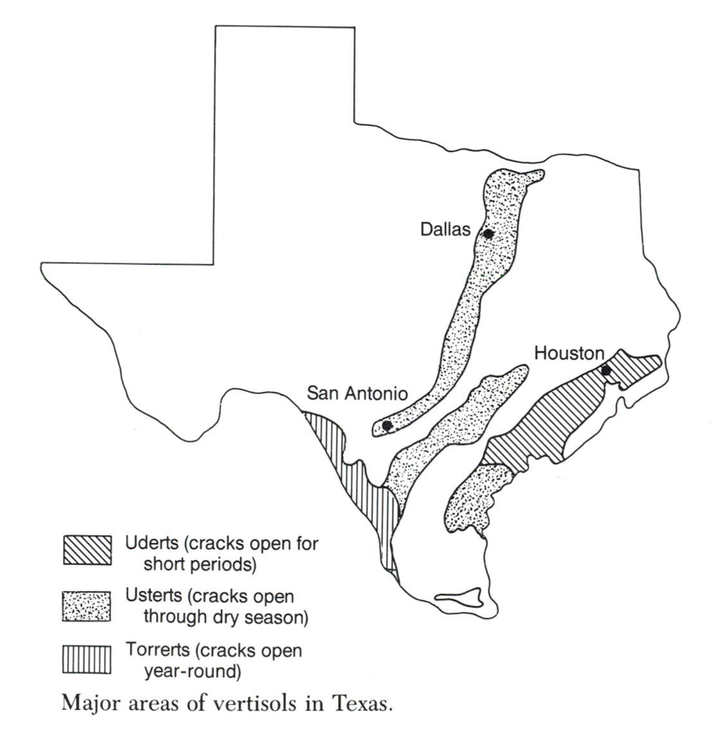

Major areas of vertisols in Texas.

Histosols These soils are dominated by a thick organic layer. Although they are widely associated with cool, humid climates, surface hydrology is the real key to their formation. Long-term saturation or shallow standing water may induce the buildup of organic litter characteristic of the gleization soil-forming regime (Fig. 14.3*a*). The chemistry of the organic mass is determined by the type of vegetation and the minerals present in the groundwater. In the subarctic and arctic zones of North America (principally in Canada and Alaska), histosols have formed extensively under northern forests that yield acidic organic litter low in plant nutrients. The resultant soils are thus low in exchangeable bases and therefore have low pHs. Histosols are used mainly for specialized agriculture depending on their chemistry and drainage; for example, truck (vegetable) farming for those with higher fertility and better drainage and cranberry farming for those with poor drainage and strongly acidic pHs. Histosols must usually be drained before they are farmed, but this in turn accelerates microfloral action and wind erosion. Thus, the histosols have a limited lifetime under agriculture—as little as 30 to 40 years for the thinner ones.

Canadian Soil Classification

The Canadian soil classification was developed by that nation's principal agricultural agency, the Canada Department of Agriculture. The Canadian system has fewer subdivisions (taxa) than the U.S. Comprehensive scheme and understandably places greater emphasis on soils of cold environments. In addition, the Canadian system has no equivalent to the suborder subdivision of the U.S. scheme.

The Canadian system recognizes nine orders of soils: chernozemic, solonetzic, luvisolic, podzolic, brunisolic, regosolic, gelysolic, organic, and cryosolic (Fig. 14.8). These orders and the great groups into which they are subdivided are listed along with their principal traits in Appendix D.

Engineering Classification of Soil

Because geographers are concerned about how soil is used, they have an abiding interest in the agronomist's interpretation and classification of it. In the past several decades, however, urban sprawl and increasingly elaborate construction schemes have compelled them to appreciate the civil engineer's outlook on soil as well.

Table of Contents

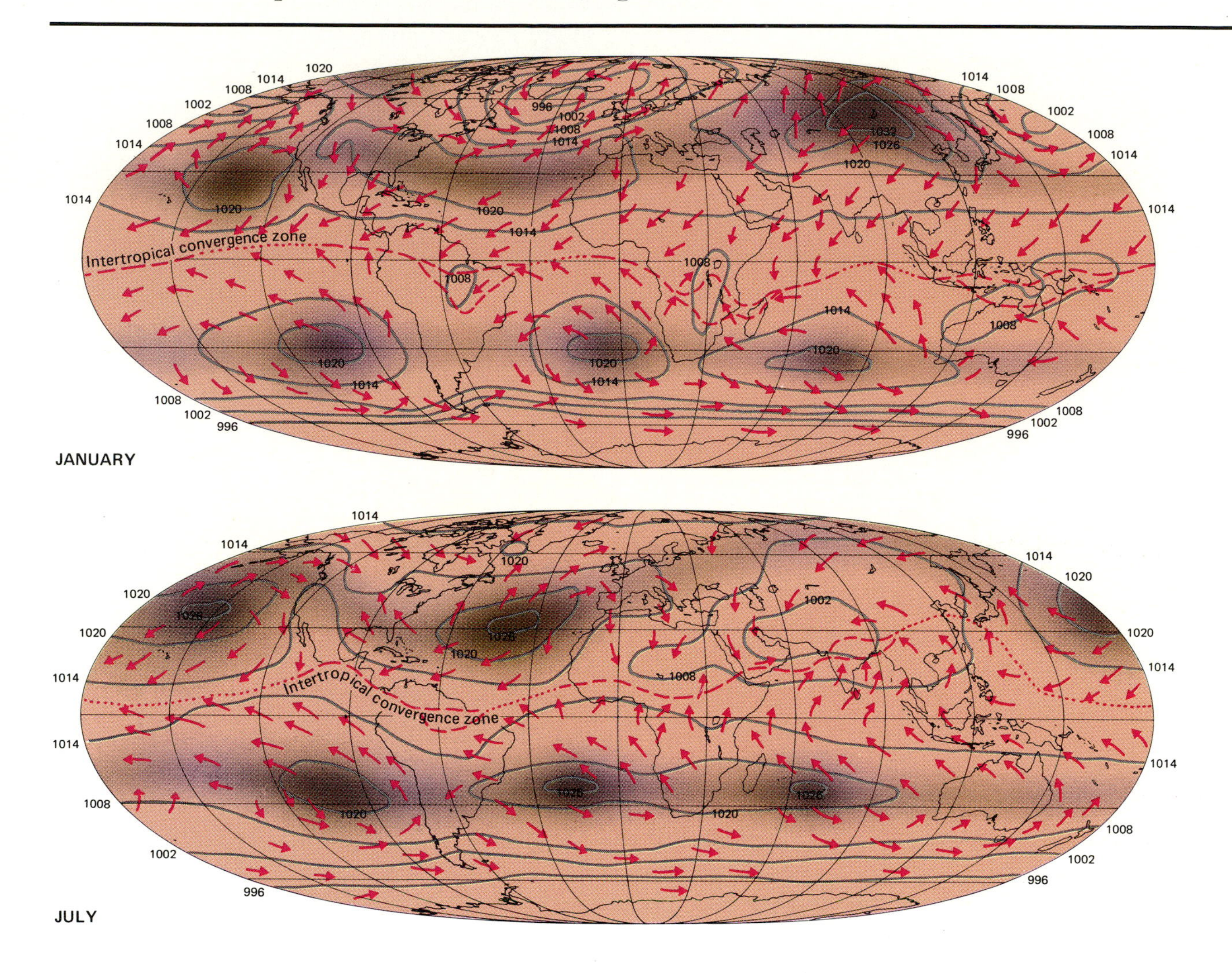
Intertropical convergence zone
JANUARY
JULY

PLATE 2 Ocean Circulation in the North Atlantic

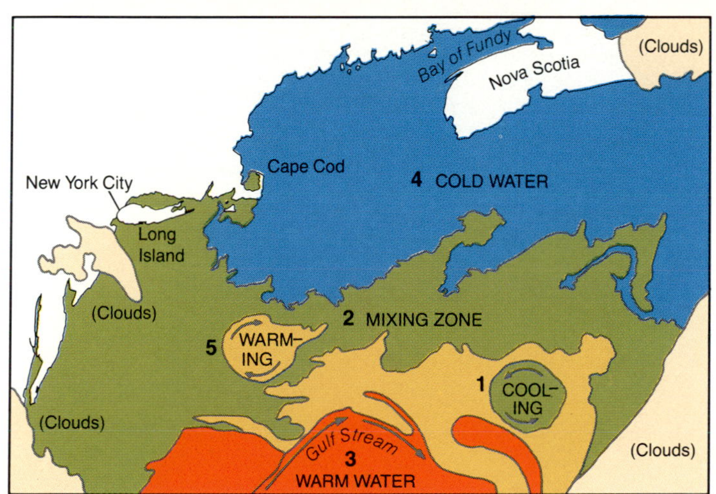

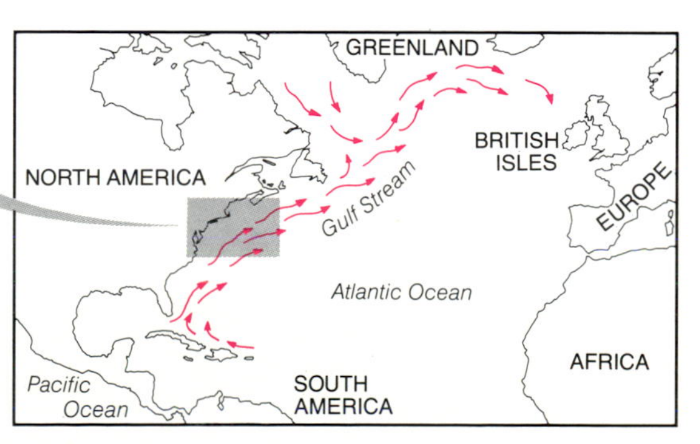

Source: NASA

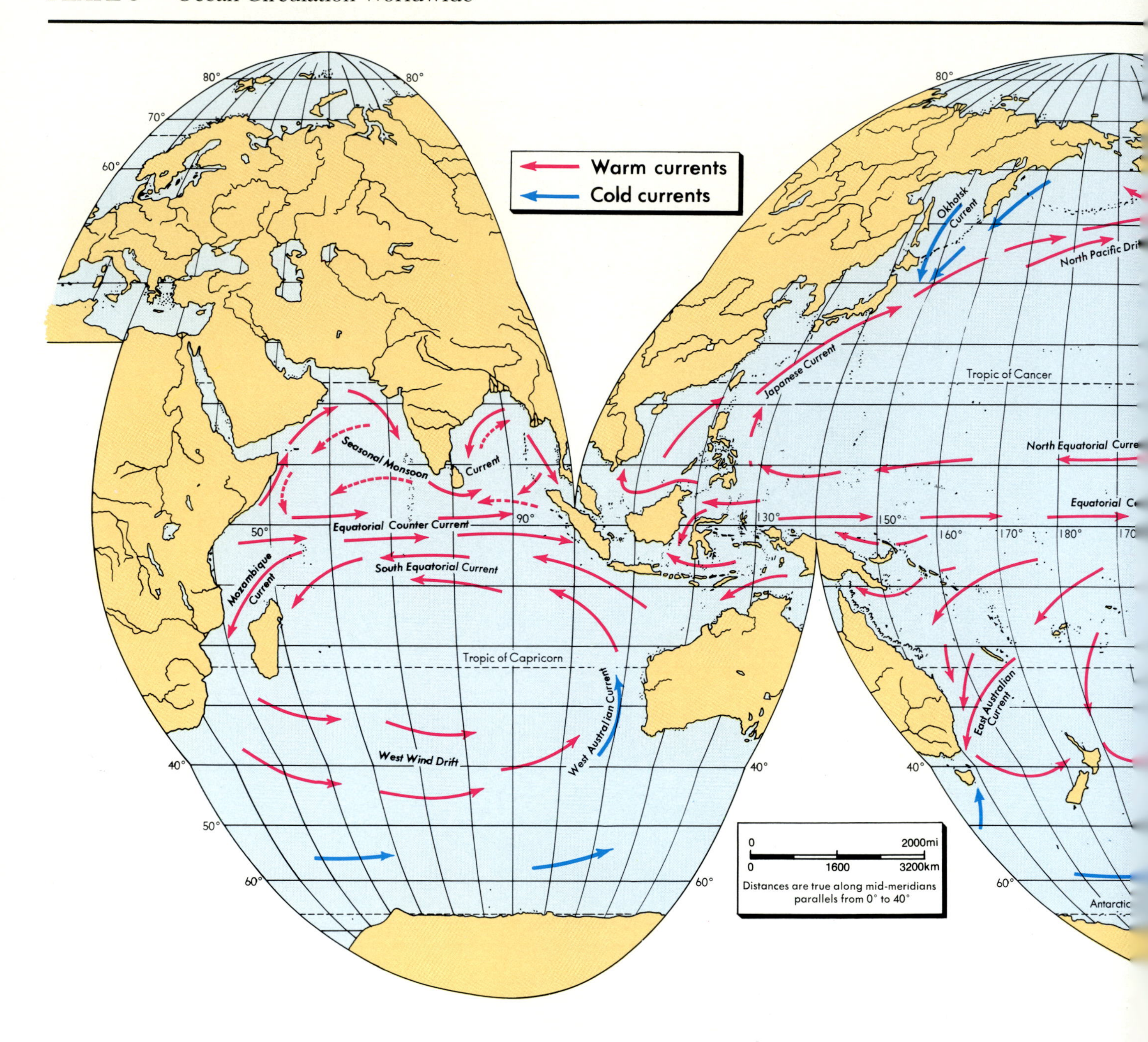

Warm currents
Cold currents
Okhotsk Current
North Pacific Drift
Japanese Current
Tropic of Cancer
Seasonal Monsoon Current
North Equatorial Current
Equatorial Counter Current
South Equatorial Current
Mozambique Current
Tropic of Capricorn
West Australian Current
West Wind Drift
East Australian Current
0 2000mi
0 1600 3200km
Distances are true along mid-meridians parallels from 0° to 40°

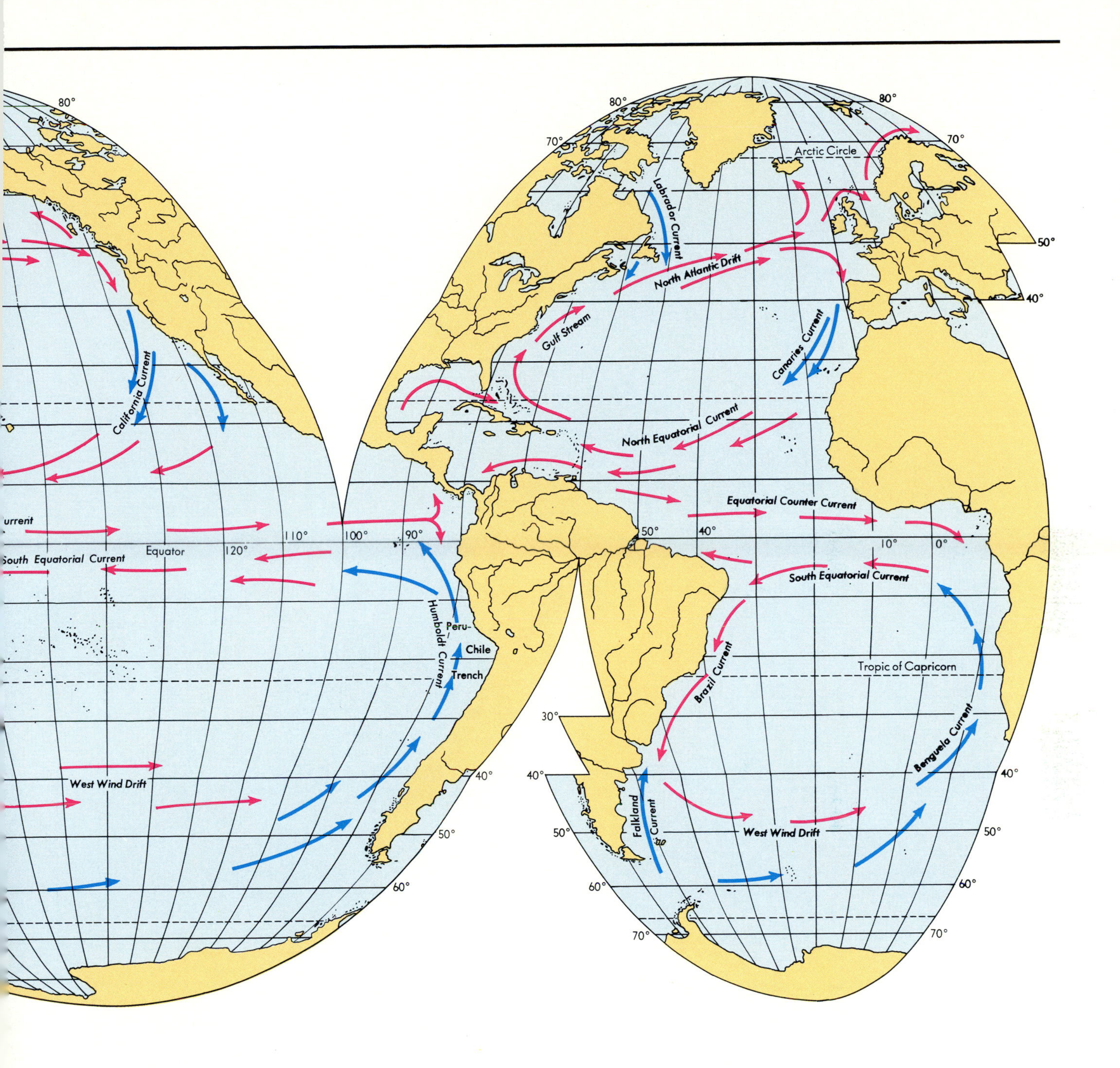

California Current
West Wind Drift
South Equatorial Current
Equator
Humboldt Current
Peru-Chile Trench
Labrador Current
North Atlantic Drift
Gulf Stream
Canaries Current
North Equatorial Current
Equatorial Counter Current
South Equatorial Current
Brazil Current
Benguela Current
Falkland Current
West Wind Drift
Arctic Circle
Tropic of Capricorn

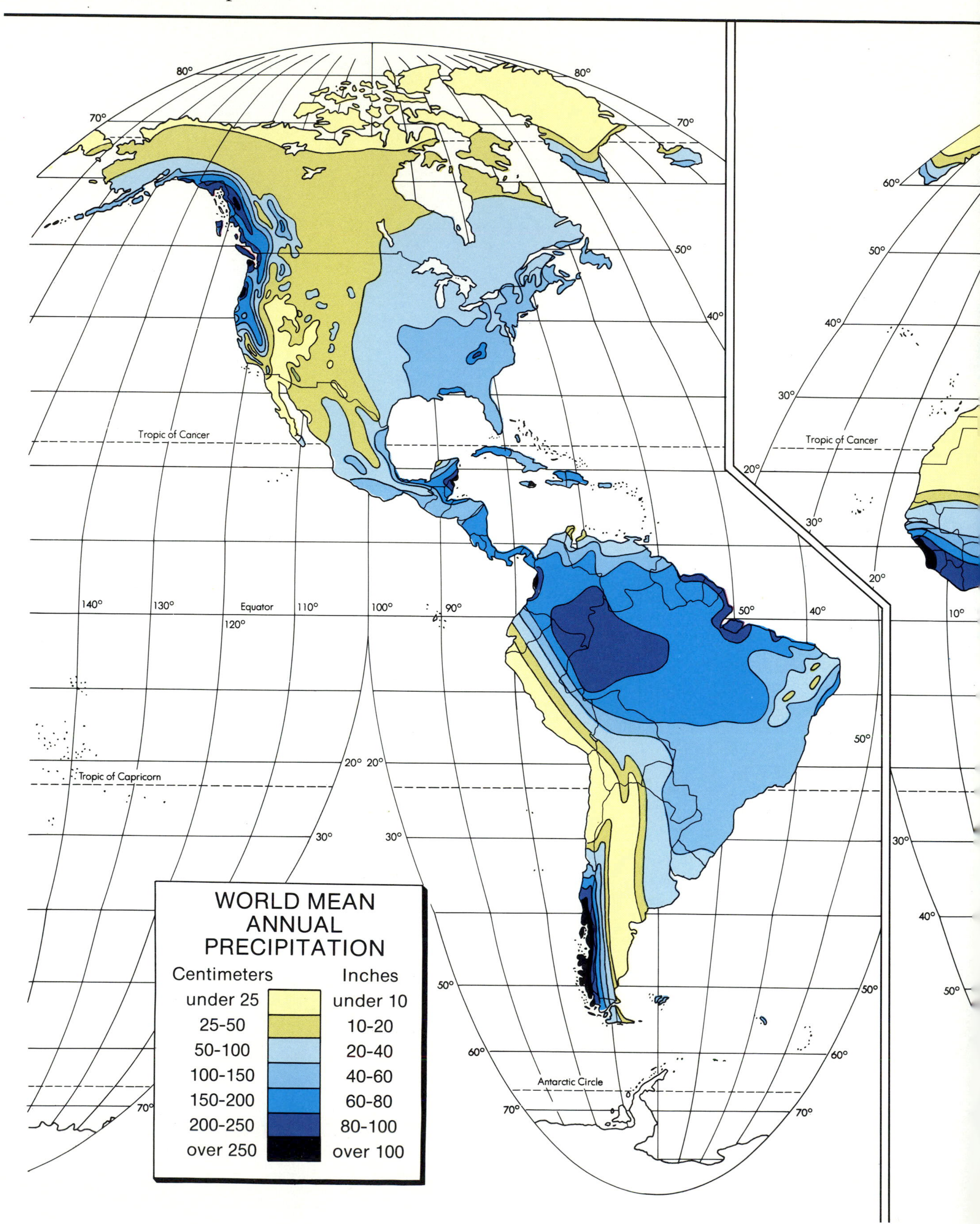
WORLD MEAN ANNUAL PRECIPITATION
Centimeters
Inches
under 25
under 10
25-50
10-20
50-100
20-40
100-150
40-60
150-200
60-80
200-250
80-100
over 250
over 100
Tropic of Cancer
Equator
Tropic of Capricorn
Antarctic Circle
80°
70°
50°
40°
30°
20°
10°
60°
140°
130°
120°
110°
100°
90°

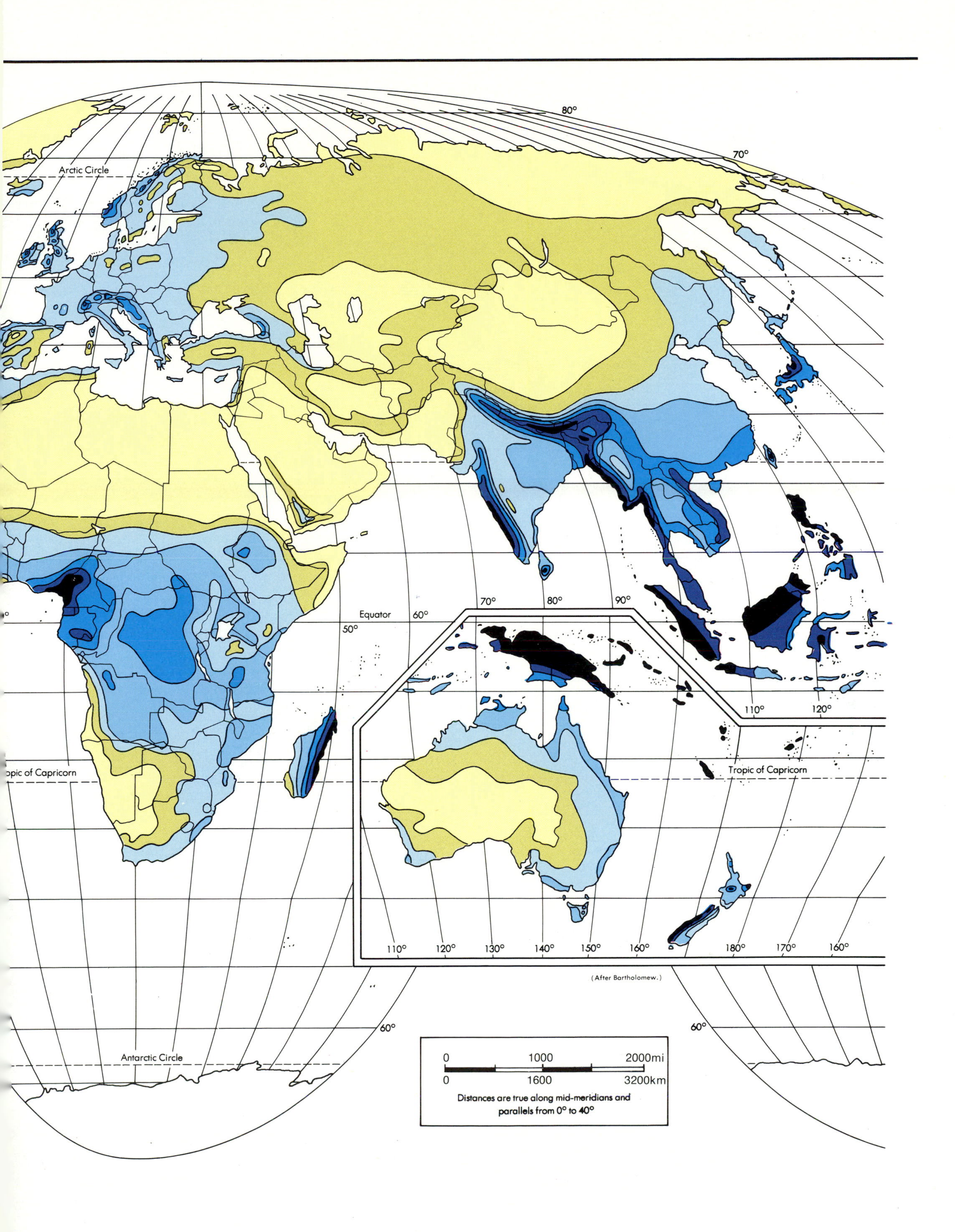

80°
70°
Arctic Circle
Equator
Tropic of Capricorn
Antarctic Circle
50°
60°
70°
80°
90°
110°
120°
130°
140°
150°
160°
180°
170°
(After Bartholomew.)
0
1000
2000mi
1600
3200km
Distances are true along mid-meridians and parallels from 0° to 40°

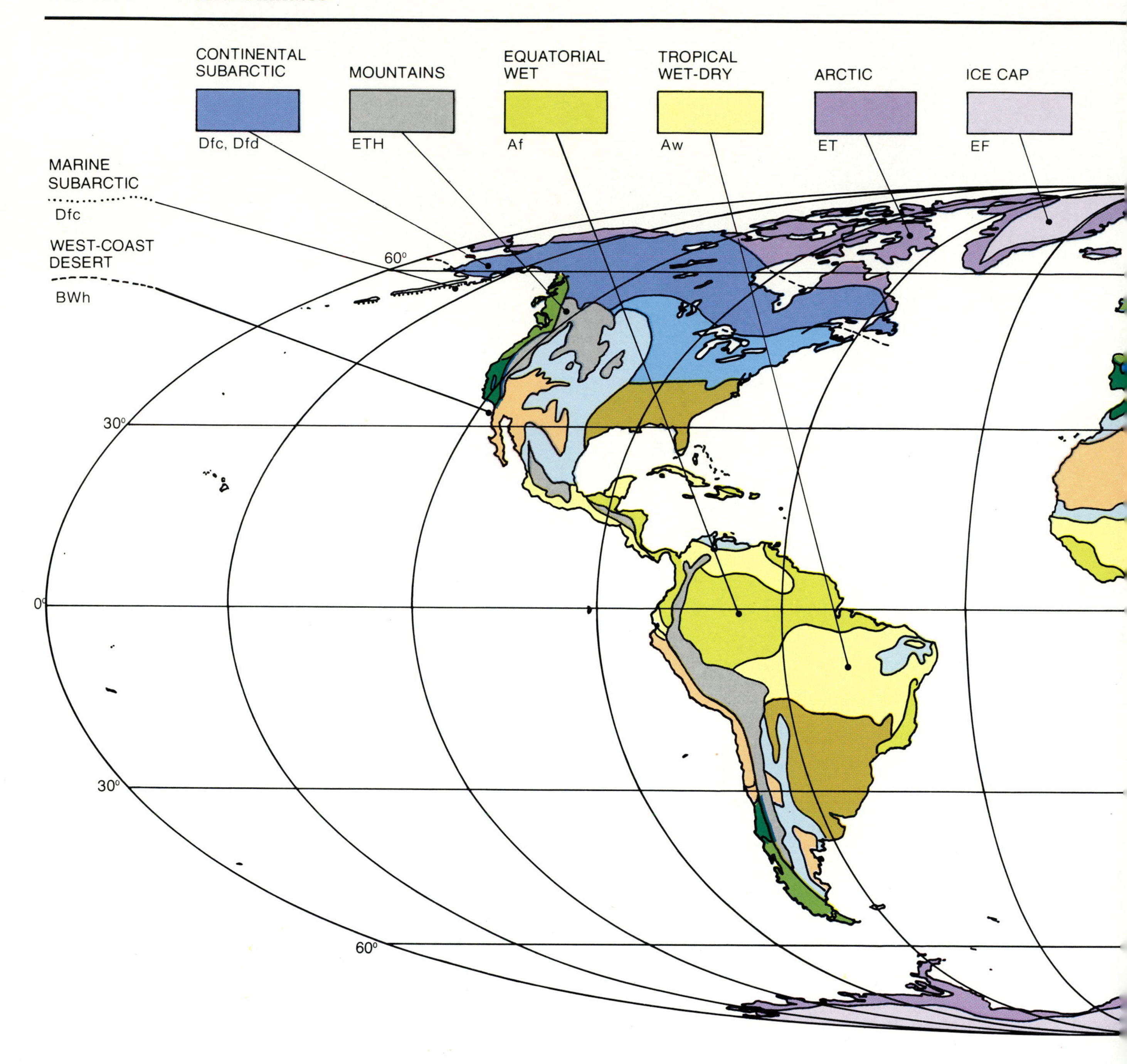
CONTINENTAL SUBARCTIC
Dfc, Dfd
MOUNTAINS
ETH
EQUATORIAL WET
Af
TROPICAL WET-DRY
Aw
ARCTIC
ET
ICE CAP
EF
MARINE SUBARCTIC
Dfc
WEST-COAST DESERT
BWh
60°
30°
0°
30°
60°

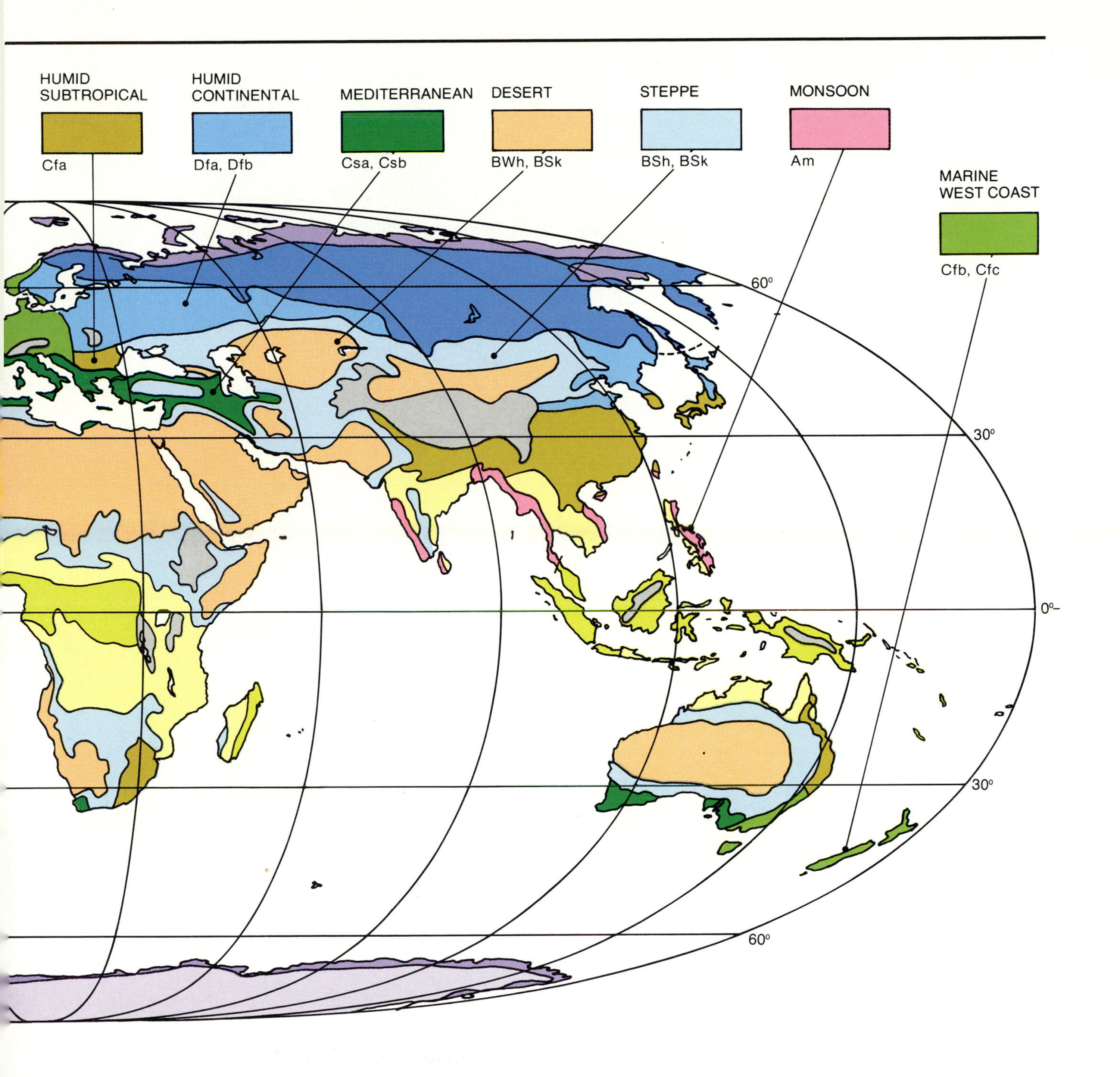

HUMID SUBTROPICAL
Cfa
HUMID CONTINENTAL
Dfa, Dfb
MEDITERRANEAN
Csa, Csb
DESERT
BWh, BSk
STEPPE
BSh, BSk
MONSOON
Am
MARINE WEST COAST
Cfb, Cfc
60°
30°
0°
30°
60°

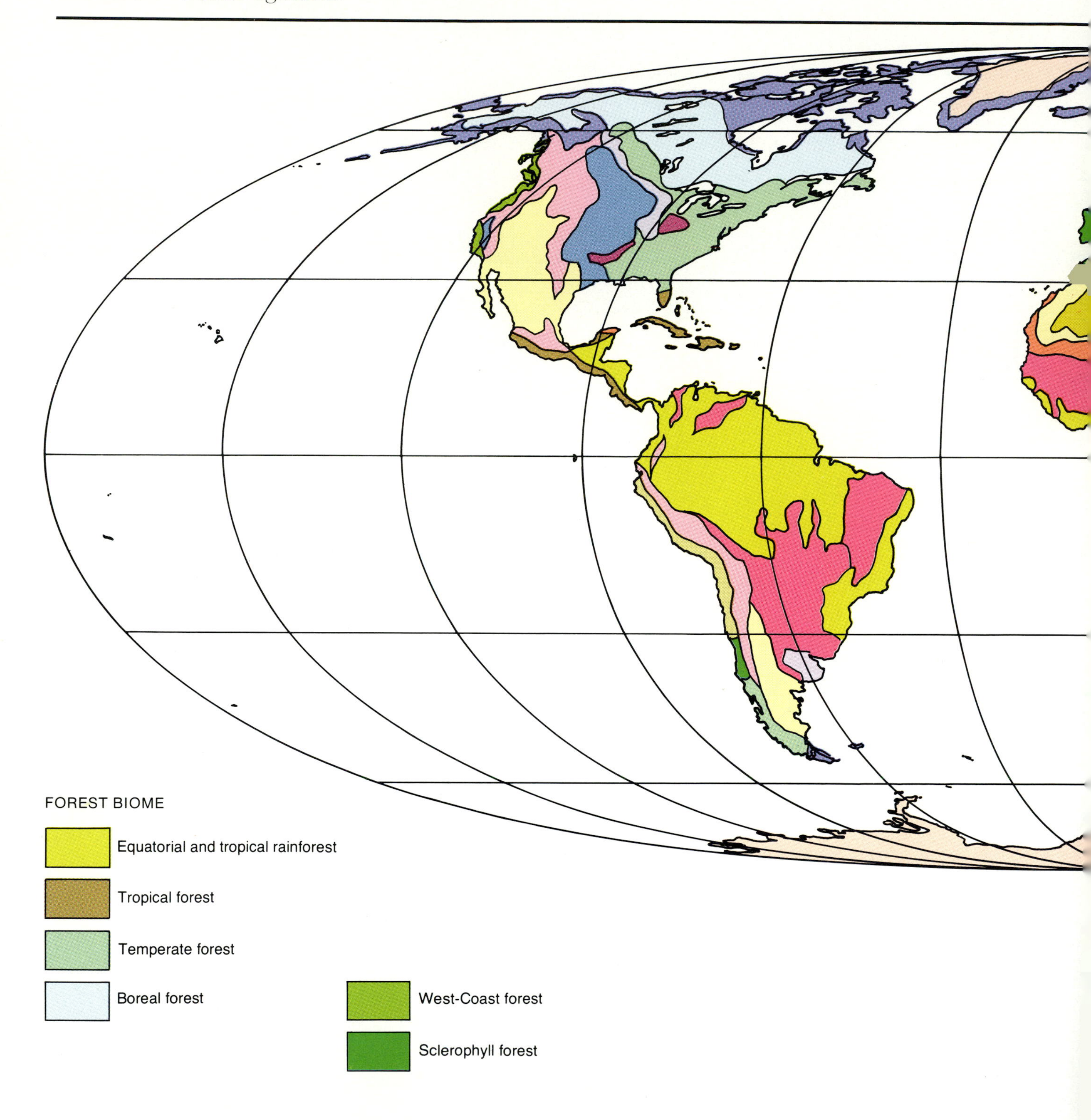
FOREST BIOME
Equatorial and tropical rainforest
Tropical forest
Temperate forest
Boreal forest
West-Coast forest
Sclerophyll forest

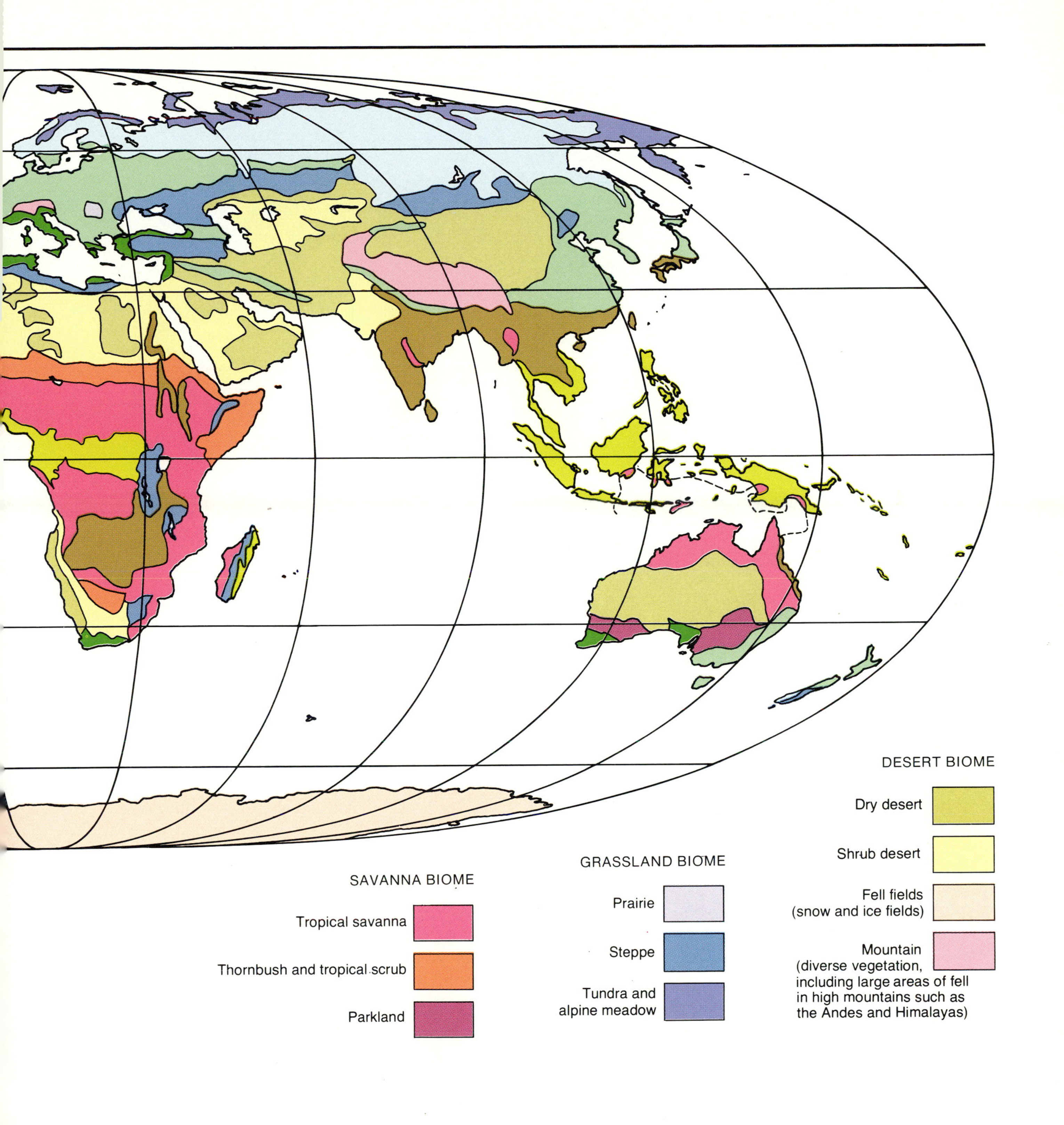

SAVANNA BIOME
Tropical savanna
Thornbush and tropical scrub
Parkland
GRASSLAND BIOME
Prairie
Steppe
Tundra and alpine meadow
DESERT BIOME
Dry desert
Shrub desert
Fell fields (snow and ice fields)
Mountain (diverse vegetation, including large areas of fell in high mountains such as the Andes and Himalayas)

PLATE 7 World Soils: USDA Traditional Classification

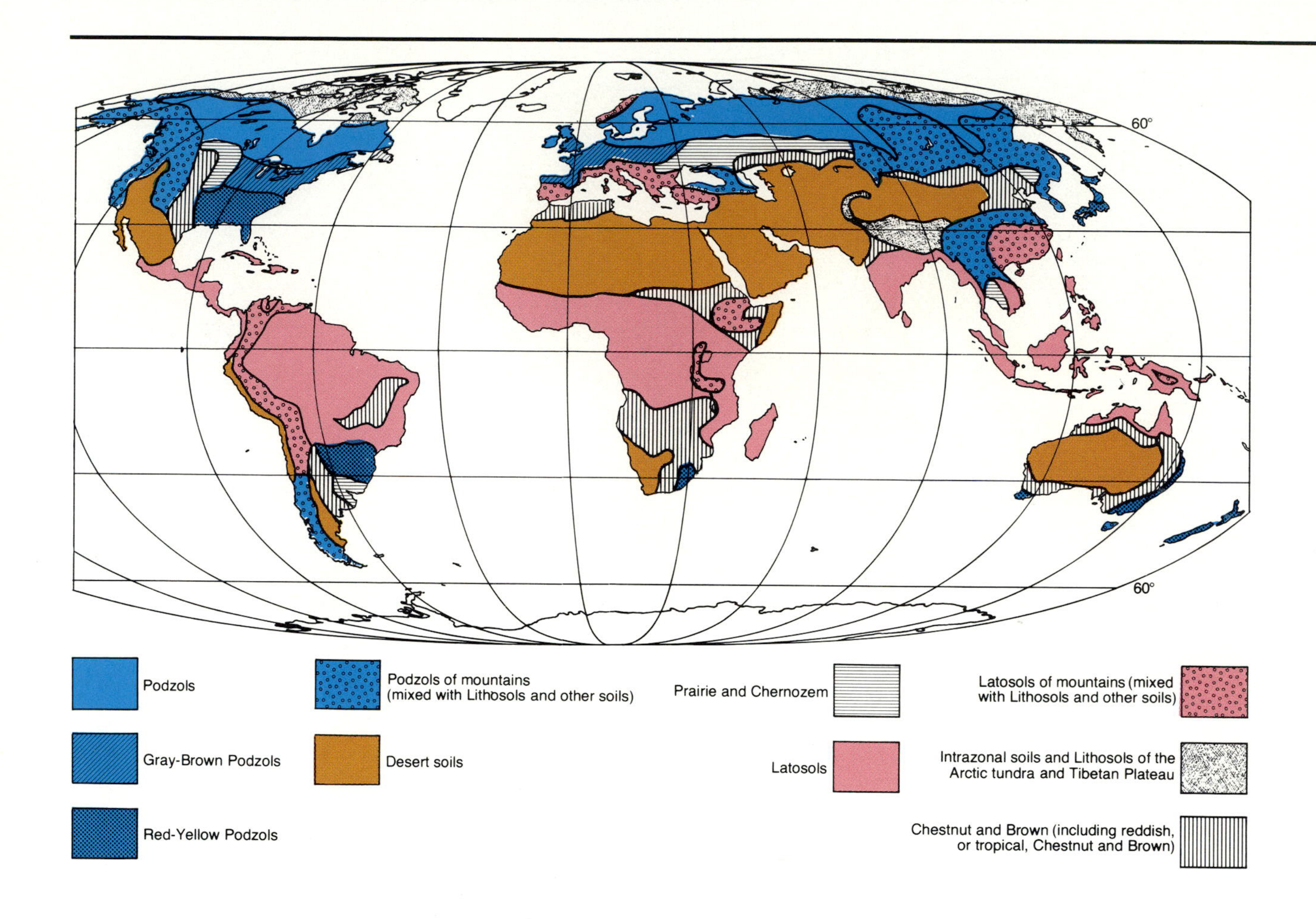

Source: U.S. Dept. of Agriculture

PLATE 8 Soils of the Western Hemisphere: USDA Comprehensive Classification

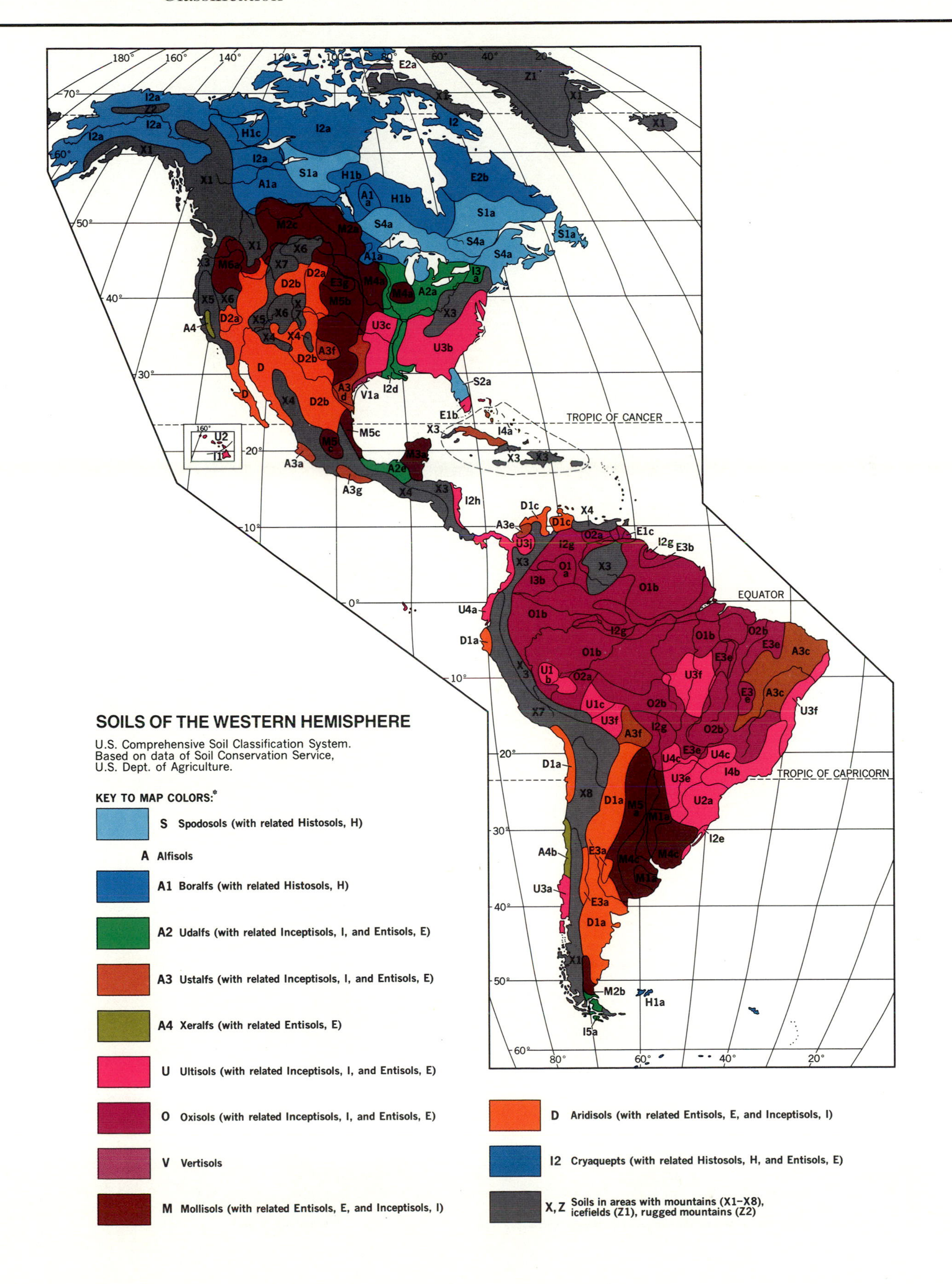

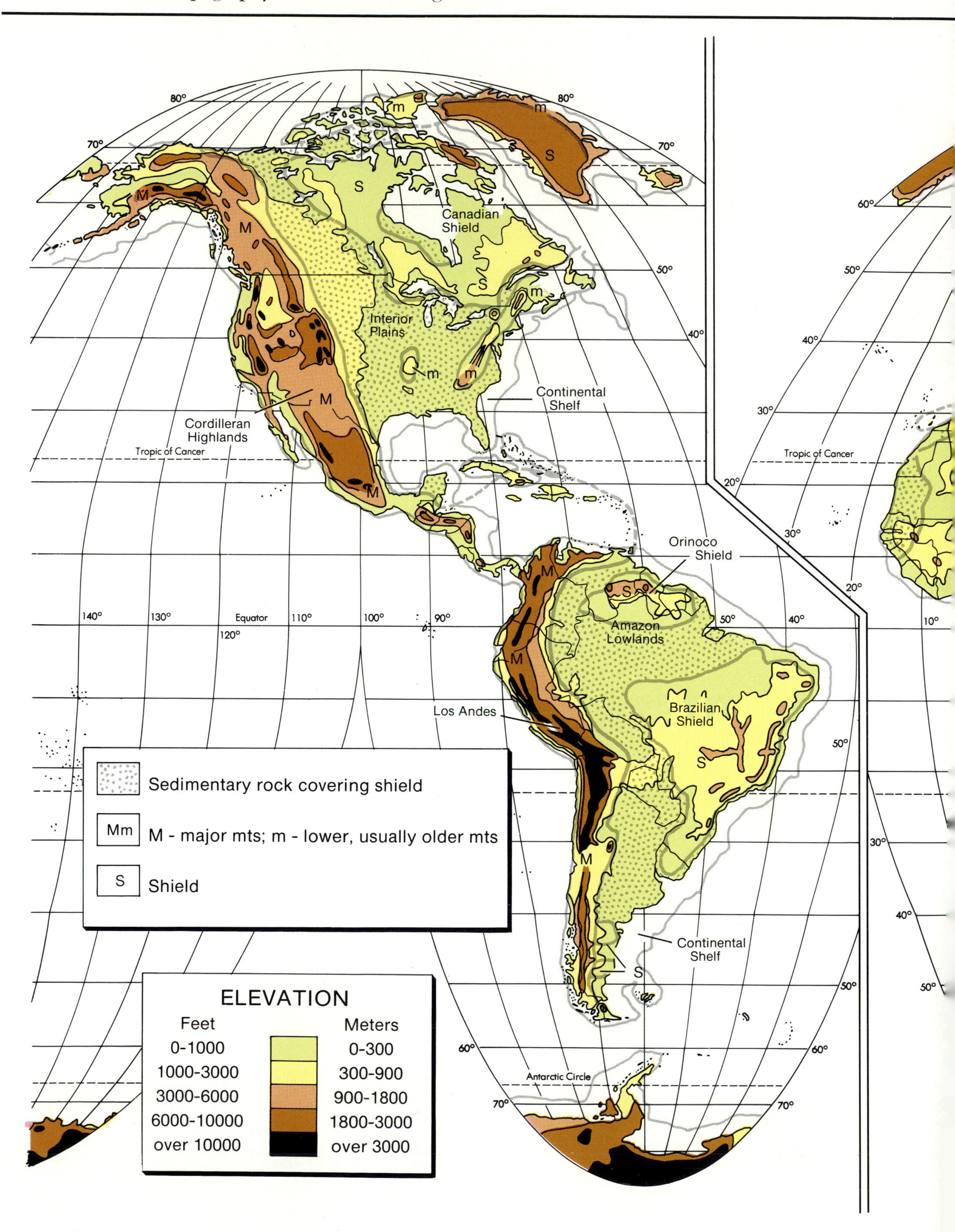

Canadian Shield
Interior Plains
Continental Shelf
Cordilleran Highlands
Tropic of Cancer
Equator
Orinoco Shield
Amazon Lowlands
Brazilian Shield
Los Andes
Continental Shelf
Antarctic Circle
Sedimentary rock covering shield
Mm M - major mts; m - lower, usually older mts
S Shield
ELEVATION
Feet
0-1000
1000-3000
3000-6000
6000-10000
over 10000
Meters
0-300
300-900
900-1800
1800-3000
over 3000

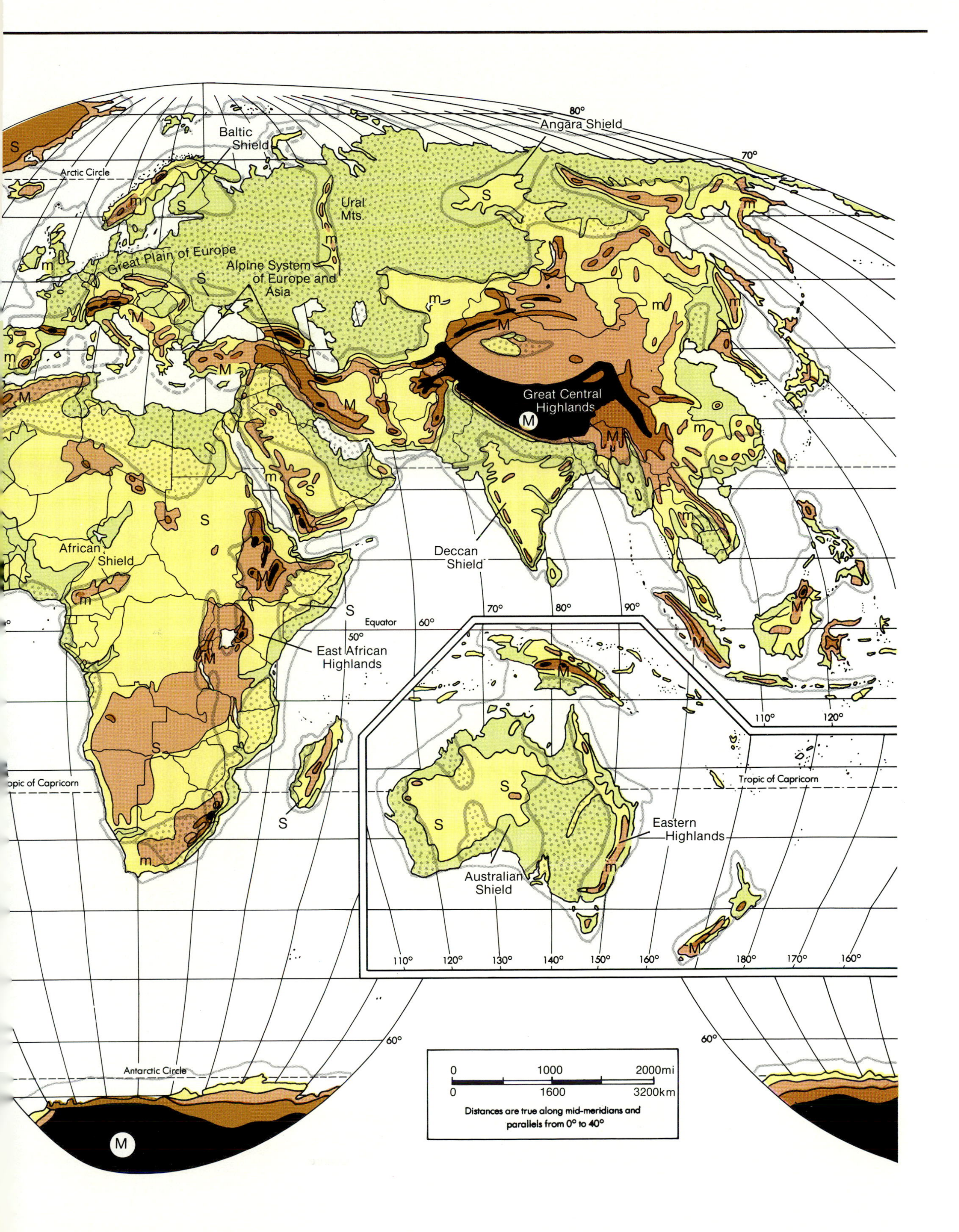

Baltic Shield
Angara Shield
Arctic Circle
Ural Mts.
Great Plain of Europe
Alpine System of Europe and Asia
Great Central Highlands
African Shield
Deccan Shield
Equator
East African Highlands
Tropic of Capricorn
Eastern Highlands
Australian Shield
Antarctic Circle
0 1000 2000mi
0 1600 3200km
Distances are true along mid-meridians and parallels from 0° to 40°

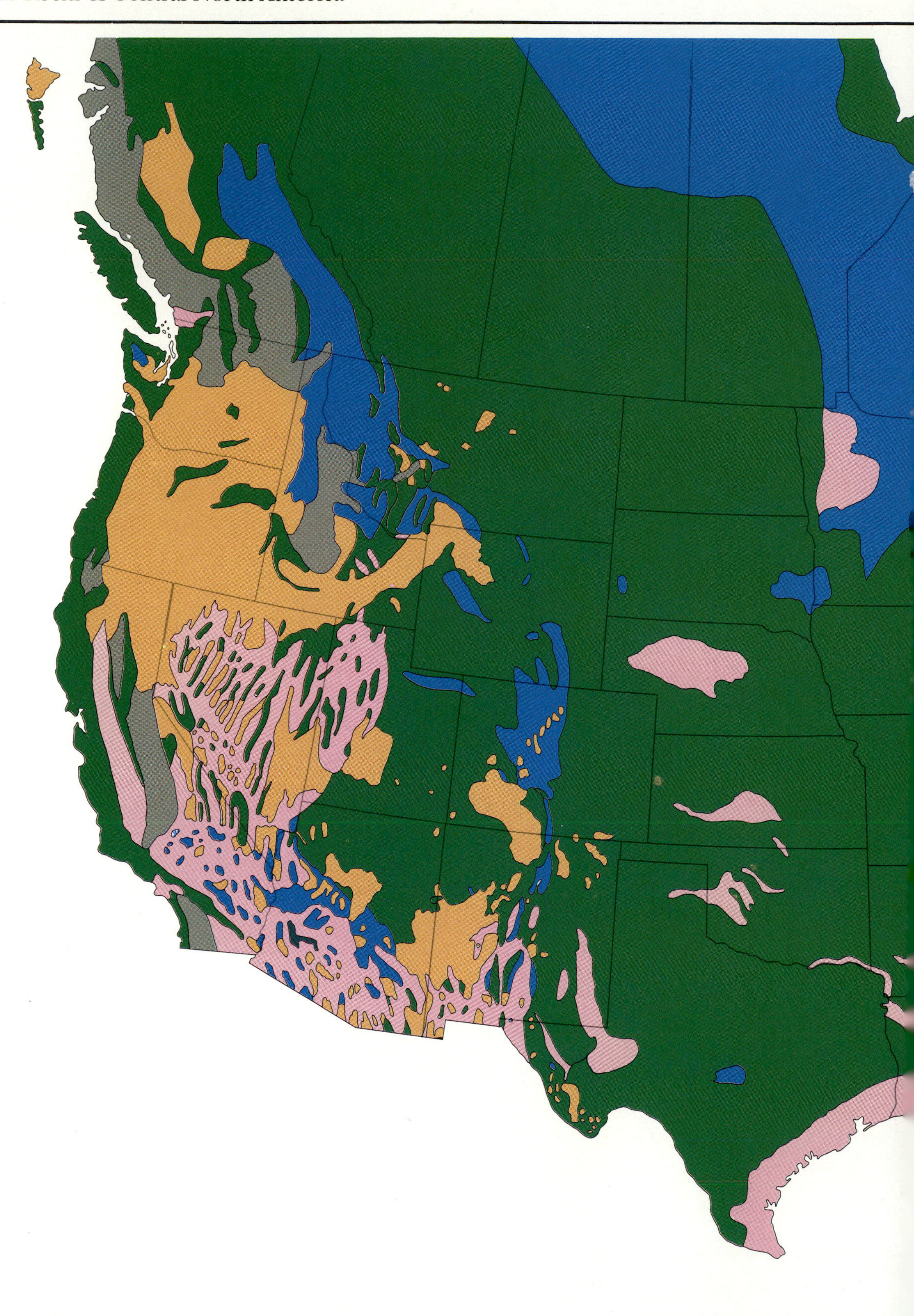

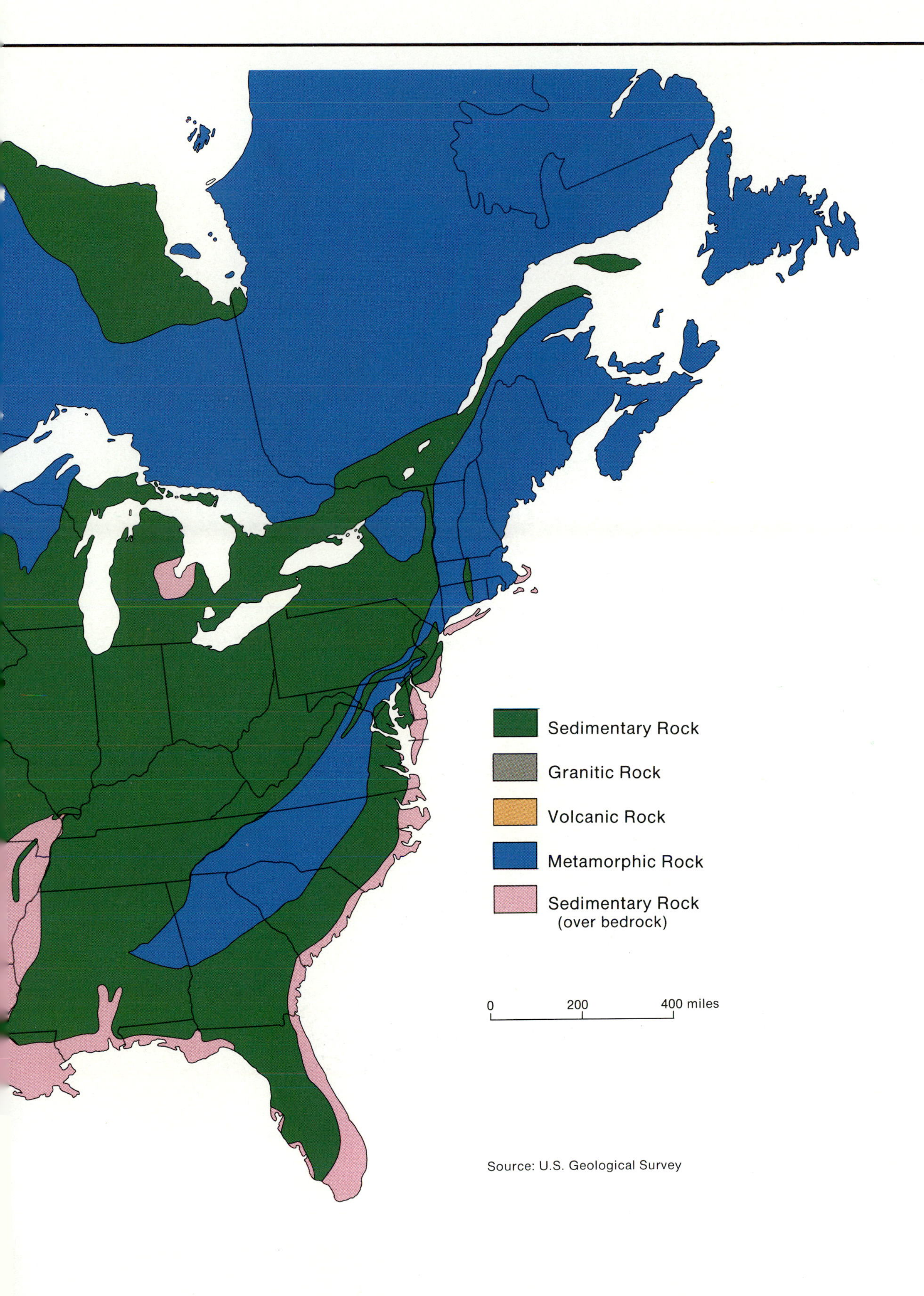

Sedimentary Rock
Granitic Rock
Volcanic Rock
Metamorphic Rock
Sedimentary Rock
(over bedrock)
0
200
400 miles
Source: U.S. Geological Survey

PLATE 11 Tectonic Plates and Border Zones

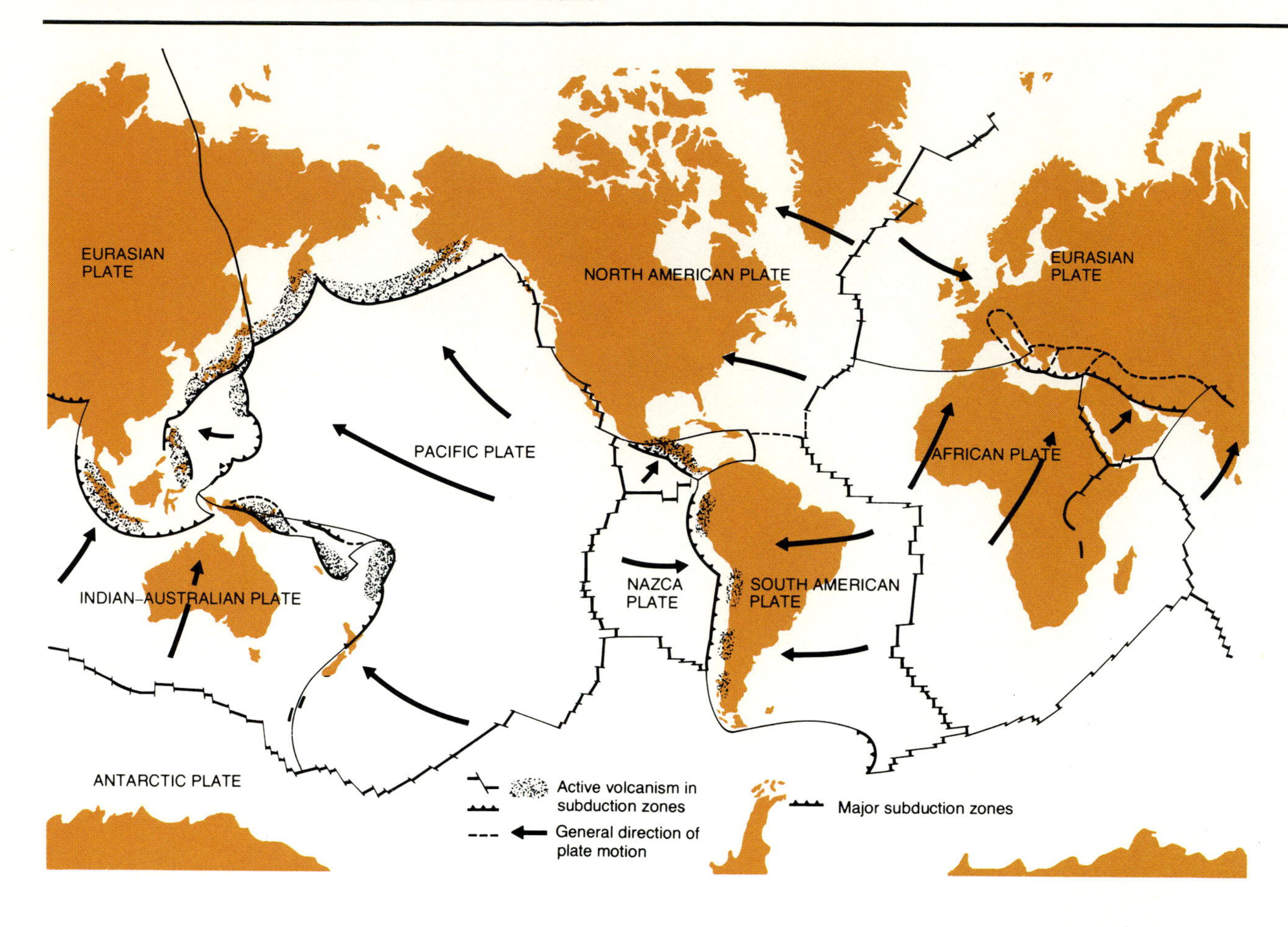

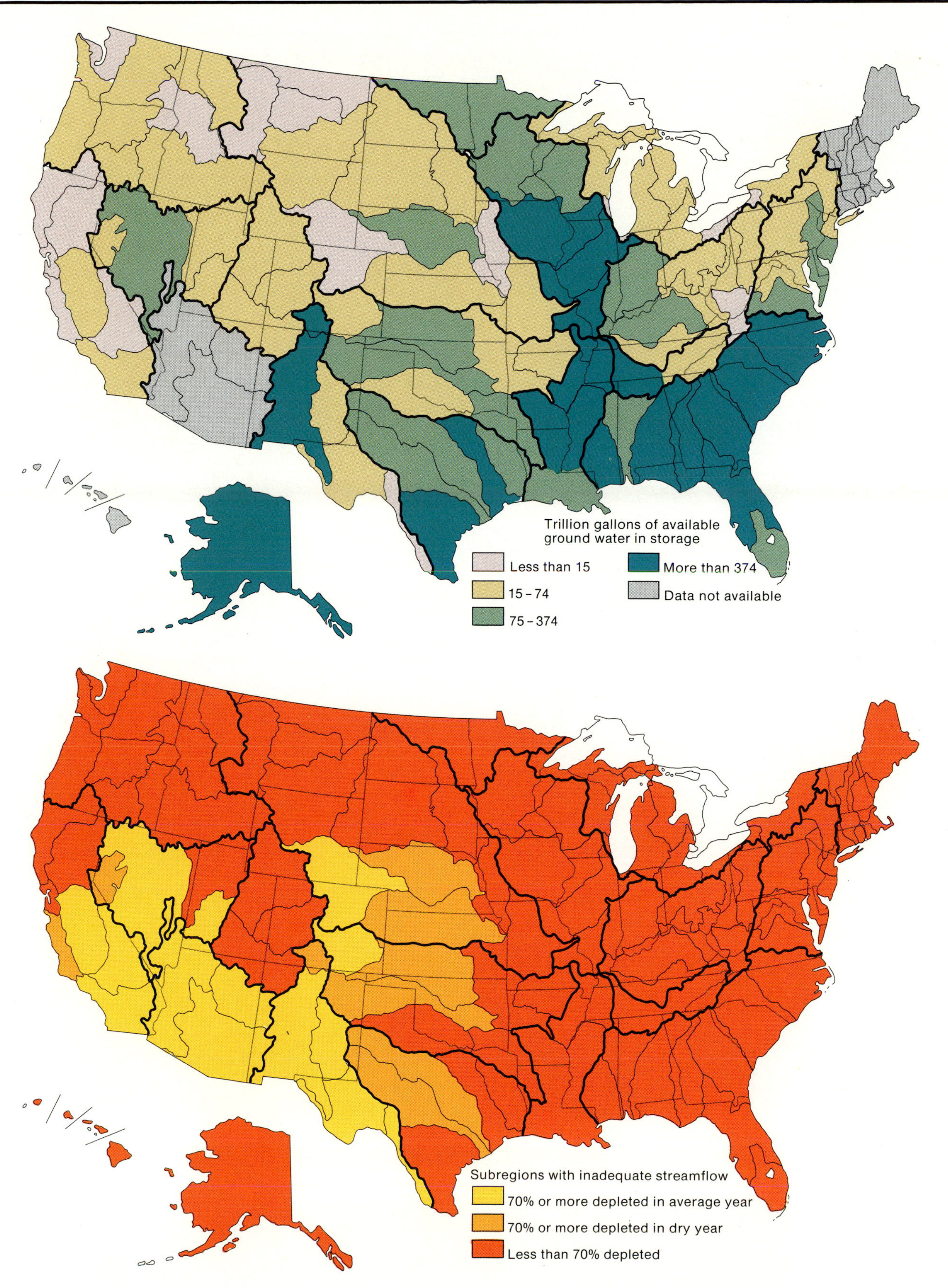

Source: U.S. Council on Environmental Quality

PLATE 13 Landsat Color Infrared Composite Image, Puget Sound (also see Fig. B-2)

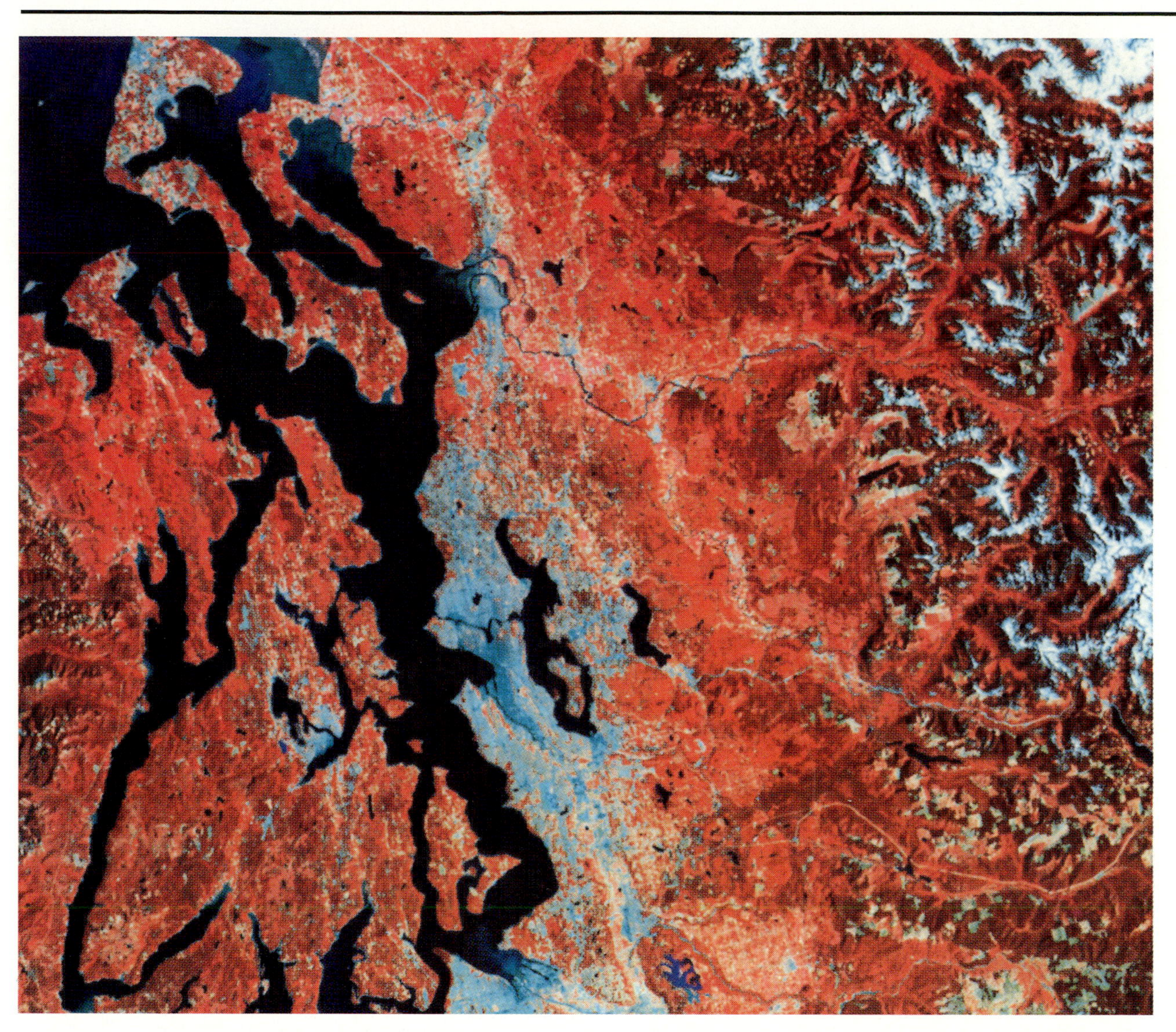

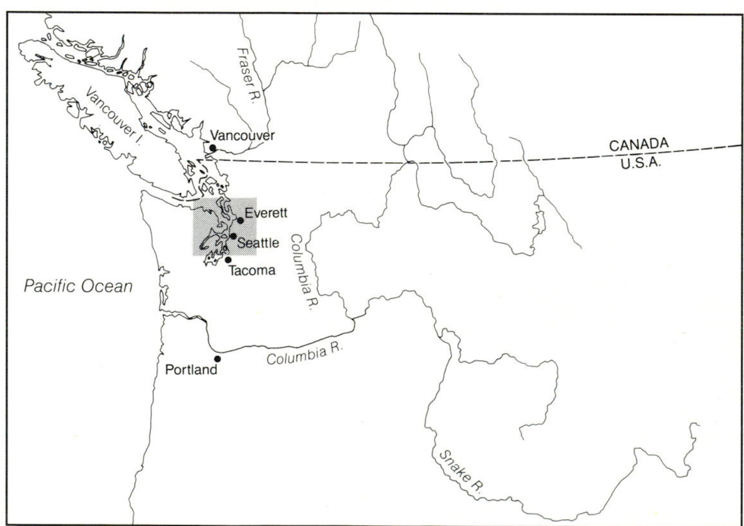

PLATE 14 Landsat Natural Color Composite Image, Cape Cod (also see Fig. B-3)

PLATE 15 Landsat Color Infrared Composite, Texas-New Mexico Border (also see Fig. B-4)

PLATE 16 AVHRR Color Infrared Mosaic, Northern Hemisphere, Summer (also see Fig. B-6)

PLATE 17 Natural Color and Color Infrared Aerial Photographs, Coastal Lake, Michigan (also see Fig. B-7)

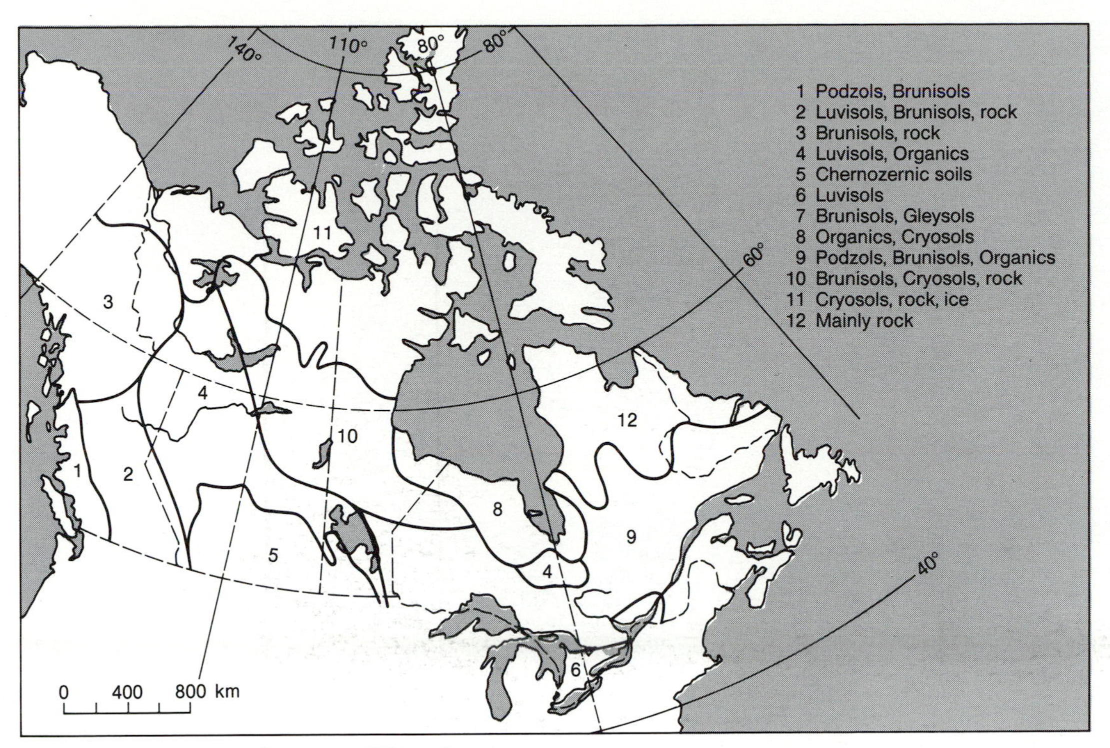

Figure 14.8 Major soil regions of Canada.

Engineers classify soil mainly according to how it performs when put to certain intensive uses, such as supporting buildings and highways or holding back reservoirs of water. As we noted in Chapter 13, the engineer is not interested in the capability of a soil to support plants; the engineering part of the soil is usually considered to begin with the material beneath the solum. For this reason, agricultural soils maps, which disregard the soil below a depth of a meter or so, are of limited value to the civil engineer. Because the weight of most structures is supported by the soil mass as much as several tens of meters below the surface, the civil engineer must be concerned mainly with what the agronomist would consider the subsoil. One widely used engineering soil classification system, the Unified System, which is based on two main criteria (texture or composition, and behavior when saturated), is given in Appendix D.

PROBLEM SOILS AND SOIL PROBLEMS

It is safe to say that all terrestrial life is dependent on soil. Virtually all plants depend on soil for water, minerals, and structural support. All higher organisms in turn depend on the plants; without them we perish. In addition, soil is the structural foundation for most of the landscape; it is what humans walk on, put their buildings on, and place their water and other pipe systems in. In both contexts, life support and structural support, we have many serious problems with soil in the world today.

Urbanization and Soil

Most of us in the industrial world live in and around cities. In the past century cities have changed enormously; the buildings are larger (many now exceed heights of 1000 feet), the developed areas are more extensive geographically, and the support systems (water, energy, communications) are more elaborate and complex. The role of soils in urban development has also changed. Many of the effects of soil on ancient cities—such as the slow sinking of parts of Mexico City and Venice, Italy, because they were unwittingly built over poorly compacted sediments—are not as apparent today because we have the means of detecting where such soils lie.

Does this mean that unstable soils are avoided in

Excavation for building foundation revealing different soil layers with different engineering properties.

modern urbanization? The answer is usually no. As cities grow and land values rise, investors are willing to spend more money for site improvements (soil, drainage, and topography) because ultimately the economic return on the land as a space for business will be high enough to amply offset the improvement costs. The problem soil may be excavated and replaced by a better one, or special building foundations may be designed to compensate for soil weakness. Therefore, modern society treats problem soils as an added expense in urban development. Only in extreme cases (where the soil limitations are too severe to overcome) or where decisions about building are based on noneconomic values, such as the need for parks and open space, habitat protection, or the maintenance of groundwater recharge areas, are areas containing problem soils set aside in the modern urban environment.

Permafrost and Development

Of the soils that pose severe limitations to development, none are more abundant than soils of the tundra (the cryosolic order in the Canadian Soil classification). These soils are usually poorly drained and high in organic content, but, more importantly, they are also underlain by permafrost (see Fig. 4.8). Soils underlain by permafrost are highly sensitive to changes in the soil-heat balance, and this balance can be upset by disturbances as slight as the damage and destruction of ground vegetation by the wheels of a truck or tractor. Roads, buildings, and airfields built here often increase soil heat enough to melt the upper permafrost. Because liquid water occupies less space than ice, the ground subsides as the permafrost melts. This can cause buildings to tilt, foundations to break, and facility lines to rupture. (See Chapter 19 for details on soil movement related to frozen ground.)

The Trans-Alaska Pipeline, which stretches 700 miles from the North Slope oilfields in northern Alaska to the ice-free Pacific port of Valdez in southern Alaska, was built over hundreds of miles of permafrost. In this cold environment, oil must be heated if it is to flow freely. Therefore, the conventional method of burying the pipe could not be used because the chances were high that burial would melt the permafrost, causing the ground to subside and rupture the pipe. After much analysis and debate among scientists, engineers, environmentalists, and government officials, it was decided to elevate the pipe above permafrost ground. The supports that hold the pipe in the air are equipped with heat shields that insulate the cold underlying ground from the hot pipeline (Fig. 14.9).

Figure 14.9 The Trans-Alaska pipeline, north of Fairbanks, Alaska.

Soil and Agriculture

Between 1975 and 1988 world population will have grown by 1 billion people (from 4 to 5 billion). The food demands of this mass of humanity are placing unprecedented stress on millions of acres of soil. This development has led to two major trends in agricultural land use: (1) more intensive use of existing cropland, and (2) expansion of agriculture into second-rate lands. The objective of the first trend is to improve annual yields per acre or hectare of land under cultivation, and it usually entails some sort of technological investment. These include new varieties of crops with better yields, fertilizer application, and new cropping techniques that employ mechanization and irrigation. In almost any case, it results in more intensive use of the land, which can result in deterioration of the soil as organic content declines, erosion increases, and fertility drops, and requires considerable financial investment.

The second trend is more common among the developing and underdeveloped countries. It is characterized by the spread of agriculture beyond the marginal lands and into lands classed as *submarginal.* These are lands with serious limitations, such as steep slopes, short growing seasons, or frequent periods of drought—lands where chances of crop failure are as great or greater in any year than success is.

But the consequences of using submarginal lands are far greater than frequent crop failure because this practice also results in (1) serious environmental damage, especially soil erosion, and thus (2) pressure to use more, even less suitable, land. The net outcome is an ever-worsening situation as farmers are pressed to produce more food for a growing population from a declining reservoir of land.

Loss of fertility and erosion take place together on most cropland. Even in the United States, where many economists consider the agricultural system to be the most advanced in the world, both of these problems are widespread. The U.S. Department of Agriculture estimates that in 1977 the average annual soil loss to erosion from runoff was 4.8 tons per acre on cropland in the United States. With good soil conservation practices, topsoil is replaced at a rate of 1.5 tons per acre. In general, a net soil loss greater than 1 to 4 tons per acre per year (depending on soil depth) cannot sustain crop production. Soil erosion has permanently damaged 200 million acres and is a serious problem on 240 million additional acres of cropland, more than half of the total cropland acreage in the United States (Fig. 14.10). Canada has experienced similar problems, and in Note 14, Geoffrey A.S. Scott of the University of Winnipeg examines the impacts of agriculture on the Canadian Prairies.

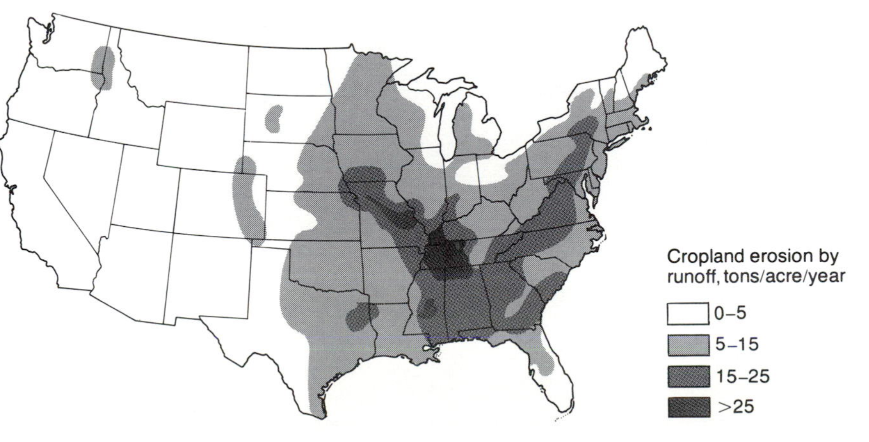

Figure 14.10 Erosion of cropland by runoff in the United States, 1975.

THE VIEW FROM THE FIELD / NOTE 14

The Impact of Agriculture on the Soils and Vegetation of the Canadian Prairies

...all the prairies were burned last autumn, a vast conflagration extending for one thousand miles in length and several hundreds in breadth. . . . If a portion of prairie escapes fire for two or three years the result is seen in the growth of willows and aspens, first in patches, then in large areas. . . . A fire comes, destroys the young forest growth and establishes a prairie once more.[1]

Fires no longer exercise a significant control on the vegetation in this region because highways, plowed fields, towns, and other factors retard their spread. As a result, in the past 100 years trees (mainly aspens (*Populus tremuloides*)) have spread rapidly into unfarmed prairie, creating vast new areas of parkland vegetation. Just how rapidly the parkland margins expanded south and west during the first half of this century can be seen in the accompanying map: Between 1905 and 1956, the aspen parkland in southwestern Manitoba spread more than 140 km (80 mi.) in some areas. But the new parkland has not been without modification either, for much of it has been altered by overgrazing and trampling by cattle. The presence of pasture sage (*Artemisia frigida*) on coarse-textured soils and snowberry (*Symphoricarpos occidentalis*) reflects the overgrazing.

As for the remaining prairie (south of the aspen parkland), the original plant cover has been greatly modified by farming and related land uses. Only relicts of the original tall grass prairie are left in the region, such as a few tens of hectares dominated by tall blue stem growing on the heavy, often waterlogged, soils of the Red River Valley in Manitoba. In addition, throughout the region the natural prairie stands have also suffered through encroachment by weeds, such as leafy spurge (*Euphorbia esula*) and sow thistle (*Sonchus arvensis*), which accompanied settlement. In drier grasslands overgrazing has encouraged a great increase in another weed, sage (*Artemisia spp.*).

Many of the actions that have altered the plant cover have also altered the soils of the Canadian Prairies. The most serious changes in soil stem directly from vegetative change: increased erosion,

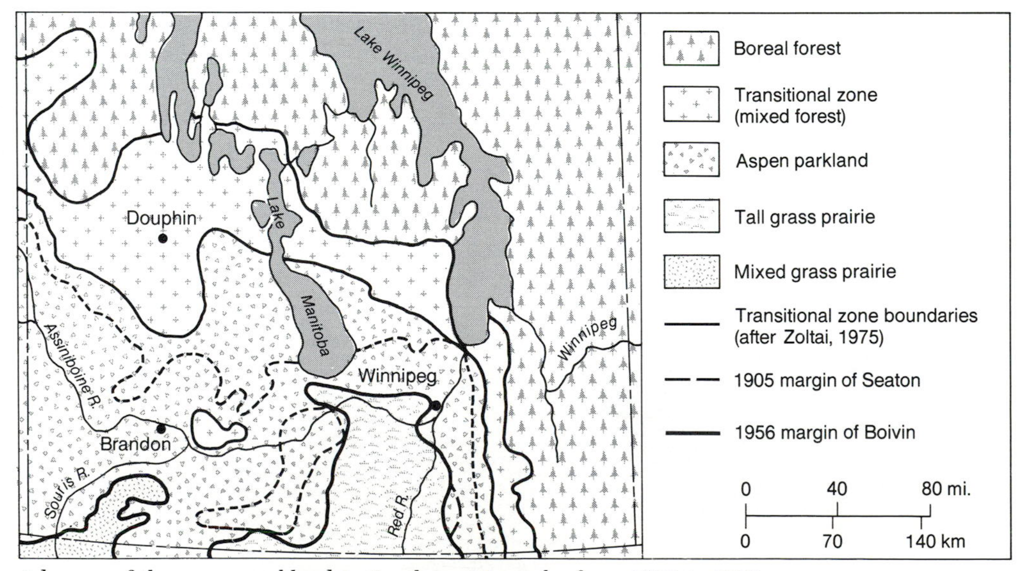

Advance of the aspen parkland in Southern Manitoba from 1905 to 1956.

[1]So reported H.Y. Hinds in 1859 after observing the effects of fires on the vegetation of the Canadian Prairies.

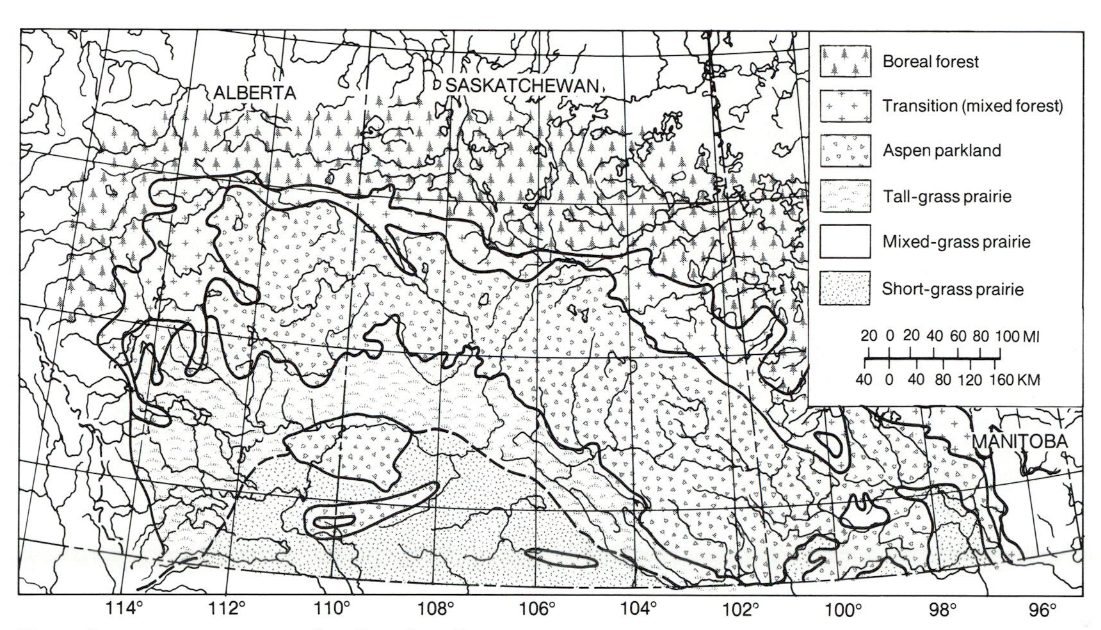

Natural vegetation cover in the Canadian Prairies.

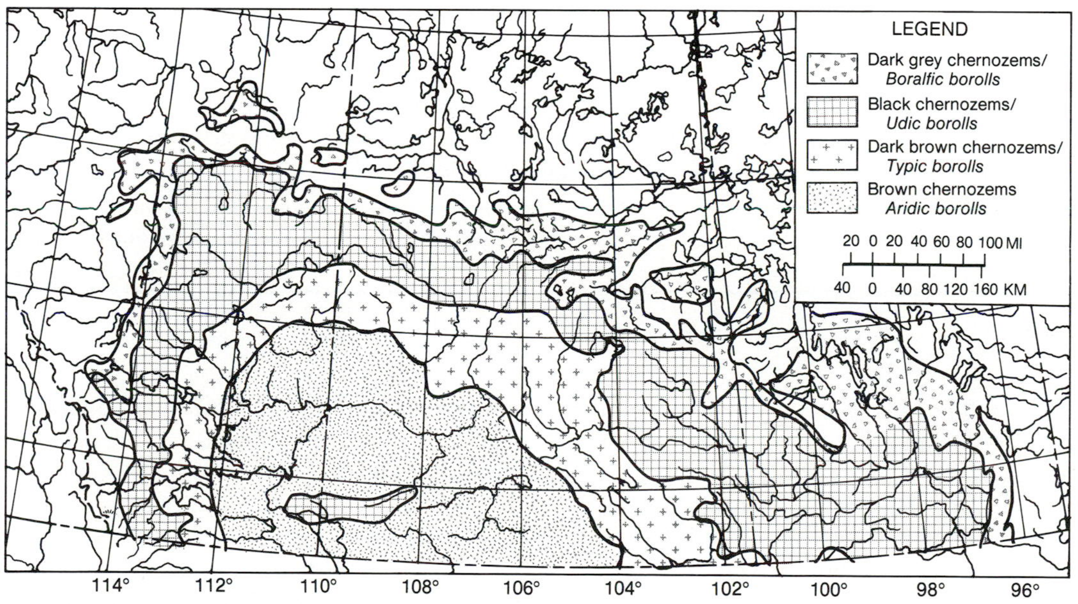

Soils distribution in the Canadian Prairies.

(Note 14 Continues)

loss of fertility, and increased salinity. With respect to fertility, for example, where prairie has given way to aspen parkland some black chernozem soils have been modified to dark-gray chernozem soils because aspen litter promotes greater leaching (being more acidic than grass litter) and less organic accumulation than do grasses.

But erosion unquestionably has rendered the greatest change in soils. Wind erosion is a persistent threat to plowed soils without protective plant covers. In southern Saskatchewan and southeastern Alberta, the chinook winds can blow away huge amounts of unprotected topsoil in a matter of hours. In the Dust Bowl era (the "dirty thirties"), drought, land abandonment, and poor farming techniques combined to induce colossal erosion rates. Estimates for some areas indicate topsoil losses alone of some 1800 tonnes per hectare (880 tons per acre) in some years. The Prairie Farm Rehabilitation Act of 1935 led to major reductions in erosion by introducing shelter belts and other conservation practices, but today 2 million hectares (5 million acres) of Canadian prairie soils still exhibit moderately to severely eroded conditions, while another 4.2 million hectares (10.5 million acres) show slight to moderate wind erosion. It is estimated that erosion problems are on the increase again today as a result of a decline in the practice of row cropping, removal of shelter (tree) belts along field borders, and the continued decline in soil organic content because of modern cropping practices.

In the drier parts of the prairies, it is common knowledge among farmers that plants deplete soil moisture faster than direct evaporation of bare soil does. This has given rise to the practice of leaving fields fallow in some summers as a way of conserving soil moisture. To be most successful, this practice requires that fields be kept clean of crops and weeds for one growing season, which means that no new organic matter is produced to make up for that lost in natural decomposition of the top soil. A lower organic content in turn reduces the soil's capacity to form aggregates (see Fig. 13.11) and resist compaction, both of which make it more vulnerable to erosion by wind and runoff.

The summer fallow practice, along with irrigation in some areas, has also led to increased soil salinity. Wheat normally uses only 50 to 65 percent of the soil water that undisturbed prairie would, and the summer fallow practice results in even lower percentages of soil water loss. Therefore, under the fallow practice, much more water is left in the soil over the long run, and part of this water seeps down to the water table, slowly raising it. As the water table rises, the dissolved salts in it are also raised in the soil. The net effect has been the gradual buildup of salts in the surface layers of the soil. Farmers have responded by changing to more salt-tolerant crops, planting fields to forage crops, or simply abandoning the land.

Geoffrey A.J. Scott
University of Winnipeg

In addition to erosion, soil fertility is also degraded by crops that draw large amounts of nutrients from the soil without replacing them. When a natural plant cover is replaced by crops, two changes usually occur in the soil-nutrient system: (1) nutrients are drawn from the soil at an accelerated rate because most crops have higher nutrient requirements than do natural plants; and (2) little or no plant organic matter is resupplied to the soil surface because the mass of the crops are removed in harvesting. In order to maintain productivity, fertilizers must be added to make up for this imbalance.

The amount of fertilizers required varies according to the crops planted and the soil conditions. Where soils are inherently weak in nutrients and the crops have high nutrient requirements, each year hundreds of pounds of fertilizer must be added to each acre in order to sustain production (Fig. 14.11). This is generally the case in the American South, where the ultisols are used for crops of corn, tobacco, and sorghum. In contrast, richer soils, such as the mollisols of the Great Plains, which are planted to low nutrient-demand crops such as wheat, may require only 25 to 30 pounds (11.4–13.6 kg) of fertilizer per acre per year, even after decades of use.

The most severe soil-nutrient problems in the world occur in the oxisols and ultisols of the tropical rainforests. Despite the high soil fertility implied by the luxuriant rainforest vegetation, these soils are deceptively low in nutrients because organic matter decom-

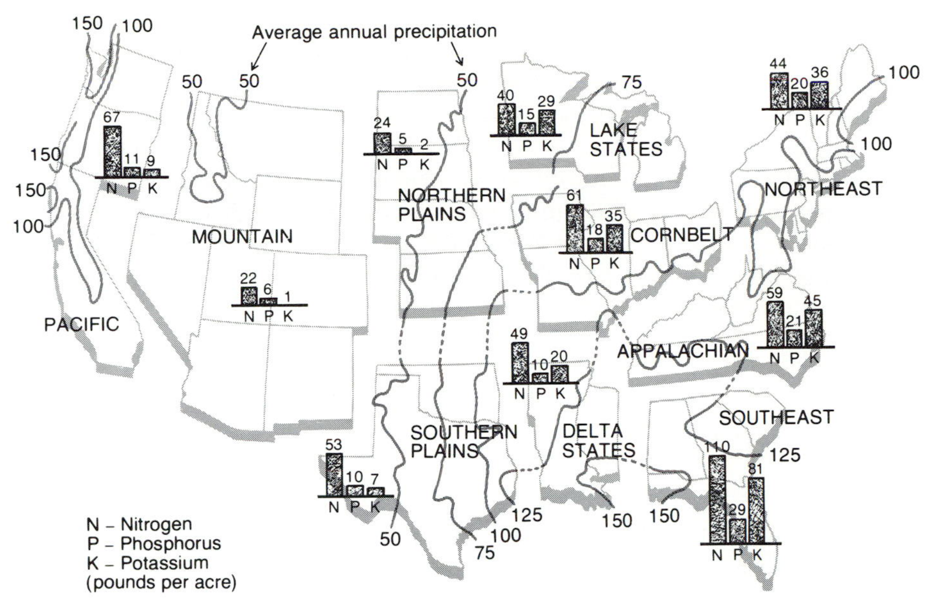

Figure 14.11 Average application of fertilizer per acre per year in the United States. Note the approximately sevenfold increases in quantities from the Northern Plains to the Southeast. This corresponds to an increase in precipitation and in turn greater soil leaching and lower soil fertility in the Southeast.

poses so rapidly here and also because heavy tropical rains so rapidly leach nutrients from the solum. Unless massive amounts of fertilizer are added to these soils after clearing and planting, they are depleted of fertility in as little as 3 to 5 years of cropping. In addition, as the organic layer disintegrates and the crop cover diminishes, the clay beneath is often baked to pavement hardness by the sun. Because the capital and technology necessary to maintain soil productivity are generally lacking among the peasants of the tropics, there is little left for the farmer to do but move to new ground when the soil declines after several years.

Under this practice of *shifting agriculture,* the abandoned plot is left to second-growth vegetation, and after many years the nutrients are restored to levels that can once again support crops for a few years. In the meantime, the farmer has cleared, used, and abandoned several more patches of soil. Clearly, the practice of shifting agriculture is very inefficient, and it requires a larger amount of land to sustain a family than is needed to sustain a family in the United States or Europe. But lacking the means to maintain soil productivity, the tillers of the rainforest latosols have no choice but to continue their soil-depleting practices. Hence, these vast regions remain among the world's great agricultural frontiers.

Arid lands are plagued with an agricultural problem of their own: salt saturation of the soil caused by excessive irrigation. Where intensive irrigation is practiced, a large amount of the water is lost in percolation; after many years the groundwater under fields is often raised to the point where the zone of capillary water immediately over the water table and the water table itself reach the surface. As the water rises through the soil column, it dissolves the salt and carries it up with it. When the water evaporates, a salt residue builds up in the root zone, eventually rendering the fields useless for cropping. This process, called *salination,* has contributed significantly to the loss of highly productive agricultural land and is cited as one of the causes of desertification, the overall degradation of land owing to problems associated with aridity (also see Note 14).

Summary

The combination of chemical, biological, and physical processes within the soil column produces differentiation of the soil into horizons. These processes are influenced by climate, parent material, vegetation and topography, and drainage. One set of soil-forming regimes is associated with bioclimatic factors, and another with factors, such as mechanical action, that are not directly tied to bioclimatic conditions.

The major soil classification schemes have come from agronomy and engineering. The traditional USDA scheme grew out of evolutionary concepts about soil formation. The USDA comprehensive scheme is based on diagnostic horizons and recognizes a broader range of "legitimate" soil-forming processes. The Canadian soil classification system has fewer subdivisions than the new USDA scheme and places greater emphasis on the soils of cold environments. Civil engineering identifies texture and behavior as the key properties in soil classification.

Soil is fundamental to the survival of most of humanity on the planet inasmuch as it is critical to agricultural productivity. Yet productive soil is being lost to urban development, and in most agricultural areas fertility is declining because of erosion, overuse, and salination of the soil.

Concepts and Terms for Review

- soil-climate relations
- soil-forming regimes
  - podzolization
  - laterization
  - calcification
  - salinization
  - hydromorphic
  - geomechanical
- soil-forming factors
- pedon and polypedon
- soil classification schemes
  - traditional USDA system
  - comprehensive system
  - Canadian system
  - engineering systems
- orders, suborders, great groups
- zonal, azonal, intrazonal
- entisol, inceptisol
- aridisol, mollisol
- spodosol, alfisol
- ultisol, oxisol
- vertisol, histosol
- urbanizization and soil
- permafrost and development
- soil erosion and agriculture
- soil impoverishment
- shifting agriculture
- salination

Sources and References

Baldwin, M., Kellogg, C.E., and Thorp, J. (1938) "Soil Classification." *The 1938 Yearbook of Agriculture: Soils and Man.* Washington, D.C.: Government Printing Office.

Berry, Wendell. (1977) *The Unsettling of America: Culture and Agriculture.* Avon.

Bridges, E.M. (1978) *World Soils*, 2d ed. Cambridge, England: Cambridge University Press.

Brown, R.J.E. (1970) *Permafrost in Canada: Its Influence on Northern Development.* Toronto: University of Toronto Press.

Bunting, B.T. (1967) *The Geography of Soil.* Chicago: Aldine.

Cunningham, R.K. (1963) "The Effect of Clearing a Tropical Forest Soil." *Journal of Soil Science* 14: pp. 334–335.

Eckholm, Eric D. (1976) *Losing Ground: Environment, Stress and World Food Prospects.* New York: Norton.

Glantz, Michael H., ed. (1977) *Desertification: Environmental Degradation in and around Arid Lands.* Boulder: Westview.

McNeil, M. (1964) "Lateritic Soils." *Scientific American* 207, 11: pp. 97–102.

Soil Survey Staff. (1964) *Supplement to Soil Classification, A Comprehensive System, 7th Approximation.* Washington, D.C.: Soil Conservation Service, U.S. Department of Agriculture.

Steila, Donald (1976). *The Geography of Soils: Formation, Distribution and Management.* Englewood Cliffs, N.J.: Prentice-Hall.

Young, A. (1976) *Tropical Soils and Soil Survey.* Cambridge, England: Cambridge University Press.

Chapter 15

Structure and Composition of the Solid Earth

The solid earth is more than a passive platform on which the landscape rests—it plays an active role in shaping the landscape. Differences in rock composition and in the energetic motions of earthquakes and volcanoes directly influence the landscape and its organisms. To set the framework for study, we must first examine the structure and composition of the earth's crust and the general nature of the material beneath it.

Only 100 years ago the frontier of research into the nature of the solid earth, the crust, was in the American West, where geographers and geologists were unfolding the story of the Grand Canyon, the building of the Rocky Mountains, and so on (Fig. 15.1). Today the frontier—or at least one of the frontiers—lies in the exploration of other planetary bodies in our solar system: the moon and our sister terrestrial planets, Mars, Mercury, and Venus.

Our probes of the moon and Mars have revealed many things that have advanced our understanding of the earth. One of the most significant findings is a sharp contrast in landscapes: the moon and Mars possess in abundance two surface features that are very scarce on earth: impact craters, and huge, dormant volcanoes. Impact craters are depressions encircled by a rim of debris caused by the impact of large meteorites with the planet's surface. Another significant finding, based on analysis of lunar rock, is that moon and earth originated at about the same time—4.5 billion years ago—and it appears that the other terrestrial planets also began then.

How did these planetary bodies form? The most widely accepted theory is that the moon, Mars, and earth originated in basically the same fashion—by the accumulation (or condensation) of interplanetary debris in the first billion years of the solar system. (See Note 15.) The bulk of each planet formed quickly, probably in the first 200 or 300 million years, but debris continued to pepper their surfaces at a heavy rate for another several hundred million years. Most of the debris probably consisted of small particles of rock and ice, but some of it was also large meteorites that bombarded the surface with great force, creating large impact craters. Thus, about 4 billion years ago the earth had to have been literally covered with pockmarks like those we see on the moon today.

What happened to these features and to the ancient

Figure 15.1 The frontiers of research into the nature of the earth's crust. Field party sponsored by the U.S. government in 1871 to investigate the geology and geography of the American West.

volcanoes that formed among them on early earth? They were destroyed by erosional and geological forces operating at or near the earth's surface. Erosional forces (principally running water) wore them down, and geological forces, principally faulting (breaking) and folding (bending), deformed and consumed the rock of the ancient impact craters and volcanoes until today scarcely any features are left from the earth's first billion years. In contrast, the moon and Mars are rich in features dating to the first 1 or 2 billions years of their existence. The lesson here is clear: although the earth, its moon, its neighbor Mars, and other terrestrial planets probably had similar beginnings, the development of the earth after the first billion years or so took a significantly different course. The platform of rock on which the landscape rests was, and continues to be, highly active on earth, a fact that geographers must consider in their analysis of modern landscapes.

Examination of rock exposed in mountain sides, in mine shafts, and in the cores extracted from oil and gas wells indicates that the earth's crust is tremendously diverse, and indeed, often suggests that it is more diverse in its form and composition than we can imagine. It is the diversity of the crust, as well as its instability, that makes it so important in landscape formation and hence of great concern to physical geography. Among the physical geographer's interests are three major questions about the crust:

- What is the configuration and composition of the crust where it meets the landscape?
- What is the influence of the crust's form and composition in the development of landforms, soil, water features, vegetation, and climate?
- What is the influence of the processes of crustal instability, such as volcanism and earthquaking, in shaping the landscape and conditions of life in different parts of the planet?

THE SHAPE OF THE EARTH'S SURFACE

In order to establish a framework for study, we begin at the broadest scale with a brief description of the major topographic, structural, and compositional units of the earth's surface and subsurface. If we disregard ocean water for the moment, two major earth surface structures are immediately apparent: the continents and the ocean basins. The ocean basins occupy nearly 65 percent of the planet's surface and have several subdivisions.

The Ocean Basins and the Continents

The largest subdivision of the ocean basins in terms of area is the *abyssal plain*, or deep ocean floor. Two smaller subdivisions are the *midoceanic ridge*, or

VIEW FROM THE FIELD / NOTE 15

Extraterrestrial Debris as a Source of Geographic Change

It is an interesting fact that each year thousands of tons of debris rain down on earth from space, gradually increasing the planet's mass. Most of this debris consists of fine particles of ash from meteors that burn up and disintegrate as they pass through the atmosphere. The flashes of light from shooting stars are visual evidence of this process.

For many years shooting stars and related phenomena were the affairs of astronomers, but geographers have grown increasingly interested in them in the past few decades because of evidence that they may be major agents of change in the earth's surface environments. As we have already indicated, in the earth's first billion years, meteorites (defined as meteors that hit the ground) bombarded the surface, creating great explosions that threw huge amounts of debris into the atmosphere. As time passed, the magnitude and frequency of the explosions declined because the supply of debris in the solar system was largely taken in by the earth and the other planets. But much debris remained in space, some (like Halley's Comet) going into orbit around the sun.

The magnitude and frequency principle tells us that the processes that shape the earth's surface operate over time at irregular frequencies and in highly variable magnitudes. As the magnitude of a process, such as the size of a volcanic eruption or a storm wave on the ocean, increases, its frequency of occurrence decreases. In other words, the very powerful events in nature do not occur very often, whereas the gentle ones occur frequently.

The change rendered in the landscape by events generally increases with magnitude, that is, the very large events typically do the most work. Small events do little work individually, but since they are many, collectively they may be capable of appreciable work in the long run. When we compare small and large events, it appears that smaller ones tend to produce more gradual change, whereas larger ones tend to produce punctuated changes at widely separated intervals.

There is no accurate way of forecasting when an event of great magnitude will happen, but we can estimate the probability of its occurrence by measuring the average interval of time between past events. In northern Arizona, there is an impact crater 4000 feet across and 600 feet deep caused by a meteorite estimated to be 130 feet in diameter. Events of this magnitude are estimated to occur every 1000 to 1500 years. However, there is also another consideration in computing the recurrence interval of meteoritic events: Many of the comets and asteroids that produce meteors are in various orbits in the solar system, and these orbits come close to earth and its gravitational field at regular intervals. Halley's Comet visits earth every 76 years and Encke's Comet every 3.3 million. An idea recently advanced proposes that episodes of major impacts occur within a 26-million-year period, corresponding to certain large comet or asteroid orbits.

What are the environmental effects of meteoritic events? In 1908 a huge blast in Siberia leveled 800 square miles of forest, but the meteor exploded before reaching the ground. For those that hit the earth, the most important effect is not the impact crater sculpted by the blast, but the debris thrown into the atmosphere. The airborne debris from a large impact can be tremendous, many times greater than that from a large volcanic eruption. The most direct effect of this debris is a reduction in incoming solar radiation leading to sudden cooling at the lower atmosphere. A decline in global temperature by 5 to 10 degrees C could drastically reduce global biological productivity and compress the ranges of many species. Extinctions would be widespread as they were at the end of the Cretaceous Period (65 to 70 million years ago) when such an event may have ended the age of the dinosaurs.

The prospect of a huge meteorite explosion is very small; the prospect of large explosions from terrestrial sources is much greater, however. Large volcanic explosions are a frequent occurrence in each century. (See Table 16.2 for a list of the major effects of the Krakatoa blast of 1883.) The spectre of global nuclear war must also be considered. A major nuclear exchange could produce many of the same physical effects as a large meteorite explosion, namely, massive amounts of airborne debris and sudden cooling of the lower atmosphere leading to what some scientists have termed nuclear winter.

mountain range, the longest continuous mountain ranges in the world; and the *ocean trench*, or deep, which occurs on the margins of the ocean basin. Where the ocean basins meet the continents are the *continental slopes*, the long inclines that lead from the abyssal plains up to the outer edge of the continents, the continental shelf (Fig. 15.2).

The continents are composed of large, somewhat elevated interiors, called *shields*, fringed by mountains and continental shelves. The shields are generally the largest of the subdivisions of the continents and are often characterized by broad dome shapes—hence the term *shield*—but in reality they are very uneven surfaces. On most continents large sections have been depressed and covered with sediments or sea; in North America, for instance, the central part of the Canadian Shield in the area of Hudson Bay forms a huge depression. On the margins of the shields are the other two subdivisions: the *orogenic belts*, which are great bands of mountains, and the *continental shelves*, which begin on land and extend underwater to the continental slope (Fig. 15.3).

Orogenic belts are extensive mountain chains, such as the Rockies of North America and the Andes of South America. The highest orogenic belt is the Himalayas of Asia, where hundreds of peaks exceed an elevation of 6000 meters above sea level. Figure 15.4 shows the distribution of shields and orogenic belts for the large land masses of the world (also see color plate 9). Note that each shield has more or less the geographic aspect of a nucleus of a continent. The shields also contain the oldest rocks of the continents (and for the earth as a whole), with large masses between 2 and 3 billion years old. The orogenic belts are significantly younger, between 100 and 500 million years. The continental shelves are younger than the orogenic belts, generally less than 100 million years old. Thus, the general pattern of rock ages suggests that the continents may have grown outward from an inner core that was probably the shield.

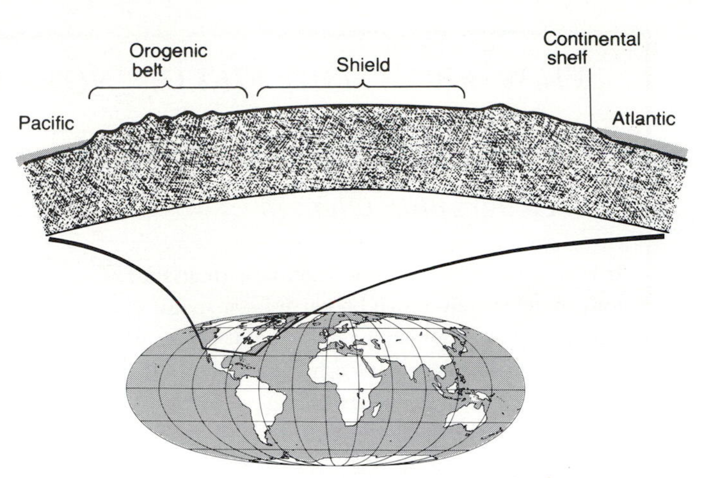

Figure 15.3 A schematic profile across North America showing the continental shelf and the orogenic belt.

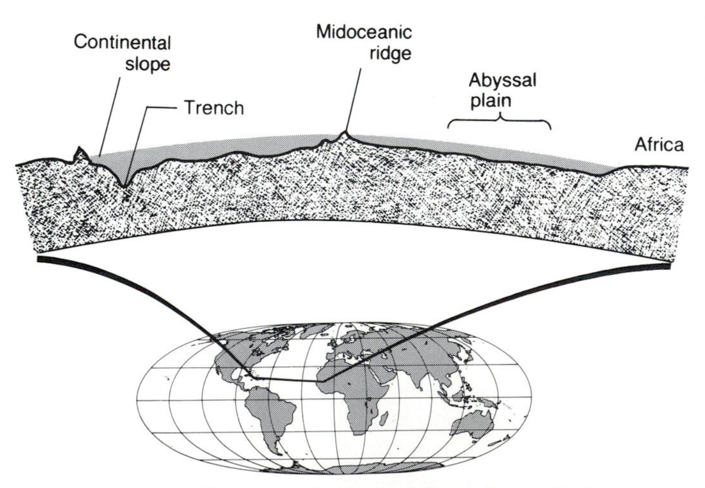

Figure 15.2 The major features of the Atlantic Basin. These types of features are found in every major ocean basin, although their relative locations and individual configurations are, of course, different.

The continental shelves, the gently sloping "shoulders" of the continents, together comprise about 5 percent of the earth's surface area. They range in width from several hundred kilometers, as in the Grand Banks near Newfoundland, to only several kilometers, along mountainous coastlines. At the outer edge of the continental shelf, the continental slope gives way to the deep ocean floor and in some locations to trenches situated near the foot of the slope. The continental shelves are becoming increasingly important as sources of food and minerals. Figure 15.5 shows their distribution in the world.

The General Topography of the Earth

Because we spend so much of our lives within 10 feet or so of the ground, the configuration of the planet's surface often seems extremely varied to us. In recent years, however, with the perspective afforded us by spacecraft and high-altitude aircraft, we have often

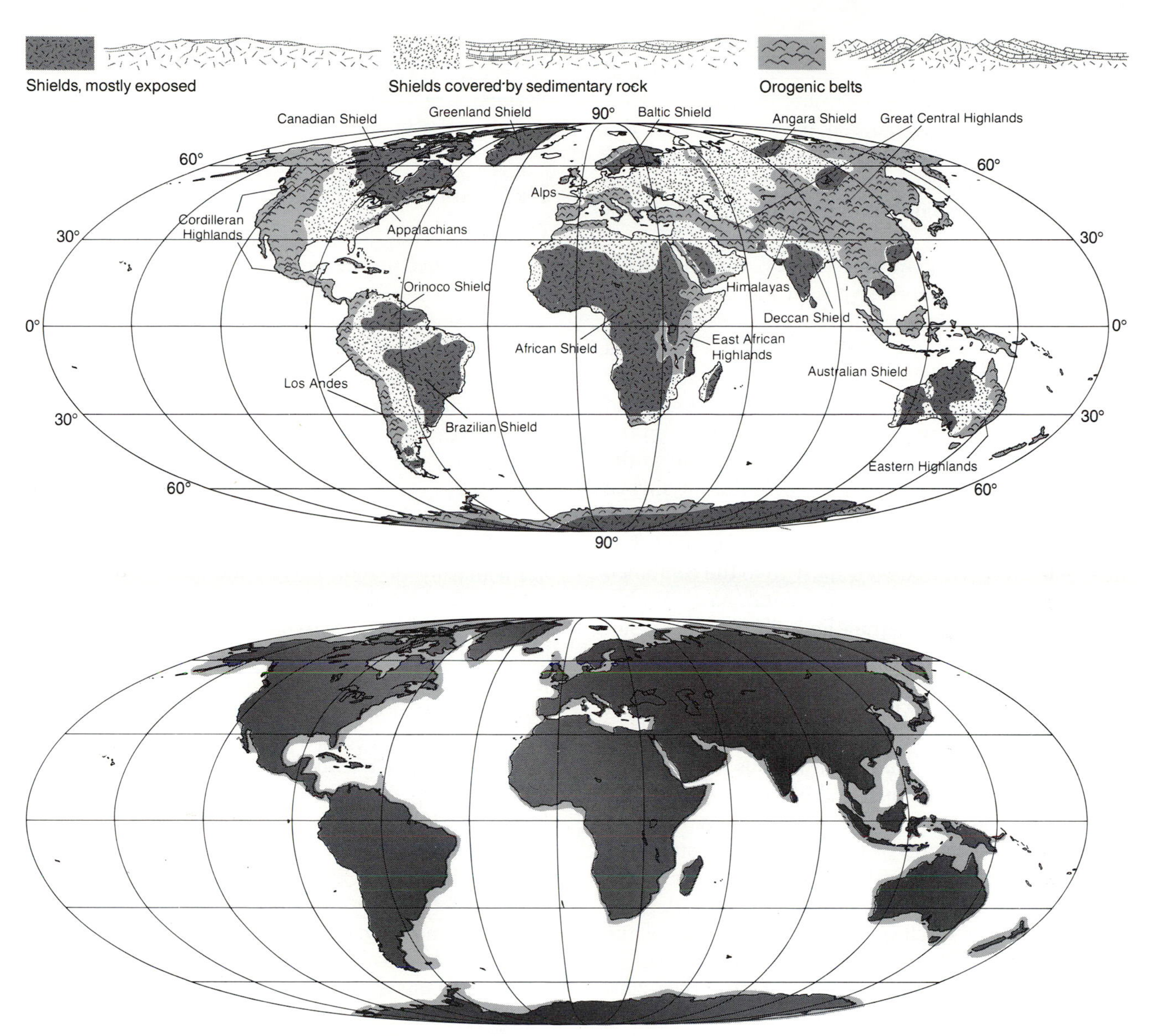

Figure 15.5 The global distribution of continental shelves. The width of the shelf varies from several kilometers to several hundred kilometers and thus may have great geopolitical implications as nations increase their reliance on the continental shelves as sources of food and minerals.

been impressed by the apparent smoothness of the earth's skin. Indeed, most of it is very smooth, more than 70 percent of it being water.

Many of the largest topographic features of the earth are obscured by the oceans and have never been seen. The ocean trenches typically reach depths of 7500 m or more below sea level. The deepest known is the Marianas Trench, near Guam in the Pacific Ocean, which reaches about 11,000 meters below sea level and about 5000 meters below the surrounding ocean floor. The most obtrusive topographic features of the ocean basins, however, are the great mountains that rise from the sea floor and the oceanic ridges.

Many mountain islands exceed a height of 6000 m above the ocean floor, and some, such as Mauna Loa of Hawaii, exceed a height of 10,000 m. Three classes of mountain islands are found in the oceans: (1) island arcs, for example, Japan, which occur in close association with ocean trenches; (2) ridge islands, for example, Iceland, which are found along midoceanic ridges; and (3) seamounts, for example, Hawaii, which rise directly from the abyssal plains. Although the seamounts that form islands are very large, most seamounts do not reach more than 3000 m or so above the ocean floor, leaving their summits well below sea level.

In terms of topographic extremes, the continents are generally less impressive than the ocean basins. The greatest continental difference is 9243 m (between Mount Everest at +8848 m and the Dead Sea at −395 m), but these points are widely separated geographically. More typically, the extremes in and around the world's major mountain ranges are on the order of 3000 to 4000 m, roughly comparable to the differences between trenches and neighboring abyssal plains (Color Plate 9).

The prize for extremes in topography for points less than 500 km apart, however, must go to the combined areas of the ocean rims and continental coasts. Within this narrow ribbon of terrain lie both ocean trenches and high continental mountain ranges. Together, these features commonly span a vertical distance of 12,000 m or more. Although these elevations look impressive, the average rate of slope between the mountain peaks and the trench floor is only about 2 to 3 m per 100 m of horizontal distance, comparable to steep inclines on U.S. interstate expressways.

The mean elevation of the earth's solid surface is 2.44 km below sea level. The mean depth of the oceans is nearly 4 km, whereas the mean elevation of the land is 840 m above sea level. About 71 percent of the earth's land area lies between sea level and a height of 1000 m, and about 13 percent is higher than 2000 m. The 13 percent figure is noteworthy because the conditions necessary to support life decrease rather sharply above 2000 m. If a larger percentage of land were at high altitude, the life-support capacity of the planet would definitely be lower than it is presently.

THE CRUST AND BELOW

The continents and the ocean basins together extend well below the surface and make up the earth's *crust*. Relative to the earth's 12,742 km (7918 miles) diameter, the crust is very thin, ranging in thickness from 8 to 65 km (5 to 40 miles). The crust is often described as the "skin" of the planet, with a thickness of only about 0.005 times the radius of the earth. It is also the surface of the earth's mantle, the massive zone of rock that surrounds the earth's core about which we will say more later. The geologic processes that change the earth's surface, such as volcanism and faulting, originate largely in the uppermost part of the *mantle*, including crust.

The Sima and Sial Layers

A traditional way to describe the crust is to divide it into two zones on the basis of rock types. The basal zone, called the *sima layer* (for the elements *Si*licon and *Ma*gnesium), is composed of a heavy, dark group of rocks called the *basaltic* rocks. The large amounts of iron and magnesium, both heavy minerals, contained in these rocks give them fairly high densities, that is, high masses per unit volume of rock. The units used for rock density are kilograms per cubic meter (Fig. 15.6).

The densities of rock in the sima generally range from 2800 to 3300 kilograms per cubic meter. (Water has a density of 1000 kg/m^3; therefore, for a given volume, these rocks are 2.8 to 3.3 times heavier than water.) The sima layer circumscribes the entire planet but is exposed only in the ocean basins. Elsewhere, it is covered by the upper zone of the crust, called the *sial layer* (for the elements *Si*licon and *Al*uminum).

The sial layer occupies about 35 percent of the earth's surface, in places reaching a thickness of 40 km, and forms the continents. The rock of sial is lighter in weight than that of the sima, with densities generally in the range of 2700 to 2800 kg/m^3. Sial is lighter because the heavier elements of iron and magnesium are less abundant in the rock of the sial, which is made up mostly of silicon, aluminum, and other light elements. Granite is the representative rock type of the sial layer, and it has been traditional to refer to sialic rock in general as "granitic rock."

The sial should not be thought of as a flat, homogeneous slab resting on the sima. Both its composition and structure are highly diversified. The lower boundary appears to be highly irregular and is characterized by a transitional rather than an abrupt change to the

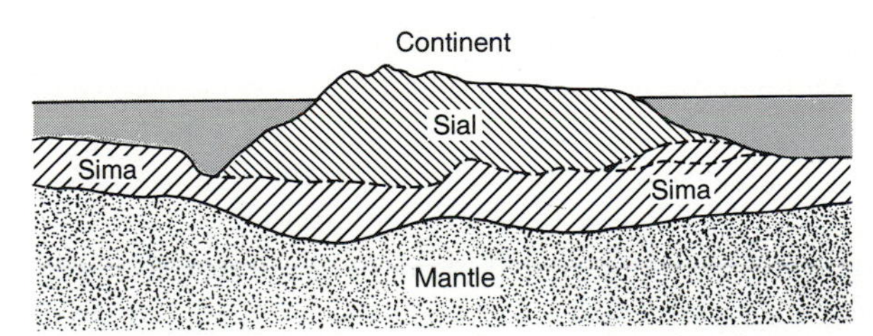

Figure 15.6 Summary diagram and data on the major features of the sial and sima.

upper sima. In composition, it includes large amounts of basaltic rocks, similar to those of the sima, that reside among the granitic rock. In sharp contrast, the sima, judging from the records of rock samples taken from the ocean basins, appears to be almost exclusively basaltic. Bear in mind that the continental crust consists of both sial and sima, whereas the oceanic crust consists of sima alone (Fig. 15.6).

Exploring the Earth at Depth

We have learned about the composition and structure of the crust and subcrust by analyzing energy waves transmitted through the earth. The concept of energy waves in the earth can be envisioned by using the analogy of a struck bell. When a bell is struck, sound waves radiate from it. At the same time, waves travel through the solid part of the bell, and if you are touching the bell, you feel these waves as vibrations. The waves are transmitted within the bell through the bumping together of trillions of atoms. As a wave travels along, successive atoms bump into their neighbors and then return to their original positions. Because the atoms always return to their original patterns after they are displaced, these waves are called *elastic waves*. The elastic waves studied in the earth are produced by sudden movement of massive quantities of rock or by artificial means such as an explosion from dynamite or a nuclear device. The branch of geophysics that studies earth waves, or tremors, is called *seismology*.

Seismic energy produces several types of waves, but two are particularly important in seismic analysis: *compressional waves* and *shear waves*. As compressional waves travel through rock, they generate a back-and-forth motion parallel to the direction in which energy is broadcast. (Sound waves—in air, or a solid such as bell metal—are also compressional waves.) (Fig. 15.7a). Shear waves cause neighboring particles to move in directions perpendicular, or transverse, to the direction of energy transmission (Fig. 15.7b). These motions can be detected with devices called *seismometers* and recorded with an associated instrument called a *seismograph*.

By studying the behavior of seismic waves as they pass through a mass of rock, we are able to learn some things about the composition and structure of the mass. For example, (1) wave velocity generally increases with rock density; (2) shear waves cannot penetrate molten masses because fluids have no strength

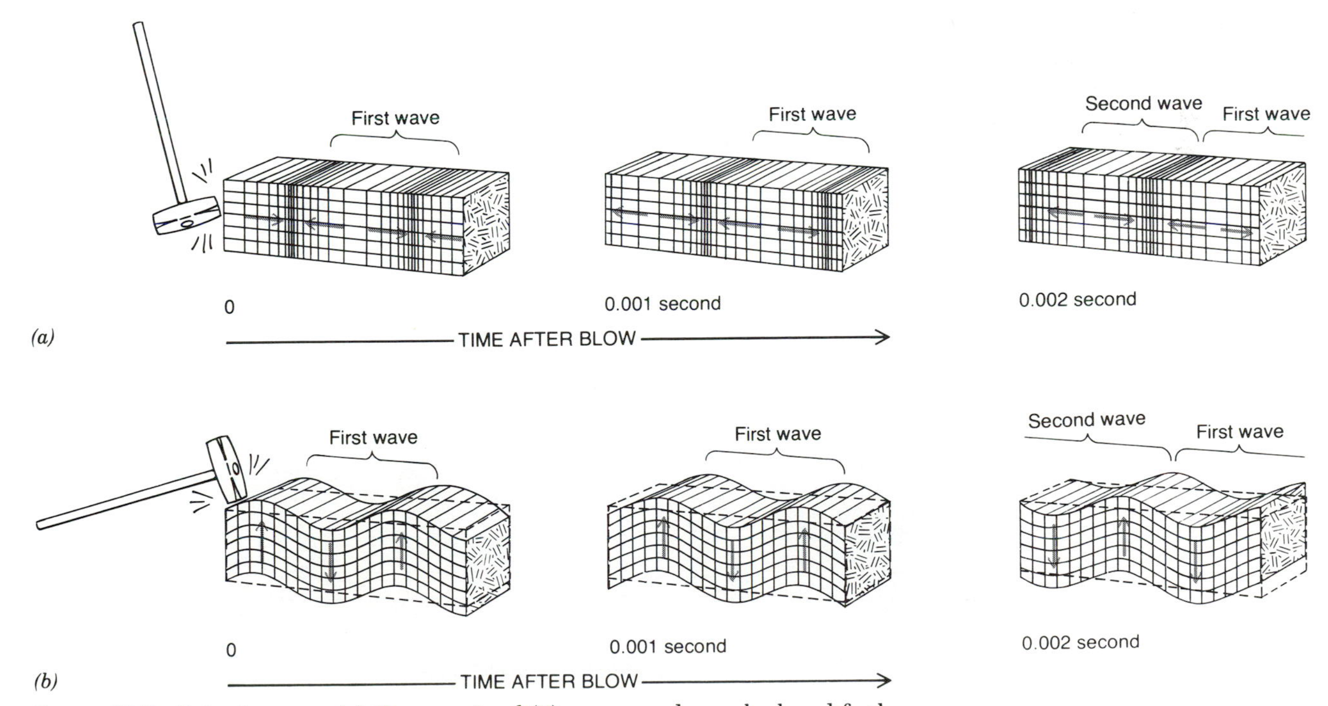

Figure 15.7 Seismic waves: (*a*) Compressional (P) waves produce a back-and-forth motion along the line of energy propagation; as the wave passes through a material, this motion compresses and decompresses particles. (*b*) Shear (S) waves move in a direction that is transverse to the energy propagation and characterized by a rolling action.

against shearing forces; and (3) when waves strike a boundary between two rock layers of different densities, part of the energy travels along the boundary and eventually back to the surface. The discovery of the base of the crust itself was accomplished through careful analysis of seismic waves obeying the latter principle. A Yugoslavian geophysicist, Mohorovičić (pronounced Mo-ho-ro-vee-chich), found that wave velocities increased sharply between what we now call the crust and the zone under it. This discontinuity has come to be known as the M discontinuity, or more commonly, the *Moho discontinuity*.

Seismic data from all parts of the world indicate that the configuration of the Moho varies with the thickness of the crust. Where the crust is thick, as under the continents, the Moho shows a broad downward flexure; the opposite appears to be true under the thin oceanic crust. Directly under belts of large mountains, the bottom of the crust forms distinct downward bulges. These, of course, are the points of greatest crustal thickness and weight (Fig. 15.8).

The variation in the depth of the Moho with the thickness of the crust is explained by the *principle of isostacy*. This principle holds that the continents,

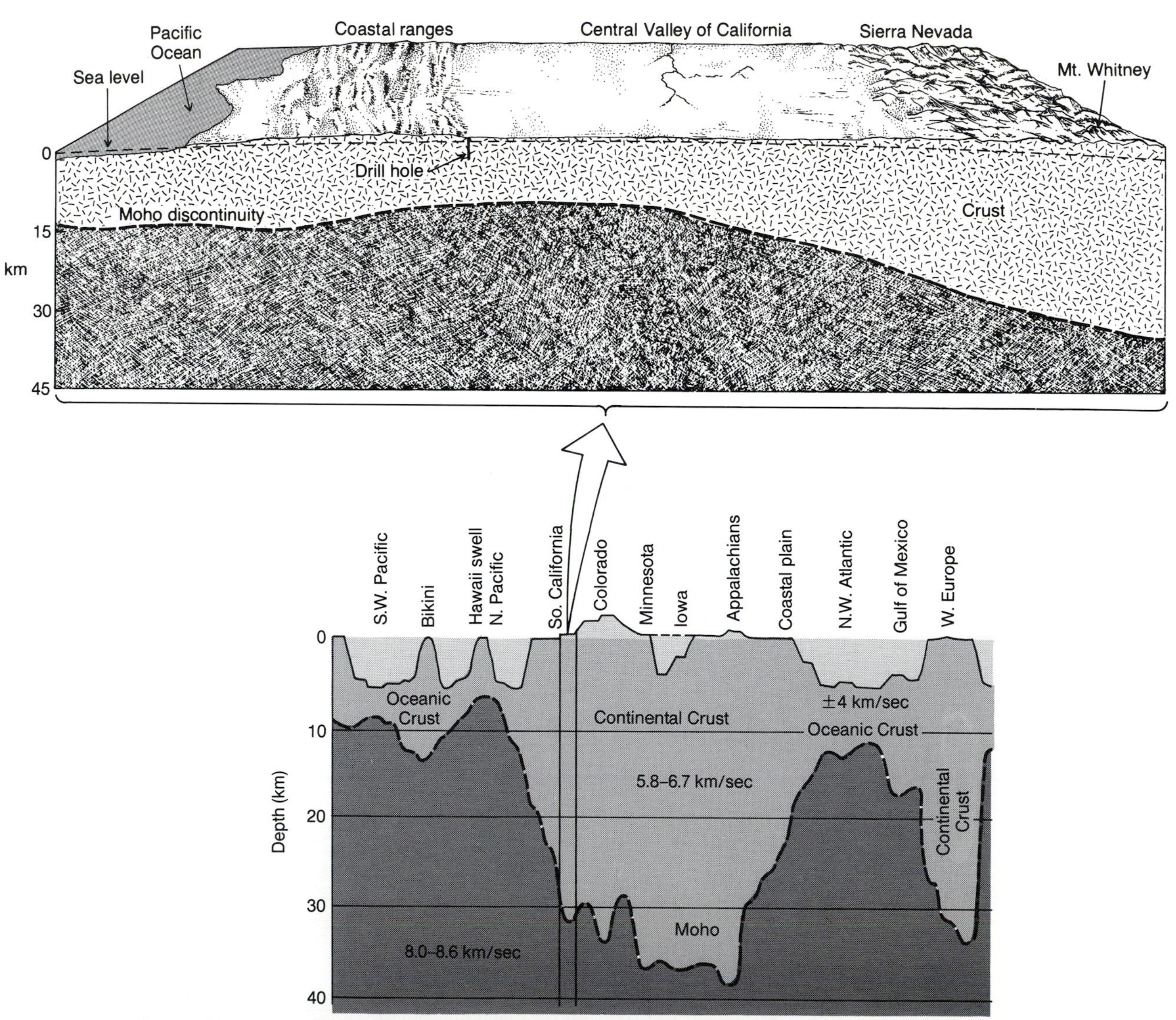

Figure 15.8 The configuration of the crust from Europe, across the Atlantic Ocean and North America, to the Pacific Ocean. The enlargement shows the crust in southern California where a large downward bulge corresponds to the Sierra Nevada range.

being the lightest part of the crust, float on the denser mantle. As mass is added to a continent, as would occur when a mountain range is pushed up, the continent will compensate by sinking deeper into the mantle directly under the mountain range. The principle of isostacy is illustrated nicely by the changes in elevation of a cargo ship as it is loaded and unloaded; in this example, the hull of the ship would be comparable to the Moho. As we will see later in this book (Chapter 22), the continental glaciers were of sufficient mass to depress the crust in central North America, which accounts for much of the Hudson Bay lowland in the Canadian Shield.

The Earth's Mantle and Core

Just as the crust is defined by a seismic discontinuity, the remainder of the earth is subdivided on the basis of the changes in the velocity of seismic waves. The crust rests immediately on (and is part of) a massive zone called the *mantle*, which extends 2700 km into the earth. The mantle contains about two-thirds of the total mass of the earth. For the most part, seismic-wave velocities increase steadily throughout this zone (Fig. 15.9*a*).

At the base of the mantle, a sudden change in wave velocities occurs. This marks the beginning of the

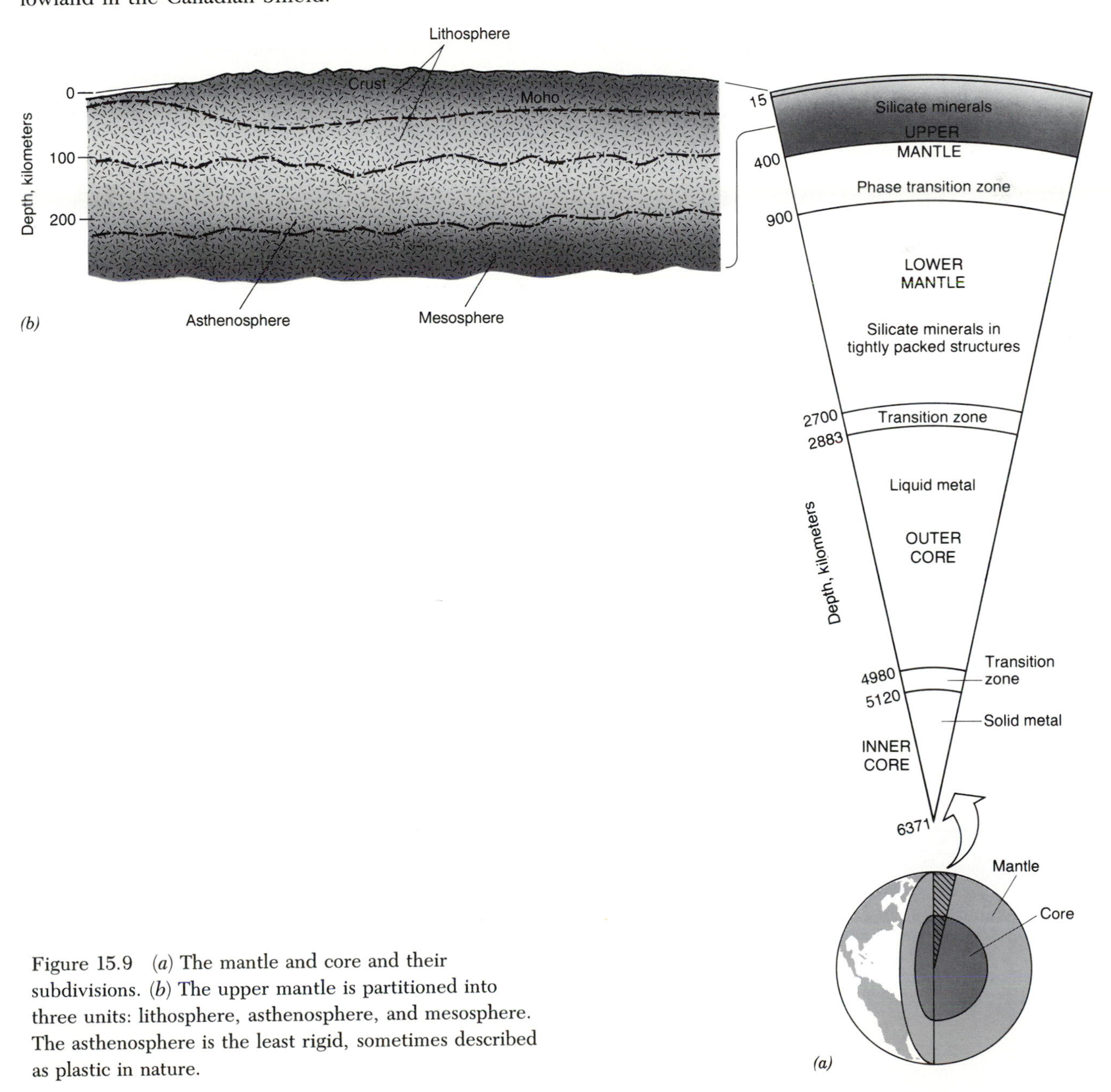

Figure 15.9 (*a*) The mantle and core and their subdivisions. (*b*) The upper mantle is partitioned into three units: lithosphere, asthenosphere, and mesosphere. The asthenosphere is the least rigid, sometimes described as plastic in nature.

core, which makes up the rest of the earth. Interestingly enough, the compressional-wave velocities drop to zero. Because shear waves cannot be transmitted through liquids, the outer portion of the core, which is about 2000 km thick, must exist in the liquid state. Seismologists have also determined that the innermost portion of the core is probably solid. Hence, the core is divided into two units, the *inner core* and the *outer core* (Fig. 15.9*a*). The core contains one-third of the total mass of the earth and is composed of an iron alloy. The magnetic field of the earth is generated in the liquid outer core.

The mantle is further subdivided into the *upper mantle* and the *lower mantle*. The upper mantle, in turn, is subdivided into three smaller zones. The uppermost 100 km, including the crust, is the *lithosphere;* the next 100 km or so is the *asthenosphere;* and the remainder is the *mesosphere* (Fig. 15.9*b*). Both the lithosphere and mesosphere are solid, but the asthenosphere appears to consist of partially melted rock. Although it may be liquid at some points, most of the asthenosphere appears to have a consistency that approximates the transitional phase between a solid and a plastic substance. The asthenosphere is also called the *plastic layer*.

EARTHQUAKES

Earthquakes are violent shaking or vibrating motions (tremors) of the ground caused by the passage of seismic waves. The seismic waves are caused by the release of energy from a sudden movement along a fault. *Faults* are fractures in the earth's crust where there has been displacement of rock. (The processes and types of faults will be discussed in Chapter 16.) Each time a displacement occurs along a fault, an earthquake of some size is produced.

The Size and Frequency of Earthquakes

Most earthquakes are small and cause little or no damage to the landscape. Seismographs record an average of more than 2000 tremors each day that are never felt by humans. Great earthquakes like the ones that destroyed Lisbon, Portugal, in 1755, San Francisco in 1906 (Fig. 15.10*a*), and T'ana-shan, China, in 1976, occur every 5 to 10 years for the earth as a whole and 50 to 100 years in California. What is a *great earthquake?* One that destroys cities and kills hundreds or thousands of people? There are two measures of an earthquake's size: *intensity,* which is based on its de-

(a)

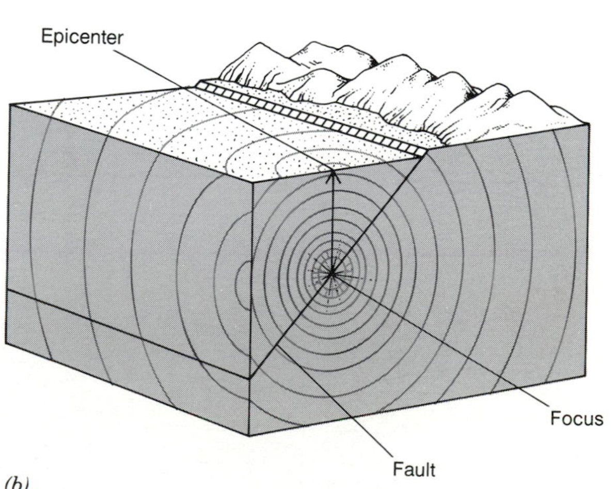

(b)

Figure 15.10 (*a*) Photograph of the damage caused by the San Francisco earthquake of 1906. (*b*) The location of the principal components of an earthquake: fault, focus, and epicenter.

structive effect; and *magnitude*, which is based on the energy it releases.

The most precise measure of the size of an earthquake is *magnitude*, which is the amount of energy released at the earthquake's focus. The *focus* is the location of the initial slip on the fault; the point on the surface immediately above the focus is the *epicenter* (Fig. 15.10*b*). A standard scale, called the *Richter Magnitude Scale*, is used to rank earthquake magnitude. This scale is based on the height of the largest wave produced on a seismograph tracing (plus an adjustment factor for weakening of the seismic wave with distance from the focus). In order to interpret Richter scale magnitudes, it is necessary to understand that the scale is logarithmic, which means that an increase of one unit represents a tenfold increase in earthquake magnitude. Thus, an earthquake that ranks 8 is 10,000 times one that ranks 4.

The largest earthquakes recorded using the Richter scale (since the 1930s) have magnitudes of close to 8.5. Great earthquakes have magnitudes greater than 8, but any earthquake of magnitude 5 or greater can produce appreciable damage in the landscape.

Earthquake Damage

Besides the magnitude of the earthquake itself, the amount of damage caused by an earthquake is also influenced by geographic factors, particularly, location, land use, and landforms. The most destructive earthquakes are those with shallow foci (0 to 100 km depth) and epicenters located in or close to (1) regions of heavy population; (2) metropolitan landscapes with buildings constructed of brittle materials such as adobe and concrete; and (c) mountainous topography where tremors can loosen materials and trigger large landslides and avalanches.

These factors were significant in the San Fernando (Los Angeles) earthquake of 1971. The magnitude of this earthquake was only 6.6, but because of its location in the Los Angeles metropolitan area, it destroyed bridges, sections of highway, and many public buildings. Total damage to property was $1 billion, and sixty-five persons were killed, mainly from building collapse. Many more people (50,000) were killed in the Peru earthquake of 1970 (magnitude 7.7), but in this instance a large proportion of the victims were inhabitants of farms, villages, and towns buried by landslides and avalanches from surrounding mountainslopes.

Earthquake Hazard Areas

The orogenic belts, midoceanic ridges, and island arcs are the areas most threatened by earthquakes (see Fig. 16.2). These are the areas of most faulting activity, which is an integral part of the mountain-building process. California, Japan, and Italy lie in these zones.

Damage from the San Fernando, California earthquake in the Los Angeles area, 1971.

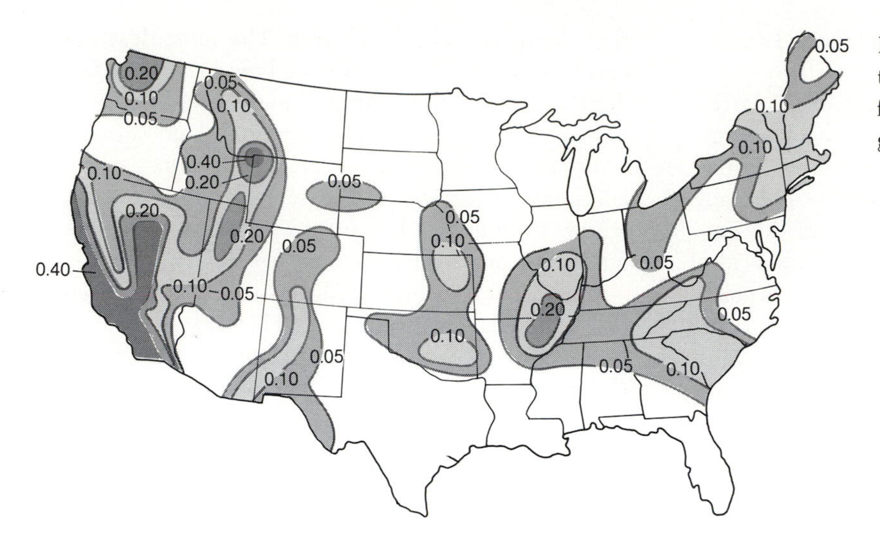

Figure 15.11 Earthquake risk areas of the United States. The magnitude and frequency of earthquakes are likely to be greater with higher numbers.

But the recorded history of earthquake activity also reveals that some nonmountainous areas have experienced powerful earthquakes. In the United States some of the most destructive earthquakes have occurred in the East: the New Madrid, Missouri, earthquakes of 1811–12, and the Charleston, South Carolina, earthquake of 1886. These earthquakes stand out on the earthquake risk map in Figure 15.11 because this map is based mainly on evidence of damage from past events, not solely on frequency of occurrence or measured magnitude. The greatest magnitude and frequency of earthquakes in the United States favors the West, although seismologists still regard the New Madrid area as having potential for high-magnitude earthquakes.

ROCKS AND MINERALS OF THE CRUST

From the question of the earth's general structure and the nature of seismic energy, we move on to the question of the earth's rock composition. Our primary concern is with the rocks that are found at or near the surface of the continents because these are the rocks that influence formation of soils, erosion of mountains, availability of mineral resources, and so on. Subsurface rocks, especially those that occur at depths of a kilometer or more, are of secondary importance in physical geography, for they do not directly affect the landscape.

Although nearly 2000 minerals and dozens of rock varieties have been identified, relatively few of these are found in appreciable quantities at the earth's surface. Only one major rock type, basalt, makes up virtually all of the ocean basins. By contrast, the continents are much more diverse, but tabulations show that only seven basic rock types occupy more than 90 percent of their total area.

Minerals

A *mineral* may be defined as a naturally occurring inorganic substance with a characteristic internal crystal (molecular) structure that is quite consistent in its physical and chemical properties from sample to sample. In general, minerals can be divided according to those that combine with other minerals to form rocks and those that do not. The most important group of rock-forming minerals are the *silicate minerals.*

The silicate minerals are composed of a basic ion of silicon and oxygen combined with one or more elements. Depending on the particular elements that combine with the silicon-oxygen ion, the mineral thus formed may vary considerably in density and color and other features. Two subgroups of silicates are recognized: *ferromagnesian* and *nonferromagnesian* (or aluminosilicates). As the names indicate, these minerals are distinguished on the basis of the presence or absence of iron, magnesium, and aluminum (Table 15.1).

In connection with our earlier remarks, it is important to note that the rocks of the continents (sial layer) are heavy in nonferromagnesian minerals, especially feldspar and quartz[1]; whereas the rocks of the

[1]Together, quartz and feldspar make up 58 percent of all rocks found on the *surface* of the continents.

TABLE 15.1 SILICATE MINERALS (BASIC COMPOUND: ELEMENT(S) + SILICON-OXYGEN ION)

Type	Important Members	
Ferromagnesian silicates (silicate ion + iron and magnesium ions)	Olivine, Hornblende, Biotite mica, Pyroxene	heavy and dark
Nonferromagnesian silicates (also called aluminosilicates) (lack iron and magnesium ions)	Muscovite mica, Feldspars: orthoclase (potassic), plagioclase (sodic, calcic), Quartz	light

ocean basins (and the lower continents) are heavy in ferromagnesian minerals. This accounts for the differences in density and color of the rocks in these zones. Below the crust, in the lithosphere and asthenosphere, the concentration of the ferromagnesian minerals is even greater, and deeper down it is even greater yet.

Rocks

Rocks are assemblages of minerals and can be classified according to several different schemes. A traditional scheme is based on the manner and environment of rock formation. Under this scheme, three major divisions may be used for virtually all rocks: *igneous*, *metamorphic*, and *sedimentary*. The distribution of each in central North America is shown in Figure 15.12 (Color Plate 10). For the continents as a whole, sedimentary rocks occupy 74 percent of the surface area; igneous rocks occupy 18 percent; and metamorphic rocks about 8 percent.

Igneous rocks Igneous rocks are composed of mineral crystals that form during the cooling of molten material. Molten material within the earth is called *magma;* it originates in places of high temperature and pressure in the lithosphere and asthenosphere (Fig. 15.13*a*). Solidification of magma can occur at any level below the surface; in fact, most magma never reaches the surface.

In the lithosphere, magma is held in vast chambers from which channels, called *veins* or *dikes*, may radiate toward the surface. Where the crust is fractured, the hot, pressurized magma may extend all the way through the crust and eventually flow forth on the surface as *lava* (Fig. 15.13*b*). If the lava builds up around the mouth of the dike, a volcano is formed (volcanic processes and forms are taken up in Chapter 16); if it spills over the ground, it is called *lava flow.* Massive lava flows emanating from many fissures in the crust are called *flood basalt,* and one of the largest such features in North America is the Columbia Plateau of the American Northwest (Fig. 15.13*c*).

Volcanic activity also produces airborne debris, called *pyroclasts* (dust, ash, and rock fragments), when water and dissolved gases in the magma explode near the surface. The gases may also permeate lava, leaving it highly porous and very low in density. Most volcanoes, including those such as Mount St. Helens in the Cascade Mountains, are composed of all three types of materials; pyroclastic, porous lava, and high-density, basaltic lava (Fig. 15.14).

Igneous rocks are divided into two main classes according to the depth of formation: (1) *extrusive* and (2) *intrusive* or *plutonic*. *Extrusive* rocks form at the surface, where the quick release of heat into the overlying air or water makes solidification rapid. Because the size of the crystals that form in the rock is inversely related to the rate of cooling, extrusive rocks are usually composed of microscopic-sized crystals and are thus known for smoothness of texture. By contrast, the magma that solidifies within the lithosphere, in dikes, veins, and related features, forms *intrusive* igneous rocks. Because the surrounding rock insulates the magma, cooling is much slower than that of the extrusives, thereby allowing crystals to grow much larger. *Plutonic* rocks develop from the aggregation of large masses of magma called *batholiths;* thus, cooling is even slower, resulting in the formation of huge crystals that may reach several centimeters in diameter.

The mineral composition of the newly formed ig-

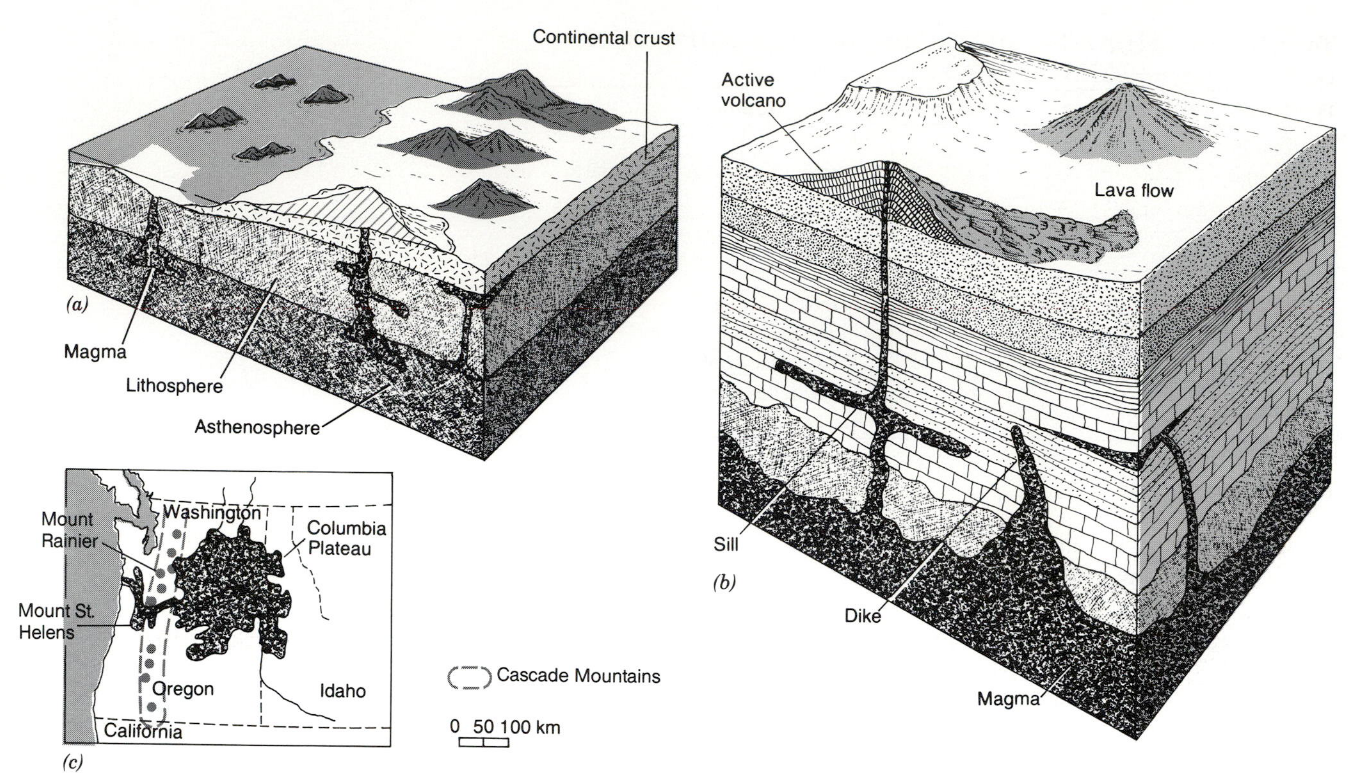

Figure 15.13 The magma of igneous rocks: (*a*) sources of magma in the lithosphere and asthenosphere; (*b*) from deep chambers, magma can make its way to the surface to form volcanoes and lava flows; (*c*) the Columbia Plateau, one of the largest areas covered by lava in North America, and the Cascade Mountains in which Mount St. Helens is located.

neous rock depends on the chemical composition of the magma, the geochemical environment through which it moved, and the heat and pressure present during its solidification. If the magma is about 60 percent plagioclase feldspar and 35 percent ferromagnesian minerals, the rocks *diorite* or *andesite* may be formed. If the proportion of ferromagnesian minerals is 60 to 70 percent, *gabbro* or *basalt* may be formed.

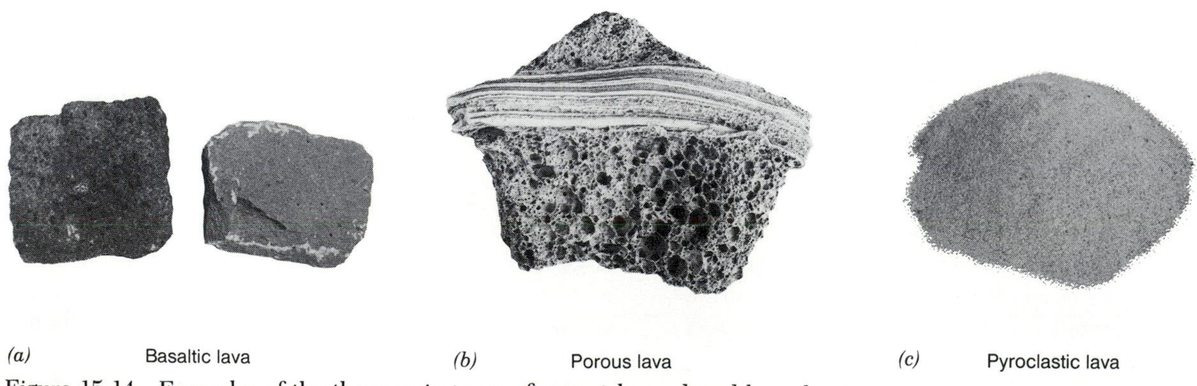

Figure 15.14 Examples of the three main types of materials produced by volcanism: (*a*) basaltic lava, (*b*) porous lava, called *scoria*, and (*c*) pyroclastic material.

Together, basalt and andesite comprise about 98 percent of all extrusive rocks (Fig. 15.12). If the magma is composed primarily of quartz and 15 to 20 percent feldspar with an admixture of the heavy minerals mica and hornblende, the rock is likely to be *granite* or *granodiorite*. These rocks are intrusive and together make up approximately 95 percent of all intrusive rocks.

In Figure 15.15 the important igneous rocks of the crust are classified according to mineral content and texture, mainly crystal size. The principal members of the silicate group are given on the left; two textural classes representing intrusive and extrusive rocks are on the right. By combining these two features, we can identify two important trends in the igneous rocks. Note that granite, diorite, and gabbro each have a fine-grained extrusive counterpart. Although the mineral content of each pair is similar, the two rocks look and feel remarkably different. Furthermore, as we go from top to bottom in Figure 15.15, in a very general way we are reading the change in rock types from the sial layer (granite and rhyolite) to the lower sial (diorite and andesite) to the sima layer (gabbro and basalt) to the lower lithosphere, represented by the very heavy rock *peridotite*.

Metamorphic rocks Because the temperature of the magma that invades the crust usually exceeds that of the crustal rock by hundreds of degrees centigrade,

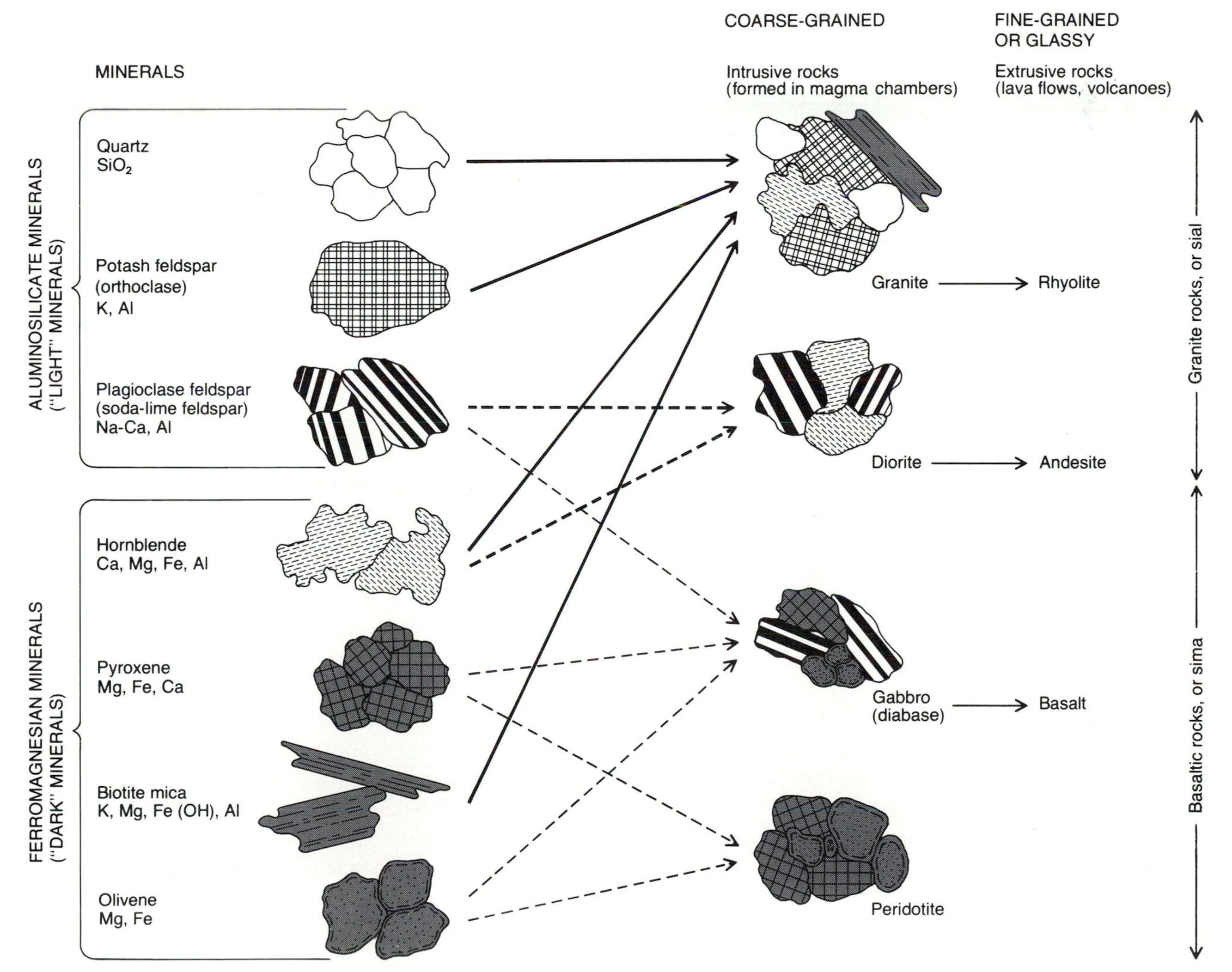

Figure 15.15 Common intrusive and extrusive rocks produced by different combinations of light and dark silicate minerals. The drawings resemble the appearance of the minerals as you would see them under the high magnification of a polarizing microscope.

a great amount of heat is transferred into the resident rock, or *country rock*, as the oldtime geologists called it. This heat may partly melt the surrounding resident rock, increase pressure in it, and superheat any water in it. Any one of these effects may change, or metamorphose, the resident rock's crystal structure, or mineral composition, or both. *Metamorphic* rocks, then, are those that have been changed by heat, pressure, and related factors. In granite, for example, metamorphism produces a realignment of the minerals into bands, and the resultant rock is called *gneiss* (pronounced "nice"). Although metamorphosed igneous rocks are common, most of the metamorphic rocks found on the surface of the continents are derived from sedimentary rock, described below.

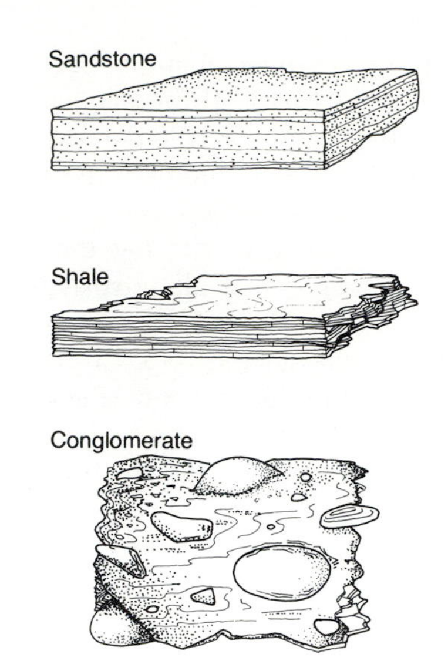

Figure 15.16 Sketches of three detrital sedimentary rocks: sandstone, shale, and conglomerate.

Sedimentary rocks Water, ice, waves, and wind as well as biological and chemical processes weaken and disintegrate the rocks on or near the surface of the crust. From this action, sediment residues (such as clay and sand), biological debris, and chemical residues are produced which accumulate in lowlands, basins, and on the continental shelves. As layers of sediment grow to thicknesses of hundreds of meters, the lower layers are compressed and in time consolidate into sedimentary rock.

Sedimentary rocks are divided into two main groups: *detrital* and *chemical*. Detrital rocks are formed of particles that have been transported to their resting place by erosional processes. Clay, silt, sand, and pebbles are the most common constituents of detrital rocks. These particles provide the key identifying feature for these abundant rocks, for example, sand in *sandstone*, clay in *shale*, and particles of all sizes in a rock called *conglomerate* (Fig. 15.16). Shale is the most abundant surface rock on the continents, occupying 52 percent of their combined areas, and sandstone is the second most abundant, occupying 15 percent of the surface area of the land masses (Fig. 15.12).

The chemical group of sedimentary rocks is produced mainly by the precipitation of minerals out of water. *Limestone* and *dolomite* are the most abundant chemical rocks, comprising 7 percent of surface rock of the land masses. Limestone (calcium carbonate) is formed mainly from either (1) animal bodies and shells composed of calcium which these creatures extracted from seawater; or (2) calcium precipitated directly from seawater. These two modes of limestone formation often occur together. The origin of dolomite is disputed, but it is probably an altered form of limestone in which some of the calcium has been chemically replaced by magnesium.

Also included in the chemical group of sedimentary rocks are rocks called *evaporites*. As the name implies, these are the residues left after the evaporation of mineral-rich water. Various kinds of salts, including rock salt, gypsum, and borax, are the most abundant evaporites.

Another sedimentary rock of the chemical group whose origins are associated with wet environments is coal. *Coal* is a hydrocarbon compound which forms largely from the buildup and burial of organic matter in ancient swamps and marshes. Coal is classified into three types according to its level of development. *Lignite*, a soft brown coal, is least developed, that is, least compact and compositionally pure. It is generally considered to represent an intermediate state between peat, the partially decomposed organic accumulations found in bogs and lakes, and true coal. *Bituminous* coal, also called soft coal, is harder than lignite as a result of compaction from deep burial. Where coal seams are subject to metamorphism in areas of mountain building, bituminous coal is transformed into *anthracite*, the hardest and purest form of coal. The coal fields of the Appalachian Plateaus, the largest body of coal reserves in the world, are largely bituminous.

Most sedimentary rocks form under the shallow waters on the continental shelves, where huge quantities of material eroded from the land are deposited by rivers. This material ranges in size from large particles, 10 cm or more in diameter, to fine clays, less than 0.002 mm in diameter, and the ions of dissolved minerals. When a river enters the ocean, it slows down and drops its sediment load, generally according to particle size. Large particles such as pebbles and sand are deposited nearest shore, whereas the clay particles, which can remain suspended in seawater for many days or weeks, are often deposited far out on the shelf, on the continental slope, or beyond (Fig. 15.17).

Throughout much of the world, the continental shelves are also areas of abundant marine life because light, heat, and various forms of plant and animal nourishment are concentrated there (see Fig. 11.11). As a result, the remains of shell creatures and other organisms in many areas constitute an important source of sediment. Coupled with the precipitation of chemicals from seawater and the inflow of detrital sediments from the land, the sediment makeup on the continental shelves is very complex.

The rate of sedimentation leading to the formation of chemical and detrital rock can be very rapid on the continental shelf. Field studies reveal that in bays along the coast, the rate can be as high as 2 meters a year. Simple multiplication tells us that even at one-

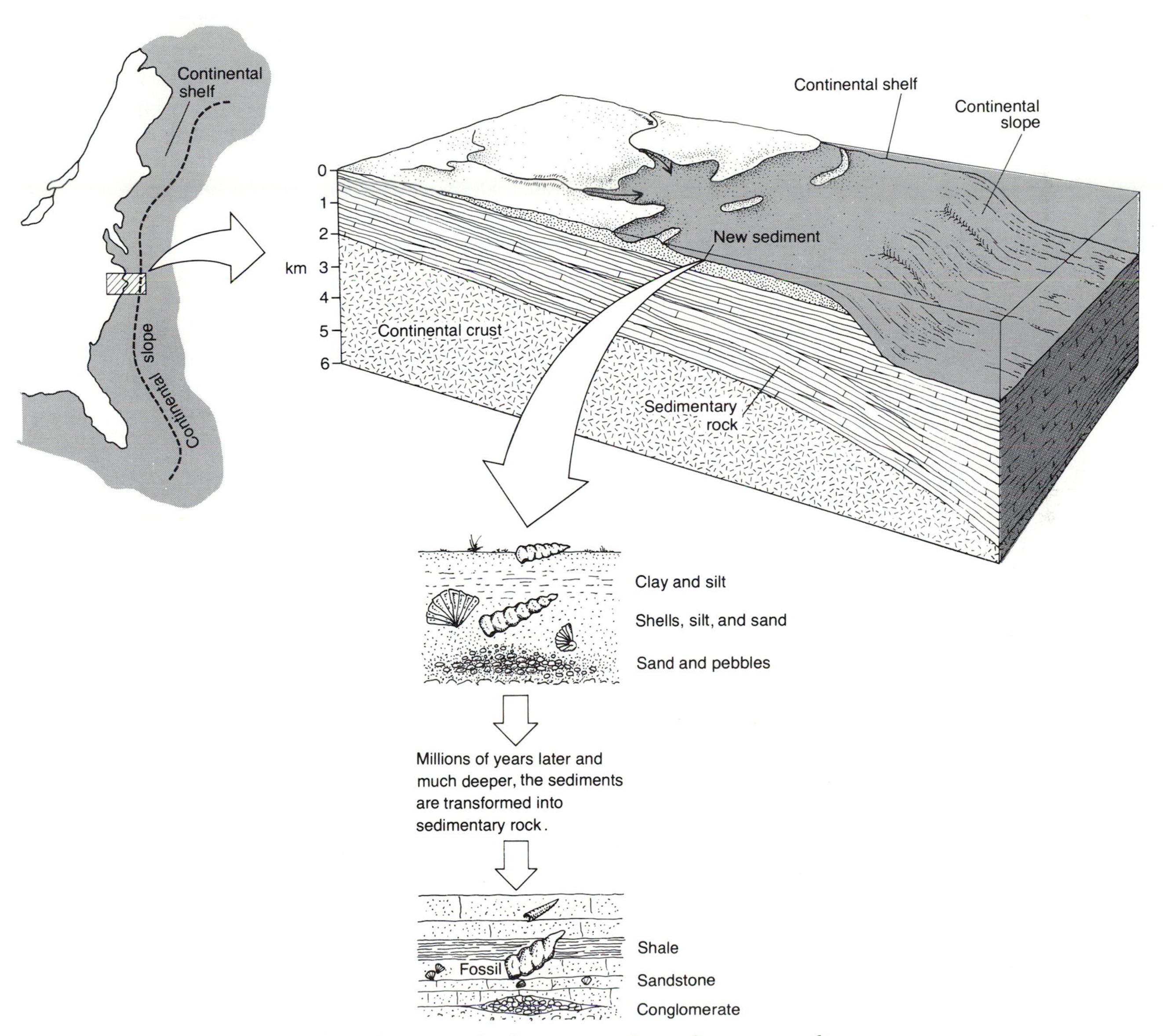

Figure 15.17 Sedimentation and the formation of sedimentary rocks on the continental shelf. Both detrital and chemical go into this sediment mass.

TABLE 15.2 METAMORPHIC COUNTERPARTS OF FOUR SEDIMENTARY ROCKS

Sedimentary		Metamorphic
Shale	⟶	Slate
Sandstone	⟶	Quartzite
Conglomerate	⟶	Metaconglomerate
Limestone	⟶	Marble

tenth this rate, it would take only 5000 years for sediment to reach a thickness of 1 kilometer. But as the sediment builds up, it is compressed under its own weight, and much of it is washed farther out on the shelf and beyond. As a result, after millions of years, the sedimentary rocks on large continental shelves, such as that off the Atlantic Coast of the United States, may grow to a total thickness of 5 to 6 kilometers (Fig. 15.17).

Sedimentary rocks are also subject to metamorphism. This usually results in an increase in density and hardness as well as a change in texture. For instance, the metamorphism of sandstone into *quartzite* results in a welding together of the sand grains to the extent that individual grains are difficult to distinguish. Table 15.2 lists the metamorphic counterparts of several common sedimentary rocks.

Rock Type and the Landscape

Earlier, we stated that surface rocks are important to physical geography. Let us close this chapter by citing a few examples of the role of rocks in shaping the landscape. Bedrock is the major source of soil particles, and variations in rock composition from place to place have in some areas produced corresponding variations in soil type. Figure 15.18 shows the relationship between sandstone and sandy soils and chalk or limestone and clay soils in a part of the Coastal Plain in Alabama. Similarly, the chemical properties imparted to soil by bedrock can have a strong influence on the distribution of plants. For example, in forested mountain regions, outcrops of serpentine rock (an intrusive igneous rock composed almost entirely of ferromagnesian minerals) are usually marked by treeless patches, or "holes," in the forest. Bristlecone pine in the upper parts of the White Mountains of California appears to prefer one rock type over others. Above an elevation of 3000 m, it grows well in soils derived from limestone, whereas adjacent areas underlain by other rock types are treeless (Fig. 15.19).

Rock type also has a major influence on erosion rates. Depending on mineral composition and the internal structure of crystals or sediments, rocks vary considerably in their resistance to weathering and erosion. Therefore, the particular topographic forms into which the land is sculptured are often closely tied to rock type. For example, in humid areas some limestones are highly susceptible to chemical breakdown. Groundwater dissolves and washes away the calcium carbonate, leaving the rock riddled with pits, caverns, and underground drainage ways. By contrast, adjacent areas of sandstone, for example, show none of these effects because the silicon dioxide that makes up the

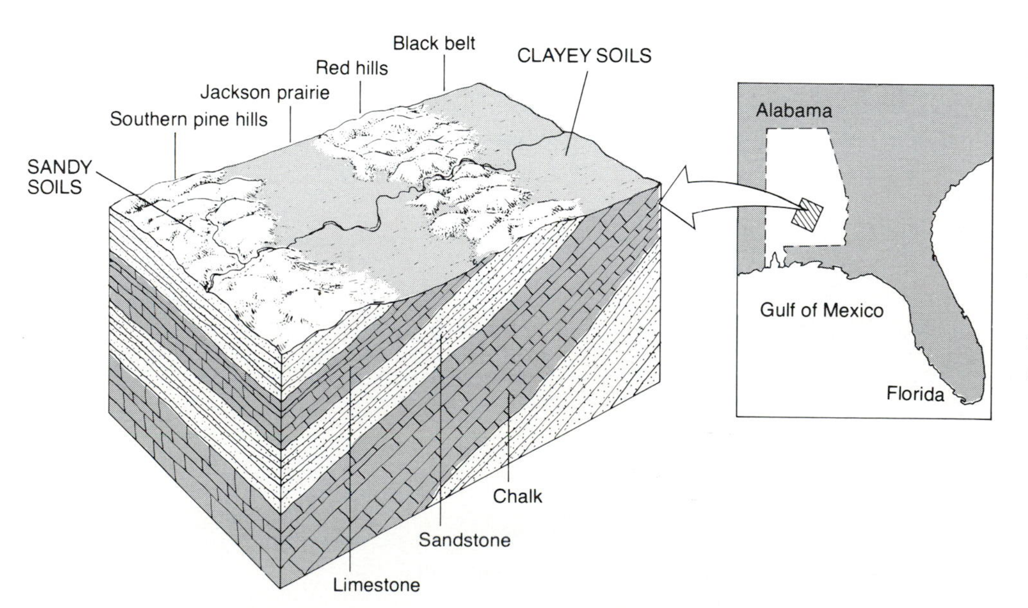

Figure 15.18 The influence of rock types on soil is evident in southcentral Alabama, where sandstone formations have given rise to sandy soils, whereas adjacent limestone and chalk formations have given rise to clayey soils.

sand particles is relatively resistant to dissolution by water. Because of these differences, limestone and sandstone will often yield markedly different landforms when subjected to the same environmental conditions (Fig. 15.20). We will examine the processes which break down rock, called weathering processes, in detail in Chapter 19.

Summary

The ocean basins and the continents are the major units of the crust. The ocean basins are composed of basaltic rocks, which are both darker and heavier than the granitic rocks of the continents. The ocean basins are made up of three main features: abyssal plains,

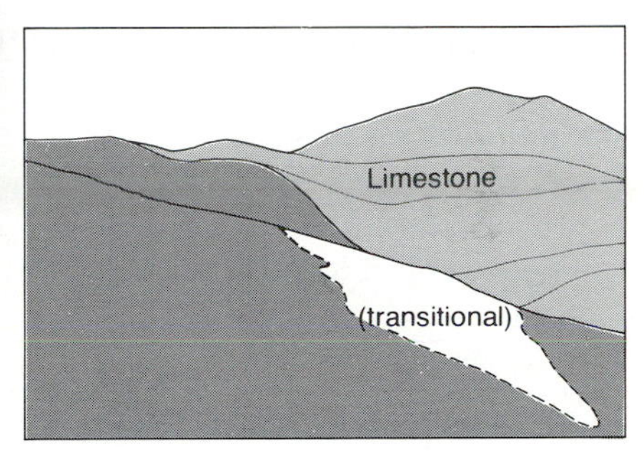

Figure 15.19 The relationship between the distributions of limestone and bristlecone above 3000 m elevation in the White Mountains of California. Note the exact boundary in the central part of the photograph, where the bristlecones grow immediately adjacent to the rock boundary.

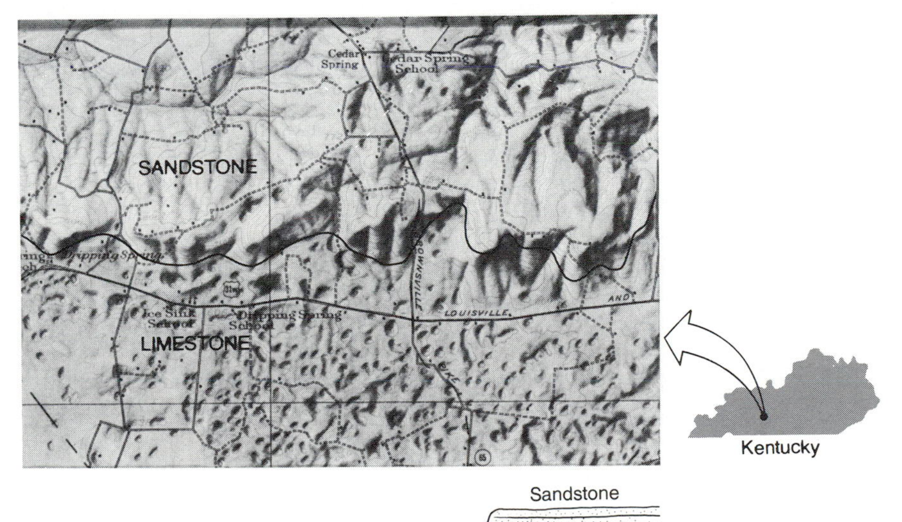

Figure 15.20 In this area in central Kentucky, near Mammoth Cave, sandstone and limestone have given rise to distinctly different landforms. The limestone not only has been lowered more by weathering and erosion, but has also developed a fine, pitted texture as a result of dissolution of the rock by underground drainage. The sandstone, on the other hand, has developed into large hills and valleys as a result of surface erosion by streams.

midoceanic ridges, and ocean trenches. The continents are also made up of three main features: shields, orogenic belts, and continental shelves. The rocks of the continents, especially those of the shields, are much older than those of the ocean basins.

The lithosphere rests on a somewhat plastic layer called the asthenosphere; both layers are a part of the upper mantle. Seismological data reveal that the mantle extends 2700 km into the earth, below which it gives way to the outer core, which is made up of liquid metal. The inner core, a mass of solid metal, occupies the center of the earth. Earthquake activity is almost constantly occurring in the lithosphere. Most are small with no influence on the landscape. Great earthquakes occur every 5 to 10 years in the world and cause major damage to the landscape.

The crust is composed of thousands of minerals, some of which combine to form rocks. Sedimentary rocks occupy nearly 75 percent of the surface of the continents, whereas virtually all of the oceanic basins are occupied by volcanic rocks, principally basalt and its relatives. Among the influences of rock types on the landscape are soil composition and vegetation distribution; for example, sandstone tends to produce sandy soils in the Alabama Coastal Plain, and at high elevations in the White Mountains of California, bristlecone pines grow only over limestone bedrock.

Concepts and Terms for Review

early impact craters
ocean basins
  abyssal plain
  midoceanic ridge
  ocean trench
continents
  shield
  orogenic belt
  continental shelf
elevation of surface features
mountain islands
crust: sima and sial
seismic exploration
  compressional waves
  shear waves
Moho discontinuity
principle of isostacy
upper mantle
  lithosphere
  asthenosphere
  mesosphere
low mantle
core
earthquake
  intensity and magnitude
  focus and epicenter
  Richter Magnitude Scale
  great earthquake
earthquake damage and hazard
minerals
silicate minerals
  ferromagnesian
  nonferromagnesian
rock
  igneous
  sedimentary
  metamorphic
surface rocks and minerals
rock-forming environments
influence of rock type in landscape

Sources and References

Bolt, B.A. (1978) *Earthquakes: A Primer.* San Francisco: Freeman.

Cargo, D., and Mallory, R. (1977) *Man and His Geologic Environment,* 2d ed. Reading, Mass.: Addison-Wesley.

Coffman, J.L., and von Hake, C.A., ed. (1973) *Earthquake History of the United States.* Washington, D.C.: U.S. Department of Commerce (NOAA).

Dalrymple, G.B., Silver, E.A., and Jackson, E.D. (1973) "Origin of the Hawaiian Islands." *American Scientist* 61, 3: pp. 294–308.

Fridriksson, S. (1975) *Surtsey: Evolution of Life on a Volcanic Island.* New York: Halsted Press.

Green, J., and Short, N.M. (1972) *Volcanic Landforms and Surface Features.* New York: Springer-Verlag.

Holmes, A. (1965) *Principles of Physical Geology.* New York: Ronald Press.

Murphy, R.E. (1968) "Landforms of the World." Map Supplement No. 9, *Annals of the Association of American Geographers* 58, 1.

Press, R., and Siever, R. (1978) *Earth,* 2nd ed. San Francisco: Freeman.

Richter, C.F. (1958) *Elementary Seismology.* San Francisco: Freeman.

Verhoogen, J., et al. (1970) *The Earth: An Introduction to Physical Geology.* New York: Holt, Rinehart and Winston.

Wyllie, P.J. (1975) "The Earth's Mantle." *Scientific American,* March: pp. 50–57.

Chapter 16

Crustal Mechanics, Rock Structures, and Related Landforms

For centuries geographers have sought to provide global perspectives on climate, vegetation, soils, and related phenomena. The way is now open via the theory of plate tectonics to extend that perspective to the earth's crust, to view large sections of the crust as part of an interconnected whole. How does this help explain the origins and locations of mountains and other large features of the crust? Beyond this question, what are the characteristics of the different classes of mountains and what sorts of deformational processes are associated with them?

It took centuries to build our theories of atmospheric and oceanic circulation but by the mid-1800s science had formulated a reasonably accurate model of global circulation. We learned that the atmosphere is made up of great belts of pressure which drive systems of prevailing winds, which in turn drive systems of ocean currents. Understanding the geographic scale of these systems was important, because it enabled us to establish cause and effect connections between widely separated regions and conditions, such as between warm, tropical water of the Caribbean Sea, and the temperate west coast climate of Northwestern Europe.

But systems of fluids that operate at or close to the earth's surface are relatively easy to observe and measure compared to systems of rock that operate at depths in the earth. For this reason, among others, theories concerning the great systems of forces and processes that rearrange the crust, that form the continents and ocean basins, and that launched the great episodes of mountain building, have been more difficult and taken longer to build. However, a substantial body of theory, called *plate tectonics,* is now in place to help us probe and illuminate these exciting problems. This theory proposes that the continents and the ocean basins are parts of great mobile sections of crust that move not only vertically but also laterally. Over geologic time (hundreds of years), the continents have moved thousands of kilometers across the earth's surface, and as they have, the oceans have grown or been squeezed out between them.

For the scientist interested in the earth's surface, plate tectonics provides the key to understanding the big structural features of the crust. For the geographer, it offers not only a unifying model for the major large-scale landforms on the continents, but also a means of understanding interrelationships among the continents themselves. Most of the traditional theo-

THE VIEW FROM THE FIELD / NOTE 16

The Origin of the Continental Drift Idea

The main motivation to propose large continental displacement across the earth's surface seems to have come from the ability to fit the continents back together as pieces of a puzzle. The remarkable coincidence between the coastlines of Africa and South America, by itself, is convincing. It is said that Francis Bacon (1562–1626) commented on this correlation in his treatise *Novum Organum.* As the subject became a matter of debate in the early twentieth century, many names came to light. Today it seems a matter of an author's nationality as to who was essential in the "drift" conception. However, it is to the German geographer and meterologist Alfred Wegener (1880–1930), who worked so prolifically and debated so cunningly, that the credit must go for the first pronouncement of the hypothesis of continental drift. When Wegener's ideas were published in his book *The Origin of the Continents and Oceans* (1915), they were, like all radical propositions, accepted by some scientists and vigorously rejected by others. Almost needless to say, the following debates were heated.

Alfred Wegener produced compelling evidence to show that the continents were once united into a single supercontinent that he called Pangaea (pronounced "pan-gee-a"). Pangaea existed 600 to 225 million years ago and gradually drifted apart in the ensuing several hundred million years to form the present continents and oceans.

The maps presented here reproduce Wegener's conception of the stages of the breakup of Pangaea. Although he relied fundamentally on the puzzlelike geometry of the continents, he carried the case of drifting continents further by drawing on diverse bodies of evidence from the fields of geodesy, geography, geophysics, geology, paleontology, biology, paleoclimatology, and physics. Among the strongest evidence he produced was that of striking similarities in rocks, fossils, and ancient geologic events in continents now widely separated geographically.

For example, working with the well-known climatologist Wladimir Köppen (who, by the way, was his father-in-law), Wegener deduced a number of climatic indicators of former geographic linkages between Africa and Antarctica, India, South America, and Australia. These indicators showed that some continents not only migrated poleward, but split apart as well. Coal and salt deposits that formed under tropical or subtropical conditions were found in climatic regions that are presently much too cold for such formations. Good evidence for Pangaea was also provided by signs of an ancient glaciation that crossed parts of South America, Africa, and India, now separated by distances of more than 5000 km. The deposits left in Africa by this glaciation are presently situated in a subtropical environment where there is no evidence of past mountain ranges of sufficient height to foster glaciation.

In the decades following 1915, advocates of the hypothesis refined and extended Wegener's ideas. Without the means to generate data on the subcrust, however, the hypothesis could not be tested, and so to most geoscientists it remained largely an interesting body of speculation. Following World War II, the intensity and sophistication of geophysical research increased greatly, and in the 1960s data were assembled that generally confirmed Wegener's hypothesis. As for Alfred Wegener, he perished on

ries concerning the development of orogenic belts assumed driving forces in the crust that were far smaller in scale than we now know to be associated with plate tectonics. In general, these traditional theories provided a poor framework for understanding the functional relations among features such as terrestrial mountain chains, ocean trenches, and active volcanoes. The case is quite the opposite with plate tectonics, and the way now appears open to further our understanding of geographic interconnections among the boldest features and geologic processes of the earth's crust, just as we have with the boldest features and processes of the atmosphere through the theories of global circulation.

What are some of the important questions that plate tectonics raises for physical geography? One concerns

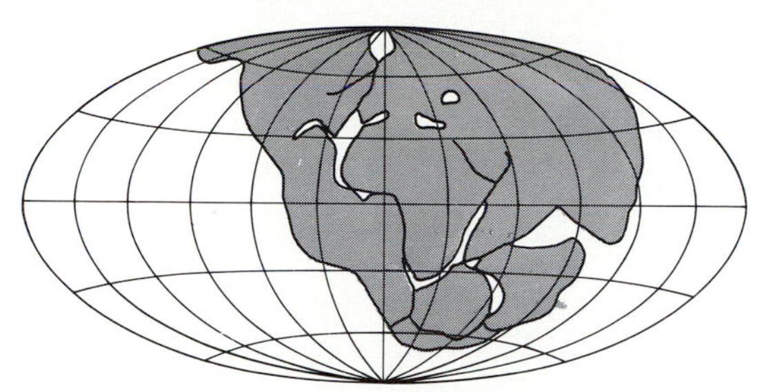

About 300 million years ago

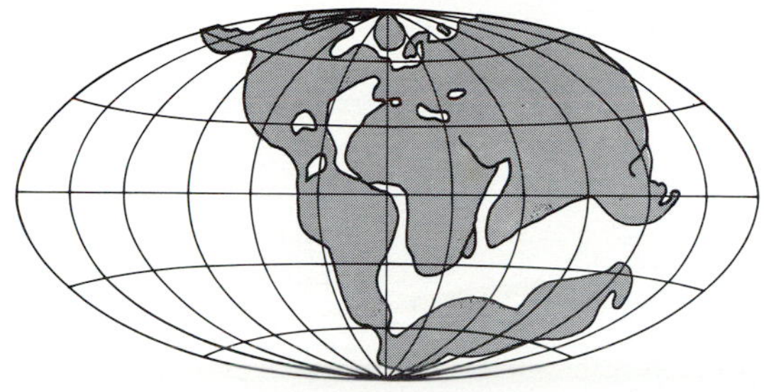

About 50 million years ago

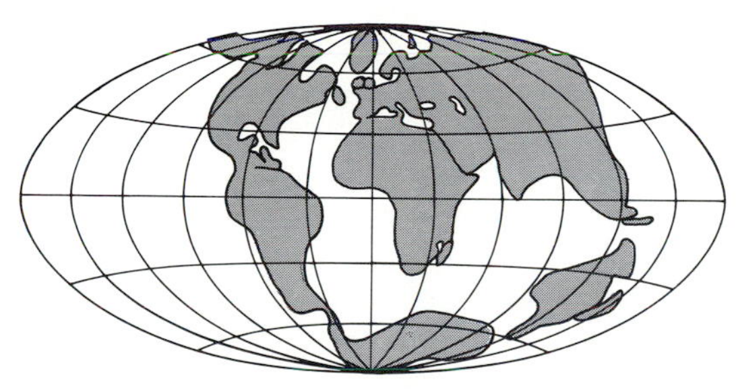

About 1 million years ago

an expedition to Greenland in 1930. Sadly, he never gained the satisfaction of seeing his ideas win the applause of scientists the world over.

Bruce D. Marsh
Johns Hopkins University

the past changes in the geographic proximity of land masses, including the formation and destruction of land bridges and how these changes have influenced the dispersal of the plants and animals from which modern populations are derived. Another relates to climatic changes. Oceanographers speculate that the oceanic circulation patterns must have changed drastically with the breakup of Pangaea, the opening of the Atlantic, and the positioning of Antarctica. Changes in the flow of warm and cold currents northward and southward would have affected the global transfer of energy, which in turn would produce climatic change. How severe were such changes? Could they have been the source of major biological changes on the planet, including extinctions such as that of the dinosaurs? And could the oceanic circulation and climatic changes have contributed to the origin of the Ice Age? We may learn the answers to some of these questions in the next decade or two as the theory of the plates is further unfolded.

The Plate Idea

The idea of migrating continents has traditionally been known as **continental drift**; however, the term is misleading in light of the body of information that has emerged on the subject. As we saw in Chapter 15, the lithosphere consists of two major units: the lighter continental blocks and the heavier basal layer that extends from the ocean floor past the Moho to the top of the asthenosphere. It is important to remember that this basal layer underlies both the continents and the ocean basins, for it is this layer, not the continental blocks, that is involved in the drift movement. Here the edges of the drifting sections, or plates, of the lithosphere are defined. Because the continental blocks rest on this layer, the continents simply go along for the ride, so to speak, when it moves. Thus, the displacement and deformation of the crust caused by movement of lithospheric plates have come to be known as *plate tectonics* instead of continental drift. The term *tectonics* refers to the deformation of the crust generally on a large scale such as that associated with the building of the orogenic belts. The development of the continental drift/plate tectonics concept is one the truly interesting stories in earth science which Bruce D. Marsh of Johns Hopkins University takes up in Note 16.

THE MOVEMENT AND DISTRIBUTION OF TECTONIC PLATES

The basic processes involved in plate tectonics are quite easy to comprehend; however, the geographic scale at which they operate is so vast that, to many people, the whole idea seems incredible. Let us, therefore, start at a much smaller scale with the assistance of an analogy drawn from another part of nature: a large ice-covered lake that is sectioned into several large sheets (Fig. 16.1).

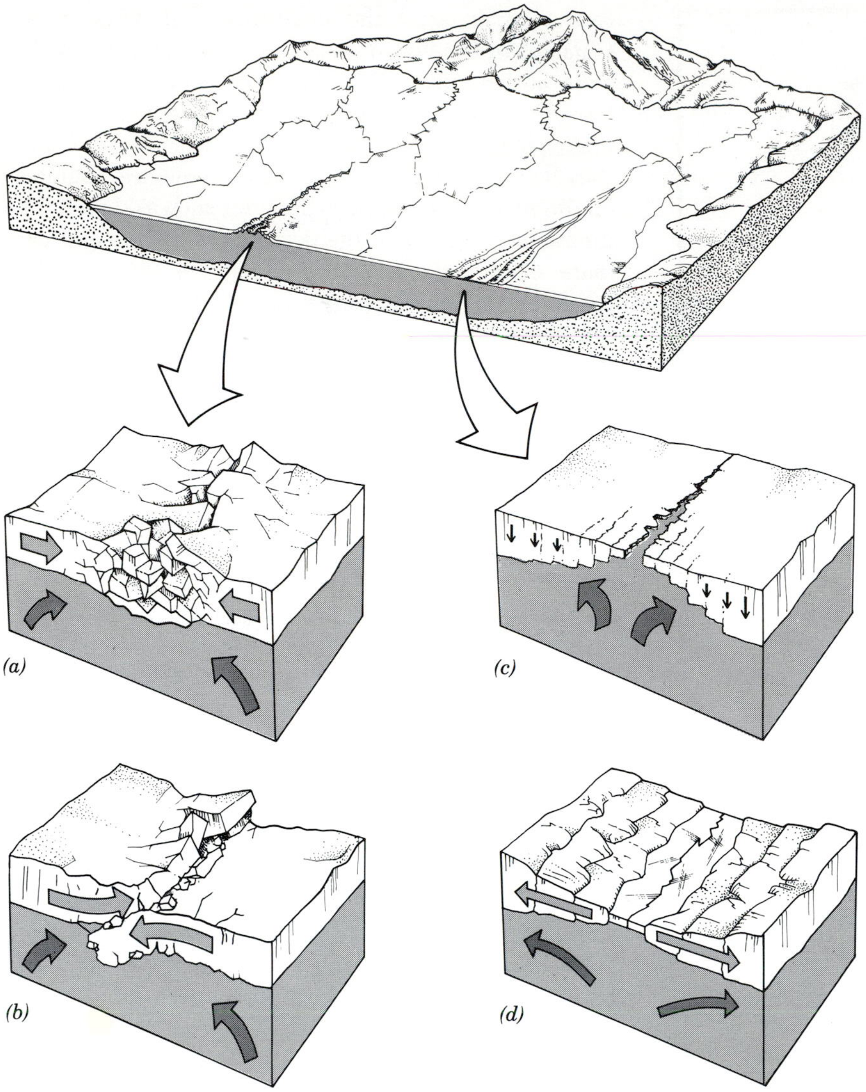

(a)

(c)

(b)

(d)

Figure 16.1 Changes that can occur at the edges of ice sheets as they move together (*a*) and (*b*), and apart (*c*) and (*d*). Lithospheric plates appear to behave similarly.

The Ice Analogy

The physical arrangement of ice over water is roughly comparable to that of the lithosphere over the asthenosphere. First, the more rigid material (frozen in both water and rock) is on top; second, the warmest material is on the bottom, thus, heat flows through the ice/lithosphere into the atmosphere; third, the ice/lithosphere is lighter (less dense) than the underlying water/asthenosphere, and therefore floats on it. We will assume that moving along the bottom of each sheet is a current of water strong enough to set the sheet into motion. Each current flows in a different direction, and, as a result, the various sheets move into, away from, or along one another.

For sheets moving against one another, compression is produced along the contact. If this force exceeds the strength of the internal forces that hold the ice in each sheet together, the edges of both sheets will crumble and buckle, forming a *pressure ridge* (Fig. 16.1*a*). The ridge, which is a crude facsimile of some types of mountain ranges, protrudes not only above the surface of the sheets but into the water below them as well.

Now let us give the sheets different densities so that one rides much higher in the water than the other. As the two sheets collide, the higher sheet will tend to override the lower one, forcing the denser ice downward several meters into the water under the pressure ridge (Fig. 16.1*b*). Surrounded by water at a temperature slightly above freezing, the submerged slab will begin to melt; if the condition is prolonged, it may melt away altogether. This example appears to be a reasonably accurate representation of the movement of an oceanic plate of the lithosphere against a plate capped by a continent. The heavier, oceanic plate plunges under the leading edge of the continental plate, sending a huge slab downward into the asthenosphere, where it is eventually reabsorbed. The entire process is referred to as *subduction*, about which we will say more later.

Back on the ice, we can find places where ice sheets drift apart rather than press together as they do at pressure ridges. As they drift apart, water fills the vacated space, loses heat rapidly to the overlying air (whose temperature is below freezing), and freezes into new ice (Fig. 16.1*c*). As the sheets are driven further apart, more new ice can be seen extending outward from the zone of separation (Fig. 16.1*d*). In the realm of plate tectonics, this example resembles certain conditions that develop along the midoceanic ridges. Here the crust separation is probably due to divergent currents in the asthenosphere, and magma is filling the vacated space. Moreover, as the crust is drawn away from the fracture zone, it grows thicker. For both crust and ice, the zone of separation and upflow has a relatively light cover of solid material and tends to be warmer than the rest of the plate or sheet by virtue of its recent formation and thinness (which allows for rapid conduction of heat). Looking at the lake surface overall, as the sheets move, old ice is destroyed along compressional zones while new ice is created in the cracks and separations. At any moment the balance between the area of ice gained and lost is equal; thus, the lake is always completely covered with ice.

Continuity of the Crust

Although the ice model has a number of shortcomings—it cannot help us understand volcanism, for example—it does help to convey the basic concept of plate tectonics. Among the concepts illustrated is *crustal continuity*. According to this idea, the production and consumption of lithosphere counterbalance each other. At any moment the lithosphere must cover the entire planet; therefore, if it is consumed or destroyed at some rate in subduction zones, it must also be produced at a similar rate along the separation zones. Understandably, the zones of production and destruction are the foci of stress in the lithosphere, where earthquaking, volcanism, and crustal deformation are most concentrated.

How quickly is the crust being consumed and produced? This can be estimated on the basis of the rates of plate movement that have been determined in various parts of the world. The fastest movement appears to be about 10 cm per year, but for most plates the rate ranges from 1.5 to 6 cm per year. If we assume that an average rate of movement for all plates is proportional to the rates of the global subduction and crustal emergence, it would take the subductable lithosphere about 160 million years to replace itself. Therefore, if the plates behaved in the past as they do today, the lithosphere under the oceans is probably nowhere more than 200 million years old. Measurements of the ages of rock from the ocean floor show this to be so. This contrasts sharply with the ages of rocks in the continental lithosphere which are as much as 4 billion years (the oldest measured to this date is 3.8 billion years). Thus, it appears that the ocean basins have been destroyed and rebuilt many times, while the continents have probably endured in some manner since their inception in the first half billion to one billion years of the earth's existence.

Plate Size and Location

The data recorded at seismic stations around the world enable us to pinpoint the locations of virtually all earthquakes that have occurred in recent years. Plotted on a world map, the epicenters tend to arrange themselves into linear patterns that reveal the borders of the tectonic plates (Fig. 16.2). From such maps, it is possible to identify as many as twenty plates in the lithosphere.

Tectonic plates tend to fall into three classes according to size. Seven major plates and four or five minor plates are widely recognized. The remaining eight or nine plates are small and are often referred to as *platelets*. Most platelets are located in the destructive zones between major plates, and several are concentrated between the Eurasian and African plates (Fig. 16.3). The area of the major plates ranges from approximately 20 million to 50 million square miles, and all but one, the Pacific plate, contains a continent.

Although we are quite certain about the locations of plates today, we are not so certain about former plate positions on the planet. This statement is based on the fact that cartographers reference geographic locations according to the international geographic grid, the global system of coordinates (intersecting lines) comprised of parallels and meridians. This system is fixed on certain landmarks, one of the most important being the Royal Observatory at Greenwich, England, which is designated 0 degrees longitude.

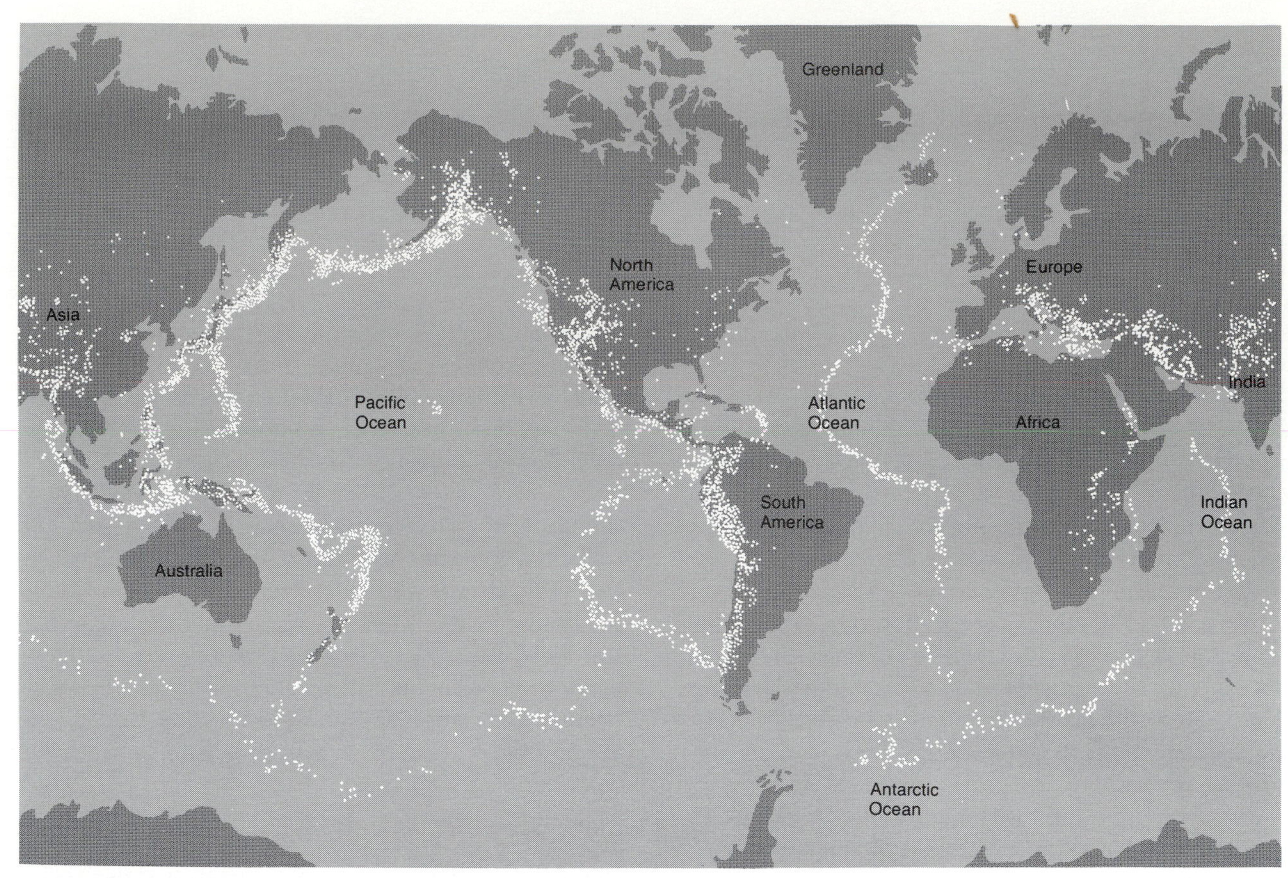
Greenland
North America
Europe
Asia
Pacific Ocean
Atlantic Ocean
Africa
India
South America
Indian Ocean
Australia
Antarctic Ocean

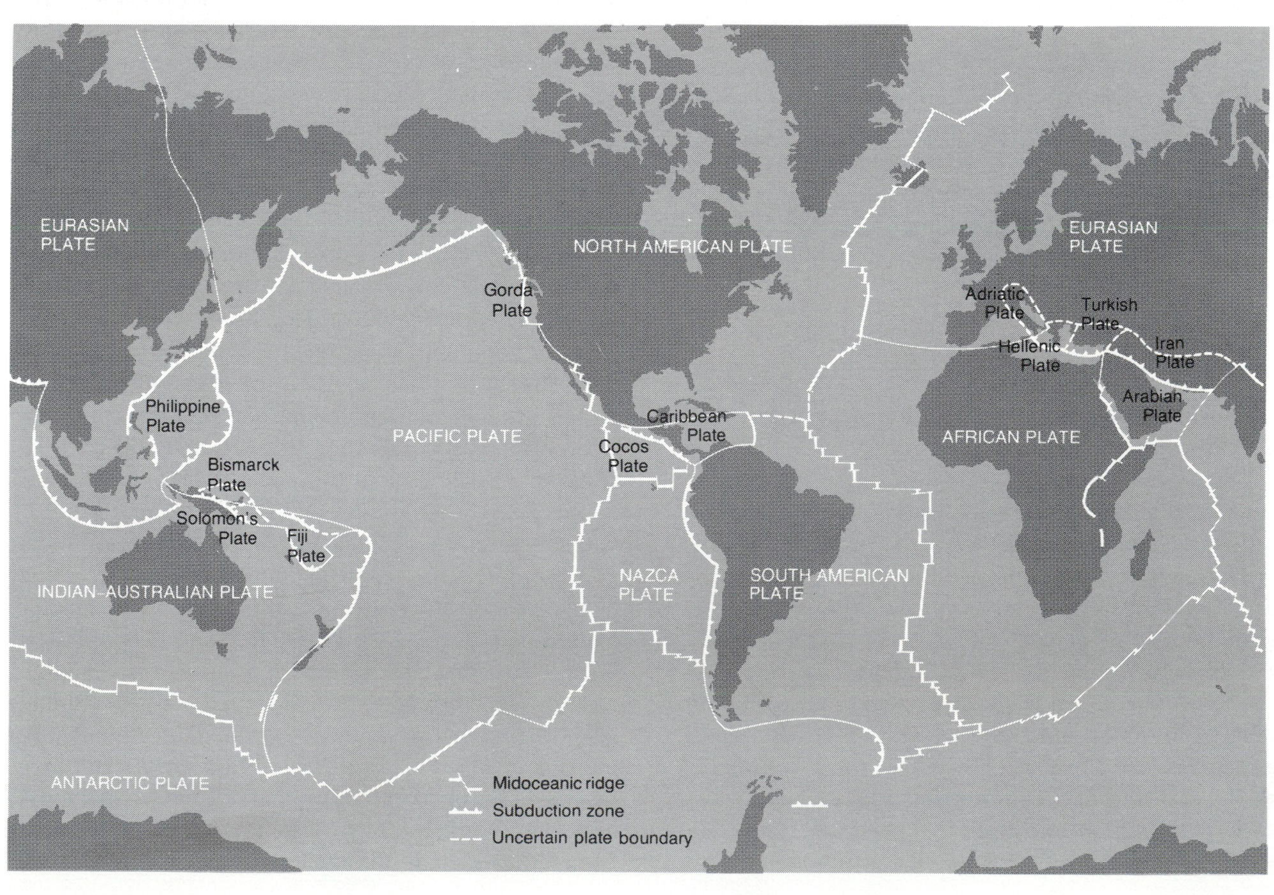
EURASIAN PLATE
NORTH AMERICAN PLATE
EURASIAN PLATE
Gorda Plate
Adriatic Plate
Turkish Plate
Hellenic Plate
Iran Plate
Philippine Plate
Arabian Plate
Caribbean Plate
PACIFIC PLATE
AFRICAN PLATE
Cocos Plate
Bismarck Plate
Solomon's Plate
Fiji Plate
NAZCA PLATE
SOUTH AMERICAN PLATE
INDIAN-AUSTRALIAN PLATE
ANTARCTIC PLATE
Midoceanic ridge
Subduction zone
Uncertain plate boundary

Figure 16.2 The distribution of earthquakes with foci at depths of 0–700 km during a recent 7-year period. The linear patterns define the boundaries of tectonic plates.

Figure 16.3 Tectonic plates of the world. Platelets are concentrated in the destructive zones between major plates, for example, between the African and Eurasian plates.

Although the founders of the global coordinate system had no way of knowing this, these landmarks are moving with the plates on which they are situated, and since the plates are moving at different rates and in different directions, the geographic grid of the world is being stretched out of shape like a huge rubber map. To plot accurately the future locations of plates on the globe, we will have to turn to an extraterrestrial reference system, probably the stars.

Projecting into the geologic past to, say, the time of Pangaea (see Note 16), we have no way of deciphering precisely where that land mass was situated on the globe. However, we can decipher the *relative* locations of the various sections of Pangaea by matching rocks, magnetic patterns, fossils, shapes of present-day land masses, and so on. Then, by differentiating the distances between the present locations of continents and their former positions in Pangaea, we can determine *net movement* and the *relative* directions of movement.

Although worldwide data on the contemporary rates and directions of plate movements are not abundant, we have learned enough to map the general patterns of movement in the major plates. One interpretation is presented in Figure 16.4. Each arrow on this map represents 20 million years of movement, and it is immediately apparent that the Pacific, Australian-Indian, and Nazca have been the most active plates in the past 20 million years.

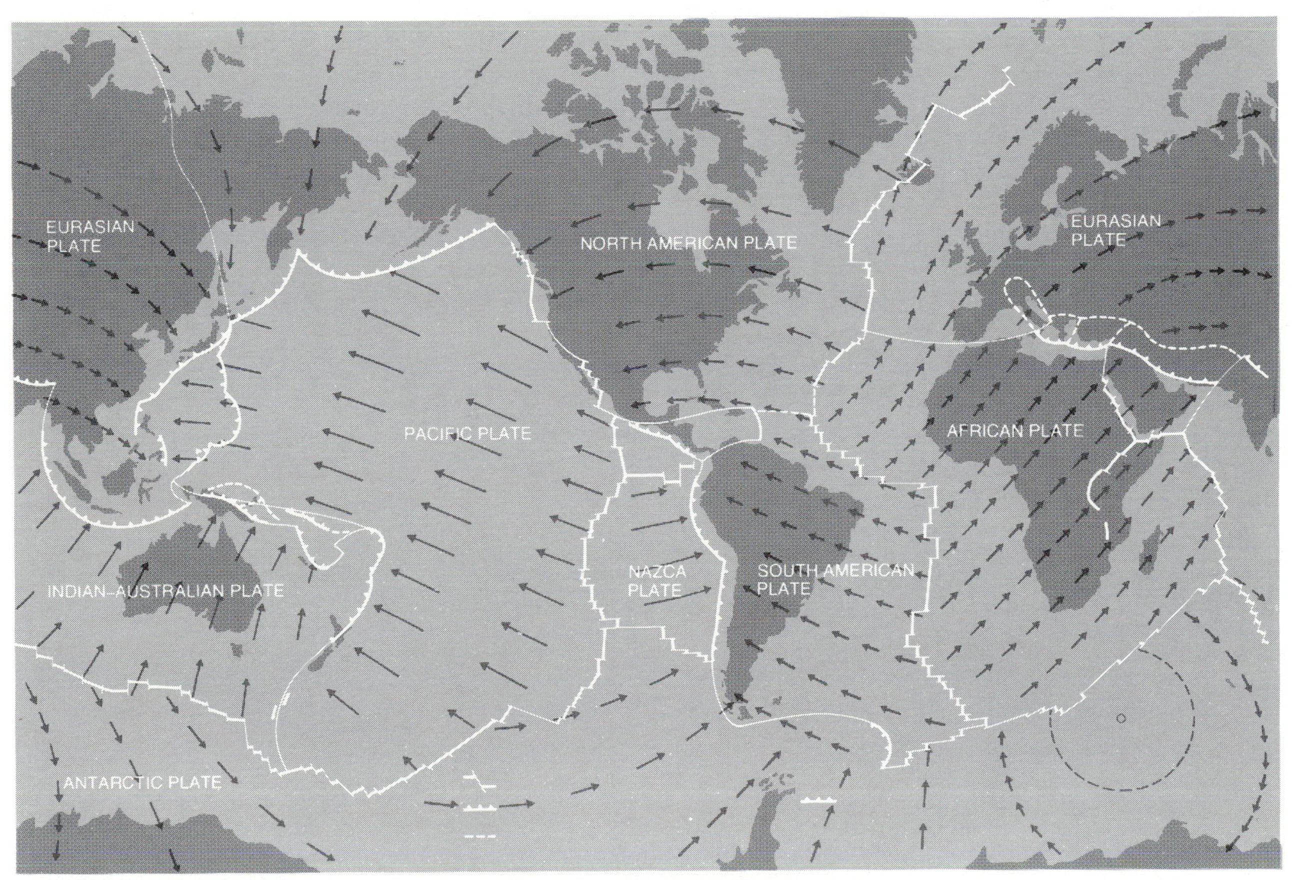

Figure 16.4 The worldwide pattern of tectonic plate movement. Each arrow represents 20 million years of movement; the longer the arrow, the greater the rate of movement.

Evidence For Plate Movement

What actual evidence is there for plate movement? Scientists have been able to marshal several bodies of evidence that reveal both the amount and the relative directions of plate displacement. One of the strongest bodies of evidence was born with the discovery of parallel bands of magnetic alignments in the rocks on the floors of the Pacific and Atlantic oceans.

Let us examine the case of the Atlantic. The mid-Atlantic ridge is a volcanically active rise that runs south from the Arctic through Iceland to near Antarctica. For millions of years, basaltic rock has flowed from the various fissures in the ridge. Such flows have been frequently observed on Iceland and other islands along the ridge. When the new igneous rock solidifies, the polarity of its magnetic minerals is permanently frozen in place and should be consistent with that of the earth's magnetic field at that time. Over periods of time ranging from tens of thousands to millions of years, however, the polarity of the earth's magnetic field undergoes reversals, that is, changes from negative to positive and vice versa. Analysis of volcanic rocks formed in various parts of the world indicates that nine major reversals have occurred in the past 3.6 million years.

The bands of magnetic deviations that have been mapped in the basaltic rock of the Atlantic floor correspond with the reversals in the earth's magnetic field which have been imprinted in volcanic rocks as they formed along the midoceanic ridge (Fig. 16.5). As the crust spread from the ridge, the rock, with its magnetic polarity fixed, was carried outward across the ocean floor. It follows that the older rock farther from the ridge shows alignments with former positions of the earth's magnetic field, whereas the newer rock nearer the ridge shows bands of alignment with relatively recent magnetic fields. If the crust were not spreading from the ridge, the newer volcanic rock would cover the older rock as it emerged, and the magnetic reversals would be piled on top of one another. Therefore, the striking geographical pattern of magnetic polarities that has been documented in the ocean floor would not be discernible. This is not the case, however, and we believe that new rock is appearing while the ocean floor is spreading from the ridges.

Along each ridge system, the rate of spreading and the emergence of volcanic rock vary markedly over time and from place to place. Along some midoceanic ridges there is no evidence of any present activity, but much evidence of former activity. Along the mid-Atlantic ridge, the sea floor around Iceland appears to be spreading at a fairly slow rate. This can be deduced from the fact that here the ridge tends to be very narrow and high in elevation, suggesting that sea-floor spreading cannot draw new rock away from the zone of emergence before it has time to pile up. The opposite interpretation can be applied to the wide, low sections of the ridge, and it explains the broad pattern of geomagnetic banding in the ocean floor.

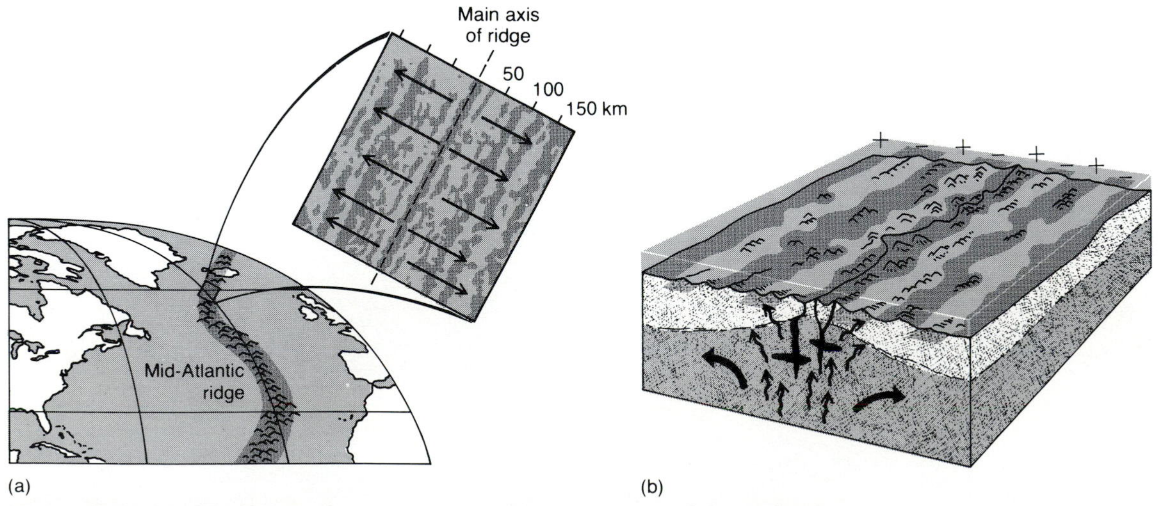

Figure 16.5 (*a*) The pattern of magnetic anomalies in a section of the Mid-Atlantic ridge south of Iceland and (*b*) a schematic interpretation of this pattern. The dark areas represent the rock with positive magnetization; the light areas, the rock with negative magnetization.

ACTIVITY ON THE EDGES OF PLATES

Three major types of contacts can be identified along the edges of tectonic plates, and all are associated with earthquakes and volcanism. Two of these, *constructive* and *destructive*, were mentioned in the context of the ice sheet analogy. Whereas these contacts tend to be located at the opposite edges of a plate that is moving essentially in one direction, the third type of contact is found along the sides of the plate that are more or less parallel with the direction of motion. Here movement is mainly lateral as the plate slips past its neighbor. Crust is neither destroyed nor created, but conserved; hence, such contacts can be termed *conservative*. Let us briefly describe how earthquakes and volcanism are related to plate movement in these three contacts.

Destructive Zones

As an oceanic plate subducts under its neighbor, the surface of the crust is drawn down to form a trench (Fig. 16.6). Beneath the trench, the subduction process is manifested in a jerking downward motion as the plates slip along the angular contact between them.

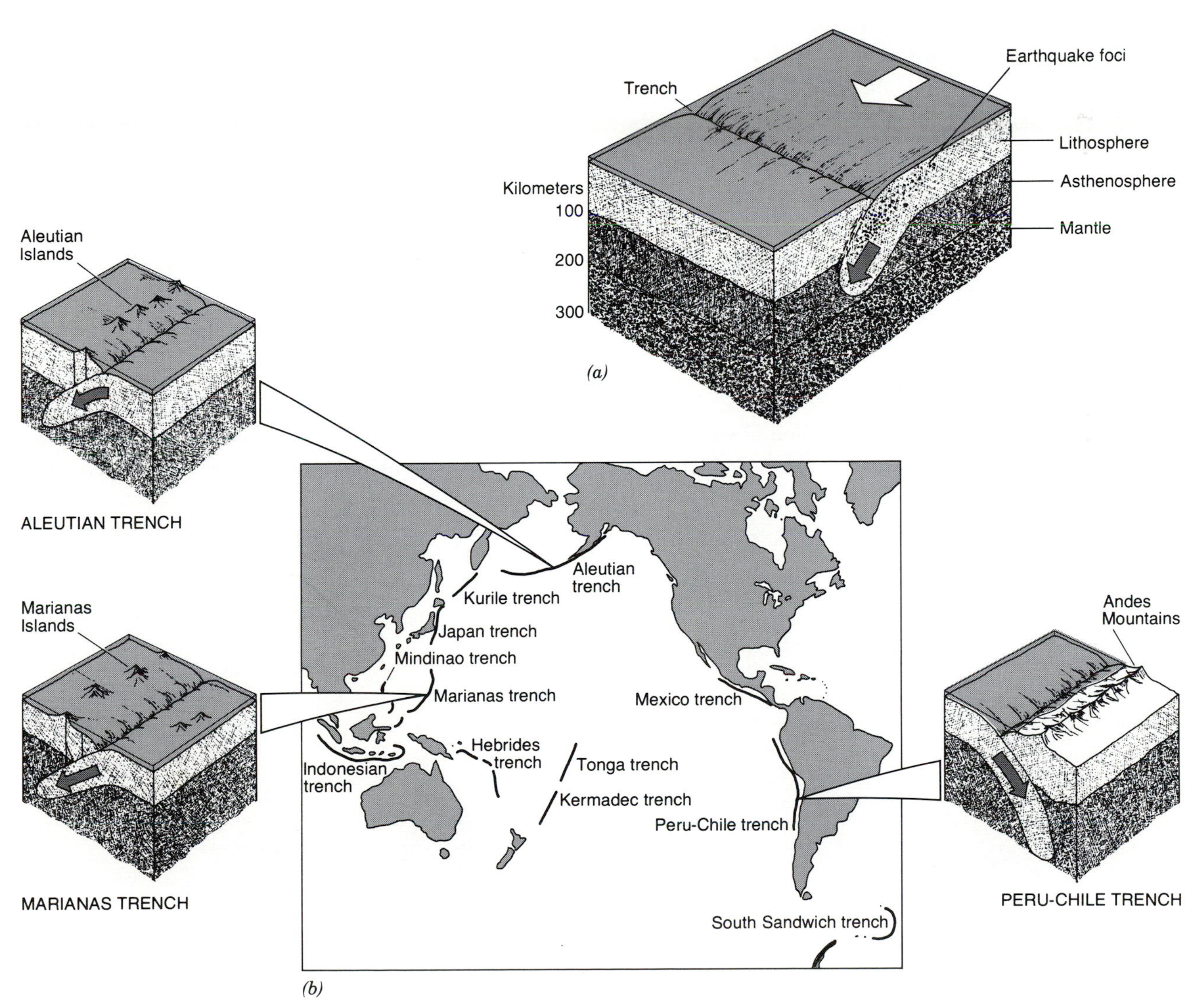

Figure 16.6 Trench formation: (*a*) the early phase in the formation of a subduction zone, showing the trench and the concentration of shallow earthquake foci; (*b*) the locations of trenches in and around the Pacific Basin.

This motion is the result of frequent small displacements of massive blocks of rock as the plate bends into the earth. Each displacement, or *fault*, releases seismic energy into the lithosphere, part of that energy traveling to the surface to produce an earthquake of some strength (see Fig. 15.10b).

From our vantage point on the surface, the size of an earthquake depends on many factors in addition to the amount of energy released at the focus of a displacement. Among these other factors are the depth at which the fault takes place and the physical nature of the rock around the zone of displacement. If the fault is shallow and the rock is rigid, the surface tremors are likely to be very strong. This is clearly the case in the upper 50 km of the lithosphere, where compressional stress in the brittle rock of the crust generates not only the strongest tremors, but also the greatest frequency of faulting. Farther down, the rock appears to grow less brittle or more plastic (because it grows hotter), and the magnitude and frequency of faulting decline markedly.

Not all of the activity in the subduction zone involves the displacement of rock (and earthquakes): volcanism is also prominent. The source of volcanic activity is found at the lower contact between the continental and oceanic plates. Here, during the early stages of subduction, a body of magma forms. Some of this magma melts its way through the overlying plates, forming conduits called *diapirs*, and then issues onto the surface to form volcanoes. According to one interpretation, initially a single line of volcanoes is formed, but as the oceanic plate subducts further, the magma body is dragged down with the plate, more diapirs develop, and a broader zone of volcanism is created, leading to more islands (Fig. 16.7). As the volcanoes grow in size and number, they may eventually coalesce to form arc-shaped island archipelagoes, such as the Japanese Islands and Sumatra (Fig. 16.8; color plate 11).

In the advanced stages of subduction, the plunging slab of lithosphere extends through the asthenosphere and into the underlying mantle. Because the slab is much colder than the mantle rock, it extends intact as deep as 700 km in some areas, where it is physically assimilated into the mantle. Although the seismic zone is extended to a corresponding depth, the magnitude and frequency of faulting decline, apparently because the slab is heating up and is less brittle Fig. 16.9.

How does the subduction process come to a halt? There are probably many contributing factors. The forces driving the plates may stop or change direction. There is really no way of knowing when this happens, but there is evidence to indicate that when continental blocks are drawn into each other, subduction slowly grinds to a halt. This appears to have been the case when the northern edge of the Indian-Australian plate subducted under the Eurasian plate. Subduction apparently continued until the Indian subcontinent lodged against the Asia continental block. In the course of convergence, a narrow ocean basin called the Tethys Sea was formed between these blocks, and over a period of tens of millions of years, it became a depository for volcanic and sedimentary rock. As the basin was compressed, these rocks were uplifted to form the massive mountain ranges of Tibet and neighboring lands (Fig. 16.10).

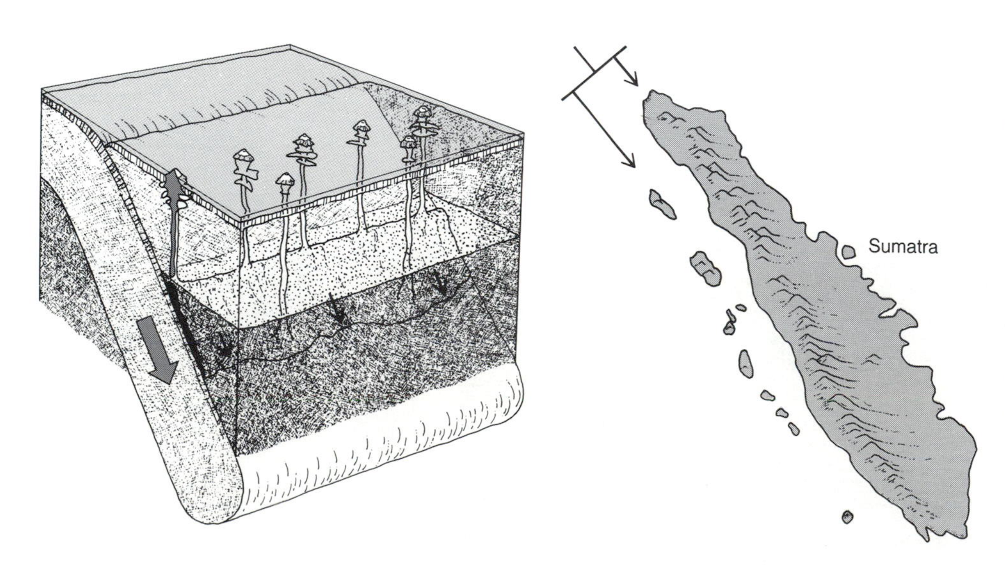

Figure 16.7 An interpretation of volcanic activity in a subduction zone leading to the formation of an island arc. The magma tube that develops in the early phase gives rise to diapirs which produce volcanoes. As subduction advances, more diapirs develop, and the island arc grows.

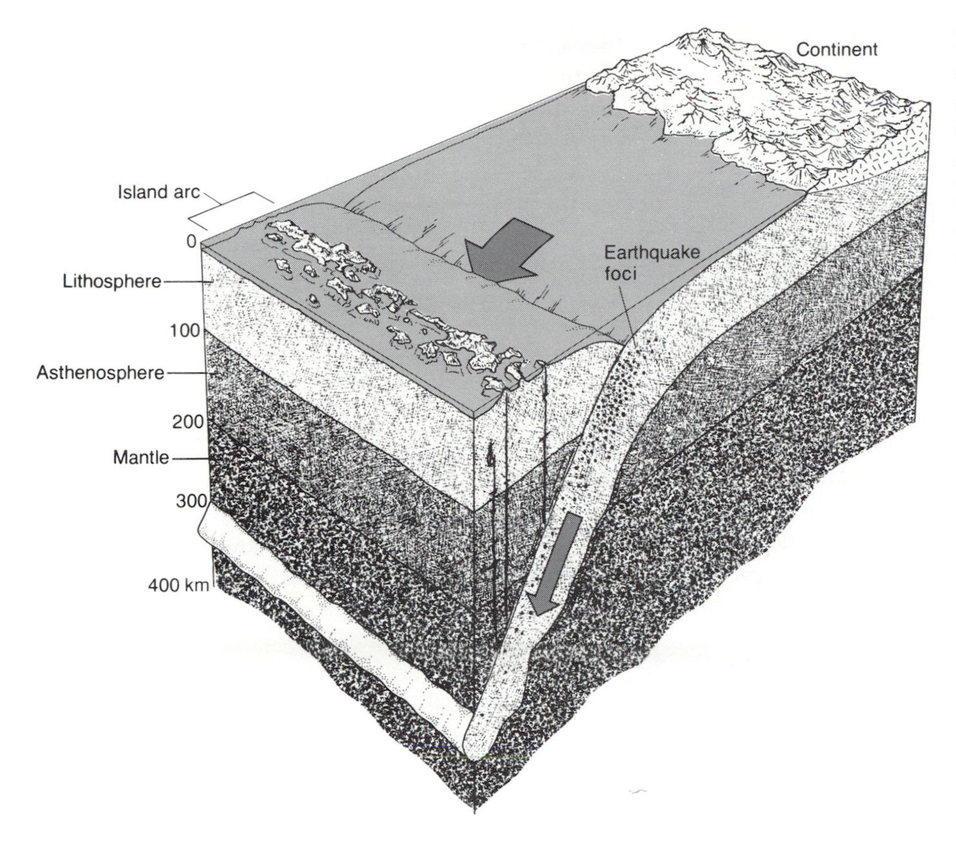

Figure 16.9 The advanced stage of subduction. The slab extends into the mantle under the asthenosphere. By this time the island arc has grown into large islands, and the zone of earthquaking may reach a depth of 700 km or more.

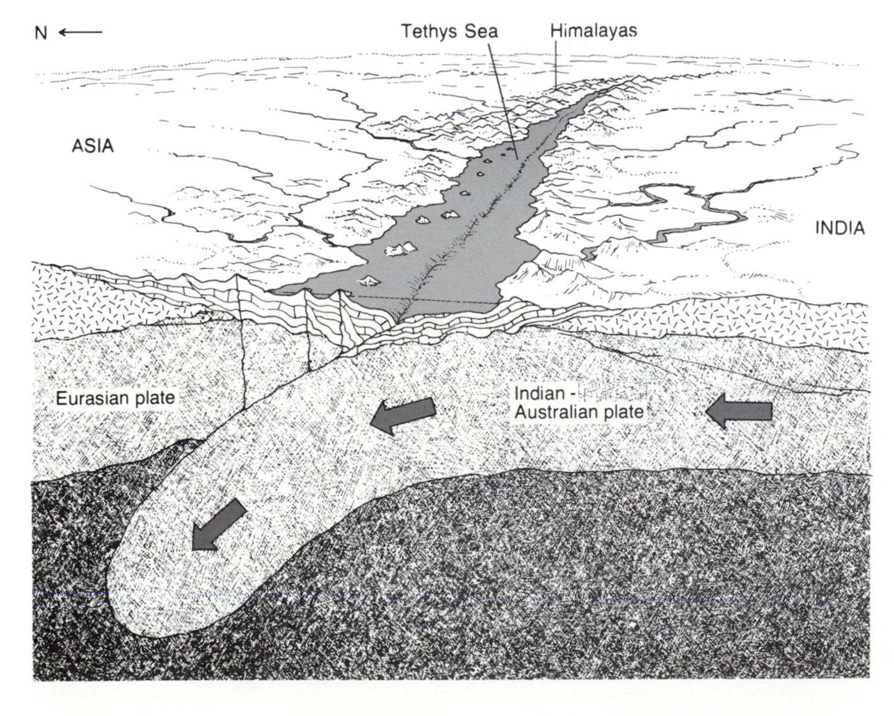

Figure 16.10 Subduction of the Indian-Australian plate along the southern edge of the Eurasian plate. The sedimentary rocks of the Tethys Sea were compressed and uplifted to form huge mountain ranges such as the Himalayas, Pamirs, and Hindu Kush.

Constructive Zones

More or less opposite the subducting side of a plate is the zone of crustal spreading where magma is emerging to form new lithosphere. Here tensional motion, which is the opposite of the compressional motion of the subducting side of the plate, is set up as the two plates move apart. As a result, the crust fractures and small blocks of ocean floor downdrop between the parting plates. These faults are usually shallow and in turn generate shallow, low-magnitude earthquakes.

As the lithosphere is pulled apart, it is reduced to a relatively thin layer that is subject to intrusions of magma from the asthenosphere. The faults provide magma with ready access to the surface, where it flows between and over the parting blocks and piles up to form a rise or mountain range. If the rate of spreading is relatively slow, say, less than 3 or 4 cm per year, the ridge may grow to great heights, 3000 m or more; otherwise, it tends to be lower and broader (Fig. 16.11).

Although the process of spreading and emergence is known to occur on the continents, it is more prevalent in the ocean basins, where it has produced the worldwide system of oceanic ridges. On the continents, there appear to be only two major zones of spreading—one in Africa and the other in Asia. The African zone, which is the better developed of the two, extends from the Red Sea down the East African Highlands and is marked by a system of rift valleys that form the basins for Lakes Albert, Tanganyika, and Malawi (Nyasa) as well as hundreds of smaller water bodies (see Fig. 16.25). We are uncertain, however, whether this zone is currently active.

Conservative Zones

In conservative zones, the movement of the plates is mainly parallel to the boundary between them, and therefore crust is neither created nor destroyed. The most famous fault system in North America, the San Andreas, is a conservative zone along the California coast. It separates the northward-moving Pacific plate from the North America plate and is the source of frequent and often strong earthquakes (Fig. 16.12). However, volcanism is uncommon here, even though it is typical of such zones.

Conservative zones are also found within active mid-oceanic ridges in the complex border configuration there which resembles a sort of interfingered pattern between the two plates (Fig. 16.13). At the ends of the fingers, the boundary is constructive, but along the sides of the fingers, the boundary is conservative. The term *transform fault* has been given to the conservative segments of the border. As the ridge pulls apart, lines of crustal weakness develop perpendicular

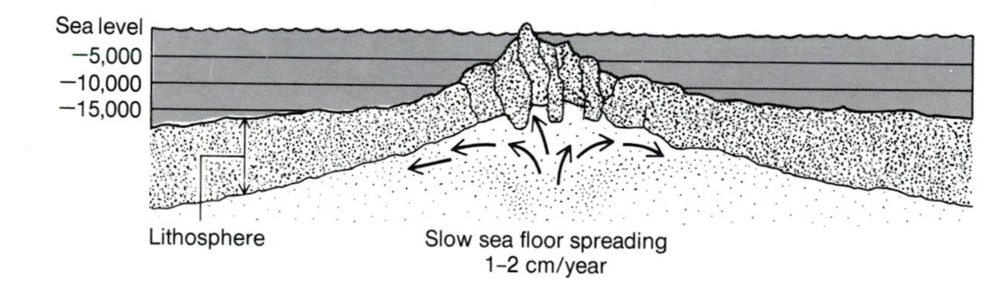

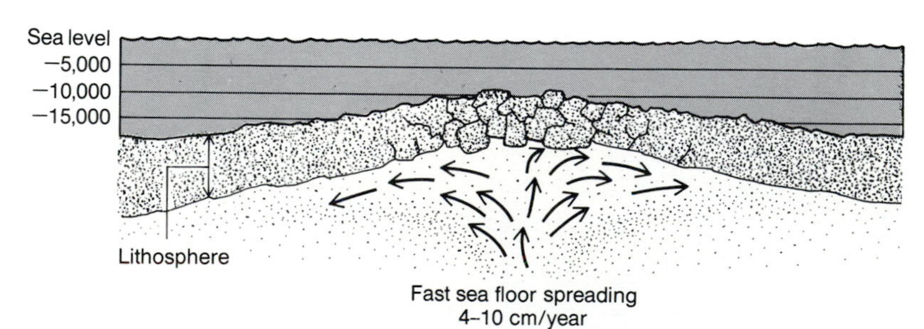

Figure 16.11 Relationship between ridge height and the rate of sea floor spreading. Ridges tend to be higher where spreading rates are low; this is the case in the North Atlantic where Iceland has formed on a high part of the ridge.

to the ridge along the paths of movement. Some of these lines become transform faults, which, when interconnected with the faults that parallel the ridge, form the interfingered border pattern. This concept is portrayed schematically in the inset of Figure 16.13.

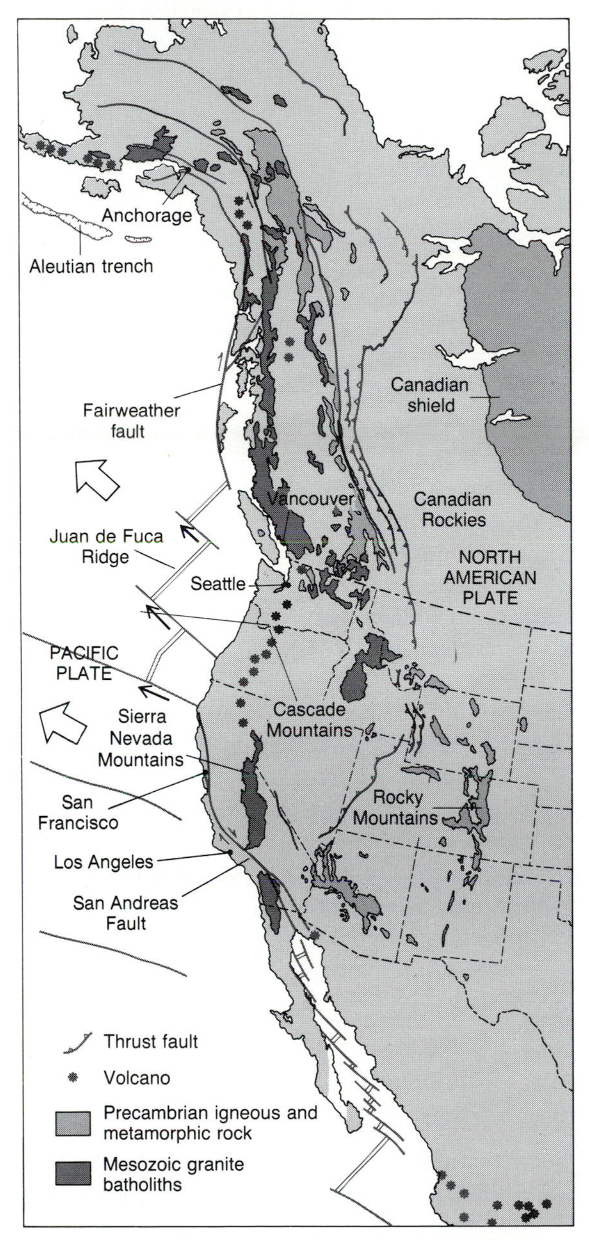

Figure 16.12 The main faults of the San Andreas system, North America's prominent conservative zone.

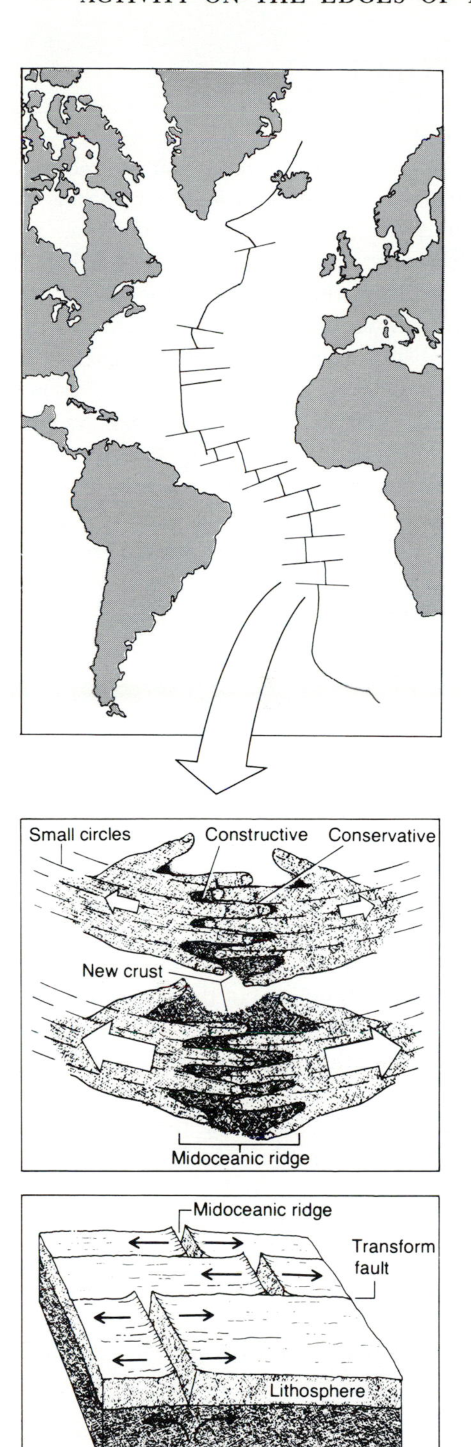

Figure 16.13 The interfingered configuration of the plate border along the mid-Atlantic ridge. The east-west-trending fault lines represent conservative borders.

DRIVING FORCE OF PLATE TECTONICS

So far in this chapter, we have described the tectonic plates and the nature of their movement. But what forces drive the plates? To answer this question, we must examine the internal energy of the earth and how it is released from the interior of the planet.

We have long known that heat is emitted from the earth's crust. Hot springs and volcanoes, in particular, are indisputable evidence of this fact. What was not known until this century, however, is that heat is being continually emitted from every square centimeter of the crust's surface. The average rate of emission is 47 calories per square centimeter per year, which is very small compared to the energy received at the earth's surface from the sun. However, when we multiply this small rate times the surface area of the earth (510 million square km), the resultant figure is enormous—ten times the total energy generated and used by humans each year.

Sources and Mechanisms of Heat Flow

What is the source of earth heat? It is thought that the interior of the earth received a great charge of thermal energy with the conversion of gravitational energy into heat when the core formed about 4 billion years ago. To this source is added the heat produced by the radioactive disintegration of the elements uranium, thorium, and potassium, which are contained in the rocks of the mantle and crust.

The heaviest concentrations of these elements are found in granite, the rock of the continents, whereas significantly lower concentrations are found in basalt, the rock of the ocean basins, and only a trace is thought to exist in the rocks of the mantle (Table 16.1). Thus, the continents are radioactively "hot," with about 35 percent of their overall heat flow contributed by the granitic rock. In contrast, the ocean basins are radioactively "cool," and only about 10 percent of their heat is provided by the basaltic rock. All the rest of the heat emitted from both oceans (90 percent) and continents (65 percent) comes from the mantle (Fig. 16.14).

TABLE 16.1 RELATIVE HEAT FROM RADIOACTIVITY IN ROCKS

Rock Group	Uranium (ppm)	Thorium (ppm)	Potassium (% wgt)
Sialic	3.0	10.0	2.2
Basaltic	0.7	3.0	1.1
Mantle	0.013	0.05	0.001

(heat output is approximately equivalent to element content in parts per million (ppm))

How does the heat get out of the mantle? In the lithosphere, conduction is virtually the only mechanism of heat transfer because, over the short-term (months and years), the rock of the lithosphere is essentially solid and immobile. But the lithosphere is thin, a mere membrane, compared to the mantle, and when we consider conduction as the main mechanism of heat flow through the mantle, we find that the rate would be impossibly slow to account for the present rate at which heat is being delivered to the crust. The reason is that solid rock is such a poor conductor of heat. Witness, for example, that it could take about 5 billion years for heat to be conducted across a slab of rock 250 miles thick. (See Chapter 2, pages 30 and 31, for a definition of heat transfer mechanisms.)

Convection, on the other hand, is a highly efficient means of heat transfer. But convection involves mass transfer (flow) of heated rock, and rock is extremely resistant to deformation by flow. However, when we recall that seismic findings show conclusively that the rock of the asthenosphere exists in a plastic state and thus may be capable of flow-type movement, the idea sounds more plausible. Furthermore, tests show that a rate of rock flow of only 2 to 3 cm per century is enough to produce a heat flow greater than that possible by conduction through stable rock. Thus, it appears that under the long-term strain of convective forces, heat is transferred through the upper mantle by an upward flow that is perhaps best described as a creeping motion. Although the nature of this motion in terms of rates and direction of movement, zones of origin, and continuity of flow is unknown, contemporary scientists are convinced of its existence (Fig. 16.15).

Mantle Convection: The Engine of the Plates

It appears that mantle convection, driven by the heat of earth's interior, is the force driving the plates. The theory of convection in the mantle not only accounts for the *rate* of heat flow from the earth, but also establishes the presence of a force of sufficient magnitude to move the huge tectonic plates. Since the early decades of this century, the debate over the meaning of the large body of evidence supporting plate tectonics often stalled on the question of the driving force responsible for the movement. Because of the size of

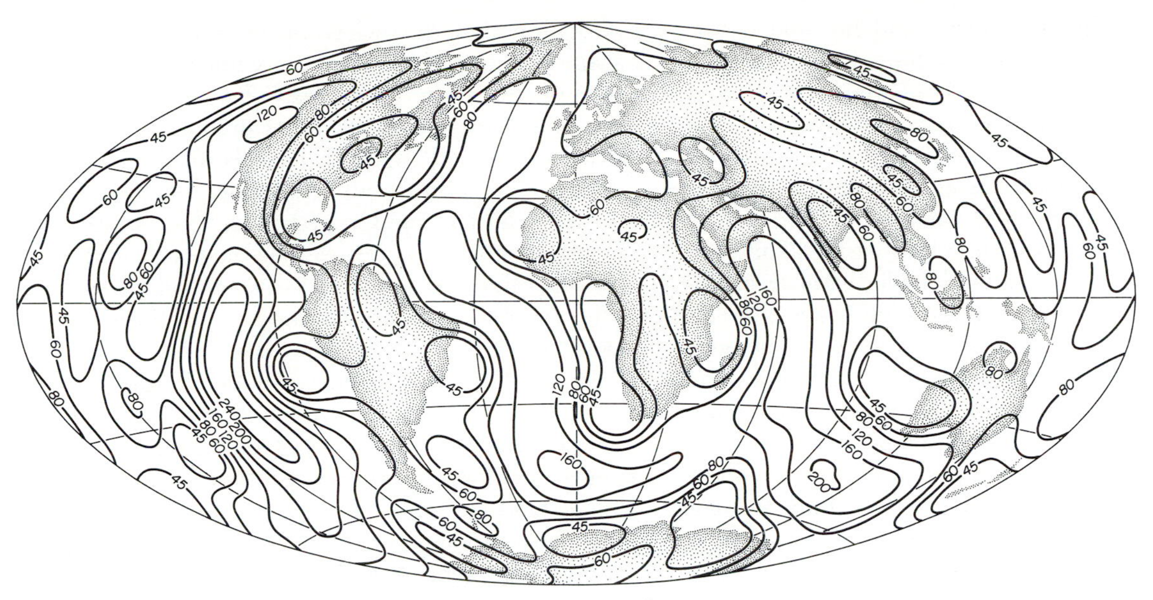

Figure 16.14 Map of global heat flow in millijoules/m²s. Enlargement represents a heat-flow profile across the mid-Atlantic ridge.

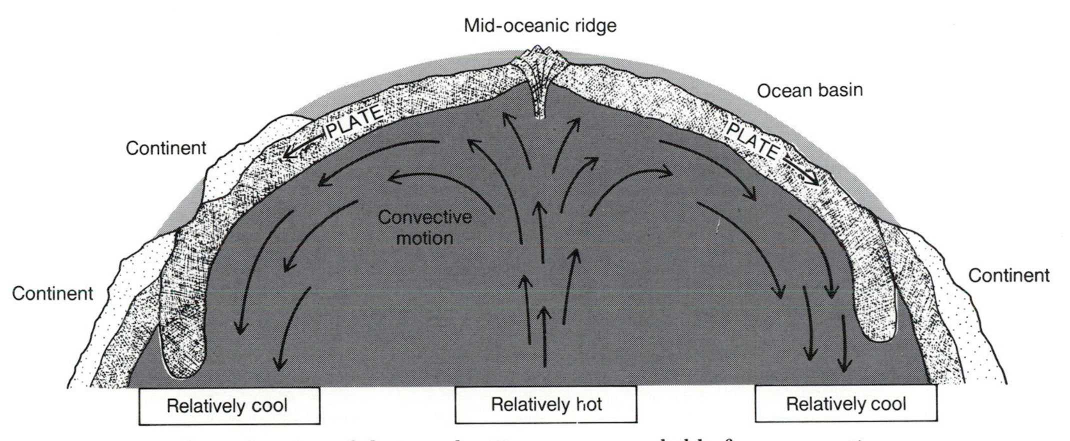

Figure 16.15 The convection of the mantle. Hot currents probably form convective chimneys that rise through the asthenosphere and generate three types of movement: upward, downward, and lateral. As a result, zones of divergence and convergence are formed near the base of the lithosphere, which in turn appears to produce corresponding motions in the lithosphere.

the crustal blocks involved, it was difficult to identify a mechanism powerful enough to produce the movement. Convective flows in the mantle has provided a plausible answer. Moving upward through the plastic layer, they not only transfer heat, but, as they spread out at the base of the lithosphere, they also exert differential stress on the lithosphere, causing it to weaken and break. When these flows cool with the loss of heat to the overlying rock, they sink back into the asthenosphere. Thus, the lithosphere can be pulled apart, pushed together, and pulled downward, depending on the pattern of flow (Fig. 16.15).

MOUNTAIN BUILDING BY FOLDING AND FAULTING

Plate tectonics is the key to understanding the general distribution of mountains and mountain-building activity. Practically all of the earth's mountains—the orogenic belts, the island arcs, and the midoceanic ridges—are located on or near the edges of the tectonic plates. Coincidentally, these are also traditional zones of heavy human settlement, especially the coastal orogenic belts and the island arcs; thus, humans have a longstanding association with mountains and mountain-building processes.

The major mountain chains of the earth fall into two main categories: volcanic and folded. Those of the ocean basins, including both arcs and ridges, are almost exclusively volcanic, whereas the orogenic belts of the continents are mainly folded mountains. The latter result from the compression of sedimentary rocks on the margins of the continents. Where neighboring continental plates move against each other, these rocks are folded into massive structures that are pushed upward thousands of meters. The folding is usually accompanied by faulting and volcanism. The ultimate result of all this deformation is long chains of complex mountains such as the Andes, Alps, and Rockies.

The Nature of Rock Deformation

In order to understand the processes of rock deformation, we should begin with a look at the behavioral characteristics of rocks when placed under stress. *Stress* is defined as a force that is acting on a body. If the force is not equal in all directions, it is referred to as *differential stress.* When rock is subjected to differential stress, such as that produced by the motion of tectonic plates, it may undergo three stages of deformation: elastic, plastic, and rupture. Under *elastic* deformation, the rock is able to return to its original shape and size if the stress is withdrawn. For each type of rock there is an *elastic limit,* beyond which the rock does not return to its original shape if the stress is released. If stress exceeds this limit, *plastic* deformation is the result. Plastic deformation of rock is called *folding,* and it is irreversible. If stress is increased still further, exceeding the plastic limit, and fractures develop, the rock is said to fail by *rupture* (Fig. 16.16). *Faults* are ruptures or fractures in rock along which displacement has taken place.

How different rocks behave under differential stress depends on their internal properties, such as mineral composition and crystal structure, as well as on the rate at which stress is applied. It is very important to understand that stress in rock builds up slowly over very long periods of time, millions of years. Because of this very long and gradual application, rocks in the crust, such as granite and limestone which appear very brittle in the laboratory, often behave as plastics in nature. We can demonstrate this idea with ice; under sudden stress it will rupture, but when stress is applied gradually, ice will bend appreciably. This accounts for the fact that the glaciers are able to bend and flow.

Mapping Geologic Structures

In order to figure out the broad patterns of crustal deformation, it is first necessary to measure individual structures. But the task is a difficult one, for the bulk of most structures is hidden underground and those portions that are exposed are often heavily altered by

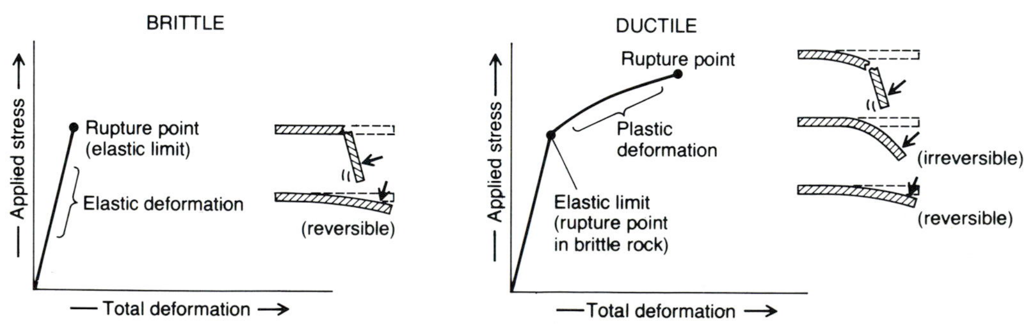

Figure 16.16 The responses of brittle and ductile rocks to stress. With the initial application of stress, all rocks behave like an elastic and deform relatively little (represented by the lower part of both graph lines). The difference between brittle and ductile rocks occurs at the elastic limit. At this point the brittle rock ruptures, whereas the ductile bends further, but as a plastic rather than as an elastic. Ultimately, of course, even the ductile rock ruptures.

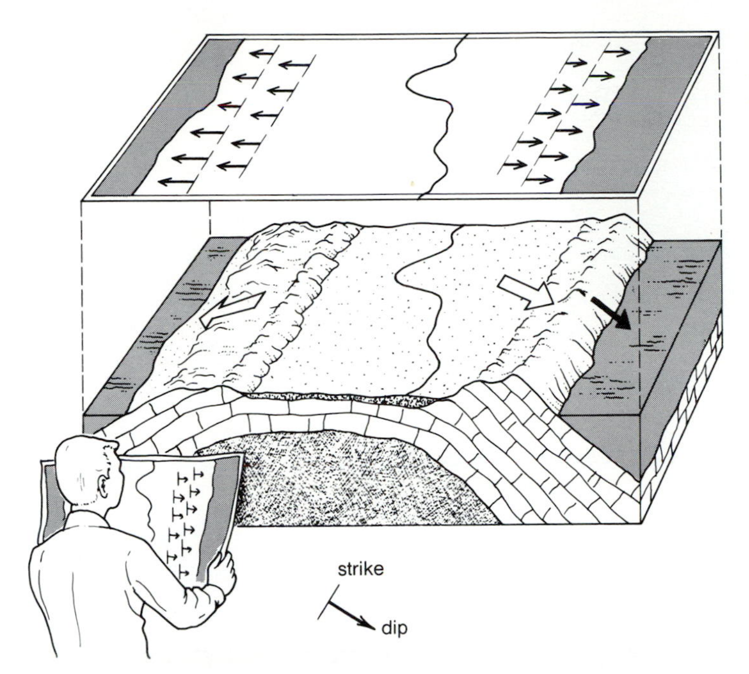

Figure 16.17 The directional properties of strike and dip are illustrated as they would appear in the field and on a geologic map.

erosion. In nonmountainous regions, moreover, practically the entire upper surface of the bedrock is buried under thick mantles of soil and surface deposits. How, then, do we piece together the evidence that leads to the identification and classification of the geologic structures of the crust?

The traditional method involves measuring the geometric attitude of the rock in outcrops. *Outcrops* are exposures of bedrock, and *geometric attitude* is their directional orientation. Outcrops range from small projections or ledges to huge exposures like those in the Grand Canyon that are several thousand meters high and many kilometers long. Inasmuch as most of the bedrock at or near the surface of the continental crust is of sedimentary origins, most outcrops will exhibit distinct layers or strata, and knowing this is important in reconstructing geologic structures.

One of the most meaningful features of an outcrop is whether the strata are horizontal or tilted. This is based on the *principle of original horizontality*, which holds that the layers of sediment that become rock strata were originally deposited in horizontal beds; thus, the bedding (strata) in undisturbed sedimentary rocks should be flat-lying. In addition, a sedimentary layer is deposited as a continuous sheet that usually ends by thinning away gradually or by changing gradually in composition. Thus, where we find an exposed face of beds—as in the walls of the Grand Canyon—it means that the sedimentary rock has been broken or eroded.

Strike and *dip* are the two directional properties measured in rock outcrops. Dip is the angle of inclination of a bed from the horizontal. Strike is the direction perpendicular to dip where the broken or eroded exposures of the dipping beds intersect the surface (Fig. 16.17). By marking the strike and dip directions of many outcrops on a map, it is possible to reconstruct individual folds and faults. They in turn can be combined with other fold and fault patterns to reconstruct larger structural patterns such as those of mountain ranges. The structural trends of mountain ranges in turn enable scientists to work out the relationship between crustal deformation and tectonic forces such as those associated with plate movement.

Properties of Faults and Folds

The displacement in a fault, relative to the fracture itself, may be up, down, back and forth, in or out, or any combination of these. Faults range in size from ruptures only a few centimeters long to those extending tens of kilometers through the lithosphere. The largest faults are formed along the edges of tectonic plates, where networks of faults such as the San Andreas can be traced hundreds of kilometers along the surface (Fig. 16.18).

A standard terminology is used to describe faults. The faces of the blocks on either side of the fault are called the *walls;* the surface separating the walls is the *fault plane.* If the fault plane is inclined, which is

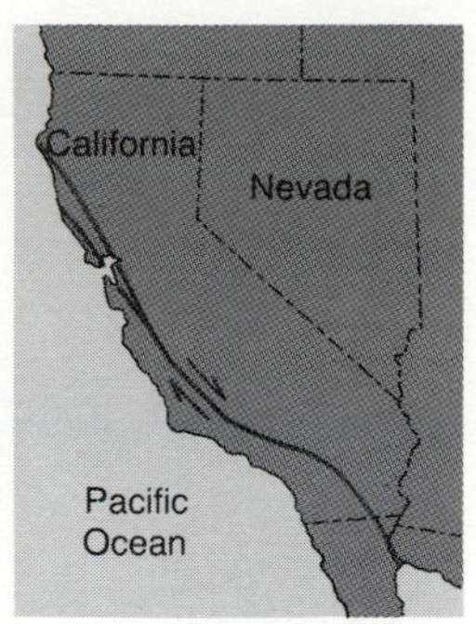

Figure 16.18 A segment of the San Andreas fault north of San Francisco. This fault can be traced for about 750 km across California.

normally the case, the upper face is called the *hanging wall*, and the lower face is called the *footwall* (Fig. 16.19). The trend of the fault along the earth's surface, as it would appear on a map, for example, is termed the *fault line*. The part of a wall exposed as a result of displacement is the *fault scarp*.

Folds are structures that resemble warps or wrinkles in rock. They are the most common geologic structure in the orogenic belts and are often elegantly displayed in exposures of sedimentary and metamorphic rocks along canyons, cliffs (Fig. 16.20), and cuts into hillsides made in the construction of roadways. Folds are also common in nonmountainous areas, for example, the midwestern United States, northern France, and southern England. The largest foldlike structures are broad upward and downward flexures called *domes* and *basins* and may be hundreds of kilometers in diameter (Fig. 16.21).

The basic terms used to describe folds are as follows.

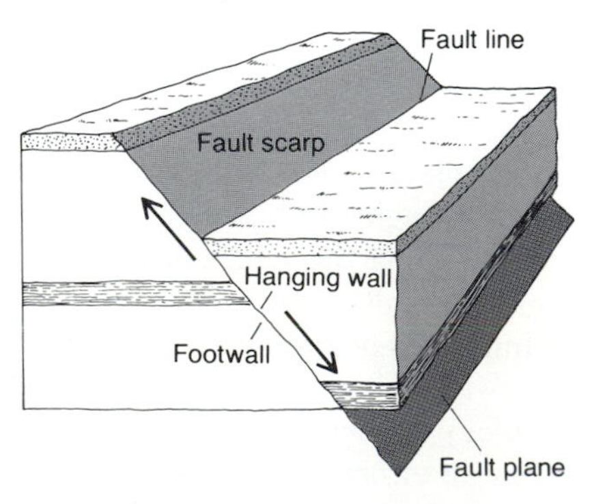

Figure 16.19 The basic features of a fault: fault scarp, fault line, footwall, hanging wall, and fault plane.

The two sides of a fold are the *limbs;* an imaginary plane drawn between the limbs, which divides the fold in half, is called the *axial plane;* and the crest of the fold is the *axis*. If the axis is inclined from the horizontal, the fold is said to *plunge*. Imagine a plunging fold to be somewhat like a submarine emerging on the ocean surface (Fig. 16.22).

Figure 16.20 Folds in sedimentary rocks exposed on a mountainside in the Canadian Rockies.

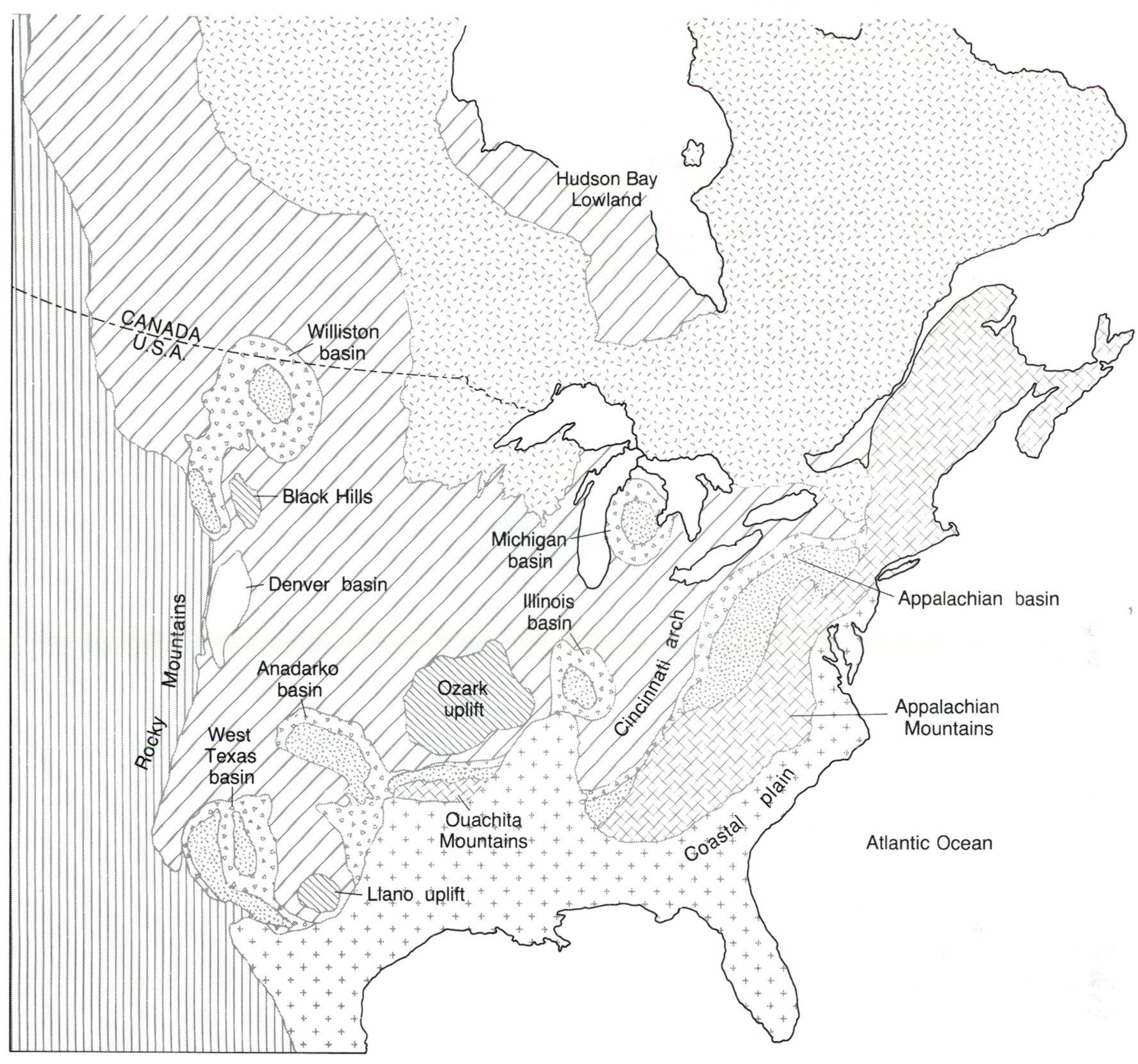

Figure 16.21 Major basins and uplifts of central North America. Uplifts such as the Ozark are essentially very broad domes.

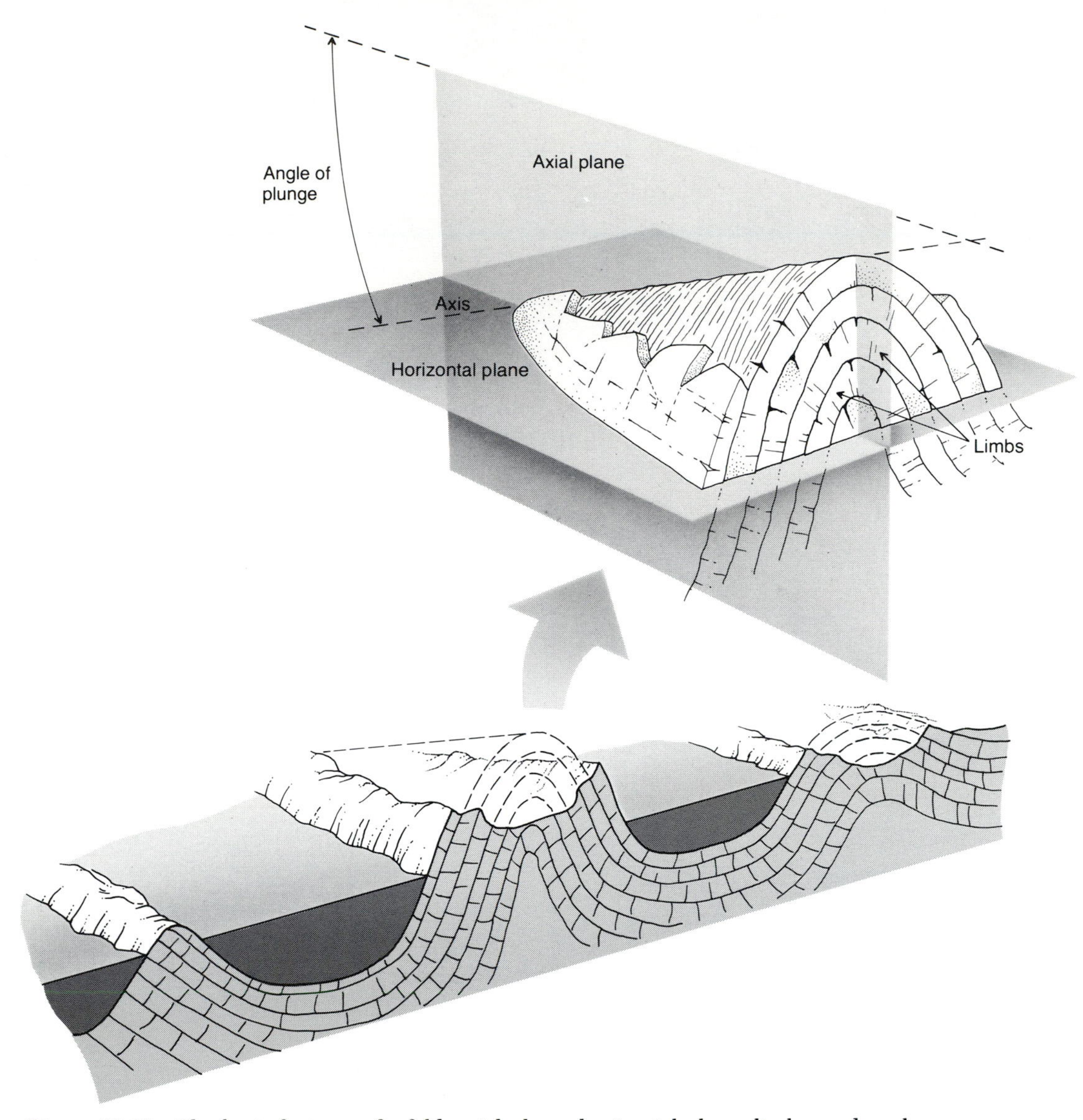

Figure 16.22 The basic features of a fold: axial plane, horizontal plane, limbs, and angle of plunge. If the axis is vertical, the fold is symmetrical; if tilted, it is asymmetrical.

Types of Faults and Folds

Earlier in this chapter we discussed faults in connection with the motion of tectonic plates. Three basic types of motion were described: tensional (pulling apart) in constructive zones, compressional (pushing together) in destructive zones, and lateral in conservative zones where the crust is displaced horizontally. The same types of motion also produce faulting on a smaller scale, and these faults are classed according to the direction of displacement of one wall relative to the other. A displacement that is up or down along the fault plane is called a *dip-slip* fault; one that is parallel to the fault line, as in the transform faults on the ocean floor, is termed a *strike-slip* fault. Displacements that combine strike- and dip-slip are *oblique-slip* faults (Fig. 16.23).

Three basic types of dip-slip faults are recognized: normal faults, reverse faults, and thrust faults (Fig. 16.24). In a *normal fault* the hanging wall is displaced downward relative to the footwall, exposing the upper part of the footwall in the form of a fault scarp. Many mountain ranges in the American West, including the Wasatch of Utah and the Sierra Nevada of California, represent normal faults. In the Sierras the fault scarp faces east; in the Wasatch the fault scarp faces west

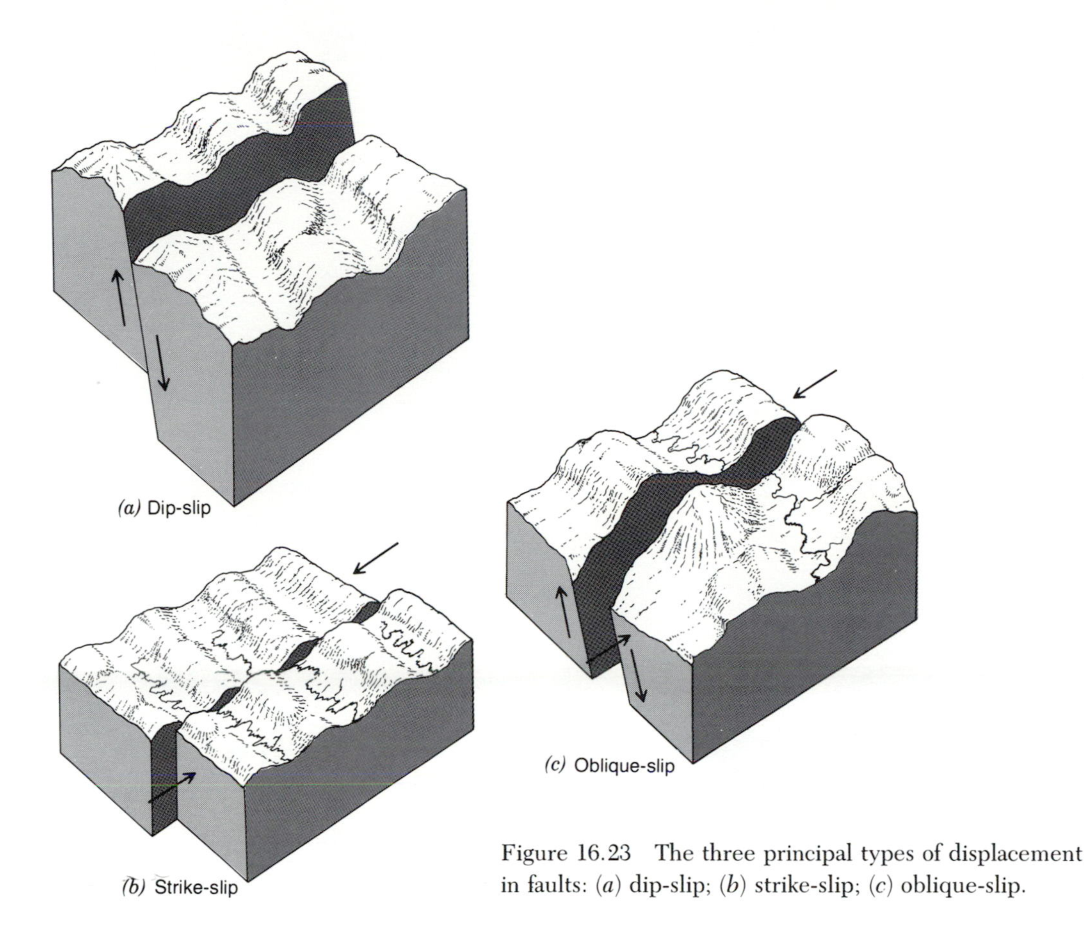

Figure 16.23 The three principal types of displacement in faults: (*a*) dip-slip; (*b*) strike-slip; (*c*) oblique-slip.

(see Fig. 16.25). Between the Sierra Nevada and the Wasatch lies an extensive area called the Great Basin, in which range after range of mountains has been formed by normal faulting.

The opposite displacement produces a *reverse fault* in which the hanging wall moves up relative to the footwall. If a reverse fault is subjected to great compressional force, and the fault plane is gently inclined, the resultant movement would be horizontal. This is a *thrust fault*. They are known to drive slabs of rock laterally for tens of kilometers over the surface. In some parts of the Rocky Mountains, rock formations were thrust great distances to form prominent front ranges along the Great Plains. The Canadian Rockies

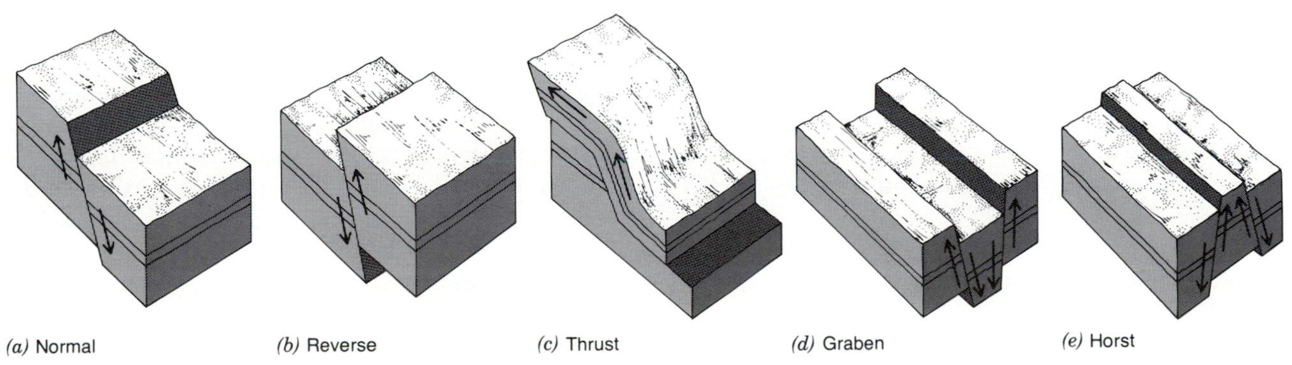

Figure 16.24 The main types of dip-slip faults: (*a*) normal; (*b*) reverse; (*c*) thrust; (*d*) graben; (*e*) horst.

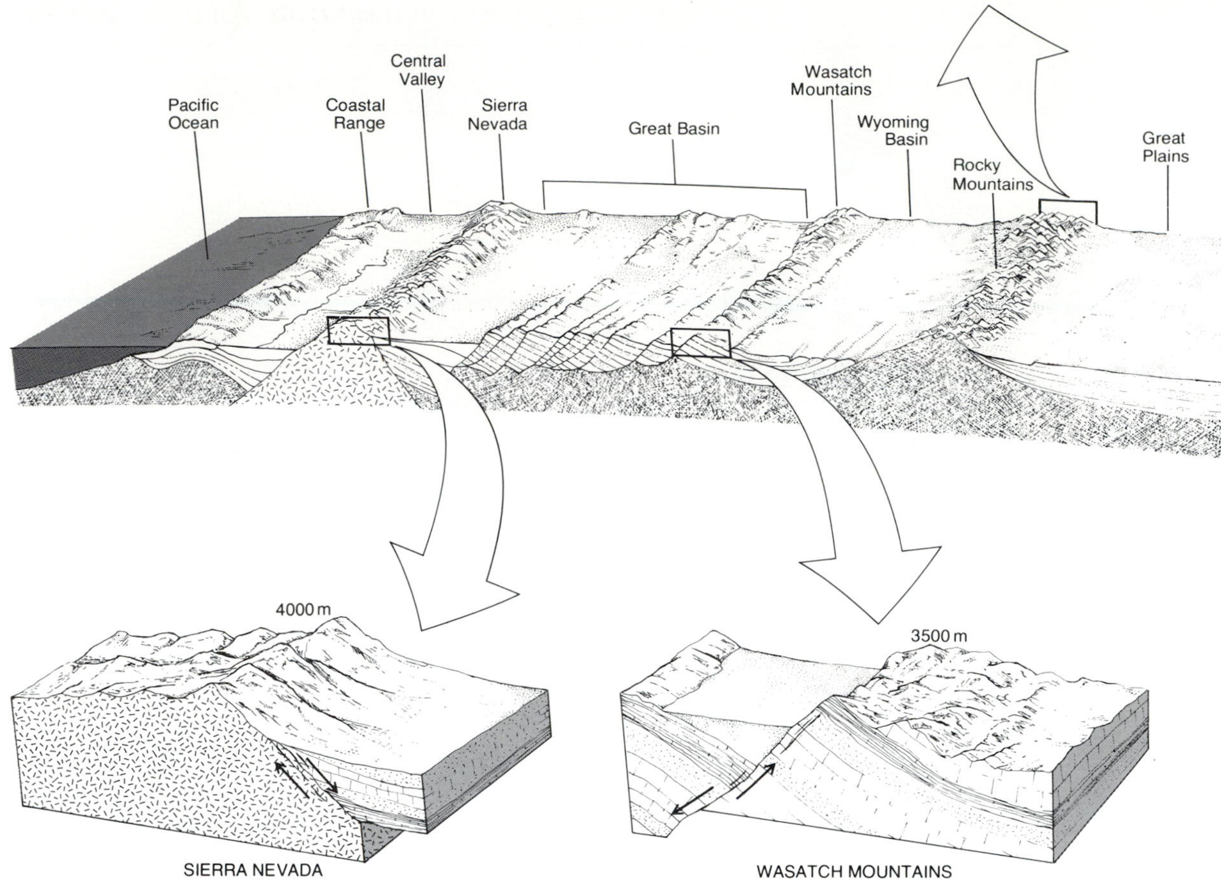

Figure 16.25 A geological diagram across North America from California to Maryland. Enlargements show the Sierra Nevada, the Wasatch Mountains, the Appalachians, and the Canadian Rockies.

in Alberta are a case in point; great slabs of rock were thrust eastward and piled against one another in an overlapping fashion (Fig. 16.25).

Finally, some dip-slip faults involve two fault planes. A rift, or *graben*, forms where a block is displaced downward between two normal faults. The opposite type of structure, called a *horst*, results in the elevation of a block between parallel fault planes (Fig. 16.24). Rifts and horsts are common features in zones of tensional motion. Today the most extensive system of rifts on earth begins in the Gulf of Aqaba on the east side of the Sinai Peninsula and extends through the basin of the Red Sea and into the West African Highlands, where it forms the basins for large inland lakes such as Lake Tanganyika and Lake Malawi (Nyasa) (Fig. 16.26).

The simplest folds can be described in two-dimensional terminology: monoclines, anticlines, synclines, and overturned folds (Fig. 16.27). A *monocline* is a single bend in an otherwise horizontal formation; an *anticline* is a double bend upward in the shape of an "A"; a *syncline* is the downward counterpart of an anticline. They may be either symmetrical or asymmetrical. An *overturned* fold represents such extreme bending that the formation on the lower part of the fold is turned upside down.

The Ridge and Valley section of the Appalachian Mountains contains spectacular examples of folding in sedimentary rocks (Fig. 16.25). The ridges and valleys trend north and south; when viewed in cross-section, they resemble a washboard terrain. On close examination, however, it is apparent that the topography produced by the folding deviates from the structural pattern of anticlines and synclines because valleys are formed from both types of folds. Synclinal valleys are easy to figure out, but in order to figure out the origin of anticlinal valleys, you must infer that the tops of the anticlines have been eroded away, exposing the underlying, often weaker, formations to erosion. The ridges are formed from the limbs of pairs of anticlines and synclines.

VOLCANISM

Volcanoes are emissions of magma, rock fragments (called *pyroclastics*), and gases, released through

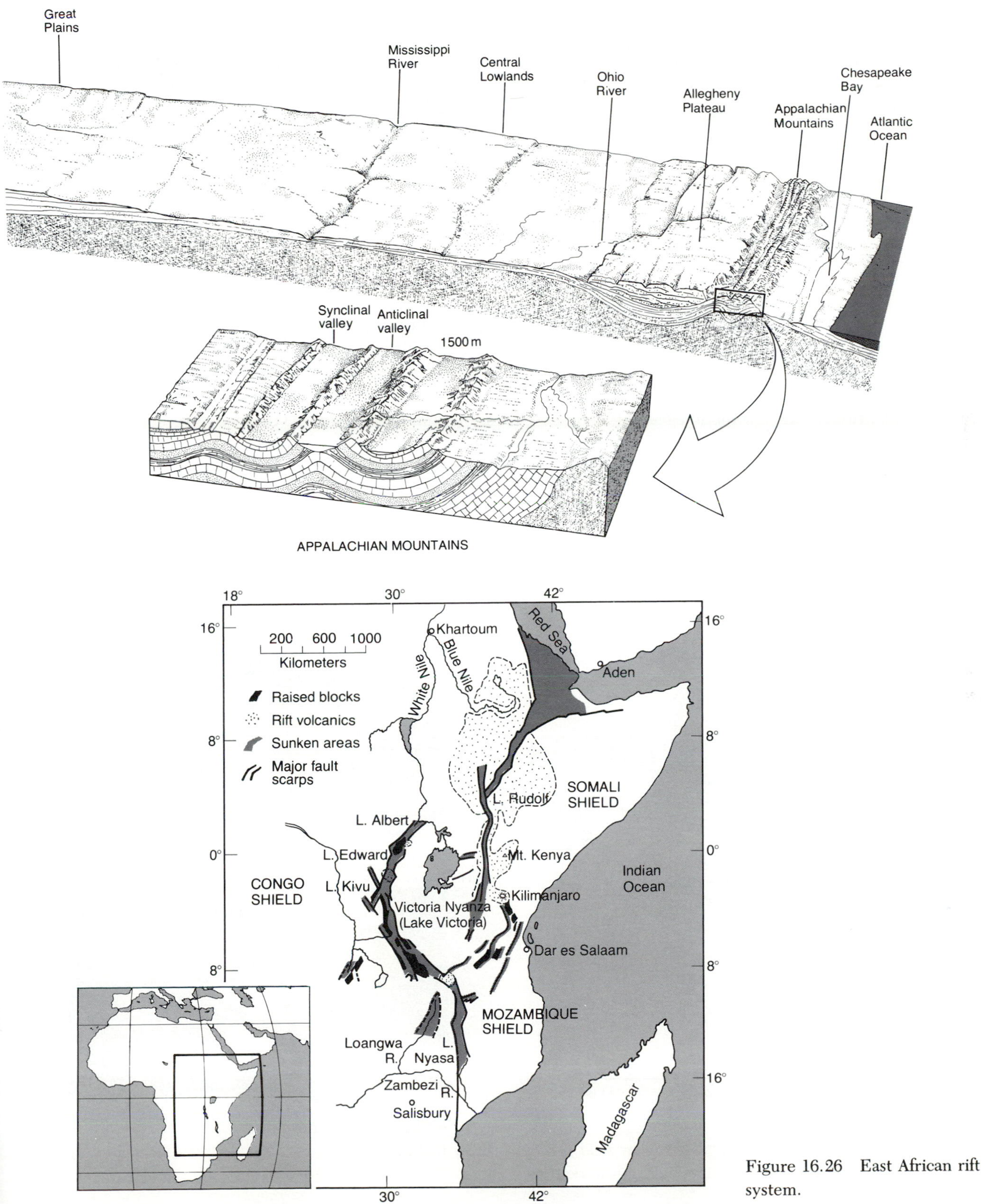

Figure 16.26 East African rift system.

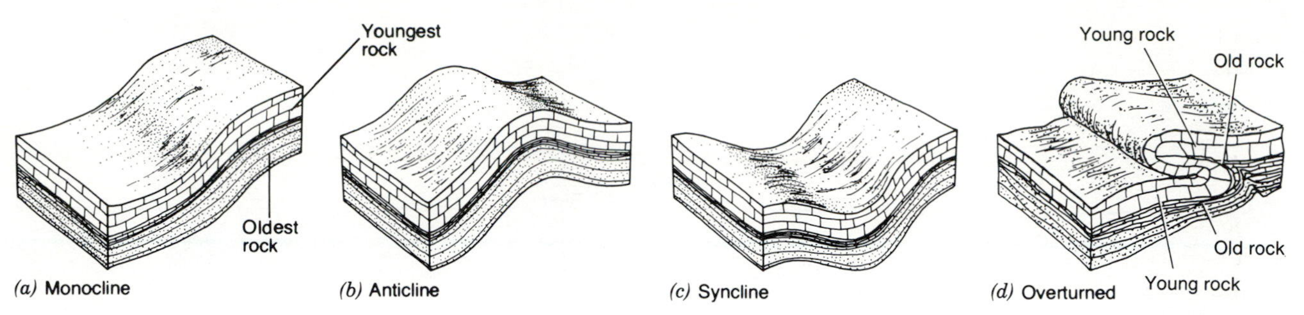

Figure 16.27 Four basic types of folds; (*a*) monocline; (*b*) anticline; (*c*) syncline; and (*d*) overturned.

openings in the earth's crust. Most volcanoes appear to be explosive in nature and produce more pyroclastics than lava. Although volcanism is common on the continents, the bulk of it occurs in the ocean basins along midoceanic ridges and subduction zones (Color Plate 11). With a few exceptions, volcanism on the continents is associated with orogenic environments.

Types of Volcanoes

Although we tend to think of volcanoes as conical-shaped mountains, they take on a great variety of forms. Some of the largest volcanoes are not mountains at all, but are more like plateaus. They are the result of massive outflows of lava, called *basalt floods*,

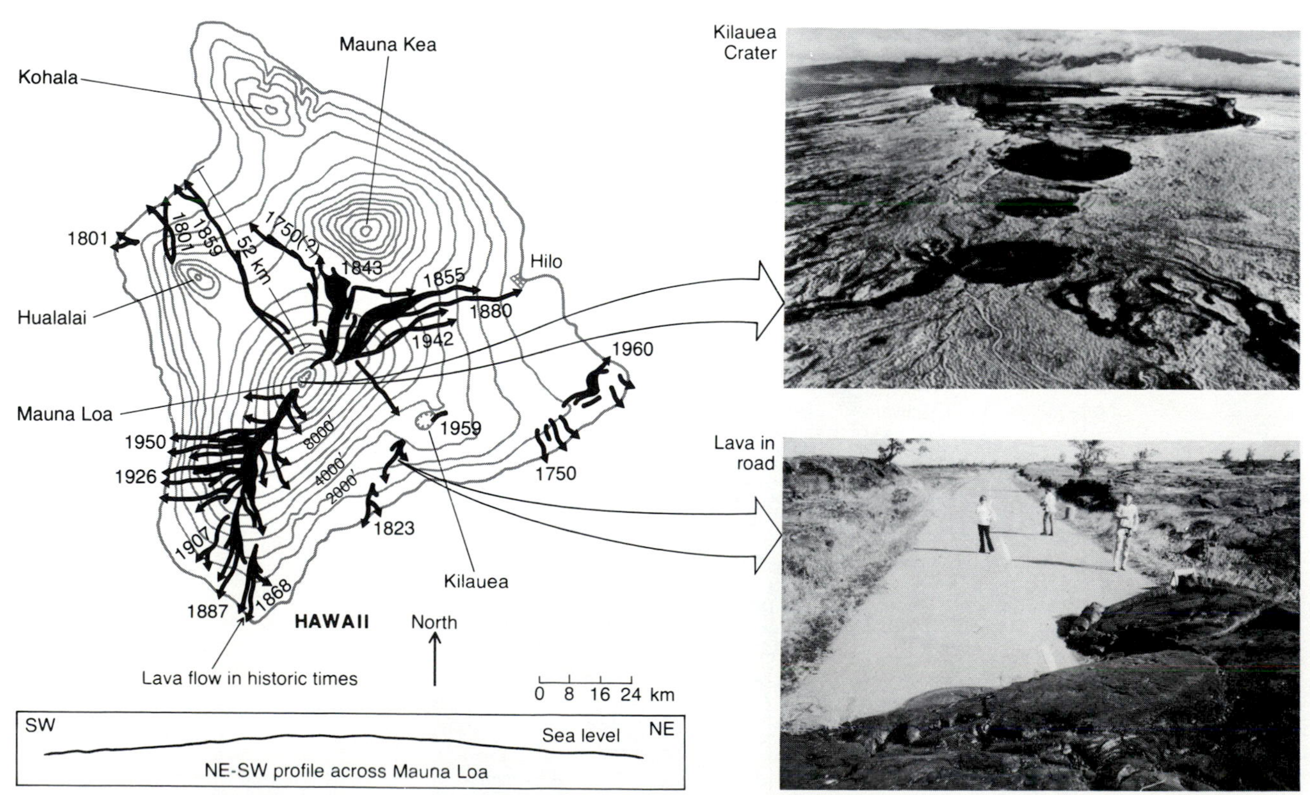

Figure 16.28 Contour map of Hawaii, showing the field shield volcanoes that form the island. The heavy lines represent lava flows during historic times.

from long, narrow openings, or fissures, in the crust. The lava spills over the landscape in a relatively thin sheet, typically covering hundreds of square kilometers with each eruption. The fluid behavior of the lava is attributed to its high temperature, often up to 1200 degrees C. On sloping surfaces it can flow at velocities exceeding 50 km per hour. The largest continuous area of flood basalts in the United States is the Columbia Plateau, which covers more than 400,000 km^2 (Fig. 15.13*c*).

If a fluid magma is released from a group of tunnels, called *vents*, which are chronically active, the lava tends to pile up, forming a broad mound called a *shield volcano*. Although shield volcanoes may grow to heights of several thousand meters, most of their growth is lateral, with lava issuing from the sides of the volcano in emissions called *flank eruptions*. Side slopes are gentle, generally less than 5 degrees, and are laced with lava flows of various sizes. The Hawaiian Islands are excellent examples of shield volcanoes. The main island, Hawaii, was built from five shield volcanoes which coalesced into a single mass as they emerged from the sea (Fig. 16.28). Like Hawaii, most shield volcanoes tend to form near the interiors of tectonic plates.

Composite volcanoes (also known as *stratocones*) are generally conical in shape and represent the popular stereotype of the "classical" volcano. Smaller and steeper than shield volcanoes, they also differ in composition; shield volcanoes are almost entirely basaltic lava, whereas composite volcanoes are made up of both lava and pyroclastic rocks. Pyroclastic material, which is described as ash and all sizes of rock particles, is commonly ejected into the air in the opening stage of the eruption. The coarse particles rain back to the surface of the volcano to form a layer of some thickness; the fine particles are carried off in the atmosphere and eventually fall out elsewhere. Later in the eruption or in a subsequent eruption, lava flows may bury the pyroclastic layer; thus, the lava and the pyroclastics become interbedded in a layer-cake fashion (Fig. 16.29). Many of the famous volcanoes belong to the composite class, including Vesuvius and Etna of Italy, Fuji-san of Japan, Ararat of Turkey, and the well-publicized Mount St. Helens of Washington (Fig. 16.30). Virtually all of the volcanoes in subduction zones are composites, which accounts for their abundance around the Pacific plate.

Composite and shield volcanoes often develop similar anatomies, characterized by a main passageway

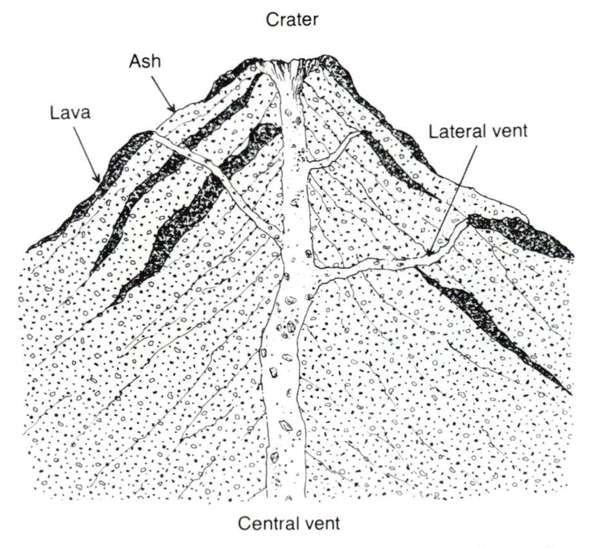

Figure 16.29 The basic structure and circulation features of a composite cone.

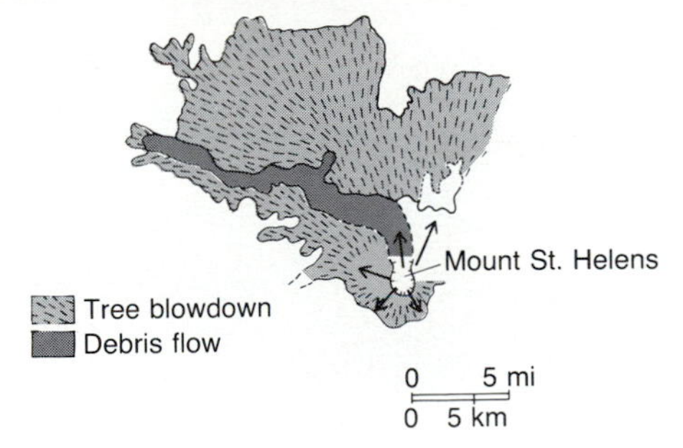

Figure 16.30 The Mount St. Helens, May 18, 1980, explosion and ash deposit, the most intensively studied and best documented volcanic eruption in history. The inset map shows the pattern and extent of trees blown down.

(called the *central vent*), multiple secondary vents (which lead from the central vent to the sides), and a *crater* at the summit (Fig. 16.29). The crater forms when magma held in the central vent is released through a lower vent, thereby causing the neck to retract. Such shifts in material can also break the hull of the volcano and create fissures that later serve as passageways for lava. A much larger depression, called a *caldera,* also forms in some volcanoes. Calderas form when a volcano collapses because of drainage of a large mass of magma through a lower vent or the expulsion of a massive amount of material in an explosive eruption. Such eruptions have been known to destroy the entire superstructure of a volcano.

Basic Mechanics of Volcanism

The underground mechanics of volcanism are in general poorly understood. Most likely, the heat that creates the magma is supplied mainly by the asthenosphere. When a pocket of magma develops at the base of the lithosphere, it begins to rise through the overlying rock like a parcel of hot air rising through the atmosphere, displacing its way upward into zones of colder material. Because of the huge quantities of heat needed to melt through rock, a great supply of magma must be available if the lithosphere is to be fully penetrated. However, once a hot conduit to the surface has been created, the ascent of additional missiles of magma appears to take place with comparative ease. With their plumping in place, individual volcanoes tend to remain active for a long time after they are born.

Most volcanic eruptions are not violent. Basalt flood eruptions, for example, typically involve little more than outpourings of lava onto the ground from fissures in the crust. Many composite volcanoes are born as modest discharges of gas and ash. The ash builds up to form a *cinder cone,* and only after months or years of activity does the eruption produce truly explosive emissions. Lava is not usually present in the early stages of a volcano's development but is more likely to appear after a sizable amount of ash has been deposited—often several years or decades after the volcano has been initiated.

Unquestionably, the most violent eruptions occur in established volcanoes that reactivate after a period of quiescence. The superstructure (hull) and the internal passageways (neck and vents) grow rigid with cooling during inactive periods, and when new magma begins to push upward, it meets resistance from these capping materials. Pressure builds, and, as in the case of Mount St. Helens, the volcano suddenly gives way at a weak point creating a great explosion. Figure 16.31 shows the ash flow created by one such explosion, Mount Katmai, Alaska, in 1912.

Figure 16.31 The Katmai, Alaska ash deposit laid down by a massive eruption in 1912.

Volcanic Events and the Landscape

Volcanoes, like earthquakes, can render sudden change in the landscape. Accordingly, volcanic events have received a good deal of attention, and the human drama associated with volcanic disasters has given rise to some of our most interesting history.

Vesuvius, Italy, A.D. 79 Vesuvius, a composite volcano located on the Bay of Naples in Italy, was, and still is, notorious for its misbehavior. One morning in the summer of A.D. 79 it erupted with a modest burst of ash, a common sight to the residents of Pompeii. Late the following day it became clear that this eruption was larger than usual, and the city's inhabitants began to flee. Before everyone could escape, however, Vesuvius suddenly released a huge mass of hot ash that literally buried people and animals in their tracks. Within several days most of the city was buried to such a depth that excavation and reuse in later years were considered all but impossible. Vesuvius has continued to remain active over the centuries; it has erupted eleven times since 1800.

Krakatoa, Indonesia, 1883 Krakatoa was a small, uninhabited volcanic island situated in an ancient caldera between Java and Sumatra. In May 1883 it began to emit ash, and on August 27 it exploded in what is generally considered the greatest natural blast ever witnessed. The energy released exceeded that of the atomic blast on Hiroshima by at least five times. The explosion was caused by the superheating (heating above the boiling point) of groundwater locked in the volcano until the volcano burst open like an overheated steam boiler. The effects of the blast are summarized in Table 16.2.

TABLE 16.2 MAJOR EFFECTS OF THE KRAKATOA BLAST, 1883

Type	Description
Atmospheric	Sound of explosion heard in Australia, more than 2000 km away
	Created air pressure waves recorded on other side of earth
	Airborne debris created almost total darkness during midday at Jakarta, 150 km away
	Dust that entered higher atmosphere increased backscattering of solar radiation, resulting in temperatures several degrees cooler over much of the earth
Oceanic	Generated giant wave nearly 40 m high which overwashed nearby coastal lands, killing 30,000 people
	Sea waves produced by explosion were recorded on tidal gauges as far away as the English Channel
Geomorphic	Discharged about 20 cubic km of debris into atmosphere
	Created a hole 300 m deep in the caldera

Mount Pelee, Martinique, 1902 The Mount Pelee eruption of 1902 also destroyed a city, but in a fashion much different from that of Vesuvius. Pelee is located on the heavily populated island of Martinique in the Caribbean. After a half century of dormancy, the volcano erupted with a series of mild explosions in the spring of 1902. Then, on the morning of May 8, it released a tremendous blast that sent a mass of incandescent gas and ash out the side of the volcano and toward the city of St. Pierre. The mass, called a *nuee ardente* (for glowing cloud), moved at a velocity approaching 200 km/hr, engulfing the city and killing its 28,000 inhabitants instantly. Only one resident survived, a prisoner in a basement cell of the jail (a convicted murderer who was pardoned and later became a missionary), but a number of sailors aboard ships in the harbor watched the event. One offered this description:

> The mountain was blown in pieces. There was no warning. The side of the volcano was ripped out and there was hurled straight toward us a solid wall of flame. It sounded like a thousand cannon. . . .
> The wave of fire was on us and over us like a flash of lightning. It was like a hurricane of fire. I saw it strike the cable steamship Grappler broadside on, and capsize her. From end to end she burst into flames and then sank. The fire rolled in mass straight down upon St. Pierre and the shipping. The town vanished before our eyes.
> The air grew stifling hot and we were in the thick of it. Wherever the mass of fire struck the sea, the water boiled and sent up vast columns of steam. The sea was torn into huge whirlpools that careened toward the open sea. . . . The blast of fire from the volcano lasted only a few minutes. It shrivelled and set fire to everything it touched. Thousands of casks of rum were stored in St. Pierre, and these were

exploded by the terrific heat. The burning rum ran in streams down every street and out into the sea.

. . . Before the volcano burst, the landings of St. Pierre were covered with people. After the explosion, not one living soul was seen on the land. Only twenty-five of those on board were left after the first blast.[1]

Summary

Plate tectonics is a model that describes the mechanics of the earth's lithosphere. It helps explain why the ocean basins, which earth scientists once thought to be the enduring, stable parts of the crust, are in fact nowhere more than 200 million years old, whereas the interiors of the continents are several billion years old. Moreover, the youngest oceanic rock is found near the ridge axis and the older rock at increasing distance from the ridge. This complements the geographic patterns of magnetic reversals in the rock of the ocean floor, which indicate that the crust is emerging from the midoceanic ridges. The plate tectonics theory holds that if the crust is emerging in certain zones, it must be consumed in others. Crust is consumed by the process of subduction. On plate borders where neither crustal emergence nor destruction is taking place, the borders are classed as conservative.

Orogenic belts are produced by massive deformation of rock near the edges of tectonic plates. In most mountain ranges, folding is the principal type of deformation, and it is usually accompanied by extensive faulting and volcanism. Folds and faults are classified according to the orientation of the structure from the horizontal and the relative direction of the displacement. The Ridge and Valley section of the Appalachian Mountains is a classic example of terrain that has resulted from the erosion of folded rock. The rift valleys of East Africa and the mountain ranges of the Great Basin of the United States are an example of terrain created largely by faulting. Terrain formed by volcanic activity is typically mountainous, and is exemplified by island arcs such as the Japanese Islands and seamounts such as the Hawaiian Islands. There are two main types of volcanoes, *composite* and *shield*, and they may produce explosive and nonexplosive eruptions at different times in their development. Violent volcanic eruptions, though not frequent, have produced some of history's most dramatic human disasters.

[1]K. H. Wilcoxon, *Chains of Fire—The Story of Volcanoes*, Chilton, 1966.

Concepts and Terms for Review

- continental drift
- plate tectonics
- continuity of the crust
- plate velocity
- plate size, location, and motion
- magnetic deviations
- plate borders
  - constructive
  - destructive
  - conservative
- subduction and volcanism
- sea floor spreading and ridge formation
- San Andreas fault
- transform fault
- earth heat flow
- heat sources
- mantle convection
- mountain building
  - volcanism
  - folding
  - faulting
- rock deformation
  - elastic
  - plastic
  - faults
- principle of original horizontality
- strike and dip
- properties of faults and folds
- types of faults
  - dip-slip
  - strike-slip
  - oblique-slip
- types of folds
  - monocline
  - anticline and syncline
  - overturned
- volcanism
- types of volcanoes
  - basalt floods
  - shield
  - composite
  - cinder cone
- mechanisms of volcanoes
  - magma movement
  - eruption
  - development
- volcanic events

Sources and References

Bird, J., and Bryan, I. ed. (1972) *Plate Tectonics.* Washington, D.C.: American Geographical Union.

Chapman, D.S. and Pollack, H.N. (1975) "Global Heat Flow: A New Look." *Earth and Planetary Science Letters* 23: pp. 23–32.

Decker, Robert and Decker, Barbara. (1981) *Volcanoes.* Freeman: San Francisco.

Elder, John. (1978) *The Bowels of the Earth.* Oxford, England: Oxford University Press.

Hallam, A. (1973) *A Revolution in the Earth Sciences.* Oxford, England: Clarendon Press.

Marsh, B.D. (1979) "Island-Arc Volcanism." *American Scientist* 67: pp. 161–172.

Phillips, Owen M. (1968) *The Heart of the Earth.* San Francisco: Freeman, Cooper.

U.S. Geological Survey, n.d. *Atlas of Volcanic Phenomena.* Washington, D.C.: Government Printing Office.

Wegener, Alfred. (1966) *The Origin of the Continents and Oceans* (translated by John Biram). New York: Dover.

Wilson, J.T., ed. (1970) *Continents Adrift—Readings from "Scientific American."* San Francisco: Freeman.

Wyllie, P.J. (1976) *The Way the Earth Works.* New York: John Wiley and Sons.

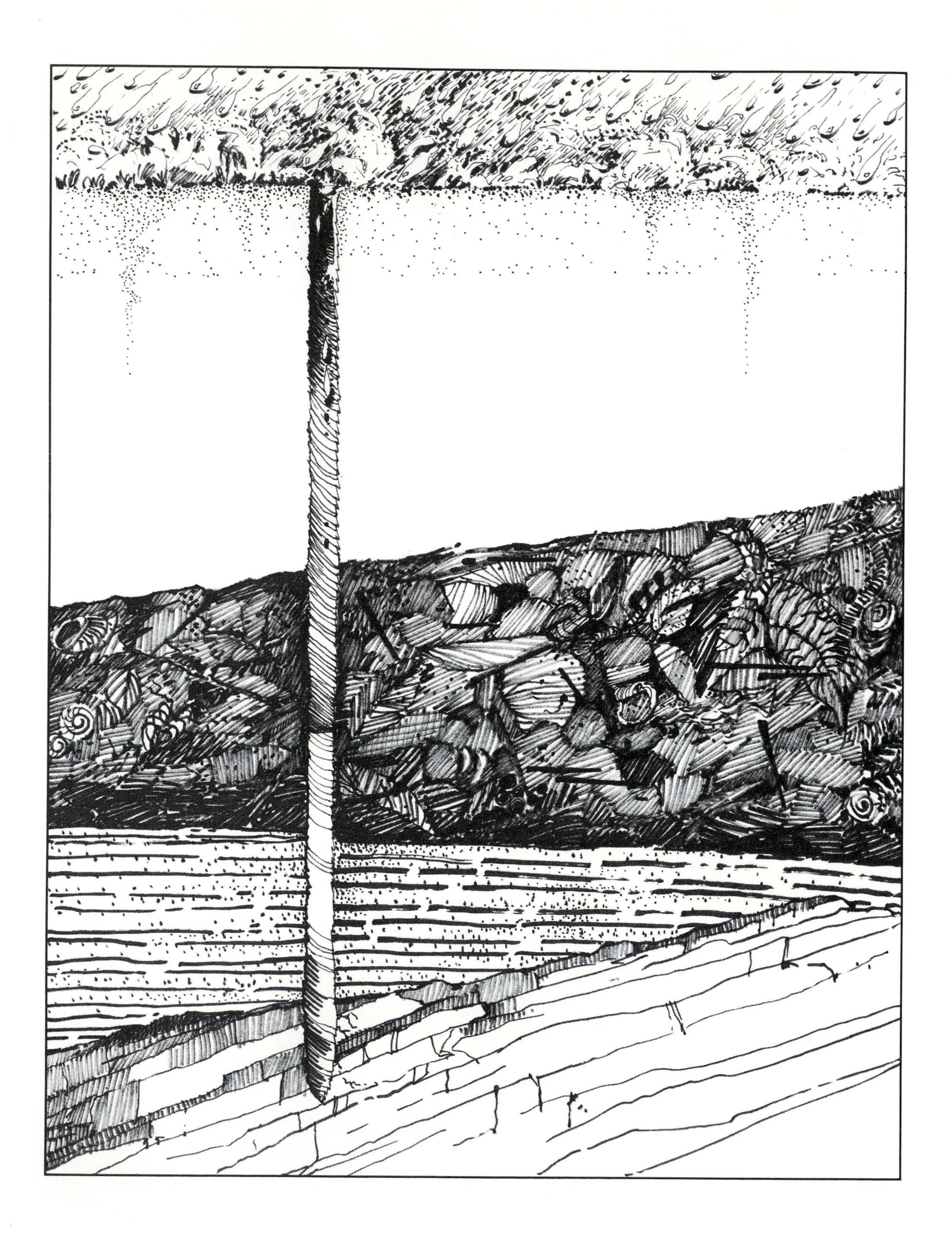

Chapter 17

Infiltration and Groundwater

An important scientific question is the destiny of the rain that falls on the land. How much rain stays on the surface and how much moves into the ground? What controls the proportions that go to each? What happens to the water that becomes groundwater? Does it build up? Does it move and is it replaced when we withdraw it for irrigation, industrial, and domestic uses?

The great platforms of rock that form the continents provide a roughly hewn foundation on which the rest of the landscape is built. If we are to understand how the other parts of the landscape—soil, vegetation, animals, water features, land use facilities, and the lower atmosphere—are arranged on this foundation and help to shape it, it is essential that we learn about the distribution processes and work of water. Water is essential to the weathering (disintegration) of bedrock, to the erosion of soil and rock, to the survival of all living organisms, and to the distribution of humans and their land uses.

Until several centuries ago the earth's crust was thought to be too impervious to allow the penetration of rainfall. The ancients thought that the water coming out of the ground into wells, springs, and stream channels came from other sources—possibly subterranean caverns linked to the sea—but definitely not from the atmosphere.

We now understand that the earth's crust is very permeable indeed. The upper crust is riddled with cracks and cavities that conduct and retain water pulled downward by gravity. On the land masses all of this water originates from atmospheric precipitation. Not all precipitation that falls on the land goes into the ground, however, because the crust is not uniformly permeable over the continents. Some water is stored on the surface in lakes and glaciers, and some runs off in streams and rivers.

Our intent in this and the next chapter is to examine the distribution and movement of water both on and within the land. In addition to learning about the principles and processes governing the movement and distribution of water, we also want to learn about water as it relates to land use and issues such as water use and flood hazards.

RECEPTION OF PRECIPITATION

Less than half (42 percent) of the water introduced to the land by atmospheric precipitation goes back to the sea as runoff. Most of it (58 percent) is lost to the atmosphere through vaporization. Many factors influence whether water is disposed of as vapor or runoff. Among the most important of these factors are various features and materials of the landscape (Fig. 17.1*a*).

As precipitation approaches the landscape, it encounters a series of barriers in the form of vegetation, land use features (buildings, etc.), soils, and bedrock. Each of these barriers dissipates, redirects, detains, or absorbs, and/or filters the water. In a heavily forested landscape, for example, most rainfall may be taken up on plant surfaces and by the soil. Any water that remains after the soil has taken its share passes on through the soil to become groundwater, which in turn slowly feeds into the streams. Runoff on the soil surface may be entirely absent.

Suppose that we cut the forest and clear the land. Now the balance between water moving into the ground and over the surface reverses: Soil-surface runoff increases and streams receive water more rapidly, whereas contributions to soil moisture and groundwater decline. Thus, we see that an understanding of water distribution, supplies, and features begins with an understanding of the landscape as a hydrologic agent.

Interception

In vegetated landscapes, plants intercept a portion of every rainfall before it reaches the soil surface. As the plant surfaces become wetted, water begins to trickle down the stems and trunks and drip from the foliage. These processes are called *stemflow* and *throughfall*, and in a heavily vegetated landscape, most precipitation gets to the ground in this manner (Fig. 17.1*b*). The difference between the amount of rain falling on the vegetation and that actually getting to the ground is termed *interception*, and it is normally expressed as a percentage of the total rainfall.

A dense forest canopy may intercept the total fall of a light rain, preventing any moisture from getting to the ground. In general, however, it appears that interception by forest canopies amounts to 10 to 25 percent of the gross annual precipitation in most forested parts of the world. Ground plants (e.g., grasses and crops) can also be effective in interception; a healthy stand of spring wheat or alfalfa, for instance, may intercept as much as 30 percent of the rainfall during the growing season.

Infiltration

What happens to precipitation once it reaches the ground? There are four possibilities: It may (1) evaporate, returning directly to the atmosphere; (2) sit on the surface in puddles, ponds, or snow; (3) soak into the soil; or (4) run off over the surface into streams, lakes, and wetlands. Which takes place depends on the infiltration capacity of the soil, the intensity of rainfall, the lay of the ground, and the atmospheric conditions, such as wind and air temperature.

The water-absorption function of the soil surface is termed *infiltration capacity* (f) and is defined as the rate at which surface water can be received by the soil. Infiltration capacity is equal to the depth of water received divided by the time taken to absorb it and can be expressed by the formula:

$$f = \frac{d}{t},$$

where d is depth of water received in inches or centimeters and t is time in minutes or hours.

The intensity of rainfall is the rate at which water reaches the ground, and like f, it is measured in centimeters or inches per hour. When the rainfall intensity exceeds the infiltration capacity, runoff is generated on the surface of the ground. This runoff is called *overland flow* and is characterized by thin sheets and tiny trickles of water that flow over the ground toward stream channels. Runoff processes are taken up in the next chapter; in this chapter we focus on the process of infiltration and the movement of water within the soil and at greater depths.

The infiltration capacity of any surface is influenced by many factors; some of these are relatively permanent, whereas others change seasonally or with weather conditions.

Soil texture One important determinant of infiltration capacity is the size and interconnectedness of spaces among the organic and mineral particles in the surface layer of soil. Large spaces that are linked together in extensive networks of open spaces tend to maximize infiltration. As any builder of sand castles knows, coarse-grained materials, such as beach sand, have high infiltration capacities, especially if the sand is well sorted, that is, of relatively uniform grain sizes. Conversely, fine-grained soils, such as clay, have low

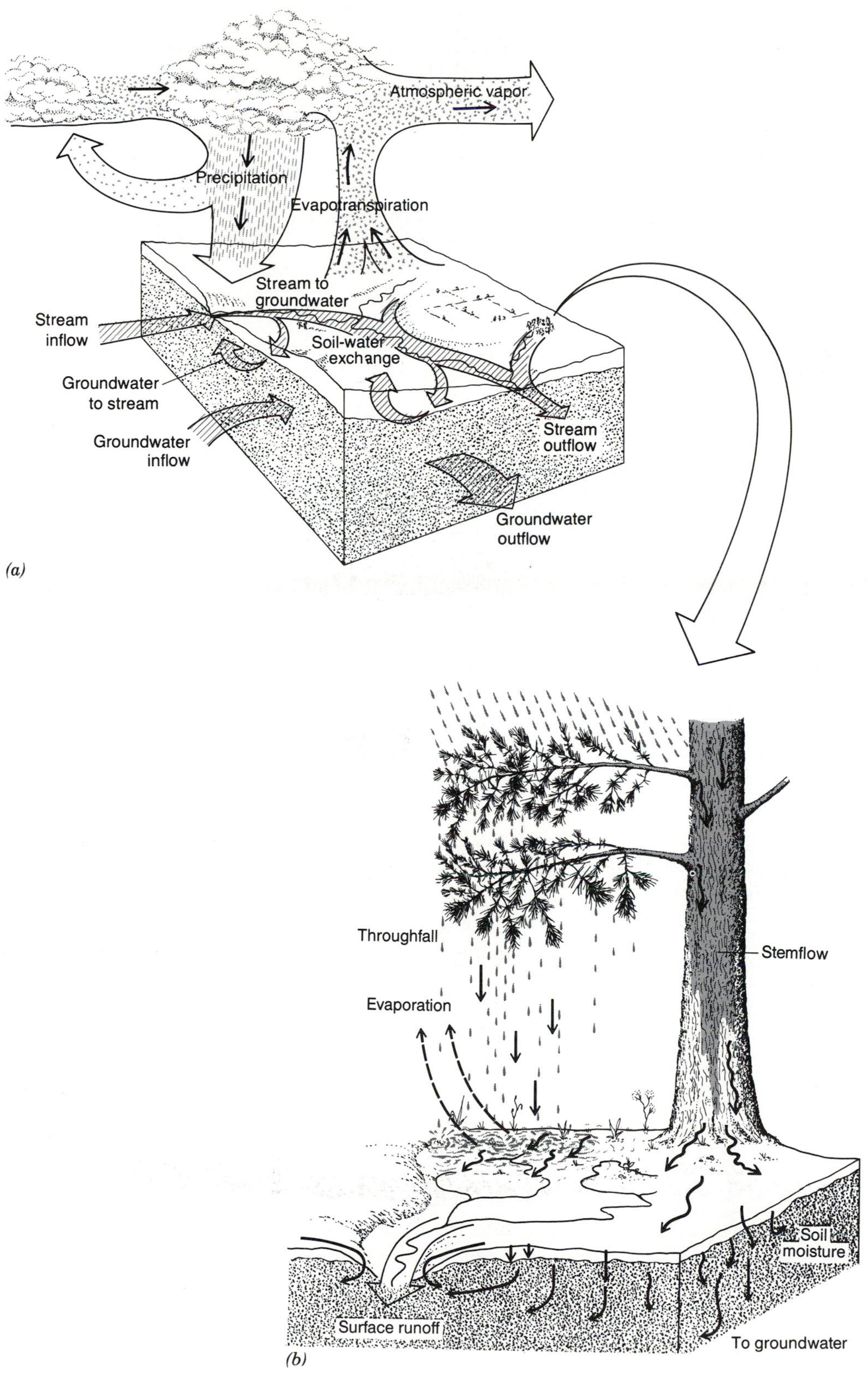

Figure 17.1 The hydrologic cycle: (*a*) the major inflows and outflows of water from a parcel of landscape; (*b*) a detailed version of these flows in a forested site within that parcel.

TABLE 17.1 REPRESENTATIVE INFILTRATION VALUES

Particle Size (Diameter)	Infiltration Rate (mm/hour)
Sand (0.05–2 mm)	10–100
Silt (0.02 to 0.05)	2–15
Clay (smaller than 0.002 mm)	0.2–7

TABLE 17.2 REPRESENTATIVE INFILTRATION CAPACITIES OF VARIOUS GROUND COVERS

Ground Cover	Infiltration Rate, cm per Hour
Old permanent pasture	6.0
Permanent pasture, moderately grazed	2.0
Strip-planted crops	1.0
Weeds or grain	0.8–10
Clean, tilled soil	0.6–0.8
Bare ground, crusted	0.4–0.6

infiltration capacities that are typically less than 7 mm per hour (Table 17.1).

Soil moisture Within the soil, moisture resides mainly in the spaces between the soil particles; therefore, when the upper soil is dry, there is generally more space available for infiltrating water. As the soil takes on water during a rainstorm (or snowmelt), however, the interparticle spaces often fill up faster than the water can be drained from the soil. As a result, with time infiltration capacity typically declines into the storm. The same holds true for several rainfalls in short succession (Fig. 17.2). As infiltration decreases, the porportion of rain converted to surface water increases correspondingly.

Vegetation Another important set of controls on infiltration is vegetation and the associated organisms that live in the soil. The roots and stems of plants loosen soil, facilitating water penetration of the surface. Burrowing creatures, such as worms and moles, make passageways that also serve as water-entry routes. This fact is demonstrated by the data in Table 17.2; for a given soil, the infiltration capacity of an herb-covered field (old permanent pasture) is nine to ten times greater than that of barren ground. Moreover, on bare soil the physical impact of the raindrops compacts the soil and reduces infiltration capacity. Vegetation virtually eliminates this effect. Figure 17.3 illustrates the influence of vegetation on the infiltration capacities of a sloping surface of three different soil textures.

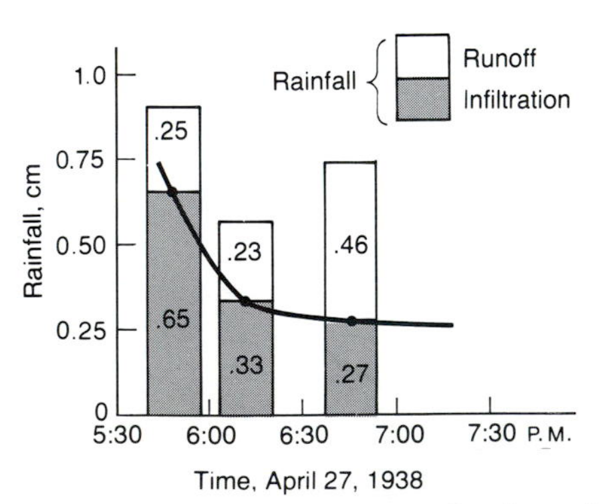

Figure 17.2 Infiltration rates for three short rain showers that fell on a 2.7-acre plot of ground within about 70 minutes. Note that the amount of water lost to infiltration decreases sharply from the first to second shower, but decreases only slightly from the second to third shower. This trend, shown by the graph line, is typical of most soil surfaces.

Topography Slope is an important control on the velocity of overland flow and in turn on infiltration. On steep slopes, overland flow moves rapidly off the surface; residence time on the surface is short, thereby limiting the chances for absorption. On gentle or flat surface, residence times are much greater, allowing more total time for infiltration.

Seasonal Factors Several seasonal factors may have a strong influence on infiltration. If freezing takes place when the upper soil is at or near saturation, infiltration can be reduced to zero. Under such conditions, interparticle spaces become blocked with ice and divert essentially all rainwater or snowmelt water into runoff (Fig. 17.4). In the Arctic, permafrost has a similar effect on percolating water. It is not surprising, then, that the soil layer above the permafrost is usually saturated or covered with ponded water in summer.

As the water content of the upper soil changes seasonally, so may infiltration capacity. In clayey soils, particularly those composed of the clay mineral montmorillonite, the addition of water may produce par-

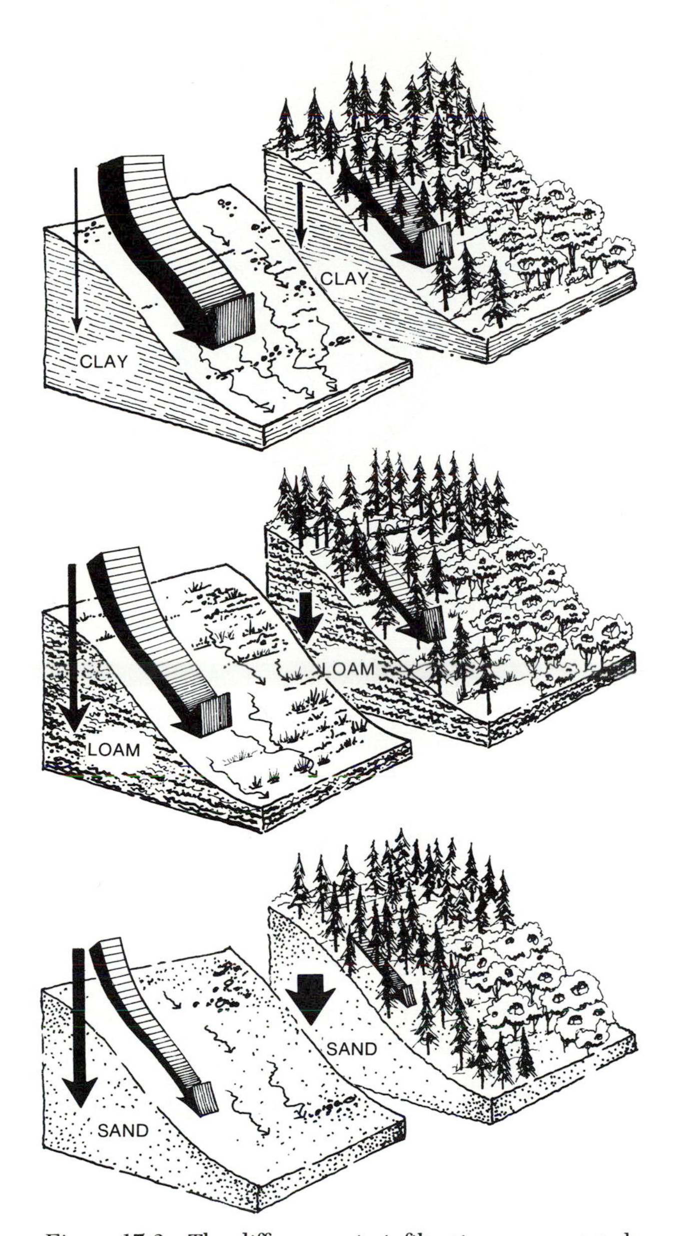

Figure 17.3 The differences in infiltration on vegetated and nonvegetated soils. The widths of the arrows indicate the relative amounts of water that infiltrate and run off.

ticle swelling and closing of interparticle spaces. In contrast, summer drying may produce contraction and cracking, although this may also be accompanied by the formation of a clayey crust on the surface. These factors, combined with the seasonality of plant growth and the ground frost of winter, characteristically render changes of 25 to 50 percent in infiltration for periods ranging from a few days to several months.

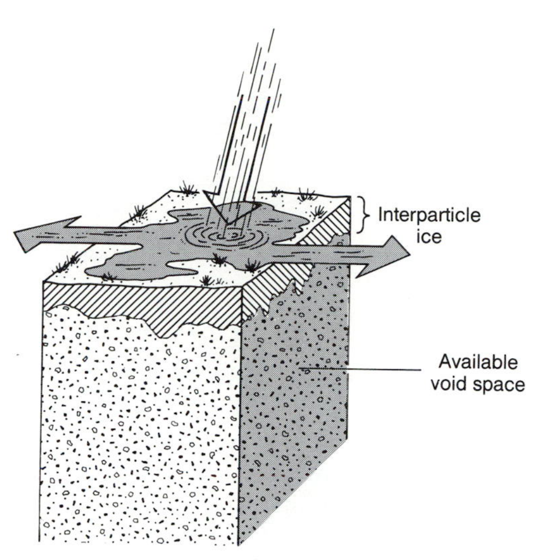

Figure 17.4 Only a thin layer of interparticle ice can reduce infiltration capacity to zero. In the northwestern United States, such conditions have been known to contribute greatly to flooding by locking out water from snowmelt and rain, thereby forcing essentially all of it into streams and rivers.

WATER IN THE SOIL

As we saw in Chapter 13, when infiltration water enters dry soil, some of it clings to the walls of particles. Microscopic examination shows that it attaches itself to an existing film of water molecules on the walls of the particles. This film is termed *hygroscopic water*, and it is held to a particle surface by molecular forces so powerful that it cannot be removed from the surface by natural processes.

Capillary Water and Field Capacity

The water that attaches to the hygroscopic film is *capillary water*. As the layer of capillary water on a particle grows in thickness, the molecular forces that hold it to the particle grow weaker. Eventually, the outer water molecules can no longer be held to the particle and yield to gravity, flowing down to the next particle.

The heaviest concentrations of capillary water occur at the contacts between soil particles (Fig. 17.5). The greater the number of particle contacts in a given volume of soil, the greater the capillary water-holding capacity. It follows that fine-textured soils can hold more capillary water than can coarse-textured soils. The capillary water-holding capacity of soil is termed

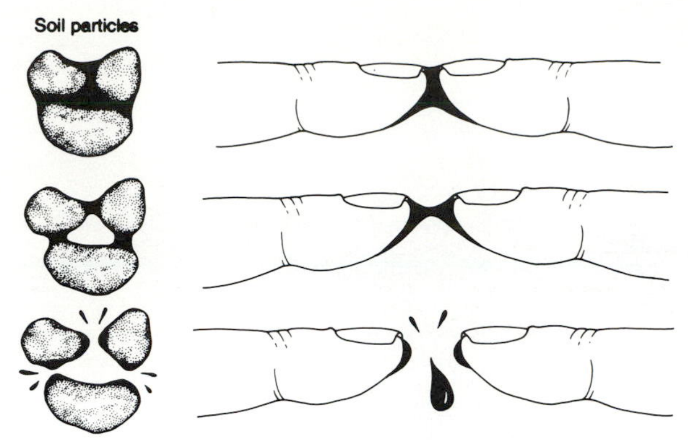

Figure 17.5 The tendency for capillary water to concentrate at the contact between particles is illustrated by this demonstration using wetted fingers.

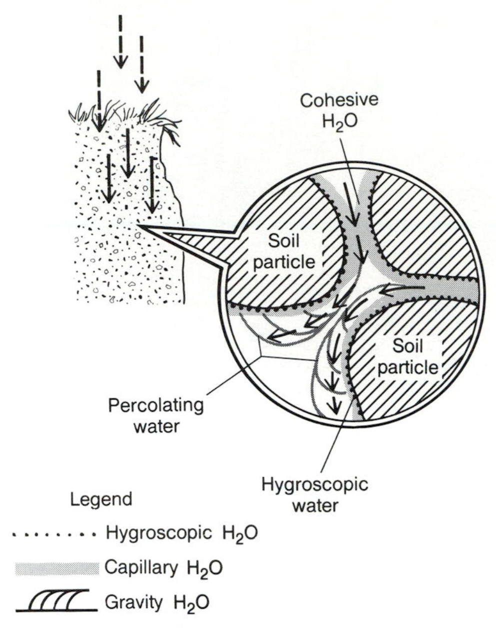

Figure 17.6 A film of cohesive water being deposited on the hygroscopic film as infiltration water percolates between soil particles.

field capacity. Table 17.3 gives the field capacities for four soil textures and an organic soil, peat. Peat has an extraordinary holding capacity because the organic particles themselves absorb water in addition to holding capillary water around and between them.

If water percolates into a soil until the capillary reservoir is full, the soil is said to be at *field capacity*. If even more water is added, it must pass through the soil because the molecular force is inadequate to hold it against gravity. This water is called *gravity water* because it flows with the gravitational gradient (Fig. 17.6).

Soil Permeability

Gravity water occurs in the liquid state and flows in the part of interparticle space not occupied by capillary water. In coarse-textured soils, where these spaces are large and well connected, gravity water is transmitted rapidly. Such soils are said to have high *permeability*. It follows that soils with high permeability should also have low field capacities, for they have fewer particle contacts and less total particle surface area than fine-textured soils do.

Permeability is determined in the laboratory by measuring the time it takes for water to pass through a specified volume of soil. A standard field measurement of permeability is called the *percolation test*. This involves first excavating a small pit and filling it with water. The water is allowed to drain, then the pit is refilled with water, and the drop in water level is timed until the pit is empty (Fig. 17.7). This test is generally used to determine the suitability of a soil for domestic (household) wastewater absorption, which is the principal means of sewage disposal in rural and many suburban areas in the United States and Canada.

TABLE 17.3 FIELD CAPACITIES OF FIVE TYPES OF SOIL

Texture	Capillary Water Capacity[a]
Sand	7
Sandy loam	15
Silt loam	25
Clay	40
Peat	170

[a]Expressed as a percentage based on the weight added to dry soil by capillary water.

Wastewater Absorption

Wastewater absorption systems are usually made up of two parts: a holding, or *septic*, tank where wastewater is collected; and a connecting *drainfield* through which the wastewater is dispersed into the soil (Fig. 17.8*a*). The soil performs two important functions: (1)

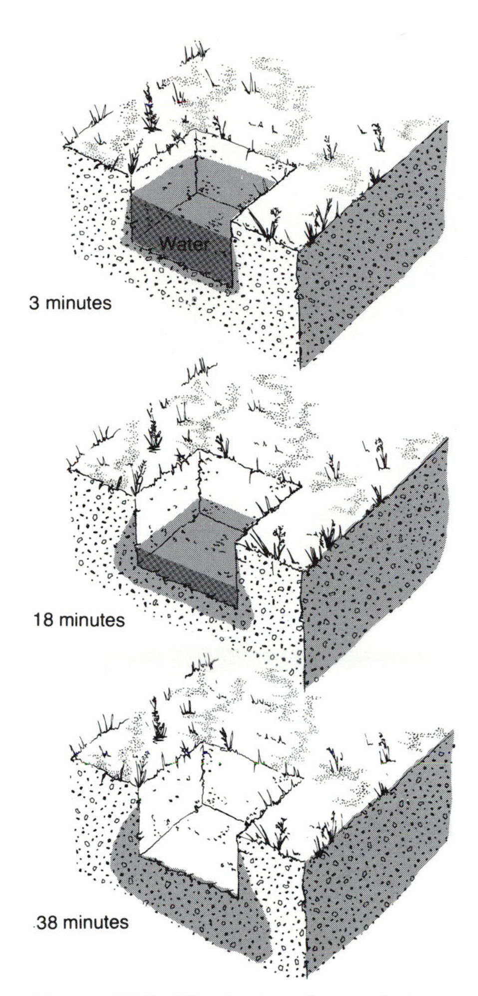

Figure 17.7 The basic soil percolation test for wastewater disposal provides a simple measure of soil permeability.

removal of solid debris and certain dissolved chemicals (such as phosphorus); and (2) elimination of certain disease-causing agents (mainly bacteria). Two soil traits are desirable for efficient wastewater disposal, namely, a good percolation rate (that is, high permeability because it is necessary to get the contaminated water below ground as fast as possible) and good filtering capacity because it is necessary to get rid of particles, chemicals, and bacteria before the water seeps into streams or groundwater. These call for two contrasting soil characteristics; coarse textures for good percolation, and fine textures (clays) for removal of contaminants. The best soils, therefore, represent a compromise between the two textures and are usually composed of intermediate textures, such as sandy loam (Fig. 17.8*b*).

GROUNDWATER

If water percolates into a soil that is already at field capacity, it flows slowly through the soil into the subsoil, eventually reaching a zone where all of the interparticle spaces are filled with water. This zone is called the *zone of saturation*, or *groundwater zone*. Although this water moves mainly downward through the soil materials, it may also move laterally where the terrain slopes steeply or impervious materials impede its downward flow. Where the soil and subsoil materials are deep, several belts of this water, each representing a different rainfall, can often be found moving in succession downward through the capillary zone.

Groundwater is an accumulation of gravity water that fully saturates the interparticle voids. The total amount of groundwater that can be held by any material (soil, surface deposits, or rock) is controlled by the material's *porosity*, which is a measure of the sum total of soil-void spaces. Porosity commonly varies from 5 to 20 percent in soils and near-surface bedrock. Figure 17.9 depicts the porosities of some representative earth materials near the surface. In general, porosity decreases with depth because the pressure of the rock overburden closes out void spaces at depth. At depths of several thousand meters, it is typically less than 1 percent.

The boundary between the soil and groundwater zones is the *water table*. In coarse-grained soils the water table approximates a boundary line, but in fine-grained soils it has more the aspect of a transition zone. This zone, called *capillary fringe*, is a layer ranging from a few centimeters to several meters in thickness across which moisture is transformed from gravity to capillary water as it moves upward in the soil during dry periods. The movements of capillary and gravity water in relation to the water table are schematically illustrated in Figure 17.10.

Groundwater Flow

If we trace the water table across the land, we will find that where the subsurface material is fairly homogeneous the elevation of the water table changes with the elevation of the land. Barring special sub-

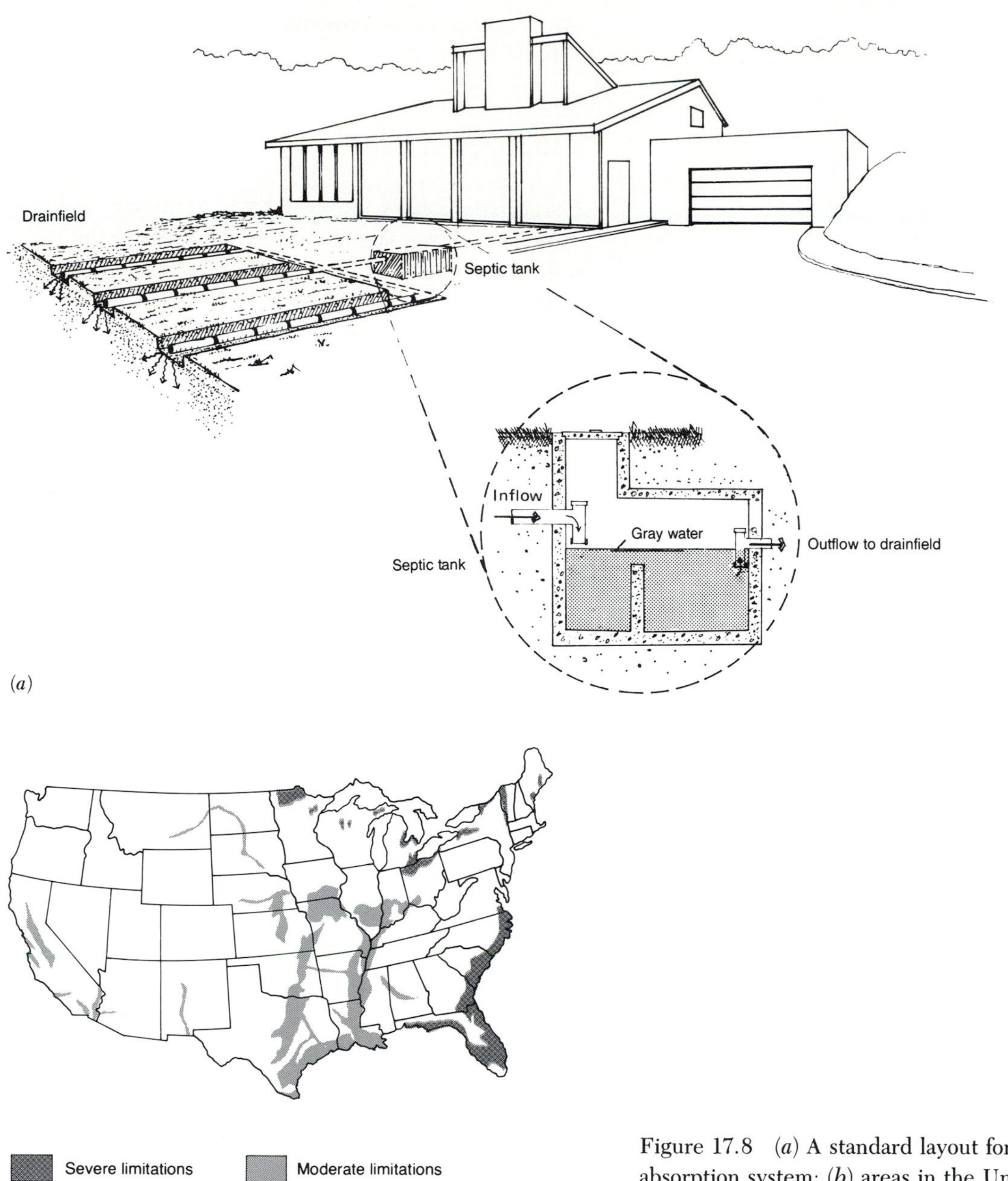

Figure 17.8 (*a*) A standard layout for a wastewater absorption system; (*b*) areas in the United States with soil limitations for wastewater absorption.

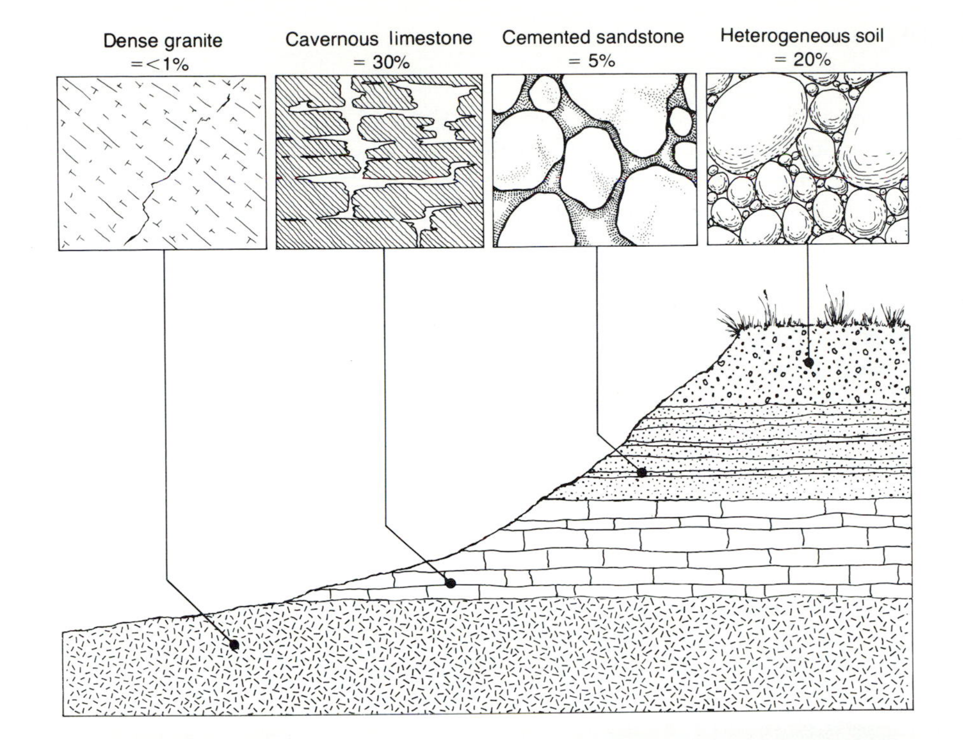

Figure 17.9 Porosities of four materials near the surface. In all materials, porosity decreases with depth because the great pressure exerted by the soil and rock overburden tends to close the voids. Porosity can be as high as 45 percent in some near-surface bedrock but it is rarely higher than 2 to 3 percent at depths exceeding 2000 m.

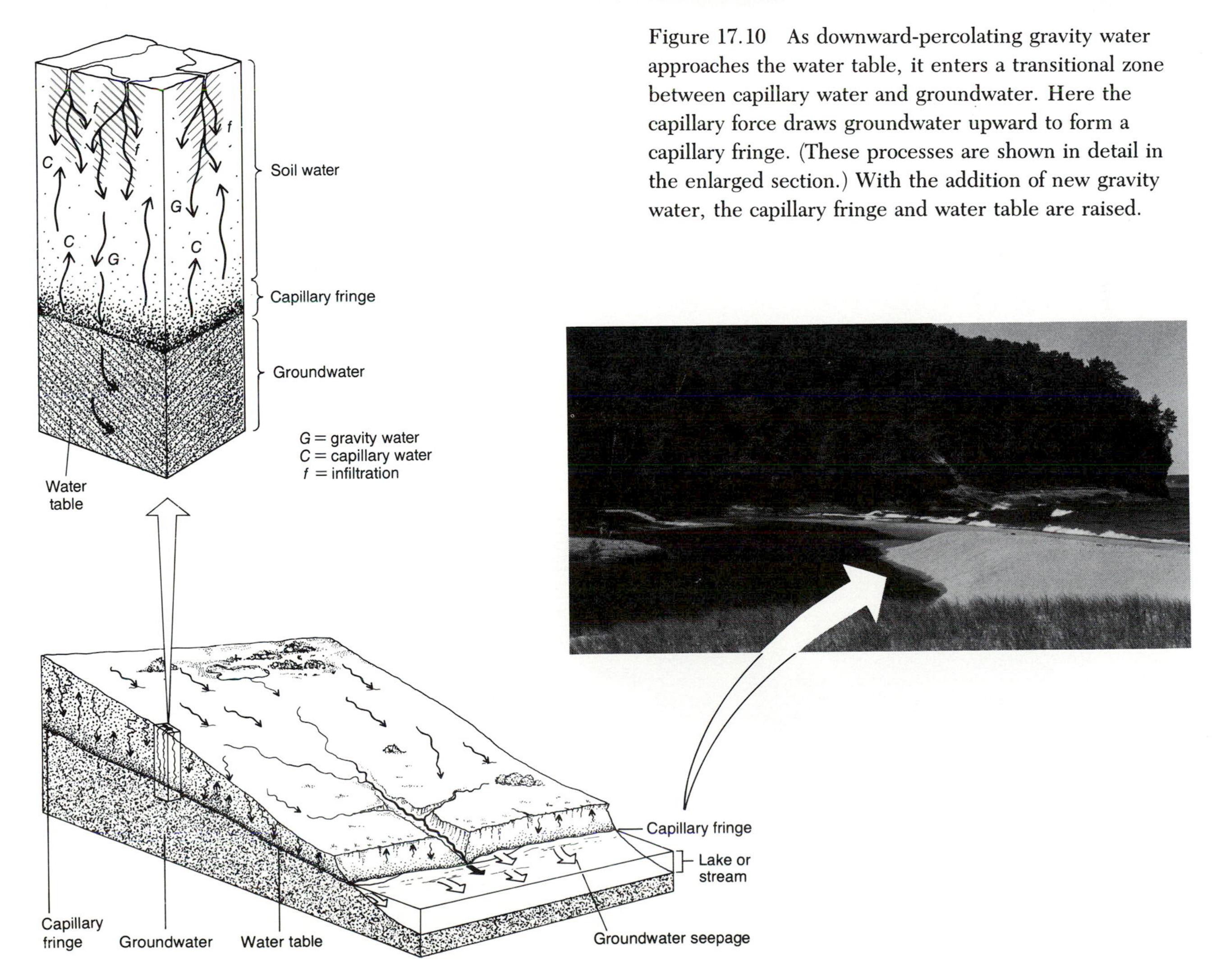

Figure 17.10 As downward-percolating gravity water approaches the water table, it enters a transitional zone between capillary water and groundwater. Here the capillary force draws groundwater upward to form a capillary fringe. (These processes are shown in detail in the enlarged section.) With the addition of new gravity water, the capillary fringe and water table are raised.

surface conditions where it may be elevated, or *perched*, on a soil stratum of low permeability, the *relief* of the water table (i.e., variations in its elevation), however, is typically much less than that of the overlying ground (Fig. 17.11). Nonetheless, the water table almost everywhere is inclined to some degree; therefore, gravitational gradients are set up beneath the landscape along which groundwater slowly flows.

These gradients are termed *hydraulic gradients* (I), and an individual gradient can be calculated by dividing the elevation difference between two points on the water table by the horizontal distance separating them. The inset in Figure 17.11 portrays the dimensions used in this calculation. The necessary measurements of the water table elevations are usually made by drilling test holes at two or more elevations on a hillslope.

Given a hydraulic gradient in some material, the maximum rate of flow which groundwater can attain is governed by the resistance the material poses to water movement. This is represented by the permeability of the material, a fact discovered more than 150 years ago by the French hydrologist Henri Darcy. The expression for the velocity (V) of groundwater flow

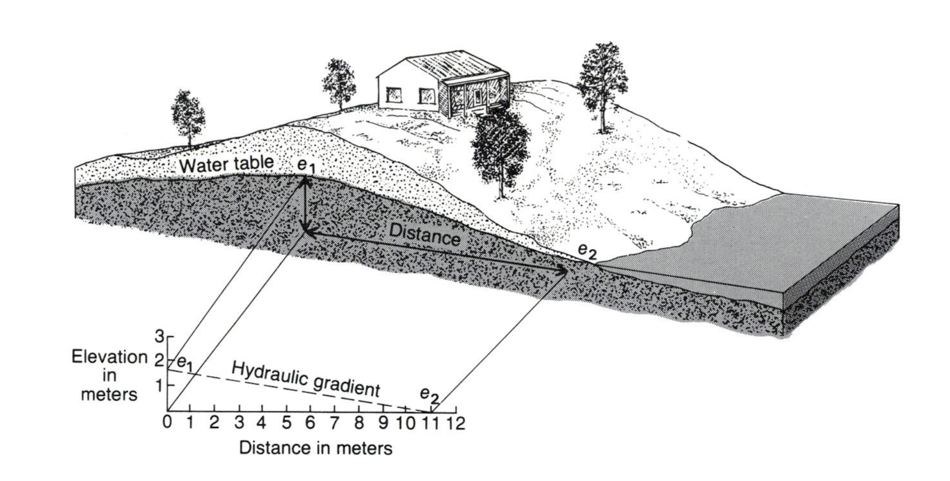

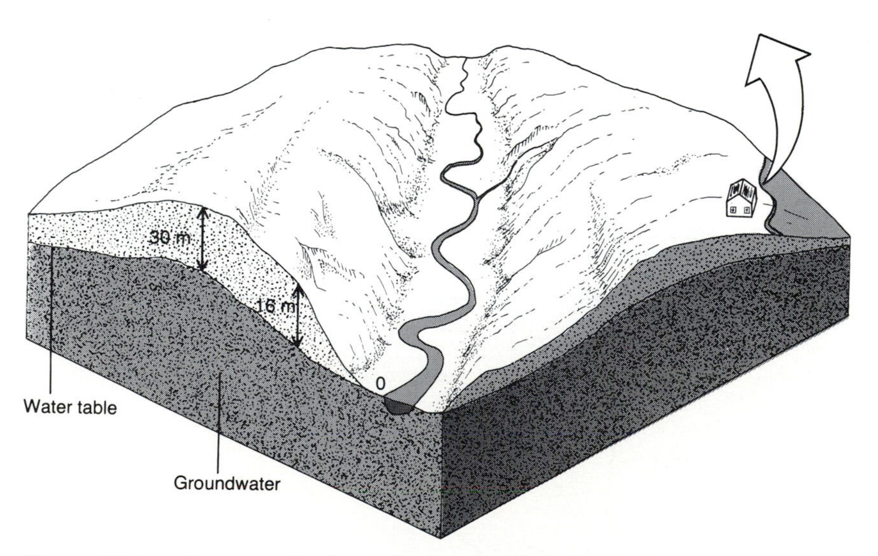

Figure 17.11 The general relationship between the configuration of the water table and that of the overlying terrain. The variation in the elevation of the water table is usually less than that of the land surface, resulting in an outflow of groundwater. The inset illustrates the measurements involved in calculating the hydraulic gradient.

has since come to be known as *Darcy's law:*

$$V = \text{Permeability} \times \text{Hydraulic gradient}.$$

How fast does groundwater flow? Compared to flow rates of surface water, flow rates of groundwater are very slow in most materials; 15 m per year is an average velocity for most groundwater. But velocities vary widely. In certain highly permeable materials, such as coarse gravels, tests reveal that groundwater can move at velocities as great as 125 m per day, but this is uncommon.

Recharge and Transmission

In all materials, the hydraulic gradient tends to vary with permeability and groundwater recharge. *Recharge* is the term given to gravity water supplied to a body of groundwater from surface sources, such as swamps and lakes (Fig. 17.12*a*). If a material of low permeability receives rapid recharge, groundwater builds up because the conduction of water is relatively restricted. As a result, the hydraulic gradient steepens, but as it does, so does flow velocity, according to Darcy's law. Under such conditions, the hydraulic gradient will continue to rise until the rate of flow, or *transmission*, is equal to the rate of recharge. Where recharge is low and permeability is high, of course, the hydraulic gradient will fall until the rates of transmission and recharge are equal (Fig. 17.12*b*).

Under these conditions, the hydraulic gradient falls to a level at which the pressure exerted by the weight of the water higher on the gradient is just adequate to sustain a balance between recharge and transmission. This is a good example of the dynamic-equilibrium principle, whereby a flow system is continuously tending toward a state of equilibrium, or what in physics is called a *steady state*. Owing to the steady-state principle, it is normal in most areas to find steep hydraulic gradients in materials of low permeability and gentle hydraulic gradients in materials of high permeability. An understanding of this concept is important to the management of groundwater supplies where groundwater is pumped for human uses.

Where groundwater flow is unconfined, that is, not bound by strata of contrasting permeabilities, the pattern of flow it assumes is characterized by broad arcs under the hydraulic gradients (Fig. 17.12*c*). This pattern is attributed to the fact that the pressure exerted by the mass of water is transferred both downward and outward, resulting in the semicircular flow lines similar, incidentally, to the flow of a glacier (see Fig. 22.5).

Aquifers

In most areas the geologic materials underground are differentiated into layers, formations, or zones of different water-holding capacities. These materials may be porous or nonporous, thick or thin, or consolidated (bedrock) or unconsolidated (deposits). Consequently, it is typical to find great contrasts in groundwater supplies at different depths in different places.

Those bodies of materials with especially extensive concentrations of usable groundwater are widely known as *aquifers* (Fig. 17.12*d*). Many different types

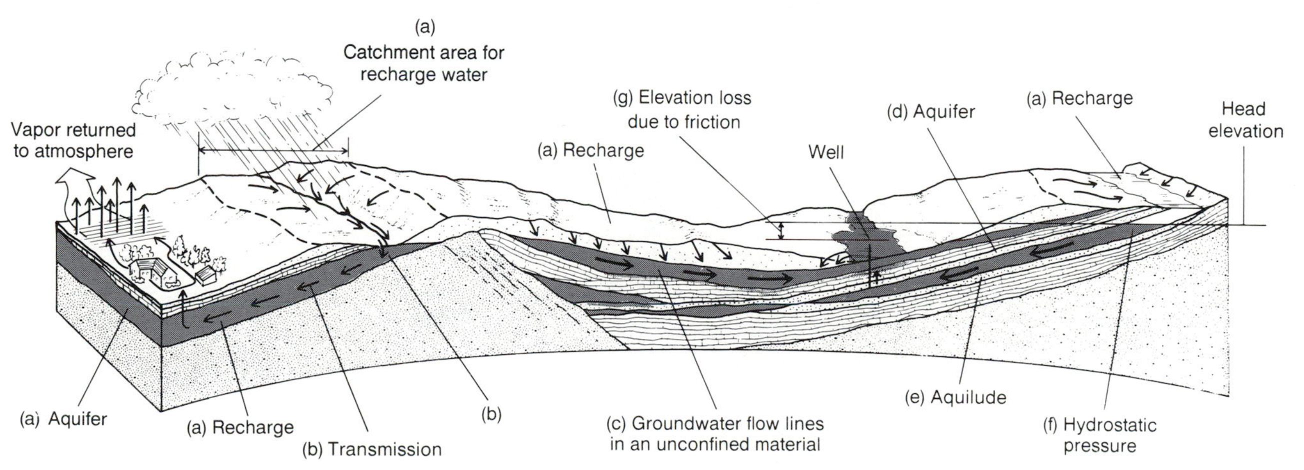

Figure 17.12 A composite diagram illustrating (*a*) recharge, (*b*) groundwater transmission, (*c*) groundwater in an unconfined material, (*d*) aquifers, (*e*) aquiludes, (*f*) hydrostatic pressure, and (*g*) elevation loss in artesian flow due to friction.

of materials may form aquifers, but porous material with good permeability, such as beds of sand and conglomerate formations, are usually the best. An aquifer is evaluated or ranked according to how much water can be pumped from it (without causing an unacceptable decline in its overall water level) and on the quality of its water. With respect to water quality, highly mineralized water, such as salt water, is not generally usable for agriculture and municipal purposes, and aquifers containing such water are usually not counted among an area's groundwater resources.

Aquifers form in two types of materials: *consolidated* (mainly bedrock) and *unconsolidated* (mainly surface deposits) (Fig. 17.13). In the central part of North America, where diverse glacial deposits lie over sedimentary bedrock, aquifers are found in both the bedrock and the surface deposits. In large river lowlands, such as the Mississippi and Missouri valleys, extensive shallow aquifers (at depths less than 100 m) are formed in the river deposits. These aquifers are recharged by river water, and their supplies fluctuate with the seasonal changes in river flow. In the American West, aquifers are found in both deposits and bedrock, but large areas are without aquifers that will yield more than small flows, say, less than 50 gallons of water per minute. In addition, many aquifers in the diverse mountainous terrain of western North America are localized in scale because they are formed in deposits associated with individual mountain slopes and intervening basins (Fig. 17.13).

Shallow aquifers have a capacity to recharge much faster than deeper ones. Although average recharge

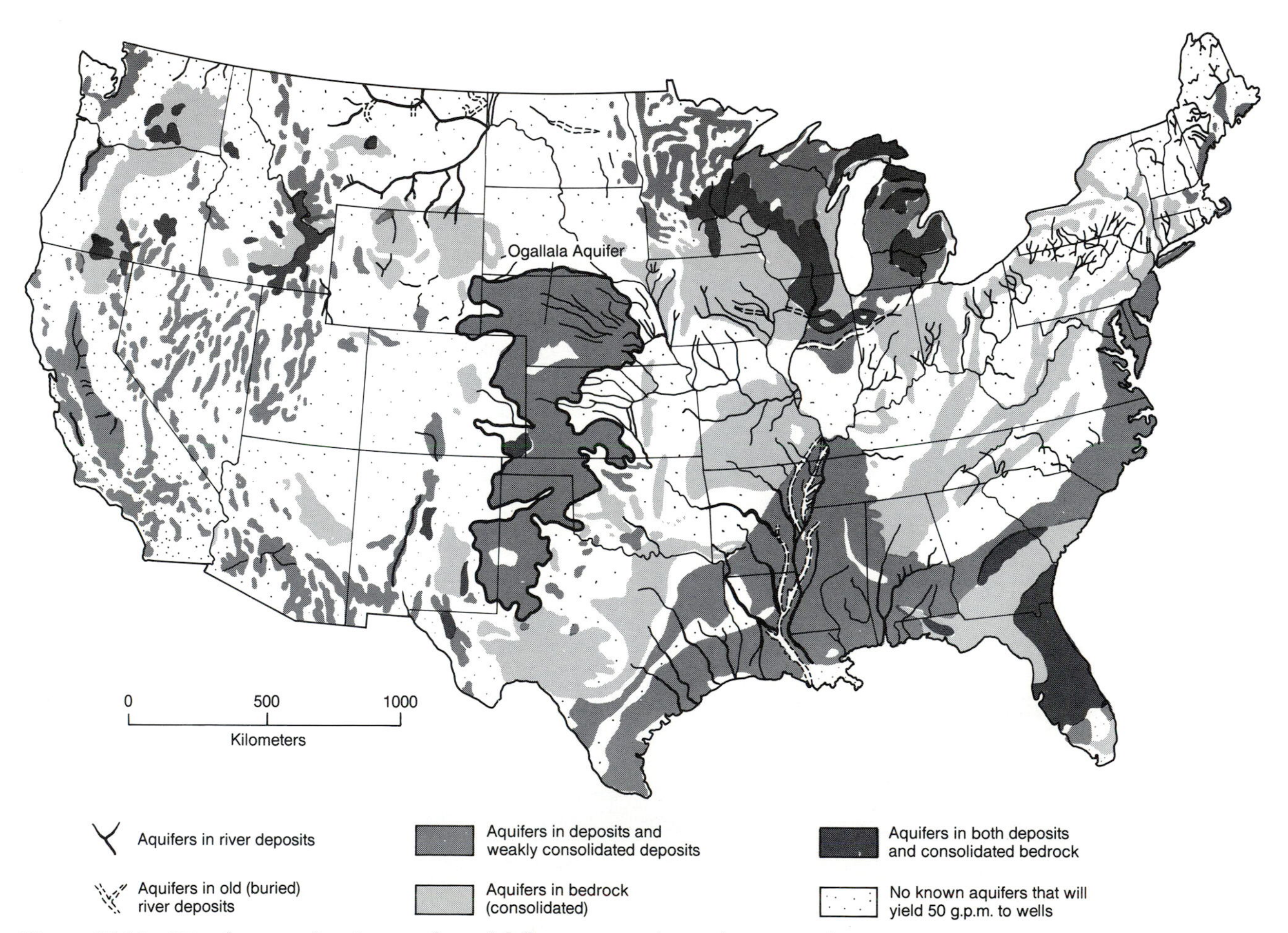

Figure 17.13 Distribution of major aquifers of different types (river deposits, other deposits, and consolidated bedrock) in the coterminous United States. A major aquifer is defined as one composed of material capable of yielding 50 gallons per minute or more to an individual well and having water quality generally not containing more than 2000 parts per million of dissolved solids.

rates decrease more or less progressively with depth, as a whole aquifers within 1000 m of the surface require several hundred years (300-year average) to be completely renewed, whereas those at depths greater than 1000 m require several thousand years (4600-year average) for complete renewal. This difference is related not only to the distance recharged water must travel, but also to the fact that percolating water moves slower in the smaller spaces found at great depths. Not surprisingly, finding groundwater is difficult in many areas. In Note 17 Chester Wilson describes some traditional folk methods and some modern scientific methods for locating groundwater.

Artesian Flow

Some aquifers produce a pressurized outflow of groundwater. This phenomenon, known as *artesian flow*, results from a geologic condition in which a tilted aquifer is sandwiched between two impermeable formations called *aquicludes* (Fig. 17.12*e*). The groundwater in the lower part of the aquifer exists under the pressure of the weight of the water upslope, called *hydrostatic pressure* (Fig. 17.12*f*). If the aquifer is tapped by a well, hydrostatic pressure will cause the groundwater to flow upward until it reaches a height *approaching* the head elevation of the water at the upper end of the aquifer. The reason that it does not reach the head elevation is that part of the energy that creates the hydrostatic pressure is lost in friction as the water moves through the aquifer (Fig. 17.12*g*).

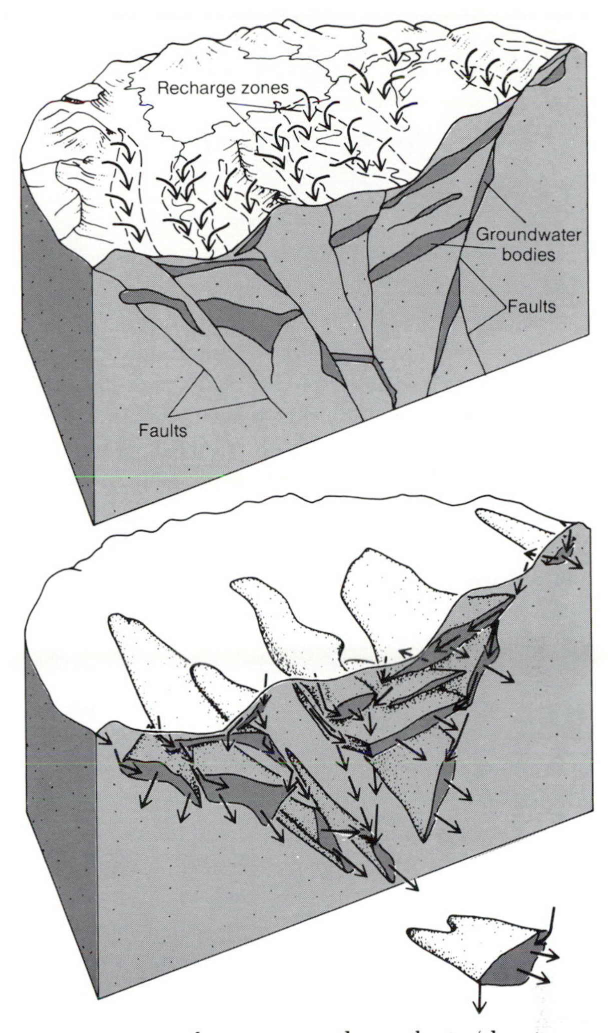

Figure 17.14 Above, a groundwater basin (showing groundwater bodies and recharge areas) in a mountainous area; below, the complex relationship among the groundwater bodies in this basin.

Groundwater Basins

To this point, we have not considered complex groundwater arrangements such as an area underlain by many interrelated bodies of groundwater. A group of groundwater bodies linked together in a large flow system is called a *groundwater basin*. Groundwater basins are typically complex three-dimensional systems characterized by vertical and horizontal flows among the various groundwater bodies—both those that would qualify as aquifers and those that would not—and between groundwater bodies and the surface (Fig. 17.14).

The spatial configuration of a groundwater basin is determined largely by regional geology, that is, by the extent and structure of the deposits and rock formations that house the groundwater bodies. Because these deposits and formations usually differ vastly in their size, composition, and shape, exactly how the various bodies of groundwater in a basin are linked together at different depths is seldom clear. This uncertainty is significant not only in forecasting safe yield limits, but also in understanding the spread of contaminants in a basin from solid and hazardous waste buried in landfills.

GROUNDWATER WITHDRAWAL

From the standpoint of water supply for human use, the best aquifers are those containing vast amounts of pure water that can be easily withdrawn without caus-

THE VIEW FROM THE FIELD / NOTE 17

Probing for Groundwater: Then and Now

The earth underground has long been recognized as a source of water. In Egypt, as early as 3000 B.C., municipal water came from shallow wells, and early writings from India indicate that groundwater was being used for irrigation purposes in 300 B.C. But the source of groundwater was unknown to our early ancestors. Plato believed that the source of all water was a great underground sea which he called Tartarus. That groundwater occurs in great underground pools and rivers is a widely held public misconception that has survived until the present time and has generated much folklore concerning its occurrence, especially in rural communities.

Because groundwater is out of sight, the aura of mystery surrounding its occurrence has been easily perpetuated. Its mysterious nature undoubtedly contributed to the successful rise of water witching or dowsing in medieval times. Water witching or dowsing is performed by a person holding the forked ends of a V-shaped willow branch or the handle of a bent metal rod in their hands and walking over a stretch of land. The willow or metal rod is pulled downward by an "irresistible force" when it is over a supply of groundwater. Faith in dowsing is believed to have persisted throughout the world since the Middle Ages. In the Far East it is called *radiesthesia,* a word of Egyptian origin connoting a science dealing with the study of radiating energy of objects. Although dowsers themselves do not have an acceptable explanation of their ability to detect groundwater, the fact is that a well drilled where the stick points usually produces water.

There may be a perfectly sound scientific explanation for the success of the doswer: probably static electricity. In recent dowsing experiments by James Trow of Michigan State University, it has been shown that many people carry small charges of electricity, with a potential of 1 to 8 volts, on their bodies and that movement of the dowsing rod occurs when these charges react with a stretch of electrically charged ground. The presence of small electrical charges associated with water and mineral deposits is well known, and electrical currents with a potential of one-half to one volt can be caused by

water moving through small pores in rocks and soil. Besides groundwater, dowsers are also able to detect mineral deposits, oil, and electrical cables. The success of the dowser depends on the polarity and voltage of the charges carried by the person and the polarity and voltage of the particular formation. Some people may be electrically neutral, which would explain why dowsing does not work for everyone.

Although the "mystery" of water witching or dowsing may have been explained (suggesting once again that some sorcerers and magicians are merely transitional scientists), the study of groundwater in conventional scientific circles has become dependent on various technical means of probing the underground. Geophysics has provided the most important tools in the development of groundwater resources. The main geophysical techniques used in these studies are *electrical resistivity* and *seismic refraction.*

The electrical resistivity method is based on the

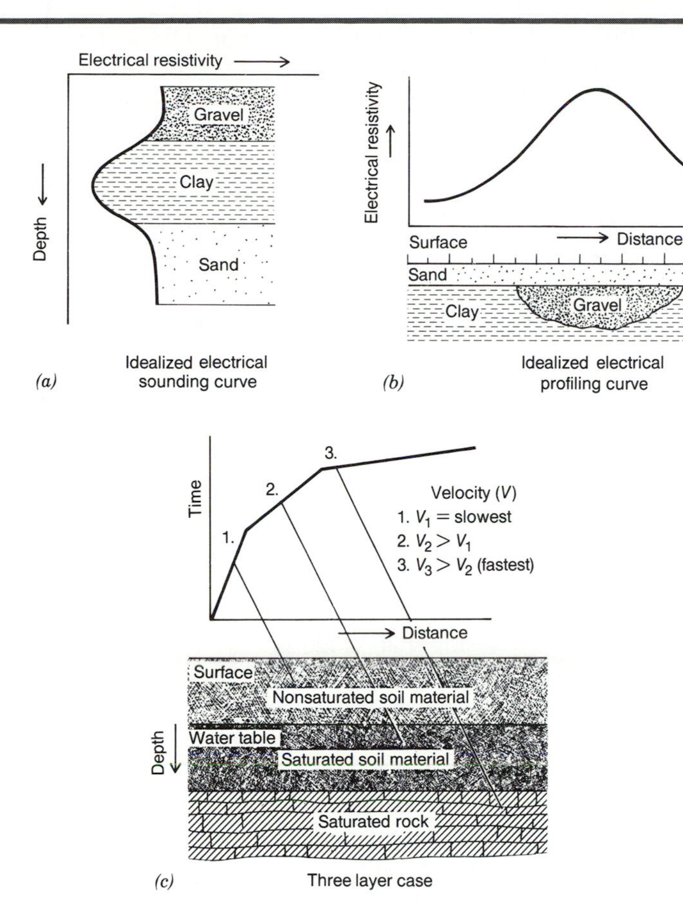

resistance of different subsurface materials to the conduction of an electrical current. The electrical current is introduced to the ground through electrodes (an electrical conductor, usually metal stakes). The depth of current penetration can be altered by changing the spacing between the stakes. From readings taken at different stake spacings, the type and thickness of subsurface materials can be interpreted, as the vertical profile in diagram (a) shows. Dry rock materials have a higher resistivity than the same material saturated with water because water increases the electrical conductivity. Salt water lowers the resistivity (improves conduction) more than fresh water. Coarse sediments have a higher resistivity than finer, clayey sediments with the same moisture content.

Resistivity can also be used to construct horizontal profiles. For this procedure, the stakes are moved to different locations along a traverse (line along the surface) or around a grid. The data can be plotted on a graph to form a resistivity profile or, in the case of a grid, to build a resistivity contour map.

This technique is good for locating buried gravel aquifers and the boundary between fresh and salt groundwater (Diagram b).

Seismic methods are used to measure the energy of seismic waves discharged into the ground from an explosive charge. As the waves pass through different materials, their velocity and direction change. These changes can be measured and used to determine the depth from the surface to bedrock and the water table, and to identify different materials and formations underground that may hold groundwater. A method called seismic refraction is used most widely in groundwater studies. It involves measuring the time it takes for refracted seismic waves (those that literally bend by curving laterally along a horizontal layer after going into the ground and then return to the surface) to travel different distances from the explosive charge. Data on travel times (called arrival times) can be used to interpret the thickness of deposits holding groundwater, the depth to bedrock, and so on (Diagram c).

With respect to groundwater, seismic wave velocities increase in saturated materials, but a great deal of caution must be exercised in making interpretations so as not to confuse a velocity increase related to the water table with one caused by a change in subsurface material. One way of overcoming this difficulty is to drill a test hole to find a representative depth for the water table and then extend the seismic survey over a larger area from that known point.

Chester Wilson
Mott Community College and
University of Michigan-Flint

ing an unacceptable decline in the head (elevation) of the aquifer. Two conditions must exist if an aquifer is to yield a dependable supply of water over many years. First, the rate of withdrawal must not exceed the transmissibility of the aquifer; otherwise, the water is pumped out faster than it can be supplied to the well, and the safe well yield will soon be exceeded. *Safe well yield* is defined as the maximum pumping rate that can be supplied by a well without lowering the water level below the pump intake. Second, the rate of total withdrawal should not outrun the aquifer's recharge rate, or else the water level in the aquifer will fall. When an aquifer declines significantly, the *safe aquifer yield* has usually been exceeded, and if the *overdraft* is sustained for many years, the aquifer will be depleted.

In arid and semiarid regions where groundwater is pumped for irrigation, the safe aquifer yield is typically exceeded and often greatly exceeded. In the American West, many aquifers took on huge quantities of water during and after the Ice Age several thousand years ago; today these reserves are being withdrawn at rates far exceeding present recharge rates. Unless pumping rates are adjusted to recharge rates, the water level will decline, wells will have to be extended deeper and deeper, and these aquifers will eventually be depleted.

In parts of Arizona, the water level in some aquifers is declining as much as 6 meters a year because of heavy pumping for agriculture and urban uses. In the Great Plains the Ogallala Aquifer, which stretches from Nebraska to Texas and is one of the largest groundwater reservoirs in the world, is declining at a rate of 0.15 to 1.0 m per year because of heavy irrigation withdrawals (Fig. 17.13). With the current shifts in the U.S. population to the Southwest, it appears that demands on these dwindling groundwater reserves will be even greater in the next several decades. At the same time, demands may decline in the water-rich Midwest and Northwest of the United States as these regions lose population (color plate 12).

Water Use in the United States

For the United States as a whole, average water use in 1975 totaled 420 billion gallons daily, of which about 150 billion gallons went to agriculture, 190 billion gallons to steam electrical generation, 45 billion to manufacturing, and 35 billion to domestic uses (municipal and residential) (Fig. 17.15). These supplies were drawn from both groundwater and surface water (rivers, lakes, and reservoirs) sources. In the West (seventeen states), groundwater currently accounts for nearly half the public (domestic) and industrial supply, whereas in the East (thirty-one states), it accounts for 29 percent of the public and only 16 percent of the industrial. In the agricultural sector, about 80 percent of the total water withdrawn (both surface and groundwater) is used by the Western states (color plate 12).

Most of the water withdrawals for agriculture in the United States are used for irrigation (see color plate

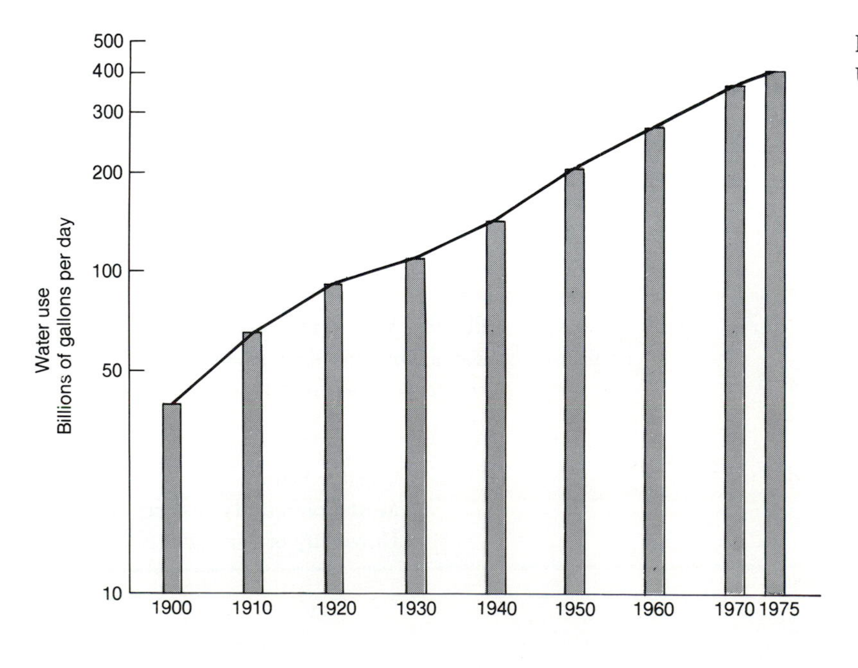

Figure 17.15 Daily water use in the United States 1900–75.

12). The bulk of irrigation water is lost in evaporation and transpiration; agriculture is therefore classed as a *consumptive user* because most of the water withdrawn is not returned to the ground or a stream in the liquid form. In contrast, residential water users (principally households) return most of the water withdrawn to a body of surface water and thus are classed as *nonconsumptive users*. However, water quality is usually seriously *degraded* by residential use because most of the water is used as a medium for discharging domestic sewage; therefore, it is contaminated with bacteria, dissolved chemicals (such as phosphorus and nitrogen), and sediments. (Canadians and Americans typically use 100 to 125 gallons of water per person daily in domestic, or home, use.) Water used in electrical generation is also degraded; in this case, the water is used for cooling, and when discharged back into the environment, it is typically 20°C warmer than its withdrawal temperature. The increased temperature reduces the water's capacity to hold dissolved oxygen, which in fresh water ecosystems results in damage or death to many organisms.

Formation of Cones of Depression

In and around urbanized areas (or agricultural areas with high concentrations of large wells), the groundwater level in aquifers may not only be greatly depressed by heavy pumping but also develop a very uneven upper surface. This happens where a high rate of pumping from an individual well is maintained for an extended period of time, causing the level of groundwater around the well to be drawn down by many meters or tens of meters. The groundwater surface takes on a funnel shape, like that on the water surface above an open drain in a bathtub, and is called a *cone of depression* (Fig. 17.16*a*).

As the cone of depression deepens with pumping, the hydraulic gradient increases, causing faster groundwater flow toward the well. If the rate of pumping is not highly variable, the cone usually stabilizes

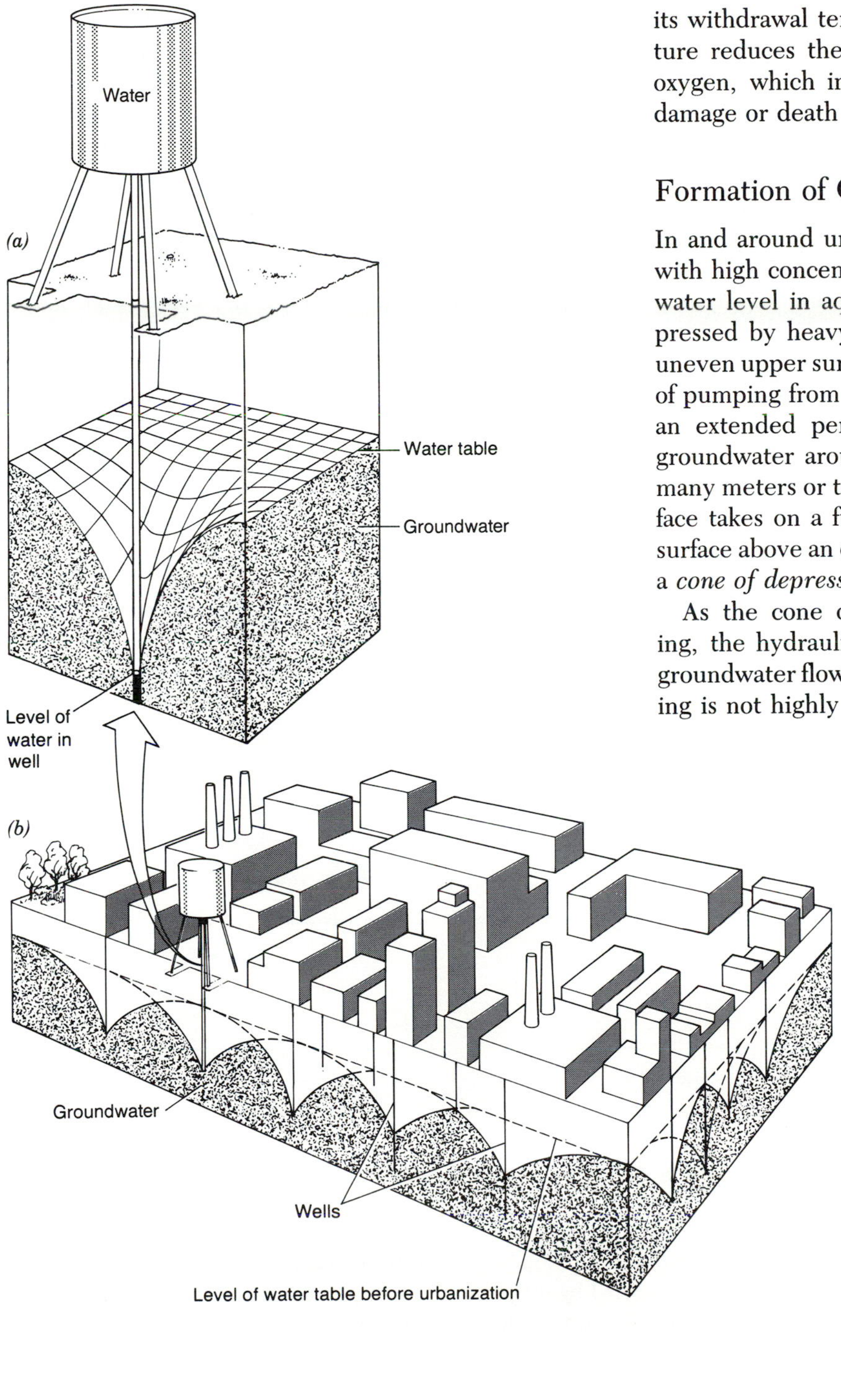

Figure 17.16 Formation of a cone of depression in a water table: (*a*) around a single well; (*b*) intersecting cones of depression resulting from many closely spaced wells in an uranized area.

in time. However, if many wells are clustered together, as is often the case in urbanized areas, they may produce an overall lowering of the water table as the tops of neighboring cones intersect each other as they widen with drawdown (Fig. 17.16*b*). If the wells are of variable depths, this may result in a loss of water to shallow wells as the cones of big, deep wells are drawn beyond the shallower pumping depths.

Drawdown of groundwater over a large area can also lead to loss of volume in the ground material in which it resides because, in certain materials, loss of water causes them to compress or shrink. This may result in subsidence in the overlying ground, as it has in Houston, Texas, for example, where much of the metropolitan area has subsided a meter or more with groundwater depletion.

Salt Water Intrusion into Groundwater

Salt water intrusion into the groundwater supply is a serious problem in some coastal areas. Under natural conditions, the groundwater under the sea is salt water. If there is no barrier along the coast, such as a ledge of impervious rock, to hold the salt water back, the salt groundwater will form a wedge under the fresh groundwater of the land. This happens because the salt water is denser. This principle is illustrated for an island in Figure 17.17: The fresh water table is above sea level, and the salt water table is below sea level.

Because it is lighter, the lens-shaped pocket of fresh groundwater floats on top of the salt groundwater. The density ratio of fresh water to salt water is 1:1.025; therefore, it takes forty-one volumetric units of fresh water to equal the weight of forty units of salt water. It follows, then, that as gravity water is added to the top of the pocket of fresh groundwater (Fig. 17.17*a*), the base of the pocket will depress the underlying salt water in proportion to the relative densities of the two fluids. For 1 m of elevation of the fresh groundwater table (above sea level), the salt water table will be depressed 40 m (below sea level).

If a well is drilled into the island, the pumping of groundwater leads to the formation of a cone of depression (Fig. 17.17*b*). Below the cone of depression, the reduced elevation of the water table leads to a *cone of ascension* in the salt groundwater as the boundary between the fresh and salt water adjusts to the reduction in mass of the overlying fresh groundwater. If the cone of depression becomes too large, the cone of ascension may intersect the well; if it does, the fresh water supply is lost to the well.

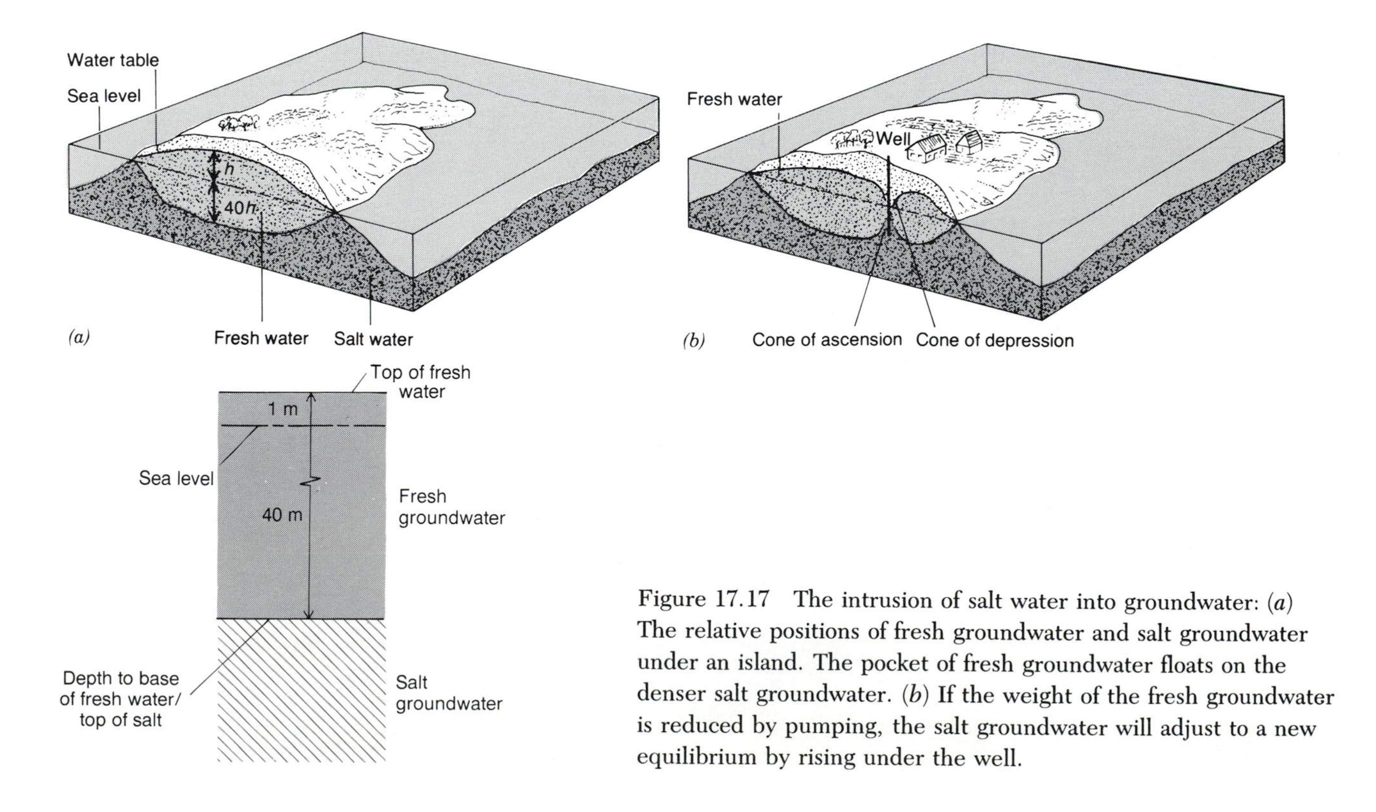

Figure 17.17 The intrusion of salt water into groundwater: (*a*) The relative positions of fresh groundwater and salt groundwater under an island. The pocket of fresh groundwater floats on the denser salt groundwater. (*b*) If the weight of the fresh groundwater is reduced by pumping, the salt groundwater will adjust to a new equilibrium by rising under the well.

Groundwater Contamination

Contamination of groundwater by chemicals leaking from buried waste is one of the most frightening environmental problems facing the United States and Canada today. The cause for concern can be demonstrated quite easily: (1) about half the population of the United States and Canada relies on groundwater for domestic water supply; (2) groundwater systems (basins) are complex, and their relationship to the landscape and surface deposits—and therefore to buried wastes—in many areas is not well understood (Fig. 17.14); (3) the locations, amounts, and types of waste buried over the past 50 years in and around the areas where we live are very poorly documented, meaning that there are all sorts of hidden buried waste sites; (4) the types of contaminants and how they behave (move and change) in the groundwater system are inadequately understood by scientists; and (5) once contaminated, aquifers may remain so for tens or even hundreds of years because of the slow rate at which groundwater renews (flushes) itself.

There are two main sources of groundwater contamination from landfills: solid waste and hazardous waste. A *landfill* is any disposal site (on land) where waste has been placed without regard to its possible influence on the surface or subsurface environment. *Solid waste* is the nonliquid and nongaseous debris discarded by cities, industry, agriculture, and mining. *Hazardous wastes* are mainly industrial residues that are toxic, flammable, or explosive. In 1980 solid waste production in the United States averaged 20 tons per person (per year), and hazardous waste averaged about 500 pounds (230 kg) per person (per year).

The U.S. Environmental Protection Agency in 1980 estimated that there were 32,000 to 50,000 landfills containing hazardous waste in the United States. Many different substances are found in hazardous wastes, including various toxic chemicals such as pesticides, cyanides, acids, and compounds of lead, mercury, and arsenic. Among the contaminants causing increased concern are manufactured organic compounds found in both solid and hazardous waste. The number of identified organic compounds concocted by humans by the late 1970s totaled about 2 million. The reason for concern is not just the large population of these contaminants (and the fact that it is increasing by tens of thousands each year), but that a significant share of them are comparatively resistant to natural degradation in the environment and can remain in surface and groundwater systems for periods of years.

The contamination of groundwater by buried waste takes place when *leachate* seeps into a groundwater body or water feeding a groundwater body. *Leachate* is water that has percolated through waste and become contaminated with toxic materials, organic compounds, and inorganic constituents such as calcium, nitrogen, and phosphorus. If the landfill is situated over permeable material, the leachate can migrate through a great volume of material in a matter of years. The landfill shown in Figure 17.18 produced a large chloride plume that migrated 700 m (2200 ft.) hori-

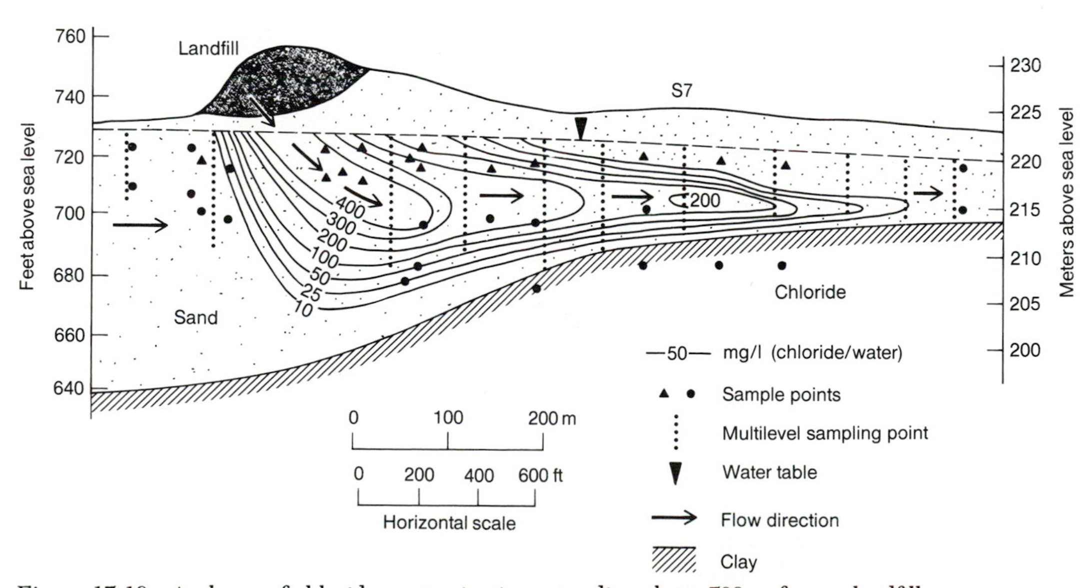

Figure 17.18 A plume of chloride contamination extending about 700 m from a landfill.

zontally through a sand layer in 35 years. Notice, however, that it did not penetrate the underlying clay which is impermeable.

Back to the Surface: Springs and Seepage

Although lots of groundwater remains in aquifers for thousands of years and some becomes locked in rock formations for millions of years, most of it remains underground for periods ranging from several months to several years. Aside from pumping, groundwater is released to the surface mainly through: (1) capillary rise into the soil from which it is evaporated or taken up by plants; and (2) seepage to streams, lakes, and wetlands from which it runs off and/or evaporates.

Many different geologic and topographic conditions can produce seepage. In mountainous areas, for example, it can be traced to fault lines and outcrops of tilted formations (Fig. 17.19). In areas where bedrock is buried under deep deposits of soil, groundwater seepage is usually found along the lower parts of slopes. Steep slopes represent "breaks" in elevation that may be too abrupt to produce a corresponding elevation change in the water table. Under such conditions, the water table may intercept the surface, especially in humid regions where the water table is high. If the resultant seepage is modest, a spring is formed, but if it is strong, a lake, wetland, or stream may be formed (Fig. 17.20).

Which type of water feature develops depends on a host of conditions, including the rate of seepage, the size of the depression, and the rate of water loss to runoff and evaporation. Most of the thousands of inland lakes of Minnesota, Wisconsin, Michigan, and Ontario, for example, are "seepage"-type lakes whose water levels fluctuate with the seasonal changes in the elevation of the water table. Most of these lakes are connected to wetlands or over thousands of years evolve into wetlands as they are filled in with organic debris.

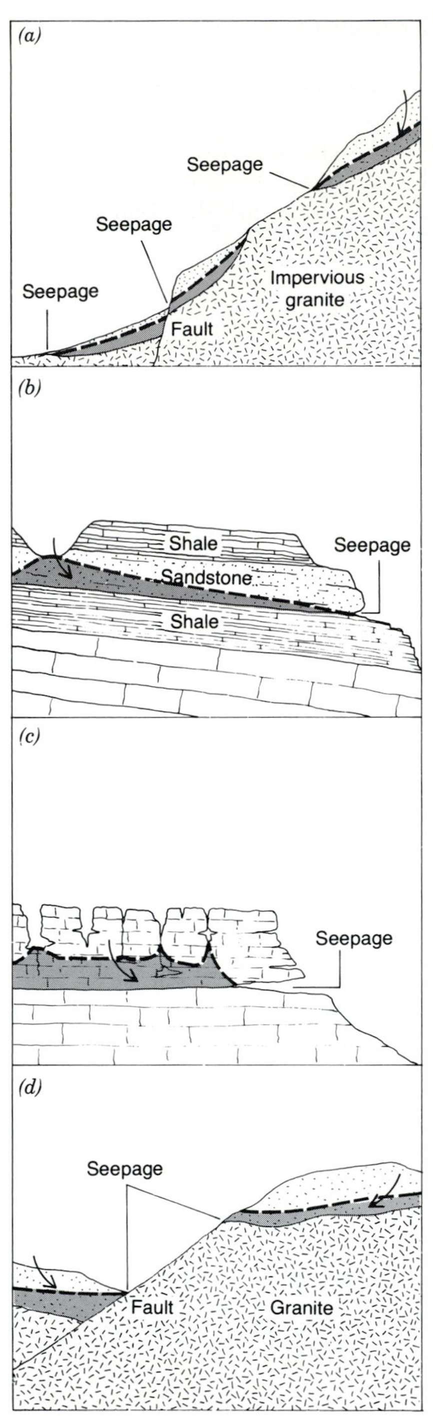

Figure 17.19 Seepage zones associated with different rock formations: (*a*) unconsolidated sediments over impervious bedrock; (*b*) inclined sandstone over shale; (*c*) cavernous limestone over unweathered limestone; (*d*) unconsolidated materials broken by a fault.

Summary

In a vegetated landscape, precipitation is intercepted by plants before reaching the ground. On the ground, water may lie on the surface, run off, or infiltrate the soil. Infiltration capacity is the rate at which water passes into the soil from the surface, and it is influenced by soil-particle size, vegetation, rainfall inten-

Figure 17.20 Groundwater seepage along a shore where the water table intercepts the surface.

sity, and other factors. Within the soil, percolating water may be taken up as capillary water or trickle on through the soil as gravity water.

Groundwater is an accumulation of gravity water in the subsoil and bedrock. The supply of groundwater varies geographically, owing to differences in climate, topography, and subsurface geology. The velocity of groundwater flow is controlled mainly by the hydraulic gradient and the permeability of the water-bearing material. Aquifers are materials with large concentrations of usable groundwater and they are valuable sources of water for agriculture, households, and industry.

Where groundwater withdrawal and transmission exceed the rate of recharge, aquifers will be drawn down. This is common in urbanized areas and agricultural districts, particularly in dry areas. Another influence of humans on groundwater is contamination from buried wastes. This is a very serious problem not only because of our dependence on groundwater for water supplies, but also because of the complex nature of groundwater systems and the uncertain nature of contaminant behavior underground. Most groundwater ultimately returns to the surface through capillary rise in the soil or seepage into streams, wetlands, and lakes.

Concepts and Terms for Review

- precipitation on the land
  - interception
  - runoff
  - infiltration
  - evaporation
  - storage
- infiltration capacity
- soil water
- soil permeability
- wastewater absorption
- groundwater
  - porosity
  - water table
  - Darcy's Law
  - recharge
  - transmission
- aquifer
- artesian flow
- groundwater basin
- groundwater withdrawal
  - safe yield
  - overdraft
  - regional water use in the United States
  - consumptive-nonconsumptive use
- groundwater problems
  - cones of depression
  - saltwater intrusion
  - contamination
  - solid and hazardous waste
  - leachate
- groundwater outflow
  - seepage and springs

Sources and References

Ambroggi, R.P. (1966) "Water Under the Sahara." *Scientific American*, 214:5, pp. 21–29.

Franke, O.L., and McClymonds, N.E. (1972) "Summary of the Hydrologic Situation on Long Island, New York, as a Guide to Water Management Alternatives." *U.S. Geological Survey Professional Paper 627-F.*

Freeze, A.R., and Cherry, J.A. (1979) *Groundwater.* Englewood Cliffs, N.J.: Prentice-Hall.

Horton, R.E. (1933) "The Role of Infiltration in the Hydrologic Cycle." *American Geophysical Union Transactions*, Vol. 14, pp. 446–460.

Leopold, L.B., and Maddock, T. Jr. (1954) *The Flood-Control Controversy.* New York: Ronald Press.

Meinzer, O.E. (1927) "Plants as Indicators of Groundwater." *U.S. Geological Survey Water Supply Paper 577.*

Pye, V.I., et. al. (1983) *Groundwater Contamination in the United States.* Philadelphia: University of Pennsylvania Press.

Ritter, W.F., and Chirnside, A.E.M. (1984) "Impact of Land Use on Ground-water Quality in Southern Delaware." *Ground Water* 22:1, pp. 38–47.

Strahler, A.N. (1957) "Quantitative Analysis of Watershed Geomorphology." *Transactions American Geophysical Union 38*, 6: pp. 913–920.

White, G.F. (1964) "Choice of Adjustment to Floods." University of Chicago: *Department of Geography Research Paper No. 93.*

Chapter 18

Runoff, Streamflow, and Flooding

At one time or another all landscapes, even the most arid, shed water from precipitation. This water is called runoff and it feeds the networks of streams and rivers that drain the land. What is the character of these networks and the different streams in them? Do they gain their flows exclusively from surface runoff? How are they affected by different precipitation events? How often are the capacities of stream channels exceeded, resulting in floods?

As we pushed along, the current got stronger and our speed less and less. By mid-April we were reduced to a crawl of no more than a knot. Our food stocks were low, we were now down to a sack of rice and a can of tea. There was not much time to fish, for all our energies were concentrated on making our way upstream. We were both suffering badly from malaria and loss of blood and lack of sleep caused by incessant insect attacks. However, we kept going, for our destination was upstream. On the twenty-fifth of April, our full supply was down to five gallons [of gasoline] and this I determined to hoard against any future emergency. Our speed slowed to a mere crawl against the seven- and eight-knot current, though, when we had any wind, we would sail like blazes. Finally, just above the small miserable settlement of Codajas, we started to *espia*, or haul the boat from one tree to the next. . . .

Day after day we slaved at the *espia*, struggling with our remaining strength against the roaring waters of the Amazon. Struggling in the stinking mist, across shore-side swamps, in the awful eternal smell of rotting vegetation, in the unceasing din of insects chirping, monkeys howling, trees crashing mightily; and always the river, the everlasting *brutal*, roaring current as it swooshed down thousands of miles to the Atlantic Ocean far away. . . .

By the fifteenth of May, after sixteen days of hard slog, we had progressed exactly 160 miles. The river was rising by the hour, threatening to overflow the banks; already vast areas of the jungle were flooded. Soon, even the *espia* would not be possible, for we would be swept into some swamp and stranded there. . . .

With aching hearts, after an hour's argument, we both conceded that we had lost this round. The Amazon had beaten us. We would have to turn back.[1]

[1]From *The Incredible Voyage* by Tristan Jones. Sneed, Andrews and McMeel, Inc., Kansas City, 1977 (1st ed.).

The Amazon River is the largest river on earth. Its average flow is some 180,000 cubic meters *per second*—ten times larger than the Mississippi's and more than sixty times larger than the Nile's (Fig. 18.1). The Amazon is fed with runoff derived from the heavy rains produced by the equatorial and tropical climates that dominate most of its huge (5.8 million km^2) drainage basin. Between the rainfall and the river, however, lies a complex landscape of rainforests, deep clayey soils, great swamps, and mountains that influence the rate and amount of water that reaches the Amazon. Our objective in this chapter is to examine the runoff processes that supply water to streams and rivers and the factors that influence the character of stream and river flows.

THE NATURE OF RUNOFF

The water that feeds streams and rivers and is carried by them is called *runoff*. There are two general categories of runoff: *surface* and *subsurface*. Surface runoff includes *overland flow*, which is characterized by thin sheets of water that move slowly over the ground during periods of rainfall or snowmelt; and *channel flow*, which constitutes the flow in all sorts of surface arteries ranging from tiny rivulets to large rivers. Subsurface flows are of two varieties: *groundwater*, which is held and moves at depth in deposits and bedrock; and *interflow*, which moves horizontally within the soil layer, usually at depths not exceeding several meters.

Sources of Runoff

Precipitation is the sole source of water for runoff. But not all precipitation ends up as runoff because most of it (nearly 60 percent for the continents as a whole) is lost to the atmosphere in evaporation and plant transpiration. The amount of runoff from any area of land represents the water left after vapor losses to the atmosphere, plus or minus any inflows or outflows from water held in storage (groundwater and soil water). We can express this concept in a formula called the *hydrologic equation:*

$$\text{RUNOFF} = \text{Precipitation} - \begin{matrix}\text{Evaporation and}\\ \text{transpiration}\end{matrix} \pm \begin{matrix}\text{Change in}\\ \text{storage water}\end{matrix}$$

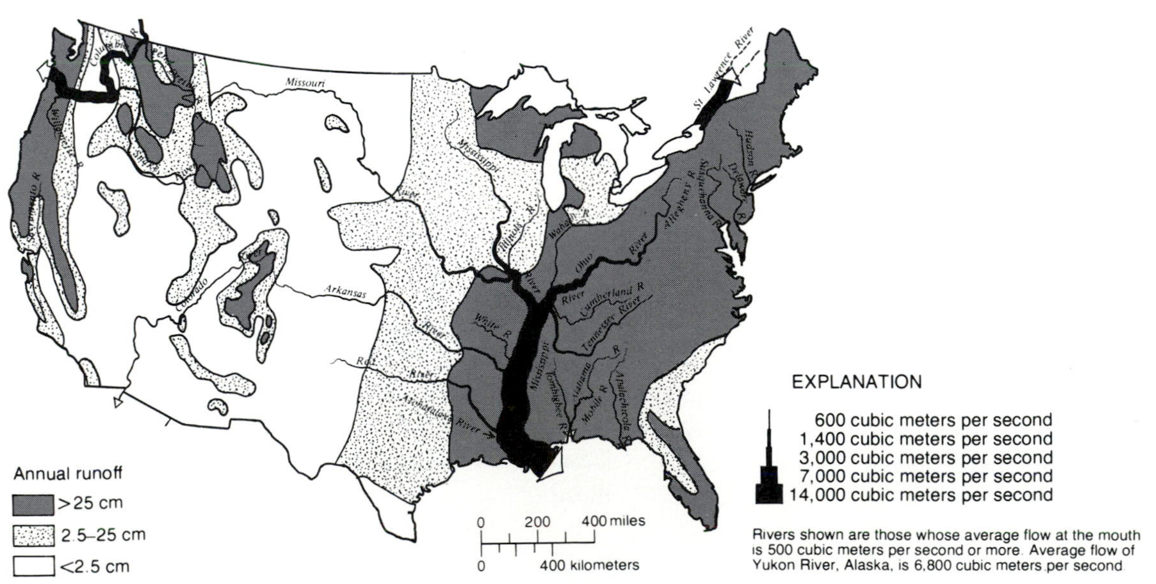

Figure 18.1 Average annual runoff in the coterminous United States, including the annual average flows of major rivers. In all parts of the country, annual runoff is less than half of the annual precipitation.

The hydrologic equation holds true over any period of time, but its components vary greatly. Precipitation varies with weather events (storm passages), the seasons in many areas, and over longer periods, ranging from several years to decades. Evaporation and transpiration usually vary most from season to season with changes in surface heating and air temperature.

Storage water also varies on a seasonal basis: we know this because the water table usually declines in the summer and rises in the winter and spring throughout most of North America. But over long periods, say, many years, the changes in storage water should balance out to zero, barring a sharp change in climate or large withdrawals of groundwater. Because much of the runoff in streams and rivers is derived from groundwater, runoff is not as irregular in time and space as precipitation is. When rainfall occurs, much of it infiltrates into the soil and percolates into the groundwater. During periods of dry weather, groundwater gradually drains into springs and streams. The enormous water storage capacity of the surface layers of the earth (unconsolidated deposits and shallow bedrock) keeps large streams flowing during several months of dry weather. Recall that, aside from the oceans and the glaciers, groundwater and soil water together represent the largest source of water on the planet (Table 7.1).

Measuring Runoff

Rainfall is measured in centimeters or inches (as depth on the surface) over time, usually per hour or day, whereas runoff is measured as the volume of flow over time. Typical units for flow are cubic meters per second (m^3/sec, or "cumec") or, in the United States, cubic feet per second (ft^3/sec, or "cfs"). The flow of water in a river expresses the rate at which water is leaving a drainage basin just as the flow of water from your home into the sewer represents the rate at which water is leaving the municipal reservoir or the storage tank in your home or town.

Runoff that flows in stream and river channels is called *discharge*. Discharge is equal to the cross-sectional area of water in the channel times the mean velocity of the streamflow (Fig. 18.2). In the United States, the U.S. Geologic Survey has the responsibility for measuring the discharge of the nation's rivers. Discharge data are gathered for the purpose of evaluating the nation's water resources and analyzing flood problems. To obtain the necessary flow data, the Geological Survey has established several thousand gaging stations on rivers where the elevation of the water surface, called the *stage*, is automatically recorded. The stage data are converted into discharge data, and, for most locations, only the peak annual discharge is recorded and published. *Peak annual discharge* is the largest discharge recorded at a gaging station in a calendar year.

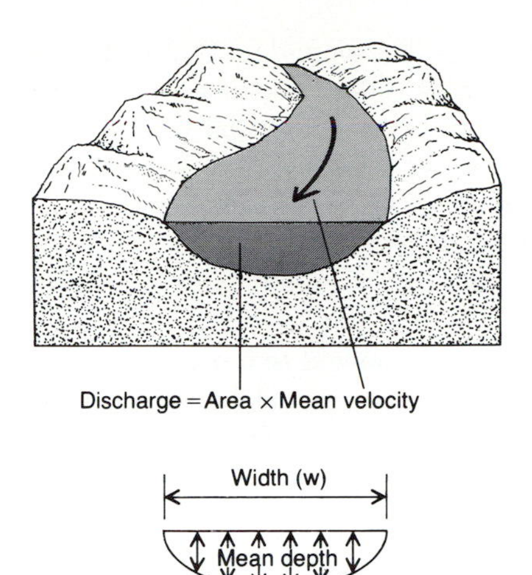

Figure 18.2 The discharge of a stream is equal to the cross-sectional area of water in the channel times the mean velocity.

DRAINAGE BASINS AND STREAM NETWORKS

An important question in many geographic problems concerns the relationship between the area or region over which something (such as water, automobiles, or people) is distributed and the system of carriers (such as channels, streets, and walks) that serves it. With respect to runoff, the area is called a drainage basin, and the system of streams and rivers that serves it (that is, drains it) is called a drainage network.

The Drainage Basin

A *drainage basin* is a discrete area of terrain which is higher on the perimeter and lower near the center or on one side (Fig. 18.3). The boundary that marks the perimeter of a basin is called the *drainage divide*. Since *all* land masses are naturally partitioned into drainage basins of various sizes, every drainage divide must separate two adjacent basins. The drainage divide functions precisely as the term implies: it is the line of separation in surface runoff between neighboring basins.

Drainage basins are organized in a pyramidal fashion inasmuch as the runoff from small basins combines to form larger basins and the runoff from these basins in turn combines to form even larger basins and so on. This mode of organization is described as a *hierarchy*, or a *nested hierarchy*, since each set of smaller basins is set inside the next larger basin (Fig. 18.4). By the same token, the streams that drain small basins combine to form larger streams, and they combine to form even larger ones, leading to the formation of a drainage network. *Drainage networks* are systems of sequentially related channels. In general the channels increase in size with lower position in the network.

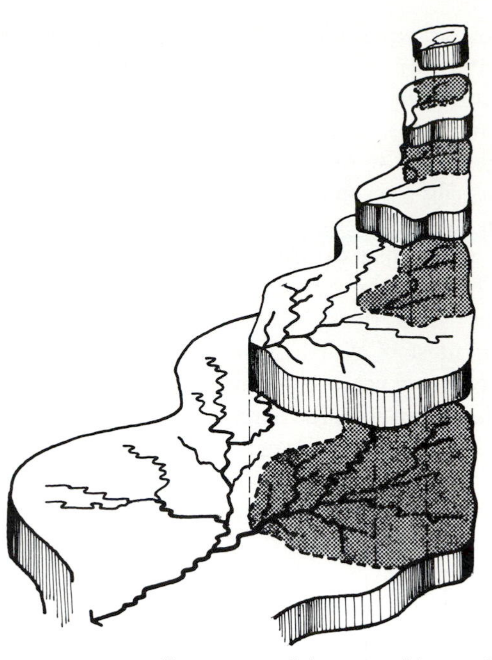

Figure 18.4 Illustration of the nested hierarchy of lower order basins within a large drainage basin.

Stream Order and Drainage Density

A stream channel and the basin it drains may be classified according to their relative position, or rank, within the drainage network. The classes of streams are called *orders*. A first-order stream has no tributaries; a second-order stream is formed by the joining of two first-order streams; a third-order stream is formed when two second-order streams join; and so on. Drainage networks (and their basins) are designated according to the highest order stream, the trunk stream; thus, first-order basins are drained by first-order trunk streams, second-order basins by second-order trunk streams, and so on.

The stream order concept is essential to understanding how a drainage network functions. If you count the number of streams in each order, you will find that first-order streams are the most numerous streams in any network, second-order are second most numerous, and so on, with the numbers declining progressively with higher orders. The relationship between stream numbers and order is known as the *principle of stream orders* (Fig. 18.5). The ratio be-

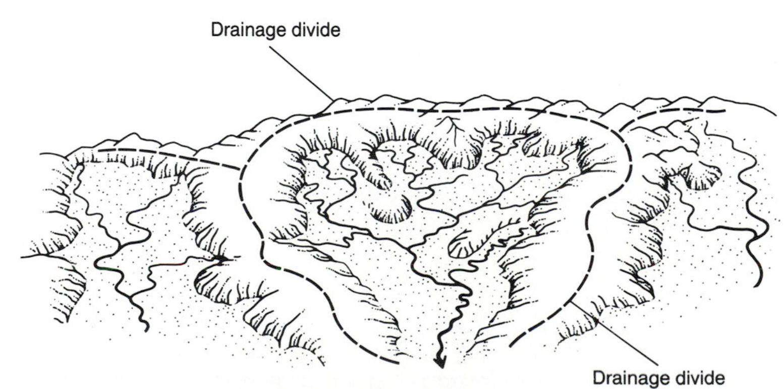

Figure 18.3 Schematic diagram of a drainage basin. The high terrain on the perimeter is the drainage divide.

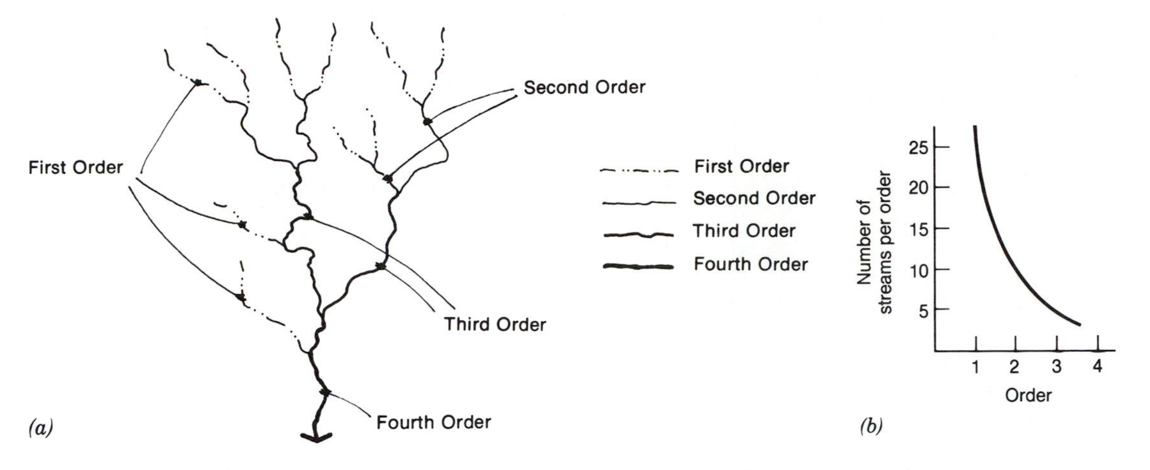

Figure 18.5 (*a*) Stream order classification according to rank in the drainage network; and (*b*) a graph illustrating the principle of stream orders.

tween the number of streams in one order and the next is called the *bifurcation ratio* (branching ratio), and it gives us an indication of the rate of increase in stream size from one order to the next. We know that when two streams merge to form one channel the size of the resultant stream will be about equal to the sum of the two contributing channels. In general, bifurcation ratios fall close to 3 for most drainage networks, meaning that streams increase in size about threefold with each higher order.

Other networks in nature also follow the stream order model. Some of the best examples are found in the root systems of plants, the branching structure of trees, and the veination system in many leaves (Fig. 18.6). It is interesting that similar branching networks are also found in the Martian landscape. Based on their branching patterns and channel forms, they were undoubtedly carved by streams, indicating that at least some parts of that planet supported liquid water in the past. On earth, drainage networks display a remarkable range of patterns that reveal a great deal about the geology and landforms of a region.

Another important feature of stream networks is *drainage density*, the total length of channels per unit area of drainage basin. Basins of high density consist of a large number of tightly interfingered channels, whereas basins of low density have relatively few and widely spaced channels. Drainage densities range from 0.5 km to 300 km of channel per square kilometer of area. Where drainage density is high, stream discharge tends to show a more direct response to rainfall than where drainage density is low because runoff gets into the channels more quickly.

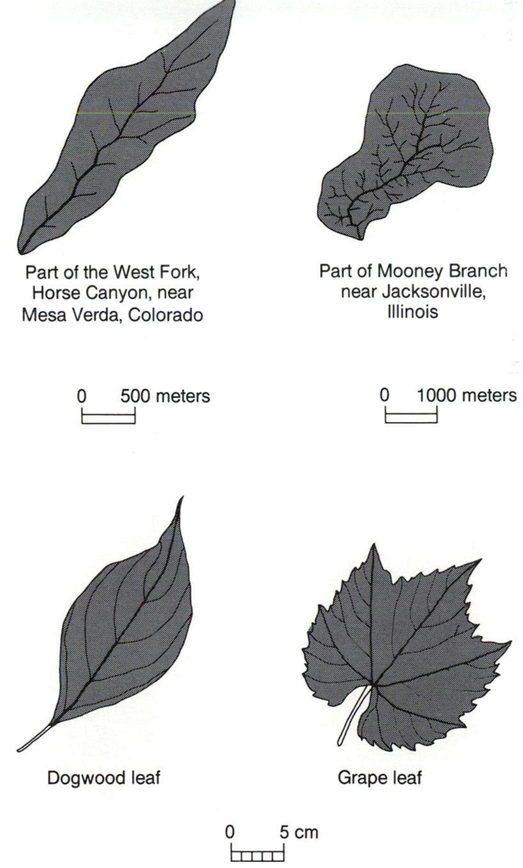

Figure 18.6 Branching patterns in nature: stream networks and leaf-vein networks.

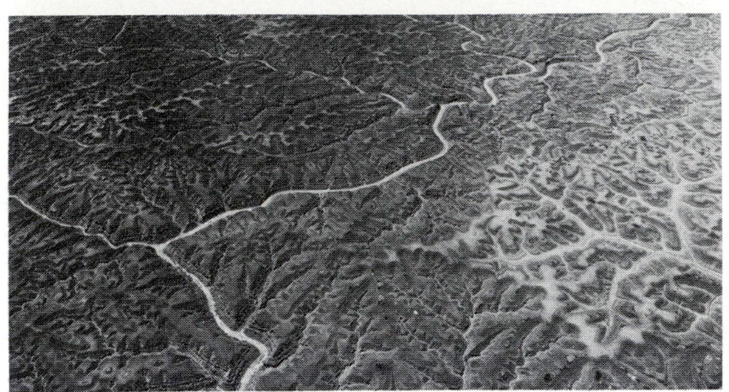

RUNOFF-PRECIPITATION RELATIONS

Precipitation is brought to drainage basins as rainfall or snow, but not all of it develops into runoff because of losses to evaporation and absorption by soil. For precipitation that does develop into runoff, its delivery to streams is highly variable over time and space, and this has an important influence on the flow characteristics of streams.

Sources of Discharge

Stream discharge is derived from four major sources: channel precipitation, overland flow, interflow, and groundwater flow (Fig. 18.7).

- *Channel precipitation* is rain or snow that falls directly into streams, lakes, or other waterways of a drainage basin. It immediately becomes discharge, but typically the amount is relatively small, unless the basin includes large areas of open water such as a lake or reservoir.
- *Overland flow* occurs when the rainfall intensity exceeds the surface infiltration capacity. This water travels over the surface, collecting in small rivulets and trickles, and flows into stream channels relatively quickly. Overland flow generally increases with precipitation intensity and lower infiltration capacities. In urbanized areas, impervious surface materials may reduce infiltration so much that 80 to 90 percent of a rainfall is typically converted to overland flow.
- *Interflow* takes place when infiltration water enters the soil more rapidly than it can percolate downward to groundwater level. Under these circumstances, the percolating water will move laterally in the downhill direction. In forested areas, where infiltration capacities are very high, interflow often constitutes a large part of the runoff

Figure 18.7 Sources of streamflow: channel precipitation, overland flow, interflow, and groundwater.

from a rainstorm; in contrast, in urban areas interflow is very slight compared to overland flow (Fig. 18.7).

- *Groundwater flow* supplies water to streams where the stream channel intersects the water table. Because the supply of groundwater is very large and its rate of flow slow (compared with the overland flow), the groundwater flow is usually very steady over long periods. In humid areas groundwater flow thus accounts for most of the steady flow of streams during dry periods.

Channel precipitation and overland flow deliver water to stream channels very rapidly after precipitation begins. Where such sources constitute a major portion of the runoff, the response of the stream to a rainstorm will be rapid and peaked. Interflow, on the other hand, is considerably slower than overland flow and tends to feed water to the stream more gradually; when rainfall occurs, it takes several hours for interflow to begin, and it may last for several days after rainfall has ended. Thus, a stream fed by interflow water will rise slowly and decline slowly compared to one fed by overland flow. Groundwater flow responds to precipitation even more slowly than interflow does because the infiltrating water must work its way through the soil to the water table, then move laterally to the streams. This may take days, weeks, or even months.

Rainfall and Basin Influences on Streamflow

Streamflow can be divided into two categories based on the rate at which it is delivered to the stream channel: stormflow and baseflow. *Stormflow* (also called *quickflow*) is the runoff that occurs during or soon after the storm; it is made up of overland flow and channel precipitation. *Baseflow* is the discharge derived from groundwater seeping into the channel. Because groundwater is released slowly, baseflow changes little over long periods of time. Interflow can contribute to either stormflow or baseflow, depending on soil and vegetation conditions in the basin.

The relationship of stormflow and baseflow can be illustrated with the aid of a diagram called a *hydrograph*, which traces the changes in stream discharge over a specific period of time, usually the time encompassing a rainstorm (Fig. 18.8*a*). If we compare the stream hydrograph to rainfall for the same time period (Fig. 18.8*b*), it is possible to measure the stream's total response to (1) different amounts and intensities of rainfall and (2) different conditions of the drainage basin.

Let us consider a few examples using different rainfalls. If a rainfall is so intensive that it exceeds the soil infiltration capacity, then overland flow will be produced and the hydrograph will show a strong peak made up of stormflow (Fig. 18.9*a*). On the other hand,

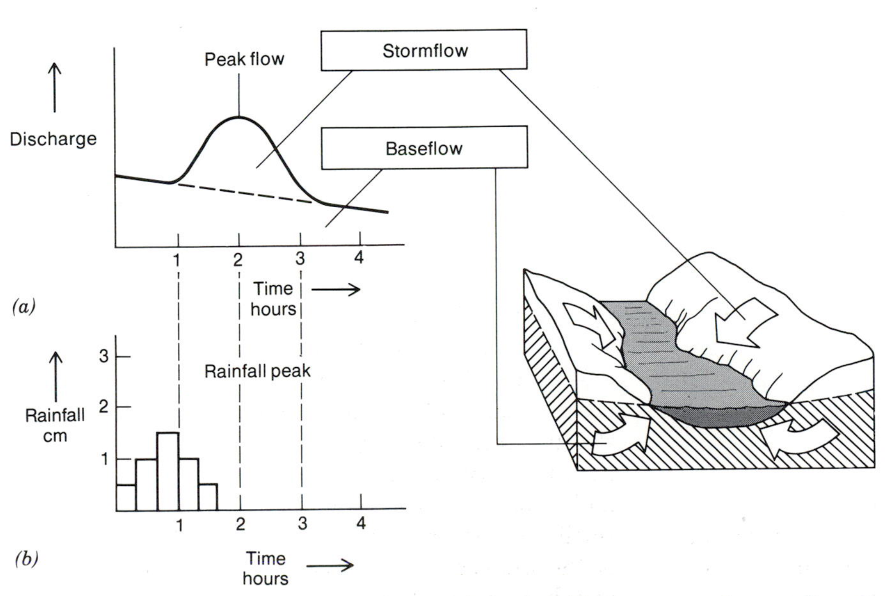

Figure 18.8 (*a*) A hydrograph illustrating the two main categories of streamflow; below (*b*), the corresponding rainstorm.

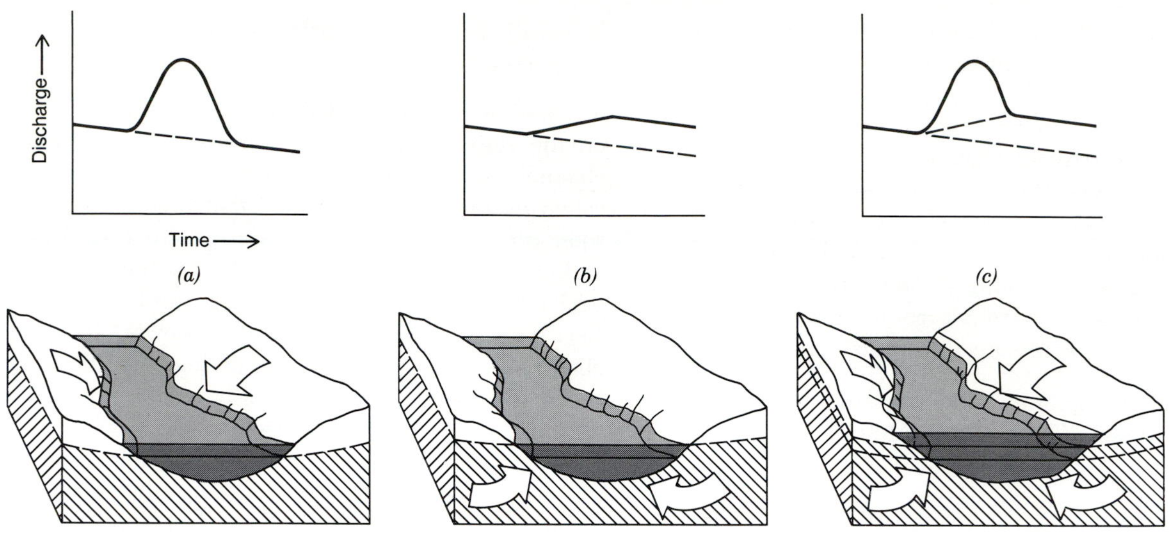

Figure 18.9 Three types of hydrographs and the types of runoff associated with each: (*a*) Stormflow produced by overland flow, (*b*) rise in baseflow produced by groundwater rise, (*c*) rise in both stormflow and baseflow.

if rainfall is less intensive, all of it may be taken up in infiltration and no overland flow will be produced. But if the water that infiltrates the soil makes its way to the water table and raises it, baseflow will be increased. This will result in a different hydrograph shape, one in which baseflow rises but there is no accompanying stormflow peak (Fig. 18.9*b*). If conditions are such, as in a long hard rain, that *both* stormflow and additional baseflow are produced, then the hydrograph will show both a peak and a baseflow rise (Fig. 18.9*c*).

We can produce essentially the same hydrographs by altering surface conditions in the drainage basin. In a fully forested basin, most heavy rainfalls will produce a hydrograph similar to the one shown in Figure 18.9*b*. Effectively all rainwater is taken up by infiltration, and as a result, baseflow increases as the water table rises. If the forest is cleared, even partially cleared, the same size rainstorms will produce stormflow because the loss of forest will reduce infiltration capacity, thereby resulting in overland flow. The resultant hydrograph will look more like those presented in Figure 18.9*a* or 18.9*c*.

Stream and River Types

An acquaintance with the concepts of drainage networks and sources of streamflow is essential to understanding the different types of streams and rivers. Streams are classified first according to whether their discharge occurs on a continuous (permanent) basis or on a temporary basis. Streams with permanent flows are called *perennial streams*. These streams are supported by groundwater which ensures continuous baseflow even during periods of little or no rainfall.

Most rivers are perennial, but some, even large ones, such as the Platt River which crosses the semiarid Great Plains, lose their flows in dry years when the water table falls below the bed of the channel. Some perennial rivers are also classed as *exotic rivers*. These are large rivers that rise in a humid zone and then flow across a desert without drying up. The Colorado River of the American Southwest and the Nile River of Africa are exotic rivers. The relationship of the exotic river to the water table is basically opposite that of rivers in humid regions: instead of receiving flow *from* groundwater, they give up flow *to* the water table beneath the streambed. Streams characterized by these contrasting modes of flow are also termed *effluent* (receiving groundwater) and *influent* (giving up flow to groundwater) (Fig. 18.10).

Streams with temporary flows usually fall into two classes: intermittent and ephemeral. *Intermittent* streams carry discharge seasonally, and *ephemeral* streams carry discharge only during rainfall or snowmelt events. Ephemeral streams are usually very small, first-order channels, situated in the upper part of the drainage nets, which receive their discharge mainly from overland flow. Intermittent streams are often the next larger set of channels in the drainage

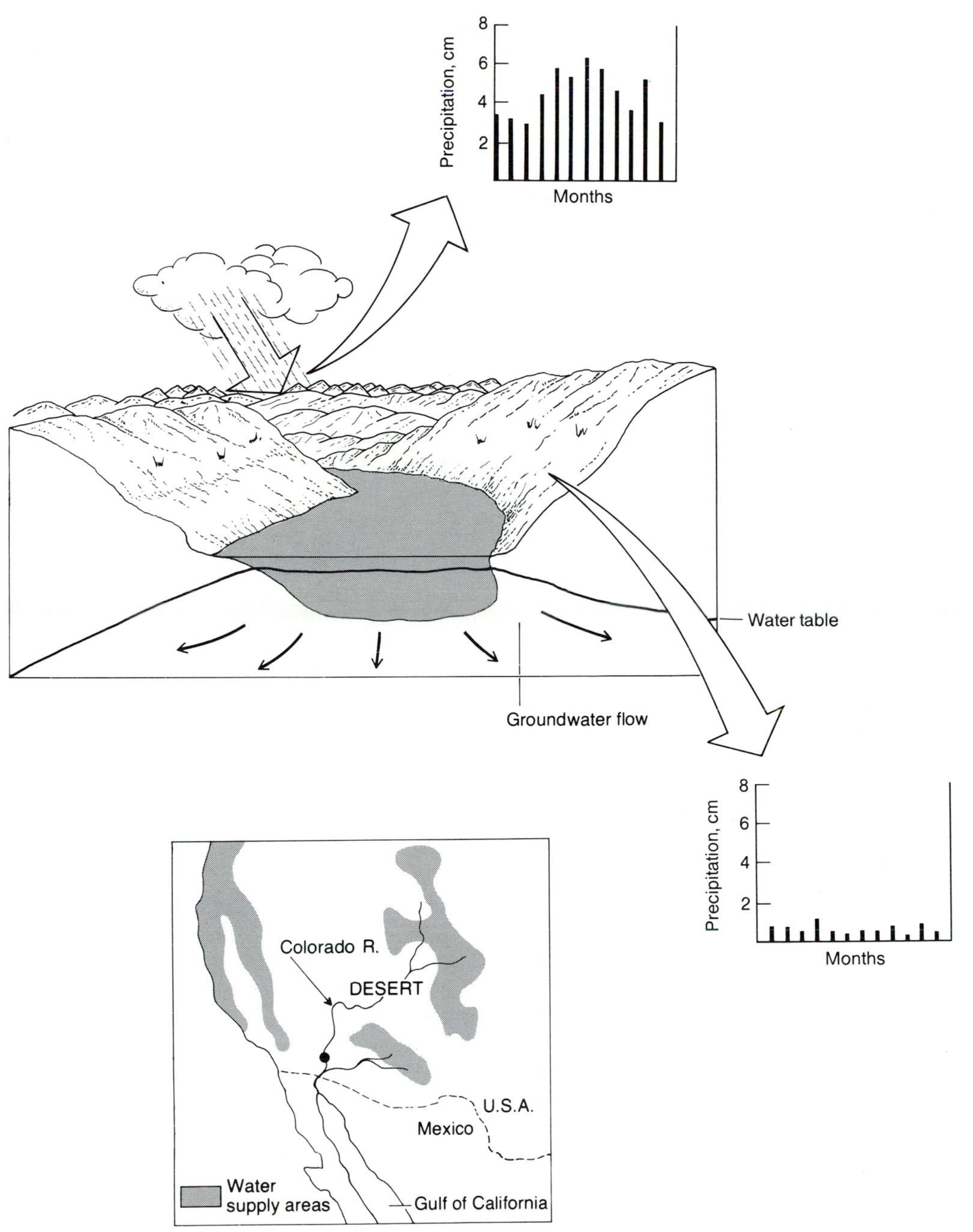

Figure 18.10 A schematic diagram of an exotic stream such as the Colorado River; flow is received from a humid region as the stream carries the water into the desert where it is taken up in groundwater recharge and evaporation.

network, and they are situated at slightly lower elevations in the drainage basins. They carry discharge only in the moist season when the water table rises and comes into contact with the streambed and when interflow is plentiful. The relationship among perennial, intermittent, and ephemeral streams in the drainage basin in different seasons is shown in Figure 18.11*a*.

Based on this knowledge, we can understand how drainage networks expand and contract with different rainfalls, snowmelts, and storage-water conditions. During an intensive rainstorm, for instance, the drainage net enlarges as the intermittent and ephemeral channels take on water (Fig. 18.11*b*). But during a summer drought, not only are the intermittent and ephemeral channels dry, but also even some of larger channels normally supported by baseflow go dry as the water table falls below their beds. In the extreme, baseflow may decline until only the trunk stream in a drainage net is left with flow. At this time the trunk stream qualifies as only a first-order stream (Fig. 18.11*c*). Thus, the order of a stream in the network is not constant over time, but may increase and decrease with the seasons or with precipitation events. However, its relative rank among the channels in the drainage network remains constant regardless of seasons and flow events.

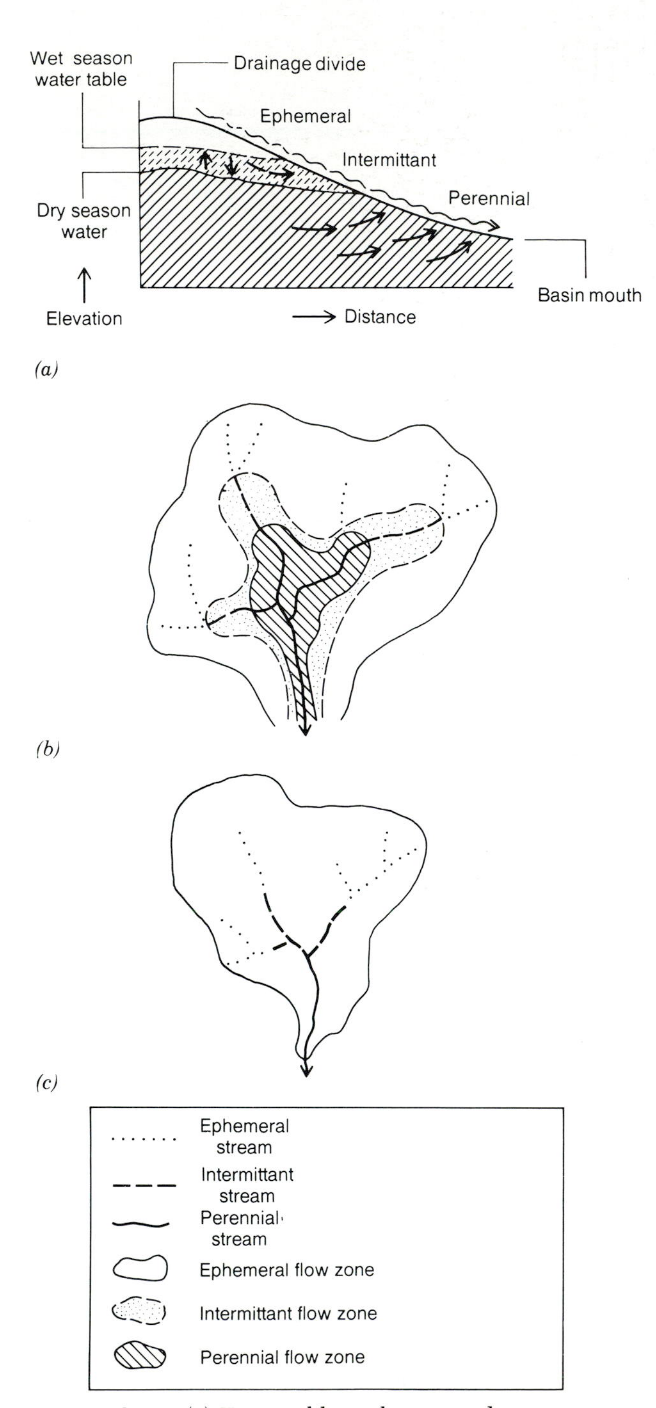

Figure 18.11 (*a*) Water table evaluation and stream types; (*b*) the flow zones in a drainage basin in relation to ephemeral, intermittent, and perennial streams; and (*c*) the character of the drainage network in a dry period when only the trunk stream carries water.

HUMAN IMPACT ON DRAINAGE SYSTEMS

Although the size, shape, and density of drainage systems are subject to natural change (as, for example, streams erode their channels), the rate of this change is typically very slow. Where humans are involved, however, this is often not the case. People have been rearranging drainage systems in major ways for thousands of years; we have only to look to the irrigation systems of the ancient world for evidence of this legacy. In the past century, the rate of drainage alteration has accelerated dramatically with urban and agricultural development.

Influences on Drainage Networks and Basins

With respect to the drainage network itself, three types of changes can be identified: pruning, grafting, and intensification. *Pruning* is characterized by the cutting back of part of the network. Certain channels, usually small ones, are mechanically filled in or diverted to another basin (Fig. 18.12). This is common with both urban and agricultural development, especially where there is a strong incentive to occupy unused ground because of high land values. *Grafting* is essentially the opposite of pruning; channels draining areas beyond the natural limits of a drainage basin are spliced into its network of streams (Fig. 18.13). Grafting is often practiced in areas of irregular terrain (and drainage), where in engineering stormwater drains for cities it is easier and less expensive to cut through a small drainage divide than to follow natural lines. This usually results in larger discharges in a trunk stream because the drainage area has been increased.

Intensification is characterized by the addition of

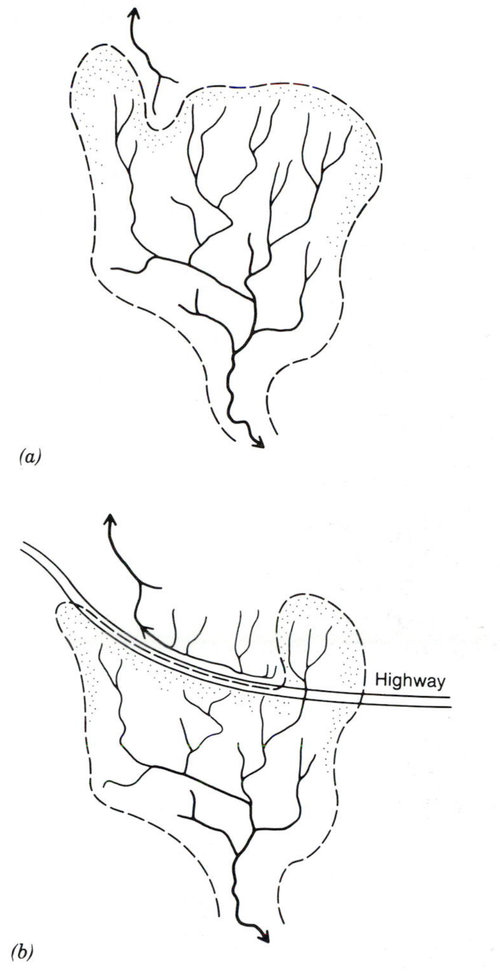

Figure 18.12 An illustration of pruning in which the upper part of the drainage network (*a*) was cut off and (*b*) diverted with the building of a highway.

new (artificial) channels to the drainage network and results in an increase in the drainage density. This change is almost always motivated by the need to hasten the release of overland flow from land use surfaces because it is a nuisance to human activities. In agriculture the need in many areas is to drain soil to facilitate plowing and planting; to do so, field drains, ditches, and tiles (small pipes buried about 2 feet under the surface) are constructed in and around fields. However, this practice seems to be prevalent only in humid regions; in subhumid and semiarid rural areas where water conservation is the objective, runoff is held on the land in ponds and reservoirs (see Note 7).

In urban areas, intensification is characterized by the construction of elaborate systems of drains composed of building roof drains, street gutters, and stormsewers. *Stormsewers* are underground pipes connected to surface drains (manholes) in streets and parking lots. Stormsewers are constructed in networks whose pipes increase in size with distance down the system (similar to natural drainage networks) and usually empty into streams and rivers. It is not uncommon for as many as 100 stormsewers, ranging from 1 foot or so to 10 feet in diameter, to empty into a one-mile stretch of river in old American cities.

As for the surface of the drainage basin itself, the major change affected by land development is the reduction in infiltration capacity. So significant is this factor that a special term, called the coefficient of runoff, has been introduced to describe this hydrologic characteristic of runoff surfaces. The *coefficient of runoff* is a number that represents the proportion of a rainfall converted to overland flow, that is, not lost to infiltration. It is given as a decimal between 0 and 1.0; the larger the decimal, the greater the overland flow. A coefficient of 0.33 means that 33 percent of any rainfall will be turned into overland flow, although strictly speaking this applies only to large rain-

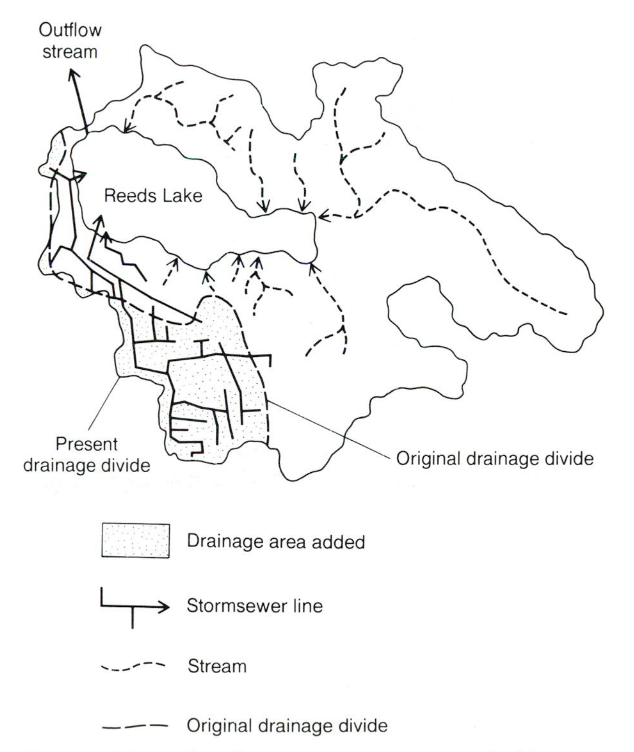

Figure 18.13 This drainage basin was extended by grafting stormsewers onto the lake which drain areas beyond the natural watershed.

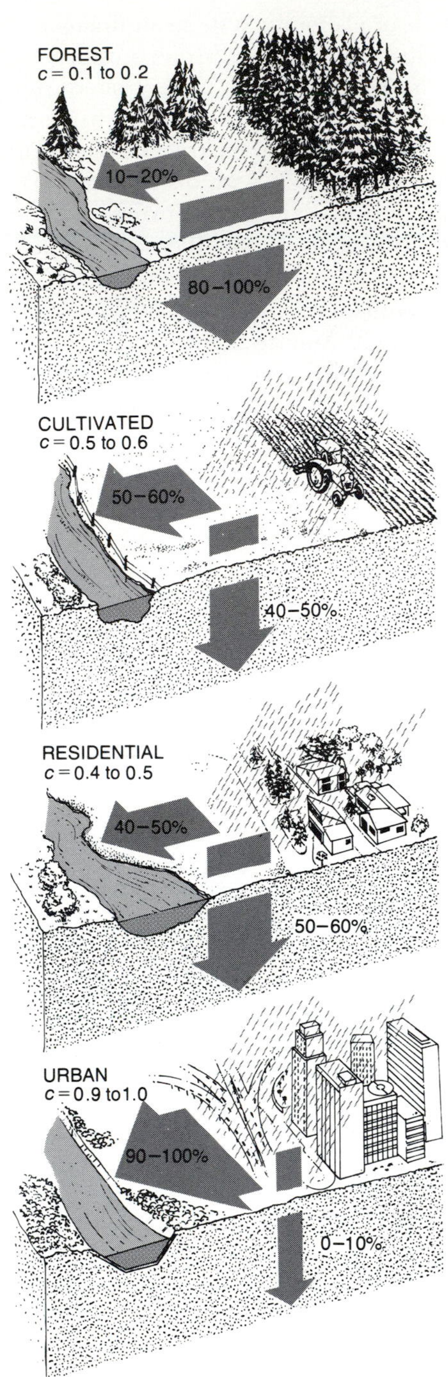

Figure 18.14 Coefficients of runoff for four types of surfaces.

storms such as those that produce several centimeters of rain per hour.

The coefficient of runoff generally increases with the density of land development, beginning with forest clearing. Residential neighborhoods made up of detached (single-family) houses have coefficients about twice those of the vegetated land they replaced. Bare soil in farmfields typically have a coefficient three times or more that of fully vegetated ground. Inner cities with continuous covers of impervious surface (concrete, asphalt, and roofs) typically have coefficients of runoff in the range of 0.9 to 1.0 (Fig. 18.14).

Influence of Development on Streamflow

When land use is examined in terms of its effects on streamflow, two significant trends can be identified: (1) A greater volume of runoff is produced for a given rainstorm; and (2) the rate of delivery of runoff to streams is much faster. The first trend is attributed to the increase in the coefficient of runoff, whereas the second is attributed to the increase in drainage density and the faster rates of channel flow in manufactured channels such as stormsewers and ditches (Fig. 18.15).

The delivery time for runoff is termed the *time of concentration* and is defined as the time it takes a droplet of water falling on the perimeter of a drainage basin to make its way to the mouth (outlet) of the basin. Stormwater moves through the basin mainly in two types of flow: overland flow and channel flow. Although overland flow moves rapidly in comparison to interflow and groundwater, it is a much slower means of water movement than is channel flow. Overland flow velocities typically range from 4 to 10 inches per second (10–25 cm/sec), whereas channel flow velocities are typically ten times greater, with the very fastest flows achieved in stormsewers and concrete-lined channels. Thus, in any drainage basin, the greater the ratio of channel flow distance to overland flow distance, the shorter the concentration time and the greater the mass of water that builds up at the basin outlet. In addition, the greater the total length of improved and artificial channels, the shorter the concentration time.

The net result of higher coefficients of runoff and briefer concentration times is that more water gets to streams faster, resulting in much larger peak discharges. This is clearly most pronounced in heavily urbanized areas, but it is also common in agricultural areas. In terms of hydrograph shape, the postdevel-

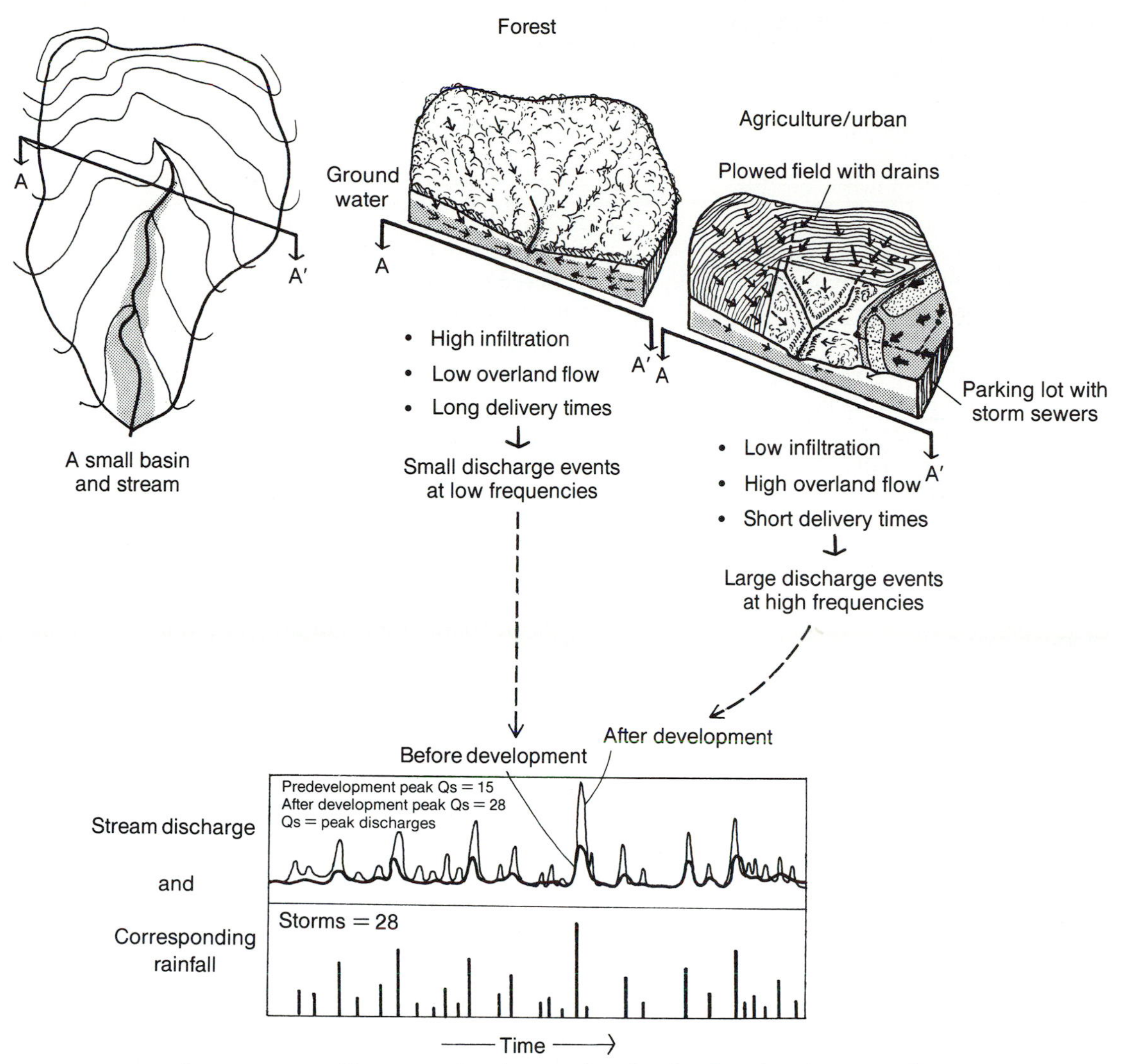

Figure 18.15 Changes in runoff associated with agricultural and urban development of a small drainage basin. The result is an increase in both the magnitude and the frequency of large flows in the small stream.

opment hydrograph is higher and its peak occurs sooner after the start (and peak) of rainfall (Fig. 18.16). Such flows produce much greater magnitudes and frequencies of flooding than the flows in predevelopment times.

In addition to increased flooding, several other changes in streams are produced by land development. Water pollution is one of the most serious of these changes because a host of contaminants are washed off urban and agricultural surfaces by stormwater. Four main types of pollutants are commonly found in stormwater: physical sediments (clay, silt, sand, and human litter); nutrients (minerals inducing rapid growth in aquatic plants); pesticide residues; and toxic materials (Table 18.1). Another serious problem is increased channel erosion, a subject covered in Chapter 20.

FLOODS

Since time immemorial, floods have ravaged human settlements in virtually every climatic region of the world. Despite the millennia of observations and folk

TABLE 18.1 STORMWATER POLLUTANTS

Category	Sources	Effects
Sediments	Cropland Construction sites Dirt roads Heavily used urban surfaces	Decreased water clarity Channel sedimentation
Nutrients	Farm and lawn fertilizers Fallout from air pollution Decaying plant litter and garbage Sewage seepage	Increased growth of aquatic plants such as algae Increased buildup in organic debris Reduced oxygen levels in water Changes in fish species
Pesticides	Cropland Lawns Pest control programs	Death in certain aquatic microorganisms Imbalances in aquatic populations Decreased recreation potential
Toxic materials	Industry Atmospheric fallout Highways	Death of various organisms Health hazard to humans Decreased recreation potential Loss of water use potential

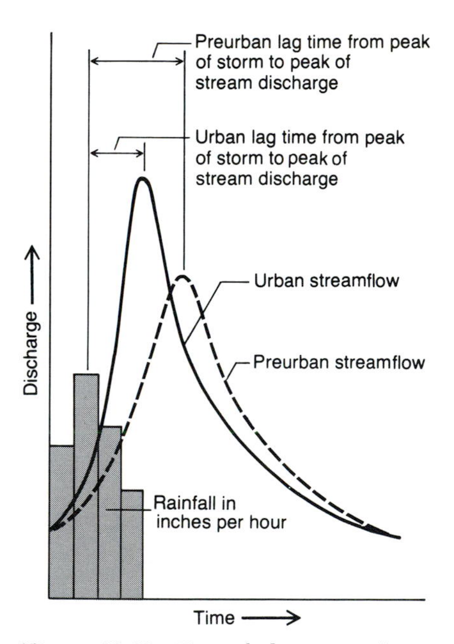

Figure 18.16 Typical changes in the response of a stream to a rainstorm as a result of urbanization of its basin.

records on floods, their often-repeated lesson has had surprisingly little effect on patterns of settlement. Indeed, floods render massive damage to America every year, and the annual averages have generally increased since the early 1900s (Fig. 18.17).

Property damage and loss of life from floods have also increased for the world as a whole, though some of the major causes are different from those in the United States. Undoubtedly, population pressure is one of the major causes of flood disasters in Third World countries. The country of Bangladesh is a good example; Bangladesh occupies a large portion of the Ganges and Bramaputra River deltas which are inhabited by tens of millions of people. These delta lands are highly susceptible to severe flooding, but as population increases in Bangladesh, people are forced to take up more and more vulnerable sites because less hazardous sites are already occupied.

In the developed world, other factors also contribute to rising flood damage. In the United States, urban sprawl has been an important factor; residential development has in many areas pushed its way into flood-prone river valleys, wetlands, and coastal lowlands. Societal attitudes and affluence have contributed to this situation inasmuch as many Americans want to live near water and can afford to move to new locations. At the same time, Americans have developed a false sense of security about their vulnerability to flooding because they tend to feel safe behind the protection of dams and other flood-control devices erected on most major rivers by various governmental agencies. Studies show that these flood protection devices have led to greater flood damage to land use in

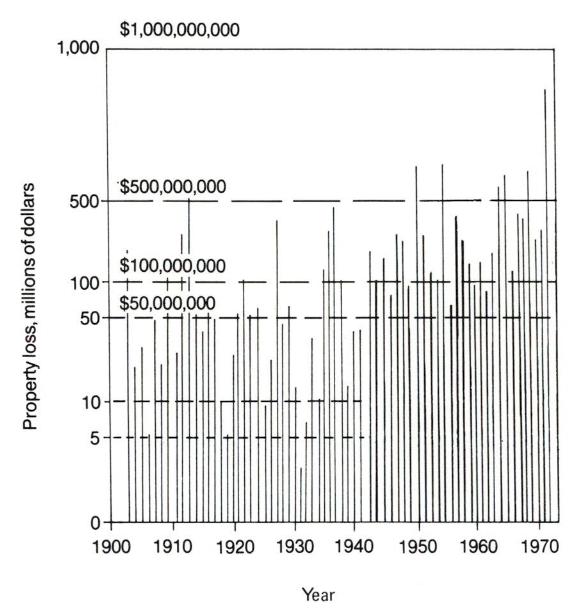

Figure 18.17 Millions of dollars in annual property damage caused by flooding in the United States since the early 1900s.

the long run. Added to all of these factors is the increase in the magnitude and frequency of stream flooding related to land development, especially urbanization.

Types of Flooding

Two different types of flooding are caused by runoff. The more spectacular type can be termed *outflooding* (Fig. 18.18), and it results when a river overflows its banks and spills into nearby lowlands, usually the river floodplain. A *floodplain* is the low, flat ground that borders the river channel (floodplain formation is discussed in Chapter 20). On large rivers with large floodplains, such as the Mississippi and the Ohio, thousands of square kilometers may be inundated, causing tremendous damage to farms, towns, and cities. Most outflooding occurs in the lower reaches of a drainage net where the floodplain is often tens of kilometers wide. Outflooding is the classical flooding reported by the ancients living in the valleys of early civilization: the Nile Valley, the Tigris-Euphrates Valley, and the Indus Valley.

Flooding also occurs in upstream areas where surface water collects in low spots before reaching a stream channel. This is called *inflooding*, and it is especially prevalent in areas of flat ground with low infiltration capacities or where drainage is closed in hilly terrain. Inflooding usually begins with the formation of "puddles" and reaches its maximum during or shortly after the rainstorm with the formation of shallow but extensive ponds.

In small drainage basins, damage caused by inflooding is often more serious than that caused by outflooding. Crop damage, delayed planting, stunted crops, septic-field malfunction, cellar flooding, and lawn damage are some of the land use problems resulting from this process.

River Flow Regimes

Although few, if any, rivers show a truly cyclical pattern of flow—that is, one with high flows and low flows occurring at the same time year after year—many do show distinct seasonal trends in their flow regimes. The strongest seasonal regimes are found among rivers fed by seasonal precipitation from climates such as the tropical wet-dry (savanna) and monsoon. Both the Ganges of India and the Indus of Pakistan, whose watersheds are dominated by the monsoon climate, produce their greatest flows in summer, the rainy season. However, since the rainy season varies from year to year, the period of peak flows often varies by a month or more.

The Nile River has a seasonal flow regime but with an interesting twist. The headwaters of the Nile are fed by the summer rains of the tropical wet-dry climate in equatorial East Africa, whereas the mouth of the river, on the Mediterranean Sea, lies in the summer-dry Mediterranean climate. The flood season on the lower Nile (in the lowlands of Egypt) occurs in late summer when the climate there is at its driest. This peculiarity in the Nile's flow puzzled the ancient Greeks, who observed the lower Nile but had no knowledge of its headwaters' climate.

In North America, the Missouri River flows through several different climatic zones with different runoff characteristics. Although much of the Missouri's drainage area lies in the semiarid Great Plains, the headwaters of most of its large tributaries lie high in the Rocky Mountains and, therefore, are strongly influenced by late-spring mountain snowmelt. As a result, the seasonal occurrence of floods on the Missouri River in the Great Plains often comes well after the peak of spring in that region (Fig. 18.19*a*). On the other hand, smaller watersheds within a single climatic zone show a seasonal distribution of floods that is more consistent with the seasonal character of the climate. The Potomac River is a case in point; most of its floods correspond to snowmelt, spring rains, and

Figure 18.18 Two types of flooding: inflooding and outflooding. In outflooding, a stream overtops its banks and spills onto its floodplain, whereas in inflooding water collects in low spots before being conducted to a channel.

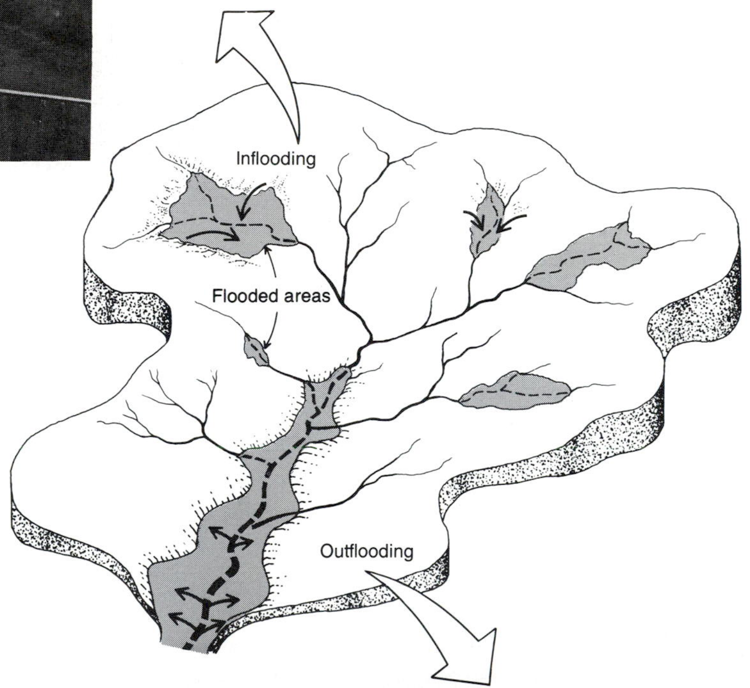

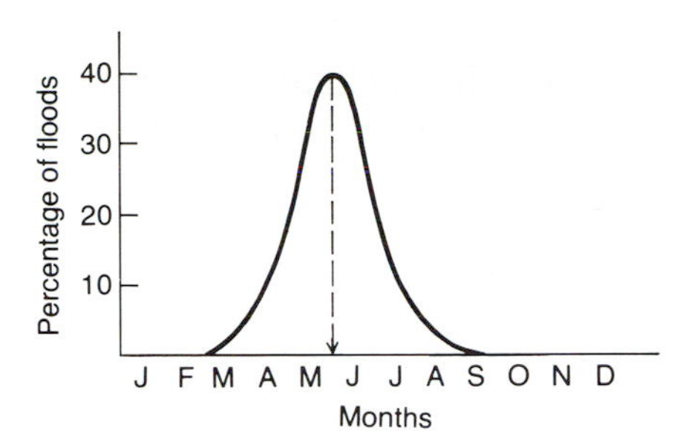

(a)

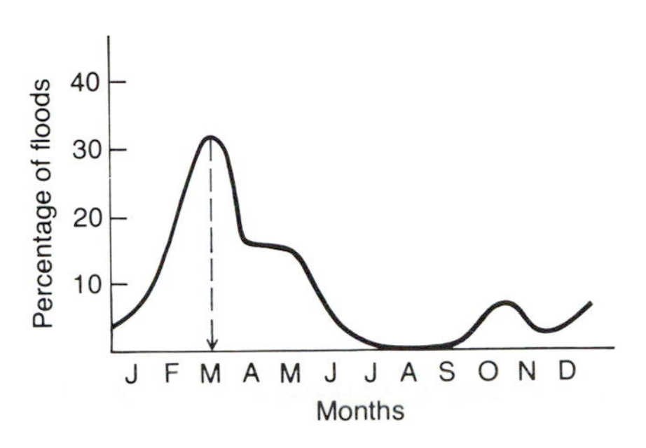

(b)

Figure 18.19 The monthly distribution of peak annual flows for rivers in different climatic zones: (*a*) the Missouri, whose peak month is 2.5 months later, which is fed by snowmelt and rainfall from the middle and northern Rocky Mountains; and (*b*) the Potomac, which is fed by rainfall and snowmelt from the middle Appalachian Mountains.

high soil water and groundwater conditions of spring (Fig. 18.19*b*).

Flood Forecasting

Forecasting flood flows is a difficult and risky task. This is not surprising considering the complex nature of the problem. Floods are caused by many factors, not just heavy rainfalls, but by rapid snowmelt, short-term rises in ocean levels caused by storm surges, and channel constrictions caused by dams of broken river ice as well. All of these factors are transient, which means their occurrence in a drainage basin tends to be sporadic. A large part of the difficulty in flood forecasting lies in not knowing when any one of these factors will occur in the basin. The problem is compounded when we consider the chances of two or more happening at the same time. Indeed, some of the worst floods in modern history have been caused by combinations of transient factors such as heavy rainfall and rapid snowmelt.

But floods are also influenced by factors that are relatively permanent (nontransient) such as the size of the drainage basins, the surface characteristics of the land, soil types, and the number and lengths of channels in the basin. A forecasting method called the *rational method* is based on two nontransient factors, basin area and coefficient of runoff, and one transient factor, rainfall intensity. Because a specified rainfall intensity cannot be forecast with much accuracy, the rational method starts out with a given intensity based on averages of rainstorm intensity (Fig. 18.20). Thus,

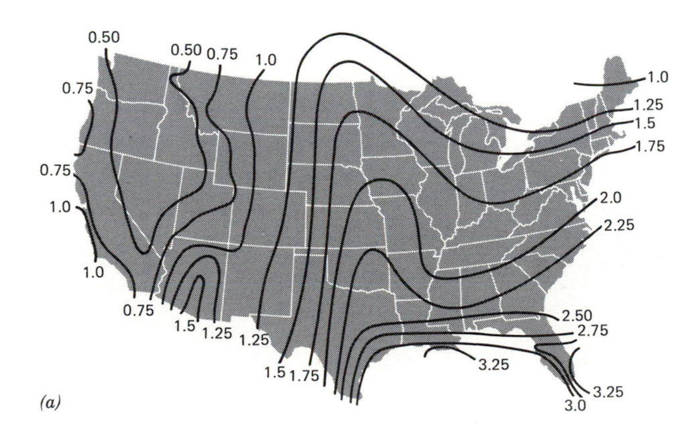

(a)

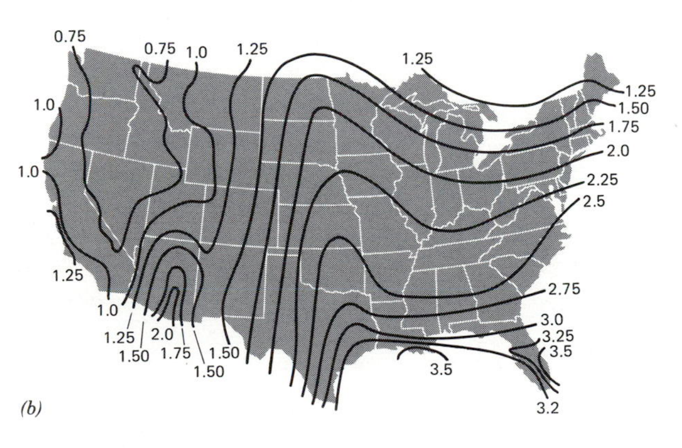

(b)

Figure 18.20 Amounts of rainfall produced by (*a*) 5- and (*b*) 10-year storms in the coterminous United States. The values represent 1-hour rainfall intensities.

in using this method, we would say, "*suppose* the 10-year, one-hour rainstorm[2] were to happen in this basin which has an area of 250 acres and a coefficient of runoff of 0.65." By multiplying one factor times another, we can estimate, or forecast, the flood flow that would be produced. Of course, the forecast is limited by the fact that we really do not know *when* the 10-year rainstorm will occur; we only know that when it does, the streamflow will reach a certain magnitude. Figure 18.20 shows the distribution for the one-hour rainstorms of 5-year and 10-year average frequency.

Another method of flood forecasting, or estimating, is called the *magnitude and frequency method.* This method is based on records of a river's past flows and is used for rivers with large drainage basins (in contrast to the rational method which is limited to small basins, generally less than 2 square miles in area). The idea behind this method could be put this way: "if we can learn how a river behaved in the past, we should be able to forecast how it will behave in the future." To do this we take records giving the sizes and years of peak annual flows and determine how frequently flows of different magnitudes have occurred on the average. This average is called the *recurrence interval* or *average return period,* and as with the rainstorms in Figure 18.20, it is expressed as the 2-year flow, the 5-year flow, and so on up to the recorded limit of peak flows, the 30-year flow.

In the magnitude and frequency method, forecasts are given as a probability. For example, the 2-year flow has a 50 percent chance of occurrence in any year (because it happens on the average of once every 2 years), and the 25-year flow, which is much larger, has a 1/25th or 4 percent chance of occurrence in any year. In addition, if a graph is constructed based on

[2]The 10-year, one-hour rainstorm is defined as the rainfall of greatest one-hour intensity that occurs on the *average* of once every 10 years (Fig. 18.20*b*).

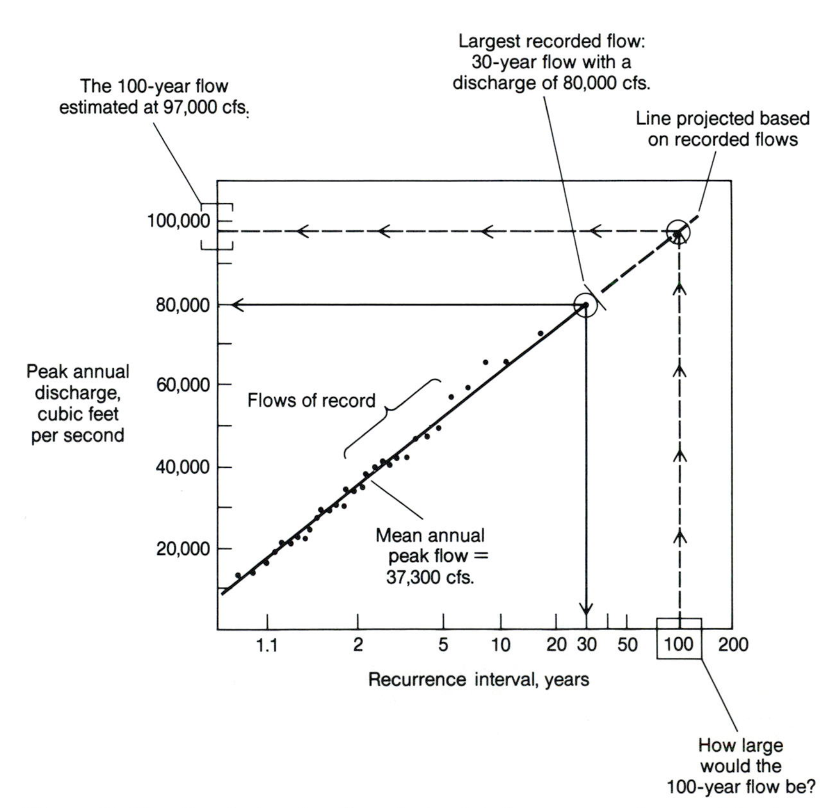

Figure 18.21 Graph showing the relationship between peak annual discharge and recurrence interval (average return period) for the Sky Komish River at Gold Bar, Washington. The broken line represents projection beyond the years of record.

a plot of recurrence intervals and their respective flow magnitudes, it is possible to make projections of flow magnitudes that have not yet been recorded, that is, to project to recurrence intervals beyond the period of record (Fig. 18.21).

Such projections are very helpful in land use planning and civil engineering because they provide a way of estimating the extent and depth of infrequent events, such as the 100- or 200-year flood, that have not yet been recorded. Such information is critical to certain land use planning programs. In the United States the federal government now requires that the limits of the 100-year flood be mapped in major river valleys for the communities in or bordering them to qualify for National Flood Insurance. This insurance program provides assistance to flood victims living in or near officially designated floodplain zones who have met certain program requirements such as flood-proofing of houses (Fig. 18.22).

Another type of flood forecasting involves *monitoring* the downstream movement of a flood crest (after the water is in the channel) and projecting when it will hit various places along the way. The U.S. Army Corps of Engineers, the U.S. National Weather Service, and other agencies provide this kind of forecasting for major rivers such as the Mississippi and for coastal areas affected by flooding from rivers and/or hurricane surges. In extreme instances, evacuation plans prepared by planning agencies are activated, and the population is required to vacate the threatened area (Fig. 18.23).

Flood Protection and Control

Two general approaches are used in flood protection: (1) control of the river by structural means such as dams; and (2) control of the distribution of population by means of government regulations and information

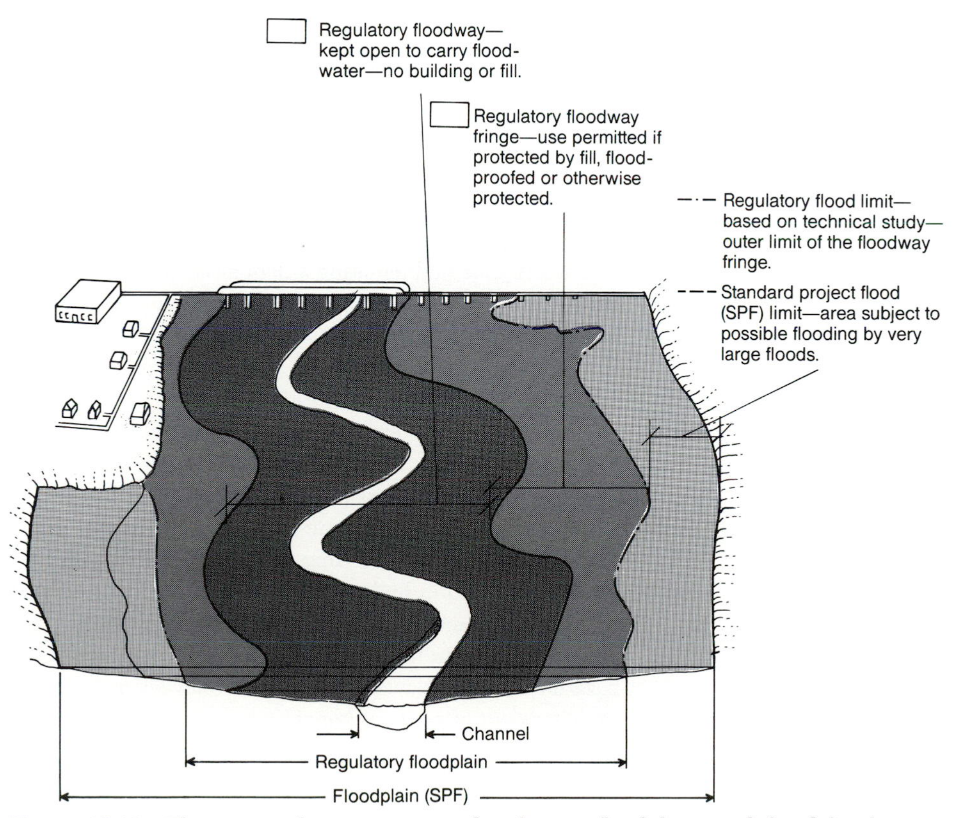

Figure 18.22 The principal zones associated with river floodplains as defined for the U.S. National Flood Insurance Program. To qualify for insurance coverage in the regulatory floodway fringe, "flood-proofing" is required.

THE VIEW FROM THE FIELD / NOTE 18

Constriction of the Mississippi River and Its Effect on Flood Magnitude

The 1973 deluge on the Mississippi River was reported by the news media as a 200-year flood, yet the discharge records show that the flow was only a 30-year event. At Saint Louis, Missouri, the flood began on March 10 and continued for 77 consecutive days, exceeding the record set in 1844, when the river was in flood for 58 days during the entire year. The river crested at Saint Louis on April 28, 1973, at a gage height of 13.18 m (13.2 ft. above flood stage) and a peak discharge of 24,100 m^3/sec. This stage (flow elevation) topped the 189-year record by 1 ft. and was 2 ft. higher than the 1844 crest; however, the discharge was about 35 percent less than the estimated peak flow in 1844. Compared to the 1908 flood, the 1973 flood had the same flow, but the peak stage was 8.2 ft. higher.

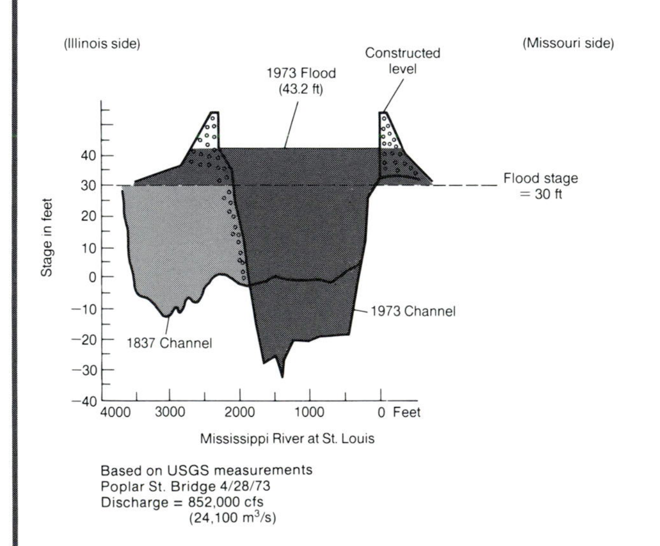

Some investigators attribute higher stages of modern flows to a combination of levee confinement and channel constriction from navigation works such as wing dikes, side channel dikes, and revetments, which reduce the cross-sectional area of the channel. Tampering with the Mississippi River in the Saint Louis area has a long history which started in 1837, when Lieutenant Robert E. Lee built the confinement dikes to remove sandbars threatening the Saint Louis harbor. Engineers narrowed the river from 1300 m (4300 ft) in 1849 to 610 m (2000 ft) in 1907 and finally to 580 m (1900 ft) in 1969.

A natural alluvial river generally widens its channel in response to a large flood, depending on the relative erodibility of its bed and banks. The width of an alluvial river channel over a long period of time is a function of average discharge, other hydrological and geomorphologic factors being equal. Between 1803 and 1860, the Mississippi widened itself mostly in response to four large floods. After 1881, it became more difficult for the river to widen itself, because of the navigation dikes and the structural bank protection. The Mississippi responded by downcutting its bed, creating a deeper channel. This is reflected in the declining level of minimum annual stages. But for peak annual flows, the trend has been toward higher stages.

Since 1837, the channel at Saint Louis has lost about a third of its volume owing to the installation of navigation works. As a result, during a flood on the thus modified Mississippi, the stages are higher for a given discharge. In some reaches of the river, deposition also occurs, causing a further rise in stages. Hemmed in by levees, flows that would otherwise be contained by the channel and lower floodplain are forced to higher elevations, creating more hazardous floods. Navigation works and levees make big floods out of moderate ones.

Charles B. Belt, Jr.
Saint Louis University

Figure 18.23 A scale model of a portion of the Mississippi River used to simulate the behavior of flood flows. The rows of cards produce an effect on flow similar to that of trees.

programs. In the United States river control has long been the favored approach (see Note 18). Typical control measures include *dams*, which are used not only for flood control but also as a source of water for agriculture and cities; *levees*, which are builtup banks designed to contain higher level flows; and *channelization*, which involves deepening and straightening channels to conduct larger flows at faster rates.

In recent decades, grave doubts have been raised concerning the long-term effectiveness of river control programs. The doubts stem from the fact that the flood damage continues to increase each decade (see Fig. 18.17) and that the *benefits* gained from the construction of reservoirs and channel "improvements" do not adequately offset the *losses* in usable farmland, woodland, natural river channels, and plant and animal habitats. As a result, alternative programs have been implemented based on (1) land use planning which guides or restricts certain land uses, such as residential, from flood-prone areas, and (2) stormwater control practices aimed at reducing the rates of runoff from developed areas which are contributing to increased magnitudes and frequencies of flooding.

Summary

There are two categories of runoff—surface and subsurface—and both are fed solely by precipitation. Outside stream channels themselves, overland flow is the principal type of surface runoff, and groundwater is the principal type of subsurface runoff. Overland flow is associated mainly with intensive precipitation and low infiltration capacities, and it is delivered rapidly to streams. Groundwater flow shows a much slower response to rainfall, and it is delivered to streams very gradually. Streamflow dominated by overland flows, such as those in arid lands, agricultural landscapes, and urban areas, produce a greater magnitude and frequency of peak flows (and floods) than those dominated by groundwater flows.

Streams and their drainage basins are altered in many ways by land development, and these alterations usually affect their flow characteristics, often severely. Among the various means of forecasting floods, the rational method and the magnitude and frequency method are used widely; however, flood forecasting remains an inexact science. Property damage from flooding increases each decade in the United States, despite major efforts to control rivers through dams, levees, and channelization. Alternative ways of reducing flood damage are now being implemented, and they include land use control in developed areas.

Concepts and Terms for Review

- runoff
  - overland flow
  - channel flow
  - subsurface sources
  - units of measure
- hydrologic equation
- discharge and stage
- drainage basins and networks
- principle of stream orders
- sources of discharge
  - channel precipitation
  - overland flow
  - interflow
  - groundwater flow
- rainfall-runoff relations
  - stormflow
  - baseflow
  - hydrograph

stream and river types
  perennial streams
  exotic rivers: influent and effluent flow
  intermittent and ephemeral
human impact on drainage systems
  pruning and grafting
  intensification
  stormsewers
  runoff rates
  water quality
floods
  human situations
  types of flooding
  river flow regimes
flood forecasting
  rational method
  magnitude and frequency method
  flow monitoring
flood protection and control
  approaches
  benefits and losses

Sources and References

Belt, C.B., Jr. (1975) "The 1973 Flood and Man's Constriction of the Mississippi River." *Science* 189: pp. 681–684.

Betson, R.P. (1964) "What is Watershed Runoff?" *Journal of Geophysical Research* 68: pp. 1541–1552.

Bruce, J.P., and Clark, R.H. (1966) *Introduction to Hydrometeorology.* Oxford, England: Pergamon Press.

Dunne, T., and Black, R.D. (1970) "Partial-Area Contributions to Storm Runoff in a Small New England Watershed." *Water Resources Research* 6: pp. 1296–1311.

Dunn, T., and Leopold, L.B. (1978) *Water in Environmental Planning.* San Francisco: W.H. Freeman.

Hewlett, J.D. (1961) "Soil Moisture as a Source of Baseflow from Steep Mountain Watersheds." *U.S. Forest Service Station Paper 132,* Southeastern Forest Experiment Station.

Kirkby, M.J., ed. (1978) *Hillslope Hydrology.* New York: John Wiley & Sons.

Kochel, C.R., and Baker, V.R. (1982) "Paleoflood Hydrology." *Science*, 215:4531.

Penman, H.L. (1961) "Weather, Plant and Soil Factors in Hydrology." *Weather*, Vol. 16, pp. 207–219.

Satterland, Donald R. (1972) *Wildland Watershed Management.* New York: Ronald Press.

Ward, R. (1978) *Floods, A Geographical Perspective.* New York: Halsted.

Chapter 19

Rock Weathering, Slope Erosion, and Landforms

If geologic processes build up the landmasses, there must be opposing processes that wear them down. Two sets of processes, weathering and erosion, are mainly responsible, and as they lower the terrain hillslopes, mountains and plains take form. How do these processes break down bedrock and remove the residues from the land? How are landforms sculpted in the process? Are these processes related to climate, vegetation, and land use in such a way that we can expect to find certain landforms in certain geographic regions?

The nature of change in the earth's surface has been a major issue in science for centuries. Before 1850 or so, the argument on how the earth changed to acquire its present landforms, water features, and organisms centered on two conflicting ideas: (1) that change took place in one or two massive episodes when the world was created; and (2) that change took place gradually over very long periods of time. The implications of these ideas have been extremely important to the development of thought on the origin of the earth and its landscapes. For one thing it has been important to geographers in evaluating the nature and rates of change in modern landscapes under the influence of massive human populations compared to those in ancient landscapes with little or no humans.

We are still a long way from a complete understanding of the nature of landscape change, but we do understand some aspects of it. First, it appears that different phenomena have different modes of change. Some features—for example, the massive bodies of crystalline rock in the continental interiors—seem to change slowly but steadily over millions of years, whereas others, such as many flowering plants, seem to undergo episodic evolution in which they go unchanged for eons and then change drastically in very short time spans—some thousands of years, which is but a moment on the geologic time scale.

The landforms of the continents—the hillslopes and valleys that we see around us—appear to result from several different modes of change. In some cases, the old stereotype of the everlasting hills may be appropriate, for some hills are changed so slowly by weathering and erosion that in several human lifetimes net variation in mass or form is effectively zero. But we also know that the forces that operate on and within hills often function like triggers that, when released, can produce massive change, such as landslides, in a matter of seconds. We also understand that the rates of change in the landscape as a whole are rising as human impact on the earth increases. In fact, the earth has probably never been changed so much in so short a time as it has in the human era.

In the great system of processes that wears down, sculptures, and generally reshapes the surface of the crust, weathering is treated as the first phase because it prepares bedrock for erosion. The erosional processes that carry away weathered material are termed *geomorphic agents:* streams (and related forms of runoff), glaciers, waves and currents, and wind. The surface forms—hillslopes, river valleys, sea cliffs, rock domes, and canyons—shaped by weathering and erosional processes are called *landforms.* The formal study of landforms, and the weathering and erosional processes that shape them, is called *geomorphology.*

In this chapter we explore first the conditions and processes of rock weathering and then the various processes that wear down and shape hillslopes. We will examine both those hillslope processes that work almost constantly at very slow rates and those such as landslides that work sporadically with infrequent episodes of high rates of work. The final section describes the common slope forms of different landscapes.

WEATHERING

By definition, weathering is different from erosion; weathering is only the breakdown of rock, whereas erosion is the removal of the debris produced by the breakdown. This distinction is purely a matter of scientific convention; in reality, weathering and erosion are usually part of a chain of related processes, the end result of which is the loss of rock materials from the land. For our purposes, we can treat weathering in two main categories: chemical and mechanical.

Weathering Environments

Weathering is most intensive at or near the earth's surface. There are two basic reasons for this fact: (1) many rocks, particularly the igneous rocks, formed under conditions of heat, pressure, and chemistry so different from those of surface environments that they tend to be unstable upon contact with surface environments; (2) the energy that drives the weathering processes is most concentrated on the earth's surface. This energy, represented by water, heat, organic matter, and some types of mechanical activity, is closely related to conditions of the bioclimatic environment. Water is the most effective natural solvent on the planet, and heat generally increases the solubility of materials because it speeds up chemical reactions. The remains of plants contribute ions to water, which may increase its efficiency. Because the availability of water, heat, and organic matter varies with climate, weathering is as much a function of surface geographic conditions as it is of rock type.

Throughout this book we have mentioned the principle of physical stability in reference to various parts of the environment. In the atmosphere, for instance, air is stable or unstable depending on its temperature structure. If cold, dense air overlies warm, light air, the two bodies will overturn, and the situation is thus said to be gravitationally unstable. This concept is the basis for understanding how the wind is produced. By the same token, the concept of chemical stability is essential to an understanding of chemical weathering.

Chemical stability refers to the tendency of the minerals that make up rock, such as feldspar and quartz, to change by chemical action into other states, as from one mineral into another type of mineral. Most minerals form within the crust in igneous and metamorphic environments that are very hot (crystallizing temperatures ranging from 800°C to 1200°C), high in pressure (ranging from 1000 bars to 10,000 bars), and low in water (less than 1 percent by volume in basaltic rocks). When exposed to the surface environment thermal, pressure, and moisture conditions are not only different, but they also come into contact with atmospheric gases and chemicals from organic sources, and chemical reactions usually take place. Whether a mineral's behavior is stable or unstable under these conditions depends primarily on the conditions under which it originally formed.

The least stable minerals are those that formed at the highest temperature and pressures in igneous environments. In the presence of water or air, these minerals are prone to chemical reactions leading to their transformation into other materials or mineral types. Calcic feldspar, for example, which forms at relatively high temperatures, is transformed by chemical action into clays and soluble ions, neither of which can be reconstituted into the original mineral. Quartz, on the other hand, crystallizes at lower temperatures and is stable compared with calcic feldspar because it can readily be reconstituted into quartz crystals. Figure 19.1 lists ten minerals according to their chemical stability at the earth's surface.

Most chemical reactions take place with greatest intensity in the presence of liquid water. The power of water to affect chemical weathering is controlled by three factors: (1) availability of water; (2) temper-

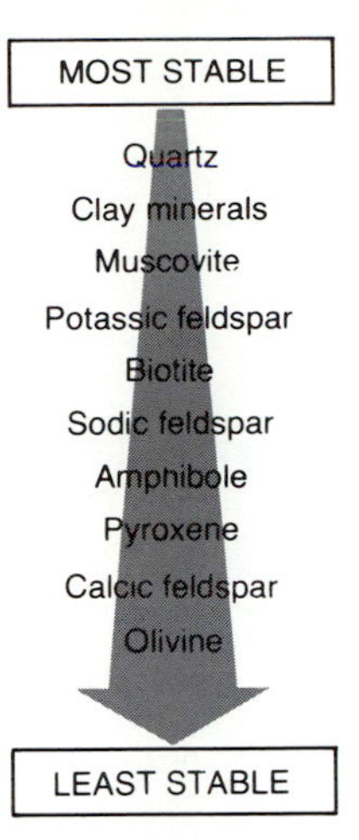

Figure 19.1 Stability of ten common minerals under various conditions at the earth's surface.

ature of the water; and (3) chemicals in the water (Fig. 19.2). The first is readily apparent when we examine rocks in deserts and wet environments. In wet areas rocks such as granite, which contain both stable and unstable minerals, appear crumbly or "rotten" because the feldspar and the biotite mica, both relatively unstable, break down around the quartz crystals when they chemically react with water (Fig. 19.3). In humid regions where the terrain has remained free of major disturbances (such as glaciation or volcanism) for millions of years, the residues from the breakdown of granite and related rocks reach thicknesses of 20 to 30 m. In deserts, granite often has the look of monumental stability because the more reactive minerals, with little water to act on them, remain intact for long periods of time.

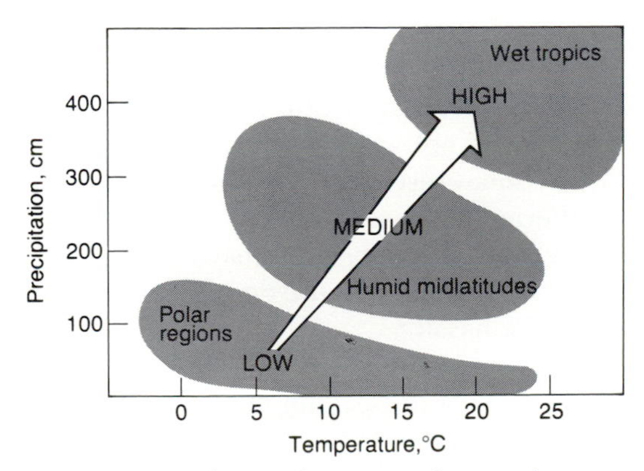

Figure 19.2 The combinations of atmospheric moisture and heat that are conducive to high, medium, and low rates of chemical weathering, respectively.

Figure 19.3 "Rotten" granite, the result of chemical weathering in the presence of moisture. The black object is a lens cap about 5 cm (2 in.) in diameter.

Temperature affects chemical weathering because heat influences the rate of chemical reactions. Most reactions double with each 10°C rise in temperature. Therefore, we can expect that, in a tropical environment with an average temperature around 27°C, chemical weathering can decompose a rock as much as four times faster than in the subarctic at an average temperature of 7°C or so.

Essentially all water in nature contains chemicals besides H_2O. For example, when water vapor condenses in a cloud, it assimilates carbon dioxide (CO_2) from the air. Through this process, the CO_2 content of rain or snow can be raised by ten to thirty times over the amount of CO_2 normally in the atmosphere (0.03 percent). Once dissolved in water, the carbon dioxide combines to form a weak carbonic acid. On the ground, additional carbonic acid can be formed if the rain infiltrates decomposing organic material. Organic matter in the topsoil is the food for bacteria, and through respiration they in turn produce carbon dioxide. In moist, aerated soils, bacterial activity may raise the CO_2 content of soil air to as much as 10 percent by volume. Added to soil water and groundwater, the CO_2 increases the concentration of carbonic acid.

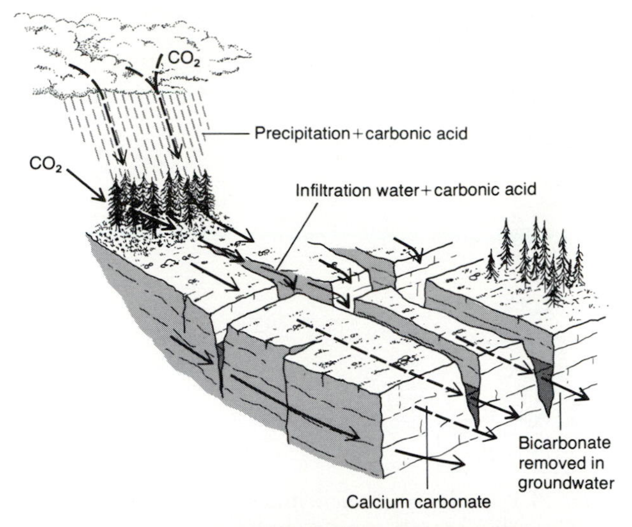

Figure 19.4 Disintegration of limestone as a result of the reaction of natural carbonic acid with the calcium. The picture was taken on an exposure in the Niagara formation, Manitoulin Island, Ontario.

Types of Chemical Weathering

When carbonic acid is introduced to limestone, it reacts with calcium. From this reaction, a bicarbonate soluble in water is formed. Thus, as acidic groundwater percolates through the limestone, the calcium carbonate, which makes up more than 90 percent of this rock, is leached away (Fig. 19.4). In humid areas this process, which is known as *solution, or carbonation, weathering*, is highly effective in the decomposition of limestone, especially if the rock is riddled with cracks. Left in place of the dissolved limestone are small residues of detrital sediments which were originally deposited with the limestone on the sea bottom. In time, these particles, which are mainly clay, accumulate along with organic material and deposits from runoff and wind to form a mantle of soil. The dissolved calcium, on the other hand, enters the groundwater and may eventually be released into a stream that ultimately discharges into the sea. In many areas of limestone, solution weathering produces rock caverns and gives rise to a type of terrain called *karst* topography, which is described in Chapter 20.

Another weathering process involving organic matter is *chelation* (pronounced key-la-tion). This process is characterized by the bonding of mineral ions to large organic molecules that are excreted from vegetation. Our knowledge of this process has developed in connection with research on agricultural fertilizers, and, although chelation is not well understood under natural conditions, some researchers regard it as an important weathering process. Lichens, for example, excrete chelation agents, and studies have shown that lichen-covered basalt is often more deeply weathered than is lichen-free basalt.

Igneous rock is also weathered by other chemical processes involving water carrying minerals. It often starts when water penetrates the rock along the contacts between crystals and is absorbed by certain minerals. This process, called *hydration*, produces no chemical change itself, but sets up a sequence of chemical reactions that alter the minerals irreversibly.

One set of these reactions is *hydrolysis,* which usually involves the reaction of water and an acid on a mineral. Hydrolysis is considered to be the most effective process in weathering granite and related rocks.

Acting on potash feldspar, a major constituent of granite, hydrolysis produces the clay mineral kaolinite as well as ions of silica in solution. Under the hot, wet conditions of the tropics, the chemical alteration of both potash and plagioclase feldspars yields bauxite, an oxide of aluminum. In some parts of the tropics, bauxite has accumulated in such quantities that it is commercially mined for aluminum ore. Bauxite is also the main constituent of laterite, the hard layer that forms in tropical soils.

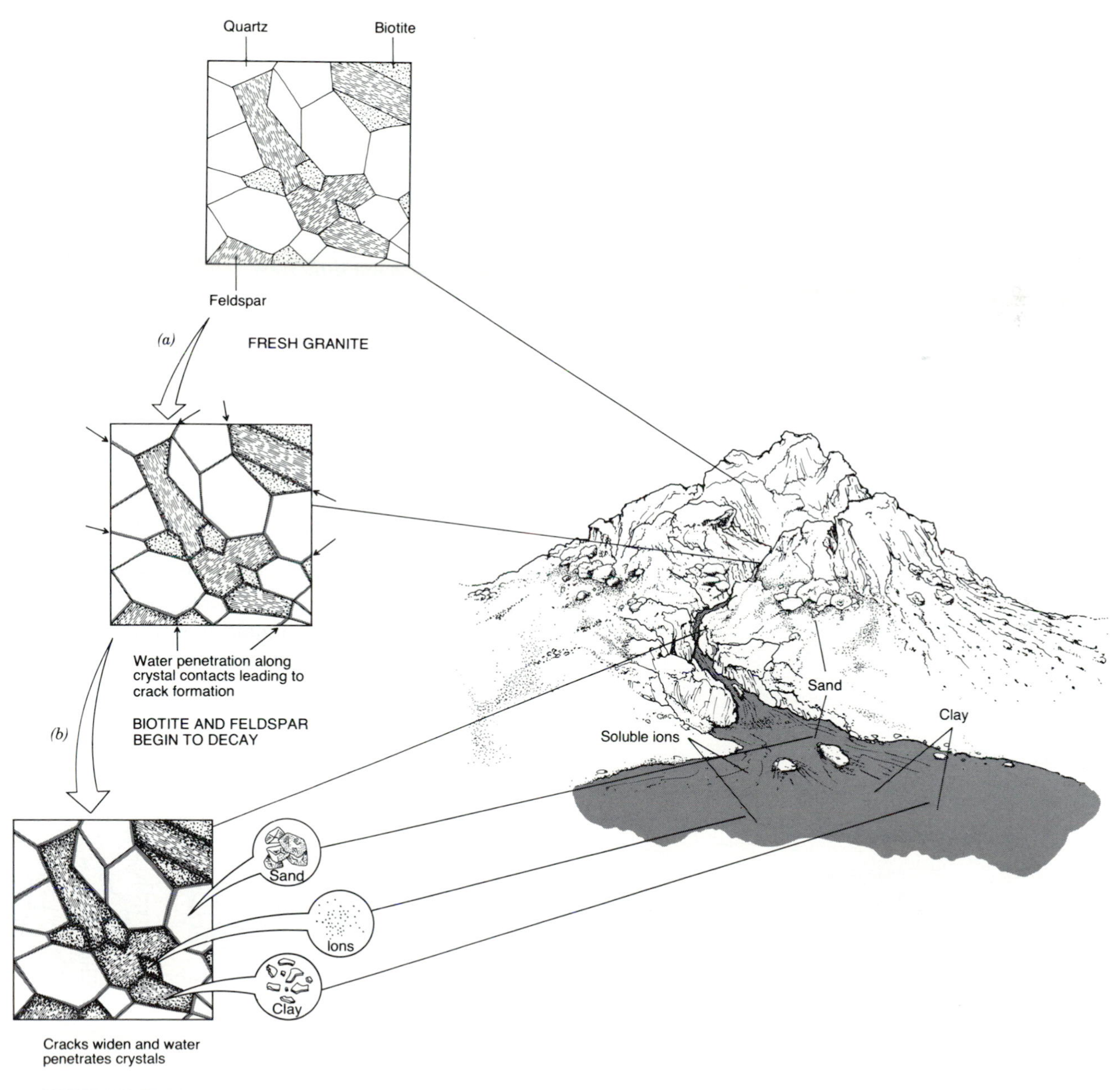

Figure 19.5 Some stages in the decay of granite. With the decay of biotite and feldspar, more cracks are opened, leading to faster and faster rates of decomposition. The prinicipal end products are soluble ions, sand (silica), and clay.

TABLE 19.1 COMPOSITION OF SEAWATER

Mineral	Parts Per Thousand
Chloride	19.3
Sodium	10.6
Sulfate	2.7
Magnesium	1.3
Calcium	0.4
Potassium	0.4
Bicarbonate	0.1
Minor constituents	0.1
	34.9

Another type of chemical weathering, *oxidation*, usually accompanies hydrolysis. It involves a number of silicate minerals but is most apparent in rocks containing iron. The iron in olivine and pyroxene, for instance, is altered in the presence of oxygen to form ferric iron, which in turn is transformed into limonite, a mineral resembling rust.

Because of the diversified mineral composition of the igneous rocks, decomposition tends to take place at very uneven rates. In the case of granite, its biotite and feldspar constituents alter rapidly, whereas its quartz is slow to change. Practically all igneous rocks, however, weather to yield four types of end products: (1) silica particles, which often end up as sand; (2) clay, which often turns out to be kaolinite; (3) soluble ions, which enter the water system; and (4) oxides of iron or aluminum, which impart reddish or brownish coloration to the soil (Fig. 19.5).

The soluble ions produced in weathering are carried off in groundwater and streamflow, eventually reaching the sea. The composition of seawater attests to this unending process (Table 19.1). Clay and quartz are left behind to form the soil cover. In time, even the quartz is corroded by organic acids, and it too is leached from the soil, but this process takes millions of years. This appears to have happened in many areas of the tropics because soils there that have been derived from granitic rocks are poor in quartz.

Mechanical Weathering

Mechanical weathering, the second major category, physically fragments rock. Virtually everywhere mechanical weathering operates hand-in-hand with chemical processes, and in most places it is impossible to ascertain how much work should be ascribed to each. By fragmenting rock, mechanical weathering increases the total surface area over which chemical weathering occurs (Fig. 19.6). Mechanical weathering is apparently more effective in cold, dry environments, whereas chemical weathering tends to be more effective in warm, wet environments. Mechanical weathering begins with the formation of cracks in bedrock. Crack formation is initiated in basically three ways: (1) differential expansion of rock masses; (2) chemical decomposition along bedding planes or contacts between different rock types; and (3) expansion within bedrock by freezing water. When cracks widen and deepen as chemical and mechanical weathering proceeds, they are called *joint lines* (Fig. 19.4). Joint lines may extend tens or even hundreds of meters into

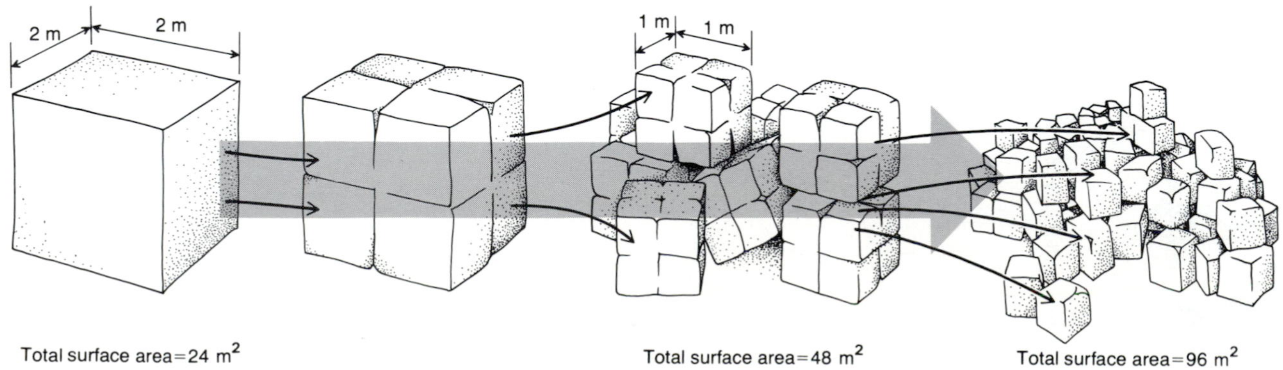

Figure 19.6 As mechanical weathering breaks rock down into smaller fragments, the amount of surface area exposed to chemical weathering increases. Here the rock is sectioned into eight pieces, and the surface area doubles with each step. Thus, the weathering of a given mass of material accelerates over time because each doubling also doubles the area exposed to chemical weathering.

Figure 19.7 Exfoliation in the Sierra Nevada, California, and the resultant accumulation of debris.

bedrock and develop lateral offshoots. When horizontal and vertical joint lines intersect, blocks of rock are freed from the solid earth.

Where massive bodies of igneous rocks are exposed, a type of mechanical weathering called *exfoliation* tends to develop. Exfoliation is thought to be caused mainly by the differential expansion of a rock body as it decompresses with the loss of heavy rock overburden. The expansion produces a system of joint lines that yield scalelike sheets or slabs of rock (Fig. 19.7). As the sheets detach, they slide downslope or disintegrate into smaller particles that are subsequently eroded away. In any case, once the process starts, it is self-perpetuating: each shedding of mass results in decompression, leading to more shedding and so on.

At a much smaller scale, individual boulders may also be subject to scalelike weathering, called *spheroidal weathering*. Although this weathering pattern looks like a miniature version of exfoliation, it is apparently due to chemical weathering and does not involve decompression and rock expansion.

Frost wedging (or *frost shattering*) appears to be a very effective weathering process in polar and high mountain environments. Unlike exfoliation, which is impossible to simulate in the laboratory, the physical process of frost wedging is easy to demonstrate. Experiments show that when water freezes, the force of crystallization can be tremendous. At a temperature of −20°C—a common surface temperature during winter in cold environments—the force exerted by the expansion of ice in a closed container is more than enough to burst steel pipes.

We know from the principles of ground-heat flow that freezing in bedrock progresses from the top down (see Fig. 4.6). Therefore, the first ice to crystallize in a water-filled crack forms a cap over the water. The cap resists some of the upward expansion produced by subsequent freezing, thereby causing a large proportion of the resultant force to be directed downward and outward against the crack (Fig. 19.8). As a result, when the water is truly confined, cracks and joint lines may enlarge, and boulders may even be split open or shattered (Fig. 19.9). Freeze-thaw expansion is also effective in breaking down weathered material, such as boulder-, cobble-, and pebble-sized particles.

Little is known about wedging caused by other substances. Crystallization of salt- or quartz-rich solutions can produce a wedging effect, although we have no information on how much work it may do. Wedging by plant roots is often cited as an example of mechanical weathering, but there is little evidence as to how effective roots are in fragmenting rock. They do cause damage to sidewalks and play an important role in chemical weathering because they help to hold moisture and organic matter in joint lines and also contribute ions to soil water.

Mechanical weathering of rock is also produced by a variety of other processes; however, we know little about them in terms of their relative effectiveness. Some of these processes include rock shattering and fracturing from the impact of a long fall, such as from the Yosemite Valley cliff (Fig. 19.7), rock plucking by ice falling from a cliff face, and breakage from lightning strikes and the intense heat of forest fires.

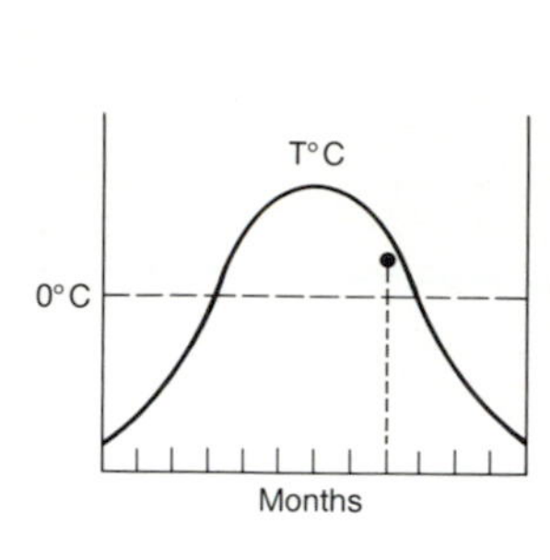

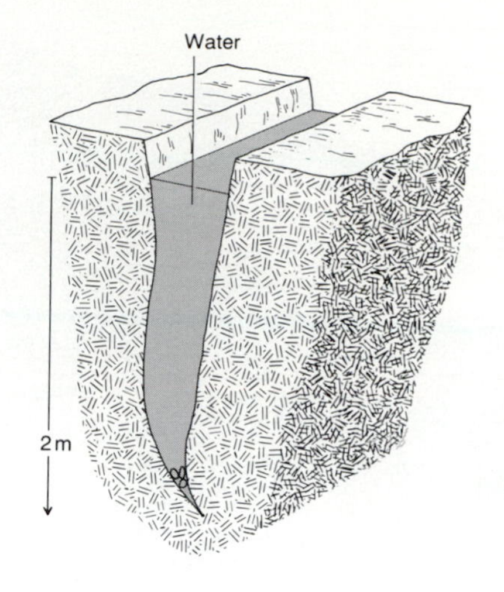

(a)

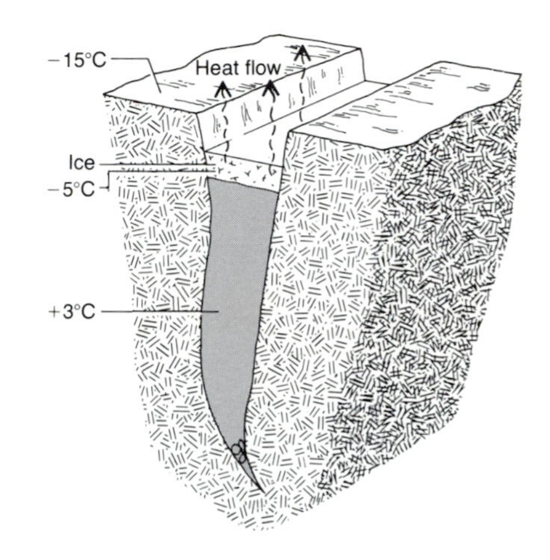

(b)

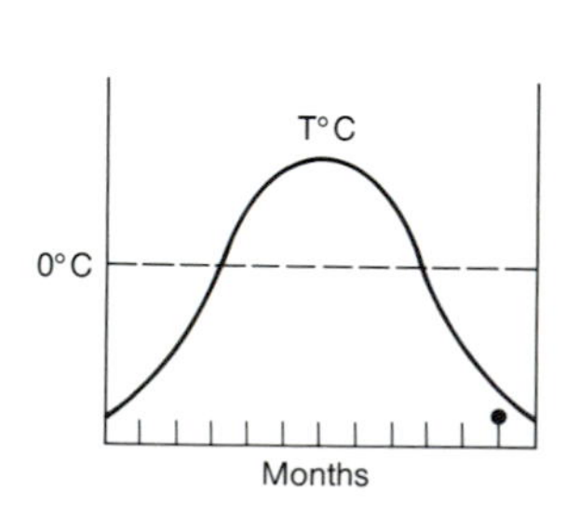

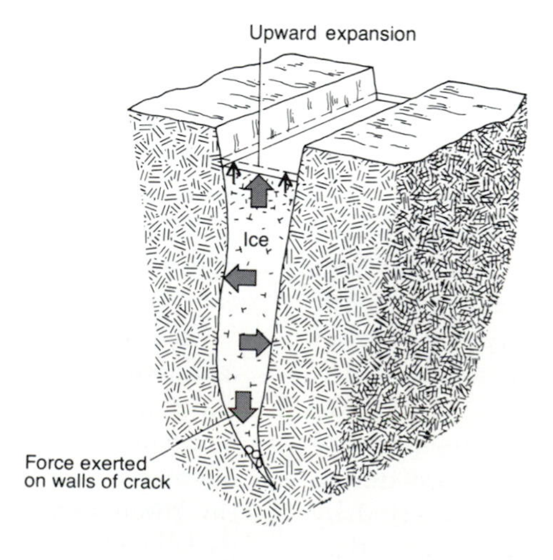

(c)

Figure 19.8 Ice formation in a water-filled crack, resulting in a form of mechanical weathering known as *frost wedging*. The force exerted by ice expansion can deepen and widen cracks and even split boulders.

Figure 19.9 Boulder fields, such as this one about 3500 m above sea level in the White Mountains of California, are considered to be strong evidence of mechanical weathering by frost action.

Weathering and Landforms

What influence does weathering have on the development of landforms? Intuitively, one would expect the least resistant rock to form the lowest parts of the terrain. But we must bear in mind that the resistance of rock varies with climatic conditions, so any comparison must be limited to a single climatic zone. Some light was cast on this problem by P.H. Rahn and his students, who gauged the susceptibility of different rock types to weathering by comparing the amounts of disintegration in century-old tombstones of different compositions. Of the four rock types examined in a cemetery at West Wilmington, Connecticut, granite proved to be the most resistant, and sandstone, the least resistant (Fig. 19.10*a*). Compared with areas of the same rock types in the New England terrain, land elevations showed corresponding variations, with granite forming the high features and sandstone the low features (Fig. 19.10*b*). A comparison of terrain developed in sandstone and limestone in central Kentucky, on the other hand, shows sandstone to be more resistant (see Fig. 15.20). This can be attributed to the fact that under the bioclimatic conditions of Kentucky, this limestone is measurably weaker than sandstone. Figure 19.11 summarizes the relative rates of chemical and mechanical weathering as a function of average annual temperature and precipitation.

SLOPE EROSION BY RAINFALL AND RUNOFF

Weathering is only the first step in the long series, or system, of processes that culminates in the deposition of detritus in the sea. The second step is the downhill movement of rock fragments from the sites where they were freed from the bedrock. The processes responsible for this movement are collectively referred to as *hillslope processes*. There are two types of hillslope processes: *erosion* by various forms of runoff, including overland flow and flow in small channels, and *mass movement*, a variety of gravitationally induced motions, such as rockfalls, landslides, and avalanches.

The term *hillslope* is taken to mean all sorts of slopes

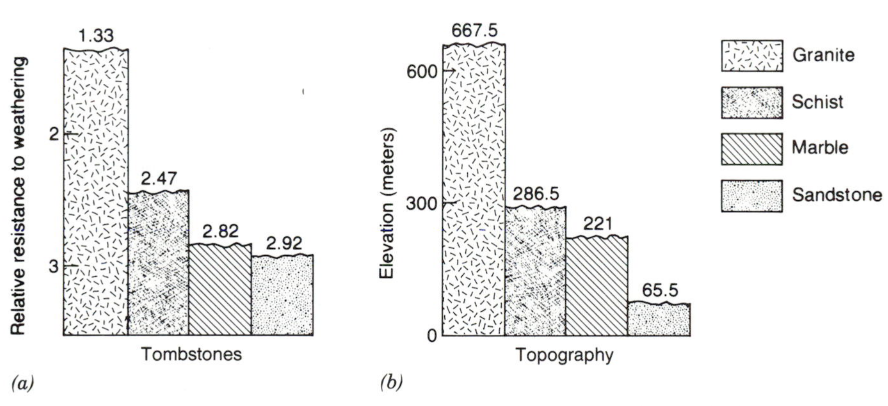

Figure 19.10 Comparison of (*a*) tombstone weathering and (*b*) land elevation near West Wilmington, Connecticut. Note that the resistance to weathering of these four rocks corresponds to the elevation of the terrain they underlie.

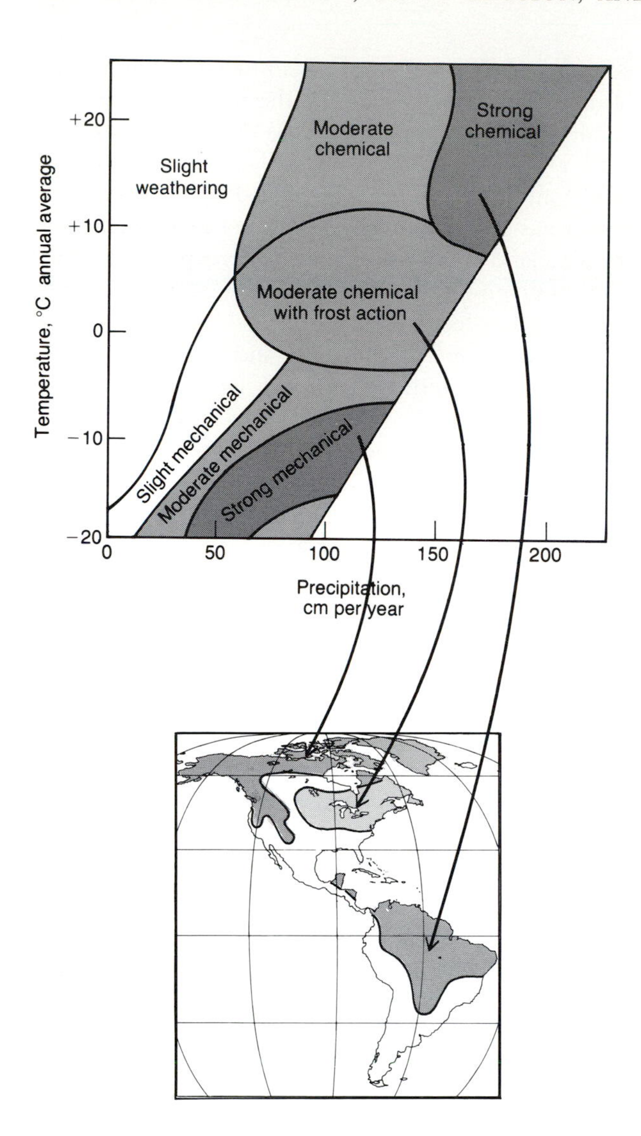

Figure 19.11 The general relationship between climate, represented in terms of average annual precipitation and temperature, and the types and relative rates of weathering. Geographical areas that exemplify the three major weathering environments are shown in the maps.

from mountain cliffs to gentle inclines in farmfields. The English language has a large variety of words to describe the various parts of hillslopes; however, only certain ones are conventional in science. Each slope is composed of three basic parts: a *crest slope*, which is the brow; a *midslope*, which is the central section; and a *footslope*, which is the base or toe. In general, the midslope is the steepest part of any slope; the surface of the midslope is also called the *slope face*.

Rainsplash and Rainwash

Erosion by water begins with the impact of raindrops on the soil. When a raindrop hits a wet surface, the impact sends out a circular splash of water and soil particles. If this occurs on a sloping surface, the downhill side of the splash travels further than the uphill side of the splash does (Fig. 19.12). The difference in the lengths of these trajectories—multiplied times the millions of raindrops hitting a slope in a single rainstorm—can account for appreciable downhill movement of soil on barren or sparsely vegetated slopes. The process is called *rainsplash*.

A related erosional process, called *rainwash* or *wash*, is produced by overland flow spilling over the slope face. Overland flow occurs when the intensity of precipitation exceeds the infiltration capacity of the surface. Near the top of the slope, it is usually slight and incapable of effecting much erosion. But as it moves downslope, it increases in volume and velocity (Fig. 19.13), and at some point generates enough force to displace small soil particles, especially organic mat-

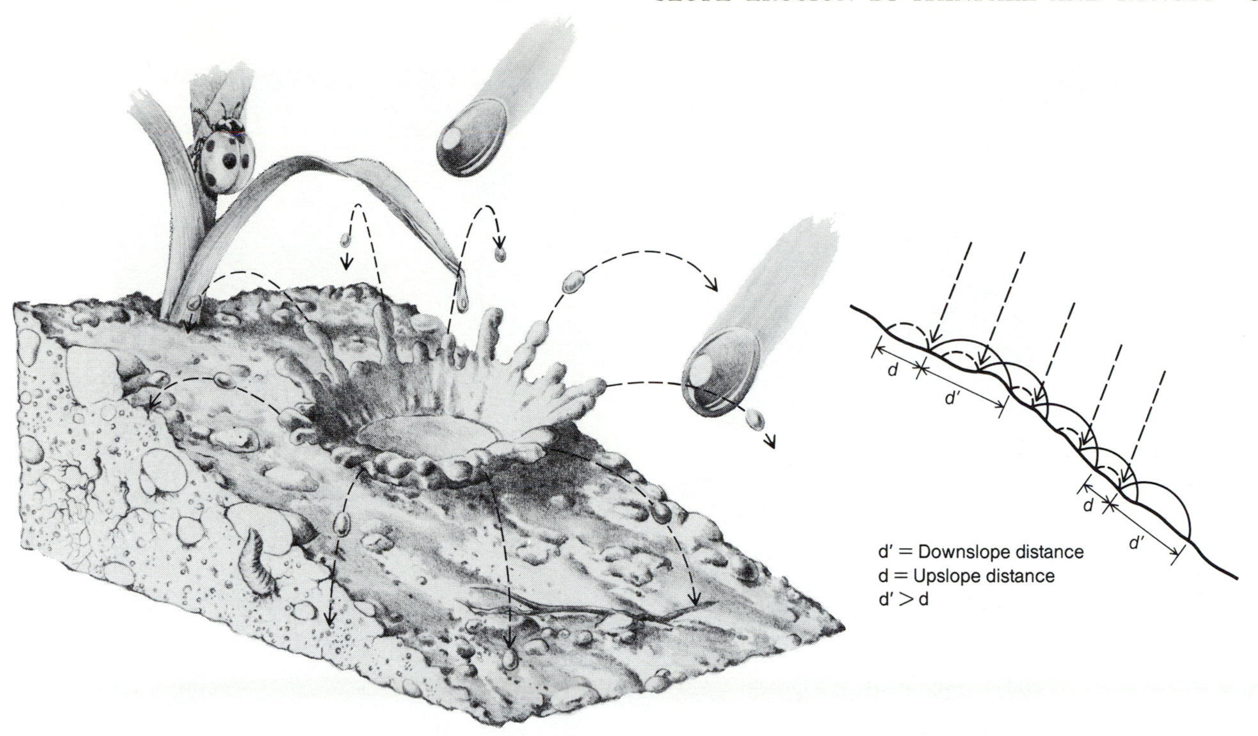

Figure 19.12 The splash created by a raindrop hitting a slope. Note the relative lengths of the downhill and uphill trajectories.

ter, silt, and small sand grains. (Clays tend to be more resistant to overland flow because of their cohesiveness.)

Together, rainwash and rainsplash can erode (lower) a slope by as much as 3 to 4 cm per year. Such rates are possible only on barren slopes, however, that is, those without the protective cover of vegetation. When vegetation is added to a slope, erosion rates decline significantly because the plants absorb part of the energy of raindrops, take up part of the rainwater through interception, increase soil intake of water through infiltration, slow the velocity of overland flow, and increase the resistance of the soil to erosion. On permanently forested slopes, erosion by these processes is so small over periods of decades or longer that it cannot be reliably measured in the field. In addition, precipitation amounts and intensities are also significant; they will be discussed in a later section.

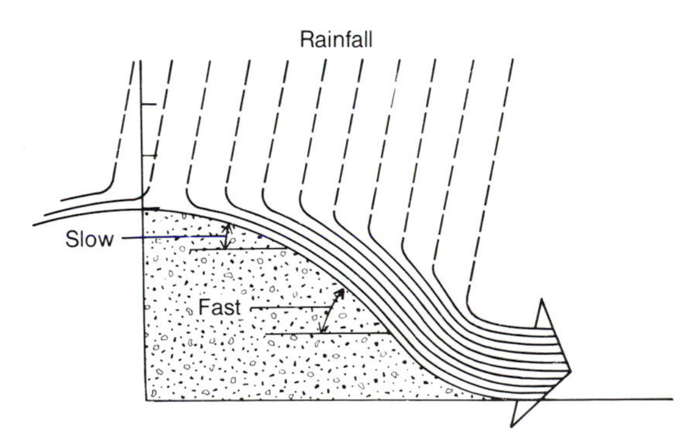

Figure 19.13 The distribution of overland flow on a hillslope. The depth of water increases with distance downslope; the velocity of flow increases with slope angle.

Gullying

As overland flow moves from the crest slope onto the midslope, the water merges into rivulets that can etch small grooves (tiny channels) into the slope. The rivulets in turn merge together a little further downslope, producing flows of sufficient force to erode great ruts or gullies into the slope face (Fig. 19.14). As the gullies deepen, they may cut into the zone of interflow water, or even groundwater in some cases, which initiates seepage into the channel.

As seepage water trickles out of the soil, it loosens and erodes fine particles. This process, which is termed *sapping*, causes the gully wall to be weakened and undercut along the seepage line. The gully walls may cave in, or slump, unless they are strong enough (as is commonly the case where there are dense mats

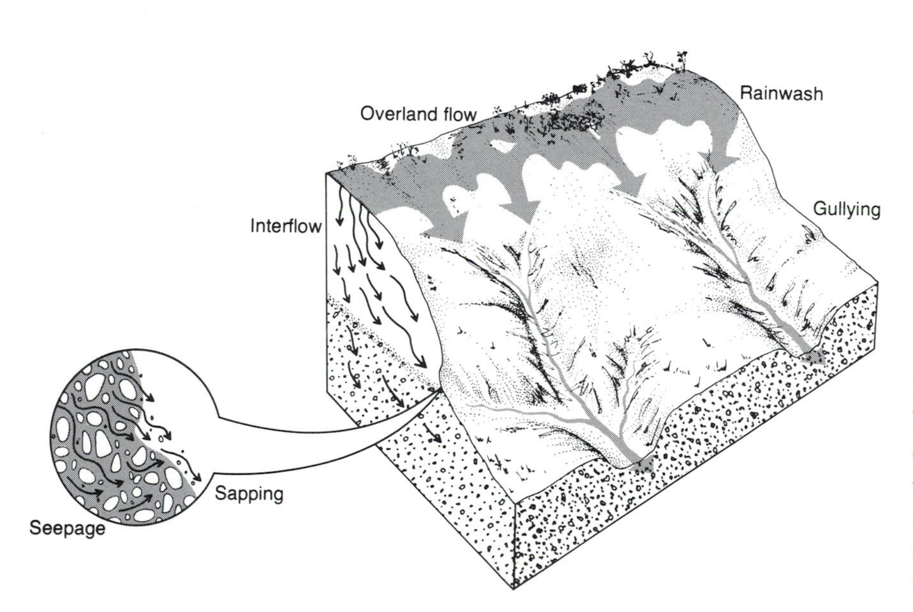

Figure 19.14 Hillslope erosion by runoff. The upper slope is eroded mainly by rainwash, whereas the midslope and lower slope are eroded by channel flow and sapping by seepage from interflow water.

of roots) to hold up against the undermining. Where they are resistant, the sapping may lead to the formation of small tunnels in the slope, a process that is known as *piping*. As the pipes advance into the slope, they increase the rate of seepage which in turn leads to further erosion and acceleration in the rate at which gullies, such as those shown in Fig. 19.14, advance into the slope. This sequence has produced some of the worst soil erosion disasters in the world, literally consuming entire hillsides in a matter of years in many places. (See Chapter 14 for more information on soil erosion.)

Sedimentation and Alluvial Fan Formation

When runoff reaches the footslope, its flow velocity declines (in response to the lower slope angle), which may cause it to deposit part of its sediment load there. The largest particles are deposited first, followed sequentially by smaller particles as the water loses speed. The finest particles, which are mainly clays carried within the water (in suspension), are not deposited near the footslope unless the water is brought to a halt. Otherwise, these sediments are carried beyond the slope to a linking stream channel and discharged into it.

Flows discharging from gullies and small streams not only lose much of their sediment to deposition as they cross the footslope, but may also lose much of their water supply to infiltration and evaporation. In arid regions, this process is so pronounced that even sizable streams may lose all of their discharge as they flow into valleys from mountain slopes. This process yields great deposits of sediment, which the streams wend their way across by breaking down into distributary channels similar to those of a river delta (Fig. 19.15). The resultant landform, called an *alluvial fan*, is a triangular- or semiconical-shaped feature; for some spectacular examples, also see Figure 19.23*b*.

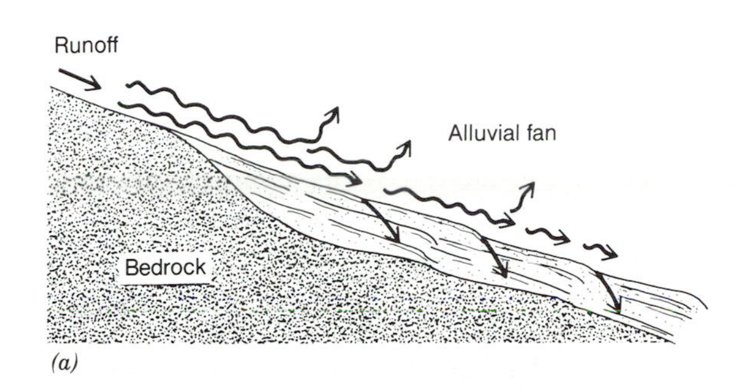

Figure 19.15 (*a*) Deposition by runoff at the foot of a slope. (*b*) Flow is lost to infiltration and evaporation, and the stream breaks down into distributary channels, which spread the sediment out in a fan-shaped deposit, or alluvial fan.

Erosion Related to Climate and Land Use

From field studies conducted in different parts of the world, we are able to sketch a brief picture of the relationship between bioclimatic conditions and sediment yield from slope erosion. Geomorphologists have found that in North America erosion is relatively low in arid and humid regions, but high in semiarid regions where precipitation averages around 30 cm a year. Semiarid lands are dry enough to limit the establishment of continuous plant covers, yet moist enough to produce substantial overland flow several times a year. In addition, the high variability of rainfall in dry areas also appears to promote greater erosion. Where there is more rainfall and it is less variable, vegetation is more abundant and it limits erosion. Where there is very little total rainfall (the deserts), low runoff rates limit erosion (Fig. 19.16*a*).

On other continents, erosion is high not only under semiarid conditions, but also in those climates with wet and dry seasons, such as the monsoon climate of India. The combination of a dry season that severely limits vegetation and a wet season that yields many intensive rainstorms appears to be effective in producing high rates of erosion. Agricultural activity in these areas (India and Southeast Asia) undoubtedly also contributes to this situation.

In those humid areas where crop agriculture is widespread, particularly where cropping is seasonal and fields are left barren for several months a year, soil erosion is also high. This is the case throughout much of the midwestern United States, for example, where it is not uncommon for farmers to leave fields cropless as much as 6 months per year which promotes average erosion rates as great as 25 to 50 tons (11.5–23 m^3) per acre per year in some areas. Without a plant cover to mitigate runoff, such fields are often severely gullied, like the one shown in Fig. 19.14, despite outward efforts at soil conservation by government agencies. With the inevitable expansion of agriculture throughout the world, we may see the balance of erosion rates shift to humid areas, if it has not already.

Finally, we can consider soil erosion related to urbanization. Urban development in North America generally followed a sequence of land use beginning in the 1800s with land clearing for agriculture. As urban areas expanded in the twentieth century, the agricultural lands gave way to suburban development. Both of these changes resulted in increases in soil erosion and stream sedimentation (Fig. 19.16*b*). The highest sediment yields have been associated with the construction phase of suburban development, when the land is devegetated and disturbed by earth-moving machines. Following suburbanization, sediment losses decline substantially as more and more of the land becomes urbanized and covered with hard-surface materials.

MASS MOVEMENT

When materials move by mass movement, they are drawn downslope under the influence of gravity rather than by a transporting agent such as a stream or a

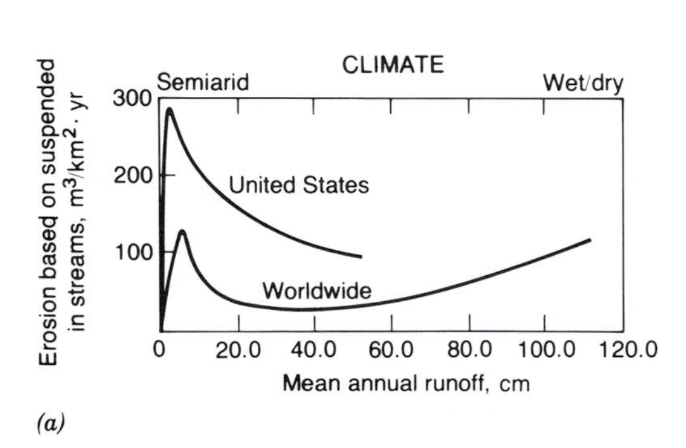

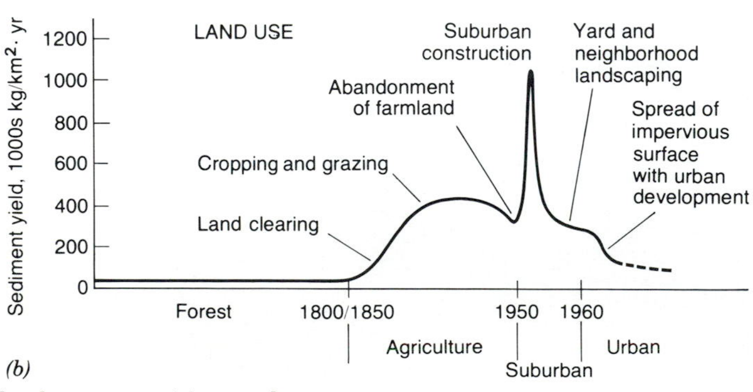

Figure 19.16 Erosion in a relationship to climate and urbanization: (*a*) A study conducted in the United States indicated that erosion is highest in semiarid areas, whereas a study using data from several continents showed erosion to be high in both the semiarid and the wet/dry (monsoon) climates. (*b*) Changes in sediment yield to streams in response to land use change.

glacier. Mass movement may be rapid, as in a landslide or avalanche, or so slow that it is barely detectable with sensitive instruments. Moreover, mass movement may involve gigantic amounts of debris, easily enough to bury towns and villages, or just individual grains of sand rolling off an anthill.

In any body of material situated on an incline, two sets of opposing forces determine whether it will move or remain in place. The forces that tend to induce movement generate *shear stress*, the magnitude of which is primarily a function of the steepness of the slope. The steeper the slope, the greater the tendency for gravity to pull objects downhill.

The opposing forces impart *shear strength* to the material. Shear strength is governed by factors such as the frictional resistance between adjacent sand particles, the cohesiveness among clay particles, and the capacity of plant roots to bind soil and rock particles together. The balance between shear stress and shear strength determines whether a slope is stable or unstable. Slope failure and mass movement occur when shear stress exceeds shear strength. When the two are equal, the slope is said to be at *critical threshold;* when shear strength is greater than shear stress, the slope is stable.

In evaluating the stability (safety) of a slope, it is important to realize that many of the factors governing shear stress and shear strength, such as groundwater, frost, and earthquakes, vary appreciably over time as from season to season. To produce a mass movement such as a landslide, it is necessary for the failure-producing forces to exceed the resisting forces for *only* an instant. Thus, in any slope the average state of affairs between these sets of forces is largely meaningless; only the extremes, particularly the coincidence of opposite extremes (low shear strength and high shear stress) really matters.

Types of Movement

Mass movements are classed on the basis of the type of motion involved. Five basic types of motion seem to cover most movements: fall, slump, slide, flow, and creep (Fig. 19.17). What type of motion prevails depends to a considerable degree on the physical properties of the materials in the slope. Bedrock, rock rubble, and sand behave as brittle elastic materials and tend to rupture (break) when they fail. The resulting movement takes the form of a fall, slide, or slump. In a *fall,* pieces of rock break free and sail and tumble over the slope face; in a *slide,* a sheet of materials slips over a slippage surface, called the failure plane. *Slumps* are characterized by a back rotational movement along the failure plane like that of someone slouching in a chair (Fig. 19.17). Slides and slumps may occur in one sudden, great movement, but it is more common for them to occur in a series of small displacements over months or years.

Materials that behave as plastics deform without rupturing and develop flowing motion when they fail. True *flows* are possible only in saturated materials that have a liquid or near-liquid consistency. Such consistencies are found in soft, wet, clayey materials or in sand or silt within which there is a flow of pressurized groundwater. Finally, there is the mass movement called *creep* which is characterized by a very slow (less than a few centimeters per year) downslope movement in the upper 0.5 to 1.0 m of soil. It is not clear what sort of motion is involved in creep; we only know that it is the slowest of mass movements and is caused by several different factors, which will be discussed later.

Factors Influencing Slope Stability

Soil water and groundwater Both soil water and groundwater have a major effect on slope stability, and both vary over time, especially seasonally. During spring and winter, for example, the interparticle voids take on water with infiltration from the surface. This increases the mass (weight) of the soil, which in turn increases the shear stress.

The addition of water also reduces the shear strength of soil materials because water changes the consistency (in clayey materials) to one that is more liquid-like. As children we learned this principle in making mudpies; beyond some critical amount of moisture, the mud simply got so runny that it would not stay in place and flowed away. In science, this critical moisture level is called the *liquid limit.* In this state, materials tend to flow downslope in a fashion similar to that of a very slow flow of sticky water. Depending on its composition, a flow is classed as either a *mudflow* or a *debris flow.* Mudflows consist of primarily clay and silt, whereas debris flows consist of clay and silt as well as a wide range of larger particles. Such flows are the source of serious property damage in the Los Angeles area, especially during wet winters (Fig. 19.18).

In stratified materials, if groundwater is added to a clay layer that underlies a stable mass, the clay may liquefy, and the entire mass may slide on the clay layer. Such slides are often caused by natural increases

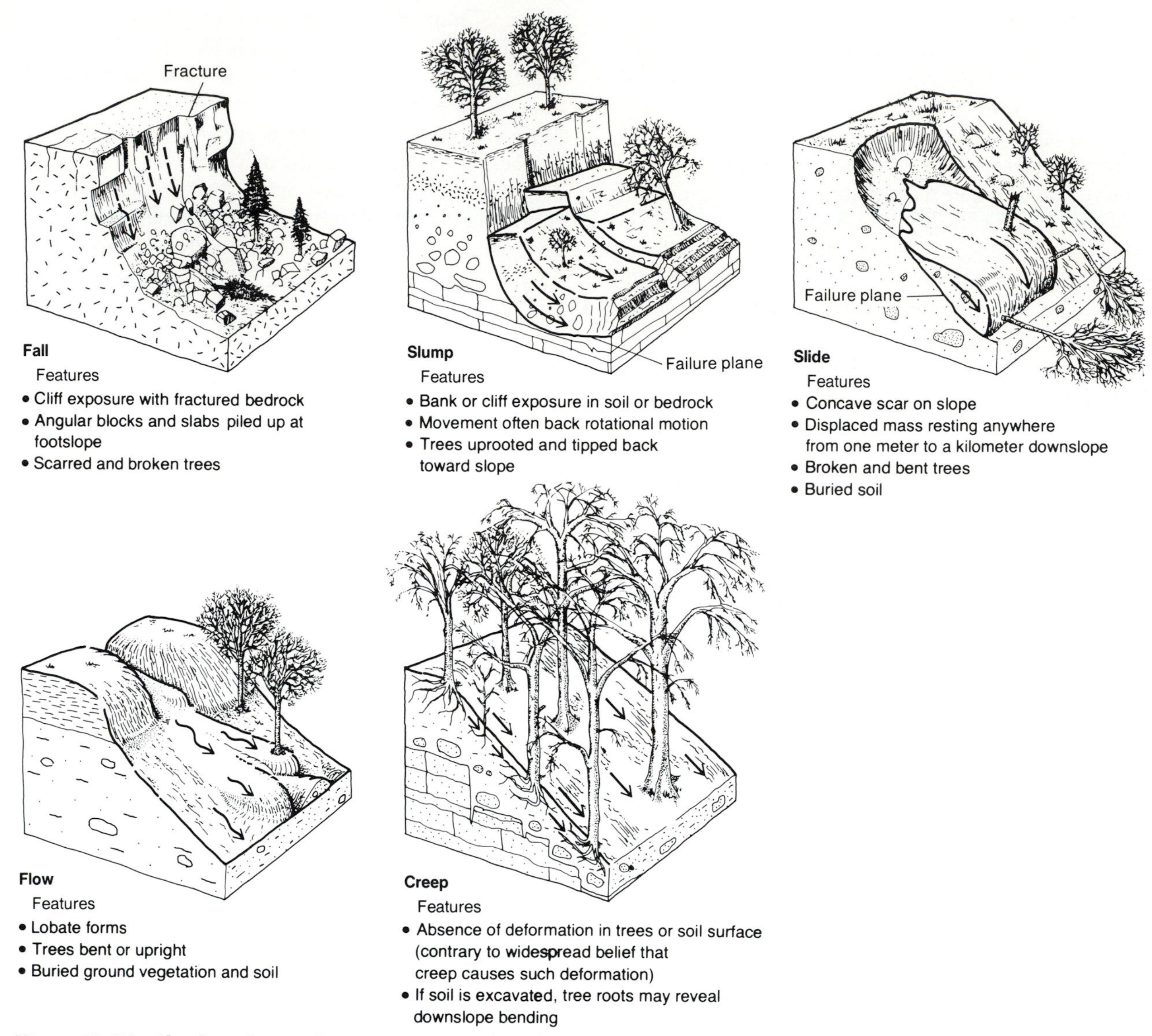

Figure 19.17 The five classes of mass movement, based on the predominant type of motion in each.

in groundwater, but in recent decades a growing number have been brought about by human-caused increases in groundwater, for example, the raising of reservoirs and seepage from irrigation and septic drain fields (Fig. 19.19).

The raising of a water table may also produce another source of slope instability: *pore-water pressure*. All groundwater exists under pressure, which explains why it is able to flow through those minute pore spaces in soil and rock materials. If the pore water pressure exerted on adjacent soil particles by groundwater is great enough to drive them apart, the particles float more or less free of one another. This reduces the shear strength of the soil, because interparticle friction and cohesion are lost and can cause failure, even in low-angle slopes. Pore-water pressure, by the way, is the cause of quicksand. Sand is normally very stable when saturated with water that is not under pressure, which is verified by the load-bearing capacity of most swimming beaches.

Figure 19.18 A localized example of destructive mass movement in Los Angeles that involved both flow- and slide-type movement.

Water can also trigger mass movements by another means. Groundwater may dissolve (leach) minerals that cement a mass together, resulting in loss of shear strength. Slope failures produced in this way often occur without a hint of warning and have produced some frightening disasters in Europe and Asia.

Ground Frost Several kinds of mass movements are caused by ground frost. One of these is rockfall produced by ice wedging on cliffs and steep hillslopes. This process can occur in fall with freeze-up and in spring, when ground ice expands and melts (Fig. 19.20*a*). Rockfall often leads to the buildup of piles of rock rubble, called *talus*, which may bury as much as half of the slope (Fig. 19.20*b*).

Frost penetration into moist soil is one of the most widespread causes of mass movement in mid- and high-latitude regions. The process involves two phases: (1) initial freezing *without* soil movement as the new ice crystals expand into the open spaces between soil particles; and (2) advanced freezing *with* soil movement as additional water drawn into the frost

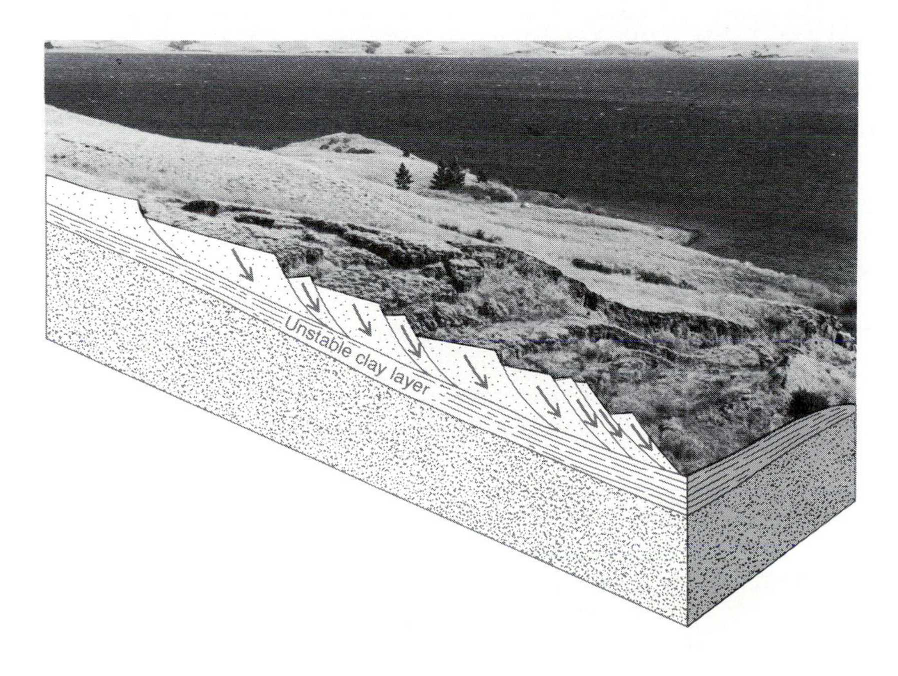

Figure 19.19 Shallow slumps and landslides over an unstable clay layer. It appears that the clay layer liquified because of an increase in groundwater caused by the raising of the reservoir in the background.

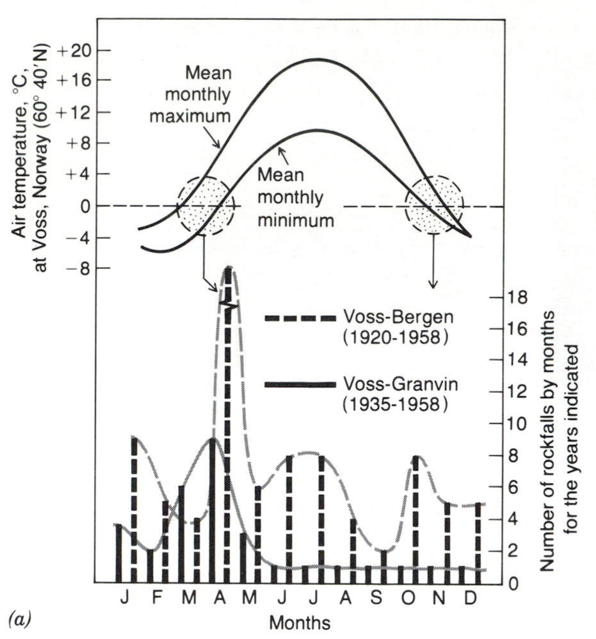

(a)

(b)

Figure 19.20 (*a*) The general relationship of rockfalls and monthly temperature along two railroads in the mountains of Norway. The period of maximum rockfalls does not occur upon freeze-up, but is delayed until ground ice expands and thaws in spring. (*b*) A large talus slope in Glacier National Park.

layer by capillary action expands beyond the availability interparticle space. As the frozen layer expands, it is pushed in the direction of least resistance, which is outward, or perpendicular to the slope face. When the ground thaws, the soil shrinks and falls, but not along the line of expansion during freeze-up. Instead, it settles a little downhill, in the direction of the gravitational force, which is perpendicular to the plane of the earth's surface (Fig. 19.21). The total downslope displacement caused by freeze-thaw activity may amount to only a few millimeters with each freeze-thaw episode, but many such episodes each winter and spring can produce appreciable mass movement.

Frost is one of the causes of *soil creep*. Creep is also known to be caused by contraction and expansion of clay with wetting and drying, although field data suggest that freeze-thaw is a more effective creep mechanism. Many geomorphologists think that creep is the dominant slope process along the crests of hills because rainwash is so slight there (see Fig. 19.13). Although it is an extremely slow process—measured rates vary from less than 0.25 cm to around 1.5 cm per year in midlatitude settings—creep appears to be highly effective in terms of work accomplished because it is so widespread, operating on virtually every hillslope in the world.

In polar and high mountain environments—which are referred to as *periglacial environments* because of the predominance of frost-related processes—freeze-thaw activity and wet soil conditions combine to produce a form of mass movement called *solifluction*. Solifluction usually takes place in the active layer above the permafrost and is distinctive because it produces lobes (tongues) of material that look as though they are flowing downslope. They actually move at slow rates, generally on the order of 5 to 15 cm per year.

Frost action and pore-water pressure (from groundwater and interflow water in the active layer) appear to be the driving forces for solifluction. Solifluction is very widespread in tundra-type environments, even on low-angle slopes of only 4 to 5 degrees. Its effectiveness in terms of total work in cold regions, however, appears to be exceeded by another process: the loss of material in solution with runoff. Other interesting surface features of periglacial environments are large concentrations of stones and patterned ground which are described later in Note 22.

Undercutting Slope undercutting by streams, waves, or other erosional agents, including humans, is a common cause of slope failure. Undercutting is a form of excavation of the slope base that produces an oversteepened slope angle, that is, one that exceeds the *angle of repose*. This is the maximum incline that materials such as sand, clay, or boulders can maintain without failing. If loose, dry sand, for example, is steepened to an angle greater than 33°, failure will result. For larger or more angular particles, the angle of repose is steeper, exceeding 40° for talus.

Earthquakes Some of the most dramatic mass movements, such as large landslides and avalanches, are limited mainly to large, steep, mountainous slopes in earthquake-prone regions. When seismic waves pass through a mass of unconsolidated material, particles tend to move differentially and even rotate somewhat, thereby breaking intergranular bonds formed by minerals, ice, and clay. As a result, the shear strength of a material can be drastically reduced, causing sudden failure. In one valley inthe Andes Mountains, for example, the landslides and avalanches triggered by the Peru earthquake of May 31, 1970, buried a city of 18,000 people. The total volume of this mass of rock, soil, and ice was estimated at 50 million to 100 million cubic meters.

The destructiveness of such events is due not only to the great mass of material involved, but also to the great distance over which it moves. The distance of horizontal movement, as across the floor of a valley,

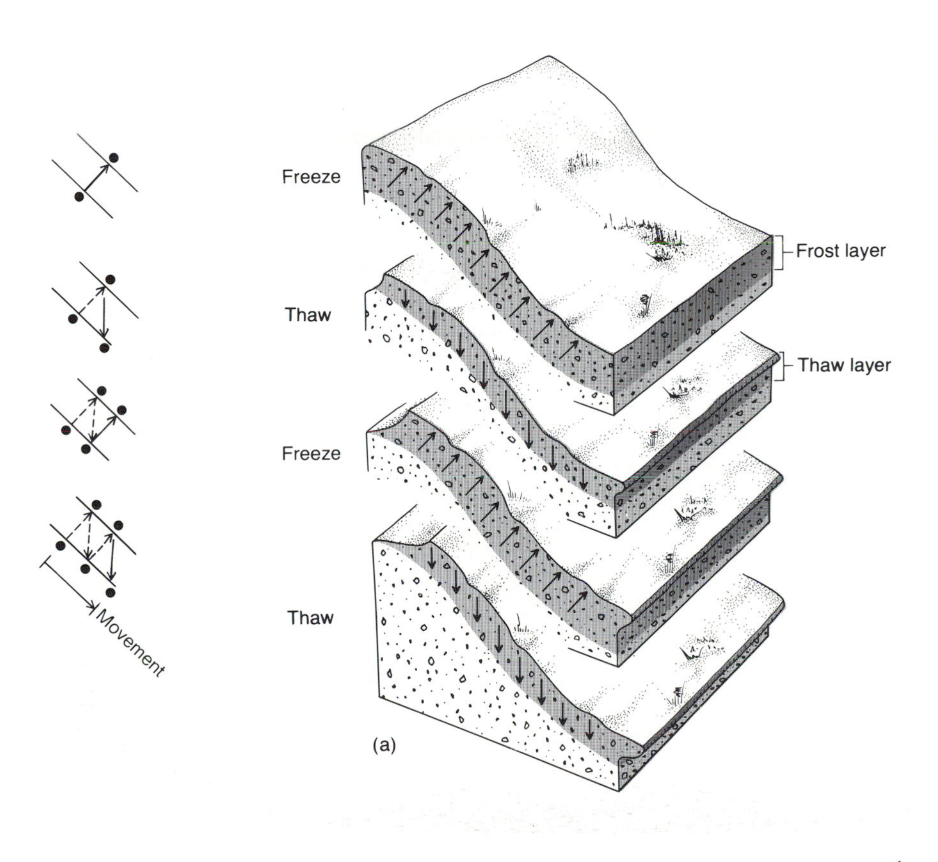

Figure 19.21 How freeze-thaw activity can produce soil creep. The amount of downhill movement suggested is greatly exaggerated.

can be so great that slides appear to defy the law of gravity. We now understand, however, that such movement is facilitated by a layer of compressed air that forms under the advancing mass. This allows the slide to move with little friction, as in the movement of a hovercraft.

Lowering of the Continents: Denudation

Hillslope processes are essential in lowering the land masses of the earth. These processes move weathered debris to a position in the landscape where it can be picked up by streams and glaciers and then transported to even lower elevations and finally to the sea. As the material is transported downvalley, it undergoes further weathering, and when it emerges on the seacoast most is in the form of sand, silt, clay, or dissolved ions (Fig. 19.22). Each year the Mississippi River dumps nearly 500 million tons of sediment into the Gulf of Mexico.

How rapidly the continents are being unloaded, or *denuded*, is equal to the total sediment load being carried to the sea by all the streams, rivers, and glaciers. Understandably, the rate of denudation for an entire continent is difficult to measure accurately, and thus our estimates are crude at best. Reasonably accurate estimates are possible, however, for sections of continents. Measurements of sediment loads of large rivers in the United States suggest that the coterminous United States is being lowered at an overall rate between 2.5 and 7.5 cm per 1000 years. Within this area, the rate varies with climate and topography; in dry, mountainous areas, such as in the Southwest, denudation may be more than 100 cm per millennium.

Could the continents ever be eroded away? This appears to be unlikely because the lowering of the land is offset by uplift of the continents as they are being unloaded. This is a response to the release of mass (weight) from the continent and represents a process likened to a ship rising in the water as it is being unloaded. Through isostatic uplift, as much as 80 percent of the elevation lost in erosion can be recovered. In other words, for every 1000 meters of land elevation lost, it appears that about 800 meters are recovered through isostatic uplift. In addition, the great masses of sediment eroded from the land may not be entirely lost from the continents, because a large share of them go into the building of the continental shelves. In the long run of geologic time, the continental shelves are in turn subject to tectonic deformation (usually in connection with the movement of tectonic plates) and are thereby reincorporated into the continents as the rocks of mountains.

FORMATION OF HILLSLOPES

The processes and rates of denudation of the land masses are important questions in natural science, but as geographers, we are also interested in the forms of the hillslopes that are created because slopes are the basic elements of landforms. Landforms everywhere are important in understanding the rest of the landscape: they influence soil formation, vegetation patterns, climate near the ground, land use, and many other features of geographic significance.

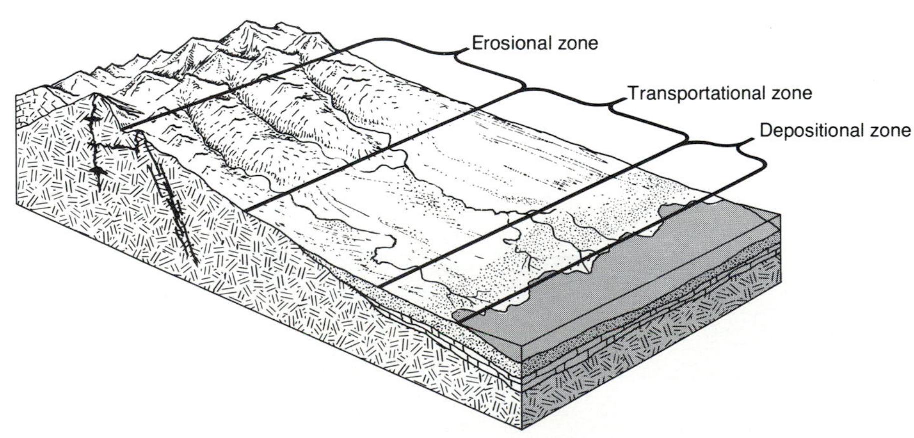

Figure 19.22 The three main zones of activity associated with the lowering of the continents. Hillslope processes are most intensive in the erosional zone. In the transportational zone, material is moved by streams to the coast, where it is deposited.

Slope Retreat

In a general way, the rate at which hillslope processes move debris over a surface can be treated as a function of slope steepness. The American geomorphologist Grove K. Gilbert observed this principle nearly 100 years ago while studying the landforms of the West. He noted that, for a given volume of runoff, the power to transport debris (particles) increases with slope inclination. If we extend this concept to include mass movement, we should find, other things being equal, that more material should be moved on steep slopes than on gentle ones.

Slope retreat is the process by which hillslopes are worked back over time. This topic has been one of lively debate in geomorphology because the long-term change—let us call it evolution—of slopes is essential to understanding the origin of major landforms such as sea cliffs (e.g., the White Cliffs of Dover), canyons (e.g., the Grand Canyon), mountain peaks (e.g., the Matterhorn), mountain ridges (e.g., the Blue Ridge), and stream valleys (e.g., the Columbia River Valley). Generally, there are two basic modes of slope retreat: parallel and nonparallel. In *parallel retreat,* the slope is worn *back* and the angle remains comparatively constant; in *nonparallel retreat,* the slope is worn *down* and the angle grows gentler (lower). Either mode is possible depending on the mass balance of the slope, but parallel retreat is generally regarded as the most prevalent mode.

Parallel retreat can be achieved in two ways: (1) when debris brought to the footslope by slope processes is transported away by another process as fast as it arrives; and (2) when a transporting agency such as a river undercuts the slope itself, causes the slope to oversteepen and fail, and then transports the resultant debris away as fast as it arrives. Sea cliffs are a prime example of the latter.

In order to maintain parallel retreat by either mode, one condition is essential: the slope must be part of an *open geomorphic system.* This system is defined by the movement of rock debris in a great train of particles—typically in the form of a drainage system such as that of the Mississippi or Colorado rivers—from the upper slopes to the edge of the continent. Through-transport is possible only if the slope producing debris is part of a drainage basin that empties into the sea. This ensures that debris released from the slope is carried away from the slope.

Where the drainage system is *closed,* on the other hand, rock debris cannot be transported out of its source area. Therefore, it builds up on valley floors and along footslopes, producing gentler angles. Over the long run, this results in nonparallel retreat on the lower parts of slopes (Fig. 19.23). On the upper parts, however, where the erosional processes are still tearing the slope down, parallel retreat may be taking place, but as slopes shift back, they grow shorter and eventually disappear at the topographic divides. Standing back from this scene over a long span of time, one would see the gross profile undergoing nonparallel retreat, as the base of the slopes built out into the valley and the upper slopes retreated toward the surrounding divides.

A surprising number of areas in the world qualify as closed geomorphic systems; most are situated in arid, mountainous regions. Death Valley and the Dead Sea Basin are famous examples of small basins where, as Figure 19.23 suggests, basin filling and lowering of footslopes are not difficult to envision. But there are large areas of closed drainage, too, for example, the Tarim Basin of northern China, the Plateau of Iran, and the Great Salt Lake Basin.

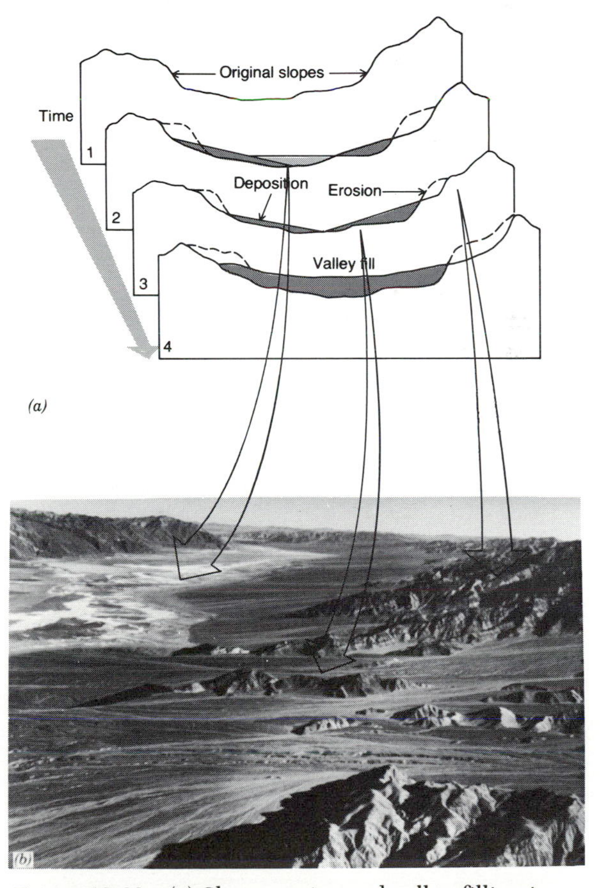

Figure 19.23 (*a*) Slope erosion and valley filling in a closed basin. The overall profile of the slope grows gentler over time. (*b*) Death Valley, a closed basin being filled by sediments, mainly in the form of alluvial fans, eroded from the mountainous side slopes.

Trends in Slope Development

It is probably rare that a particular trend in slope development would continue undisturbed for millions of years. The earth is too dynamic in most places for this to happen. In areas of active mountain building, as in parts of western North America, faulting can reestablish slopes that were originally worn down by erosion and thus trigger a new episode of erosion and retreat (Fig. 19.24).

Another source of alteration in slope development is climatic change. Increased aridity in the American Southwest in the past 10,000 years, for example, caused the reduction of many lakes and streams and thus the loss of an erosional agent with sufficient power to remove debris from footslopes. As a result, former wave and river-cut slopes in many areas have tended to grow gentler in the past several thousand years. The opposite is also possible, of course; the walls of stream valleys can be steepened with increased erosion from greater discharge. Sometimes we see examples of increased streamflow and erosion where people have diverted water into a stream or where storm discharge has been increased by urbanization or agricultural development.

Common Slope Forms and Their Origins

If we examine the landforms in most localities, we are sure to notice some distinct similarities in the shapes of different hillslopes. Similarly, slopes of a particular form tend to be common to different geographical regions as well, and geomorphologists have diligently sought to provide explanations. The best we can do at the present, however, is to say that the causes appear to be multiple; in some regions, they appear to be related to rock structure and type; in others, to

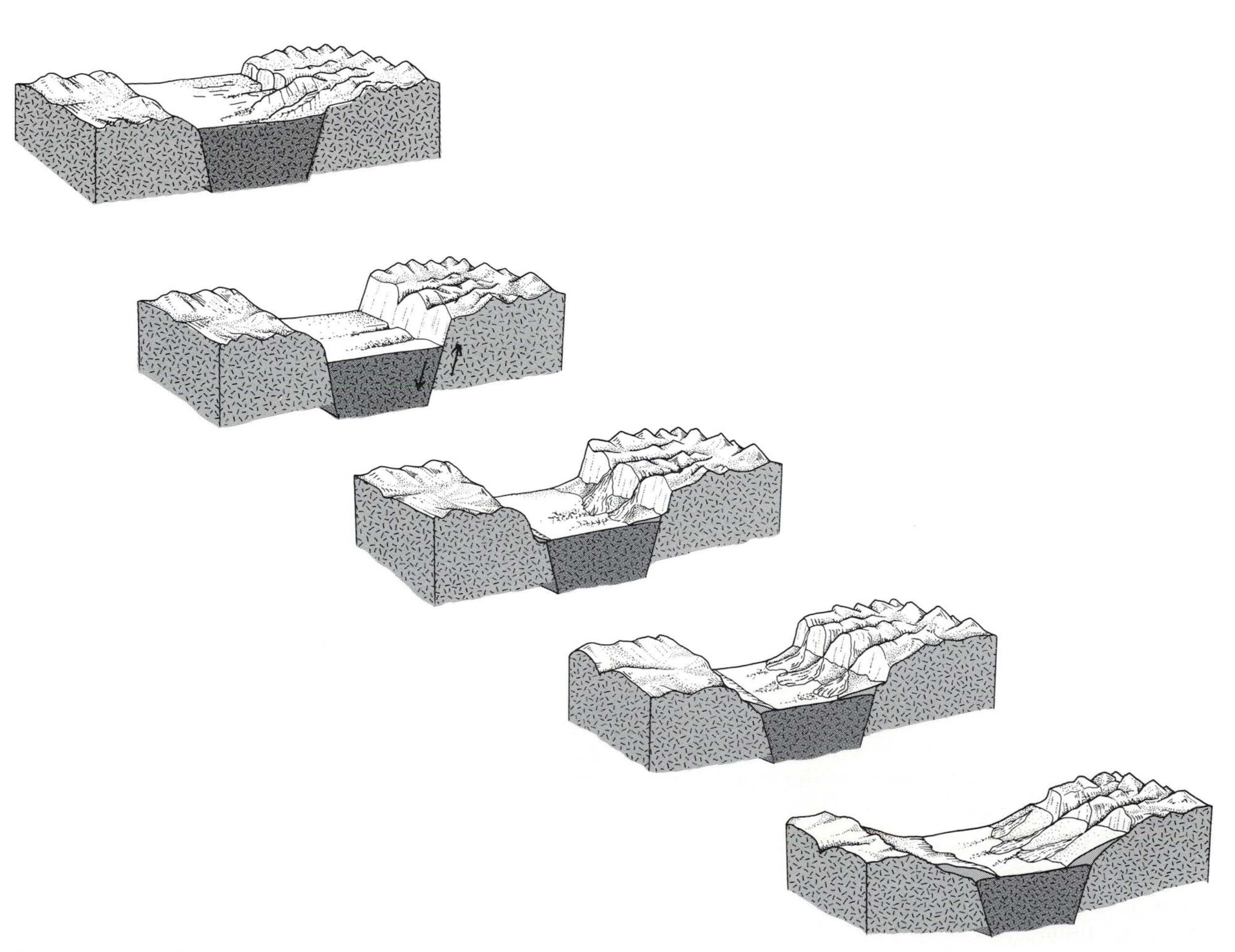

Figure 19.24 Slope development along a fault scarp. Each of these phases can be observed along many different mountain fronts in the American West.

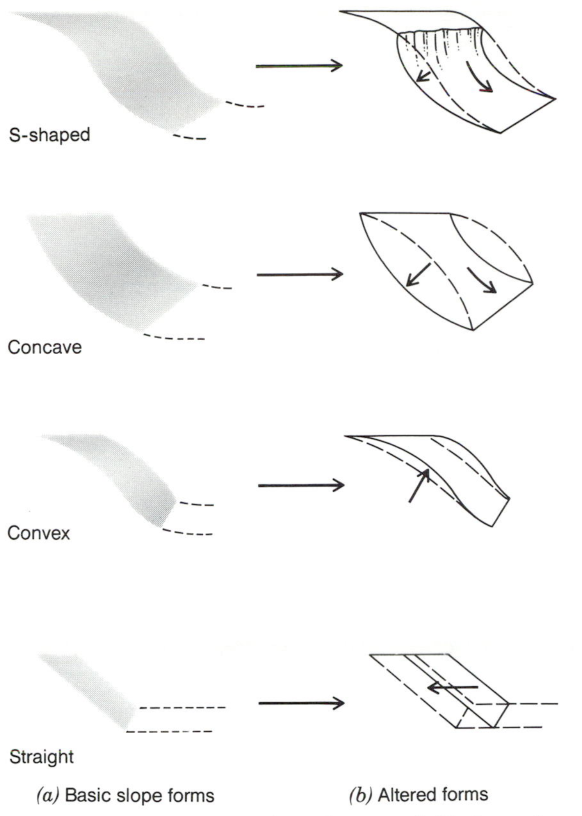

Figure 19.25 (*a*) Basic slope forms and (*b*) their altered forms.

vegetation and hillslope processes; and in yet others, to erosion and deposition by certain agents.

There are four basic hillslope forms: S-shaped, concave, convex, and straight (Fig. 19.25*a*). Gentle *S-shaped slopes* are very common in nonmountainous regions where there is a thick soil overburden and heavy vegetative cover. Such areas include much of the United States east of the Mississippi River, northwestern Europe, and the wet tropics of South America, Africa, and southeastern Asia.

The origin of the *S form* appears to be tied to a particular combination of hillslope processes, notably runoff and soil creep, working on an erodible soil held in place by a heavy plant cover. The role of plants is very important and is especially evident when these slopes are devegetated for land development or farming, and accelerated erosion sets in. The slope is quickly transformed from the S shape into a concave shape as gullies are cut into it and large chunks of material are removed by slumps and slides (Fig. 19.25*b*). See Note 19 for some remarks about slope ordinances in land development.

Concave slopes also develop under natural conditions, though typically in areas of rough terrain. This slope form is usually the work of a powerful erosional agent or a mass movement such as a landslide or slump. Glaciers, for example, are able to scour the sides of mountains into deep concavities called *cirques*, and a mountain glacier flowing down a stream valley carves both side slopes into broadly concave forms. The inclination of such slopes increases upward and may approach 90° near the midslope (Fig. 22.11).

Convex slopes, by contrast, are usually related to resistant rock formations or to uplift of the slope itself, and can be found in any climatic region. In a slope composed of sedimentary rock, for example, resistant formations in the midslope may protrude to form a convexity. Convex slopes may also result from the uplift of the mountains by tectonic forces, as in domes being unloaded by exfoliation (Fig. 19.25*b*).

Finally, there are *straight* slopes. They form under various conditions and one of the most studied examples are pediments. These are broad, gentle slopes, usually with only several degrees inclination, which lie at the foot of cliffs, alled *free faces*. Pediments are composed of bedrock over which is strewn a light veneer of gravelly rock debris and appear to form in the wake of a retreating free face. Because they are widespread in arid lands, many geographers and geologists reason that pediments are unique to dry climates (Fig. 19.26).

Landforms and Climate

There is a widespread belief among students of the landscape that different climates give rise to different

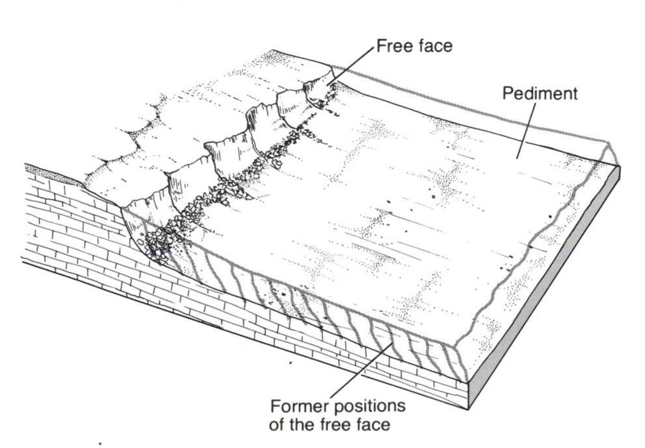

Figure 19.26 The broad, gentle footslope leading to the cliff, or free face, is called a pediment. Pediments result from parallel slope retreat.

VIEW FROM THE FIELD / NOTE 19

Slope Limitations in Land Development

Throughout the United States and Canada communities have come to recognize the pitfalls of uncontrolled land development. Not only are valued landscapes such as stream valleys, wetlands, and farmlands lost or damaged, but also improper placement of buildings and related facilities often increases the likelihood of disasters from flooding, earthquakes, and slope failure. To avoid these problems, many communities have enacted ordinances that set guidelines or rules about which settings and

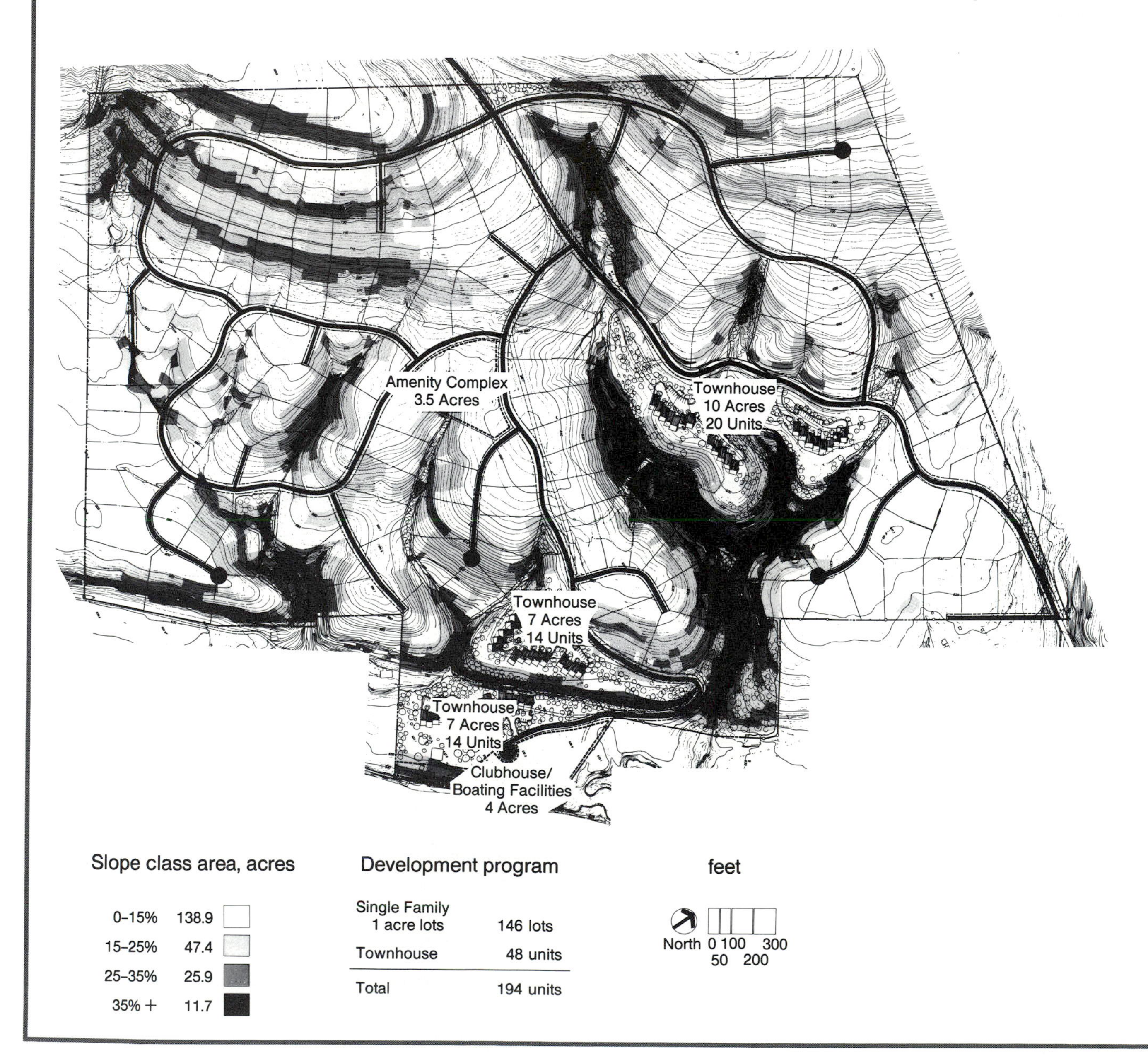

features are to be preserved or put to special uses such as park and recreation.

One of the challenges with such ordinances—from the geographer's standpoint—is measuring and mapping the pertinent environmental features and conditions. This can be a herculean task when dozens of features and conditions are involved. Instead of mapping each one, communities are often advised to map a few essential features, especially "indicator" features. Indicator features are those whose presence is representative of other features or conditions. Slopes are a good indicator feature. In addition to being important in their own right, they are often indicative of soil types, runoff rates, vegetation associations, and habitat types. Steeper slopes tend to be more limiting—owing to, among other things, higher runoff rates and thinner soils—and, as a result, ordinances usually require lower development densities on steeper slopes.

In Austin, Texas, development is pushing into the rugged hill country northwest of the old city. Concerned about the loss of water quality, groundwater recharge, and stream valley habitats, Austin has enacted a comprehensive watershed ordinance that requires planners to map slopes in four classes according to percent inclination: 0–15 percent, 15–25 percent, 25–35 percent, and greater than 35 percent. Percent slope is measured by dividing elevation change by map distance from the toe to the top of the slope:

$$\text{Percent slope} = \frac{\text{change elevation}}{\text{map distance}} \times 100$$

Development is limited to the 0–15 percent slope class. The maximum density for commercial land use, measured on the basis of impervious groundcover (concrete, asphalt, etc.) is 40 percent. For residential land use, maximum density is one unit per acre for cluster development and one unit per two acres for detached, single-family development. Development in steeper slope classes is possible only on a special case basis and requires a formal variance from the city.

The accompanying map shows the distribution of slopes and proposed land use plan for residential development on a 224-acre parcel. Although the lots extend into areas with slopes greater than 15 percent, building sites would be limited to ground of less than 15 percent; the rest would be yard space.

Figure 19.27 Rock towers such as the Captains of the Canyon in the Canyon de Chelly National Monument, Arizona, are examples of landforms often associated with the American West rather than the East.

landforms. Most Americans who have thought about the geography of their nation would be inclined to think that way, if only because the landforms of the arid West look so different from those of the humid East. And indeed certain kinds of landforms are found almost exclusively in the West (Fig. 19.27). But is this due to differences in climate or to differences in the original geology of the two regions?

Finding an answer to the question of climate and landforms has proven difficult largely because it is difficult to find areas in different climatic regions where the rock type, rock structure, and geologic history are similar enough to allow reliable comparison. Moreover, there is the problem of climatic change. In general, landforms respond to climatic conditions less acutely than do vegetation and soils. Thus, it is likely that the existing landforms in most regions are products of past as well as recent climatic conditions. Finally, such a range of factors are involved in slope formation—rainfall intensity, vegetation, rock type, rock structure, and soil formations—that some geomorphologists argue that it is possible for virtually any kind of landform to develop in any climatic region.

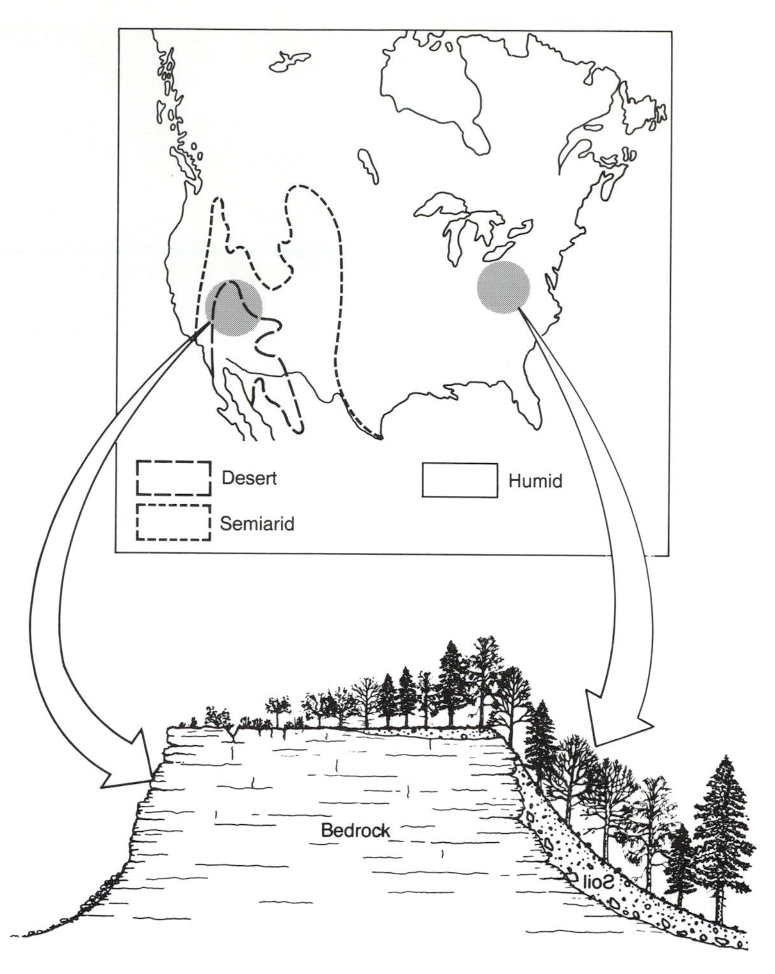

Figure 19.28 Top, a mesa in which the right side has developed under humid conditions and the left side under arid conditions; below right, the Appalachian Front, an escarpment in a humid region; below left, an escarpment on the edge of the Grand Mesa in Arizona.

The only exception is glacial areas, where certain landforms are uniquely associated with glaciated or frozen ground.

Barring glaciated areas, there are, nevertheless, some observable differences in landforms of contrasting climatic regions, but they tend to be more relative than absolute. For example, the landform called a *mesa* is a common feature in parts of southwestern

United States. Mesas are flat-topped "islands" of relatively resistant, flat-lying sedimentary rock. The sides are usually steep free faces at the base of which are debris deposits leading onto the gentler pediments, the lowermost slopes (see Fig. 19.26).

Are mesa forms found in humid areas of comparable geology? Not exactly, but similar features are there if we look through the forests and heavy mantles of soil. The humid-region versions are not as bold because the soil and vegetation tend to soften the angular form of the bedrock. If we had to put our finger on one factor to explain the difference, it would have to be vegetation. Plants hold weathered debris in place and minimize erosion by runoff. As a result, a heavy soil cover often forms on the side slopes, which may cause them to be a little gentler than their nonvegetated counterparts along the arid mesa (Fig. 19.28).

Summary

Water is the most essential ingredient in weathering. Both chemical weathering and frost action, the two most effective types of weathering, are water-dependent. Dry climates, whether warm or cold, generally produce low rates of weathering. In contrast, in moist environments the collective rates of weathering tend to be high under both warm and cold temperatures.

Hillslope processes move weathered material downslope and in so doing shape the slopes of the landscape. Erosion on hillslopes involves a variety of processes, including rainsplash, sapping, and gullying. Mass movements are gravitationally induced displacements of materials in the form of processes such as soil creep, solifluction, rockfalls, and landslides. The conditions that induce mass movements are related to a variety of factors including groundwater, earthquakes, and groundfrost.

The rate at which hillslope processes perform work is related in a general way to climate inasmuch as the magnitude and frequency of rainfall and the density of vegetation are climatically controlled. Overall, the highest rates of denudation are found in semiarid and seasonally dry climates. The manner in which slopes retreat and the forms they assume are related to many factors, including slope mass balance, rock structure, and vegetation.

Concepts and Terms for Review

- nature of landscape changes
- geomorphic agents
- landforms
- weathering
- chemical stability
- chemical weathering
  - solution
  - chelation
  - hydrolysis
  - oxidation
- mechanical weathering
  - exfoliation
  - frost wedging
- weathering and landforms
- hillslope processes
  - rainsplash and rainwash
  - gullying
  - sapping and piping
  - sedimentation
  - alluvial fan formation
- erosion, climate, and land use
- mass movement
  - shear stress and shear strength
  - fall, slide, slump, flow
  - soil creep
  - solifluction
- slope stability
  - liquid limit
  - pore-water pressure
  - ground frost
  - undercutting
  - earthquakes
- denudation
- slope retreat
  - parallel and nonparallel
  - open and closed geomorphic systems
- slope forms
  - origins
  - relation to climate

Sources and References

Carson, M.A., and Kirkby, M.J. (1972) *Hillslope Form and Process.* Cambridge, England: University Press.

Cooke, R.U., and Doornkamp, J.C. (1974) *Geomorphology in Environmental Management.* London: Oxford University Press.

Douglas, I. (1967) "Man, Vegetation and the Sediment Yield of Rivers." *Nature* 215: pp. 925–928.

Gilbert, G.K. (1909) "The Convexity of Hill Tops." *Journal of Geology* 17: pp. 344–350.

Peltier, L. (1950) "The Geographical Cycle in Periglacial Regions as it is Related to Climatic Geomorphology." *Annals of the Association of American Geographers* 40: pp. 214–236.

Rapp, A. (1960) "Recent Developments of Mountain Slopes in Karkevaage and Surroundings, Northern Scandinavia." *Geografiska Annaler* 42, pp. 2–3, 65–200.

Schuster, R.L., and Krizek, R.J., eds. (1978) *Landslides: Analysis and Control.* Washington, D.C.: National Academy of Sciences.

Strahler, A.N. (1952) "Dynamic Basis of Geomorphology." *Bulletin of the Geological Society of America* 63: pp. 923–938.

Terzaghi, K. (1950) "Mechanism of Landslides." Berkey Volume, *Geological Society of America:* pp. 83–123.

Varnes, D.J. (1958) "Landslide Types and Processes." In E.B. Eckels, ed., *Landslide and Engineering Practice.* Washington, D.C.: U.S. Highway Board, Special Report 29, pp. 20–47.

Washburn, A.L. (1956) "Classification of Patterned Ground and Review of Suggested Origins." *Geological Society of America Bulletin* 67: pp. 823–886.

Williams, P.J. (1982) *The Surface of the Earth: An Introduction to Geotechnical Science.* New York: Longmans Green.

Chapter 20

The Work of Streams in Shaping the Land

When we realize that streams and rivers carve their own valleys—as opposed to having them formed by some other agent—it is apparent why small streams are found in small valleys and large ones in large valleys. What processes enable streams to remove such large amounts of material from the land? What landforms are created as valleys take shape? Working together, is it possible for drainage systems to lower entire land masses?

By the second half of the 1800s, much of the attention in geomorphology had turned from the question of the age of the earth to which of the erosional agencies—the sea (waves and currents) or runoff (streams and rivers)—was the earth's primary erosional force. Charles Lyell, the eminent British geologist, argued that the sea at higher levels was mostly responsible, a view initially supported by his colleague Charles Darwin. But other scientists, especially in North America, were observing and measuring runoff and sediment transport by streams, and the results were pointing to runoff as the primary erosional agent of the land.

After the Civil War, the U.S. Congress funded several expeditions to explore the lands beyond the American Great Plains. Their main charge was to provide information on the condition of the land, its resources, and its potential for settlement. Part of their work also involved describing and interpreting the landforms of the West, which until this time had not been examined scientifically.

John Wesley Powell, a former Union Army major, led three expeditions (float trips) through the Grand Canyon and reported the amazing fact that the Colorado River had eroded this huge cataract, nearly 2000 m deep in places. Grove K. Gilbert and other scientists demonstrated that runoff in the forms of rainwash, rivulets, and gully flow eroded massive amounts of sediment from uplands, plateaus, and valleys alike. It became increasingly clear that sediment moved by such runoff, together with sediment moved by streams and rivers, far outweighed the sediment moved by waves, currents, wind, and glaciers. This in turn showed that in the long term runoff lowers the land masses more than winds, glaciers, and waves do—indeed, more than winds, glaciers, and waves together do.

This chapter is concerned with streams and rivers

as geomorphic agents. The focus is on the movement of sediment down the stream system to the sea. We open the discussion with an examination of flow in open channels and the principles governing stream velocity, erosive power, and sediment transport. This is followed by a description of the various landform features built by rivers as they move their sediment loads and sculpture their valleys.

FLOW CHARACTERISTICS OF STREAMS

Velocity

Water must itself move if it is to move any particles on a streambed; therefore, the velocity of a streamflow is a major determinant of erosion. The flow velocity of a stream is governed by three factors: the slope or

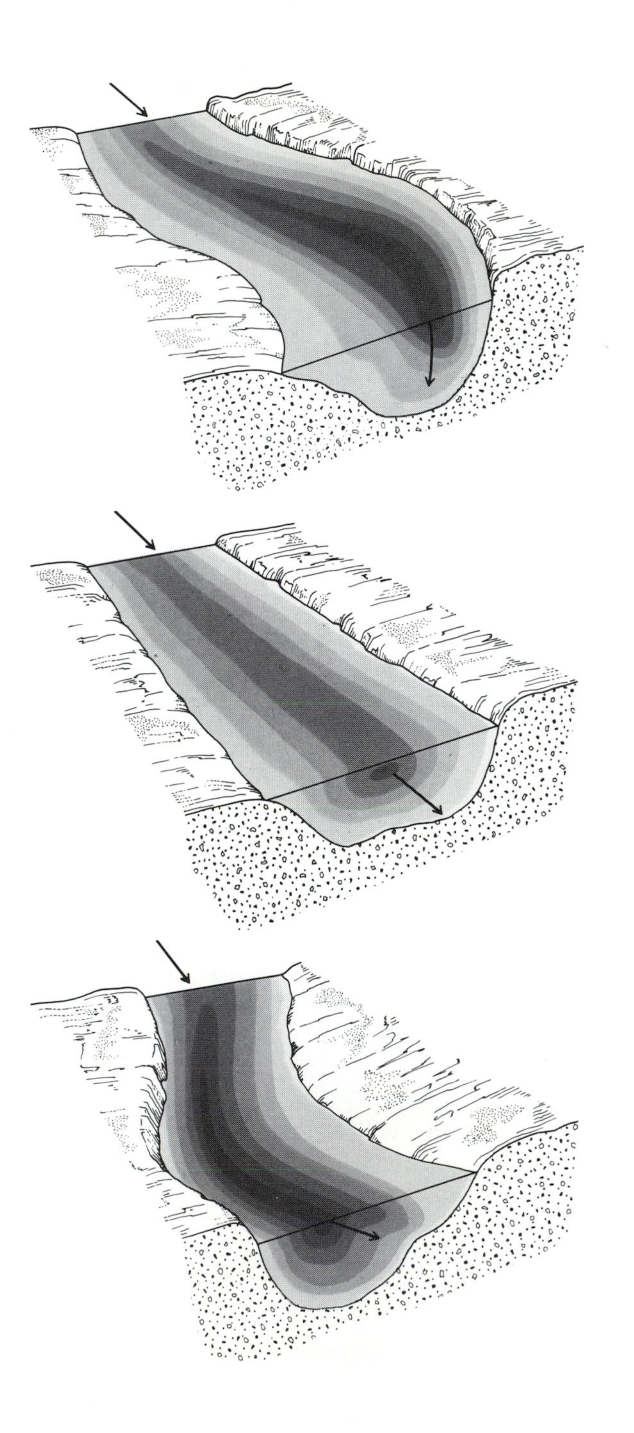

Figure 20.1 Relative velocities of flow in different sections of a stream channel. In straight sections the zone of fastest flow is located in the middle, just beneath the water surface. In the bends, velocity increases and slides to the outside of the channel.

gradient of the channel, the depth of the water, and the roughness of the channel. Velocity increases with greater depth and slope, and decreases with greater channel roughness.

We tend to think of steep mountain streams as having the fastest flows, but this is not so. The fastest flows on the average are found in the large, deep rivers that flow across the plains and lowlands. The reason is that water depth has a greater influence on velocity than channel gradient does. Near the mouth of the Amazon River, for example, where the channel gradient is only a few inches per mile but water depth is 50 m (150 ft.), average velocity is about 2.5 m per second. This is more than twice the velocity of the Grand Canyon River in Yellowstone where the gradient is 200 ft. per mile and the water depth is only 1 to 2 m on the average.

Within the stream, velocities vary with water depth and location between the channel walls. The fastest flow is usually found just below the water surface near midstream. Velocity decreases toward the streambed, and near the bottom where friction is greatest, it falls to zero. The midstream zone of highest velocity also shifts sideways with bends (meanders) in the stream (Fig. 20.1). When the water enters a bend, centrifugal force tends to throw fast-moving water to the outside. This is very significant to understanding the pattern of erosion and deposition in stream channels, a topic discussed later in this chapter.

Flow Types in Streams

When we mention the flow type in a fluid, we are referring to the nature of its motion during movement. Two types of flow are possible in streams: *turbulent* and *laminar*. Laminar flow is characterized by tiny sheets of water sliding over one another within the "skin" of water (only a few millimeters thick) that covers the streambed. This flow is very slow and represents an insignificant portion of the total water in the channel. Turbulent flow is characterized by mixing motion, and it dominates the flow of all streams. In turbulent flow, whirls of water, called *eddies*, cause mixing between the slower and faster moving parts of the stream. Turbulent flow increases with the roughness of the streambed and the velocity of the stream (Fig. 20.2).

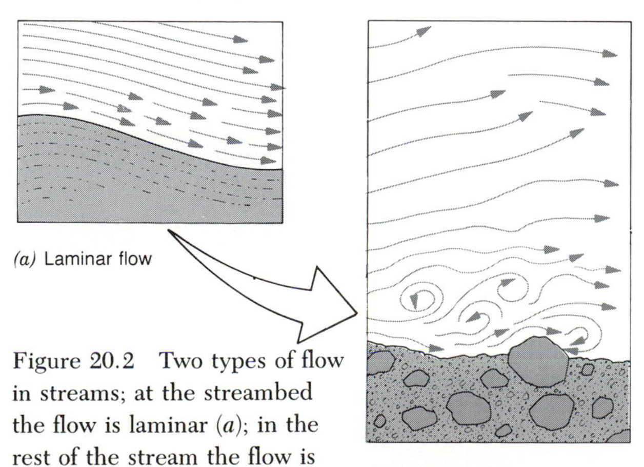

Figure 20.2 Two types of flow in streams; at the streambed the flow is laminar (*a*); in the rest of the stream the flow is turbulent (b).

The Forces Governing Streamflow

We all have an intuitive understanding of why streams flow: a stream is a mass of not-very-sticky fluid that is easily pulled downhill by gravity. The water in a stream channel represents potential energy because it is a quantity of mass raised above sea level, the base elevation for most stream systems. The mass is set into motion by the force of gravity, thereby converting the potential into kinetic energy, the energy of motion.

The total force driving the water down the channel is equal to the mass of water times the slope (downstream incline) of the channel times the force of gravity (or gravitation acceleration). As the water moves down the channel and the stream loses elevation, potential energy is constantly being converted to kinetic energy; therefore, we might expect the stream to accelerate with distance downslope, but it does not.

As with any moving object, there are forces that pose resistance to its movement. With an automobile the principal source of resistance is the friction imposed by air. With a stream it is mainly the friction imposed by its channel, that is, the streambed that lies in contact with the water. The more contact stream water has with the streambed and the rougher the streambed, the greater the resistance to flow. The principal reason stream velocity does not accelerate progressively downstream—as a rollercoaster does on a short incline—is that the force driving the water down the channel is counterbalanced by the resisting force of the streambed.

This resisting force is called *bed shear stress.* It represents that portion of a stream's driving force that is given up in order to get the water over the streambed. The bed shear stress generated by a stream is also proportional to the stream's erosive force, the force exerted by the moving water on the material making up the streambed.

The average erosive force in a channel is equal to the total bed shear stress *relative to* the area of contact between the stream water and its channel. The water-

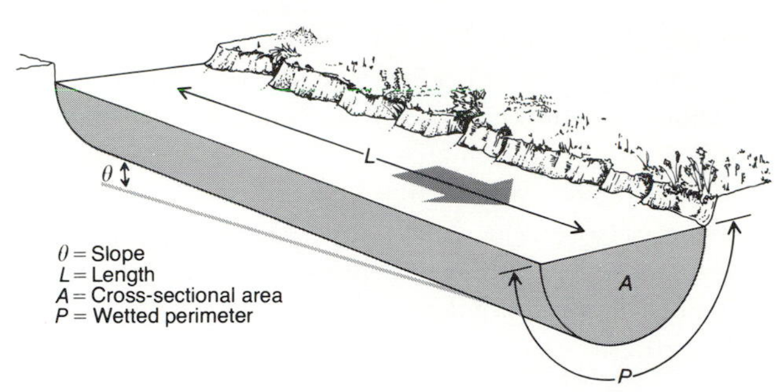

Figure 20.3 The essential dimensions of a stream channel. The force generated by the stream (bed shear stress) is exerted over an area of channel represented by the wetted perimeter over the section of channel.

line of contact in a cross-section of channel is called the *wetted perimeter*; the larger the wetted perimeter per square meter of water in the cross-section, the lower the average magnitude of the stream's erosive force at any point on the channel (Fig. 20.3). Broad, shallow channels tend to produce low levels of erosive force compared to narrow, deep ones because shallow channels have higher ratios of channel contact to water; in other words, they are less energy-efficient. This difference accounts for the fact that streams typically produce more channel erosion at high stages of flow when the water is deep than at lower stages of flow when the water is shallow.

If the stream dissipates the energy gained from the conversion of potential energy, what becomes of the energy? Most of it is converted to heat, but a small amount is also devoted to the work of moving sediment. The energy devoted to sediment movement is ultimately converted to heat by the time the sediment finally comes to rest in a delta or on the continental shelf. The generation of heat is the principal reason why streams do not freeze up completely in winter and why a stream can melt its way into the surface of a glacier.

Channel Erosion

The sediment carried by a stream comes from two sources: (1) the discharge received from tributaries and (2) direct erosion of its own channel. Although most streams transport huge amounts of sediment, the amount derived from channel erosion is relatively small for many streams because the material making up the channel is so resistant to moving water. Whether a stream can erode its channel depends on

(*a*)

(*b*)

Figure 20.4 The channel in (*a*) was scoured into volcanic rock (such as scoria and rhyolite) over a period of 70 years; individual scour marks are shown in detail in (*b*).

(1) the magnitude of bed shear stress it generates *relative to* (2) the resistance of the material on the streambed.

Generally, bedrock is the most resistant channel material. Erosion of bedrock channels occurs through three mechanisms: (1) *chemical* reactions between minerals in the water and in rock; (2) *cavitation,* whereby sudden losses in pressure as water flows around a sharp rock corner cause the formation of small air bubbles, which in turn explode, creating large, localized pressures; and (3) *scouring* by material carried in the bed load, whereby heavy particles bump and skid along the bottom, loosening and freeing material. The effectiveness of scouring is evidenced by the fact that it took only 70 years for the channel in Figure 20.4a to be cut 10m into rock. The features in Figure 20.4*b* were scoured relatively rapidly, probably in a matter of several years, and the scouring agent was particles of sand and volcanic ash.

Most stream channels are composed of less resistant materials, mainly deposits of various types made by the stream itself. The erodibility of these deposits depends mainly on the size of the particles in them. The largest (pebbles) and smallest particles (clay) have the greatest resistance to erosion because of the massiveness of the pebbles and the cohesiveness of the clays. Intermediate-sized particles, principally sand, tend to be most erodible, lacking both sufficient mass and cohesiveness to resist erosion (Fig. 20.5).

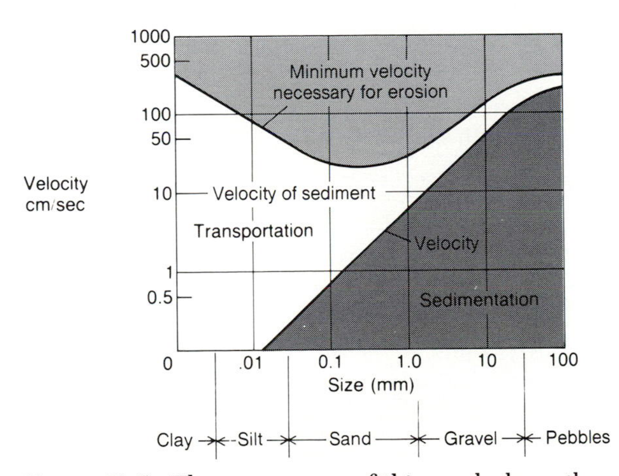

Figure 20.5 The upper curve of this graph shows the relationship between the velocity of streamflow (for a given water depth) and the erosion of different-sized particles. Clay is relatively resistant because of its cohesiveness. Sand is least resistant. The middle zone of the graph gives the velocities at which various-sized particles are transported; the lower line represents the velocity of particle deposition.

TABLE 20.1 RELATIVE RESISTANCE OF CHANNEL MATERIALS TO STREAM EROSION

Material	Relative Resistance[a]
Fine sand	1.0
Sandy loam	1.13
Grass-covered sandy loam	1.7–4.0
Fine gravel	1.7
Stiff clay	2.5
Grass-covered fine gravel	2.3–5.3
Coarse gravel (pebbles)	2.7
Cobbles	3.3
Shale	4.0

[a]Resistance relative to fine sand, the least resistant channel material.

The resistance of materials in the streambed is influenced by many other factors. Vegetation is unquestionably one of the most important because it takes a much greater force to erode stream banks protected by stems and roots than it does to erode barren ones. Compaction of clayey materials and cementing of particles by minerals precipitated from groundwater and soil water also add to the resistance of channels (Table 20.1).

Sediment Transport and Land Types

Most of the work of streams is devoted to transporting sediment. The sediment being moved by a stream is called its *load,* and three classes of load are defined: bed load, suspended load, and dissolved load. *Bed load* consists of coarse particles (sand and larger) that roll and bump along the streambed. The movement of these particles, which can be as large as boulders 1 to 2 feet in diameter, is related to two forces associated with turbulent flow near the streambed, *drag* and *lift.* Lift is an upward force capable of pivoting particles upward, whereupon drag, a lateral force, pushes them downstream. The maximum size particle that can be moved as bed load is a function of bed shear stress and is referred to as a stream's *competence.* Because competence varies greatly with different magnitudes of flow, the makeup of bed load can also vary greatly. As flow is rising, increasingly larger particles will be set into motion; and as flow is declining, progressively smaller particles will fall to rest and be deposited on the streambed. This explains why deposits in streambeds and river deltas are often

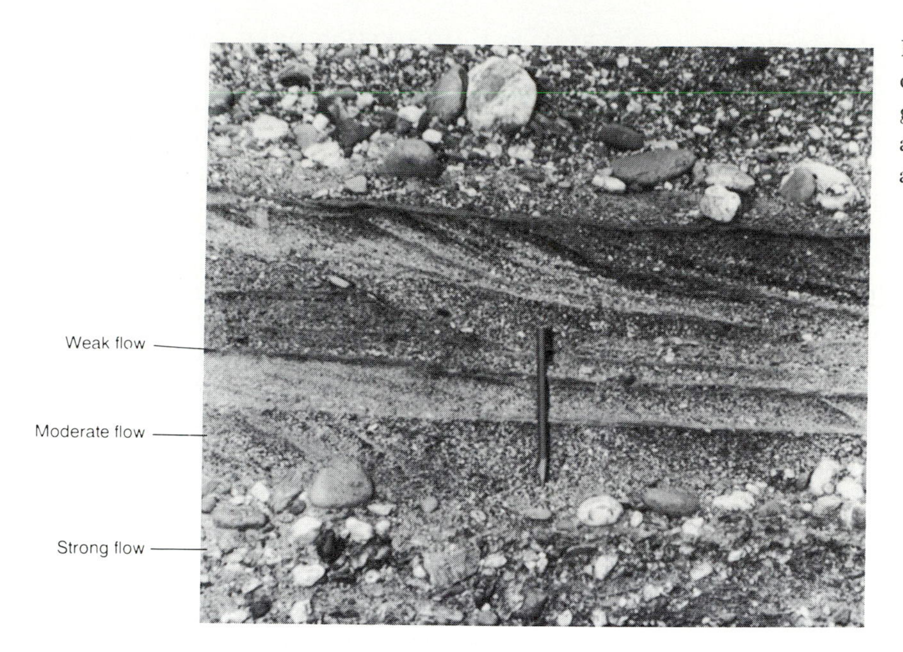

Figure 20.6 Layers of sediment on a streambed showing the gradation of particle sizes associated with strong, moderate, and weak flows.

graded upward from large to small particles (Fig. 20.6).

Suspended load consists of small-sized particles (mainly clay and silt) that are held aloft in the stream by the turbulent motion of the water. Unlike bed load, suspended load does not respond acutely to changes in bed shear stress. Once suspended load is introduced to a stream, it remains in the water a long time because streamflow is essentially always turbulent. Therefore, the amount of suspended load we see in stream water at some point along the channel is not usually derived from that part of the channel but from tributaries and erosion upstream.

A third type of load, *dissolved*, is that carried in solution. Ions of minerals produced in weathering are released into streams through groundwater and runoff. Total ionic concentrations in stream water are generally on the order of 200 to 300 milligrams per liter (200 to 300 parts per million), but in humid areas with low relief and high rates of soil leaching, as in sections of the American East and South, concentrations may reach several thousand milligrams per liter (ppm). In

TABLE 20.2 TOTAL AND DISSOLVED LOADS REPRESENTATIVE OF RIVERS IN DRY AND HUMID CLIMATES

River	Climate	Average Discharge, CFS	Total Load (tons/yr · mi^2 of drainage basin)	Dissolved Load as Percent of Total Load
Little Colorado (Arizona)	Arid	63	199	1.2
Green River (Utah)	Arid/semiarid	6,737	530	12.0
Iowa River (Iowa)	Humid	1,517	510	29.0
Delaware River (New Jersey)	Humid	11,730	270	45.0
Juniata River (Pennsylvania)	Humid	4,329	265	64.0

Source: Adapted by permission from L. B. Leopold, M. G. Wolman, and J. P. Miller, *Fluvial Processes in Geomorphology* (San Francisco: Freeman, copyright © 1964).

Figure 20.7 A plume of suspended sediment pouring from the Flathead River into Flathead Lake, Montana. This process is the same as that which occurs on a much larger scale at the mouths of large rivers emptying into the sea.

dry areas dissolved load may be much lower than the general level of concentration, typically less than 10 percent of the total stream load. This is so because rates of chemical weathering are low in dry areas, whereas suspended sediment loads are higher (Table 20.2).

Deltas: The End of the Line

All rivers must come to an end; thus, the sediment load they carry must be deposited somewhere. The end of the line for most rivers is the sea, but the termination process is not an abrupt one. Rather, the momentum of the river's forward motion usually carries it far beyond the shore, and, as it slows down, the sediment load drops out. The deposits created by this process are called *deltas*.

Because the river loses its velocity gradually, its sediment load is usually deposited in a sequential manner, beginning with the coarsest (bed load) material nearest shore. The suspended load is carried out much farther and can often be seen as a cloudy plume extending many kilometers into the sea (Fig. 20.7). The dissolved load mixes with the receiving water and has no direct impact on delta formation. Simple deltas are characterized by three types of layers or beds: *foreset* beds, which are inclined layers dipping seaward; *topset* beds, which are horizontal layers deposited over the foreset beds; and *bottomset* beds, which are thin horizontal beds of finer material deposited in advance of the foreset beds (Fig. 20.8).

The deltas of major rivers, such as the Mississippi and the Nile, are extremely complex and not easily described using the model of a simple delta. These

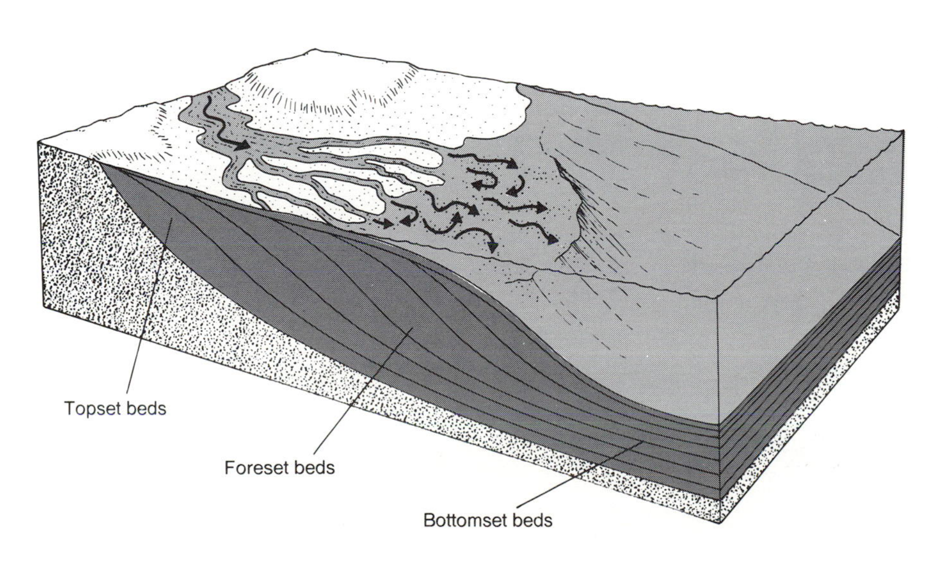

Figure 20.8 The structure of a simple delta showing the distributaries on the surface and the three types of beds that form underwater: topset beds, foreset beds, and bottomset beds.

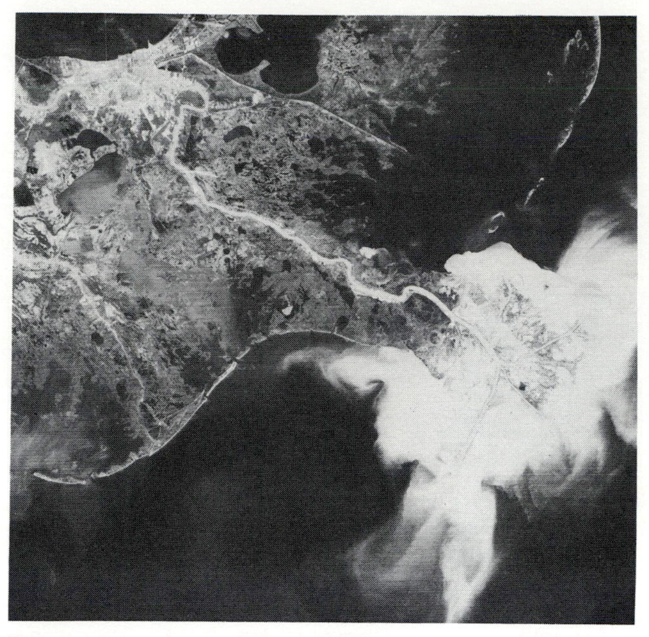

Figure 20.9 The western part of the Mississippi River Delta, about 70 miles southwest of New Orleans. This illustrates the complex nature of large deltas.

large deltas are composed of many small deltas that form at the ends of distributary channels. Distributaries are subdivisions of the main channel, and they funnel water and sediment to different parts of the delta (Fig. 20.9). The Mississippi River has four main distributaries and hundreds of smaller ones. The Mississippi Delta is one of the largest and oldest in the world; over the past 60 to 70 million years, it has grown southward about 1600 kilometers from the point where Cairo, Illinois, is located today. Chapter 21 describes several major types of deltas.

CHANNEL AND VALLEY FEATURES

Stream channels are the work of a multitude of flows or flow events. Generally, however, the moderately large flows do the greatest amount of work in moving sediment and shaping the channel. These are flows such as bankful discharges which occur once to several times per year on the average. Smaller flows—such as those we are apt to see on an occasional visit to a stream—occur with much greater frequency, of course, but their erosive power is so small by comparison that the total work they accomplish is slight overall. By contrast, very large flows, such as the 25- or 50-year discharges, do a huge amount of work, but their occurrence is so infrequent that in the long run total work adds up to less than that of the moderately large flows.

Channel Shape

Most streams can change their channels enormously in response to changes in discharge and sediment supply. When discharge rises, both velocity and water depth increase, producing scouring of the streambed (Fig. 20.10). Conversely, when discharge falls, much

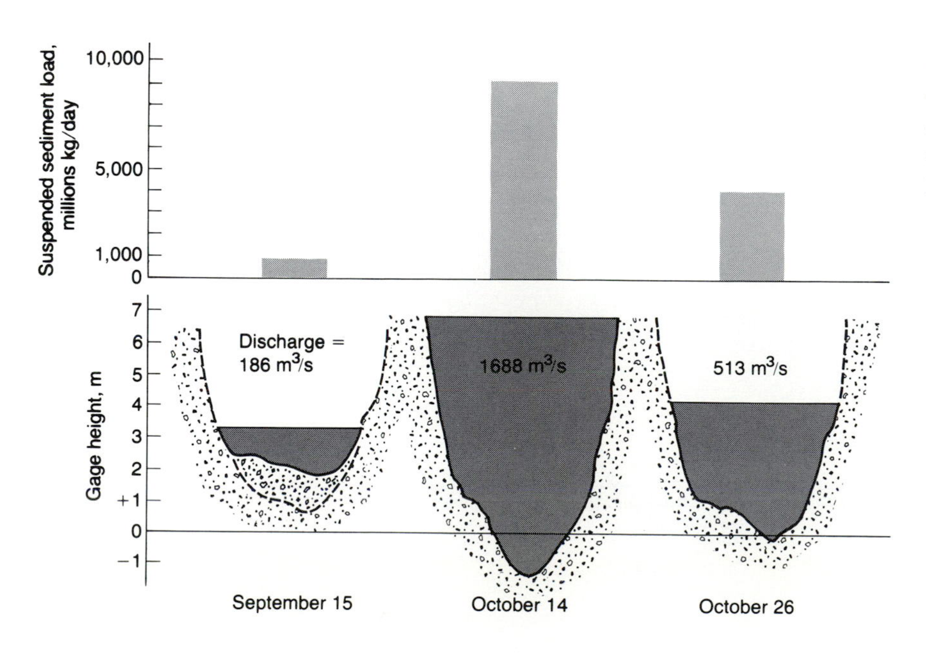

Figure 20.10 Channel scouring and changes in suspended sediment load with changes in discharge, San Juan River near Bluff, Utah.

of the channel bed fills in with sediments. These two channel processes, termed *degradation* and *aggradation*, respectively, are also associated with land use changes in drainage basins.

When land is cleared for farming, soil erosion and sediment input to streams increase, resulting in increased suspended load and some channel aggradation. Later, when farmland gives way to suburban development and the land is torn up for construction, soil erosion rises dramatically for several years. Unless controlled by special drainage facilities, this results in huge sediment loads and heavy channel aggradation. If urbanization follows, sediment input to streams declines below old agricultural levels because soils are now covered with hard-surface materials such as concrete and asphalt. At the same time, however, these materials and the stormsewers that connect them to the stream channels promote greater and faster runoff, resulting in greater magnitudes and frequencies of flows. Together, the increased flows and reduced sediment supplies produce channel degradation in the urban streams (Fig. 20.11).

Channel Patterns and Dynamics

One of the most distinctive aspects of streams is the pattern of their channels as they appear on maps. Two channel patterns occur in natural streams: braided and single-thread. *Single-thread channels*, the most common, are almost always curving or sinuous. Strongly sinuous channels such as those with loops are said to be *meandering*. The size of the meanders is a function of stream size; as average discharge increases, the width of the stream meander belt also increases (Fig. 20.12*a*).

Braided channels are characterized by multiple

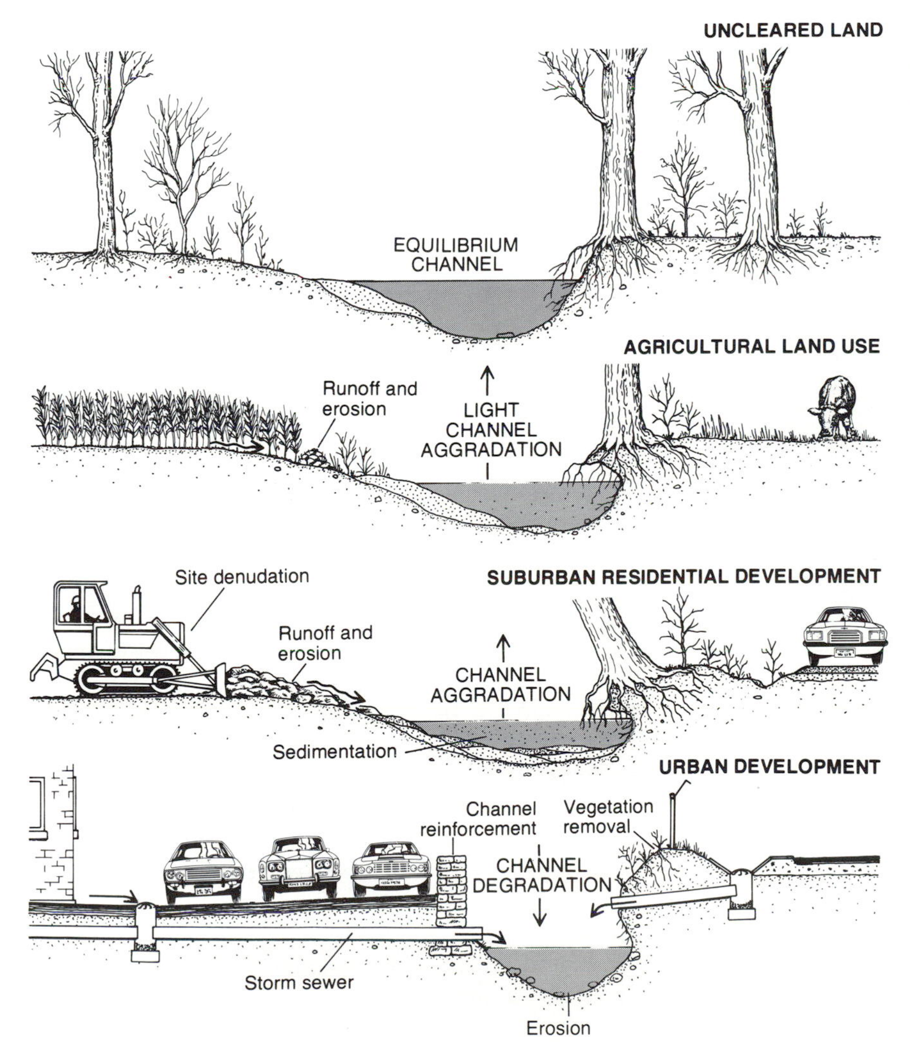

Figure 20.11 Sequence of channel changes related to land use changes. Heavy aggradation is associated with land clearance for residential development. Degradation is often brought on by urbanization.

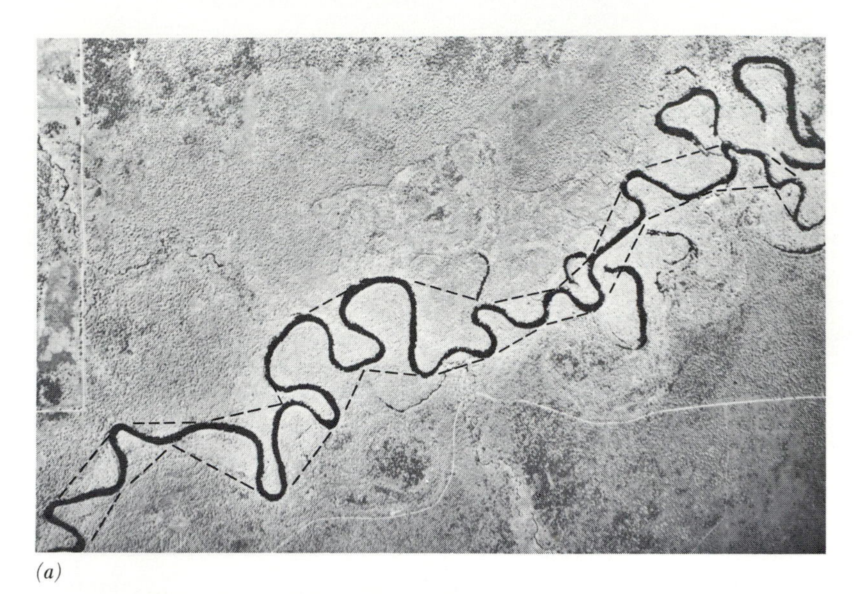

(*a*)

(*b*)

Figure 20.12 A meandering single-thread channel and its meander belt (between the broken lines) in (*a*); a braided channel in stream in the Canadian Rockies (*b*).

threads of water that wind in and out among each other. Braided channels form in streams that flow over loose, coarse materials (usually sand and gravel) and whose discharge is characterized by radical fluctuations. These conditions are common in streams draining from mountain glaciers or building construction sites. When the discharge rises, the threads begin to merge into larger flows and sediment is scoured from the bed. As flow declines, the sediment is deposited in the form of sand and gravel bars. The bars in turn split the shrinking flow, forcing the stream back into a multiple thread arrangement (Fig. 20.12*b*).

Meanders are essential to understanding the features we see in stream channels and their valleys. Like the sled and riders in a twisting bobsled run, the main body of a stream is always thrown by centrifugal force to the outside of a meander. Therefore, the stream's erosive force (bed shear stress) is always concentrated on the outside of the meanders. If we examine the flow pattern carefully, we will find that the fast-moving water is focused a little downstream of the axis of the meander. (The axis is a line that bisects the meander from base to top.) On the inside of the meander, on the opposite bank, things are reversed and velocity is the slowest of any zone of the channel (Fig. 20.13).

These contrasting zones of flow produce erosion on one bank and deposition on the other. Erosion occurs on the outsides of bends and slightly downstream, and it is manifested in an *undercut bank* (Fig. 20.14a). Deposition occurs on the insides of bends, forming features called *point bars*. Each year or so a new increment is added to the point bar while the river erodes a similar distance into the opposite bank. In

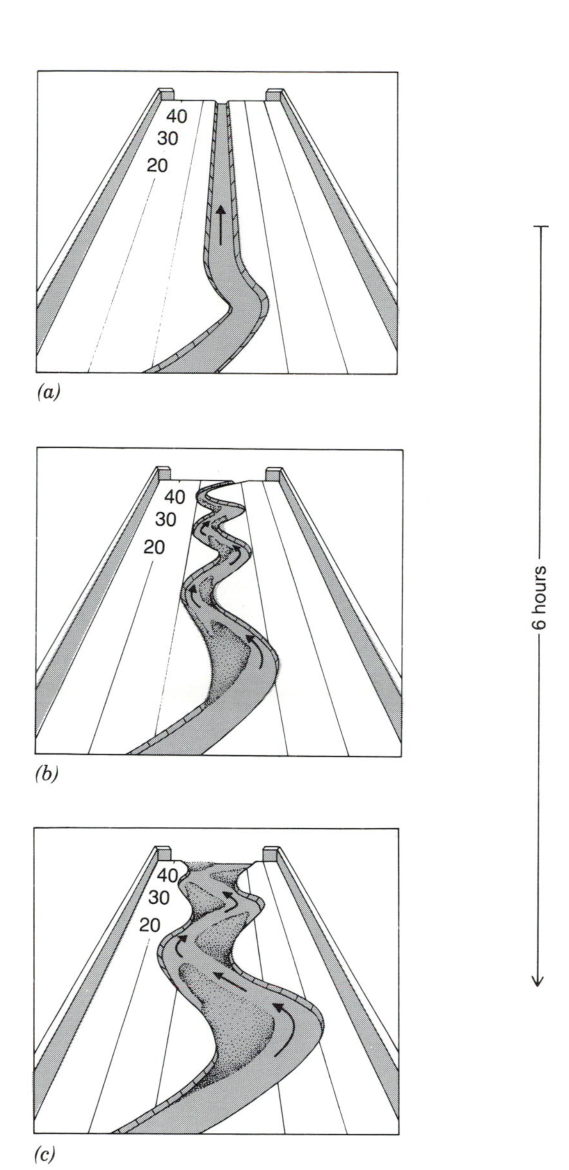

Figure 20.13 The formation of meanders over six hours in an experimental model of a stream flowing in sandy sediments. In nature, the process is much slower because of the influence of vegetation, bedrock, and other factors in holding the channel against erosion.

this way the river shifts laterally, gradually changing its location in the valley (Fig. 20.14b).

Rivers can also undergo sudden changes in location. This often occurs when a stream erodes new segments of channel and abandons old ones. Abandonment is most common where a meander forms a large loop and the river erodes toward itself from opposite sides of the loop, eventually breaching the meander. The old channel is abandoned because the new route is steeper and thus more efficient in terms of the energy of flow. The old channel forms a small lake, called an *oxbow*, but in time it fills with sediment and organic debris, becoming a wetland. On aerial photographs of floodplains, such features are often easily identified by the swamp or marsh vegetation over them (Fig. 20.15).

Floodplains

The flat ground that forms the floor of most stream valleys is the floodplain, one of the most distinctive landforms of stream and river valleys and one of the most important from the standpoint of human settlement. The ancient centers of Middle Eastern and Indian (Asian) civilization were located in floodplains, and today hundreds of millions of humans who live in floodplains struggle against the hazard of inundation by floodwaters.

Formation Floodplains form mainly as a result of lateral shifting by the river. The process works as follows. When the river flows against the high ground at the edge of its valley, called the *valley wall*, it undercuts the wall, which fails and thereby retreats. At the same time, new ground is being formed on the opposite bank in the form of a point bar (Fig. 20.14). The new ground forms at a low elevation, near that of the river. In time, the valley walls are cut back so far that a continuous ribbon of low ground is formed along the valley floor. This is the floodplain. When the river floods, this low ground receives the overflow, hence, the term *floodplain*. Although floods alter the surface of the floodplain by eroding it and leaving deposits on it, they are not the main cause of its formation, and in this respect the term is a little misleading.

When the river is not in contact with the sides of the valley, hillslope processes continue to work on the valley walls, bringing new material to the floodplain. Studies of floodplain materials reveal that material brought down by mass movements, such as slumping and soil creep, may make up an appreciable part of a small floodplain. Added to this are deposits from overland flow and small streams that drain the sides of the valley. Coupled with the channel and flood deposits, these materials help to produce a very diverse composition in floodplains.

Features Meander scars, oxbows, and point bars (described earlier) are present in virtually every floodplain. In addition, natural levees, scour channels, backswamps, and terraces are common features. *Lev-*

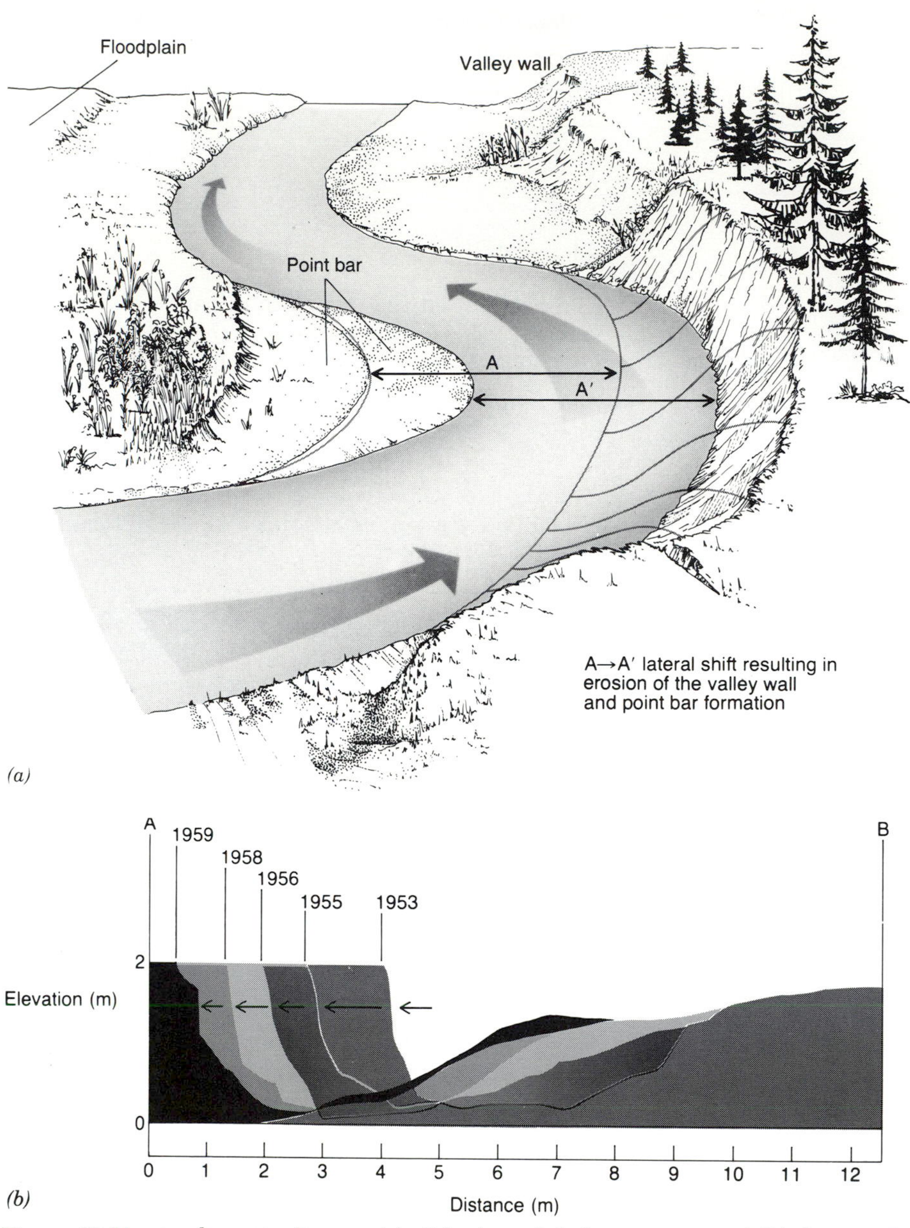

Figure 20.14 A schematic diagram (*a*) of the lateral shift in stream; and (*b*) the actual record of the lateral shift of a channel over a period of 6 years, 1953–59.

Figure 20.15 A portion of the lower Mississippi River floodplain showing the diversity of features typically found in the valleys of large rivers, including backswamps, levees, point bars, undercut banks, scour channels, filled oxbows, oxbows, and future meander beaches.

ees are mounds of sediment deposited along the river bank by floodwaters. They occur on the bank because this is where flow velocity declines sharply as the water leaves the channel, causing part of the sediment load to be dropped. In the low areas behind levees, water may pond for long periods, forming *backswamps*. *Scour channels* are shallow channels etched into the floodplain by floodwaters. They often form across the neck of a meander loop and carry flow only during flood periods (see Fig. 20.15).

Terraces are elevated parts of a floodplain that are formed when a river downcuts and begins to establish a new, lower floodplain elevation. Downcutting may be induced by many factors, including increased flow, loss of sediment load, and uplift of the land. The terraces found in many stream valleys in the American West resulted from the uplift in sections of the Rocky Mountains.

KARST PROCESSES AND TOPOGRAPHY

In addition to the work of streams on the surface, we must also recognize the work of runoff underground. In areas of carbonate rocks (limestone and dolomite), groundwater has the capacity to remove huge volumes of bedrock in solution. If the rock removal is concentrated along cracks, joint lines, and bedding planes, cavities and caverns will form. These features and the surface depressions, tunnels, and stream valleys that result from them are called *karst topography*. The term *karst* comes from an area of Yugoslavia where this kind of terrain is prominent.

Cavern Formation

As we saw in Chapter 17, infiltration water receives carbon dioxide from the atmosphere and vegetation which gives the water a weak charge of carbonic acid. The acid causes a chemical reaction that dissolves carbonate rock. In order for caverns to form, however, there must be considerable water movement through the rock because the solution processes will cease when the water becomes saturated with ions of calcium or when carbonate needs to be drained away. Therefore, from the standpoint of climate, humid regions are most conducive to cavern formation because ion-saturated groundwater is rapidly exchanged with unsaturated water from the surface.

In addition, the character of the bedrock itself is important to cavern formation. Large caverns can form only where the bedrock is strong enough to support itself against undermining. Limestone formations composed of thick strata that are well consolidated are ideal for cavern formation. Weak bedrock, such as that which is heavily fractured, poorly cemented, or made up of thin beds, is prone to sag as weathering advances. This is one of the ways in which sinkholes form.

Collapse Features

Sinkholes, or *dolines*, are surface depressions caused by the collapse of bedrock. The mechanisms of formation range from gradual subsidence to sudden collapse of cavern roofs (Fig. 20.16). Those that form from cavern collapse are typically steepsided pits, with a pile of rock and vegetative rubble at the base. Deep sinkholes (typically 50 to 100 m in depth) often extend below the water table, giving rise to small lakes or ponds. The lakes in Florida's Lake District have formed in sinkholes (Fig. 20.17).

Entire landscapes can be dominated by karst processes and features, and several levels or phases of landform development can often be recognized. The first phase is characterized by the formation of solution cavities and/or caverns and isolated sinkholes. In the second phase more sinkholes form, and old ones enlarge and begin to merge with neighboring sinkholes, revealing part of the groundwater flow and cavern

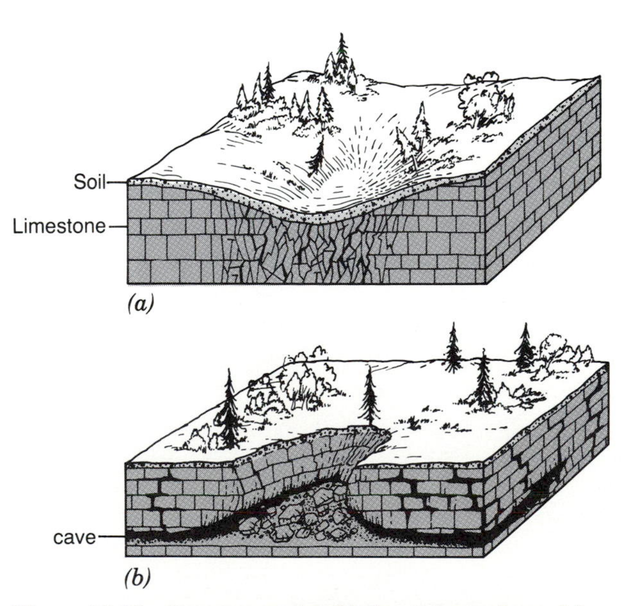

Figure 20.16 Two types of sinkholes: (*a*) one formed by sagging of bedrock; and (*b*) the dramatic variety formed by cavern collapse.

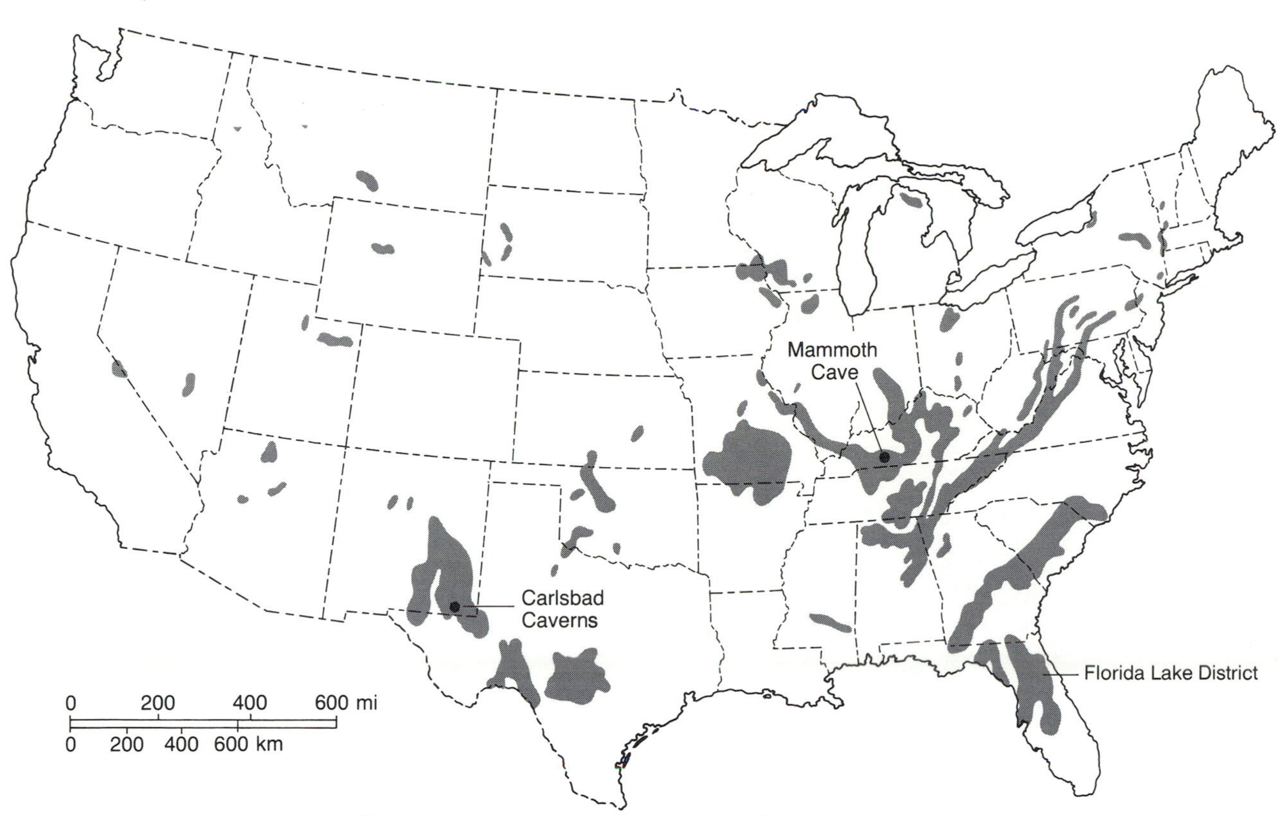

Figure 20.17 The major areas of karst topography in the coterminous United States.

network. The third phase is characterized by overall lowering of the terrain and the formation of large valleys that may contain rivers.

FORMATION OF RIVER-ERODED LANDSCAPES

How entire landscapes form and change as a result of the work of streams and rivers has been a major challenge for geographers and geologists since the late 1800s. Several significant theoretical contributions have been made to explain how streams lower whole land masses and contribute to the formation of landform regions such as the Rocky Mountains and the Great Plains. Here we examine two of these big ideas: the geographic cycle and the dynamic-equilibrium concept.

The Geographic Cycle

The first big idea on the formation of landscapes by streams and rivers was formulated before the turn of the century by a prominent American geographer named William Morris Davis. He proposed that the landscape evolves through a series of developmental stages as rivers deepen and widen their valleys. He envisioned three main stages of development, which, following biological terminology, he named youth, maturity, and old age (Fig. 20.18). During the old-age stage, the landscape is reduced to a low, rolling plain called a *peneplain*. Davis called his idea the "geographic cycle" because he thought that when the old-age stage was reached, the land mass would be uplifted and valley formation would be "rejuvenated," that is, begin over again.

The geographic cycle was immensely popular among educators and scientists in the first half of the twentieth century. It has fallen from favor, however, because of several scientific shortcomings. Among these is the question of when in the cycle uplift of the land takes place. Geophysical evidence indicates that uplift can take place at any time—and not necessarily at the end of the cycle—and that it is as much a response to tectonic forces as it is to unloading of the land per se. Thus, where uplift is sporadic, an orderly sequence of landscape stages is quite unlikely; where uplift is rather steady, the landscape could for long periods of time remain in one "stage" while in a state of continual change.

On the other hand, several of the concepts on which the geographic cycle was based are still valid. One of these is *base level,* a concept introduced by John Wesley Powell in the late 1800s. Base level is the lowest elevation to which a river can downcut its channel. Because a river that flows into the sea cannot deepen its channel much below sea level, the ocean sets the base level for most major rivers. For streams that do not terminate in the ocean, base level may be set by a lake, another stream, or even a resistant rock formation that intersects the streambed. Essentially all base levels, whether set by the ocean or features on the continents, are temporary because their elevations are continually changing with tectonic, isostatic, erosional, and water-level fluctuations.

Another concept used by Davis that is still valid is *grade* or *graded profile.* If we plot an elevation profile of a stream from headwaters to mouth, we find that

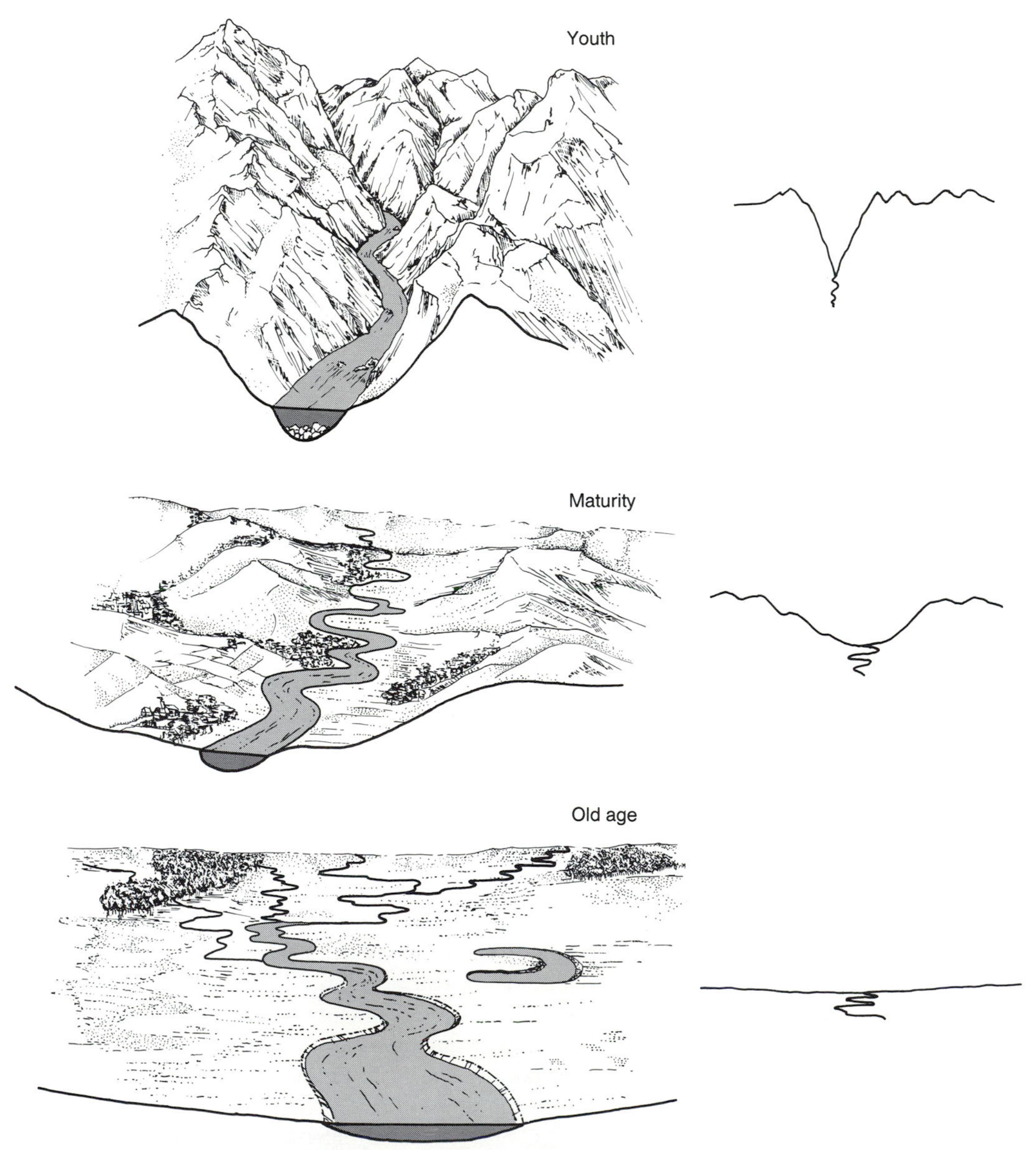

Figure 20.18 The three main stages of landscape development proposed by William Morris Davis in the geographic cycle.

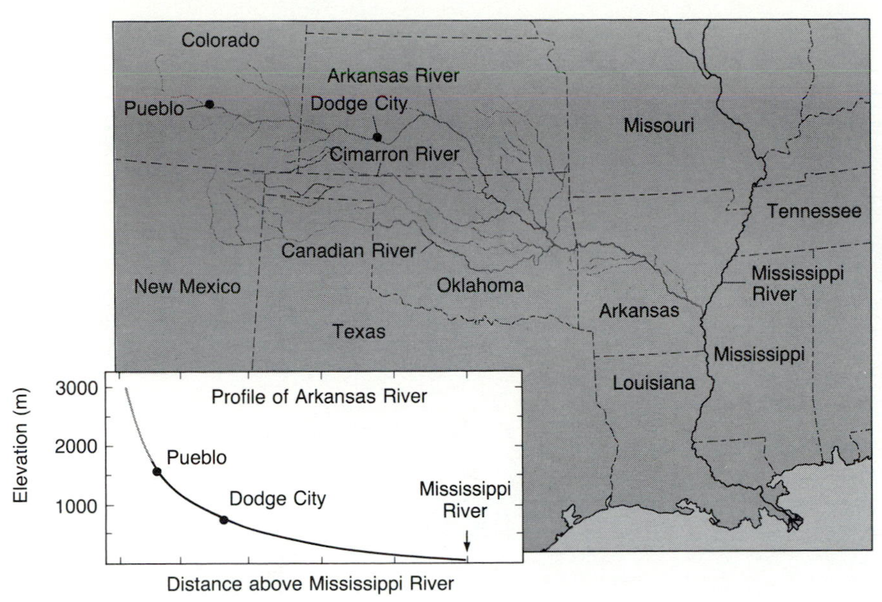

Figure 20.19 The profile of the Arkansas River from the Rocky Mountains to the Mississippi.

the overall slope of the channel flattens downstream (Fig. 20.19). Although the profiles of individual streams vary in detail, all tend to assume this long concave shape. Grove K. Gilbert termed this the graded profile, suggesting that it represents an equilibrium condition to which a stream adjusts itself in response to its discharge and sediment load. Like base level, however, the graded profile is a concept, not an absolute condition, because the stream is continually adjusting to changes in discharge and sediment load. In addition, base-level changes and changes in the inclination of the land owing to tectonic action can affect the graded profile. It seems, then, that streams *trend* toward a graded profile, and although most approximate it, they never truly achieve it.

The Dynamic-Equilibrium Concept

The observation that streams trend toward an equilibrium was put to use in another major idea about river-sculptured landscapes. Around 1950 John T. Hack, a geomorphologist with the U.S. Geological Survey, offered a serious counterview of Davis's geographic cycle. Hack argued that the landscape is in a state of "dynamic equilibrium," meaning that it is continually trending toward an equilibrium, or steady-state condition, but because of the changeable nature of the stream's energy system (driven by climate, runoff, and uplift), rarely achieves it. This is in sharp contrast to the assumptions underlying Davis's stagewise development, in which the energy system winds down as the cycle proceeds, approaching, in the language of systems theory, a state of entropy. *Entropy* is the state of minimum available energy in a system.

According to the dynamic-equilibrium concept, if the climate were to remain steady while the land uplifted in response to denudation (unloading), the stream system would maintain itself, because the water available for runoff (mass) and elevation or slope (which set the stream's energy gradient) would hold steady with the passage of time. Furthermore, if the rock types and geologic structures remained constant as the stream system cut its way into the land, the valley forms would also remain constant. Thus, stagewise development of stream valleys is untenable according to this concept.

Summary

Streamflow is governed by many factors including channel shape, roughness, and downstream slope. Velocity is greatest in large, deep rivers. Most of the stream's energy is expended in frictional resistance as the water moves over the streambed. The bulk of the stream's geomorphic work is devoted to moving the

The James River at flood stage near Richmond, Virginia, as a result of record rainfall from Hurricane Camille in 1969.

sediment and debris brought to the valley by tributaries and hillslope processes. Bed shear stress determines the size and rate of bed load moved by a stream. Suspended load transport, however, is determined largely by the supply of sediment input by tributaries upstream. Streamflow is highly variable, but it appears that moderately large flows, such as bankfull discharge, do the most work in the long run.

The features and landforms of river valleys, such as floodplains and terraces, are built over long periods of time as the river shifts laterally and downcuts its way into the landscape. In areas of carbonate rock, karst topography may dominate the landscape and its landforms contrast with those of river-eroded landscapes which dominate most of the earth's landmasses. Among the major ideas about river-eroded landscapes, the geographic cycle was, and probably still is, the most widely known; however, it does not adequately account for the complex interplay of land uplift and unloading by denudation. The dynamic-equilibrium concept argues that river systems function as energy systems, always trending toward an equilibrium condition in response to changes in land elevations and climatic conditions.

Concepts and Terms for Review

- Western expeditions
- streamflow
  - velocity
  - flow type
- forces of streamflow
  - potential and kinetic energy
  - bed shear stress
- channel erosion
  - chemical reactions
  - cavitation
  - scouring
- sediment transport
  - bed load
  - competence
  - suspended load
  - dissolved load
  - delta formation
- channel dynamics
  - aggradation
  - degradation
  - lateral movement
- channel patterns and features
  - braided and single thread
  - meandering
  - undercut bank
  - point bar
- floodplains

formation and dynamics
levee, backswamp
oxbow, scour channel
terrace
karst topography
solution weathering
cavern
sinkhole
geographic cycle concept
stages of development
peneplain
base level
graded profile
dynamic equilibrium concept
steady-state trend

Sources and References

Brice, J.C. (1969) "Evolution of Meander Loops." *Geological Society of America Bulletin* 85: pp. 581–586.

Chorley, R.J., Dunn, A.J., and Beckinsale, R.P. (1964) *The History of the Study of Landforms*, Vol. 1. London: Methuen.

Hack, J.T. (1960) "Interpretation of Erosional Topography in Humid Temperature Regions." *American Journal of Science*, Bradley Volume, 258-A: pp. 80–97.

Hjulstrom, F. (1939) "Transportation of Detritus by Moving Water." In *Recent Marine Sediments: A Symposium.* ed. P. Trask. Tulsa, Oklahoma: American Association of Petroleum Geologists.

Know, J.C. (1977) "Human Impacts on Wisconsin Stream Channels." *Annals of the Association of American Geographers*, 67:3, pp. 323–342.

Leighly, J. (1936) "Meandering Arroyos of the Dry Southwest." *Geographical Review*, April, pp. 270–282.

Leopold, L.B., Wolman, M.G., and Miller, J.P. (1964) *Fluvial Processes in Geomorphology.* San Francisco: Freeman.

Leopold, L.B., and Wolman, M.G. (1960) "River Meanders." *Bulletin Geological Society of America*, 71, pp. 769–794.

Petts, Geoffrey and Foster, Ian. (1985) *Rivers and Landscape.* London: Edward Arnold.

Williams, G.P., and Wolman, M.G. (1984) "Downstream Effects of Dams on Alluvial Rivers." *U.S. Geological Survey Professional Paper*, 1286.

Wolman, M.G. (1967) "A Cycle of Sedimentation in Urban River Channels." *Geografiska Annaler*, 49A, pp. 385–395.

Wolman, M.G., and Leopold, L.B. (1957) "River Flood Plains: Some Observations on Their Formation." *U.S. Geological Survey Professional Paper 282-C.*

Chapter 21

The Formation of Shores and Coastlines

On a planet dominated by water, coastlines are a prominent geographic feature. Here the motion of the sea, represented by waves and currents, exerts force against the land, eroding rock and soil and redistributing the sediment brought to the sea by streams. By what processes do waves and currents dislodge and transport particles? What landforms are shaped where the coast is being eroded and where sediments are accumulating?

People have a great investment in the world's coastlines. Millions of Europeans and North Americans of all ages and backgrounds each year are drawn to the water's edge for the sheer pleasure they derive from this environment. Besides these visitors, hundreds of millions of people live on the world's coasts for economic reasons—because that is where many of the major urban centers are located.

In New York City, Tokyo, Rio de Janeiro, and Houston, for instance, a total of more than 30 million people live less than 20 m (66 ft.) above sea level. In New Orleans and Alexandria, Egypt, 4 million people live less than 2 m (6 ft.) above sea level. Weigh these figures against the behavior of the seas. Virtually every day the oceans produce tides that exceed 3 m (10 ft.) over much of the earth, storm waves that reach a height of 10 m (33 ft.), and storm surges that elevate the water surface by 3 to 5 m (10 to 16 ft.). Over hundreds of years, wave erosion can drive a shoreline back by thousands of meters, and over several thousand years, sea level can rise and fall by 100 m or more.

In this chapter, we examine the motion of the sea and its capacity to change its shorelines by erosion and deposition. The motion of the sea is limited principally to the surface layer of water; therefore, the movement of sediment is restricted mainly to the shallower parts of the basin, mainly the continental shelves. Indeed, the growth of the continental shelves themselves can be attributed in large part to the work of the sea, inasmuch as waves and currents erode material from the shallow water near shore and deposit it in deeper water farther offshore.

As waves and currents move material around, the shape of the boundary between land and water, the *shoreline*, is altered, being cut back at some points and filled in at others. Over the short run, the overall trend of shoreline change is toward a smoother configuration geographically (Fig. 21.1). Over the long run, however, changes in elevations of land or water and tectonic activity such as volcanism can reverse this trend. One of the major causes of changes in elevation in the past 20,000 years has been changes in the volume of glacial ice on the continents. When glaciers grow, the oceans lose water; when they recede, the oceans rise. During the glacial maxima, sea level was 50 m to more than 100 m lower than it is today; and if today's glaciers were to melt away, sea level would be high enough—almost 80 m higher than it is at present—to drown all the cities mentioned earlier.

Anyone who has observed the assault of the sea on the land during a severe storm knows that its force is awesome. A powerful storm can cut a shoreline back 20 to 30 meters overnight; the waves of a hurricane can destroy a coastal settlement in a matter of hours. Indeed, Charles Lyell and Charles Darwin, insightful scientific observers both, once believed that the sea at higher levels was responsible for most of the erosion that led to present landforms of the continents. Today we know that the sea is not as effective as runoff in lowering the land. But the oceans are nonetheless remarkably effective in shaping the margins of the continents and islands, for each year waves and currents move great trains of sediment of hundreds of thousands of cubic meters along many coastlines of the world.

WATER IN MOTION

Waves are the principal form of motion in bodies of standing water. Most waves are generated by wind, but some are also generated by geophysical forces, such as earthquakes, by pressure differentials associated with large air masses, such as hurricanes, and by the gravitational attraction of the moon and sun.

Causes and Types of Waves

The most common geophysical waves are produced by faulting and volcanic eruptions in the ocean floor. These large waves, called *tsunamis*, commonly reach heights of 10 m and lengths of 150 km, but they are infrequent. The largest known wave of geophysical origin was created when an earthquake dislodged a mass of rock that fell 1000 m into the head of Lituya Bay, Alaska. The resulting wave reached a height of 530 m (1700 feet) above sea level.

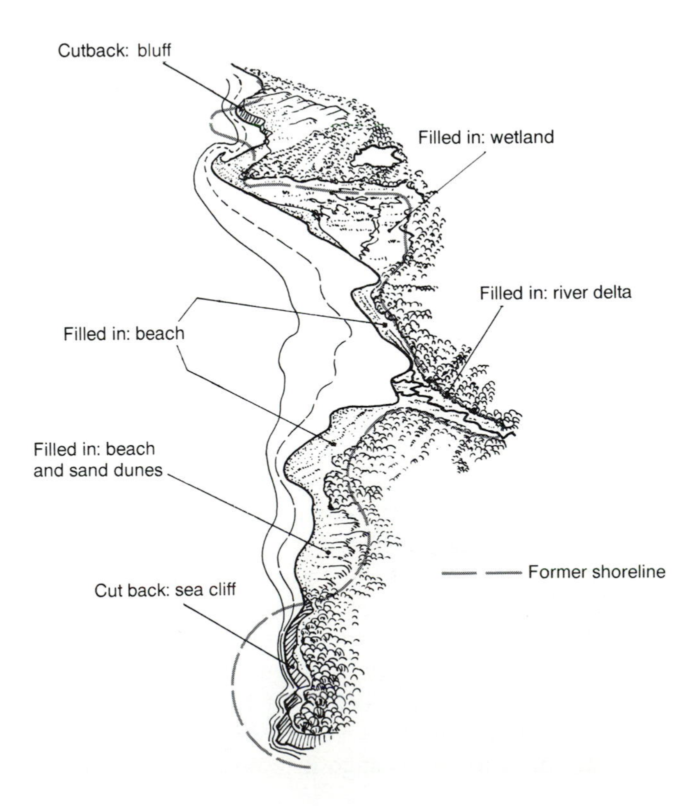

Figure 21.1 Short-term trends in shoreline changes in which an irregular shoreline is cut back at certain points and filled in at others, resulting in a more uniform coastline.

Surges are large waves caused by atmospheric pressure and strong storm winds pushing up the water surface. Hurricanes are the most common cause of surges, although they can be produced by any storm system that exerts great stress on a water surface. Hurricane surges often reach 10 m or more above sea level and in low-lying coastal areas can produce extensive flooding.

The broadest and longest waves are *tides,* which are great bulges formed in the sea in response to lunar and solar gravitational forces. The most frequent tide is associated with the moon's revolution about the earth. This tide is governed by the rates of lunar revolution and earth rotation; on the average, it rises and falls every 12 hours and 26 minutes (Fig. 21.2*a*). The tidal effect of the moon is about twice that of the sun; the largest tides, called *spring tides,* form when the sun and moon are aligned with the earth (Fig. 21.2*b*). Spring tides occur during full or new moons and, like all tidal fluxes, tend to be greatest near the equator. When the moon and sun are positioned at a right angle relative to the earth, the effect is the opposite, and the tides, called *neap tides,* tend to be lowest.

The wave motion created by tides is very complex, owing to the obstructions posed by the land masses and differences in the friction resistance of the ocean basins related to variations in water depth. The result is an irregular pattern of tides worldwide, ranging from massive fluctuations of 10 m or more in certain bays—such as the Bay of Fundy in eastern Canada—to modest fluxes of only a meter or so in polar regions. Where fluctuations are great, large flows of water, called *tidal currents,* may surge in and out of bays, estuaries, and river mouths, moving large quantities of sediment in the process.

Figure 21.2 The nature of the moon's gravitational attraction on the earth in creating daily (or twice daily) tides (*a*); in (*b*) the moon and the sun are aligned, and their combined gravitational pull creates spring tides, the large tides.

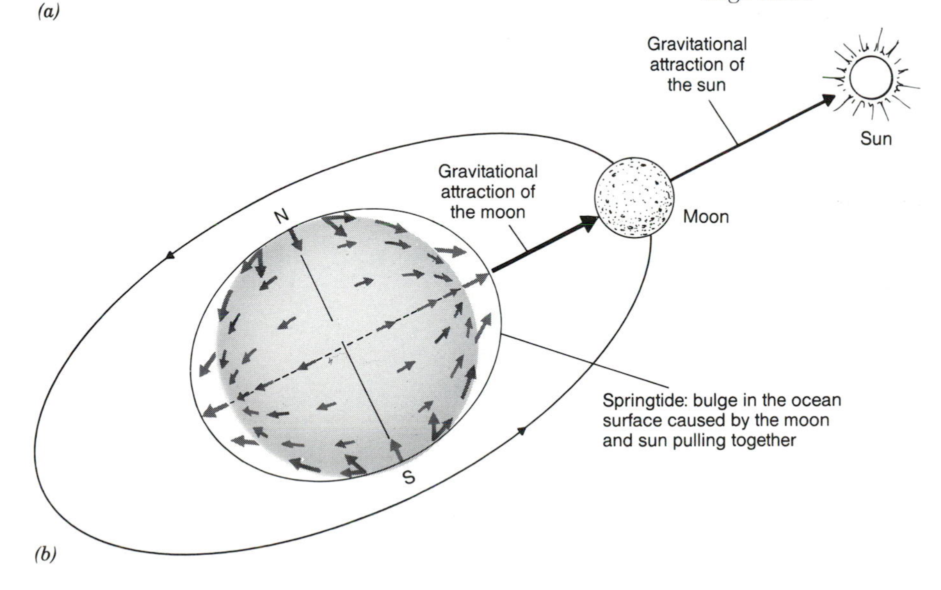

Figure 21.3 The choppy wave motion characteristic of *seas* in areas of wave generation.

Generation of Wind Waves

The mechanisms involved in wave generation by wind are not well understood. They involve the transfer of momentum from moving air to the water surface, but we are uncertain how the wavey surface is produced. It is probably related to air-pressure variations associated with wind gusts that differentially depress the sea surface. In any case, where waves are being generated, the water surface is characterized by a very choppy sort of motion, called a *sea* (Fig. 21.3). Ultimately, however, the drag of the wind on the water surface sets up a downwind direction to the wave motion. As the wave travels from the area of generation, it develops a more symmetrical wave form and may travel distances of hundreds of kilometers with little loss of size. Such waves are termed *swells*.

The size that a wave can attain is controlled by four factors: wind velocity, wind duration, water depth, and *fetch* (the distance of open water in one direction across a water body.) Large values of each are necessary to generate the largest waves: in other words, fast wind blowing from one direction for a long time over a great expanse of deep water. For any combination of these factors, there is a maximum wave size that can be generated.

Wave Forms and Motion

The description of a wave must include both its surface form and the fluid motion beneath it. Let us start with the basic terms used to describe the surface form. The

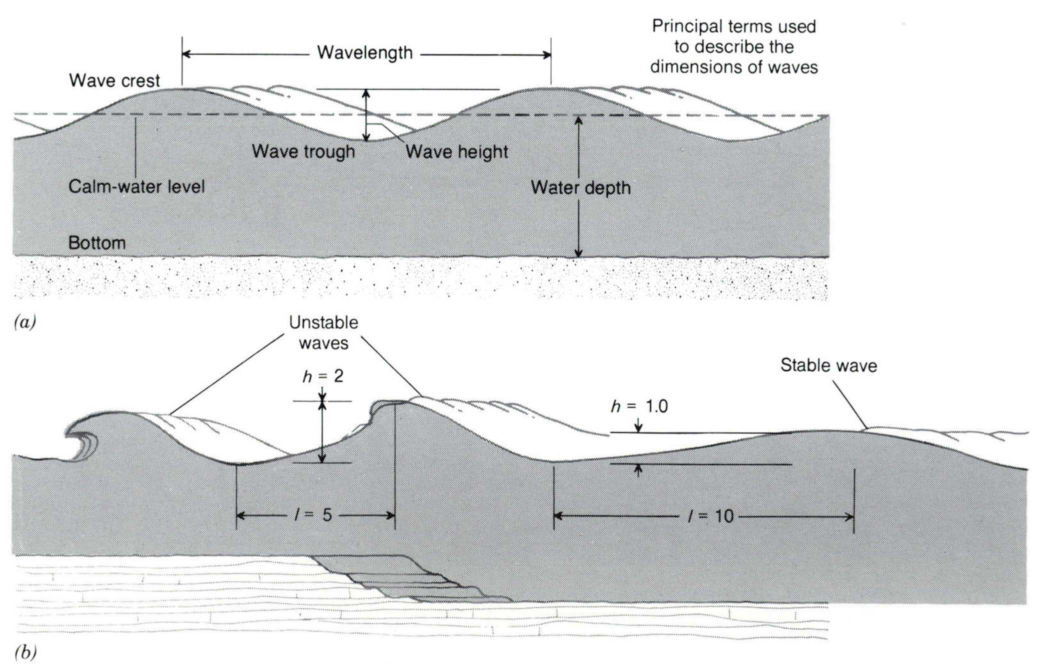

Figure 21.4 The principal dimensional properties of a wave (*a*). In shallow water (*b*), the wavelength shortens and the height increases, steepening the wave slope to an unstable angle.

part of the wave that extends above the calm-water level is the *crest*, and that below the calm-water level is the *trough*. The *wavelength* is the distance from crest to crest, or trough to trough, and the *waveheight* is the vertical distance between crest and trough (Fig. 21.4*a*). The typical slope of a wave, expressed as the ratio of waveheight to wavelength, ranges from 1:25 to 1:50; above a slope of 1:7, a wave is unstable and falls over itself, or *breaks* (Fig. 21.4*b*).

Unlike the passage of water in a river, the passage of a wave in deep water does not result in the transfer or flow of water. Rather, it is only the *wave form* that travels. Moreover, although water particles *do move* when a wave passes, the motion is only circular under the wave form. This can be demonstrated by tracing the motion of a floating object such as a cork: with the passage of each wave it returns to the same point—if it is not influenced by wind. Such waves are termed *oscillatory*, or *nearly oscillatory*. The time it takes a wave to travel a distance of one wavelength is called the *wave period*. The velocity of a wave is equal to the distance traveled by a wave in one wave period.

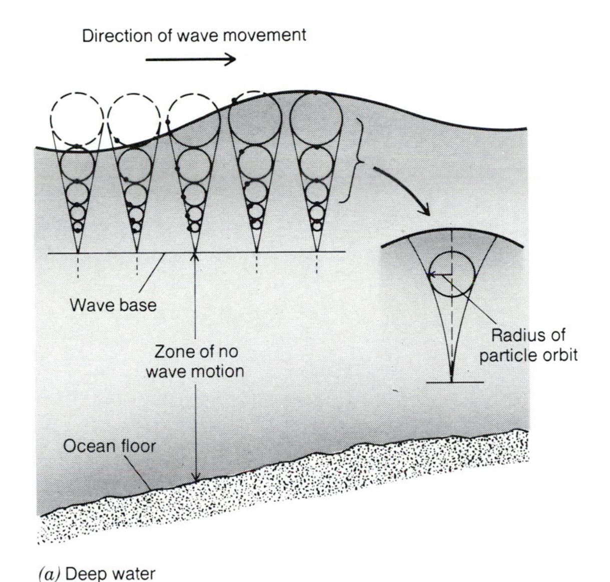

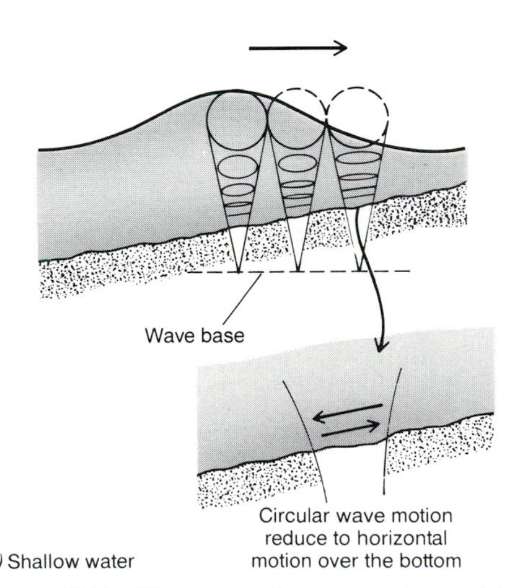

Figure 21.5 Character of wave motion in (*a*) deep water and (*b*) shallow water.

The fluid motion of an oscillatory wave can be described as a series of circular orbits of water particles (Fig. 21.5*a*). The size (radii) of the orbits is greatest at the surface and decreases with depth. For larger waves, radii are larger and the wave motion extends to greater depths than for smaller waves. For all waves in deep water, there is a depth below which they influence no water motion; at depths below this, the *wave base*, a submarine or diver will feel no motion from waves passing overhead.

As a wave approaches the coast, it reaches a depth offshore where the lowermost part of the wave touches bottom. When this happens, the shape of the orbit of rotation of the particles in the wave base changes from a circle to an ellipse. Closer toward shore, the elliptical motion is compressed into a linear motion; the water "slides" back and forth over the bottom with the passage of each wave (Fig. 21.5*b*). This produces friction, which slows down the base of the wave. At a depth of about 1.3 times wavelength, bottom drag becomes so great that the upper part of the wave outraces the lower part, and the wave becomes unstable and breaks. In contrast to oscillatory waves, the water in breaking waves is displaced (transported) toward shore; such waves are called *waves of translation*.

WAVE ENERGY AND ITS DISTRIBUTION

The energy of a wave consists of both potential and kinetic forms. Potential energy is represented by the mass of water displaced above the wave trough, and kinetic energy, by the combined velocities of the water particles associated with wave motion. Remember that waves gain energy from wind; therefore, in crossing an area of sea, they may grow if winds are rising or they may fall if winds are falling. The energy trend, or balance, of a group of waves can be estimated by comparing their total energy upon entering an area of water with their total energy upon leaving the area. A reduction in wave velocity and/or wave size indicates a reduction in total energy. From our standpoint, the most important energy reduction occurs when waves enter shallow water and dissipate energy in erosion, in friction against the bottom, and in the turbulent motion of wave action. Erosion represents the conversion of wave energy into work, whereas friction and turbulence (dissipation of force without erosion taking place) result in the conversion of wave energy into heat.

Circulation Near Shore

The energy of a 3-m wave is equivalent to a row of full-sized automobiles approaching shore side by side at full throttle. If the shore slopes away gradually under water (in the form of a broad ramp), most of this energy is spent in crossing the shallow-water zone. The force of the wave is spread over the ramp from the point where the wave first "feels" bottom to the point where it finally washes up on the beach (Fig. 21.6*a*). We can see evidence for this effect on swim-

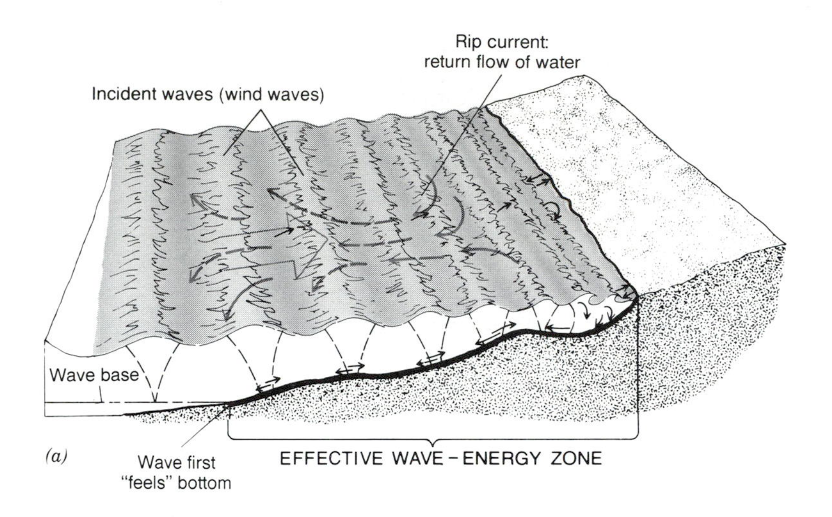

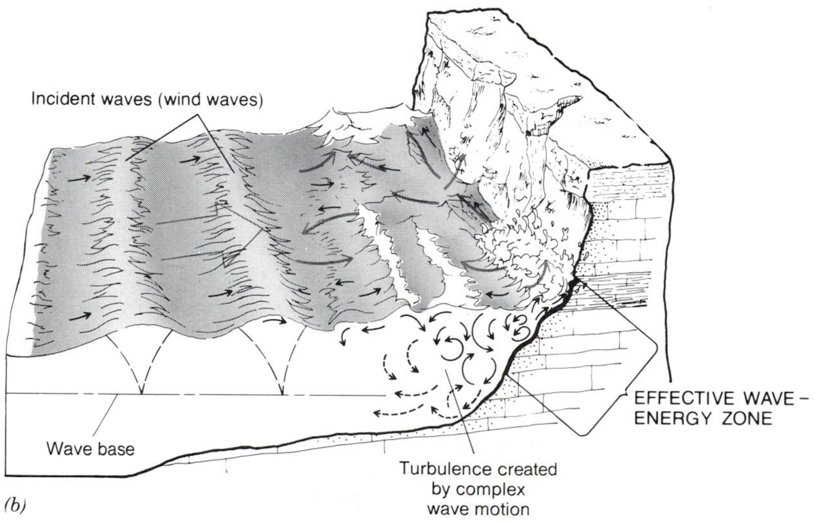

Figure 21.6 Wave-energy distribution near shore along: (*a*) a shallow-water shoreline; and (*b*) a deep-water shoreline. The effective wave-energy zone is much broader on the shallow-water shoreline; however, the magnitude of force exerted on shore is much greater per unit area on the deep-water shoreline.

ming beaches: as wavelength becomes shorter nearer the shore, the greater the energy loss to friction, turbulence, and erosion.

At some point near shore, the wave breaks and part of its water mass is thrown toward shore. As wave after wave adds its water to the shoreline, the water level there literally becomes elevated, forming a seaward slope on the sea surface. This sets up a gradient that forces the water back seaward, against the incoming waves. An important mechanism of this seaward flow is a current called a *rip current*. This is a narrow jet of water that shoots seaward through the base of the breaking waves (Fig. 21.6*a*). (Surfers often use rip currents to propel themselves seaward against large waves.)

The picture is quite different on rocky coasts where deep water runs close to shore. Here wave energy undergoes little attenuation before reaching shore, and waves tend to smack the shore nearly full-force, somewhat like the row of automobiles striking a cliff head-on. As a result, a massive amount of energy can be focused on a relatively narrow zone at the shoreline. Because the surface area of this zone is so small, not all of a large wave's energy can be dissipated in the usual manner, and there is often a lot left over. This excess energy is converted into new wave forms that move along the shore or back toward sea (Fig. 21.6*b*).

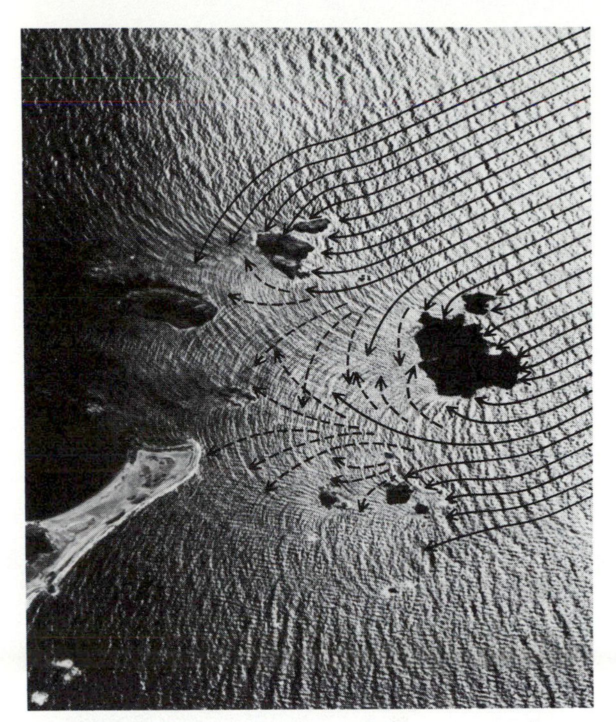

Figure 21.7 Wave refraction and the corresponding orthogonals around some small islands along the coast of Rhode Island.

Wave Refraction

From a geographical standpoint, we need to know why certain parts of a coastline are the focus of erosion and other parts the sites of deposition. This leads us to the question of the areal distribution of wave energy in shallow water. To begin with, no coast has truly uniform offshore topography; therefore, as a wave approaches shore, some segments touch bottom before others do, and those that touch first lose energy first. Therefore, the velocity of the wave is reduced over shallow water, but not over nearby deep water. This produces a bend in the axis of the wave, and because a wave always travels in a direction perpendicular to its axis (crest line), this results in a reorientation of wave energy relative to the shoreline (Fig. 21.7).

The process of wave bending, called *wave refraction,* is analogous to refraction in light and sound waves. The specific offshore location at which the wave begins to refract depends on wave size and water depth. At depths greater than one-half of the wavelength, the wave functions as a deep-water wave, and no refraction takes place. Where water depth is between 0.5 (1/2) and 0.04 (1/25) of the wavelength, conditions are considered to be transitional; velocity begins to decline, and the wave refracts correspondingly. Maximum wave refraction can be expected where water depths are less than 0.04 of the wavelength. Wavelength is greatly shortened in this zone; velocity is proportional to the square root of water depth, and the direction of advance is nearly perpendicular to the contours of bottom topography.

The redistribution of wave energy as a result of refraction can be defined by drawing *orthogonals* (lines perpendicular to the wave crest) for a group of waves traveling across the shallow zone (see Fig. 21.7). The relationship to the shape of the coastline is plainly evident: wave energy is *convergent* on promontories and *divergent* in embayments, meaning that wave energy is greater than average where the land protrudes into the sea and less than average where the sea protrudes into the land (Fig. 21.8*a*). Along straight coasts, wave energy is more or less evenly distributed, but oriented in the direction of the approaching waves (Fig. 21.8*b*). Although refrac-

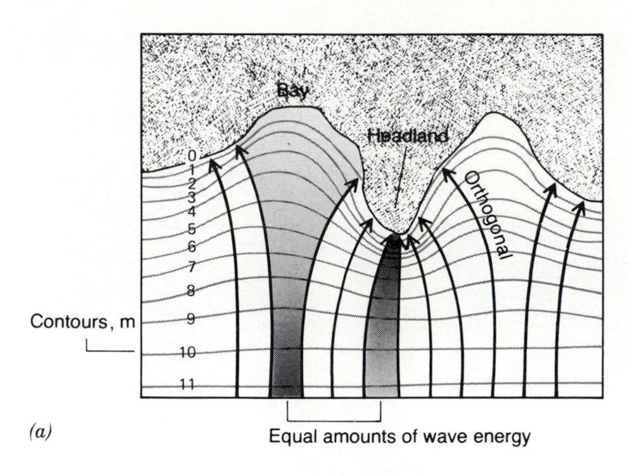

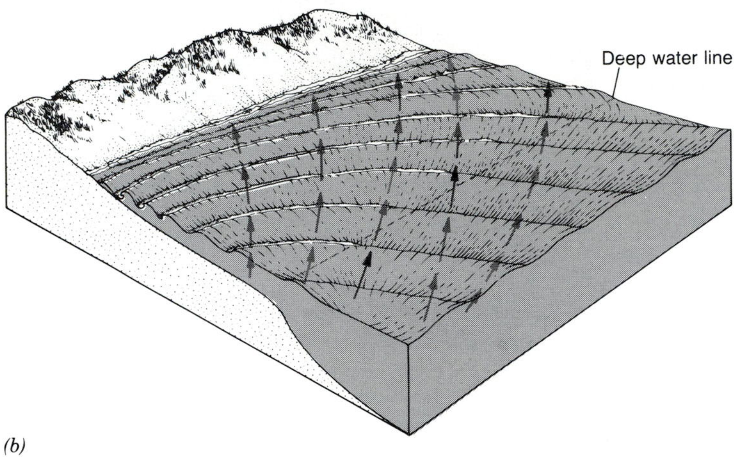

Figure 21.8 Redistribution of wave energy by wave refraction: (*a*) Seaward of the 10-m contour, wave energy, represented by the area of the cells formed between the orthogonals and the contour lines, is equal all along the wave front. Landward, the cells compress or enlarge with wave refraction around the headland and in the bay. (*b*) Along a straight coast, waves refract evenly; thus wave energy is evenly distributed.

tion can reduce the angle between a wave crest and the shoreline to as little as 10 degrees, rarely does a wave approach the shore straight-on. The fact that the force of waves is exerted in a direction oblique to the shore is very important in terms of sediment transport because when particles are lifted off bottom by wave turbulence, they tend to be carried along the shore with the flow of wave action.

Longshore Currents

Currents are also an important component of water movement and sediment transport. Currents are driven by waves, and when waves move in one direction along the coast, a gentle flow of water is set up parallel to the shoreline. This *longshore current* moves with the waves at velocities between 0.25 and 1.0 m/sec.

Along coasts where winds limit waves to one approach direction in all seasons, the flow of longshore currents is consistently in one direction, and only velocity varies with wave energy. Where winds shift seasonally, longshore currents may reverse from summer to winter. Along some north-south-trending coasts in the midlatitudes, for example, longshore currents change from northflowing in summer to southflowing in winter as wind systems shift with the seasonal change in air masses. In other areas, owing to the orientation of the coast and the variable direction of winds, longshore currents are highly variable, changing over a matter of days with different weather systems.

If we combine the various wave and current movements we have discussed, a circulation system can be described for shorelines called *nearshore circulation cells* (Fig. 21.9). These cells are made up of waves bringing water toward shore, rip currents carrying water seaward, and longshore currents moving water along shore.

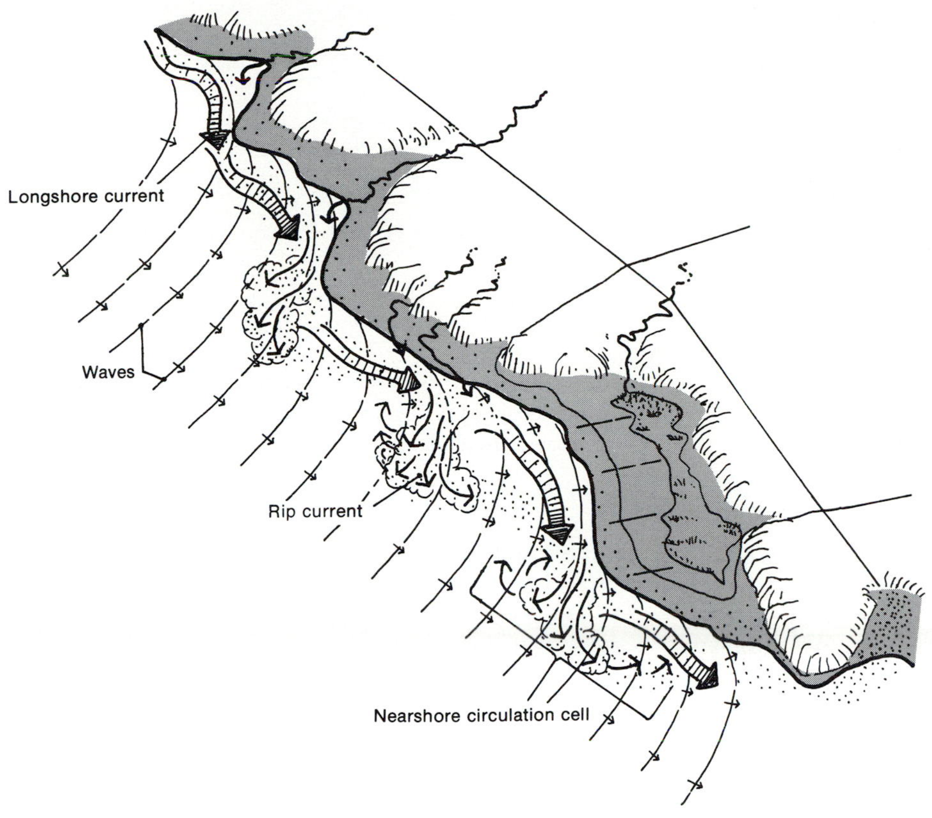

Figure 21.9 Nearshore circulation cells consisting of waves, rip currents, and longshore currents.

WAVE EROSION, SEDIMENT TRANSPORT, AND SHORE FEATURES

Waves driven by wind are the primary erosional agent on shorelines. Wave erosion removes material from the shore. It may take place underwater, at the waterline, or above it, and it usually results in *recession* of the shoreline. (Recession is also caused by a rise in water level, a subsidence of land, or both.) Erosion is measured in terms of the volume of material removed, whereas recession is measured in terms of the distance of landward displacement of the shoreline.

Wave Erosion and Erosional Features

Wave erosion takes place when the hydraulic pressure of the moving water is sufficient to move shore materials. The applied hydraulic pressure of waves is greatest where the velocity and mass of the waves are greatest, which is in the zone where waves break. How much erosion a given wave can produce depends not only on wave force but also on the resisting strength on the material it strikes. Generally, bedrock is most resistant to waves, and loose (unconsolidated) sediment is least resistant. This difference is evident along shorelines composed of cliffs (bedrock) and banks (unconsolidated material) that are being attacked by storm waves. The cliff may be cut back little, if at all, whereas the bank can be cut back rapidly—as much as several feet per hour in big storms.

When a bank is eroded by waves, it is undercut at the toe, causing it to collapse onto the beach. Subsequent waves overwash the heap of debris, quickly removing all but the heaviest particles, usually the large boulders. The fine particles are moved both downshore and offshore into deeper water. For the 100 m stretch of the shoreline shown in Figure 21.10 to retreat a distance of 2 m, 1000 cubic meters of material would have to be eroded.

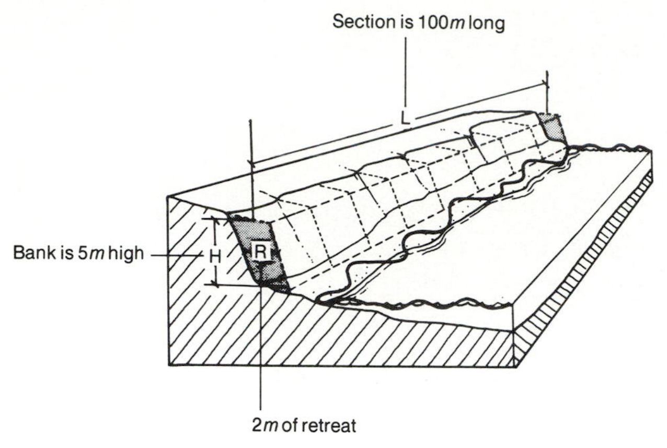

Figure 21.10 Wave erosion of a bank; retreat of this 5 *m* bank a distance of 2 *m* over a 100 *m* segment of shoreline produces a total of 1000 m^3 of erosion.

Erosion of bedrock is not only slow, but also is usually related to a combination of processes. In addition to hydraulic pressure, *corrasion* and *solution weathering* also contribute to erosion. Corrasion is abrasion by stones which are rolled, bounced, and hurled by storm waves against solid rock, breaking fragments free. Solution weathering takes place when minerals are dissolved into seawater.

The shoreline features formed by wave erosion are familiar to most of us (Fig. 21.11). Erosion of bedrock often results in the formation of a *sea cliff*, which may be undercut near water level to form a *wave-cut notch*. Where bedrock has variable resistance to wave erosion, an assortment of interesting features may form, including *sea caves* and *sea stacks*. Sea stacks are pillars of resistant rock left standing offshore.

Along nonbedrock shorelines, most of the features are formed in sediments that are in transit along the coast. Only the *backshore slope*, the steep bank or bluff landward of the shore, is composed of *in situ* (in

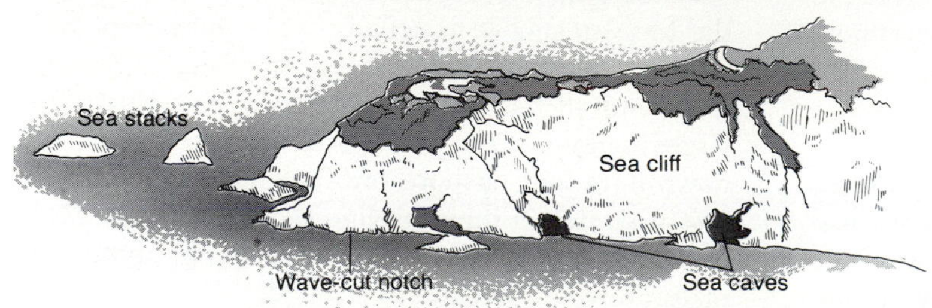

Figure 21.11 Features of a shoreline formed by wave erosion of bedrock near the Big Sur, California. The diagram pinpoints and names the features.

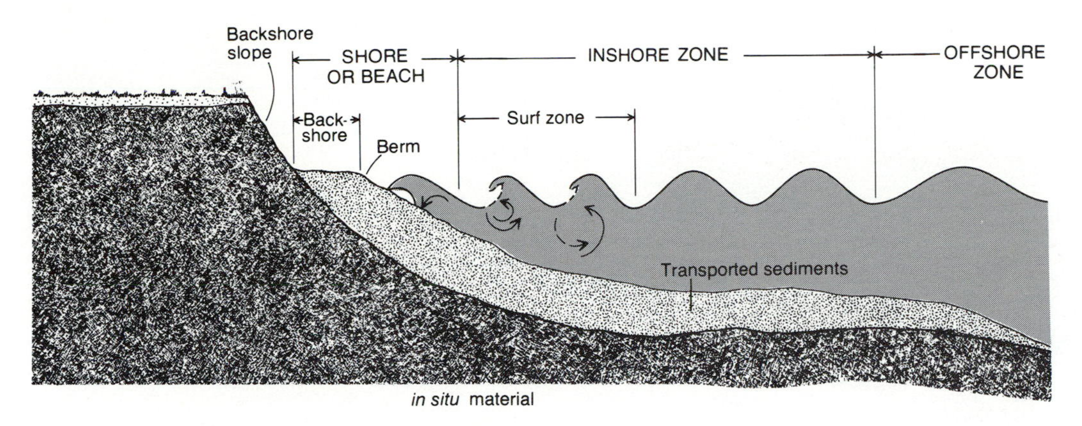

Figure 21.12 The principal zones and features of a coast comprised of unconsolidated materials and sediments.

place) material. The *shore,* or *beach,* stretches from the foot of the backshore slope to the shallow water just beyond shore; along erosional shorelines the beach may be only meters wide. Near the water, wave action forms a small ridge called a *berm.* On wide shores the area between the berm and the foot of the backshore slope is designated the *backshore.* Seaward of the shore is the *inshore zone,* the inner portion of which is the *surf zone,* where waves break. Beyond the inshore is the offshore zone, where waves behave like deep-water waves (Fig. 21.12). Longshore currents are strongest in the inshore zone, but they extend into the adjacent part of the offshore zone as well.

Sediment Transport

The material eroded by waves is incorporated into a train of sediment that moves along the shore with the flow of wave and current action. The source of more than 90 percent of this sediment is the debris brought to the coast by streams and rivers. Each year the world's land masses contribute around 15 billion m^3 of sediment to the coastlines. This material is moved by waves and currents, and some of it is deposited in *sediment sinks,* environments along the coast that are favorable to massive sediment accumulation.

The movement of sediments along a coastline is called *littoral transport,* and the material that is moved is known collectively as *littoral drift.* Littoral transport is made up of two components: longshore transport and onshore-offshore transport. The longshore component consists of sediment movement associated with wave and current action mainly in the surf zone.

Part of the longshore transport also takes place on the beach itself, driven by wave action in the form of *swash* and *backwash.* Swash is a thin sheet of wave water that slides up the beach face; backwash is its counterpart in return flow. Because waves strike the shore at an angle, swash flows obliquely onto the beach. Backwash, on the other hand, flows more perpendicularly to the shoreline. Together, swash and backwash produce a ratchetlike motion of water and sediment, resulting in a net downshore movement (Fig. 21.13). Much of the sediment, especially the pebbles and larger particles, moved in this fashion is rolled along the beach and is referred to as *bed load* (a term which, as we saw in Chapter 20, is also used to define the particles transported along the bottom of a stream channel).

Breaking waves in the swash zone.

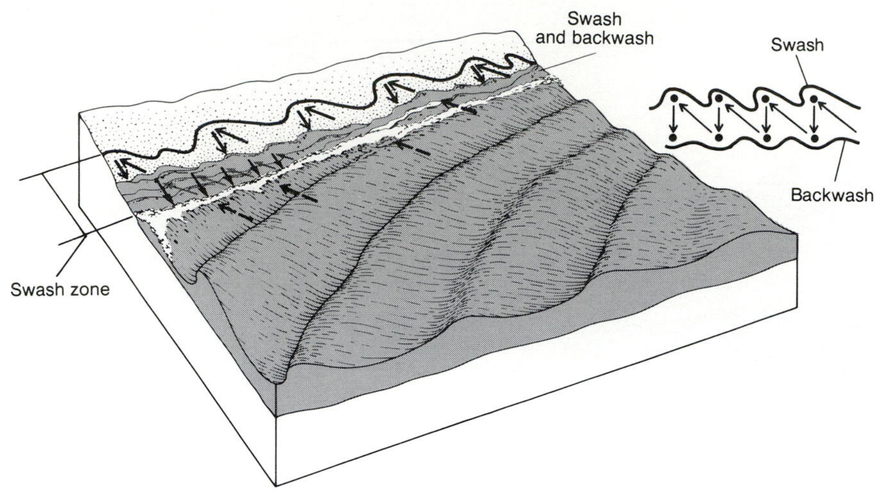

Figure 21.13 The nature of swash and backwash. Sediment moves by a rachetlike motion, rolling up the beach with the swash and down the beach with the backwash.

Beyond the beach, longshore transport is most concentrated where waves are breaking (Fig. 21.14). The turbulence created by the breakers lifts sediment high into the water whereupon it is carried downshore by longshore currents. Sediment moved in this fashion is called *suspended load* (because it is carried in suspension), and it constitutes the bulk of the littoral drift along most coasts. Most of it is sand; the fine sediments (silt and clay) raised by wave action or introduced by rivers are generally carried into the offshore zone, as illustrated in Figure 21.14.

Rates of Longshore Transport

Rates of longshore transport vary with wave energy, the angle at which waves approach the shore, the size and availability of sediments, and certain other factors including coastal ice and vegetation. Along a sandy shoreline that is free of ice, bedrock, and other controls, longshore sediment transport is directly proportional to the flux of wave and current energy in the longshore system. In measuring annual transport rates at various places on the coasts of the United

Figure 21.14 The relative distribution of longshore sediment transport.

States, the Army Corps of Engineers has found, not surprisingly, considerable differences on shores located only a few hundred kilometers apart. One of the most striking differences is found in southern California; at Oxnard Plain Shore, north of Los Angeles, longshore transport is ten times greater than it is at Camp Pendleton, south of Los Angeles (Fig. 21.15).

On the East Coast, rates range from 100,000 to 380,000 m^3/year except for sheltered locations, such as Atlantic Beach, North Carolina, where longshore transport amounts to less than 25,000 m^3 per year. In the upper Great Lakes, longshore transport averages between 50,000 and 100,000 m^3 per year for most sandy shorelines. Along arctic shorelines it is even less, on the order of 5,000 to 10,000 m^3, owing to the presence of grounded ice in the shallow-water zone for much of the year.

The longshore transport figures given above represent the sum of sediment transport in two directions along the coast: one the primary direction, the other the secondary direction. This is also known as *gross sediment transport.* A more meaningful statistic in terms of coastal trends is the *net sediment transport,* which gives the balance between the two directions of movement:

$$\text{Net sediment transport} = q_p - q_s,$$

where q_p = sediment moved in the primary direction and q_s = sediment moved in the secondary direction. A large net transport indicates that sediment is being removed from a source such as a river delta and transported to a sediment sink such as a submarine canyon or the mouth of a bay. Where net sediment transport is small but gross transport is large, sediment is merely being shifted back and forth; in other words, the same sediment mass is being reworked year after year.

Onshore-Offshore Transport

Sediment not only undergoes longshore transport but also onshore-offshore transport, that is, transport that is perpendicular to the shore. The amount of sediment moved in this fashion is very small compared with longshore transport, but it is important to beach topography. The onshore movement is brought about by low-energy waves, usually the summer wave regime, and is characterized by a shoreward migration of sand bars (Fig. 21.16). Sand is added to the beach in the swash zone, from which it may be blown and washed farther landward. This trend usually reverses with high-energy waves, typically the winter wave regime in the midlatitudes. As the storm waves of winter erode sediment, the beach face steepens, rip

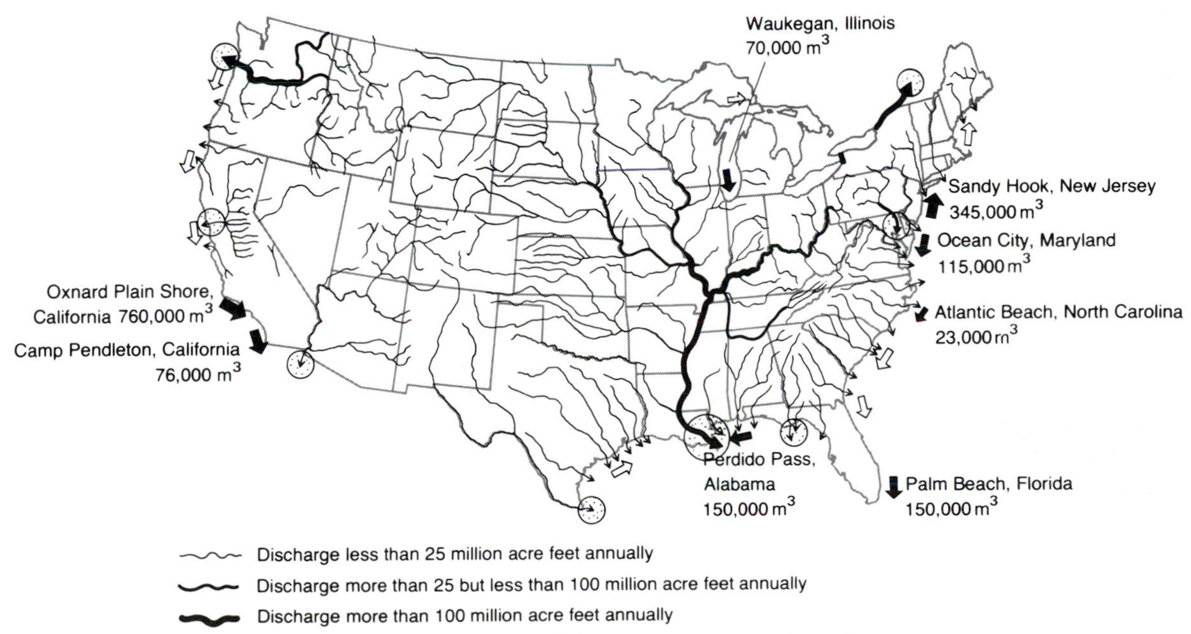

Figure 21.15 The general pattern of average annual longshore transport along the continental United States. Longshore transport rates for selected shorelines are represented by the heavy arrows. The river outlets represent points of sediment input to the coasts.

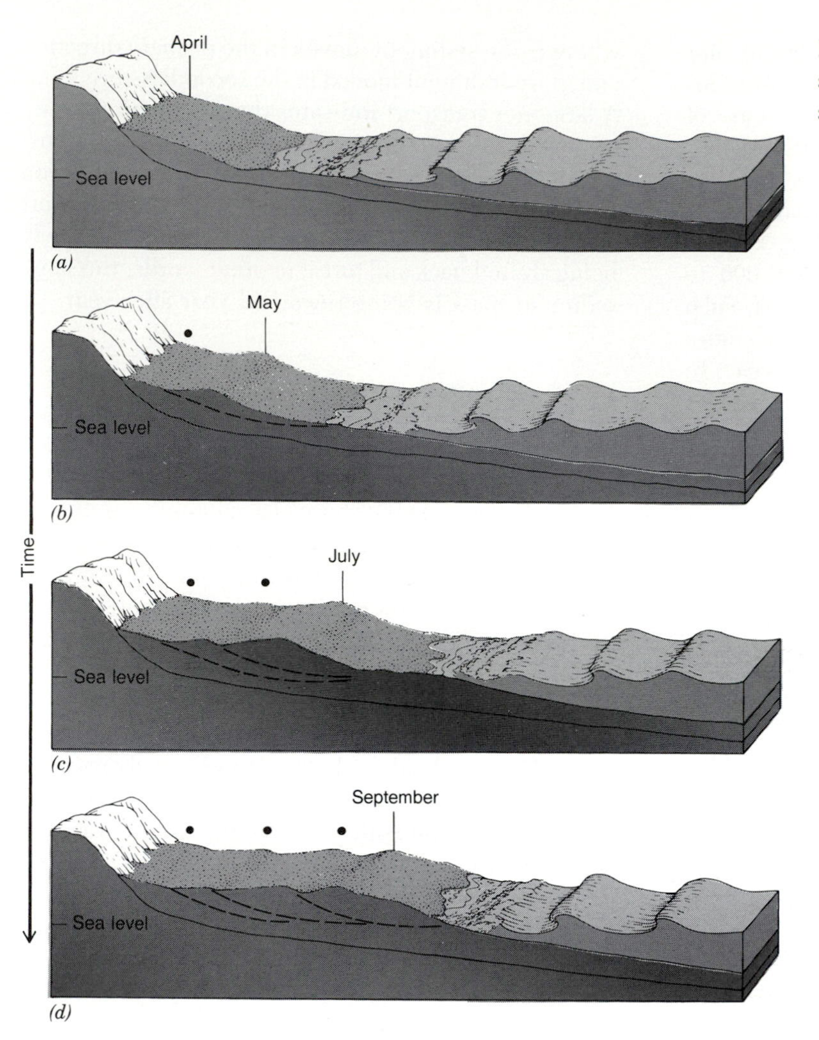

Figure 21.16 Growth of a beach in summer by the onshore accretion of sand.

currents intensify, and sand bars migrate seaward. Longshore transport continues during all seasons.

Growth and Loss of Beaches

To help us understand how sediment transport relates to trends and rates of change in a segment of coastline, it is helpful to discuss sediment transport in terms of the balance of the sediment mass. For any parcel of coastline, sediment is almost constantly being brought to it and taken away. Whether the body of sediment that makes up the beach in Figure 21.17*a* grows or shrinks over a year, a decade, or longer depends on the relative balance between inputs and outputs by waves and currents. If output exceeds input (negative balance), the body of stored sediment shrinks, and the beach begins to "dry up" (Fig. 21.17*b*). Should this condition be prolonged, the body of sediment may be drawn down to such a small amount of sediment that the *in situ* material becomes exposed, and waves go to work directly on it (Fig. 21.17*b*, *c*). In this case, the backshore slope would be cut back, and a new shore profile would be established seaward from the beach (Fig. 21.17*d*). In time the profile would smooth out, and wave energy would become rather uniformly distributed across it, resulting in what is termed an *equilibrium profile*. Each time sea level changes or the sediment mass balance changes appreciably, the trend toward a new equilibrium profile is initiated.

Erosion of the body of beach sediments can be caused by many factors. In southern California, beaches began to decline when rivers were dammed for water supply, thereby trapping within reservoirs sediment that would otherwise feed the beaches. In many parts of the world, breakwaters have been constructed at harbor entrances to intercept the longshore drift. As a result, the beaches downshore dry up because the flow of the sediment train has been broken. Moreover, once the sediment is gone, wave energy erodes *in situ* material. In the Great Lakes, high water levels are the most common cause of beach erosion. In the past decade, lake levels have been as much as 1 m above average, which has allowed storm waves to reach the beaches with greater force. As a result, sediment output has increased without a commensurate increase in input, and beaches have declined.

The opposite condition is one in which total sediment input exceeds total output and the body of stored sediment grows. The beach widens, beach ridges develop, and sand dunes begin to form. Where this trend is long term, the coastline builds seaward and is said to be *progradational*. The particular forms and causes of progradation are discussed in the next section.

SHORELINE DEPOSITION

At the point along a coast where wave and current energy declines, part of the train of sediment it is pushing along may be deposited. A common cause of energy decline is the divergence of waves with refraction in a bay or harbor. Sediments transported down the sides of the bay will accumulate at the bayhead to form sand bars, a broad beach, and other depositional features. An offshore barrier, such as an island or a detached breakwater, can also effect a reduction in wave energy by shielding the shore behind it from the full force of onshore waves. The zone behind the barrier forms a sort of energy shadow. Sediment is dropped in this shadow, and owing to wave refraction around the barrier, deposits may build out from both the shore and the barrier. If the two are not too far apart, a neck of land called a *tombolo* may form between them (Fig. 21.18*a*)

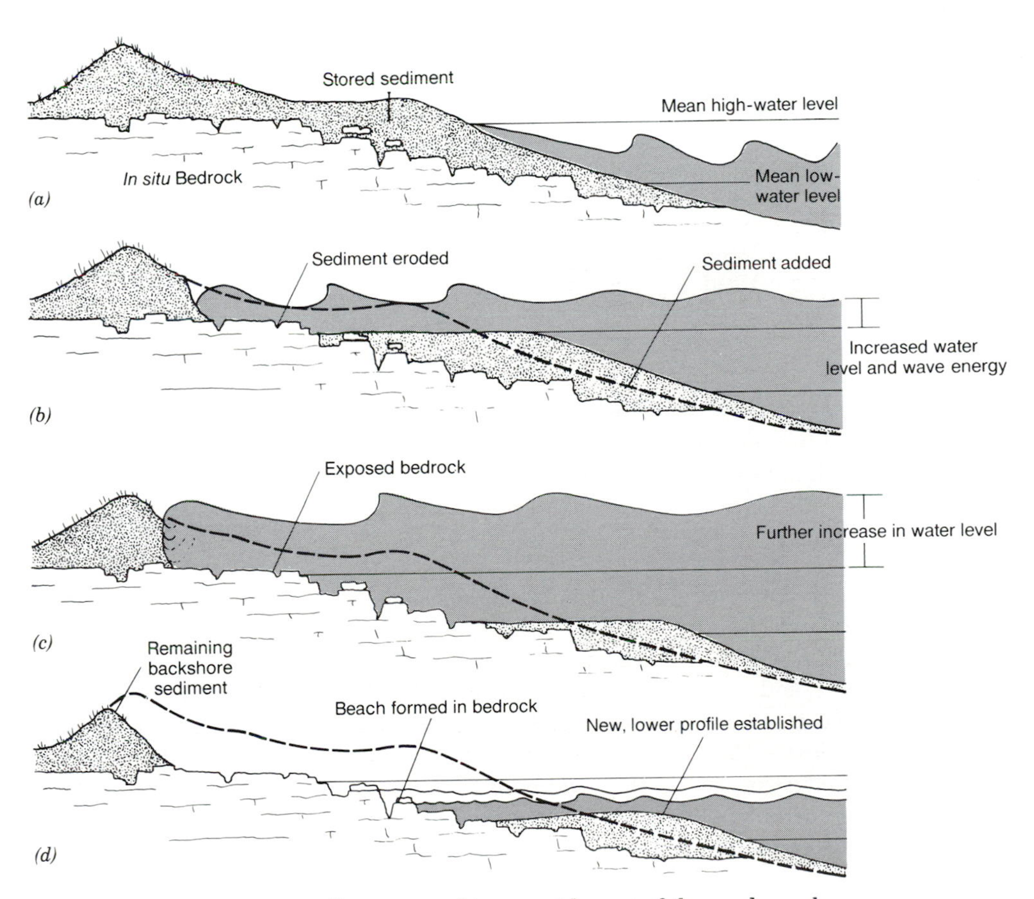

Figure 21.17 A sequence of erosion and impoverishment of the sand supply on a beach, resulting in exposure of the underlying bedrock and establishment of a new beach profile.

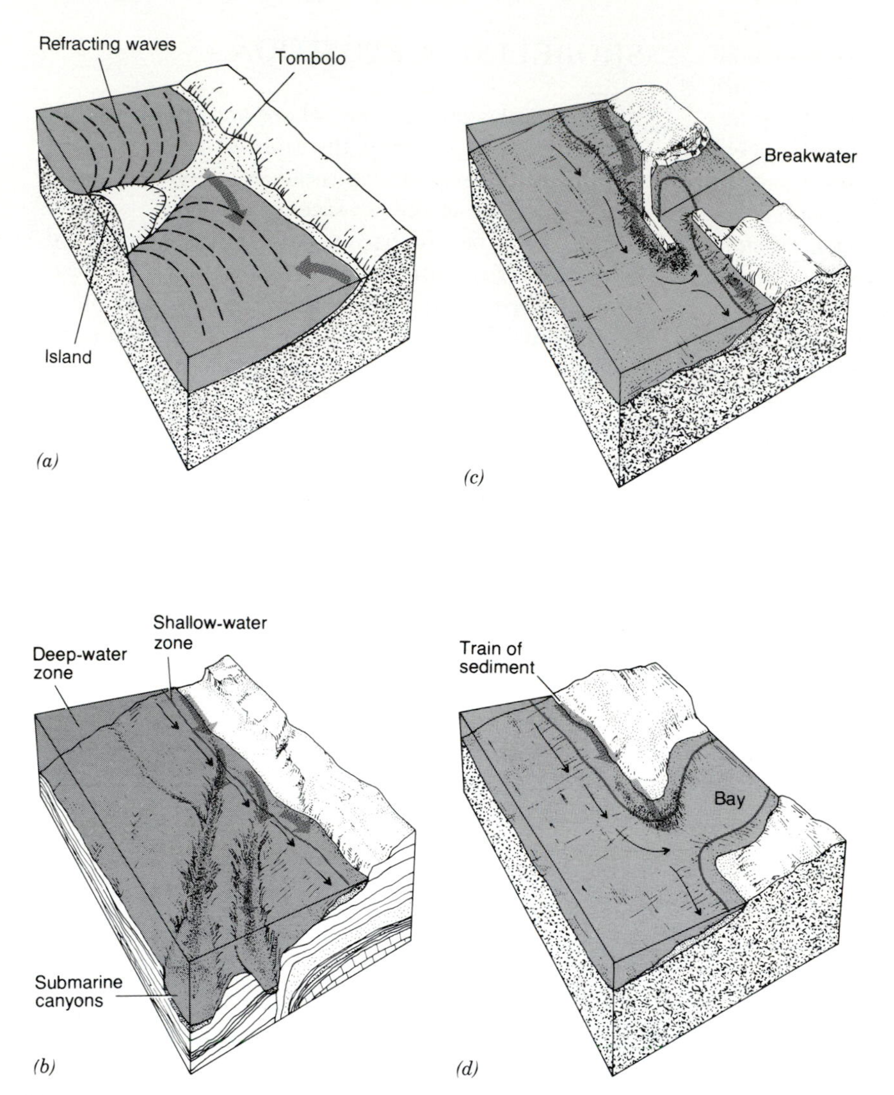

Figure 21.18 Causes of deposition: (*a*) reduction in wave energy behind an island; (*b*) interception of the sediment train by submarine canyons; (*c*) diversion of the sediment train by a breakwater; (*d*) an abrupt change in the orientation of the shoreline.

The other major cause of deposition is a sudden increase in water depth in the path of the longshore sediment train. As the sediment train passes into deep water, sediments fall below the wave base and are deposited. This commonly occurs in three types of settings: (1) where submarine canyons run close to the shore (Fig. 21.18*b*); (2) where the sediment train is diverted into deep water by an opposing current or by a reef, pier, or breakwater (Fig. 21.18*c*); and (3) where the shoreline abruptly changes its orientation, and the momentum of the waves and currents carries the sediment train into deep water, as at the mouth of a bay (Fig. 21.18*d*).

Depositional Features

The bulk of the sediments deposited along any coast are laid down under water and are not at all evident from the shore. This is especially so with those sediments that slip into submarine canyons and flow down the continental slope into very deep water (Fig. 21.18*b*). Near San Diego, California, where the continental shelf is very narrow and dissected by long canyons, much of the longshore drift is lost in this fashion.

In other settings, nearer shore, the sediments accumulate incrementally year after year and eventually form such a heap that a narrow ribbon of sand appears at the water surface. At the mouth of a bay, such a ribbon often grows from the shore itself in the form of a slender, sandy finger called a *spit*. As the spit grows, it may bend into the bay (Fig. 21.19*a*). Depending on the directions of sediment transport and the sand supply, spits may grow from one or both sides of a bay mouth. A ribbon of sand extending across the bay mouth is called a *bay-mouth bar*. These bars are often breached where storm waves overwashed them or where tidal flow and/or river discharge erodes through them (Fig. 21.19*b*).

Spits and bars are common features along coasts with irregular configurations, that is, where there are many embayments to serve as sediment sinks. The segment of Massachusetts coastline shown in Figure 21.19*a* is such a coast, and it is similar in many ways to much of the rest of the East Coast of the United States. The headlands, which are composed of unconsolidated material, are readily eroded and thus serve as sediment sources. The trend of change is toward a smoother shoreline as the headlands are cut back and the bays are filled in. (Also see color plate 14.)

This contrasts with the Maine coastline shown in Figure 21.20. Both coasts receive comparable amounts of wave energy and are approximately the same age, dating from the last glaciation, but the Maine coast exhibits none of the alteration from erosion and deposition. The reason for this difference is compositional; the Maine coast is comprised of resistant bedrock (mainly igneous and metamorphic rocks) and very little loose materials. Thus, the supply of sediment is too small to build spits and bars.

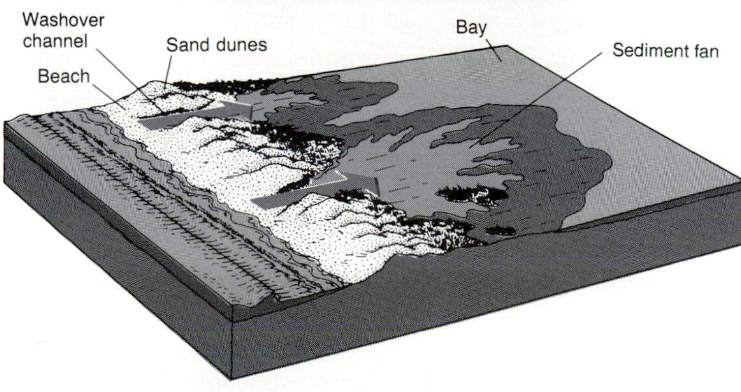

Figure 21.19 Spits and bay-mouth bars along the Massachusetts coast near Falmouth (*a*) and a diagram showing the features associated with a breech in a spit or bay-mouth bar by storm waves (*b*).

Figure 21.20 The rocky Maine coast near Brunswick. Note that very little modification by wave action is evident.

PRINCIPAL TYPES OF COASTLINES

A coastline acquires its character from its geologic structure, its initial topographic configuration, and the subsequent modification of these by wave and current action. Also important in shaping the character of a coastline is the trend toward submergence, emergence, or depositional development. In general, *submergent coasts* are heavily indented owing to flooding of the river valley mouths by seawater; *emergent coasts* are often terraced because wave-cut profiles have been elevated above sea level; and *depositional coasts* are composed of abundant depositional features such as deltas, spits, and bars.

Submergent Coasts

Ria Coasts "Ria" is a Spanish term used to describe coasts marked by prominent headlands and deep reentrants (embayments) (Fig. 21.21). Most ria coasts appear to result from partial submergence owing to a rise in sea level, subsidence of the land, or both. If this elevation change includes a river mouth, flooding of the lower valley takes place, resulting in an elongated embayment called an *estuary*. Chesapeake Bay is an extreme example of valley drowning which produced a main estuary and many tributary estuaries as well. If ria coasts are composed of erodible material, they are readily modified, as illustrated in Figure 21.19.

Fiord Coasts These coasts resemble ria coasts except that the valleys have been glaciated (Fig. 21.21). As a result, the drowned valleys are deep, steepsided, and often very long (see Fig. 22.11). Norway's fiords are world-famous, but equally spectacular ones can be found in British Columbia, southern Chile, and New Zealand. Because fiords are so deep (usually hundreds of meters), it is virtually impossible for spits and bay-mouth bars to form at their mouths.

Depositional Coasts

Barrier Island Coasts Barrier islands begin as offshore bars in shallow water where there is an ample supply of sand. As the bars grow, a beach ridge emerges, small sand dunes begin to form, and vegetation becomes established. The area of water between the island and the mainland is called a *lagoon*. Owing to the shallowness and relative calmness of this environment, rates of biological productivity and sedimentation are high in the lagoon (Fig. 21.21).

Barrier islands parallel much of the Texas shoreline of the Gulf Coast, where some islands reach lengths greater than 200 km. The Sea Islands along the Atlantic Coast of Georgia are also barrier islands, although they are more segmented than the ones of the Gulf Coast. (See Fig. 24.3).

Sandy Coasts These are coasts along which the input and output of longshore drift are balanced over long periods of time. As a result, net change is not appre-

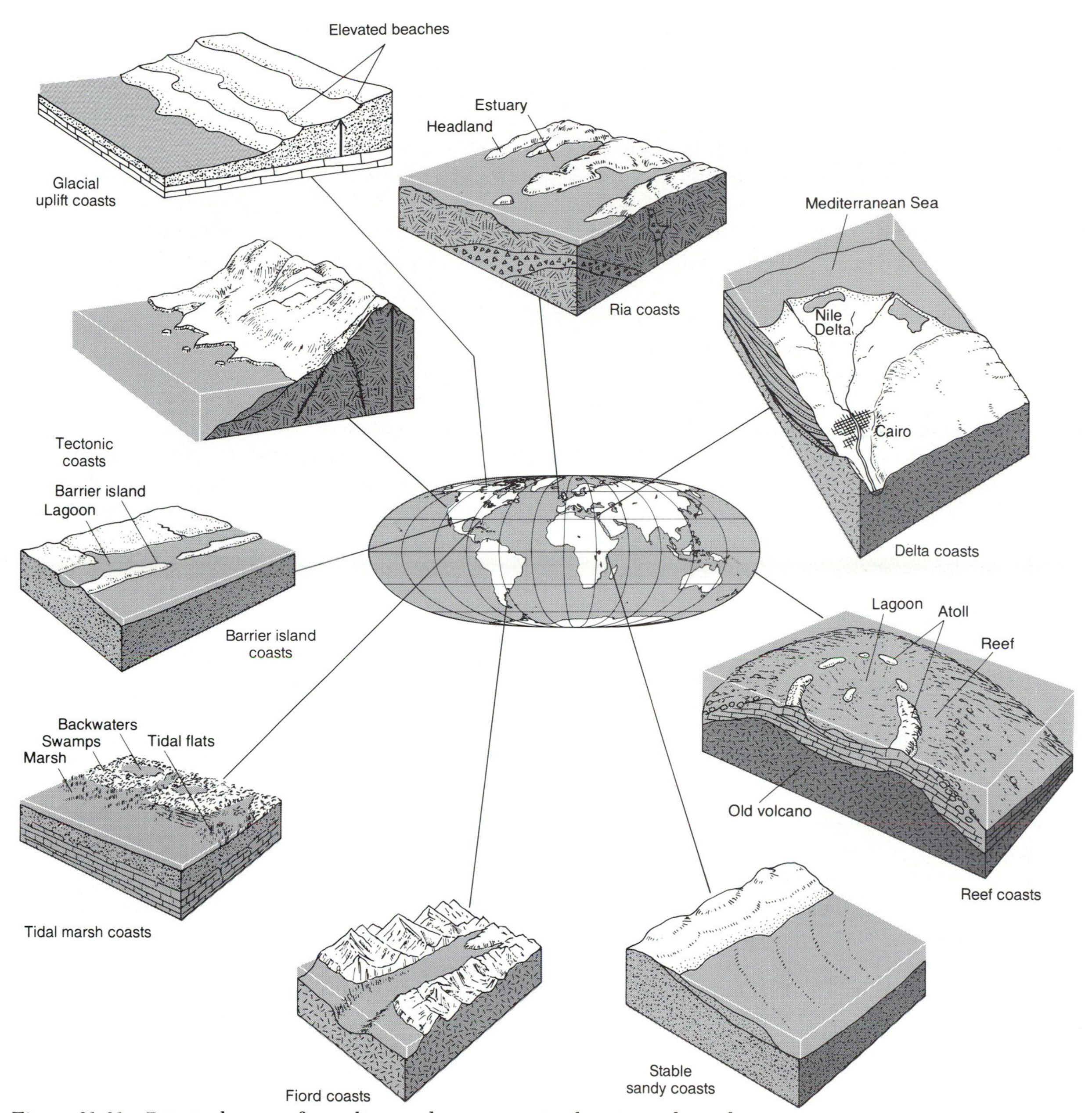

Figure 21.21 Principal types of coastlines and representative locations where they are found in the world.

ciable, although different features and configurations may form from time to time. Large scallops and cuspate forelands are common features (Fig. 21.21). Stable sandy coasts are often located between a sediment source and a sink.

Tidal Marsh Coasts On shorelines where fine sediments are abundant, the influence of bedrock and topographical relief are inconsequential, and there is a sizable tidal flux, tidal marshes are often the predominant coastal type (Fig. 21.21). These gently sloping environments are truly transitional between sea and land because they are daily covered and uncovered by the sea. Vegetation plays an important role in the tidal marsh inasmuch as it secures the environment against wholesale erosion by tides and storms. In the tropics, tidal marshes often give way to swamps of mangrove, a small tree that grows on stiltlike roots.

Delta Coasts Deltas form at the mouths of rivers where the rate of river deposition exceeds the capacity of the littoral processes to carry the sediment away. Coarse sediments are deposited near the margin of the delta, whereas clays are dispersed into deeper water, often far out to sea. (See Fig. 20.8.)

The shape into which the delta is built depends on the configuration of the river mouth, the pattern of the distributaries, and the rate of sedimentation. Four basic types of deltas are recognized: *estuarine*, which forms in the head of an estuary; *arcuate*, which forms in a bay and is fan-shaped; *cuspate*, which is pointed because of strong reshaping by waves; and *bird-foot*, which has long fingers built into the sea by the distributaries. The Mississippi Delta is a bird-foot; the Nile Delta, which is shown in Figure 21.21, is an arcuate form; and the Seine Delta fits the estuarine class (see Fig. 20.9).

Reef Coasts Coral reefs form in tropical seas where the water is less than 20 m deep. Reefs grow from the buildup of massive amounts of coral, an organism that secretes calcium carbonate. The calcium cements the coral and the remains of other organisms together into a wave-resistant body of limestone.

Early scientists were mystified by the presence of coral-reef islands in the deep ocean until they discovered that the islands had formed on old volcanoes, which through subsidence and erosion had been lowered to sea level or slightly lower. The coral builds up on the shoulders of the volcano to form a concentric ring of islands, called an *atoll*, and associated reefs, called *fringing reefs*. In the area of the crater, a small basin or lagoon forms (Fig. 21.21). Some reefs have formed on *guyots*, which are flat-topped, submerged volcanoes that have been eroded down to sea level and then submerged with subsidence of the ocean floor.

Much larger accumulations of calcium carbonate sediments, similar to the ancient limestones found on the continents, also make up reeflike formations in tropical seas. These *carbonate platforms* form extensive shallow-water areas, such as the Great Bahama Bank southeast of Florida.

Emergent Coasts

Tectonic Coasts A tectonic coast is dominated by mountain-building processes. Not surprisingly, volcanic and faulted coastlines can be considered as tectonic coasts. Volcanic coasts are dominated by lava and ash deposits. The lava is usually fairly resistant to wave attack, but the ash easily gives way to wave action; therefore, the resultant shorelines are usually irregular. Faulted coasts, on the other hand, may be fairly straight, especially if the shoreline is formed by a fault scarp. Coasts that are elevated as a result of mountain building are often terraced where wave-cut platforms have been raised above sea level. (See Figs. 21.21 and 21.11).

Glacial Uplift Coasts Emergent coasts are also found outside tectonic regions. In areas that have been glaciated and depressed isostatically, the coasts emerge after the ice has melted away. This isostatic emergence is very prominent in the Baltic Sea and Hudson Bay where terraces and elevated beaches are the major coastal features (Fig. 21.21).

SAND CONSERVATION AND BEACH PROTECTION

Coastal real estate is valuable property, and people are understandably concerned about maintaining it. Unfortunately, our love for the seaside and the natural trends of change in shorelines often create incompatible arrangements. Structures are typically placed close to the water's edge; when the shore recedes, they are subject to damage and destruction. The management alternatives are either to suffer the loss and attempt to relocate the structures on the shore; abandon the area entirely (a solution some conservationists favor); or reduce or stop the recession by altering the sediment mass balance. The answer depends on politics and economics, and it seems that changing the sediment mass balance is the favored solution in most areas.

Groins are generally the preferred means of reducing the loss of beach sand in most areas. A groin is a wall built perpendicular (or nearly so) to the shore for the purpose of slowing and capturing some of the longshore drift. The drift accumulates on the upshore side of the groin and builds outward, eventually spilling around the end of the groin. A number of groins constructed in succession can slow the longshore transport sufficiently to build and maintain a beach in some locations. In other places, however, they have been known to increase erosion in beaches downshore because they interrupt the natural flow of sediment in the same manner as breakwaters often do.

THE VIEW FROM THE FIELD / NOTE 21

Breakwaters and Sand Bypassing

Breakwaters are barriers constructed at harbor entrances to reduce wave energy and make navigation safer. Because they reduce wave energy, breakwaters also interrupt the longshore sediment transport. Sediments accumulate behind or on the upshore side of breakwaters, depriving the beach downshore of its sediment supply. This often results in beach recession downshore because, lacking sediment to move, wave energy is spent in the erosion of *in situ* material, while the breakwater area becomes clogged with sediments.

To solve this problem, a method called sand bypassing is used to maintain the flow of sediment past the harbor entrance. Three techniques may be employed. The first is hydraulic bypassing, which involves sucking up the sand and water mechanically and then pumping this mixture (slurry) past the entrance to the downshore beach. Dredging and barging, the second technique, involves excavating sand by crane or hydraulic techniques, transporting it downshore, and offloading it from the barge into shallow water. The third technique involves excavating a storage pit on the upshore side of the breakwater and piling the excavated sediment on the downshore side of the breakwater. The pile feeds the beach downshore while the pit collects sediment upshore. The sizes of the pit and pile are scaled to the expected annual sediment transport and the desired number of years between renewal operations.

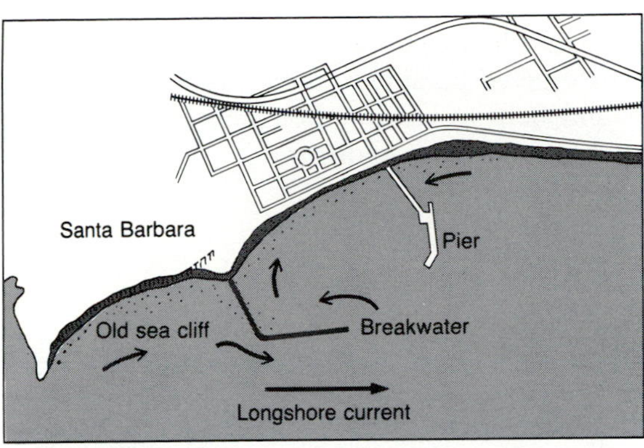

(a) Construction 1928

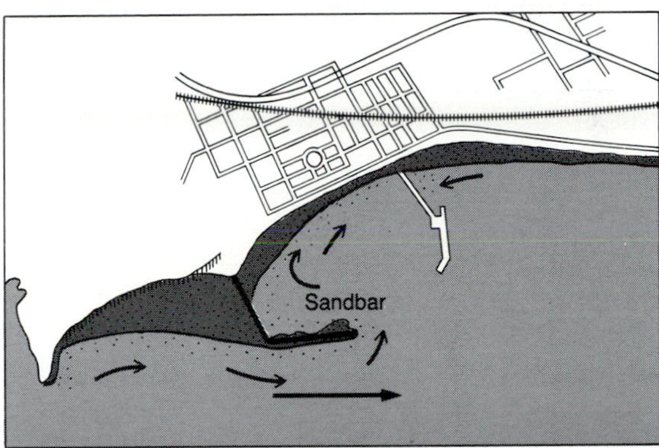

(b) 1948

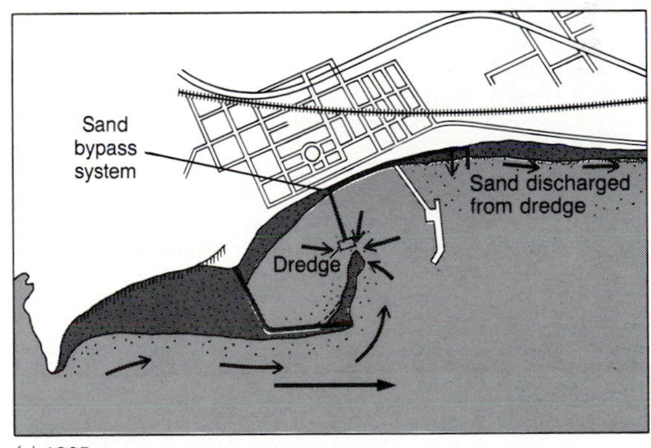

(c) 1965

Shore changes associated with the harbor and breakwater at Santa Barbara, California. Sand is now bypassed with the aid of a hydraulic dredging system.

Another method used to maintain beaches is *artificial nourishment*. Sand is trucked or hydraulically piped to the beach in order to sustain a body of sediment. This method is expensive and must be carried out frequently to replace the sediment transported away. Where groins or breakwaters block the sediment train, it is sometimes necessary to artificially nourish the beaches downshore by a technique called *sand bypassing*. Note 21 offers some additional information on this shoreline-management technique.

Where erosion has reached or threatens to reach critical proportions, the beach or the backshore slope can be protected by building structures resistant to wave attack. *Riprap* (boulders or broken concrete) placed along the shore helps to dissipate wave energy and hold materials in place. Walls constructed of concrete blocks called *revetment* serve the same purpose. *Seawalls* erected parallel to the shore intercept large waves and help to reduce their energy and in turn their capacity to move sediment.

Summary

Wind waves are the principal form of motion in bodies of standing water. Wave size increases with wind velocity, wind duration, and fetch. The rotational motion of water particles in a wave extends well below the water surface, and in shallow water this motion exerts force against the bottom. This force results in the expenditure of work.

The work of waves and currents includes both shore erosion and sediment transportation. Longshore transport is concentrated in the surf zone, but also occurs in the deeper water as well as in the swash zone on the beach. The direction of longshore transport along the coast follows that of the transporting waves and currents, and on many coasts the prevailing direction of flow changes from season to season. For any parcel of space along a shoreline, the sediment mass balance is equal to the relative inputs and outputs of sediment by waves, currents, runoff, and wind.

Among the depositional features built by shoreline processes are spits, bars, and cusps. Land use has given rise to serious problems along changeable shorelines, and management programs have traditionally favored installation of various kinds of structures to protect against erosion and recession.

Concepts and Terms for Review

- waves
  - tsunamis, surges, tides
  - wind waves: seas and swells
- wave form and motion
  - wave height and wavelength
  - oscillatory wave
  - wave base
  - breaking wave
  - wave of translation
- wave energy
- near shore circulation
- wave refraction
- longshore current
- shore erosion and recession
- wave erosion and features
  - hydraulic pressure
  - chemical weathering
  - corrasion
  - sea cliff, sea caves, sea stacks
  - shore, backshore, backshore slope
- sediment transport
  - sediment sources and sinks
  - littoral transport
  - bed load and suspended load
  - gross and net sediment transport
  - onshore-offshore transport
  - equilibrium profile
- depositional environments
- depositional features
  - tombolo
  - spit
  - bay-mouth bar
- coastlines
  - submergent: ria and fiord
  - depositional: barrier island, sandy, tidal marsh, delta, reef
  - emergent: tectonic, glacial uplift
- beach protection
- sand conservation

Sources and References

Bascom, W.N. (1964) *Waves and Beaches.* Garden City, N.Y.: Doubleday.

Bird, E.C.F. (1985) *Coastline Changes: A Global Review.* New York, Wiley–Interscience.

Bowen, A.J. (1969) "The Generation of Longshore Currents on a Plane Beach." *Journal of Marine Research* 37: pp. 206–215.

Davies, J.L. (1977) *Geographical Variation in Coastal Development.* London: Longman.

Dolin, R., et al. (1980) "Barrier Islands." *American Scientist.* Vol. 68: pp. 16–25.

Inman, D.L., and Brush, B.M. (1973) "The Coastal Challenge." *Science* 181: pp. 20–32.

Komav, P.O. (1976) *Beach Processes and Sedimentation.* Englewood Cliffs, N.J.: Prentice-Hall.

Marsh, W.M., Marsh, B.D., and Dozier, J. (1973) "The Geomorphic Influence of Lake Superior Icefoots" *American Journal of Science* 273: pp. 48–64.

Meade, R.H., and Parker, R.S. (1984) "Sediment in Rivers of the United States." In *National Water Summary 1984*, U.S. Geological Survey, Washington, D.C., pp. 49–60.

Zenkovich, V. (1967) *Processes of Coastal Development.* Trans. D. Fry. Edinburgh: Oliver and Boyd.

Chapter 22

Glacial Processes and Landforms

Ice ages are fairly common to earth. The present one has sent great ice sheets over much of North America and Eurasia and is likely to do so again. By what processes do glaciers move and rearrange the terrain beneath them and at their margins? What is the character of the landscapes that emerge after glaciation and do the glaciers have much influence on the environment beyond their immediate areas?

A time comes when creatures whose destinies have crossed somewhere in the remote past are forced to appraise each other as though they were total strangers. I had been huddled beside the fire one winter night, with the wind prowling outside and shaking the windows. The big shepherd dog on the hearth before me occasionally glanced up affectionately, sighed, and slept. I was working, actually, amidst the debris of a far greater winter. On my desk lay the lance points of ice age hunters and the heavy leg bone of a fossil bison. No remnants of flesh attached to these relics. The deed lay more than ten thousand years remote. It was represented here by naked flint and by bone so mineralized it rang when struck. As I worked on in my little circle of light, I absently laid the bone beside me on the floor. The hour had crept toward midnight. A grating noise, a heavy rasping of big teeth diverted me. I looked down.

The dog had risen. That rock-hard fragment of a vanished beast was in his jaws and he was mouthing it with a fierce intensity I had never seen exhibited by him before.

"Wolf," I exclaimed, and stretched out my hand. The dog backed up but did not yield. A low and steady rumbling began to rise in his chest, something out of a long-gone midnight. There was nothing in that bone to taste, but ancient shapes were moving in his mind and determining his utterance. Only fools gave up bones. He was warning me.

"Wolf," I chided again.

As I advanced, his teeth showed and his mouth wrinkled to strike. The rumbling rose to a direct snarl . . . his eyes were strained and desperate. "Do not," something pleaded in the back of them, some affectionate thing that had followed at my heel all the days of his mortal life, "do not force me. I am what I am and cannot be otherwise because of my shadows. Do not reach out. You are a man, and my very god. I love you, but do not put out your hand. It is midnight. We are in another time, in the snow . . . the big, the final, the terrible snow. . . .[1]

We are creatures of the Ice Age, Loren Eiseley suggests. Less than 10,000 years have passed since the last glaciers melted from North America and Eurasia. Prior to that time humans, with mental and physical capabilities equal to our own, lived near the fronts of the great ice sheets, and they no doubt understood that the ice moved, pushed down forests, discharged huge quantities of meltwater, and was associated with a cold climate. That valuable knowledge about this planet somehow slipped away from humans in the centuries after the ice sheets melted from populated areas. Not until the nineteenth century did we rediscover the Ice Age, so to speak. Since then we have learned not only about the geographic extent of the great ice sheets, how they moved, and how they eroded the land, but also that we are today living in one of the warm interludes of the Ice Age. Humans are likely to see another great advance of the ice.

[1]Loren Eiseley, *The Unexpected Universe* (4th ed), 1969. Harcourt, Brace and World, New York.

GLACIERS ANCIENT AND MODERN

The earth has seen two or three, perhaps many, ice ages. The last ice age, which is also the one we are currently in, began 1 to 2 million years ago and is referred to as the Pleistocene Epoch.

The Ice Age

In North America, the Pleistocene glaciers formed in central and eastern Canada and spread both southward and northward. The sheets that moved southward covered the St. Lawrence Lowlands, New England, the Great Lakes region, and the upper Great Plains, an area in which today more than 100 million people live. On the west, the ice sheets met another sheet of ice coming from the Canadian Rockies, and

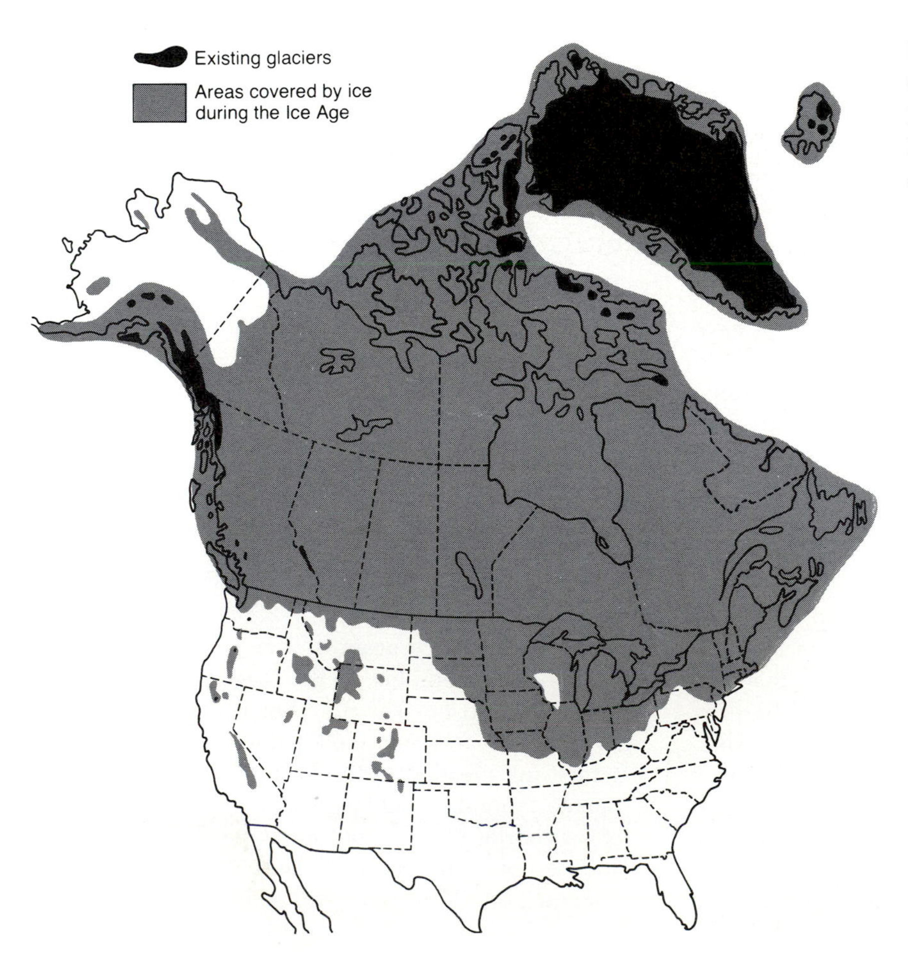

Figure 22.1 The shaded area shows the maximum extent of glaciations in North America during the Pleistocene Epoch. The dark areas show the areas of modern glaciers.

for a time glacial ice stretched completely across North America (Fig. 22.1).

A similar pattern of glaciation has also been documented in Europe. In both North America and Europe, the periods of glaciation were interrupted by *interglacials*—relatively warm, dry periods when the bulk of the continental glaciers apparently melted away. Together, the glacial stages and interglacial stages constitute the *Pleistocene* Epoch, but there is conflicting evidence on how many glacial stages there were and how long ago the Pleistocene began. Deposits on land indicate there were four glacial stages; ocean sediments indicate there may have been as many as eighteen.

The Pleistocene is of special interest to both natural and social scientists because it was not only a time when much of the landscape we see today was shaped, but also a time of rapid biological change on the planet. Many large mammals, for example, the woolly mammoth, became extinct near the close of the Pleistocene Epoch. Of paramount importance is the fact that the rise and geographic dispersal of the forerunners of modern human populations and societies took place during the Pleistocene. Around 10,000 to 12,000 years ago—near the end of the last glaciation and coinciding with the extinctions of the large mammals—humans invented agriculture. In the ensuing millennia, it spread across the world and fostered the growth of billions of humans.

The Cause of Glaciation

The search for the cause of the ice ages has occupied scientists for more than a century. Scientists generally agree that the cause—whatever its origins—must produce a lowering of global or hemispheric temperature by 4 to 5 degrees C, especially in the summer, and a substantial increase in snowfall in subarctic and arctic regions. Although lots of plausible ideas—including reductions in solar radiation and changes in the geographic locations of land masses with plate tectonics—have been advanced, there is still no widely accepted explanation for the rise of the ice ages, including the Pleistocene. On the other hand, several scientifically acceptable explanations have been advanced to account for the glacial stages and interglacials of the Pleistocene.

The idea most seriously entertained at the present is that the warming and cooling in the northern latitudes are caused by eccentricities in the earth orbit about the sun. Eccentricity refers to the degree of deviation of the shape of the orbit from a perfect circle. The earth's orbit is elliptical (oblong), but over periods of 100,000 years its shape varies enough to cause significant changes in the receipt of radiation in the upper northern latitudes. These variations correlate with evidence derived from ocean sediments which reveals that the period of glacial stages in the Pleistocene was also 100,000 years. Moreover, the troughs and peaks of the cooling and warming correspond fairly well with the dates of the glacials and interglacials.

Occurrence and Types of Glaciers

Currently, glacier ice covers about 10 percent of the land areas, compared with about 30 percent during the glacial maxima of the Pleistocene. The most extensive glaciers are now found in Antarctica and Greenland. (The North Pole ice pack, unlike the ice of which glaciers are composed, is formed of seawater which has frozen on the ocean surface. It rarely exceeds 5 m in thickness.) Elsewhere, glaciers are confined to mountain ranges, but the elevations at which they occur vary with latitude. Near the Arctic Circle alpine glaciers form as low as 500 m above sea level; at 45 degrees latitude they form only above 3000 m; and in the tropics mountains less than 5000 m cannot support glaciers (Fig. 22.2). In addition cold environments, such as northern Siberia, with little snowfall do not sustain glaciers.

Glaciers are generally classified according to size. The largest, of course, are the *continental glaciers*, which at their maxima covered land areas of the midlatitudes and subarctic on the order of 5 million km^2. Although they are long gone, the continental glaciers are not to be ignored because they not only were

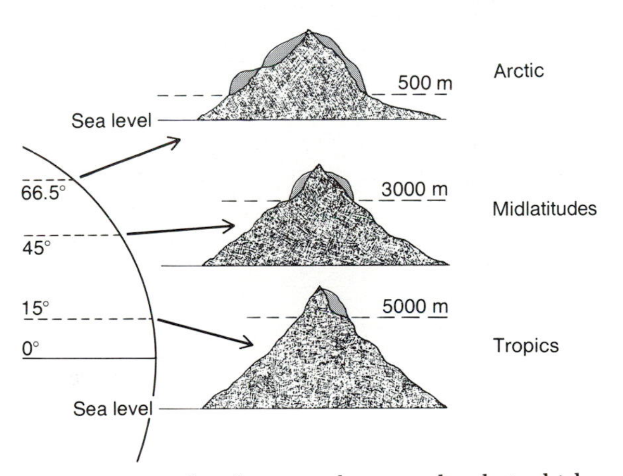

Figure 22.2 The elevation above sea level at which glaciers can form decreases from the tropics to the poles. Three representative elevations of glacier formation are shown here.

important in shaping the present landscapes of North America and Eurasia, but also are likely to appear again on the planet. The polar ice sheets of Greenland and Antarctica can be considered contemporary versions of continental glaciers.

Piedmont glaciers, the next class, are formed by the merger of many mountain glaciers. Piedmont glaciers range from several thousand to several tens of thousands of km^2 in area and are found mainly along the lower slopes of major mountain ranges, such as the Himalayas or the Andes of southern Chile.

Mountain, or alpine, glaciers are the third class; these are the most widespread glaciers in the world. They may range in size from a patch of ice several hundred meters across to rivers of ice many kilometers long. Mountain glaciers are found in all latitudes, including the tropics, where they occur only on the highest mountains, such as Kilimanjaro and Mount Kenya in East Africa or the Andes of South America.

Glaciers can also be classified as temperate or polar, based on their thermal characteristics. *Temperate glaciers* can be thought of as relatively warm because the internal ice is at or very close to the melting threshold. These glaciers are often characterized by profuse melting both within and beneath the ice, which apparently facilitates their slippage on the ground. *Polar glaciers* are frozen throughout and are distinctive for the absence of meltwater. The base of the ice is solidly frozen to the ground which prohibits slippage at the base of the ice.

GROWTH AND MOTION OF GLACIERS

Most of what we have learned about the growth, nourishment, and behavior of glaciers has come from studies of temperate mountain glaciers. These glaciers are small enough that they serve as field laboratories of sorts, rising as they do in one climatic zone and then descending a distance of only a few kilometers to a lower and warmer climatic zone where they terminate.

Snow to Ice

Ice that forms on rivers, lakes, and the open seas originates directly from liquid water, whereas the ice that forms glaciers originates as snow. New-fallen snow is composed of loosely packed lacy crystals (Fig. 22.3*a*). The density of new snow ranges from about 50 to 300 kg/m^3 (the density of pure water is 1000 kg/m^3). After the snow falls, the crystals are reduced by (1) *ablation*, which includes both melting and sublimation; and (2) physical compaction. As ablation and compaction proceed, the density of the snow increases. Additional changes take place as a result of melting and refreezing, and the original snowflakes have now been transformed into rounded crystals. This partially melted, compacted snow is called névé and has a density exceeding 500 kg/m^3 (Fig. 22.3*b*).

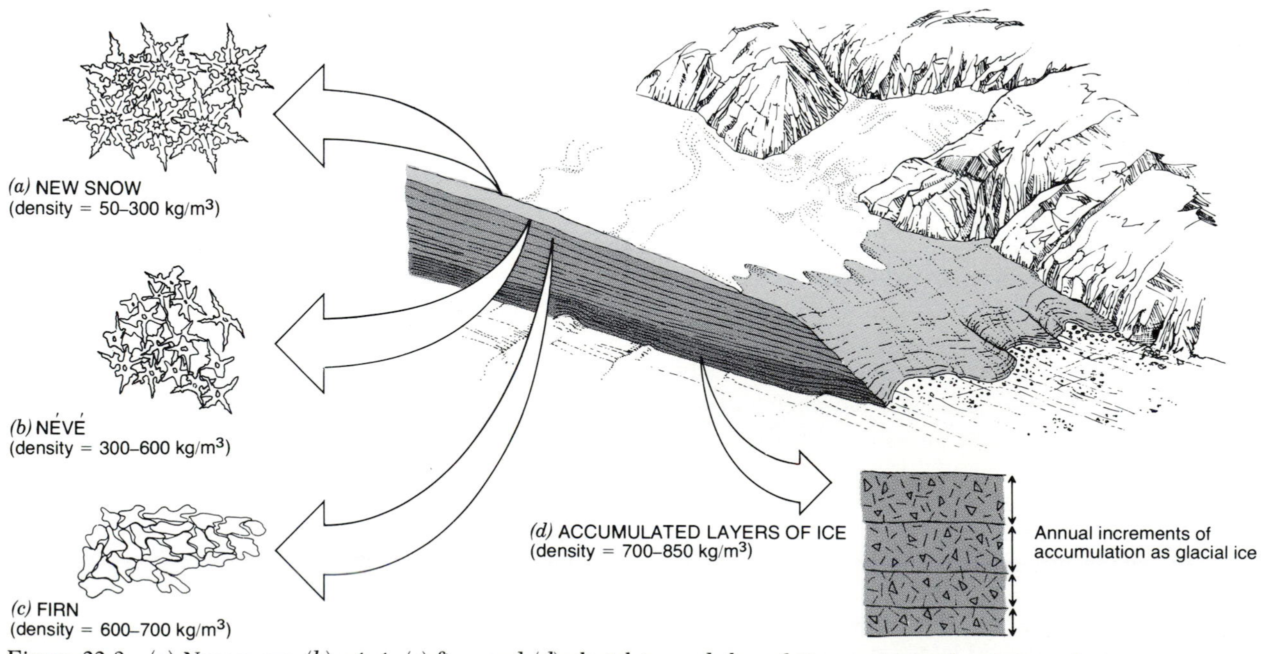

Figure 22.3 (*a*) New snow, (*b*) névé, (*c*) firn, and (*d*) glacial ice and the relative position of each in a glacier.

In most areas of the world where snow falls, all of the névé melts before the end of the ablation season.

If the névé survives the ablation season, it is called *firn* (Fig. 22.3*c*). Where this happens year after year, each year's layer of firn accumulates on top of the previous year's layers, and the layers below the surface are compressed. This causes a further increase in density, and the firn is changed into *glacial ice*, with typical densities of 850 kg/m^3. This is somewhat lower than the normal density of ice (917 kg/m^3) because some air is trapped in the glacial ice. The transformation of névé into glacial ice may take 25 to 100 years, and the exact nature of the processes involved is not fully understood. In the ice each year's increment of accumulation can often be seen as ice strata in the glacier (Fig. 22.3*d*).

Glacier Movement

To qualify as a glacier, a mass of ice must be capable of movement by plastic deformation. Glacial movement occurs as an ice mass grows too heavy to maintain its shape and literally squashes out in the lower portions (Fig. 22.4*a*). In mountain glaciers, this begins when the ice reaches a thickness of 20 m. On a mountain side, the movement is downslope and along routes of lowest resistance; as a result, mountain glaciers typically flow in existing stream valleys or valleys cut by previous glaciers (Fig. 22.4*b*).

Once set in motion, the ice develops zones of different rates of flow. If we look down on the surface of a glacier, the central zone appears to move fastest; the margins, adjacent to the valley wall, move more slowly. Near the bottom of the glacier, velocity drops off to rates comparable to those on the sides (Fig. 22.4*c*).

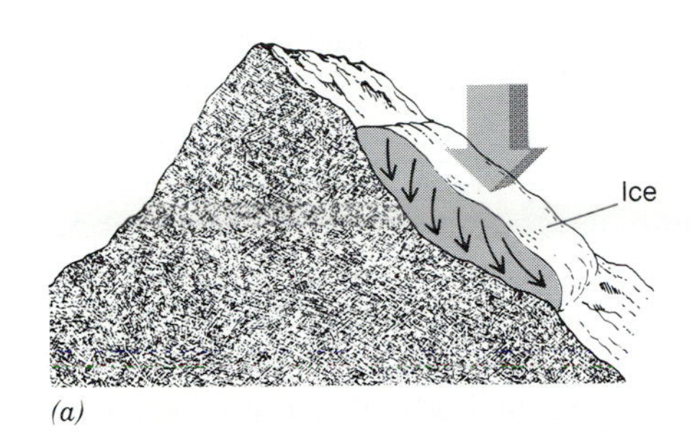

(*a*)

(*b*)

VELOCITY PROFILE
v = Velocity
Surface of glacier
Ice
Depth
Flow streamline
Base of the glacier

(*c*)

Figure 22.4 (*a*) The weight of a thick ice deposit exerts sufficient stress to cause the ice to deform as a plastic. (*b*) Deformation is most pronounced on the downslope edge of the ice. Within the glacier (*c*) velocity is greatest near the surface and declines to zero at the base.

The overall pattern of ice flow in a typical valley glacier is as follows. In the upper part of the glacier, flow lines tend to converge downslope, forming a zone of compressed flow in the middle part of the glacier. At the lower end of the glacier, the ice decompresses, and the flow lines spread out (Fig. 22.5*a*). There is a corresponding flow pattern in the interior of the glacier; flow lines in the upper part move from the surface toward the center of the glacier, and in the lower part from the center toward the surface (Fig. 22.5*b*).

The velocity of flow of the glacier is influenced by many factors. Generally, it appears that velocity increases with the steepness of the valley floor, the thickness and temperature of the ice, and the constriction posed by the sides of the valley. In a temperate glacier, the movement of the ice over the ground is similar in principle to the mechanism involved in ice skating. The immense pressure at the contact between the runner and ice causes sudden meltout and reduction of friction. This facilitates slippage of the ice over the land and allows some glaciers to move as fast as 50 m per day. Most, however, are much slower, moving at rates of less than 1 m a day. In a polar glacier the movement takes place by internal slippage because the base is frozen to the ground. Rates of movement are much slower than those of temperate glaciers.

Glacier Mass Budget

In order to understand how glaciers can grow and shrink, advance and retreat, and erode and deposit, we must examine their energy balances, or what in glaciology is termed *mass budget*. In brief, the mass budget of a glacier includes three main components: (1) accumulation of snow on the upper glacier; (2) forward movement of the glacier; and (3) ablation of the lower glacier. If over a year accumulation and ablation are equal, the glacier has neither lost nor gained any mass. If accumulation, forward movement, and ablation are all equal, the glacier has gained and pushed ahead an amount of ice equal to the amount lost. Given a year in which more ice accumulates than ablates, the glacier mass grows, and the terminus of the glacier, called the *snout*, may advance downslope.

Although some modern glaciers are advancing, most are currently in a state of retreat (Fig. 22.6). This indicates that their mass balances are negative, which suggests that climate in this century has grown either warmer or less snowy, or both. But this trend can

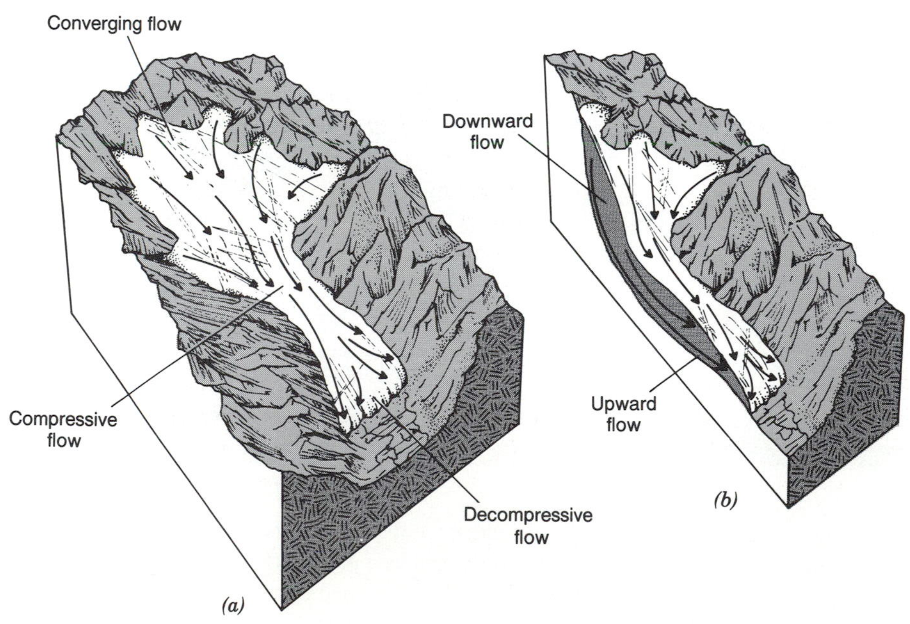

Figure 22.5 Pattern of flow in a valley glacier. (*a*) From the surface, flow lines converge in the upper region and diverge in the lower region. Flow is most constricted in the center, where flow lines closely parallel the long axis of the glacier. (*b*) In the interior of the glacier, flow lines tend to move downward in the upper region and upward in the lower region.

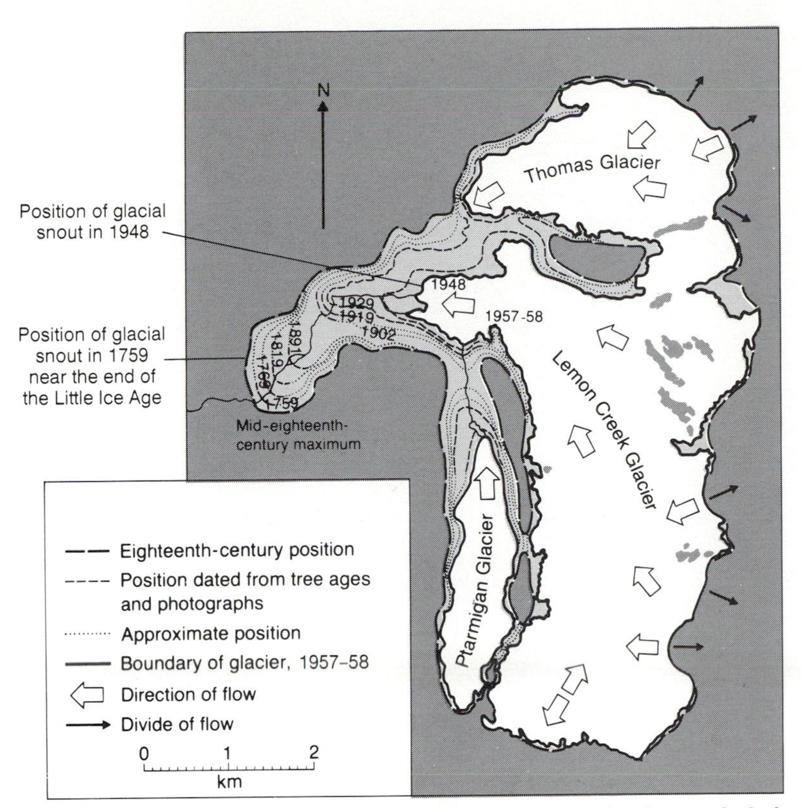

Figure 22.6 Variations in the limits of the Lemon Creek Glacier of Alaska over 200 years. During the Little Ice Age of the eighteenth century, the ice extended 2 km farther down the Lemon Creek Valley than it did in 1957–58.

reverse itself, as it did in the 1700s, when glaciers over much of the world made such strong advances that the period is referred to as the Little Ice Age (Fig. 22.6). (For additional comments on the Little Ice Age also see Chapter 9.)

To understand the true character of a glacier, we must realize that forward flow continues while accumulation is taking place at one end and ablation at the other (Fig. 22.7). The rate of flow should be controlled by net accumulation, but the relationship is complex because of the time lag between one season's accumulation and the corresponding movement of the glacier. The lag may amount to several decades in large glaciers, and for this reason it is not uncommon for their behavior to be out of phase with climatic trends. This may help to explain the phenomenon known as *glacial surges,* in which a glacier bursts forth, driving ahead at rates of 10 to 20 m per day, partially in response to a massive input of energy (net accumulation) some years before.

Although forward movement is an important control on the behavior of the glacier, it is not the only control. The rate of ablation at the snout is also important; if equal to the rate of forward motion, the snout may remain stationary while the glacier continues to move. The glacier acts in this way much like the top side of a conveyor belt. However, when the ablation rate exceeds forward movement, the snout retreats; the reverse can be expected when movement exceeds ablation. These motions—stability, advance, and retreat—are critical to the following discussion concerning the erosional and depositional work of glaciers.

Eventually, all glacier ice is destroyed. Most is lost to ablation by melting and sublimation, but some is also lost to the sea. The process by which ice breaks off the glacier and floats to sea is called *calving.* Great icebergs are produced by calving of the Greenland and Antarctic glaciers, and they can last months, or even years, in cool ocean waters. In the North Atlantic, icebergs were traditional menaces to transoceanic travel; it was such a piece of ice that sunk the *Titanic* in 1912. Today icebergs can be detected with shipboard radar, making North Atlantic sailing much safer, though not failsafe (Fig. 22.8).

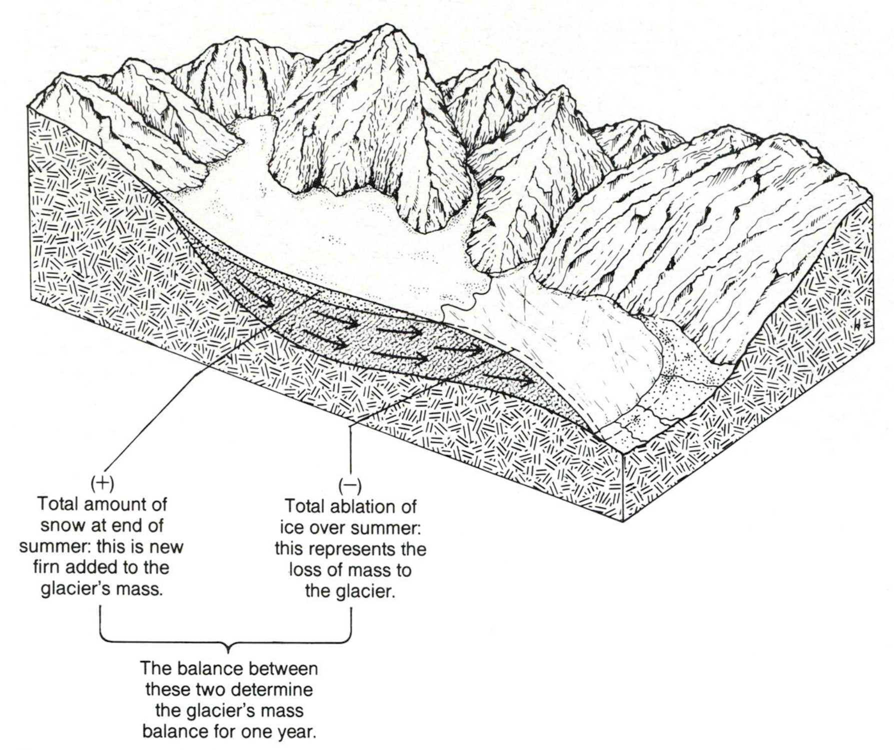

Figure 22.7 The basic idea of the mass budget of a glacier: accumulation of firn, ablation of ice, and delivery of ice from the accumulation zone to the ablation zone by forward movement.

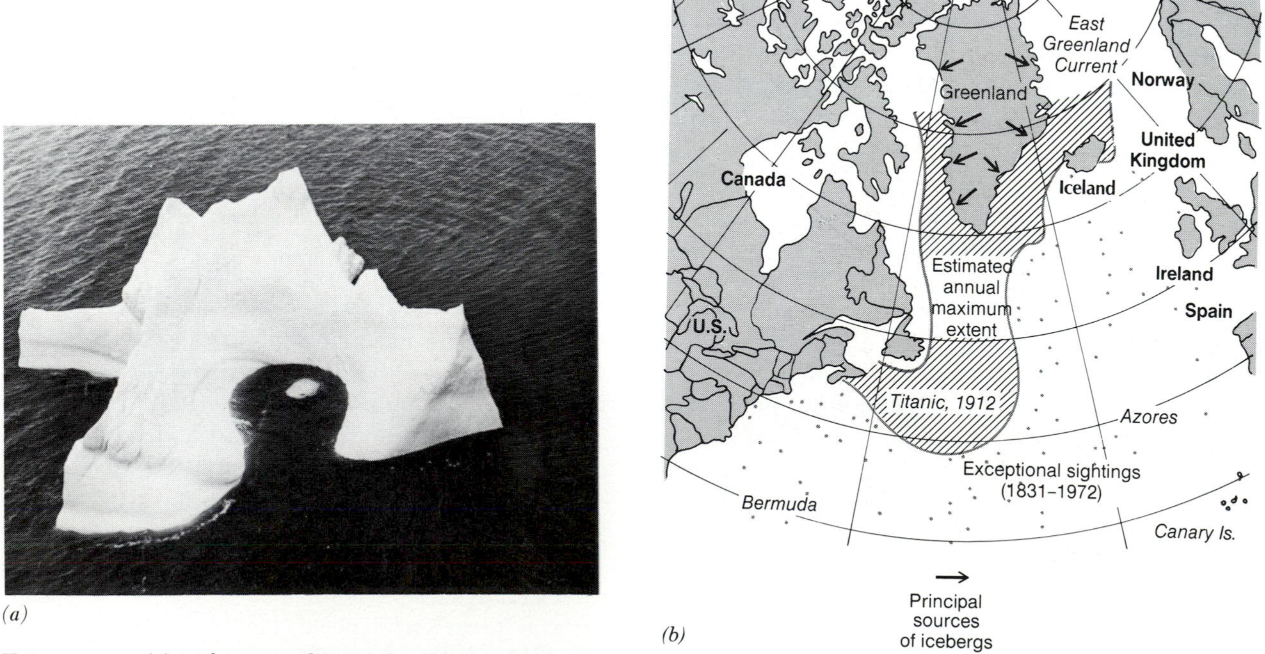

Figure 22.8 (*a*) Icebergs in the North Atlantic. These huge chunks of ice, about 90 percent of which are underwater, have calved from the snouts of large glaciers. (*b*) The map shows the distribution of icebergs in the North Atlantic.

THE WORK OF GLACIERS

Glaciers are important components of the environment—inasmuch as they influence modern climate, vegetation, and water supplies—but our principal concern is with their capacity as a geomorphic agent. Their ability to erode soil and rock, transport sediment, and create deposits is remarkable. When we take into consideration that they recently—geologically speaking—worked on nearly 50 million square kilometers of land surface, we can appreciate their geomorphic significance all the more. The following discussion focuses on temperate glaciers. Much of the work of glaciers takes place at the base of the ice where it slips over the ground. In polar glaciers little if any such slippage takes place; therefore, they accomplish very little geomorphic work.

Glacial Erosion

Two major erosional processes take place on the bed of a glacier. One is *scouring* as the ice, armed with rock debris it has incorporated from various sources, skids over the land. The rocks in the ice act as an abrasive agent, literally rasping the bedrock under them. Scouring creates a variety of features, the most notable of which are scour lines, called *striations*, and finely abraded surfaces called *glacial polish*. Polish is usually limited to well-consolidated igneous rocks, especially granite (Fig. 22.9). The abrasive action also yields a fine sediment, usually clay-sized, which often ends up in meltwater. As a result, glacial meltwater often has a light, cloudy appearance, which has led to the name *glacial milk* or *glacial flour*.

Figure 22.10 The craggy aftermath of glacial plucking; the glacier moved from right to left over this granite knob on the southern edge of the Canadian Shield.

Figure 22.9 Glacial polish on a granitic dome in the Sierra Nevada. Polishing is usually most pronounced near the crests of domes and knobs composed of tightly consolidated igneous rock.

The second erosional process on the glacier bed is known as *plucking*. It appears that the ice at depth melts and refreezes in response to pressure changes and other factors; when it does, it may refreeze around blocks of jointed bedrock. When the ice moves, the block is drawn out and carried away. The plucking process occurs with greatest intensity on the down-ice side of rock knobs (Fig. 22.10). Combined with scouring action occurring on the other side of a knob, the result is a curious, asymmetrical feature called a *roche moutonnees*, which is smooth on one side and steep and jagged on the other.

Glaciers flow over the land along the paths of least

resistance. When a mountain glacier enters a stream valley, it advances rapidly forming a distended tongue, called a *lobe*. As it moves down the valley, the lobe often merges with other lobes emanating from nearby mountain sides. The glacier grows with each tributary lobe; thus, to a limited extent a network of mountain glaciers follows the principle of stream orders.

Owing to the deep and steepsided nature of mountain river valleys, the glacier must flow in highly confined space. As a result, a great deal of stress is exerted against the walls of the valley. Talus and other footslope deposits that line the walls of stream valleys, such as the one shown in Figure 22.11*a*, are readily removed and carried downvalley in the initial advance of the ice. The glacier then begins to erode the bedrock of the valley walls and floor. The valley is both widened and deepened, eventually emerging with a beautifully symmetrical U-shape (Fig. 22.11*b*). This shape contrasts sharply with the original V-shape of the river valley, and for this reason valley shape is a good indicator of how far downvalley glaciers once extended.

Another way in which glaciated valleys differ from

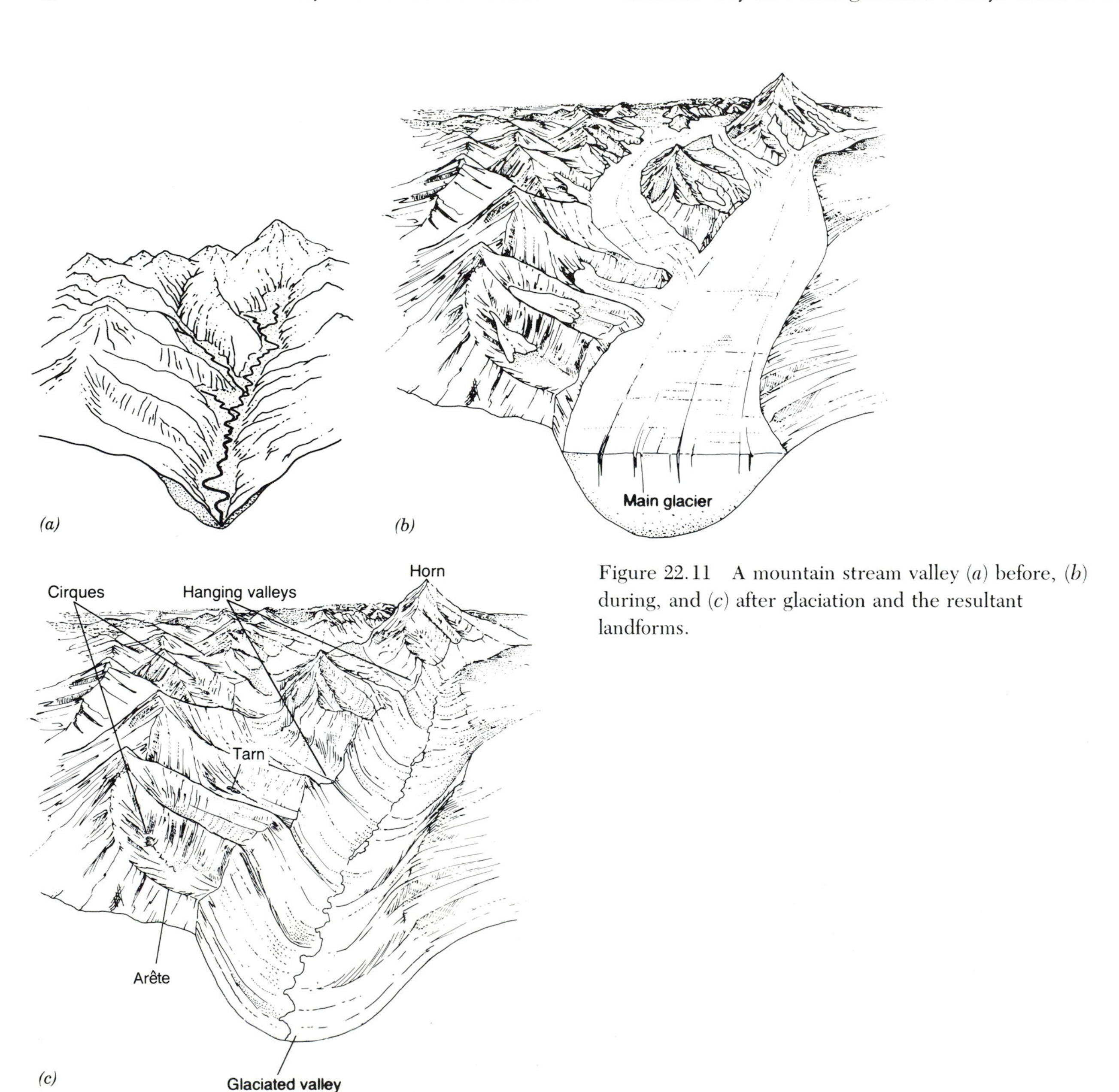

Figure 22.11 A mountain stream valley (*a*) before, (*b*) during, and (*c*) after glaciation and the resultant landforms.

nonglaciated stream valleys is in the elevation at which small valleys join larger ones. In river networks small valleys enter large valleys at essentially the same elevations, whereas in glacier networks small valleys enter large ones at much higher elevations. This is because large glaciers have so much more erosive power than small glaciers that they are able to lower their valleys much faster. Thus, in glacier systems side valleys are left high above the floor of the main valley. After deglaciation, these side valleys appear to hang above the main valley and are called *hanging valleys* (Fig. 22.11*c*). They are often the sites of spectacular waterfalls, and nowhere is this better illustrated than in the famous Yosemite Valley in the Sierra Nevada of California. (Note that hanging valleys are glaciated valleys, too.)

Other features associated with erosion by alpine glaciers include cirques, horns, and arêtes. Where an ice field works on the side of a mountain for a long time, it often sculpts out a small basin called a *cirque*. Many cirques that formed during the Pleistocene Age no longer contain glaciers, and in their bottoms small lakes, called *tarns*, have formed. When several separate glaciers develop on different sides of a large mountain, they often erode the mountain's shoulders, transforming it from, say, a dome or box shape into a conical shape. Each glaciated side usually develops a faceted aspect, giving the mountain the shape of a broad Eiffel Tower. Such mountains are called *horns*, the most famous of which is the Matterhorn of Switzerland. Each side, or facet, is called a *face*, and faces are generally separated by sharp ridges termed *arêtes* (Fig. 22.11*c*).

The erosional features created by the continental glaciers are in general less distinct than those created by mountain glaciers. It appears that the continental glaciers also followed river lowlands, particularly where the trends of valleys coincided with the direction of ice movement. The continental ice sheets, however, were so thick, probably 4000 to 5000 m, that they completely filled and overflowed most valleys. Nevertheless, in some places, such as the Finger Lakes region of New York, U-shaped valleys were created from more V-shaped stream valleys. In most areas, however, erosion by the continental sheets produced shallow basins. It is generally agreed that the basins of the large lakes on the fringe of the Canadian Shield (including those of the Great Lakes) were sculptured from preglacial lowlands (Fig. 22.12).

Glacial Deposition

Visitors to glaciers are often surprised to find that the ice surface is covered with a mantle of rock debris and that there may be few places where one can see the ice itself. This is especially common near the snout of

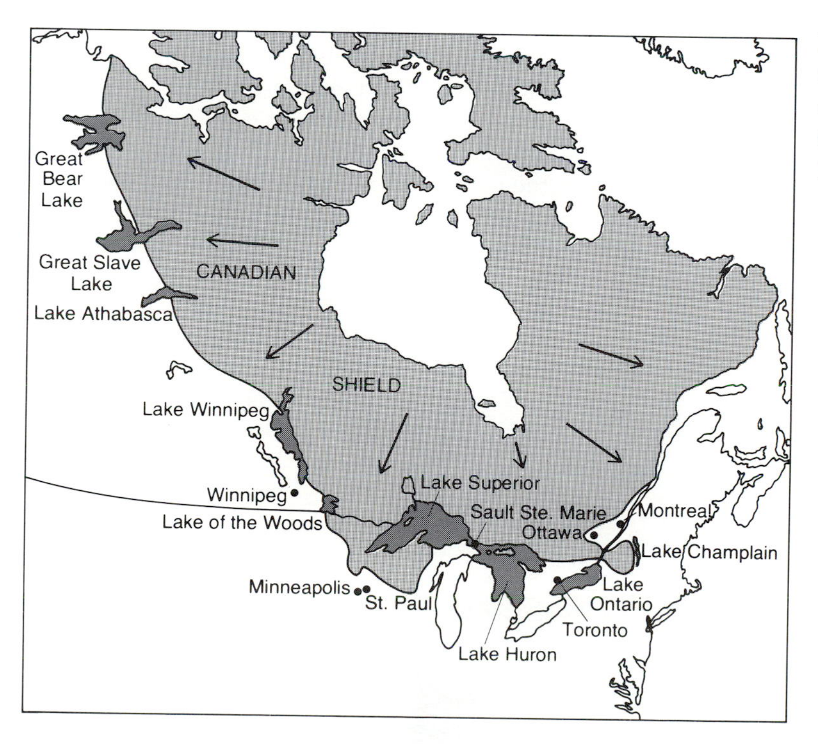

Figure 22.12 Lakes on the margin of the Canadian Shield. The basins of these lakes were sculpted by the continental glaciers as they moved off the shield onto the interior of North America.

Figure 22.13 Rock debris on the surface of the Bandaka Glacier, Hindu Kush, Afghanistan. The large pedestal in the center is more than 3 m high.

the glacier, where most of the ice has been lost to ablation. What are the sources of the rock debris in a glacier? We have already mentioned one important source: the material picked up by plucking at the base of the ice.

A second important source of rock debris is the slope at the margin of the ice. In mountainous areas steep slopes are subject to mass movement, such as rockfalls, landslides, and avalanches, which deposit large amounts of debris on top of the glacier (Fig. 22.13). Some of this material falls into crevasses, but most is buried by snow and eventually incorporated into the firn. Because the debris falls from side slopes, most of it is concentrated in a narrow belt along the margins of the ice. Thus, when two glaciers flow together, the two debris belts on the inside merge to form an interior belt of debris, called a *medial moraine* (Fig. 22.14).

Because a glacier functions like a conveyor belt, all of the debris in the ice is eventually delivered to the terminus, where it is deposited. (Contrary to the idea some people have, glaciers move relatively little material by "bulldozing.") There are two major modes of deposition: (1) direct from the ice, or (2) meltwater flow from the ice. Meltwater deposits are called *glaciofluvial* deposits; the material deposited by the ice is called *till* or *moraine*. All glacial deposits are collectively known as *glacial drift*.

Till is a heterogeneous mixture of unstratified materials ranging in size from massive boulders to clay. The particle composition of the till and the manner of its deposition are highly variable. Till deposited along the edge of the ice often forms irregular hills and mounds called *moraines* (Fig. 22.15). Moraines deposited at the point of farthest advance of the glacier are called *terminal moraines;* those deposited during halts in the retreat of the glacier are called *recessional moraines*. Other types of moraines include *lateral moraines*, which are laid down on the margins of ice, and *ground moraines*, which are deposited beneath the ice. *Till plain*, which resembles ground moraine, is formed when a sheet of ice becomes detached from the main body of the glacier and melts in place. The debris in the ice falls to the ground directly under it. Should the ice contain large boulders, such as the one shown in Figure 22.16, they too are set down. When such boulders are deposited far from their place of origin, they are termed *erratics*.

In contrast with till, glaciofluvial materials are usually stratified and less diverse in particle size. Generally, the most extensive of the glaciofluvial deposits in a glaciated area is *outwash plain*. Outwash deposits are composed mainly of sand eroded by meltwater from the ice and nearby till deposits. As streams of meltwater leave the ice, they cross the moraines and

Figure 22.14 A medial moraine in the Kaskawalsh Glacier, Alaska.

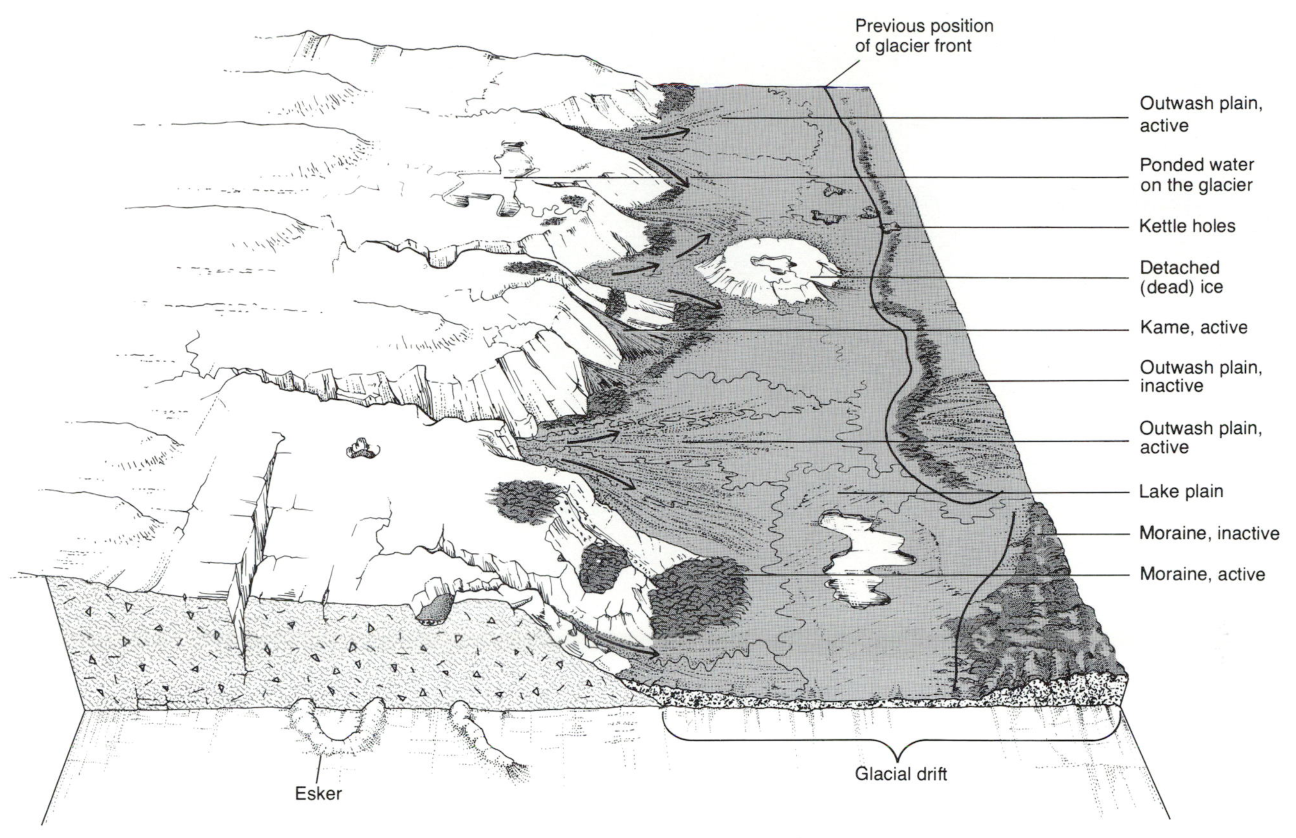

Figure 22.15 An interpretation of conditions at the front of a continental glacier, including most of the depositional landforms that typically form there.

spread over the ground beyond them. Here the streams break down into distributaries, much as a stream does on an alluvial fan, spreading sediments in a broad fan or apron. The outwash plain is formed when the aprons deposited by numerous streams of meltwater coalesce into a single feature (Fig. 22.15).

Glaciofluvial features are also deposited in contact with the glacier. Where sediment-laden water pours off the snout of the ice, a conical-shaped pile of sediment may build up. Such deposits, called *kames*, are

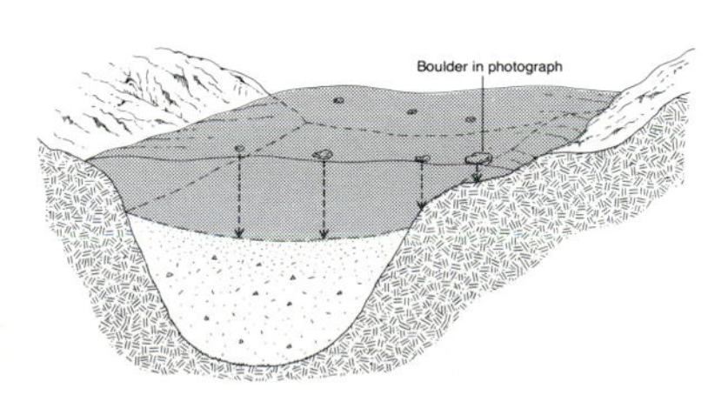

Figure 22.16 A glacial erratic perched on the edge of Little Yosemite Valley. Compare this boulder with the ones on the glacier in Figure 22.13. The accompanying diagram shows how the erratic was set down here.

Figure 22.17 An example of a modern glacial lake at the snout of the South Cascade Glacier, Washington, August 1961.

often situated on or at the edge of moraines (Fig. 22.15).

Although most meltwater flows off the surface of the glacier, some also flows in ice caverns at the base of the ice. The beds of such streams often build up with sand and gravel; if the ice melts from around them, the streambeds are left as curious winding ridges called *eskers.* Eskers are often prime sources of gravel for construction.

What happens to all of the water emitted from a massive continental glacier? Most of it found its way into stream systems that were taking shape on the newly formed terrain. The bulk of these drainage systems are still active, although the discharges they carry are considerably lower today. Some of the largest drainage features disappeared with the ice, however. In particular, large channels called *sluiceways* and vast areas of ponded meltwater, called *glacial lakes,* which formed along the ice front, drained away with changes in the position of the ice and the formation of new courses of drainage (Fig. 22.17). This is well illustrated in the Great Lakes Basin, where numerous glacial lakes rose and fell as the glacier advanced and melted back (Fig. 22.18).

Pitted Topography As the various deposits are forming, the ice disintegrates not only by melting, but also by breaking up into blocks that become detached from the main body of the glacier. These ice blocks are sometimes referred to as *dead ice* because they no longer respond to the movement of the glacier. If drift is deposited around them and the blocks melt away, a depression, called a *kettlehole,* is left in the surface. Both moraine and outwash plain may develop pitted

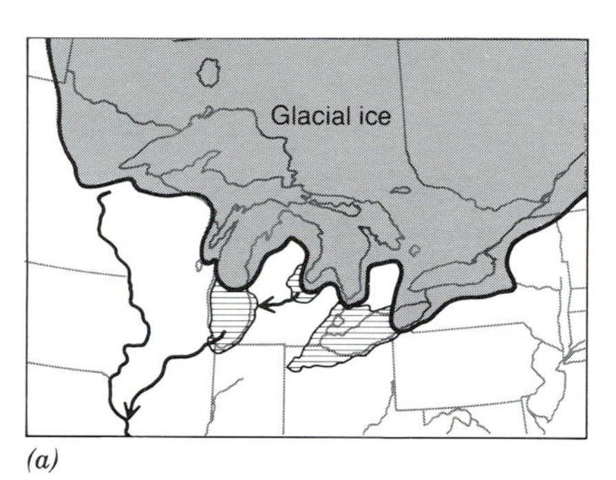

(a)

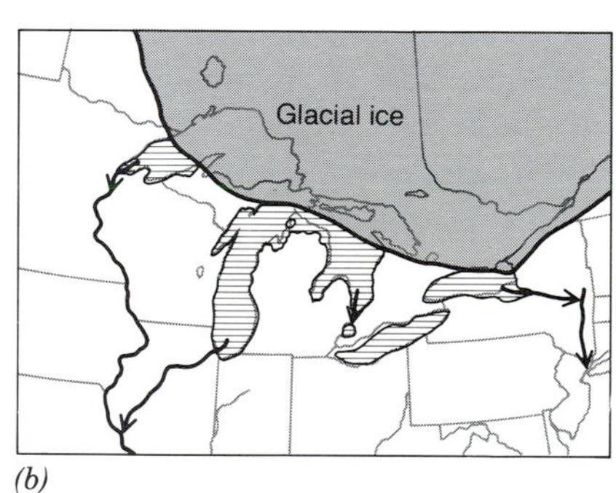

(b)

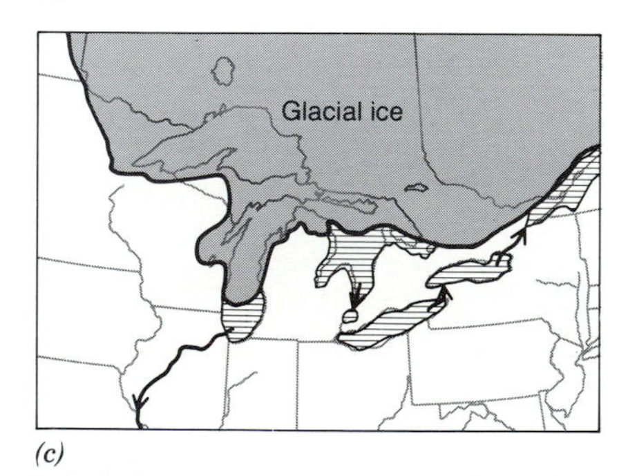

(c)

Figure 22.18 Glacial lakes and sluiceways in the Great Lakes Basin associated with advances and retreats of the Wisconsin ice sheet: (*a*) 12,500 years ago; (*b*) 12,000 years ago; (*c*) 11,400 years ago.

surfaces as a result of this process. Large kettleholes that reach below the water table become flooded with groundwater, forming lakes. Shallow lakes may, in turn, fill with organic material and thereby be transformed into wetlands such as bogs, swamps, and marshes.

CONTINENTAL GLACIATION AND THE LANDSCAPE

Probably no natural event in the time of humans on earth has rendered as much change in the landscapes of the Northern Hemisphere as the continental glaciers have. At the beginning of this chapter we mentioned that the present epoch of continental glaciation began about 1 to 2 million years ago, and based on land deposits at least four stages of glaciation are discernible. Each stage was characterized by a number of substages in which the ice advanced after a period of meltback. In North America the last stage of glaciation was the Wisconsinan stage; its maximum extent in central North America is shown in Figure 22.19.

During the Wisconsin substages, the ice sheet melted back as far as the Canadian border and then readvanced southward. In the Great Lakes region, the pattern of ice movement and retreat in each substage followed the contour of the lake basins and associated lowlands, such as Saginaw Bay in Michigan and Green Bay in Wisconsin. Not surprisingly, the landforms left behind, especially the moraines, also follow the configuration of the basins, resembling a series of collars around the Great Lakes (Fig. 22.20).

How have the landforms left by the glaciers influenced the modern landscape? The influences may be classed into two broad categories: direct and indirect. The direct influences, mainly those already discussed in this chapter, are the erosional and depositional alterations of the land associated with the movement and wastage of the ice sheet. The indirect influences are not as easily described and are not very well understood in many respects. These influences stem from changes in climate, sea level, and land mass elevation that either accompanied or were brought on by continental glaciation. They are discussed later in this chapter.

Let us begin with a few words on some of the influences of glacial landforms on various aspects of the landscape. First, the landforms built by the continental glaciers are not very large compared to the great size of the ice sheets. Practically anywhere you can find hills and valleys of similar proportions which were formed by less spectacular processes. But despite the lack of topographic magnificence, glacial landforms generally have a pronounced influence on the development of the landscape. This influence is usually striking in freshly glaciated terrain but declines with time; in older glaciated landscapes it is quite subtle or not even evident to the casual observer.

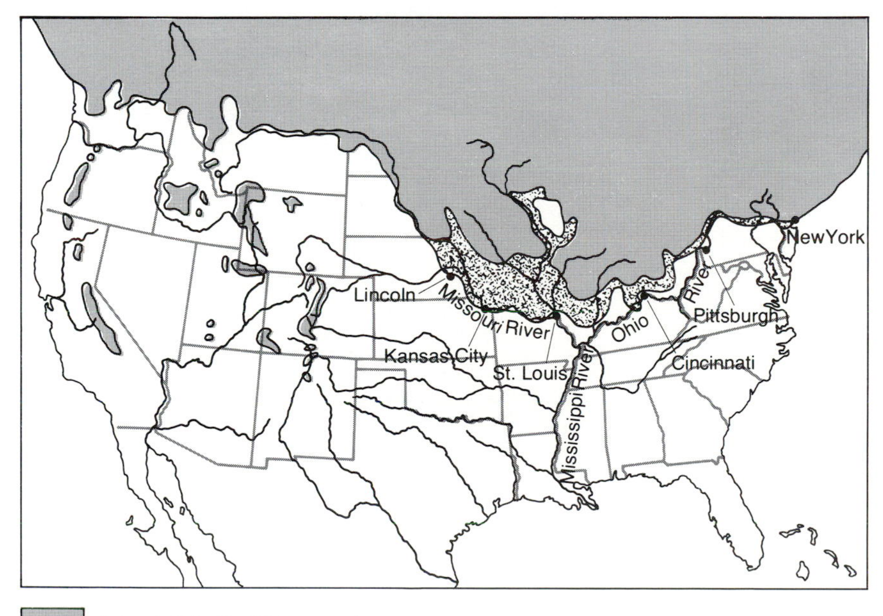

Figure 22.19 The extent of Wisconsin and earlier glaciations in the United States and southern Canada.

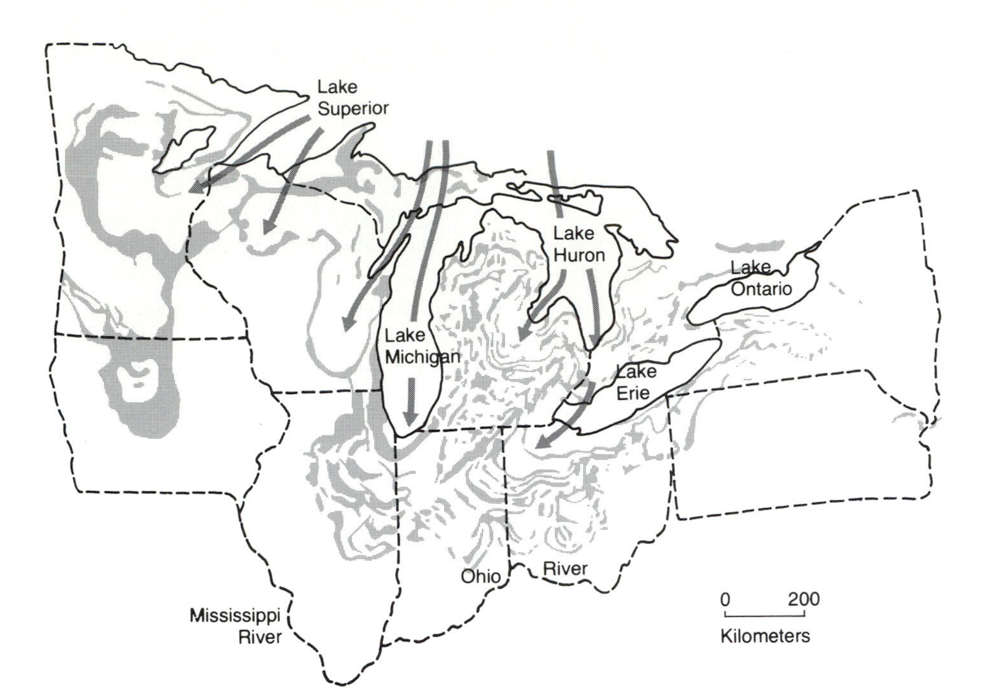

Figure 22.20 The major moraines of the Wisconsin glaciation in the Great Lakes region. The arrows show the pattern of ice movement from the Great Lakes' basins. Compare this map with the one of the glacial lakes in Fig. 25.18.

Soils and Vegetation

Perhaps the most important influence of glacial landforms is on soil formation and drainage. Of course, glacial deposits vary radically in composition, ranging from those that are mainly sand and gravel to those that are practically all clay. Glaciofluvial deposits such as outwash plains and kames are usually sandy and well drained. The soils of these deposits are inherently poor in nutrients such as nitrogen and phosphorus, compared with those of till deposits. Under the cool, moist midlatitude climates of Europe and North America, these conditions are ideal for the establishment of conifer forests and the formation of podzol (or spodosol) soils. Indeed, the correlation among conifers, podzols, and glaciofluvial landforms is so strong that in some areas the occurrence of one can be taken as an indication of the presence of the other two. The great forests of white and red pine that were found in New England and the northern Midwest in the 1800s reflected these conditions. Today forests of jack pine are often found where the great pine forests once grew. Farther north, in the Canadian Shield, drainage of sandy soils is often poor because of the irregular configuration of the underlying bedrock. As a result, the forests on outwash plains are often dominated by spruce and tamarack instead of pines (Fig. 22.21).

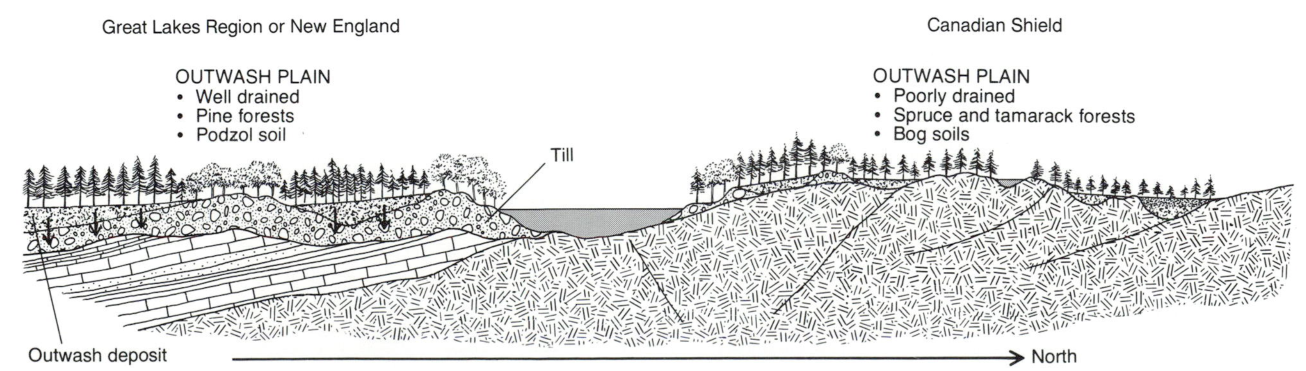

Figure 22.21 A schematic diagram showing the association of forest types with well-drained outwash plains south of the Canadian Shield and poorly drained outwash plains on the Canadian Shield.

THE VIEW FROM THE FIELD / NOTE 22

Patterned Ground in Periglacial Environments

The geographer's inquiry into spatial patterns on the earth's surface ranges from the global scale to the microscale of investigation. At all scales, the first level of concern is usually with describing the dimensions, distributions, and spatial relations associated with a particular pattern or set of patterns. The second level of concern is finding scientific explanations for the patterns, that is, discovering their origins.

In periglacial environments concentrations of stones are a conspicuous surface feature in some areas not only because of their abundance, but also because they are often arranged in curious geometric patterns. The patterns may be hexagonal, rectangular, or circular and range in size from a cell diameter or a meter or so to tens of meters or more. The origin of such patterned ground is poorly understood, but appears to be related to frost action involving heaving (upward movement) and thrusting (lateral movement) of stones.

One explanation holds that each fall these stones are lifted up with the surface layer of soil as it expands upon freezing. When the layer thaws in spring, the stones are unable to settle back into their original positions because small soil particles have fallen into the cavity created when the stone was lifted. Another explanation holds that because stones have higher thermal conductivity than the surrounding soil material, frost penetrates through them rapidly, inducing the buildup of a small pocket of ice at their base, which forces the stones upward. When the ice melts in spring, the fine soil particles adjacent to it fall into the cavity, thereby limiting the depth of settlement of the stone.

Evidence also shows that expansion, contraction, and cracking of the soil mass because of changes in soil moisture contribute to stone sorting and the formation of patterned ground. Other forms of patterned ground are manifested in the distribution of vegetation or water features rather than stones. Patterned ground can also be found outside permafrost regions, but rarely does it take the form of sorted stones. In cases where stone patterns are found in a nonperiglacial setting, it is considered to be good evidence that a periglacial landscape once existed there, and climate changed toward a warmer state without permafrost.

(a)

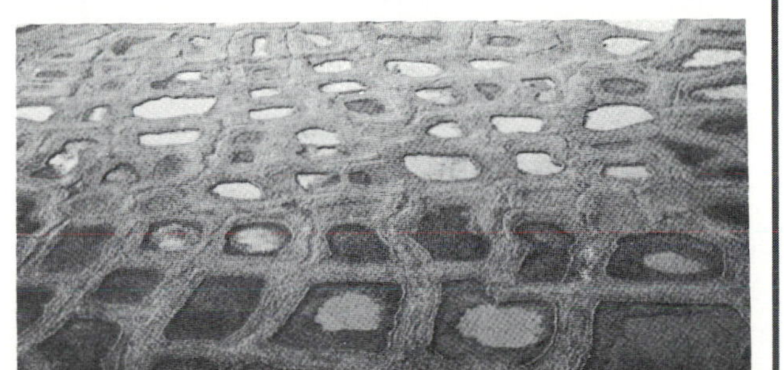

(b)

(a) Stone polygons, Cambridge Bay, Victoria Island, Northwest Territory, Canada. *(b)* Large polygons marked by vegetation and ponded water, near Prudhoe Bay, Alaska.

Drainage

The influence of glacial landforms on drainage can be illustrated at several scales. In both North America and Europe the continental ice sheets originated in shield areas where the geology is extremely diverse. When subjected to the erosional forces of the ice, the surface of the shield yielded differentially, resulting in a highly irregular topography. The drainage lines that developed took on equally irregular patterns, referred to as *deranged* drainage (Fig. 22.22*a*).

Outside the shields are areas where the courses of rivers are controlled by belts of moraines. The rivers tend to flow in the lowlands between the moraines. But moraines are rarely continuous for great distances, and for this reason the pattern is usually broken where a river breaches a morainic belt and joins other rivers. Intermorainal drainage is especially pronounced in northern Germany, where lengthy segments of the Aller, Elbe, and Netze rivers flow in east-west-trending lowlands between terminal moraines (Fig. 22.22*b*).

Wetlands are also a characteristic drainage feature of glaciated terrain. They form not only in shallow kettleholes and in lakes that have filled with sediments, but also in the valleys of large drainage channels that once carried massive discharges of glacier meltwater. Today the floor of the old channel is often occupied by wetlands linked by small streams. The Minnesota River Valley, which joins the Mississippi River at Minneapolis, is a good example of such an area of wetlands (Fig. 22.23).

Indirect Influence of Continental Glaciation

Because the earth has a fixed supply of water, about 98 percent of which is held in the oceans, we reason that the water that went into building the continental glaciers (which amounted to some 55 million cubic kilometers of ice) had to lower the sea level the world over. From underwater surveys conducted on the continental shelves, it appears that worldwide sea level during the glacial maxima was about 130 m lower than it is today. Where the continental shelf is very wide, as along the East Coast of North America, the ocean shoreline during the Pleistocene glaciations extended as much as 75 kilometers beyond its present location.

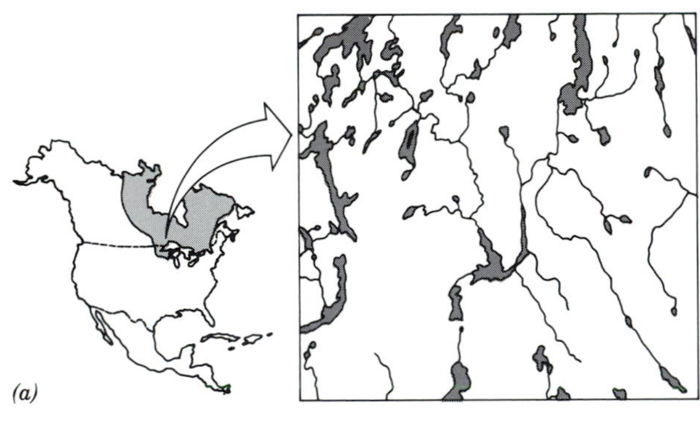

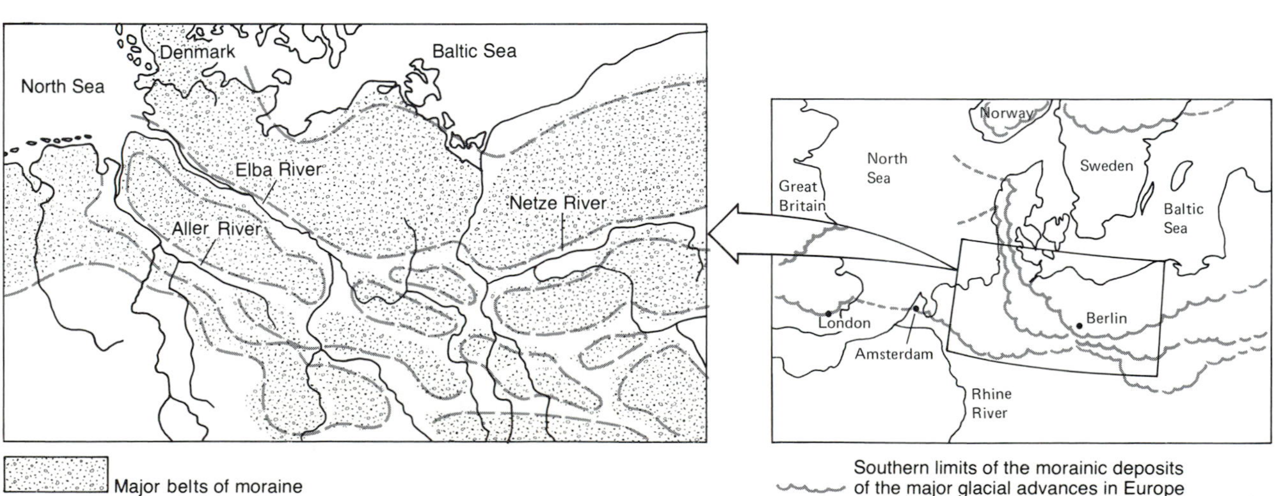

Figure 22.22 Two examples of the influence of glaciation on drainage: (*a*) deranged drainage patterns on the Canadian Shield; and (*b*) intermorainal drainage patterns in Northcentral Europe.

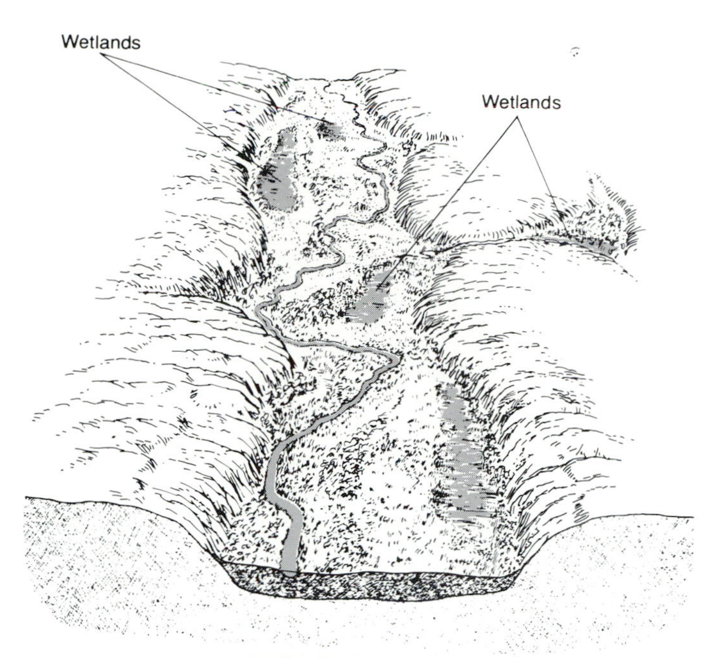

Figure 22.23 Wetlands are common features of river valleys, such as the Minnesota River, which served as a glacial drainage channel.

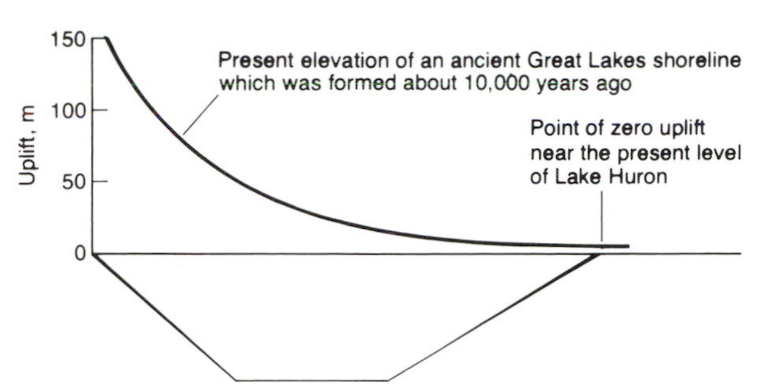

Figure 22.24 Profile of an unwarped shoreline in the Great Lakes Basin. The original elevation is represented by the southern end of the profile; the northern end represents the total uplift in the past 10,000 years.

Between Alaska and Siberia, most of the Bering Sea disappeared, creating a land bridge between North America and Asia. Some anthropologists feel that this land bridge was an important route of early human migration from Asia to North America.

Another effect of the continental ice masses was to depress the earth's crust. Owing to the enormous weight of the ice masses and the elastic nature of the crust, the land subsided under the continental glaciers. This *isostatic depression* was greatest where the ice was thickest and lasted the longest. In North America one of these areas was Hudson Bay, where crustal depression under the ice may have exceeded 1000 m. Wherever depression occurred, however, it was not permanent because when the ice melted away, the crust began to rebound. This is referred to as *isostatic rebound*, or glacial uplift. In the past 10,000 years *isostatic rebound* has recovered most of the depression caused by the Wisconsin ice sheet in the northern Great Lakes.

Northward from the Great Lakes, rebound is still in progress, and in the Hudson Bay region it amounts to several centimeters per year. The best measure of rebound is found in the ancestral shorelines of large water bodies, such as the Great Lakes and Hudson Bay. If we trace one of the Great Lakes' ancient shorelines from south to north, it begins to increase in elevation (Fig. 22.24). Because the entire shoreline of any water body had to have formed at a uniform elevation, such warping is solid documentation of isostatic rebound (also see Fig. 21.21).

The Pleistocene was also a time of accelerated wind erosion, judging from the wind deposits dating from that time. On and around the shorelines of glacial lakes and on outwash plains, sand dunes formed before vegetation could stabilize new surfaces. Farther away from the glacial front, wind deposited silt over extensive areas. The deposits, referred to as *loess*, are composed of fairly uniform silt that was apparently winnowed from sluiceways, lake beds, and barren deposits near the ice, though no one is sure of all of its sources. In any case, vast areas of loess are found in the central part of the United States, China, and in European Russia, well beyond the southern limits of glaciation (see Figs. 13.3 and 23.23).

Summary

Glaciers presently cover about 10 percent of the earth's land area, but during the Pleistocene Era, they covered about three times as much land. Glaciers are classified according to size and thermal characteristics. The smallest are mountain glaciers, and the largest are continental glaciers. Thermally, glaciers are either temperate (warm) or polar (cold). Glaciers are fed by snow, which over a number of years is transformed from névé into firn and then ice. When the thickness of the ice exceeds 20 m, it deforms as a plastic, and the glacier begins to flow.

The erosional work of glaciers involves mainly scouring and plucking. In mountainous regions glacial erosion produces distinctive landforms including U-shaped valleys, cirques, and horns. The debris transported by the ice is deposited as till and glaciofluvial material in the form of till plains, moraines, outwash plains, and related landforms. Although glacial deposits are not large as landforms, they have a pronounced influence on the development of drainage, soils, and vegetation patterns, as is evident in the landscapes of northwestern Europe, Canada, and the Great Lakes region.

Concepts and Terms for Review

- Ice Age
  - glacial stages
  - interglacial
  - Pleistocene Epoch
  - causes of glaciation
- types of glaciers
  - continental
  - piedmont
  - alpine
  - temperate
  - polar
- glacial ice
  - snow, névé, firn
  - ablation
  - movement
  - velocity
  - mass budget
  - calving
- glacial erosion
  - scouring
  - plucking
  - U-shaped valley
  - hanging valley, cirque, tarn
  - horn, arête
- glacial deposition
  - glacial drift
  - moraine
  - glaviofluvial deposits
- glacial landscapes
  - outwash plain
  - till plains
  - pitted topography
  - lake basins
  - soils and vegetation
  - drainage patterns
- indirect influences of glaciation
  - sea-level changes
  - isostatic depression and rebound
  - loess deposits

Sources and References

Donn, W.L., and Ewing, M. (1966) "A Theory of the Ice Ages, III." *Science* 152: pp. 1706–1712.

Embleton, C., and King, C.A.M. (1975) *Periglacial Geomorphology.* New York: Halsted Press.

French, H.M. (1976) *The Periglacial Environment.* Longman: New York.

Marcus, M.G. (1964) *Climate-Glacier Studies in the Juneau Ice Field Region, Alaska.* Chicago: University of Chicago, Department of Geography Research Paper 88.

Nye, J.F. (1952) "The Mechanics of Glacier Flow." *Journal of Glaciology* 2: pp. 82–93.

Price, R.J. (1973) *Glacial and Fluvioglacial Landforms.* New York: Hafner.

Ritter, D.F. (1986) *Process Geomorphology.* Dubuque, Iowa: Wm. C. Brown.

Sugden, D., and John, B. (1976) *Glaciers and Landscape.* London: Edward Arnold.

Washburn, A.L. (1973) *Periglacial Processes and Environments.* New York: St. Martin's Press.

Young, G.J., and Ommanney, C.S.L. (1984) "Canadian Glacier Hydrology and Mass Balance Studies; A History of Accomplishments and Recommendations for Future Work." *Geografiska Annaler,* 66A:3, pp. 169–182.

Chapter 23

Airflow and the Work of Wind on the Land

Air is in constant motion over the earth. At sea it generates waves; on land it erodes and transports soil particles, mostly sand and silt. But the work of wind on the land is highly uneven geographically. What governs the capability of wind as an erosional agent? What sorts of landforms are created by wind erosion?

By the second decade of the twentieth century, it seemed that the earth's remote corners had all been discovered and mapped, and that only scientific questions remained to be investigated. In one sense this was true because both poles had been reached by humans, the Nile's headwaters had been found, the Grand Canyon had been traversed, and Mount Everest had been discovered (though not climbed). Early explorers had been interested mainly in determining "what is where." They opened the way for another era of geographic exploration centered more on the question of "why something is where we find it."

In the 1930s Ralph Bagnold, a young officer in the British Royal Engineers, was stationed in Cairo, Egypt. Life in the British garrison proved dull, and Bagnold and his companions sought relief from the boredom of garrison duty by taking trips into the desert along the ancient caravan routes. West of Cairo lay the great sand deserts, which were impenetrable by traditional travel, the source of many folk tales and strange beliefs. The dunes were a challenge that Bagnold could not avoid. With the aid of specially adapted vehicles, he explored this sea of sand to study its forms and movements.

Bagnold's tour of duty ended in 1935. He returned to England, where he brought his investigation of wind-blown sand to the laboratory. To answer the questions of the mechanics of sand movement, he constructed a wind tunnel that enabled him to measure in detail the processes of sand movement. From this work emerged a book, *The Physics of Blown Sand and Desert Dunes* (1941), in which Bagnold set down the principles of sand transport and dune formation. To this day this book remains the single best work on the subject.

In 1941 Bagnold found himself back in North Africa participating in the war effort against the Axis powers. His knowledge of dunes and vehicular travel through dune fields enabled him to carry out highly successful raids against Italian supply lines, which helped cripple the Italian Army. Thus, the story of Bagnold and the dunes turned full circle, as the scientific knowledge he had acquired first in field exploration and later in laboratory analysis was applied to an urgent wartime problem in the very setting of his original field explorations.

In this chapter we will follow in Bagnold's footsteps and examine the role of wind as a geomorphic agent in the landscape. We begin with the motion of air as it moves over the land. Next, we focus on wind energy, the erosive power of wind, and the transportation of sand and silt particles by wind. Finally, we describe the principal landforms created by wind action.

AIRFLOW NEAR THE EARTH'S SURFACE

The atmosphere is an enormous ocean whose bottom is the landscape and the surface of the seas. The currents of the atmosphere, that is, winds, are governed by the same principles of fluid motion as govern liquids, notably water. But there are some important differences, chief among which is that the density of air (1.29 kg/m^3) is much less than that of water (1000 kg/m^3); therefore, wind cannot move as large a particle as a water current can.

The Boundary Layer

In considering wind as an erosional agent—as opposed to a climatic agent—we need to examine only the lowermost zone of the atmosphere, called the *boundary layer*. This layer represents the transition between the atmosphere and the earth's surface. The boundary layer is about 300 m thick and more or less follows the contour of the earth's surface over both land and water (Fig. 23.1).

Recall from the discussion of mechanical convection in Chapter 4 that wind speed increases with elevation above any surface (see Fig. 4.12). At an elevation of 300 m or so, however, the increase ends, and this marks the top of the boundary layer. The bottom of the boundary layer usually lies very close to the ground but never directly on it. It is defined as the level of zero wind speed below which there is a thin zone of air termed the *laminar sublayer*, within which there is effectively no measurable air movement. The laminar sublayer covers all surfaces, natural and synthetic, but it is very thin, only a few millimeters deep at the most.

If a surface is roughened by adding pebbles or vegetation to it, the base of the boundary layer is pushed up well above the ground level. This deeper zone of relatively calm air ranges from several centimeters in grass to several meters in heavy forest. The thickness of this larger layer is called the *roughness length*. It is very important in considering the effectiveness of wind as an erosional agent because it defines how close fast-moving currents of air reach to the ground (see Fig. 4.13 and Table 4.3).

Two types of flow can be detected in moving air: laminar and turbulent. In laminar flow, a particle of air moves horizontally to the surface in essentially a straight-line route. The airflow in the upper part of the boundary layer often approximates laminar flow, but near the ground it becomes decidedly turbulent. Turbulent flow is characterized by an irregular, swirling motion over the ground that moves both vertically and horizontally. The vertical component accounts for as much as 20 percent of the total movement.

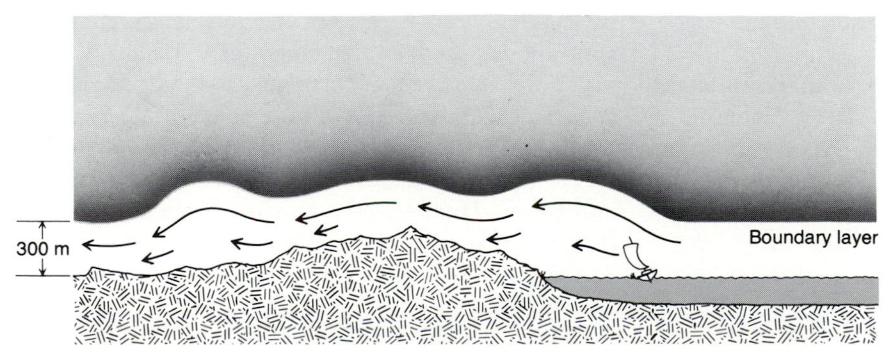

Figure 23.1 A schematic diagram of the boundary layer of the atmosphere, a zone about 300 m deep, within which wind velocity decreases toward the earth's surface.

Direction, Velocity, and Power

Wind direction and velocity at most places are highly variable and not easy to summarize in a brief statement. It is important to understand that average wind speed (based on airflow in all directions) or average wind velocity (based on single direction) tends not to be very meaningful because the range of wind speeds and velocities is so great. This is because near the ground, wind is very gusty, changing direction and velocity in a matter of minutes or even seconds (Fig. 23.2).

A second, and very important, point is that wind speed is not a good indicator of wind power (which you can think of as the force it exerts on a surface) because power and speed are not directly proportional. Rather, power increases according to the cube of speed or velocity. Therefore, fast winds, which on the one hand account for comparatively little total time in a sampling period, may on the other hand be the most important class in terms of power because of the power/velocity relationship. Experts are quick to point out that estimates of the electrical-generating capacity of wind based on average speed are consistently in error by more than 100 percent. The same would be true for estimates of erosion based on average speed or velocity. The upper curve in Figure 23.3 shows the power equivalent for the wind velocities represented by the lower curve.

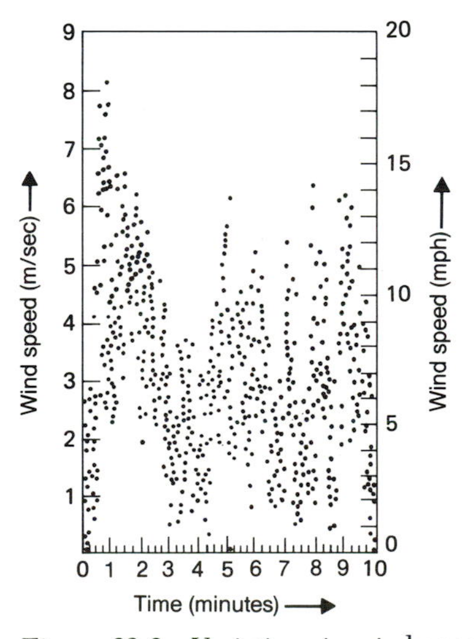

Figure 23.2 Variations in wind speed over a 10-minute period at a height of 2 m. Each dot represents an individual reading.

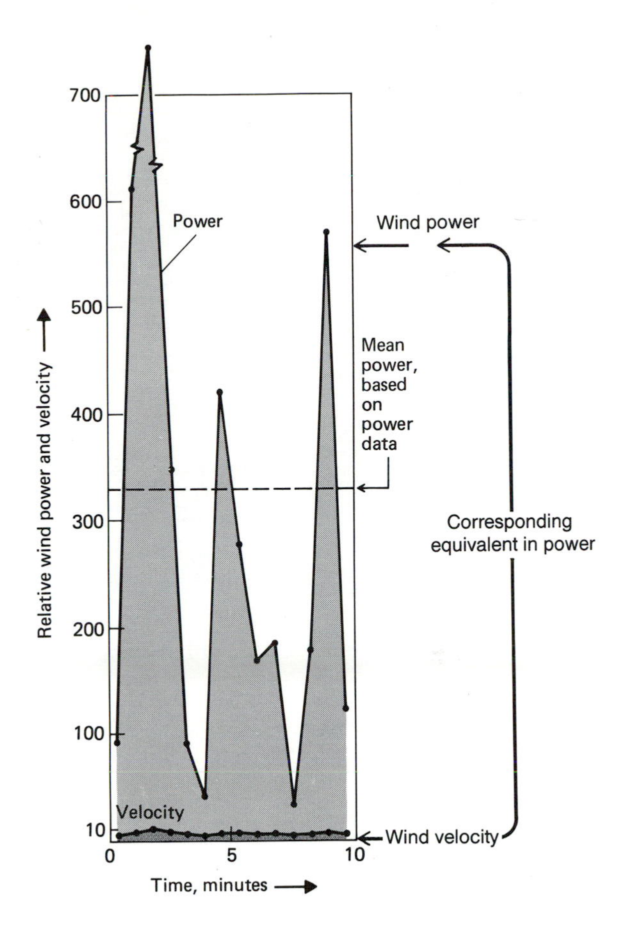

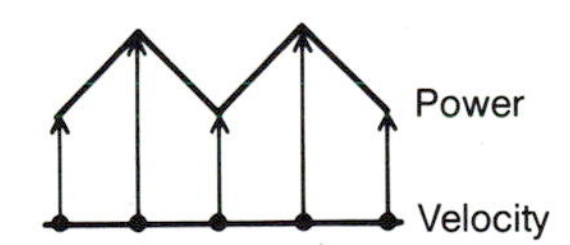

Figure 23.3 The relationship between wind velocity and wind power. Because power increases so rapidly with velocity, the curve of wind power appears extremely exaggerated compared to the corresponding curve of wind velocity.

SOME GEOGRAPHIC CHARACTERISTICS OF WIND

The energy crisis renewed our interest in wind as a resource and improved our understanding of both its aerodynamics and geographic characteristics in North America and Europe. From our standpoint as geographers, it is important that we understand the meaning of geographic scale in observing and measuring wind because wind velocity varies so much over spaces ranging from a few centimeters to hundreds of

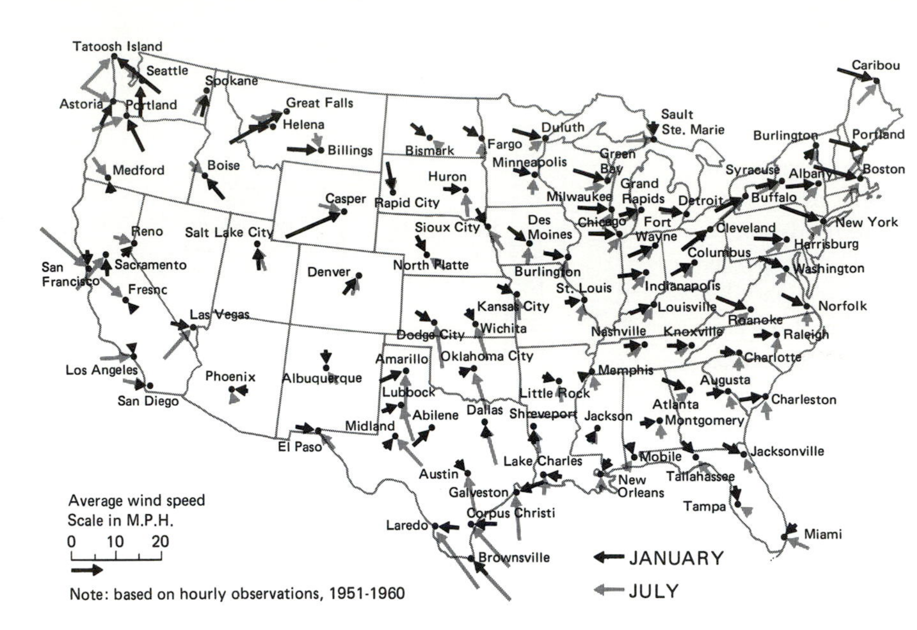

Figure 23.4 Mean wind speeds and directions for January and July in the coterminous United States. The greatest seasonal changes in speed occur along the Texas Gulf and West coasts.

kilometers, depending on terrain and atmospheric conditions.

Regional Variations

In the coterminous United States, winter wind speeds are higher for the northern half of the country, where airflow is predominantly westerly (Fig. 23.4). This corresponds to the movement of air masses, especially midlatitude cyclones, across the middle of North America. In summer the overall pattern changes considerably, with the strongest winds coming off the western Gulf of Mexico and the Pacific Ocean and blowing onto the continent. This is a response to the thermal differences between land and water. As the land heats up, low pressure develops over it, which draws in air from the ocean.

Wind and Topography

Wind velocity and flow patterns are influenced by topography in many places, especially along mountain fronts, canyons, and around isolated obstacles such as a building or a prominent hill (Fig. 23.5). January winds are especially strong in mountain settings such as Great Falls, Montana, and Casper, Wyoming, in part because the chinook blows down the valleys in which they are located. The influence of topography is also apparent where wind speeds up as it moves through a constriction in a valley, an opening in a forest, or between buildings. As it moves over an isolated hill, wind at first accelerates and then decelerates on the downwind side. At this point flow can become turbulent, and large eddies may form which

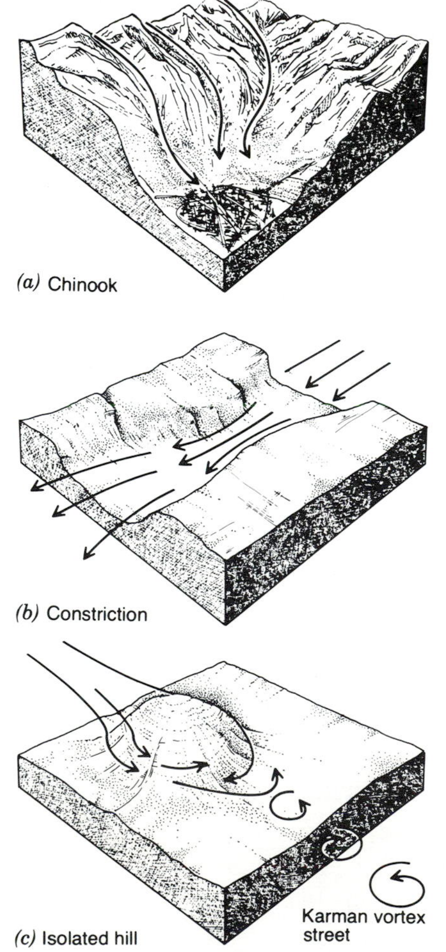

Figure 23.5 Topographic influences on airflow; (*a*) chinook setting, such as Casper, Wyoming; (*b*) constriction in a valley; (*c*) isolated hill with a Karman vortex street.

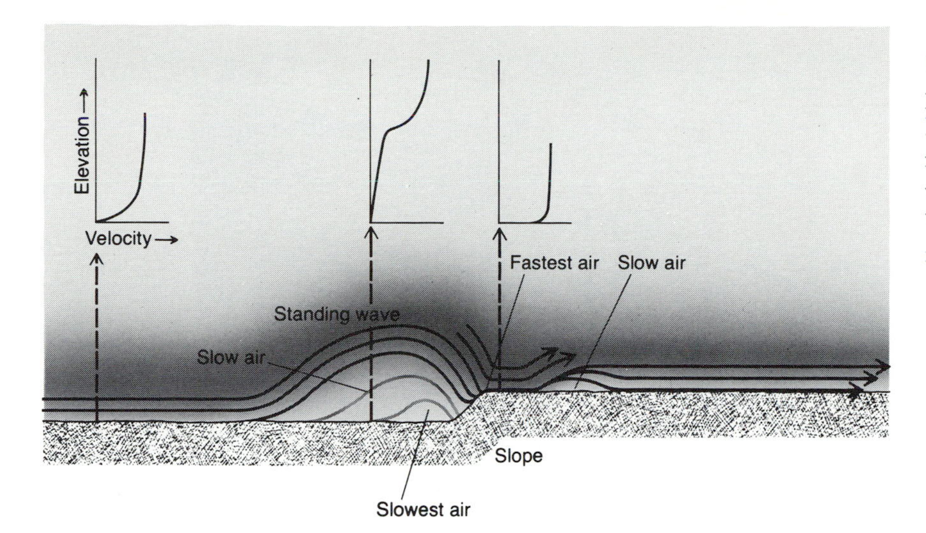

Figure 23.6 The approximate pattern of airflow over a slope. A large standing wave develops in front of the slope, from which fast wind descends on the upper slope. Velocities on the lower slope are relatively slow.

drift downwind in the wake of the hill. A train of such eddies, called a *Karmen vortex street* (named for Theodor von Karman, a pioneer in aerodynamic research), has been observed in the lee of some islands.

Figure 23.7 Conditions near the crest of a sea cliff on a windy day. The force of the wind is so great that a person standing on the brow of the cliff is easily knocked down.

One of the most important influences of topography involves airflow over a ridge or a long hillslope. Such topographic rises form an obstacle to airflow; for the air in the boundary layer to maintain its flow, velocity must increase as the air crosses the rise. This condition is portrayed in the pattern of streamlines (lines of equal wind velocity) as they approach and cross a broad slope. At some distance in front of the slope, the fast streamlines rise to form a large standing wave and then descend toward the crest slope (Fig. 23.6). Beneath the wave, near the foot of the slope, a low-velocity zone develops. At the crest of the slope, velocity increases substantially, reaching levels two to four times that near the foot of the slope (Fig. 23.7).

What happens to velocity in the area downwind of a slope? In general, it declines as the streamlines spread out. If the topography itself drops off sharply behind the slope, as along the back of a ridge or a sand dune, a calm zone may form. Wind-tunnel experiments show that the flow may undergo separation, meaning that it divides into zones of different velocities. This is manifested in a distinct low-velocity zone at some distance behind the crest slope.

The increase and decrease in wind velocity with topography are important to our understanding of the erosional and depositional effects of wind. They tell us that wind stress is greatest on windward slopes, particularly near the crests, and given that conditions on such slopes are conducive to erosion, that is where erosion will be greatest. By the same token, they tell us that wind stress is lightest on downwind slopes (called leeward slopes), and that is where deposition should take place. Evidence for this pattern can be seen in the formation of sand dunes, in the erosion of coastal slopes, and in the formation of snow cornices along mountain ridges (Fig. 23.8).

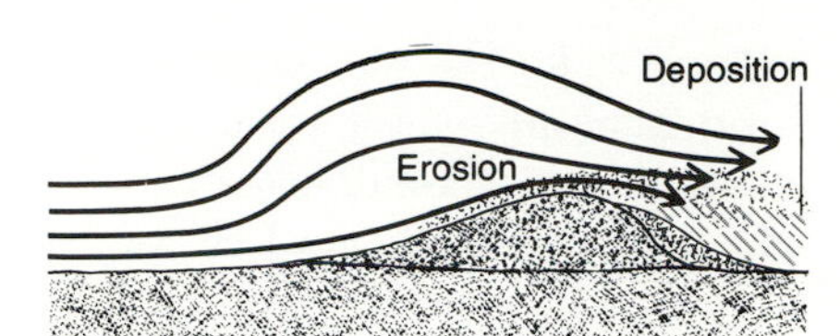

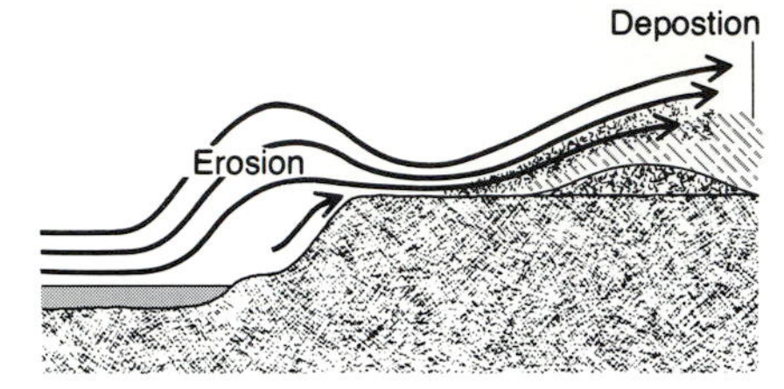

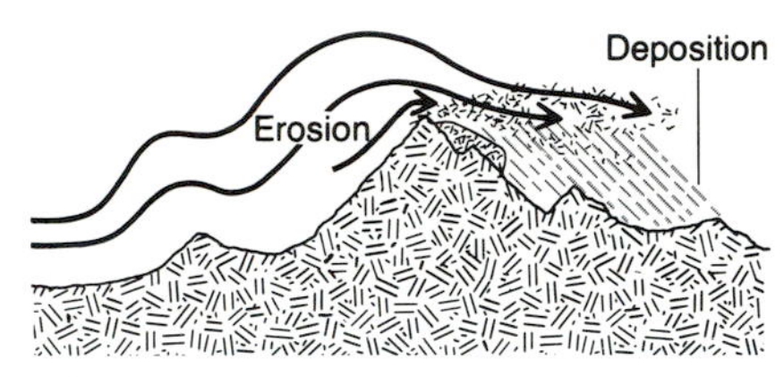

Figure 23.8 Examples of airflow over landforms that result in a zone of wind stress and erosion on one side and a zone of deposition on the other.

EROSION AND TRANSPORTATION BY WIND

The power of wind to erode is governed by two factors: air density and wind velocity. Air density has only a minor influence on the erosive force (shear stress) of wind. Moreover, it varies relatively little within several hundred meters of sea level. Velocity, on the other hand, has an acute influence on the erosive force of wind. Erosive force increases sharply with increases in wind velocity. For example, a velocity increase from 2 m/sec to 4 m/sec produces an eightfold increase in erosive power, whereas a velocity increase from 2 to 10 m/sec produces a 125-fold increase in erosional power. Thus, fast winds are capable of doing far more work than are slow winds (see Fig. 12.2). But let us not forget that slow winds occur much more frequently than fast ones do. Thus, the question of which winds do the most work in the long run in terms of sand movement can be difficult to answer.

Sand Transport

Most of the credit for our present understanding of wind erosion and transportation of sand goes to Ralph A. Bagnold, whose exploits were mentioned at the beginning of the chapter. In his many experiments dealing with sand movement, he discovered most of the key principles governing sand erosion and transport. In particular, Bagnold identified the critical relationship between wind velocity and total sand transport over a dune surface as one in which transport increases with cube of velocity (Fig. 23.9).

The mechanisms of sand transportation by wind are very interesting and not difficult to understand. When a wind begins to rise over a sand surface, the first particle movements are discernible at a velocity of about 4.5 m/sec (10 mph). The initial movement of particles is characterized by a rolling motion called *traction* (also called *creep*, which is not to be confused with soil creep). In light winds, only small particles move in traction, but as a strong windstorm reaches its peak, particles as large as small pebbles are moved in this mode.

A second mode of sand transport involves particles lifting off the ground, becoming airborne, and sailing in short trajectories several centimeters downwind. When a falling grain strikes the surface, part of its momentum is transferred to another grain, and it in turn goes sailing off. This mode of transport, called

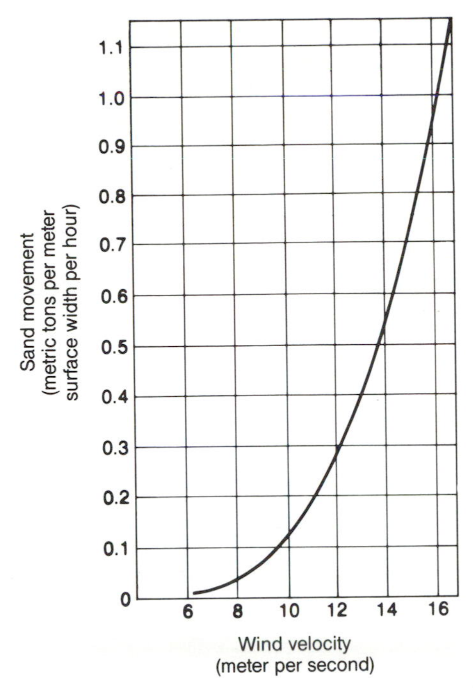

Figure 23.9 Sand transport (across a 1-m line in one hour) relative to wind velocity.

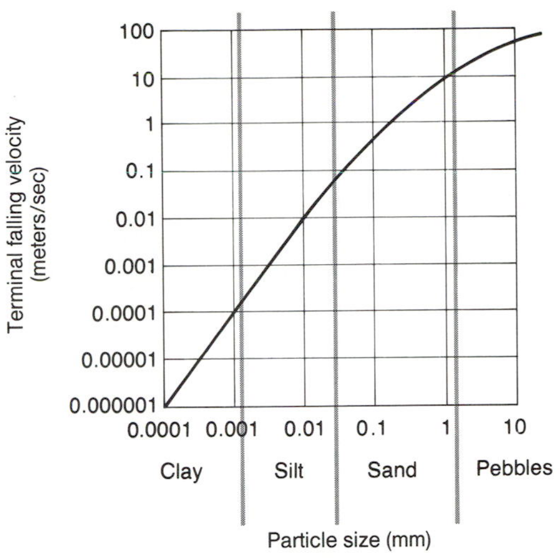

Figure 23.11 Falling velocities of different-sized particles. Note that sand falls at least 1000 times faster than clay does.

saltation, accounts for 75 to 80 percent of the sand transport over sand dunes. (The remaining 20 to 25 percent is moved by traction.) Most sand grains travel within several centimeters of the surface, but very strong winds are able to lift grains as high as 2 m and to carry them 10 m or more downwind (Fig. 23.10).

Saltation is a very effective mode of sand transport, for two reasons. First, air, which has a low density, offers little resistance to dense airborne missiles (except for those that are microscopically small); second, the air turbulence near the ground has a strong lifting component. Once a particle is wafted into the air, it tends to settle back toward the surface but can do so only if it is heavy enough to overcome the force of the uplifting currents. If a particle's terminal fall velocity is more than 20 percent of the velocity of the updraft, it cannot be held aloft. Note in Fig. 23.11 the great difference in the terminal settling velocities for sand, silt, and clay particles.

Figure 23.10 Sand transport by saltation over a sandy slope.

Selective Sorting of Particles

Light particles—principally silt and clay—which lack the necessary mass to settle rapidly out of moving air are lifted well above the zone of saltation and during very strong storms may be carried in suspension thousands of meters into the air and hundreds of kilometers downwind. Therefore, on the basis of particle size (mass) alone, we can account for the separation, or sorting, of sand from pebbles and larger particles and silt and clay from sand. This helps to explain why sand dunes are made up exclusively of sand and why deposits of wind-blown silt, called *loess*, are located at great distances from areas of wind erosion (Fig. 23.12).

The magnitude of silt transportation by wind was documented in the 1930s when silt from the Dust Bowl in Kansas, Oklahoma, and Texas so darkened the sky that it elicited responses such as the following from a Texas schoolboy:

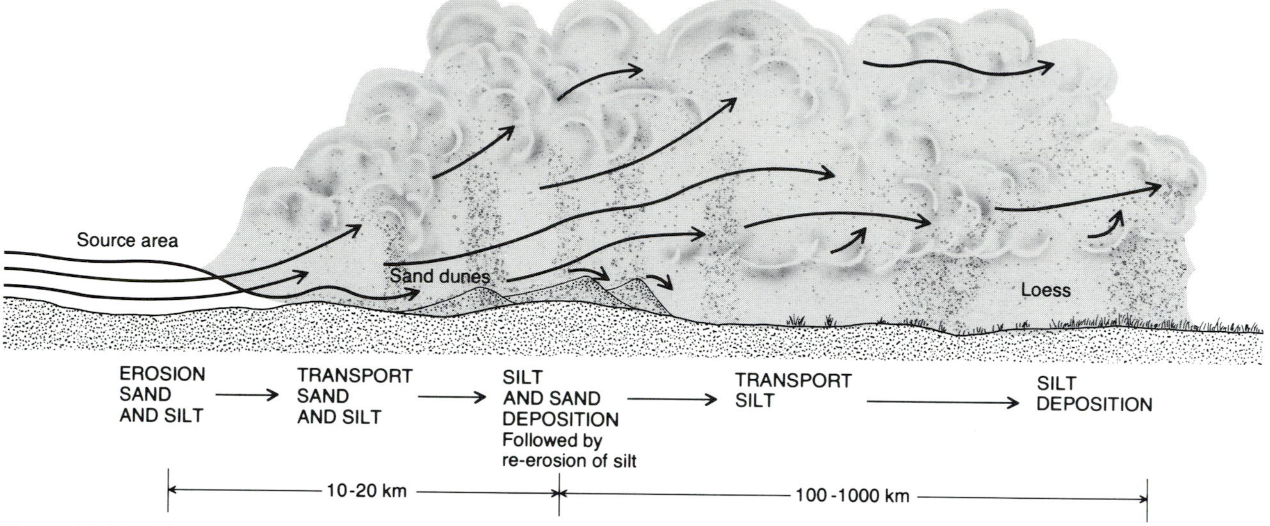

Figure 23.12 The sequence of sand and silt deposition downwind from an erosion zone.

Figure 23.13 The leading edge of a dust storm sweeping across Baca County, Colorado, in 1935.

> These storms were like rolling black smoke. We had to keep the lights on all day. We went to school with the headlights on and with dust masks on. I saw a woman who thought the world was coming to an end. She dropped down on her knees in the middle of Main Street in Amarillo and prayed outloud: "Dear Lord! Please give them a second chance."[1] (Fig. 23.13).

[1]Patrick Hughes, *American Weather Stories* (Washington, D.C.: U.S. Department of Commerce, 1976), 116 pp.

Sand Dune Formation

Sand dunes originate in aerodynamic environments that favor the deposition of sand. These are usually places where flow separates, giving rise to a zone of slow-moving air under much faster moving air. Such zones form behind obstacles, on the leeward sides of slopes, or beneath standing waves (Fig. 23.14). As the fast air slides over the calm zone, saltating grains fall out of the air stream and accumulate on the ground.

At first, the pile of sand that forms is stationary,

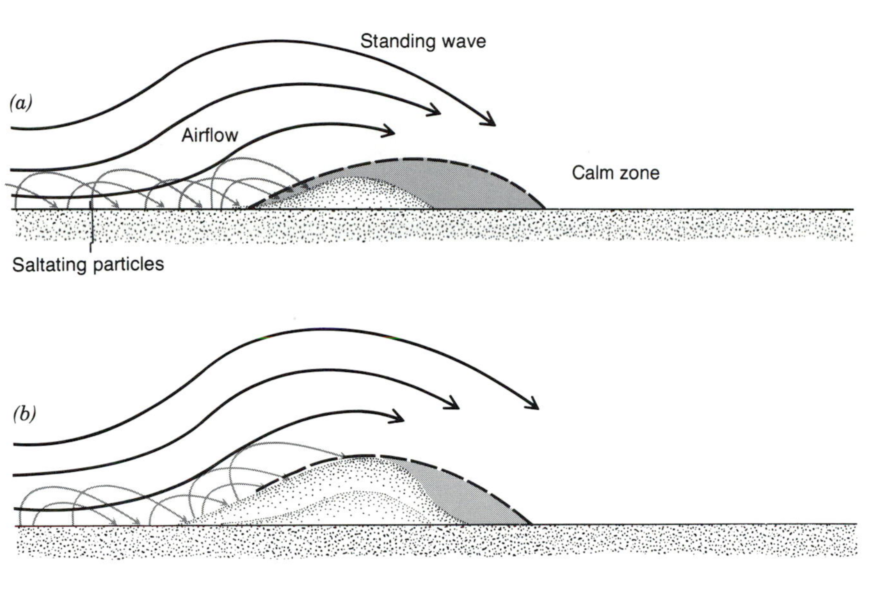

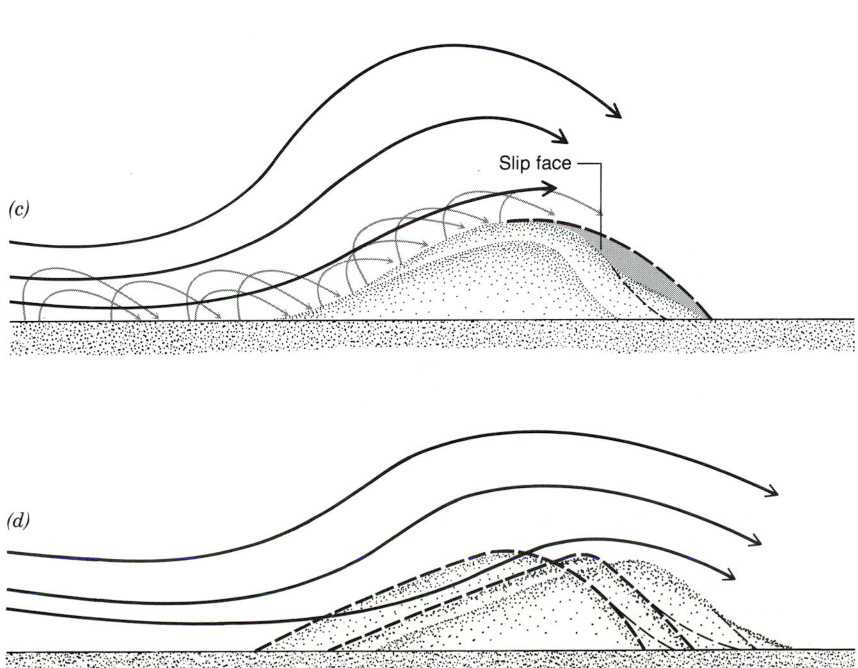

Figure 23.14 The formation of sand dunes: (*a*) saltating sand falls into low-velocity zones; (*b*) pile grows both vertically and horizontally, gradually filling low-velocity zone; (*c*) pile fills low-velocity zone, and streamlines are displaced upward, resulting in horizontal transport of sand over the crest of the pile; (*d*) the pile migrates downwind, thereby taking on the designation sand dune.

but as it grows higher, it reaches the level of faster and faster streamlines that move the sand across the top. Grains are moved from the windward to the leeward side of the pile up just over the crest. As the upper leeslope steepens to an angle around 33 degrees, the angle of repose for sand, it becomes unstable and small slides break loose, moving sand to the lower part of the slope. The sand pile has now become a dune: a mobile heap of sand with a low-angle windward slope and a steep leeward slope, called the *slip face* (Fig. 23.15). The dune migrates over the ground as sand is eroded from one side and deposited on the other.

The height of a dune is limited by the velocity of the wind above ground level. As a dune grows, two conditions change: It reaches faster streamlines with higher elevation, and streamlines become focused near the crest with the development of the windward slope. At some point, therefore, vertical growth must cease because wind stress becomes too great to allow further deposition. The maximum height is variable but usually falls in the range of 10 to 25 m. The largest sand dunes in the world are found in Saudi Arabia and measure more than 200 m high. These are not individual dunes, but rather are massive complexes of sand dunes that grow when smaller, faster-moving dunes migrate onto them.

Controls on Wind Erosion

Theoretically, every particle on the earth's surface is subject to wind stress, but whether the stress is sufficient to move it depends on the balance between the driving force of the wind and the resisting force of the particle. Because the driving force of wind is mainly a function of velocity, any control on velocity is a control on erosion or the potentiality for erosion. At ground level, the roughness length is the critical control on wind velocity. Large boulders, trees, and buildings can raise the roughness length to 1–2 m. Shrubs produce a somewhat lower roughness length, and grasses produce one of only a few centimeters. Yet all of these materials produce virtually the same effect because they raise the effective frictional surface of the wind above the particles.

The picture is quite different for smooth, unprotected ground, especially soils without a vegetative cover. As a general rule, the first places to look for substantial amounts of wind erosion are areas with little or no vegetative cover. There are basically three such environments in the world: (1) the deserts; (2) environments with fresh deposits of sediments, such as beaches and alluvial fans; and (3) cultivated farmland (Fig. 23.16).

Wind erosion is not a certainty in these environ-

Figure 23.15 Slip face of a small barchan sand dune (shown by the light area to the right of the crest line). Airflow is left to right. The tree stubs are part of an exhumed forest, buried earlier by a much larger sand dune.

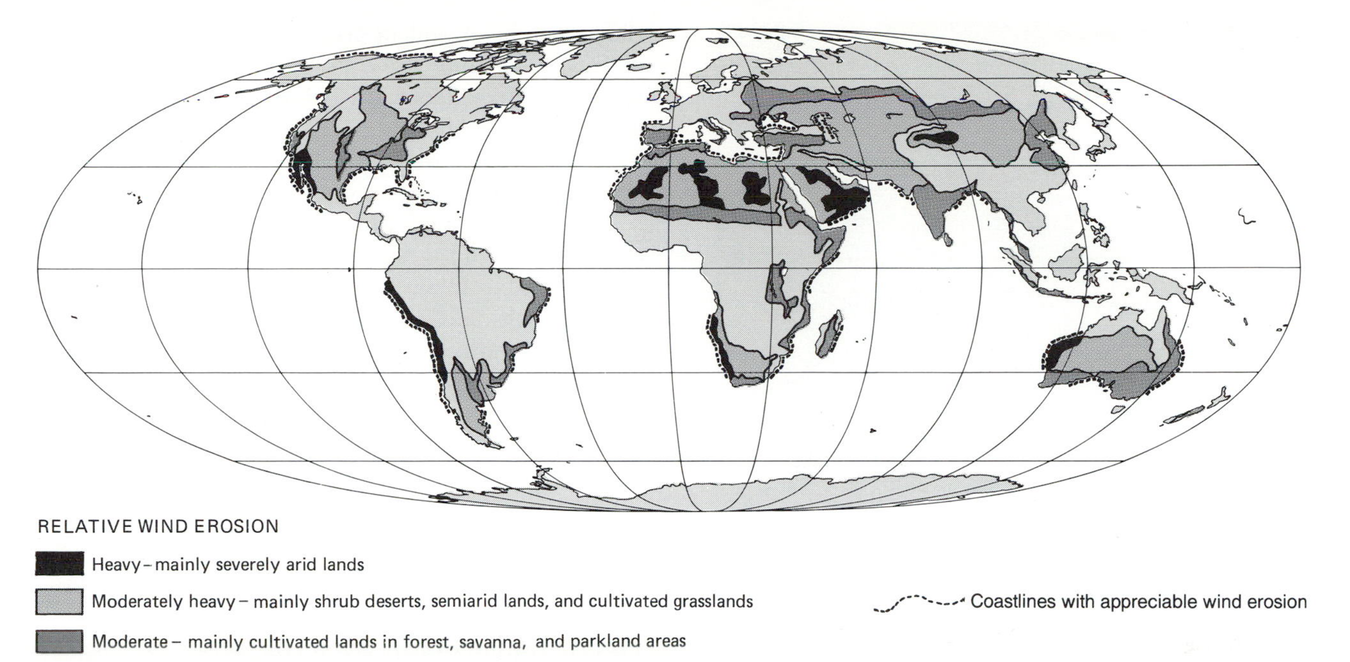

Figure 23.16 Major zones of wind erosion in the world, including coastlines.

ments, however, because several factors other than the absence of vegetation play a part. Particle size is one of the most important factors. For any sized particle, a certain wind velocity, called the *threshold velocity*, is necessary to move it. The larger the particle, the higher the threshold velocity.

Figure 23.17 A lag surface of pebbles. This surface formed as the finer particles were blown away. Note that the lag functions as armor to protect the fine material under it from being eroded.

In some deserts where fines have been winnowed from soils of mixed particle sizes, only pebbles and larger particles are left behind. Surfaces laden with such particles are called *desert lag* and sometimes resemble a cobblestone street when they become worn, polished, and cemented together by chemical residues (Fig. 23.17). The threshold velocity required to move these lag particles is exceedingly large, probably 100 meters per second or more. The fastest naturally occurring wind ever measured on the earth's surface was 103 m/sec (231 mph), recorded at the summit of Mount Washington (2000 m elevation), New Hampshire, in April 1934.

Another factor that increases the resistance of particles to wind erosion is the cohesiveness of a soil material. Clay tends to be highly cohesive, of course. Therefore, the threshold velocity required to move a clay particle is considerably greater than its small particle size suggests. Ground frost can function in much the same way; unlike salts, however, crystals of ice can be sublimated away to dry winter air, thereby releasing the particles to the wind.

LANDFORMS PRODUCED BY WIND ACTION

How effective is wind as an erosional agent? It is hard to say. Compared with runoff, it is decidedly second-

Figure 23.18 A rock face on which wind works in combination with runoff and various weathering processes to produce a striking microrelief.

ary in terms of the total amount of erosion produced worldwide. But it is always important in weighing such questions to recognize that geomorphic processes always work in combination. On the rock face, such as the one shown in Figure 23.18, wind undoubtedly works in combination with weathering and mass movement to etch grooves and pits into the face of the sandstone cliff. The same can be said for many coastal slopes, where wind combines with waves, seepage, and runoff to erode and shape sea cliffs and backshore slopes. Our current state of knowledge does not allow us to say how much work can be ascribed to wind directly, especially in complex settings.

Erosional Features

Unlike streams, wind can move sediment uphill as well as downhill. Where wind stress is focused on a spot in the landscape, it is possible, therefore, for wind to sculpt out a pit. In dune fields and sandy coastal areas, small cavities, called *deflation hollows*, are formed in this manner. They range from several to a hundred meters or so in diameter and may form in a matter of several days or seasons. Much larger depressions are found in the arid regions throughout the world, but only a few of these appear to be eroded by wind; the majority are believed to have been formed from faulting or from the collapse of limestone caverns. Broad, shallow depressions (called *pans*) are thought to be of aeolian origin. In the Lybian Desert of Egypt, an impressive group of such depressions shows strong evidence of wind erosion, although their formation was probably initiated by other processes. The largest of these, the Qattara Depression, covers around 15,000 km^2.

As we mentioned earlier, nonvegetated areas of sand- and silt-sized particles are most susceptible to wind erosion. Deserts contain the most expansive tracts of such areas, and, not surprisingly, this is where the largest surface areas of wind erosion are found. The sand eroded from such areas accumulates in large dune fields. Contrary to popular impressions, however, sand dunes are not the predominant landscape in the deserts; rather, rocky landscapes (called *hamadas* or *regs*), salt flats, and dry lake beds (called *playas*) together are more widespread. Nevertheless, there are huge tracts of sand dunes in most deserts.

Depositional Features

Sand dunes are the most prominent landforms created by wind action. The largest dune fields are found in the Middle East and North Africa and are so vast that they are called *sand seas*. One of the most spectacular sand seas, called the Rub'al Khali, or the Empty Quarter, is located in the southern part of the Arabian Peninsula, where dunes cover about 400,000 km^2 of land (Fig. 23.19). Another vast area of sand dunes is situated in the Lybian Desert, about 350 km south of the Mediterranean Sea where Ralph Bagnold roamed about and harassed the Italian Army.

Many large fields of sand dunes apparently function more or less as closed systems: once sand enters them, it does not leave. Although the dune fields may shift about considerably, especially with seasonal changes in wind direction, the net movement over many years is often insignificant. Thus, most of the work of wind is secondary; that is, it is devoted to erosion and transportation of sand that has already been eroded by wind many times before. In some instances, however, the dune fields may migrate great distances and end up at the sea or a river, where it loses part of its sand supply.

Settlements and agricultural land can be plagued by invasions of sand, and in some instances it is severe enough to destroy local land uses (Fig. 23.20). Some scholars point to dune encroachment as a contributing factor in the decline of some ancient cities in Asia and North Africa. Whether such behavior in sand dunes was related to a change in climate, such as increased aridity, we do not know.

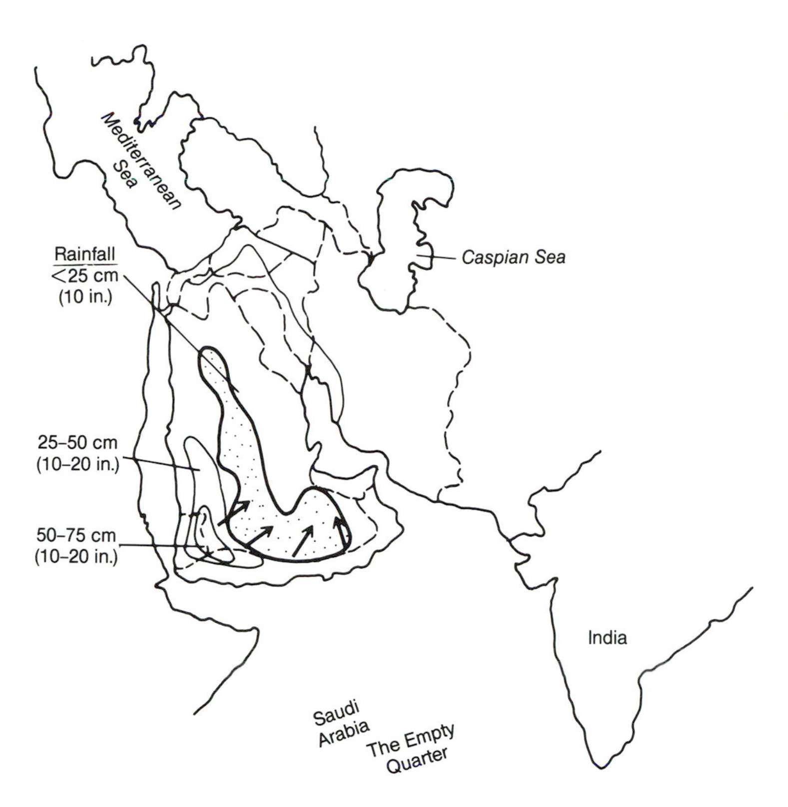

Figure 23.19 Arabian Peninsula and neighboring desert regions. Most of the area is rocky desert; the great sand dunes of the empty quarter occupy the very dry south-central part of the Arabian Peninsula.

Figure 23.20 After a long battle against enroaching sand dunes, which involved all sorts of attempts to stop the sand, including the erection of the strange-looking board barrier on the right, the village of Biggs, Oregon, was abandoned in 1899.

Desert Dunes

Desert sand dunes are free-moving heaps of sand that occur in an amazing variety of forms. They are generally considered to be among the most beautiful landforms on earth and are often used as a setting or backdrop in literature and photography. The movement of desert dunes is unimpaired by vegetation, and this fact, combined with their geographic prevalence and aesthetic magnificence, suggests that the term *classical* dunes may be appropriate for them.

Although the particular conditions that give rise to desert dune forms are not well understood, it appears that sand supply and wind directions are important factors. A *barchan* is a crescent-shaped dune whose long axis is transverse to the principal sand-moving wind. The points of the crescent, called the wings of the barchan, are curved downwind, partially enclosing the slip face (see Fig. 23.15). Barchans usually form where there is a limited supply of sand, reasonably level ground, and a fairly steady flow of wind from one direction.

As sand is added to a barchan, the wings may spawn smaller dunes, which migrate much more quickly than the parent dune does. The smaller barchans may, in turn, overtake and merge with a larger dune farther downwind. This seems to account for the tendency of barchans to migrate in schools, integrating and disintegrating into a variety of configurations as they move.

Where the sand supply is large, as in the sand seas, and formative winds come from more than one direction, sand dune forms are complex and not easy to classify. Among these are *seifs*, elongated dune masses that resemble great windrows of sand. Seifs lie parallel to the general direction of wind flow but are clearly a product of two or more winds blowing oblique to their main axis. These winds blow sand back and forth across the axis of a dune, giving rise to a faceted topography on the flanks. Most of the gigantic dunes in the sand seas are seifs (Fig. 23.21).

Figure 23.21 A large seif in the Western Sahara. The formative winds in a seif flow at an angle to the main axis of the sand dune. The segment of dune shown here is several kilometers long.

Coastal Dunes

Outside of the deserts, the only appreciable areas of active sand dunes are found in coastal dunes. Coastal dunes form wherever there is an ample supply of beach sand and strong onshore winds to blow it off the beach. In most cases, the beach must be broad and sufficiently agitated by wave action to keep it free of vegetation. Given these attributes, coastal sand dunes can form in any region, including the wet tropics, the humid midlatitudes, and the arctic.

Most coastal dunes form in association with blowouts in beach ridges or berms. *Blowouts* are open-ended deflation hollows at the end of which a sand deposit builds up. As the deposit grows, it begins to migrate inland in the form of a sand dune. Because the deflation hollow focuses wind and sand on the middle of the dune, the flanks do not move much. As a result, they are suitable habitats for many beach plants, especially dune grass, sea oats, and sand cherry, which are tolerant to sand deposition (see Fig. 12.9).

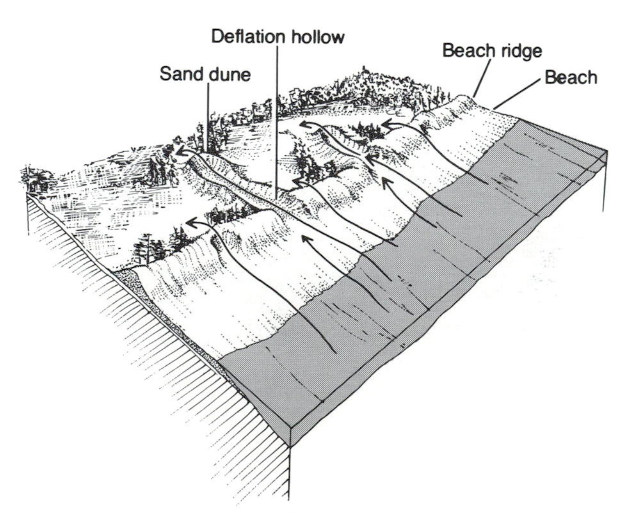

Figure 23.22 Setting and features associated with the formation of coastal sand dunes. The dune is the hairpin-shaped feature extending inland from the deflation hollow.

The plants help to stabilize the flanks, and in time the dune takes on a *parabolic* or *hairpin* shape as the mobile midsection moves farther inland (Fig. 23.22). The basic parabolic shape is sometimes difficult to identify in the field, however, because individual dunes often migrate onto one another, resulting in complex forms. Near the landward margin of dune penetration, where the coastal dunes often reach their greatest heights, they sometimes merge to form an irregular band of sand hills parallel to the coastline, which in some areas is called the *barrier dune*.

Coastal dunes are unlike desert dunes not only in their shapes, but also because they are at least partially covered by plants. The presence of vegetation substantially alters the dune environment by changing the patterns of airflow, reducing sand erosion, and stabilizing dune leeslopes. Thus, coastal dunes are not truly free-moving heaps of sand and in this regard are in a different class from desert dunes.

Loess Deposits

Loess is the only other major deposit produced by wind. Though less prominent than sand dunes topographically, loess is very widespread and important in the landscape for its role in soil formation (see Fig. 13.3). Most loess deposits appear to have been winnowed from Pleistocene glacial deposits, but the deserts also appear to have yielded appreciable amounts of the material.

Loess is principally silt, and when airborne it is carried in suspension. Observations indicate that loess can be dispersed high into the atmosphere and carried great distances by wind. This accounts for not only the widespread nature of loess deposits, but also the fact that, unlike dune deposits, loess deposits tend to blanket the landscape, covering hills and valleys alike. Once deposited, however, loess is subject to erosion, especially by runoff, which may fragment the original distribution pattern.

Loess is the parent material for large areas of prairie and chernozem soils (or molosols) and contributes significantly to the agricultural value of these soils. Soils formed in loess are friable, are often rich in minerals such as calcium carbonate, and possess good capillarity (Fig. 23.23).

Figure 23.3 Loess deposits in the Mississippi River Valley. These deposits were laid down during the Ice Age.

Summary

Airflow over the ground is mainly turbulent. The roughness length defines a zone of calm air immediately over the surface, which increases in thickness with the height of vegetation and other surface materials. Roughness length, particle size and cohesiveness, vegetation, and wind velocity are the main controls on wind erosion. The power of wind increases with the cube of velocity: therefore, fast winds of short duration may have a much greater capacity to do geomorphic work than do slow winds of long duration.

Sediment transport by wind occurs in three modes: traction, saltation, and suspension. Sand is moved mainly by saltation, whereas silt and clay are moved in suspension. Sand dunes and loess are the principal wind deposits. The largest accumulations of sand dunes are found in the deserts, where they may form great sand seas; outside arid regions, sand dunes are most prevalent in coastal zones where vegetation often plays an important role in dune formation.

Concepts and Terms for Review

boundary layer
laminar sublayer
roughness length
velocity-power relationship
geographic characteristics of wind
  regional patterns
  variations with topography
erosion and sediment transport
  threshold velocity

traction
saltation
suspension
sand and loess
sand dune
landforms produced by wind
deflation hollows and pans
desert lag
sand sea
barchan
seif
parabolic dune
barrier dune
loess deposits

Sources and References

Bagnold, R.A. (1973) *Blown Sand Desert Dunes.* London: Chapman and Hall.

Bowen, A.J., and Lindley, D. (1977) "A Wind-Tunnel Investigation of the Wind Speed and Turbulence Characteristics Close to the Ground Over Various Escarpment Shapes." *Boundary-Layer Meteorology* 12: pp. 259–771.

Chepil, W.S., and Woodruff, N.D. (1963) "The Physics of Wind Erosion and Its Control." *Advances in Agronomy* 15: pp. 211–302.

Cooke, R.U., and Warren, A. (1973) *Geomorphology in Deserts.* London: Batford.

Hack, J.T. (1942) "Dunes of the Western Navajo Country." *Geographical Review* 31: pp. 240–263.

Haff, R.K. (1986) "Booming Dunes." *American Scientist*, 74:4.

McKee, E.D., ed. (1979) "A Study of Global Sand Seas." *U.S. Geological Survey Professional Paper* 1052.

Olson, J.S. (1958) "Lake Michigan Dune Development 1: Wind-Velocity Profiles." *Journal of Geology:* pp. 254–262.

Prospero, J.M., et. al. (1981) "Atmospheric Transport of Soil Dust from Africa to South America." *Nature.*

Chapter 24

The Physiography of the United States and Canada

Physical geography's principal aim is to understand the nature of the earth's surface by examining the processes that shape it. One of our purposes is to account for the differences in the physical character of the landscape from place to place, what geographers traditionally call "physiography." What are the different types of physiography in the United States and Canada? What landscape features exert the greatest influence on the physiographic character of different regions? Is there a relationship between physiography and land use?

Since the nineteenth century, when physical geography emerged as a field of study in American, Canadian, and European colleges and universities, one of its chief accomplishments has been to describe and map landscape features over broad regions. Drawing on the results of geographic exploration and scientific investigations, physical geographers produced maps and reports on the distributions of landforms, rocks, mineral resources, water features, biota, soils, and climate.

For certain regions, this information was combined to form large composite maps. By virtue of the associations that appeared mainly among landforms, soils, climate, and vegetation, natural regions were identified. Known as *physiographic regions*, these became a major focus of study in physical geography in the first half of the twentieth century. Today physiographic regions are often used as a framework for regional geography, environmental science, and regional planning programs.

While physiographic regions are defined by the composite patterns of landscape features as they appear on maps, it is important that we do not lose sight of the fact that the physiography of any region represents the product of a host of processes that operate at or near the earth's surface. As we noted in earlier chapters, these processes are arranged in various systems: drainage systems, climatic systems, tectonics systems, ecosystems, and land use systems that overlap and interact at different geographic scales. These systems have developed over various periods of time in North America ranging from hundreds of millions of years in the case of the tectonic systems that built the Appalachians and the Rockies to only several centuries in the case of agricultural and urban land use. The physiographic patterns and features we see today

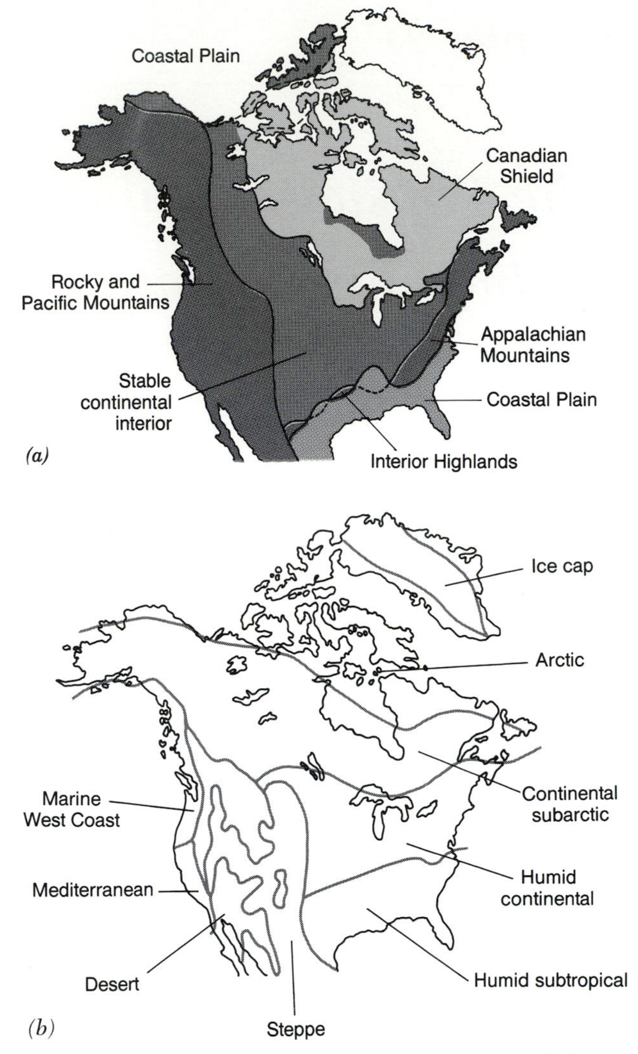

Figure 24.1 (*a*) The major structural divisions and (*b*) climates of the United States and Canada.

represent an evolving picture—at this moment a mere slice of geography at the intersection of the time lines of many forces and systems.

Generally regional geology provides a useful framework for describing the physiography of North America—in particular, some combination of the gross geologic structure, surface and near-surface rock types and deposits coupled with the geomorphic trends of the different parts of the continent (Fig. 24.1*a*). In addition, climate is also a significant influence on physiography because of its relationship to vegetation, soils, runoff, permafrost, and water resources (Fig. 24.1*b*).

Our discussion will be limited to the combined areas of the United States and Canada. Ten major physiographic regions are traditionally defined for the United States and Canada, and they are broken down into *provinces* (Table 24.1).

TABLE 24.1 PHYSIOGRAPHIC REGIONS AND PROVINCES OF THE UNITED STATES AND CANADA

Region	Provinces
Canadian Shield	Superior Uplands
	Laurentian Highlands
	Laborador Highlands
	Hudson Platform
Appalachian Mountains	Blue Ridge
	Piedmont
	Ridge and Valley
	Appalachian Plateaus
	Northern Appalachians
Interior Highlands	Ozark Plateaus
	Ouachita Mountains
Atlantic Coastal Plain	Outer Coastal Plain
	Inner Coastal Plain
	Mississippi Embayment
Interior Plains	Central Lowlands
	Great Plains
	St. Lawrence Lowlands
	Arctic Lowlands
Rocky Mountain Region	Canadian Rockies
	Northern Rockies
	Central Rockies
	Southern Rockies
Intermontane Region	Colorado Plateau
	Columbia Plateau
	Basin and Range
Pacific Mountain System	Alaska Range
	Coast Mountains
	Frazier Plateau
	Cascade Mountains
	Coast Ranges
	Sierra Nevada
	Central Lowlands
	Puget Sound-Willamette Lowlands
Alaska-Yukon Region	Brooks Range
	Yukon Basin
Arctic Coastal Plain	North Slope
	MacKenzie Delta

THE CANADIAN SHIELD AND THE EASTERN HIGHLANDS

The Canadian Shield

The Canadian Shield is a large physiographic region centered on Hudson Bay (Fig. 24.2*a*). It is composed of the oldest rocks in North America (older than a billion years) and is the geologic core of the North American continent. Structurally, the Canadian Shield is extraordinarily complex, with intersecting belts of highly deformed rocks throughout. These rocks have been subjected to not one or two, but many, early episodes of deformation; thus, most of the rocks are metamorphosed, tightly consolidated, and diverse in mineral composition.

Most of the Canadian Shield has been relatively stable (free of mountain-building activity) for the past 500 million years or so. During that time, erosional

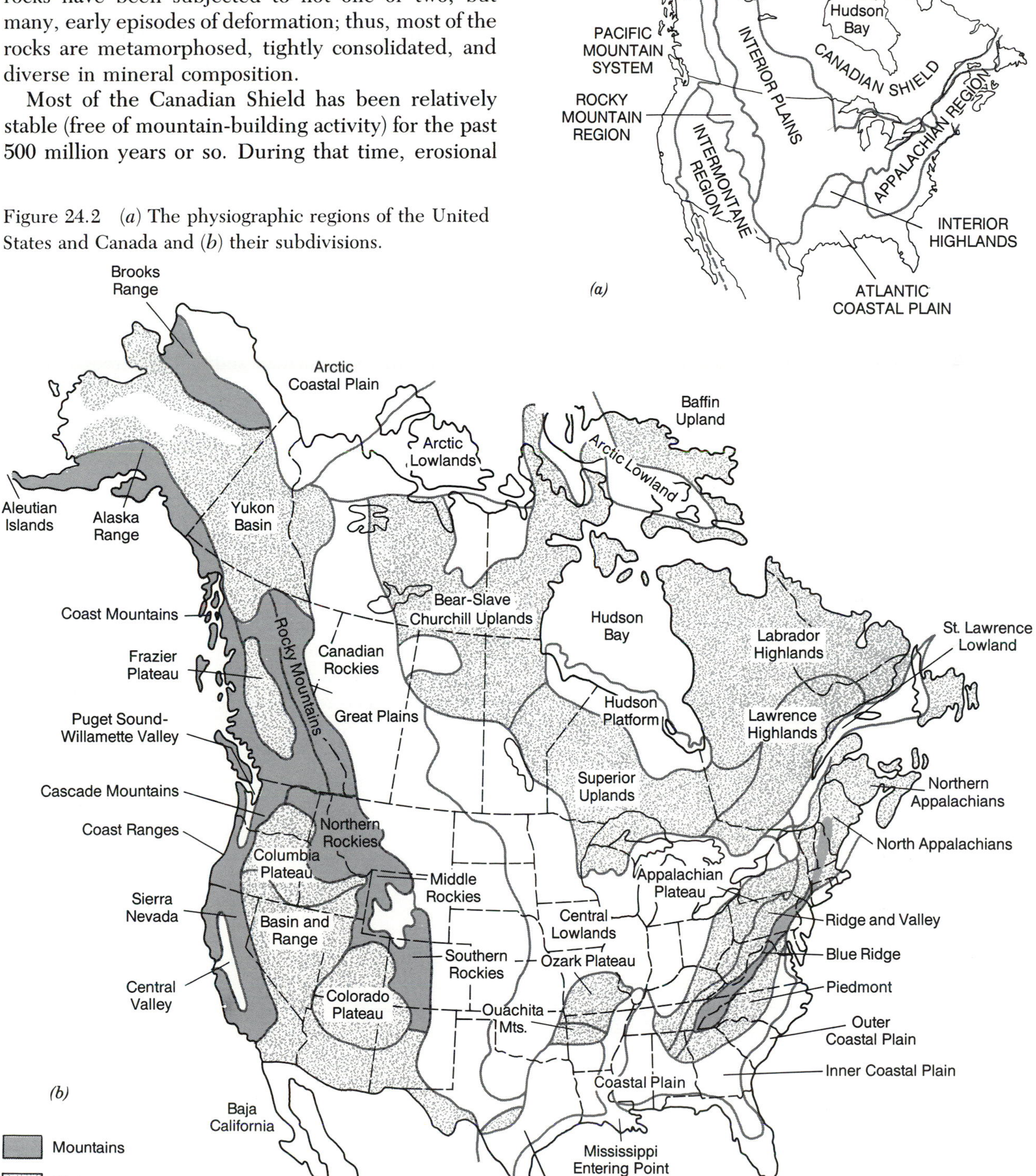

Figure 24.2 (*a*) The physiographic regions of the United States and Canada and (*b*) their subdivisions.

forces have worn the rocks down to irregular (hilly) plateau surfaces of 300 to 500 m elevation. There are five major provinces of such surfaces: the Superior Uplands, the Laurentian Highlands, the Laborador Highlands, the Baffin Upland, and the Churchill Uplands.

In addition, large sections of the shield lie at lower elevations and are covered with sedimentary rocks. The largest of these areas—though it does not belong to the Canadian Shield region—is the broad interior of the continent stretching from the Midwest through the Canadian Interior Plains northward to the Arctic Lowlands of northern Canada. Within the shield *itself* are three major sections of sedimentary rocks. The largest is found along Hudson Bay and is called the Hudson Platform. The other two lie in the Churchill Uplands (Fig. 24.2*b*).

A significant recent chapter in the long and complex physiographic development of the Canadian Shield was the glaciation of North America (see Fig. 22.1). Great masses of glacial ice formed in the Hudson Bay and Laborador Highlands regions and, in at least four different episodes in the past 1 to 2 million years, spread over all or most of the Canadian Shield. (The last ice sheet melted from the shield only 6000 to 8000 years ago.) The advancing ice sheets scoured the shield's surface, removing soil cover and rasping basins into the less resistant rocks. As a result of this action, much of the shield was left with an irregular and generally light soil cover interspersed with swampy low areas. Add this to the already diverse surface geology, and it is easy to see why the Canadian Shield is one of the most complex landscapes in North America (see Fig. 22.22*a*).

The Canadian Shield lies principally in the continental subarctic climate zone, which is characterized by fiercely cold winters and short cool summers (Fig. 24.1*b*). A great proportion of the shield is underlain by permafrost, though the coverage is discontinuous. The vegetative cover is characterized as mainly northern forests (spruce, fir, and birch, for example) mixed with wetlands in the more rugged sections and tundra in the northern zone (see Fig. 10.6). Throughout much of the shield are extensive outcrops of barren rock. Drainage patterns are very irregular; lakes and wetlands are abundant, and settlement is light. Land use is sparse, and where settlements are found they are usually related to an extractive economic activity, usually mining, forestry, or fishing.

The Appalachian Region

The Appalachian Region is an old mountain terrain that formed on the southeastern side of the Canadian Shield several hundred million years ago (Fig. 24.2*a*). Evidence indicates that the origin of the Appalachians was associated with an episode of plate tectonics prior to the breakup of Pangaea. Typical of mountains on the margins of continental platforms, they are composed of highly deformed sedimentary rocks, metamorphosed sedimentary rocks, and some intrusive igneous rocks (see Color Plate 10).

Although the Appalachians date from around the same time as the Rocky Mountains, they have developed appreciably different terrains. The Appalachians are lower, less angular, and less active tectonically. Whether these differences have always existed, or whether they are due to a long period of inactivity in the Appalachians during which erosional forces have worn them down, we cannot say. In any case, the Appalachians are characterized by rounded landforms that are generally forest-covered in their natural state and rarely exceed 2000 m in elevation (see Fig. 16.25).

The Appalachian Region stretches from northern Alabama in the American South to Newfoundland in the Canadian Maritime Provinces. It is subdivided into five provinces on the basis of landforms (Fig. 24.2*b*). The highest and smallest province (in area) is the Blue Ridge. It is composed of folded metamorphic rocks that form the narrow, central axis to the Ap-

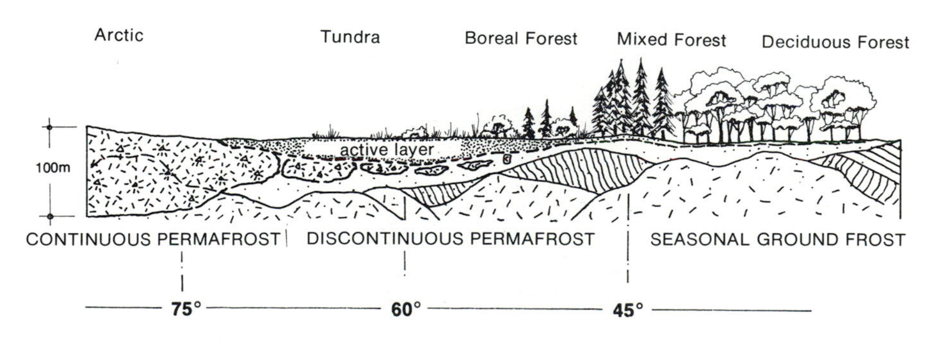

Ground frost zones between 45° and 80°N latitude, and the vegetation associated with each in the Canadian Shield.

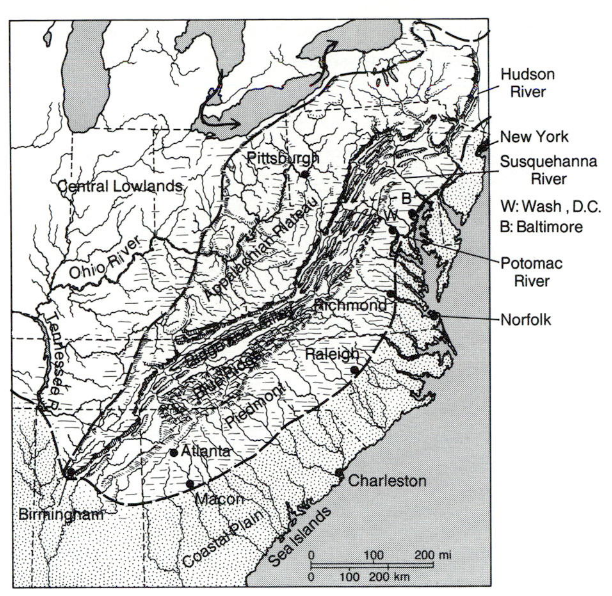

Figure 24.3 The Southern Appalachians, its provinces, and bordering regions.

palachians in the southern part of the region. The southern Appalachians lie south of the Hudson River Valley (in New York State).

East of the Blue Ridge is the Piedmont Province of the Appalachians (Fig. 24.3). Like the Blue Ridge, this province is composed mainly of metamorphic rocks. From the Blue Ridge, the Piedmont slopes gradually eastward until it disappears under the sedimentary rocks of the Coastal Plain, which borders the Atlantic Ocean. The Piedmont lies principally in the humid subtropical climate of the American South (Fig. 24.1*b*). Soils tend to be heavily leached and generally poor in nutrients (they belong to the Ultisol soil order) (see Fig. 14.7). Early in the development of Southern agriculture, the Piedmont was a favorite area for cotton and tobacco farming, but these activities waned as the soils declined and cotton farming shifted to the Mississippi Valley and later into Texas.

Drainage in the Piedmont is toward the Atlantic and Gulf of Mexico. Stream and river gradients are relatively steep, but where they cross onto the Coastal Plain, they decline somewhat. For early settlers moving up rivers from the Atlantic, the first fast water (rapids) they would encounter started with the Piedmont. The eastern border of the Piedmont became known as the Fall Line, and on some rivers it became a place of settlement: Richmond, Virginia, Raleigh, North Carolina, and Macon, Georgia, are Fall Line cities (Fig. 24.3).

West of the Blue Ridge, and stretching from middle Pennsylvania to middle Alabama, is the Ridge and Valley Province. One of the most distinctive terrains in North America, the Ridge and Valley is characterized structurally by folded and faulted sedimentary rocks that have been eroded into long ridges separated by equally long valleys. Both ridges and valleys may be continuous for several hundred kilometers, broken only by narrow gaps that certain rivers appear to have cut through ridges (see Fig. 16.25).

Drainage lines generally follow the trend of the landforms, with trunk streams flowing along the valley floors and their tributaries draining the adjacent ridge slopes. But there are notable exceptions, for some of the large rivers in the north, specifically the Susquehanna and the Potomac, drain across the grain of the ridges and flow into the Atlantic. In the southern part of the Ridge and Valley, most streams drain into the Cumberland and Tennessee rivers, which are part of the Mississippi System. Settlements and farms in the Ridge and Valley are concentrated in the valleys; the ridges have been left mostly in forest, being too steep for much else (Fig. 24.3).

West of the Ridge and Valley Province is a section of elevated sedimentary rocks into which rivers have cut deep valleys. This province is referred to as the Appalachian Plateaus (Allegheny Plateau in the north and Cumberland Plateau in the south). The rocks are flat-lying for the most part, and besides the deep valleys and flat-topped uplands, this section is famous for its great coal deposits (Fig. 24.3).

The Northern Appalachians are made up of several separate mountain ranges composed principally of crystalline rocks. The most prominent of these are the Green Mountains and White Mountains of Vermont, New Hampshire, and Maine, the Notre Dame Mountains of the Gaspé Peninsula of Quebec, and the Long Range Mountains of northern Newfoundland (Fig. 24.4). (The Adirondack Mountains of northern New York appear to be another prominent range of the Northern Appalachians, but they are a southern extension of the Canadian Shield.) The whole of the Northern Appalachians was glaciated, leaving it with abundant rock exposures and a generally thin cover of glacial deposits. These characteristics combined with northern conifer forests and abundant lakes and wetlands give much of the Northern Appalachians a character similar to the uplands of the Canadian Shield. The lakes, by the way, have been the focus of the acid rain phenomenon attributed mainly to the sulfur dioxide emissions from power plants and industrial sources in the Midwest. In addition, this section of the Appalachians borders on the Atlantic Ocean, producing an especially rugged coastline (see Fig. 21.20).

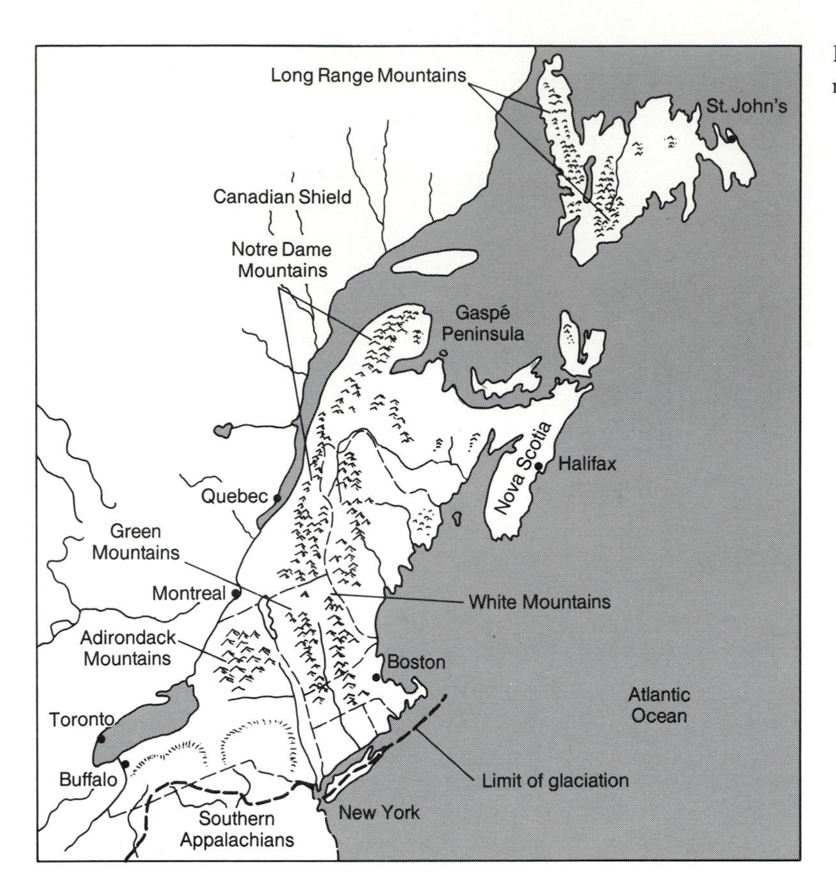

Figure 24.4 The Northern Appalachians and neighboring regions.

The Interior Highlands

West of the Cumberland Plateau, in southern Missouri and northern Arkansas, lies a small region of low mountainous terrain that closely resembles the Appalachians. This region is called the Interior Highlands, and it is made up of two main provinces: the Ozark Plateaus, whose landforms are similar to the Appalachian Plateau; and the Ouachita Mountains, whose landforms are very similar to those of the Ridge and Valley section except the grain of the terrain is east-west trending (see Fig. 16.21). The highest elevation in the Ozarks, as this area of plateaus is commonly called, lies around 1000 m; those in the Ouachita Mountains lie close to 1500 m. Most ridges in the Ouachita Mountains are forested and too steep for settlement; in the Ozarks, however, ridges are often flat-topped and cleared for farming.

THE PLAINS AND LOWLANDS

The Atlantic Coastal Plain

Seaward of the Piedmont Province of the Appalachians and continuing in a broad belt around the Gulf of Mexico is the Coastal Plain (Fig. 24.2*b*). Geologically, this region is the landward extension of the continental shelf; consistent with the submarine part of the continental shelf, the Coastal Plain is composed of sedimentary rocks that dip gently seaward (see Color Plate 10). Where these rocks outcrop within the Coastal Plain, they are often marked by a band of hilly topography and soils that are compositionally similar to the bedrock as is illustrated in Figure 15.18. Overall, however, the topographic relief of the Coastal Plain is very modest, and the highest elevations (mainly along the inner edge) reach only 100 m or so above sea level.

Near the sea, in the province called the Outer Coastal Plain, the land is generally low and wet. Swamps, lagoons, and islands are abundant, and these are subject to periodic overwash by storm waves and hurricane surges (see Fig. 6.20 and Fig. 21.19*b*). Inland, on the Inner Coastal Plain, the land is generally higher and better drained, but it is not without lowlands. Large lowlands are found along all the major river valleys, and most contain large areas of wetland. The largest river lowland is the Mississippi Embayment, which stretches from the Mississippi Delta northward to the southern tip of Illinois (Fig. 24.5). To describe this area as a river lowland is somewhat misleading because it comprises many river lowlands, some modern (active) and some ancient (inactive).

Flooding is frequent and widespread in the Mississippi Embayment and the other large river lowlands. The damage it wreaks on property and life is enormous and has increased steadily throughout the twentieth century (see Fig. 18.7). The cause of the problem is twofold: (1) increased development in high-risk areas with the growth and spread of residential population; and (2) engineering (structural) changes made in river channels, as is illustrated in Note 18, which increase the levels of floodwaters for many large flows.

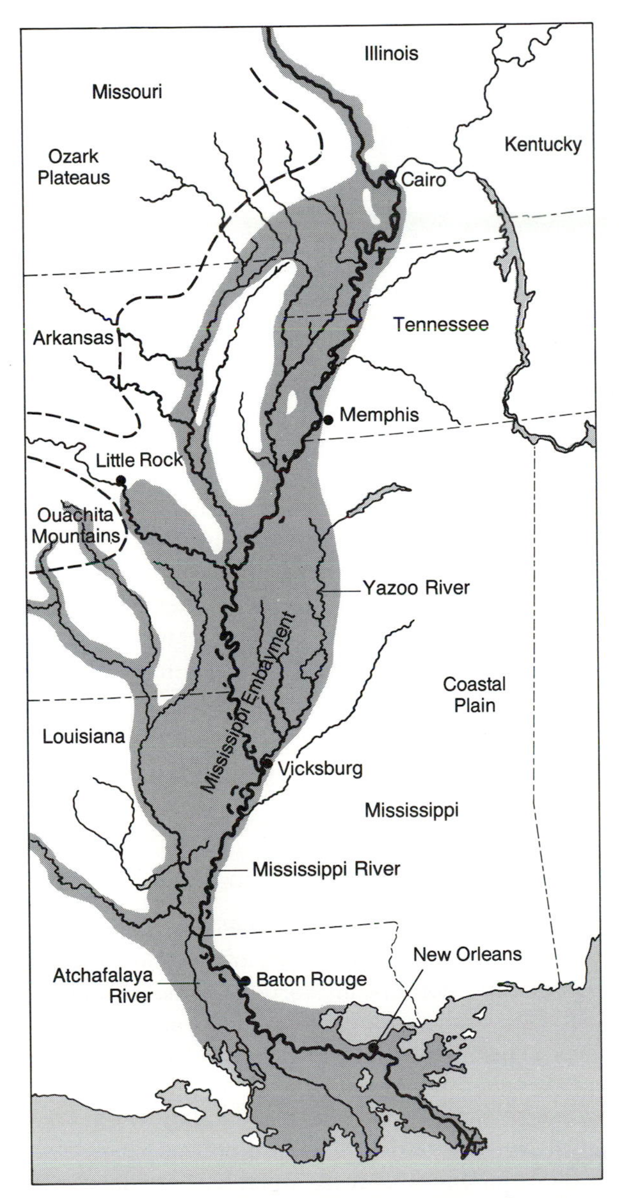

Figure 24.5 The Mississippi Embayment and connecting lowlands.

The Coastal Plain lies almost entirely within the humid subtropical climatic zone (see Fig. 24.1*b*). Soils tend to vary depending on the underlying bedrock, nearness to the coast, and river valley location. Vegetation patterns tend to follow the patterns of soil and drainage, with pines generally on the sandy higher ground and trees such as tupelo, gum, and bald cypress in the wet lowlands. Land use is diversified, ranging from recreation and specialized agriculture (vegetable, fruit, etc.) along the coast to cotton farming, forestry, and pulp and paper manufacturing inland.

The Interior Plains

Between the Appalachians on the southwest, the Coastal Plain on the south, the Rocky Mountains on the west, and the Canadian Shield on the northeast lie the Interior Plains (Fig. 24.2*a*). This region is made up of two large provinces; the Central Lowlands and the Great Plains, and one small province, the St. Lawrence Lowlands. All three are underlain with sedimentary rocks that are covered with diverse deposits of varying thicknesses. In the Central Lowlands, north of the valleys of the Ohio and Missouri rivers, these deposits are mainly glacial, but in Illinois, Iowa, northern Missouri, Kansas, and Nebraska loess covers glacial deposits and is the dominant surface material (see Fig. 13.3).

The Interior Plains are drained by three major watersheds: the Mississippi, which covers most of the region; the St. Lawrence, which drains the Great Lakes area; and the Nelson, Churchill, and Mackenzie drainage basins, which drain the Canadian Plains and small portions of North Dakota and Minnesota (Fig. 18.1). Water is generally abundant in the Central Lowlands, especially the Great Lakes area, but it declines westward into the Great Plains (see Color Plate 12). The Great Plains range from subhumid to semiarid conditions and since 1950 modern grain farming there has grown more and more dependent on irrigation (see Color Plate 15).

Soils in the Central Lowlands are extremely diverse, especially in the Great Lakes states and southern Ontario, owing to the extraordinary mix of glacial deposits left there during the last glaciation of the region, 10,000 to 20,000 years ago (see Fig. 22.20). In the northern part of this region, in the newer glacial terrain, lakes and wetlands are abundant. Westward,

Figure 24.6 A characteristic landscape in the Great Plains of Nebraska.

this diverse terrain gives way to a more uniform terrain, the prairies, and farther west, to the Great Plains (Fig. 24.6). In the prairies and plains, the terrain tends to fall into two classes: river lowlands (for example, those of the Illinois, Mississippi, Missouri, and Iowa rivers), and broad, level, or gently rolling uplands between the river valleys, which make up the bulk of the landscape.

The Interior Plains is perhaps the richest agricultural region in the Western world. The Midwest (Ohio, Michigan, Indiana, Illinois, Wisconsin, northern Missouri, Iowa, and Minnesota) and southern Ontario are dominated by corn and livestock farming, whereas the dryer western part of the Central Lowlands and the much dryer Great Plains are dominated by wheat farming—spring wheat in the north and winter wheat in the south. In Canada, the northern limits of the wheat belt extend to southcentral Saskatchewan, southern Alberta, and southwestern Manitoba (see Note 14). Beyond this border, the grasslands give way to northern forest which extends to the Arctic Circle near Great Bear Lake. Above the Arctic Circle, the Interior Plains extend northeastward across the large islands of the Northwest Territories, an area of tundra, permanent snowfields, and glaciers.

Weather is an especially integral part of life and land use in the Interior Plains. The Great Plains, Midwest, and adjacent areas of Canada are characterized not only by strong seasonal contrasts but also by extreme weather events in most seasons. Situated as these areas are in a broad corridor stretching from the Arctic to the Gulf of Mexico, the region is prone to incursions of arctic and polar air masses, the development of fierce blizzards and tornadoes, and even the invasion of an occasional hurricane from the Gulf Coast (see Fig. 6.9).

THE WESTERN MOUNTAINS AND PLATEAUS

The Rocky Mountain Region

The western border of the Great Plains is formed by the Rocky Mountain front, one of the most distinct physiographic borders in North America. From the Great Plains (at an elevation of 1500 m) the terrain abruptly rises 1,000 to 2,000 m, and we enter the rugged region of the Rockies. This region stretches, in a band ranging from 150 to 650 km in width, from central New Mexico to near the northern border of British Columbia. It is made up of four provinces: the Southern, Middle, and Northern provinces in the United States and the Canadian Province in Canada (Fig. 24.2*b*).

The Rocky Mountains are considered to be relatively young mountains because they are still active—though many of the rocks are of ages comparable to those in the Appalachians. Geologically, they are most diverse in the American provinces, where volcanic, metamorphic, and sedimentary rocks are all very prominent (see Color Plate 10). The Canadian section of the Rockies, though extremely rugged, tends to be more uniform in rock types, being composed mainly of sedimentary rocks and their metamorphic counterparts (see Fig. 16.25). The highest terrain in the Rockies, which is around 4000 m to 4300 m (13,000–14,000 ft.) elevation, is found in Colorado, Wyoming, and along the British Columbia/Saskatchewan border. The largest areas of low terrain are the Wyoming Basin (1500 m to 2000 m elevation), a structural basin composed of sedimentary rocks, and the Rocky Mountain Trench, a remarkably long, narrow valley stretching for 800 km (500 mi.) along the western edge of the Canadian Province (Fig. 24.7).

Americans have traditionally recognized the Rocky Mountains as the "continental divide"—the high ground that separates drainage between the Pacific and Atlantic watersheds. To the Atlantic go the Missouri, Platte, Arkansas, Rio Grande, and many other large rivers that rise in the Rockies; to the Pacific go only two large rivers that rise in the Rockies: the Colorado and the Columbia. In the northern part of the Canadian Province, by contrast, the Rockies do not form the continental divide. The Peace River, which drains into Hudson Bay, and the MacKenzie River, which drains into the Arctic Ocean, both rise west of the Rockies in British Columbia.

Owing to their diverse geology and topography, the Rocky Mountains are highly varied in soils, vegetation, and climate. The American Rockies lie generally in an arid/semiarid zone, but moisture conditions are often quite different from range to range within each province. West slopes usually receive greater precipitation than east slopes because the Rockies produce an orographic effect as the westerly flow of wind and air masses passes over them. Moisture conditions also improve with elevation because precipitation tends to be greater higher up and evapotranspiration rates are decidedly lower at cooler temperatures (see Fig. 12.7*a*). This helps explain the concentration of heavy forests (pines, firs, spruce, for example) at elevations between 2300 m (7500 ft.) and 3000 m (10,000 ft.) and grasslands at lower elevations. In the Canadian Rockies, forest elevations are lower, and glaciers and snowfields are more abundant at higher elevations owing largely to the higher (subarctic) latitude of this area.

Settlement in the Rocky Mountains is light; there are no large cities within the region. (Denver, Salt Lake City, and Calgary lie on the borders of the region.) Most land is publicly owned, and in both Canada and the United States, extensive tracts have been set aside as national parks, national monuments, and national forests. Mining for metals such as gold, silver, and copper is a traditional activity in the Rocky Mountains; more recently, coal, petroleum, and natural gas extraction has increased substantially. Crop agriculture is limited to selected small areas, and grazing is widespread on both public and privately owned range lands.

The Intermontane Region

Between the southern province of the Rockies and Pacific Mountain Region to the west lies an elevated

Figure 24.7 A view of the Rocky Mountain Trench looking southward.

region of plateaus and widely spaced mountain ranges. There are two major plateaus in this region: the Colorado, which lies at elevations around 2000 m and is composed of sedimentary rocks; and the Columbia, which lies at elevations around 1500 m and is composed of basaltic rock. The remainder of the region, about half its area, is the Basin and Range Province (Fig. 24.2*b*).

The Basin and Range (also called the Great Basin) is characterized by disconnected, north-south trending mountain ranges formed by faulting and tilting of large blocks of rock (Fig. 24.8). The basins between the ranges are filled with thick deposits of sediment (hundreds of meters deep) eroded from the adjacent mountains. The heavy accumulation of sediment is partially attributed to the region's dry climate, which generates insufficient runoff to support continuous-flowing streams to transport the sediment away. Indeed, the Basin and Range is the only large area of North America with internal drainage, that is, where streams dry up before reaching the sea (see the Humboldt River in Fig. 24.8).

In the past several decades, population has grown tremendously in the southern part of the Intermontane Region, principally in Arizona. This has resulted in both urban and agricultural development, which has placed great demands on limited water supplies in this arid region. Groundwater aquifers are being

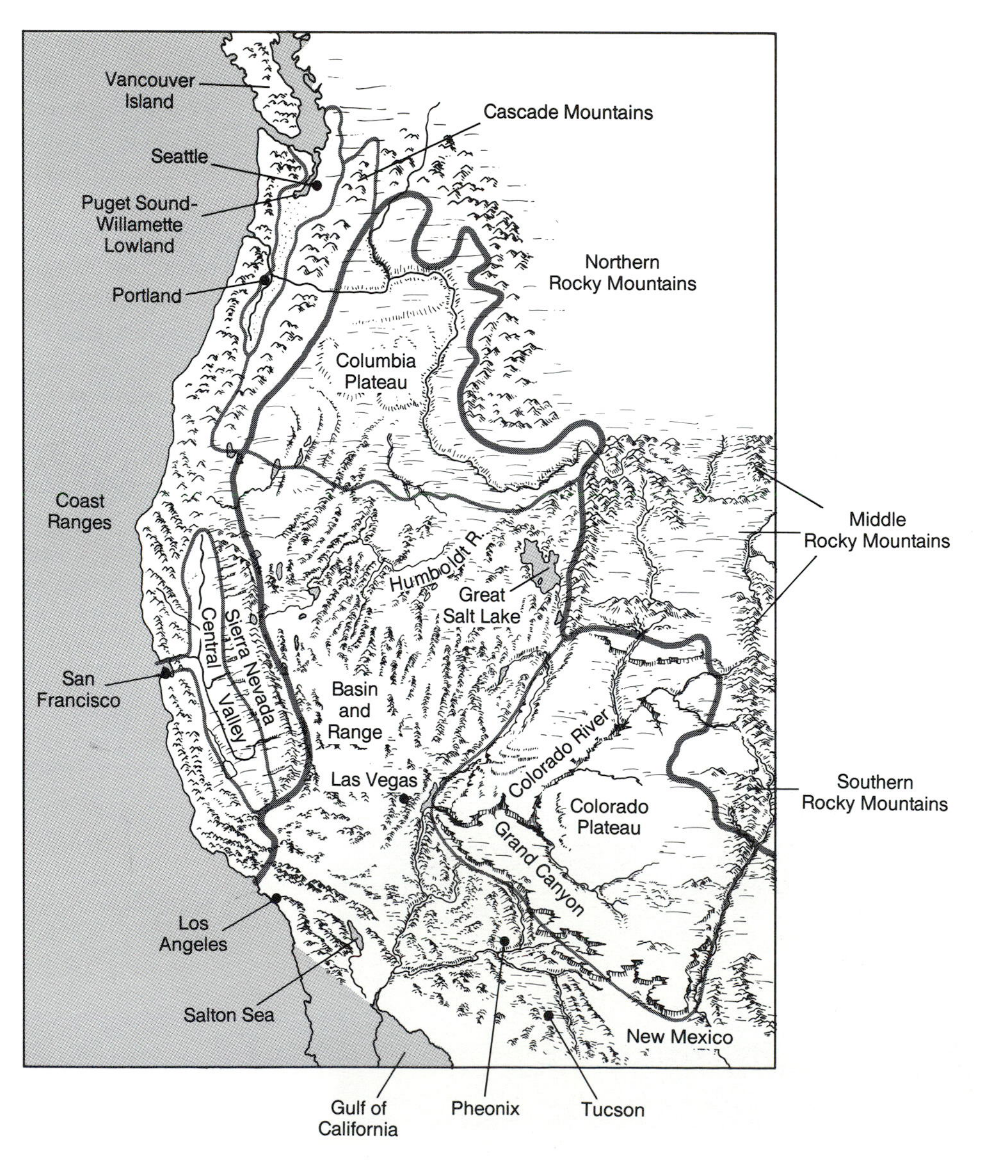

Figure 24.8 The Intermontane Region, its provinces, and the southern part of the Pacific Mountain System.

rapidly depleted, and the region is sure to face increasingly serious water-supply problems in the next decade (see Color Plate 12).

The Pacific Mountain Region

Along the Pacific margin of North America is a complex region composed of many mountain ranges (Fig. 24.2). This region, referred to as the Pacific Mountain Region, contains just about every geologic structure and mountain type imaginable, and by virtue of its location on or near the contact between the Pacific and North American plates, it is the most tectonically active part of the continent. Earthquakes are commonplace throughout the region, and volcanism is active in the Cascade Range (which contains Mount St. Helens), in the Alaskan Peninsula, and in the Aleutian Islands.

Some of the more prominent mountain ranges include the Coast Mountains of British Columbia and Alaska, the Sierra Nevada of California, the Cascades of Washington and Oregon, the Coast Ranges of California and Oregon, and the Alaskan Range of southern Alaska. The Coast Mountains and the Sierra Nevada are mostly great masses of granitic rock uplifted by faulting. The Coast Ranges, on the other hand, are mainly sedimentary rocks wrenched into a series of ridges and valleys by movements associated with the great fault system paralleling the plate border (Fig. 24.8). The Alaska Range, the northernmost part of the region, is a complex body of mountains built by folding, faulting, and volcanic activity.

Two major lowlands are found in the Pacific Mountain Region. The larger of the two is in the Central Valley of California, which lies between the Sierra Nevada and the Coast Ranges (see Fig. 16.25). The other is the Puget Sound-Willamette Valley lowland which lies between the Coast Ranges and the Cascades in Oregon, Washington, and British Columbia (see Color Plate 13). In addition, two large interior plateaus, the Frazier of British Columbia and the Yukon Basin of the Yukon Territory and central Alaska, take up a large share of the interior of the region in Canada (Fig. 24.2*b*).

The climate, vegetation, and soils of the Pacific Mountain Region are as diverse as its geology. In Alaska and Canada, heavy precipitation in the Coast Mountains—the annual average exceeds 250 cm (100 in.)—nourishes large glaciers at elevations above 2000 m and great fir forests at lower elevations. These forests extend down the coast to northern California, where redwood forests become the dominant coastal forests. Precipitation declines sharply farther down the California coast, and between San Francisco and Los Angeles, the redwoods give way to chapparal. Still farther south, in extreme southern California and the Baja California of Mexico, the chapparal gives way to grass and shrub desert.

In California, the combination of warm dry lowlands, seasonal rainfall, water supplies from mountain streams, and subtropical thermal conditions has produced one of the most diverse and productive agricultural regions in the world. Grains, vegetables, grapes, and fruits are grown extensively in the southern two-thirds of the state, mostly with the aid of irrigation. Agriculture is prominent in only one other

Volcanic mountains on the Alaska Peninsula.

area of the Pacific Mountain Region—that which is composed of the valleys of the Columbia and the Snake rivers and the Puget Sound-Willamette lowland—but here the total output is much lower than that of the California area. The forest industry flourishes in the area between San Francisco and the Alaska Panhandle; in this area, as in the Rocky Mountains, public-owned forest and parks occupy large tracts of land in both the Canadian and the U.S. portions. Population is concentrated in the coastal border (including the two large lowlands) between southwestern British Columbia and southwestern California. More than 25 million people inhabit this area, and the population is steadily increasing.

THE NORTHLANDS

The Alaskan-Yukon Region

South of Alaska's North Slope lies the Brooks Range (and its eastern limb in Canada, the British Mountains), a low, east-west trending mountain range. South of the Brooks Range and occupying the large interior of Alaska and the adjacent portion of the Yukon Territory is the Yukon Basin (Fig. 24.9). This large basin is drained by the Yukon River, which flows westward into the Bering Sea. Fairbanks, the principal city of central Alaska, is located near the center of the Yukon Basin.

South of the Yukon Basin are the northern ranges of the Pacific Mountain Region. These are the highest and most active mountains in North America: Mount McKinley, Alaska, is 6194 m (20,320 ft.) high, and Mount Logan, Yukon Territory, is 6000 m (19,850 ft.) high; volcanic activity in the Aleutian chain is virtually a continuous occurrence on one island or another.

Because of the marine influence of the Pacific Ocean, the southern coast of Alaska has a very moderate climate for its latitude (55 degrees to 62 degrees north); it is mainly of the cool marine west coast and arctic marine type. Inland the climate grows colder rapidly. The Yukon Territory and Interior Alaska are classed mainly as continental subarctic, and the Brooks Range and the Arctic Coastal Plain are classed as tundra climate. Permafrost underlies most of Alaska and the Yukon Territory; in the Yukon Basin and most of the Yukon Territory, its distribution is discontinuous (see Fig. 4.8).

Figure 24.9 The Yukon Basin north of Fairbanks showing a section of the Trans-Alaska Pipeline.

Figure 24.10 Adaptation to permafrost; utility lines (water, heat, sewer) are placed above ground to avoid disturbing the permafrost.

The Arctic Coastal Plain

On the northern fringe of North America is a coastal plain similar in topography to the Coastal Plain of southern United States (Fig. 24.2). The Arctic Coastal Plain, however, is narrower and more desolate, being extremely cold and locked in by sea ice most of the year. The North Slope of Alaska, now famous for its oil reserves, belongs to the Arctic Coastal Plain; so does the MacKenzie River Delta (just east of the Canada-Alaskan border) and the plain that fringes the northern islands of Canada. Virtually the entire Arctic Coastal Plain is underlain by permafrost, which in some areas extends offshore under the shallow waters of the Arctic Ocean (Fig. 24.10).

Summary

The physiography of any region is the product of many geographic systems that operate at or near the earth's surface. For North America, the regional geology sets the basic framework for the physiography of the continent. Each physiographic region has a particular combination of structures, rock types, landforms, soils, drainage, vegetation, and often land use. The Canadian Shield is the old, diverse inner region of the continent. The mountainous regions are the Appalachian Mountains, Rocky Mountains, and Pacific Mountain Region. They are physiographically complex, especially the Rocky Mountain and the Pacific Mountain regions. Between the Rockies and the Pacific Mountain Region in the United States lies the Intermontane Region, a dry region of plateaus and block-faulted mountain ranges. Two coastal plain regions are found in North America, the Atlantic and the Arctic. The broad interior of the continent is a region underlain by sedimentary rocks, the Interior Plains; it is widely recognized for its agricultural prominence.

Concepts and Terms for Review

- physiographic regions
- provinces
- Canadian Shield
  - geologic diversity
  - Pleistocene glaciation
  - subarctic climate
  - northern forest
- Appalachian Region
  - Piedmont and Blue Ridge provinces
  - Ridge and Valley Province
  - Appalachian Plateaus
  - Northern Appalachians
- Interior Highlands
  - Ozark Plateaus
  - Ouachita Mountains
- Atlantic Coastal Plain
  - continental shelf geology
  - river lowlands; flooding
  - humid subtropical climate
  - varied soils and vegetation
- Interior Plains

flat to gently rolling terrain
glacial and loess deposits
humid, subhumid, and semiarid climates
corn, wheat, livestock
Rocky Mountain Region
Southern, Middle, Northern, and Canadian Provinces
diverse soils, geology, vegetation
low population density
public owned lands
Intermontane Region
plateaus and mountain ranges
arid and semiarid
urban and agricultural development
Pacific Mountain Region
Coast Mountains
Sierra Nevada
Cascade Mountains
Coast Ranges
Alaska Range
Central Valley
Puget Sound-Willamette Valley
Northlands
Alaskan-Yukon Region
Arctic Coastal Plain

Sources and References

Atwood, W.W. (1964) *The Physiographic Provinces of North America.* New York: Blaisdell.

Bretz, J.H. (1965) *Geomorphic History of the Ozarks of Missouri.* State of Missouri, Jefferson, City, Missouri.

Hunt, C.B. (1974) *Natural Regions of the United States.* San Francisco: Freeman.

King, P.B. (1959) *The Evolution of North America.* Princeton, New Jersey: Princeton University Press.

Menard, H.W. (1974) *Geology, Resources and Society.* San Francisco: Freeman.

Sauer, C.O. (1981) *Selected Essays 1963–1975.* Berkeley, CA.: Turtle Island Foundation.

U.S. Council on Environmental Quality. (1984) *Environmental Quality: 15th Annual Report.* Wash. D.C.: U.S. Government Printing Office.

Appendix A

Maps and Map Reading

Maps are models, or miniature replicas, of the landscape. As models, they are abstractions of reality and highly selective in the phenomena they portray or represent. They are the most widely used form of graphics in geography, both for recording spatial phenomena and displaying the result of analysis. In order to use maps effectively, one must be familiar with some of their basic properties, such as direction, location, and scale.

DIRECTION: VARIATIONS ON NORTH

In order to be read, a map must be oriented; that is, it must be placed in its correct relation to the earth. This is a simple matter; in essentially all maps, north is at the top of the sheet; south, at the bottom; east, at right; and west, at left. Practically all of the maps in this book are oriented with north at the top. The orientation is shown by an arrow or similar symbol pointing north. On some maps two arrows are shown—one pointing to true north and one to magnetic north; the map should be oriented to true north.

True north represents the straight-line direction to the North Pole, whereas magnetic north is the direction of the compass needle as determined by the earth's magnetic field. The locations of the North Pole and the magnetic pole do not agree, the latter being situated in northern Canada; thus, there is a deviation between north on a compass and north on a map. This deviation is known as *magnetic declination*, and it is read as degrees east or west of the 0 degree declination, which is the meridian where true north and magnetic north coincide. For the coterminous United States and southern Canada, magnetic declinations range from 0 degrees to 25 degrees east or west. Because the earth's magnetic field shifts somewhat from

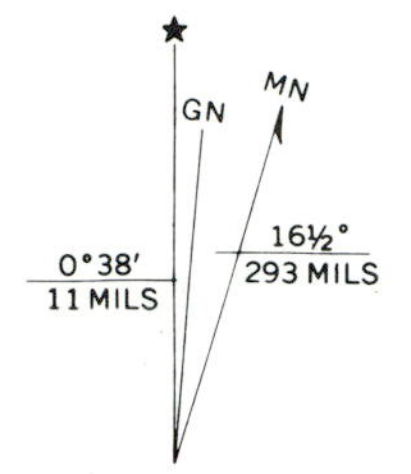

Figure A.1 North arrows that appear on U.S. Geological survey topographic maps; magnetic north, true north, grid north.

year to year, accurate surveying and detailed field mapping require that the most recent magnetic declination readings be used. In the United States these are prepared by the National Ocean Survey.

A third arrow, representing *grid north,* will also be shown on certain maps. This arrow defines north according to the grid lines used in different mapping systems. Each mapping system has its own geometric bias that yields a north that is rarely the same as magnetic north or true north (Fig. A.1).

LOCATION: GRID COORDINATE SYSTEMS

Location is a second important consideration in map reading. Several standard location and grid systems are used for the identification of an area covered by a map. The principal geographic coordinate system consists of a rectilinear network of orthogonally intersecting lines. At the global scale, this network is comprised of parallels and meridians, which are designated in degrees, minutes, and seconds. The characteristics of this system are described in Chapter 1. The standard parallel and meridian system always appears on globes, maps of very large sections of the earth, national maps, and many maps of small areas such as topographic maps.

On many maps reference is also made to other grid coordinate systems. In the United States, the National Ocean Survey has designed a grid system for each state called the Plane Coordinate System. The basic unit or cell of this system is a square measuring 10,000 feet on a side. The topographic maps prepared by the U.S. Geological Survey give state plane coordinates, township and range (where applicable), longitude and latitude, as well as reference to UTM coordinates (Fig. A.2).

The letters UTM stand for the Universal Transverse Mercator grid, a system based on square grids with 100,000-meter spacing arranged within 6 degree sections of longitude extending between latitudes 80 degrees south and 80 degrees north. The UTM grid is drawn on topographic maps prepared by the U.S. Army Topographic Command (formerly the Army Map Service), whereas longitude and latitude are referenced by tick marks on the map margins (Fig. A.2).

Throughout much of North America, the grid system referenced on maps is the township and range system. Originally devised for the division of the landscape in the old Northwest Territories of the United States, the township and range system has since been extended to the majority of the United States as well as to northern Ontario and the western provinces of Canada. This modified rectangular grid is based on a set of selected meridians, termed *principal meridians,* and parallels, called *baselines,* which intersect at an initial point (Fig. A.3).

Distances are measured in the four cardinal directions from the initial point, and locations are identified at 24-mile intervals along the baseline and principal meridian. However, owing to the earth's shape, meridians converge toward the poles, making it impossible to fit an exact square to the earth's surface. Consequently, the ideal planimetric grid that forms the basis of the township and range system is intentionally and necessarily distorted at certain points in order to conform to the earth's curvature.

Within each set of 24-mile-wide strips, 6-mile strips are defined. The strips oriented east-west are defined by the parallels and are termed *townships;* those oriented north-south and bounded by the principal guide meridians are termed *ranges.* Each township and range strip is assigned a number to indicate its position vis-à-vis the initial point. Thus, each small square, generally referred to simply as a township and measuring 6 miles on a side, is easily identified by a notation such as T4N, R3W ("township 4 north, range 3 west"). This notation identifies the township that is formed by the convergence of the fourth township strip north of the baseline and the third range strip west of the principal meridian (Fig. A.3*a*).

Every township is subdivided into thirty-six units, termed *sections,* each measuring one mile on a side. Each section within a township is given a number designation, beginning with section 1 in the northeast corner and proceeding sequentially westward to section 6, then dropping down to the next tier and proceeding back to the east, and so forth, as shown in Figure A.3*b*.

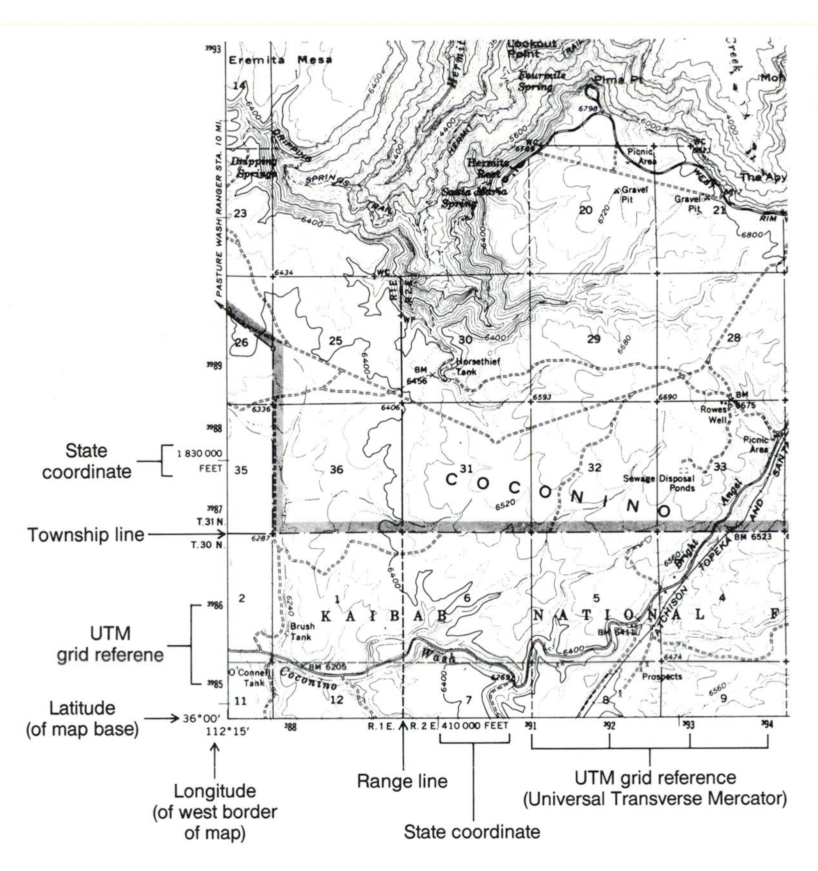

Figure A.2 For much of the United States, four types of coordinate systems are referenced on maps prepared by the U.S. Geological Survey: longitude and latitude, state coordinate, UTM, and township and range.

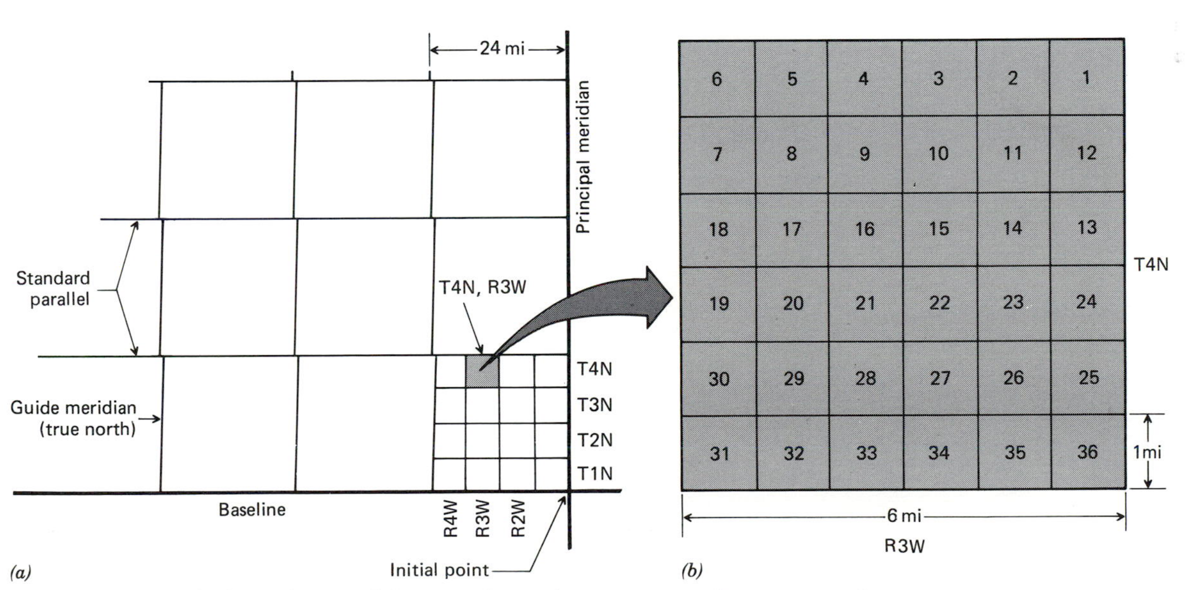

Figure A.3 (*a*) The basic layout of the township and range system, beginning with a principal meridian and baseline; (*b*) within the large grid formed by the township and range lines, each township is subdivided into thirty-six sections, and each section is one square mile (640 acres) in area.

The errors resulting from the convergence of the meridians toward the poles are accumulated along the eastern and northern column and row in each township. Thus, sections 1, 2, 3, 4, 5, 6, 7, 18, 19, 30, and 31 are often a fraction less than 640 acres (one square mile) in area.

SCALE

Scale is defined as the relationship between distance on the map and the corresponding distance on the earth's surface. Maps of small areas, such as planning sites, are called *large-scale maps*, whereas those of large areas are called *small-scale maps*. The level of detail on a map varies with scale; the smaller the scale, the less detail possible.

Scale is generally indicated on a map in either a graphic or an arithmetic form, and occasionally both are included as part of the map legend. The simplest scale indicator employed is the graphic, or bar, scale (Fig. A.4). This consists of an actual line or bar calibrated to indicate a precise map distance and labeled to indicate the corresponding ground distance. Any linear measurement on the map can be compared directly to the bar scale to determine the actual ground distance.

The arithmetic scale represents a ratio of units on the map to like units on the ground and is called a *representative fraction*. A representative fraction of 1:50,000, or 1/50,000, indicates that one unit on the map is equivalent to 50,000 of the same units on the earth's surface. Because the scale is expressed in terms of a ratio, the proportion between the two distances (map and ground) is constant. Thus, the representative fraction is applicable to all systems and all units of measurement simultaneously. Hence, 1:50,000 can be read as "1 map inch to 50,000 ground inches," or 1 map centimeter to 50,000 ground centimeters. Similarly, any other unit of measurement can be substituted. The need for conversion factors between measurement systems, for example, U.S. customary and metric, is thus avoided.

MAP TYPES

The symbols you see on maps depend on the type of map and the phenomena portrayed by the map. Three basic types of maps are used in geography: choropleth, dot, and isopleth. A *choropleth map* may be used to portray either numerical or nonnumerical phenomena. The key feature of this map is its patchlike appearance. Each patch (area) represents a different class or category, and any class may abut against any other. Sections or counties are often used as the mapping units, and both natural and human features can be portrayed with choropleth maps.

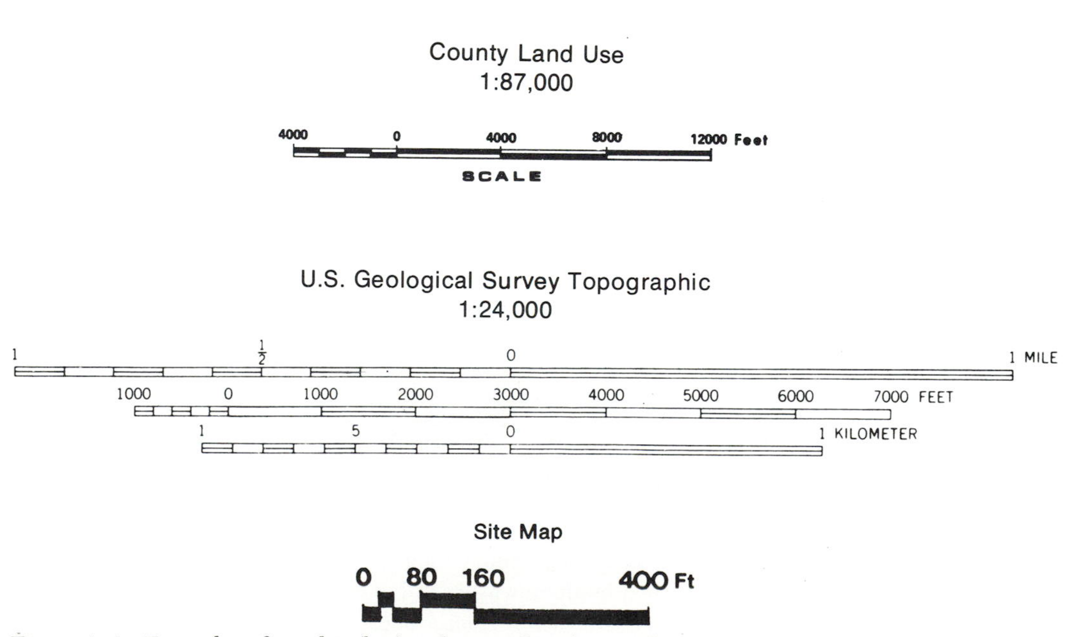

Figure A.4 Examples of graphic (bar) scales employed in modern maps.

A *dot map* is generally used to portray numerical phenomena such as population or crop production. The placement of the dots is usually intended to be representative of the location of the phenomena being portrayed. Each dot may represent a fixed value, or the dot may be sized in proportion to different values. Dot maps are useful in portraying such things as water use by state or country or weather events such as the occurrence of tornadoes by state.

An *isopleth map* is designed to show the pattern or trend of numerical values over an area. This type of map utilizes lines, called *isolines*, to connect points (places) of equal value. If a value is not known, the location of the line is interpolated on the basis of the nearest known values. Among the rules governing isopleth maps are: (1) a given isoline must have the same value over the entire map; (2) isolines cannot cross each other; and (3) the change in value from one line to the next must not exceed the iso-interval, that is, the specified difference in value of adjacent lines in a sequence.

Isopleth maps are used extensively in physical geography, especially for regional phenomena such as solar radiation, precipitation, and soil erosion rates. One of their most common uses is in portraying the topography of the earth (Fig. A.5).

Topographic contour maps are perhaps the most widely used maps in the world today because they are so valuable in terrain analysis, land use planning, and agricultural development. In the United States, the

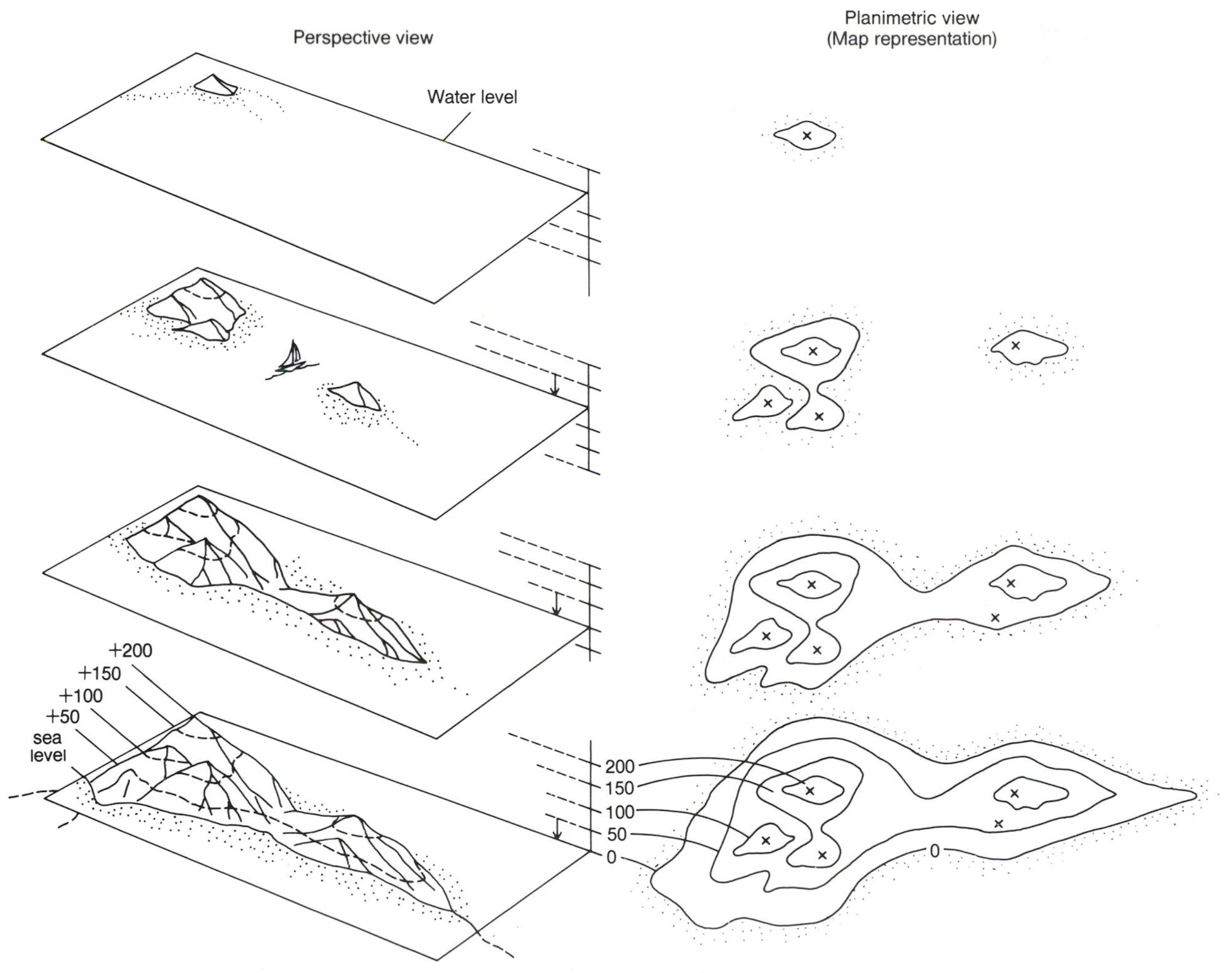

Figure A.5 The concept of a topographic contour map is portrayed here using a water plane (level) around an island. The planimetric view shows how the contours representing each water level appear on a map.

U.S. Geological Survey is charged with the task of preparing topographic maps for the nation. These maps, called *topographic quadrangles*, are prepared at a variety of scales, for example, 1:24,000, 1:62,500, and 1:250,000, and are available to anyone at a relatively low cost. In Canada, topographic maps are prepared by the Department of Energy, Mines and Resources under a program called the National Topographic System.

In addition to contours and elevation data, the U.S. Geological Survey quadrangles provide a great deal of other information about the land. This includes drainage features, forested areas, wetlands, roads, highways, urbanized areas, and even individual structures such as homes and schools in rural areas (Fig. A.6).

The most common variety of geologic map is basically a complex choropleth map showing the distri-

TOPOGRAPHIC MAP SYMBOLS

Primary highway, hard surface
Secondary highway, hard surface
Light-duty road, hard or improved surface
Unimproved road
Trail
Railroad: single track
Railroad: multiple track
Bridge
Drawbridge
Tunnel
Footbridge
Overpass—Underpass
Power transmission line with located tower
Landmark line (labeled as to type) TELEPHONE

Dam with lock
Canal with lock
Large dam
Small dam: masonry — earth
Buildings (dwelling, place of employment, etc.)
School—Church—Cemeteries Cem
Buildings (barn, warehouse, etc.)
Tanks; oil, water, etc. (labeled only if water) Water Tank
Wells other than water (labeled as to type) Oil Gas
U.S. mineral or location monument — Prospect
Quarry — Gravel pit
Mine shaft—Tunnel or cave entrance
Campsite — Picnic area
Located or landmark object—Windmill
Exposed wreck
Rock or coral reef
Foreshore flat
Rock: bare or awash

Horizontal control station
Vertical control station BM ×671 ×672
Road fork — Section corner with elevation 429 58
Checked spot elevation × 5970
Unchecked spot elevation × 5970

Boundary: national
State
county, parish, municipio
civil township, precinct, town, barrio
incorporated city, village, town, hamlet
reservation, national or state
small park, cemetery, airport, etc.
land grant
Township or range line, U.S. land survey
Section line, U.S. land survey
Township line, not U.S. land survey
Section line, not U.S. land survey
Fence line or field line
Section corner: found—indicated
Boundary monument: land grant—other

Index contour
Intermediate contour
Supplementary cont.
Depression contours
Cut — Fill
Levee
Mine dump
Large wash
Dune area
Tailings pond
Sand area
Distorted surface
Tailings
Gravel beach

Glacier
Intermittent streams
Perennial streams
Aqueduct tunnel
Water well—Spring
Falls
Rapids
Intermittent lake
Channel
Small wash
Sounding—Depth curve 10
Marsh (swamp)
Dry lake bed
Land subject to controlled inundation

Woodland
Mangrove
Submerged marsh
Scrub
Orchard
Wooded marsh
Vineyard
Bldg. omission area

VARIATIONS WILL BE FOUND ON OLDER MAPS

Figure A.6 Index of symbols that appear on U.S. Geological Survey topographic maps.

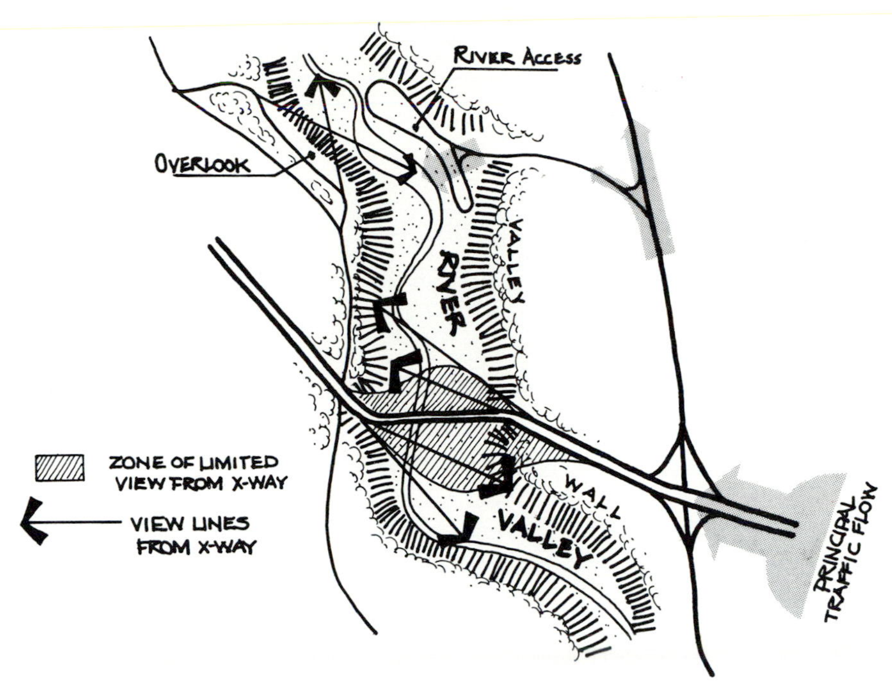

Figure A.7 An example of one type of map used in planning which shows the directional components of circulation and views.

bution of rock types and rock units of different ages. Additional symbols, such as lines representing major faults and arrows showing the directions of strike and dip, are often superimposed on the choropleth base.

Various types of isopleth maps are also used in geology, including *isopach* maps, which show the surface configuration of a subsurface rock formation, topographic contour maps, geothermal heat-flow maps, and maps showing the thickness of deposits such as volcanic ash or glacial drift.

Most of the maps used in planning are of the conventional types. Land use, land ownership, and zoning, for example, are usually portrayed with choropleth maps, whereas topography and rainstorm intensities are portrayed with standard isopleth maps. Conventional maps are not used in some planning problems. Many of these maps are special purpose maps used in the formulation of land use and site plans. Vector maps, for instance, are designed to show the directional aspect of a process or feature such as traffic flow, pedestrian views, noise, wind, and sunlight, as they relate to a specific point or area (Fig. A.7). Linkage maps are used to show the spatial relations and interactions between certain features, land use activities, and/or systems within a prescribed area. In both vector maps and linkage maps, quantities may be portrayed symbolically or noted directly on the map.

Appendix B

Remote Sensing And Image Interpretation[1]

Gathering data on land, water and atmospheric conditions over large areas has always been one of geography's principal tasks. Given the size, complexity and changeable nature of the earth's surface this is also a monumental task that cannot be accomplished very effectively if an observer's perspective is limited to the ground. Remote sensing expands our perspective on the earth's surface from the ground to a distant vantage point such as an aircraft or an earth orbiting satellite, and the data gathering task begins to assume much more manageable proportions.

All earth surface materials reflect and emit radiant energy known as electromagnetic radiation (see Chapter 3), but different materials emit and reflect different amounts and types of radiation. If we understand the nature of the energy flow from materials like water, rock and vegetation, it is possible to measure it from a remote perspective. This in turn extends our capacity to gather geographic data in a revolutionary way, for among other things, we are able to map broad reaches of the earth's surface in, as it were, a single glance.

Aerial photography has traditionally been the most widely used means of remote sensing. In the last twenty years, however, the capabilities of remote sensing technology have expanded significantly. Developments have been made in two fundamental areas. First, a greater variety of sensors have become available so it is now possible to measure electromagnetic radiation more precisely both within and beyond the limits of aerial photography. In addition these new sensors record data electronically so that images can be tailored to particular problems through enhancement and other data manipulations.

[1]Prepared by Richard Hill-Rowley, University of Michigan, Flint, and John Grossa, Jr., Central Michigan University.

The second fundamental development is the ability to mount these new instruments on platforms that fly in space. Satellite remote sensing has given us the ability to view the earth from a totally new perspective; to look at processes operating over much larger areas than was previously possible. We can now map in a precise way the geographic limits of events such as floods or crop failure due to drought. Crop estimates can be made during the growing season to predict final production totals for whole agricultural regions and complex natural vegetation can be evaluated to determine the potential for commercial forestry development or preservation as wildlife habitat.

The remotely sensed image also provides a record of the environment at a specific time from which change can be measured and calibrated, and to which an observer can return for new perspectives at a later date. Sequential image coverage allows the observer to go from geographic inventory to environmental monitoring. This expands our potential to understand geographic change and allows us to do it in a cost effective and timely way. Satellite remote sensing in particular has opened up a new era in the gathering of geographic data. It is an era in which capabilities are expanding rapidly and science is still discovering just how great the potential is.

ELECTROMAGNETIC RADIATION AND SENSOR CAPABILITY

In order to understand remote sensing systems it is necessary first to briefly examine electromagnetic radiation (see Chapter 3, especially sections on, *The Nature of Radiation* [page 37], *Absorption and Reflection by the Atmosphere* [page 38], *Surface Reflection* [page 44] and *Longwave Radiation* [p. 48]). Electromagnetic radiation occurs as a continuum of wavelengths, which we breakdown into various classes (see Fig. 3.1). The broad band of radiation between the ultraviolet and microwave regions is most important in remote sensing. Within this band three more specific zones or areas of radiation can be identified where remote sensing measurements are most useful: (1) visible, near and middle infrared; (2) thermal infrared, and (3) microwave.

The first zone is the most important. This radiation (as incoming solar radiation) is *reflected* off earth surfaces, and except for two narrow bands (1.45 μm and 1.95 μm) which are absorbed by water vapor, is readily transmitted upward through the atmosphere. Thus spectral (radiation) information about earth surface features can be detected from a distant point above the earth without much interference by the atmosphere. Recording this radiation can be done with aerial cameras using different types of film, and electronic sensors mounted in aircraft or satellites.

The second zone is the thermal infrared (between 3 μm and 14 μm) with two particular areas at 3 μm − 5 μm and 8 μm − 14 μm, where again sensing systems can operate without atmospheric interference. With this radiation sensors are measuring *emitted* as opposed to reflected energy. This is relatively longwave radiation generated by heated surface materials; consequently the resultant information tells us about the thermal character of the earth's surface such as warm and cool areas related to differences in soil moisture or vegetation coverage. Thermal infrared radiation is recorded with electronic sensors mounted in aircraft or satellites.

The third zone is the microwave which covers a broad band of longer wavelengths. It is used by radar sensors which are "active" sensors in that they broadcast their own radiation, sending it to the earth's surface in pulses, and then collect the returning radiation reflected off landscape features. A variety of wavelengths can be used in microwave remote sensing systems, but the most commonly used are around 3 cm wavelength for aircraft systems (known technically as X band), and another around 23.5 − 25.0 cm for satellite systems (known technically as L band).

REMOTE SENSING SYSTEMS

The various technological devices used in remote sensing are called *systems*. In geographic data gathering, there are three groups of remote sensing systems:

- Earth resources satellite systems, designed for detection of visible and reflective infrared radiation, and experimental satellite radars.
- Meteorological satellites which gather reflective and the thermal infrared data used primarily for weather forecasting and secondarily for mapping certain surface features.
- Aerial photographic systems which record visible and reflective (near) infrared. They remain the most widely used remote sensing system.

Landsat Systems

The era of earth resources satellites was initiated in 1972 with the launch of the first Landsat satellite (known at that time and until 1975 as ERTS). Two other Landsat satellites launched in 1975 and 1978 completed a series of three similar earth resources satellites. The satellites were launched into a polar (north-south) orbit at an altitude of 918 km (560 miles). The orbit is designed so that the earth's surface is completely imaged every 18 days. Furthermore, for every location on the earth, each pass of the satellite throughout the year will occur at the same time of the day, because the satellite is in sun-synchronous orbit around the earth.

The primary sensor on Landsats 1, 2 and 3 was a multispectral scanning system (MSS). This type of system measures radiation from the earth's surface along scan lines that run perpendicular to the satellite path. Each scan line is 79 m wide which the scanner in turn splits into units 79 m long, creating a unit of ground resolution of 79 m × 79 m (250 × 250 ft.). The total length of an MSS scan line is 185 km, and the data from 2342 adjacent scan lines are framed into scenes 185 km by 185 km in area. Radiation received from a scan line is directed by a rotating mirror to an instru-

Figure B.1 Examples of two standard types of aerial photographs: (*a*) black and white, and (*b*) infrared, or photographic infrared. Notice the differences in the way water, vegetation, and farmland appear on the two.

ment which breaks it down so it can be detected and recorded in digital form. The Landsat MSS is designed to record reflected energy in four wavelength bands: two in the visible spectrum, green (0.5 to 0.6 μm) and red (0.6 to 0.7 μm), and two measuring reflected infrared energy at 0.7 to 0.8 μm and 0.8 to 1.1 μm.

Since MSS data are recorded in digital form they can be analyzed using a computer and converted into image formats. Images of single MSS bands present distinctly different views of the landscape (Fig. B-1). The green band image, though of relatively low contrast, usually provides more information about water conditions than the other bands. The red band is often useful for land use classification because the shapes, tones, textures, and patterns of agricultural land use stand out clearly in contrast to the relatively dark toned, irregularly shaped forest areas. Major transportation arteries can be seen on both the green and red bands. While no details of water features can be seen, the sharp contrast between land and water on the infrared bands facilitates the precise location of the shorelines and the identification of smaller streams, lakes, ponds, and wetlands. Differentiation between deciduous and coniferous forest stands is often possible on the infrared bands due to differences in infrared reflectance from the two tree types. The individual MSS bands can also be combined to produce a false-color infrared composite image. The green, red and second reflective infrared bands are assigned blue, green and red colors respectively to produce an image in which healthy vegetation is red, bare soil is light blue tones, clear water is black and cities are a variety of light blue tones.

The Landsat color infrared composite image in Figure B-2 (color plate 13) shows the Puget Sound area of Washington State. The eastern side of the image is a section of the Cascade Mountain Range; dark red tones indicate the coniferous forest, and the small, square blocks indicate clear-cut timber harvesting. The central section of the image shows the Puget Sound. Along the eastern shore of the Sound is a large belt of urban development represented by the cities of Tacoma, Seattle and Everett, Washington, which appears in blue tones. Between the urban belt and the Cascade Mountains is a zone dominated by dark and bright red tones representing coniferous and deciduous forests. Across the waters of Puget Sound the western part of the image shows the Olympic Mountains in the north, also covered with coniferous forest and snow.

In 1982 a fourth Landsat was launched followed by a fifth in 1984. Landsats 4 and 5 still carry the MSS system; but changes have been made to improve satellite performance and data transmission, and also to accommodate a new sensor, the Thematic Mapper. The orbit of Landsat 4 and 5 has also been changed so that these satellites fly at somewhat lower altitudes, and global coverage is now acquired in 16 days. Otherwise continuity with Landsats 1, 2 and 3 has been maintained.

Thematic Mapper

The Thematic Mapper (TM) is a scanning system which operates much like the MSS, but it is technically more advanced. Briefly, seven spectral bands are available (compared with four in MSS) with refined

TABLE B-1 THERMATIC-MAPPER SPECTRAL BANDS

Band	Wavelength, μm	Characteristics
1	0.45 to 0.52	Blue-green—no MSS equivalent. Good penetration of water. Useful for distinguishing soil from vegetation and deciduous from coniferous plants.
2	0.52 to 0.60	Green—refined MSS band 4. Strong vegetation reflectance; is useful for assessing plant vigor.
3	0.63 to 0.69	Red—refined MSS band 5. Matches a chlorophyll absorption band that is important for discriminating vegetation types.
4	0.76 to 0.90	Reflected IR—coincident with portions of MSS bands 6 and 7. Useful for determining biomass content and for mapping shorelines.
5	1.55 to 1.75	Reflected IR. Very moisture sensitive. Good contrast between vegetation types.
6	10.40 to 12.50	Thermal IR. Nighttime images are useful for thermal mapping and for estimating soil moisture.
7	2.08 to 2.35	Reflected IR. Coincides with an absorption band caused by hydroxyl ions in minerals. Good geological discrimination.

After: Sabins 1986 and Curran 1985

green, red and reflective infrared bands, new data capabilities in two reflective infrared bands and a thermal infrared band (Table B-1). The Thematic Mapper's spatial resolution has been improved to provide for a 30m × 30m ground cell, and it is capable of measuring radiometric data to a more refined level of detail.

Several types of color images can be produced with Thematic Mapper data by combining various bands. A natural color image is created by combining the blue, green and red bands and a color infrared image is created by combining the green, red and the first infrared bands (Fig. B-3; color plate 14). Vegetation appears in various tones of green, the ocean is dark blue over deep water and light blue over shallow water or water carrying sediment; beaches, sand dunes, and settlements appear in the lightest tones.

Substitution of other TM bands is also possible in order to improve detection of certain features in a scene. Figure B-4 (color plate 15) is a portion of an enlarged scene of an agricultural area in the high plains on the Texas-New Mexico border, prepared using both the visible and infrared bands. The cropland appears as rectangular and circular fields, both heavily irrigated, and many are clear of crops, indicated by the blue tones associated with bare soil. Cropped fields exhibit a range of tones from a deep red through orange. These colors represent different crops, at different stages of development. The large blue-green strip extending east-west across the lower part of the image is an uncultivated area of sand hills into which agriculture, with the aid of irrigation, is encroaching. Two cities appear in the scene: Cloris, New Mexico in the upper left and Farwell, Texas in the upper right.

Other Earth Resources Satellite Systems

The Landsat systems were designed and operated by the National Aeronautics and Space Administration (NASA) until 1984 when operational responsibility passed to the National Oceanic and Atmospheric Administration (NOAA). Responsibility has recently been passed from NOAA to a private sector company called EOSAT that will implement future earth resource satellite systems for the United States. The French have entered commercialized remote sensing market with their SPOT (Systeme Probatoire de Observation de la Terre) satellite launched in 1986.

SPOT employs a new type of sensor called an HRV (high-resolution visible-range instrument), which use an array of detectors to image a complete line on the ground and consequently avoids the mechanical scanning element of both the MSS and TM sensors (Table B-1). SPOT flies at an altitude of 832 km. in sun-synchronous orbit, has a swath width of 60 km, and provides repeat coverage every 26 days. A revolutionary feature of SPOT is an ability to point its HRV sensors to either side of the orbit track so that the observer can "look" at the same ground area on successive orbits, a useful capability for environmental monitoring.

The Japanese are also about to enter the earth resources satellite arena. The Marine Observation Satellite (MOS1) is planned for the late 1980s and will include a resolution capability of 50m. A second system, the Earth Resource Satellite (ERS1) which will carry special camera and radar sensors, will be launched in the 1990s.

Radar Systems

Radar systems are different in several respects from other remote sensing systems. They generate their own energy, broadcast it to the earth's surface, and then measure its return to the system. Radars operate with microwave energy made up of wavelengths that are not interrupted by droplets of water that make up clouds. As a result radar systems can penetrate clouds to collect data in almost any weather conditions, day or night. The relationship between image tone and the earth surface is also very different on radar from that of reflective sensors and this has been one of several factors limiting the use of radars in surface mapping. Radar sensors do nevertheless highlight linear trends in the landscape and have found particular utility in geological mapping.

The cloud penetration capacity of radars is important for mapping in cloud shrouded areas such as the tropical rainforest. Measuring the destruction of rainforest over vast areas of South America, for example, cannot be reliably accomplished with Landsat based satellite sensors. Satellite-based radars, however, offer this potential (Fig. B-5). Similarly in cloudy subarctic and arctic regions such radars are necessary to effectively monitor movements of sea ice in the northern shipping lanes.

At present (1986), there is no continuing radar sensor in operation that can provide data to perform these mapping and monitoring functions. Three experimental systems, however, have been deployed; and they illustrate the advantages of satellite-based radar systems. The systems were Seasat, launched in 1978 and planned for an extended mission, but which failed due to technical problems within several months; SIR-A

Figure B.5 Radar image of the transition from the Andes Mts. to the Amazon rainforest in Columbia. Radar can penetrate clouds, and is thus highly valuable in mapping in overcast regions.

the first shuttle based radar which flew in 1981; and SIR-B a second shuttle based radar which flew in 1984. Each of these systems use the 23.5 cm. wavelength band and gather data with a spatial resolution better than Landsat MSS (Fig. B-5). Only limited parts of the earth surface were imaged but the quality of the images produced have confirmed plans for future shuttle-based radars and operational systems to be launched by the European Space Agency, the Japanese and the Canadians.

Meteorological Satellites

Meteorological satellites were first launched in the 1960s. They fall into two major groups: (1) those with polar, sun-synchronous orbits that image different points on the earth's surface at the same time each day and night and (2) those with geostationary orbits that are synchronized to the rotation of the earth and image the same area of a continuous basis. The main purpose of these satellites is to generate images of cloud patterns to assist in weather forecasting and monitoring. Cloud-free images, on the other hand, can be used in geographic analysis of land or sea surfaces.

The most recent generation of NOAA polar orbiting meteorological satellites use the Advanced Very High Resolution Radiometer (AVHRR) to image the earth's surface. This instrument records data in five wavelength bands: red (0.55–0.68 μm), near infrared (0.73–1.1 μm), middle infrared (3.55–3.93 μm) and two thermal infrared bands (10.5–11.5 μm) and (11.5–12.5 μm). Two satellites are usually operating from an altitude of 850 km with each acquiring global coverage every twenty-four hours. Their orbits are staggered so that imagery is collected for any area twice daily. The spatial resolution of the system is low but large area coverage makes analysis at continental or global scales feasible. Figure B-6 (color plate 16) is an infrared composite of AVHRR images for the Northern Hemisphere in July. The darker red tones are the boreal forests surrounding the white tones of the Arctic ice. The vast extent of the Saharan desert belt extending across North Africa and into the Arabian peninsular is also clearly identifiable. The frequency with which AVHRR data are gathered makes it possible to prepare global composites for different seasons and to analyze annual vegetation changes. Thermal bands of AVHRR data have been used for other environmental application such as mapping sea surface temperature, estimating the moisture content of snow and even locating forest fires.

The world's major geostationary satellites (the Geostationary Operational Environmental Satellites (GOES)) are also operated by NOAA. GOES East and GOES West are positioned at 35,000 km above the equator at 75° west longitude and 155° west longitude respectively. When functional, and their record has been very good providing hemispheric images which together cover the Americas along with the Atlantic and Pacific oceans. Since the GOES sensor sees the whole hemisphere, repetitive coverage is governed by image acquisition time, and this is every 30 minutes.

Images are generated in a visible band (0.55–0.7 μm) which operates in daylight hours and a thermal band (10.5–12.6 μm) which operates both day and night. GOES images also have a low spatial resolution, but the data have important uses. Analysis of weather patterns is preemiment. Cloud patterns and hurricane development can be monitored using the visible band. Snow cover analysis, freeze warnings and the development of severe convective storms are all applications for the thermal band. As with AVHRR data important information can also be obtained through the analysis of GOES data for land and water surfaces particularly in the thermal band. Soil moisture and sea surface temperature studies are examples.

REMOTE SENSING SYSTEMS

Aerial Photography

Aerial photography, the most widely used remote sensing technique, normally focuses reflected sunlight from the earth surface through a camera lens onto photographic film. Although most photographic film is sensitive to ultraviolet and visible light of 0.3 to 0.7 micrometer (μm) wavelengths, and is converted into black and white or natural color prints for interpretation, specially prepared emulsions have extended sensitivity into the near infrared portion of the spectrum to 0.9 (μm). While this infrared film *does* record infrared energy, it is rereflected solar radiation in the near or photographic infrared wavelengths (Fig. B.7; color plate 17). The energy recorded in infrared photographs is *not* emitted by the earth and does not represent different surface temperatures.

Image Interpretation

Aerial photographs and other remote sensing images usually provide a highly detailed record of some portion of the landscape. Such detail may evoke the question, "How do I begin?" or "What do I look for first?" Although no single starting point or approach may be best for all image interpretation projects, you must be guided mainly by the objectives of the problem or project. Initially, you might find it helpful to scan the image to get a general overview of the landscape itself before more precise or site-specific analyses are undertaken. Even before the image interpretation process begins, you should note the characteristics of the imagery such as the date and year, the scale, and the characteristics of the image system used. Remember, the way the landscape is portrayed on the image is determined mainly by these factors. In addition, take advantage of any supplementary sources of information that may aid in the interpretation task. Topographic maps produced by the U.S. Geological Survey are available for most areas. They facilitate interpretation by providing baseline information such as road locations, stream courses, spot elevations, topographic contours, and significant land survey boundaries in a planimetrically correct format. Standard soil maps prepared by the Soil Conservation Service are also helpful. Other thematic maps that show land use, vegetation, or watershed information may also be available locally. Finally, other types of remote sensing imagery of the area may be used to augment or corroborate interpretation. Even imagery from very high altitudes or from space can support the interpretation effort by showing the study area in its surrounding regional context.

ELEMENTS OF AERIAL PHOTOGRAPH INTERPRETATION

Aerial photographs are interpreted by examining and assessing the specific properties or elements of the features that make up the image. Although these elements may not always be consciously considered in the interpretation process, it is the collective assessment of them that leads to interpretation. What are some of the elements of interpretation and how are they used?

Tone or Color There must be an apparent difference in tone or color among objects in order to identify them as individual entities. Because the earth's surface is portrayed in gray tones on black and white photographs, it is often difficult to distinguish between surfaces that have only slight reflective differences (Fig. B.8). For example, areas covered by herbaceous vegetation, barren soil, or concrete may appear in similar tones of gray on images using visible light but in distinctively different tones on images using reflected infrared wavelengths. On a color photo, however, these same surfaces would likely appear in different colors (Fig. B.7; color plate 17).

Shape Some features can be identified on aerial imagery by their characteristic shape. Circular fields clearly seen even on imagery obtained from orbiting satellites are diagnostic of center pivot irrigation sys-

Figure B.8 Tonal variations related to soil types, soil moisture crop types and land use.

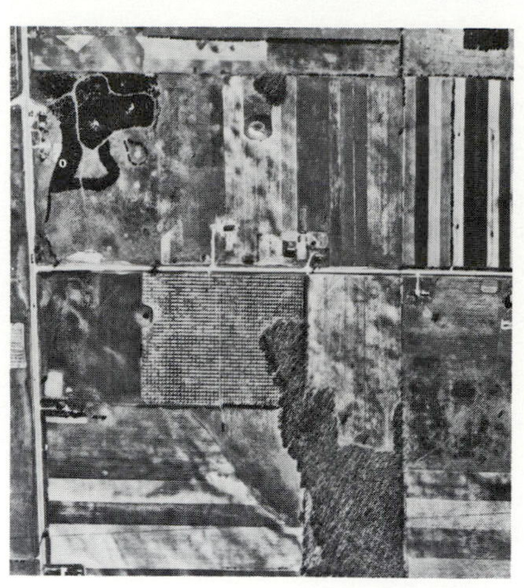

Figure B.9 Shapes and patterns in an agricultural landscape strip-cropped fields, orchards, tree plantations, natural forest.

tems, as are the cloverleaf patterns at some freeway interchanges (see color plate 15). Although features with geometric shapes are normally associated with human activity, there are unusual natural landscape features that have linear, curvilinear, or polygonal shapes. Even when not diagnostic, shape can often be used along with other elements of interpretation to identify landscape features (Fig. B.9).

Size It is often helpful to compare the relative size (height, area, etc.) of a known feature that can be seen on the image with the size of those to be identified. Some features have standard dimensions that can be used as a basis for size comparison. For example, on large-scale aerial photos, football or soccer fields or tennis courts can sometimes be seen, as can railroad tracks which also have a standard width (4.71 ft. or 1.46 m in the United States). On medium- and small-scale imagery, road systems that follow land survey section lines are one mile or 1.6 km apart. In other cases, the actual size of the feature may have to be calculated using techniques described later.

Shadow The shadows cast by objects can be both an advantage and a disadvantage to the image interpreter. On a vertical aerial photograph, the shape or silhouette of a vertical feature as seen in its shadow may often be diagnostic. Since the lengths of shadows are proportional to the height of the objects casting them, the relative height of objects can quickly be determined.

Pattern The repetition of certain features over the landscape may produce a characteristic pattern on remote sensing images. Quite often they reflect human use of the land such as the patterned arrangement of orchards, contour plowing, and strip cropping which typify modern agricultural land use practices (see color plate 15). The patchwork of rectangular fields and road networks oriented in the cardinal directions produces a landscape pattern that reflects the gridlike township and range survey system used in many parts of the United States (Fig. B.9). On large-scale aerial photographs, gullies and stream channels, the smallest components of drainage networks, are usually detectable, whereas on images acquired from space entire drainage networks may be evident (Fig. B.5). Similarly, longshore currents can be plotted from satellite images that reveal plumelike patterns of sediment transported along coastlines or into sediment sinks (color plate 14).

Texture The visual impression of texture is produced when individual features or objects are too small to be clearly discerned on the image. Like pattern, texture will often provide information on land use, vegetative cover, and plant conditions. On aerial photographs, planted vegetation will often appear to have smoother texture than natural vegetation. Actively cultivated farmland will normally have a finer, and more uniform, texture than abandoned fields with their diverse mixture of herbaceous and shrub vegetation. Mature stands of deciduous trees will exhibit

Figure B.10 Contrasts in texture between forest and open fields.

a rough, mottled texture, whereas dense stands of younger trees will appear fine-textured. Woodlots in which trees have been selectively cut may also display a very coarse, cobbled appearance due to the fuller crowns of the remaining trees and by their shadows cast on the forest floor or understory (Fig. B.10).

Site and Association The geographic associations that can be discerned on aerial photographs are also a helpful source of information. In river valleys, the boundary between valley walls and floodplain can usually be located in a stereomodel, as can the relief and slope characteristics in hilly terrain. As sites vary from uplands to bottomlands, from hilltops to swales, and from gentle to steep slopes, the soil and vegetation assemblages found at these sites will likely change as well. Many features of the cultural landscape are also site-specific. Certain manufacturing industries and electricity-generating plants, for example, are commonly located along navigable waterways, while agricultural practices such as contour plowing and terracing are indicative of sloping terrain.

Resolution In order for a feature to be considered in the interpretation process, it must be discernible on the image. The capacity to resolve fine image detail is referred to as its *resolving power*. The resolving power in photographic systems depends on the characteristics of the lens, the type of film and filter, the exposure time, and film processing. With cameras and films designed for high-resolution photography, very small objects may be seen when examined under a microscope.

In order for even a sizable feature to be seen on a photograph, it must appear in a different tone or color than its background or adjacent objects. On a black and white or color photograph, the exact shoreline of a lake might not be apparent owing to a lack of tonal difference between the shallow water and beach. A similar lack of resolution may exist between coniferous and deciduous trees growing in a mixed stand because of their similar reflectance characteristics. Both the shoreline and the different tree species, however, would be apparent on either a color or black and white infrared photograph because of sharp tonal differences between the water and land and the coniferous and deciduous trees (Fig. B.7).

The solution of most interpretation problems relies on the use of several elements of interpretation, with each providing evidence required to solve the problem. Remember that accurate interpretation involves the synthesis of evidence drawn from the careful and systematic study of the remote sensing image itself, coupled with the use of supplementary information sources such as topographic maps and soils data.

STEREOSCOPIC VIEWING

When a pair of overlapping vertical aerial photographs is viewed stereoscopically, the landscape appears in a three-dimensional perspective. Such a perspective is achieved by simultaneously viewing two photographs taken from slightly different camera positions but showing the same ground features. As a result, each eye is viewing the same scene from the position of the aerial camera when each of the two photographs making up the stereo model was acquired. When this model is viewed through a simple instrument, called a *stereoscope* (which allows each eye to focus independently on each component photograph), a three-dimensional perspective of the terrain results.

Because aerial photographs are usually acquired so that each photo overlaps the next by 60 to 70 percent, a stereoscopic model can be produced from successive photographs in a flight line. This is accomplished by locating the center of each photograph, called the *principal point*. The principal point (PP) is located at the intersection of straight lines drawn from the fiducial marks on opposite sides of the photo. The fiducial marks are printed on the margins of the photographs (Fig. B.11). After locating and marking

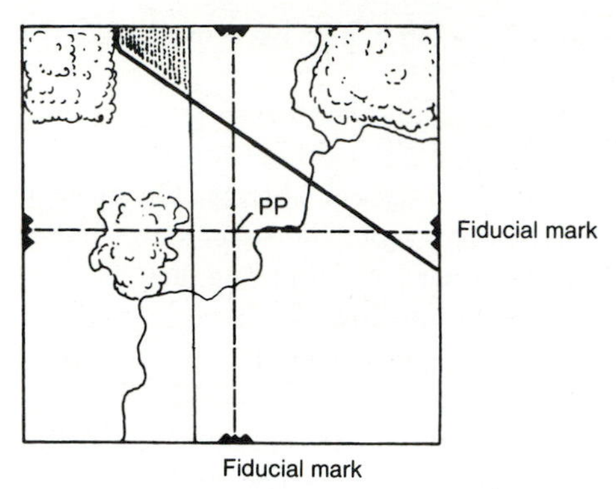

Figure B.11 The features of a standard aerial photograph: fiducial marks and principal point (PP).

the principal points on both photos (PP1, PP2) with a pin prick, transfer and mark the position of each principal point on the other photographs. These are called the *conjugate principal points* (CPP). Next, draw a straight line between the CPP and PP on each photograph. Each line represents the aircraft flight line between photo exposures and is known as the *air base*.

Having located the principal and conjugate principal points, the two photographs can be arranged for stereoscopic viewing (Fig. B.12).

1. Orient the two photographs on the viewing surface so that the aircraft flight line and the long axis of the stereoscope are parallel and so that shadows on the photographs fall toward the viewer.
2. After taping down the edges of one photograph, place the second photograph so that the corresponding CPP is about 60 mm (2.3 in.) from the PP of the first photo and secure it.
3. With the lenses of the stereoscope about 60 mm apart, place the stereoscope so that the left lens is over the left photo and the right lens is over the right photo, ensuring that the corresponding images on each photo will be viewed separately.
4. Looking through the stereoscope, adjust its position slightly until the stereoscopic effect is achieved. Remember that the stereoscope must be oriented so that its long axis is parallel to the flight line in order to see in stereo.

The perspective of the stereo model may seem somewhat abnormal as elevation changes and the height of objects appear to be exaggerated. Although this characteristic may at first be somewhat misleading, it soon becomes useful in discerning not only slight variations in height or elevation, but also the overall interpretation process.

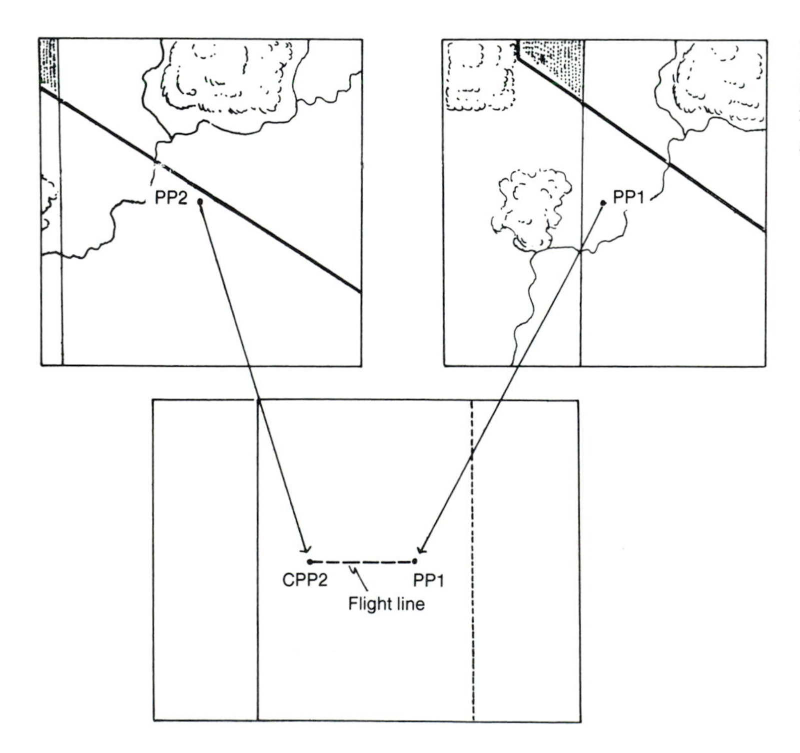

Figure B.12 The basic stereographic model (below) in which two successive photographs in a flight line are arranged with the principal points about 60 mm apart.

Appendix C

Climatic Maps of the United States

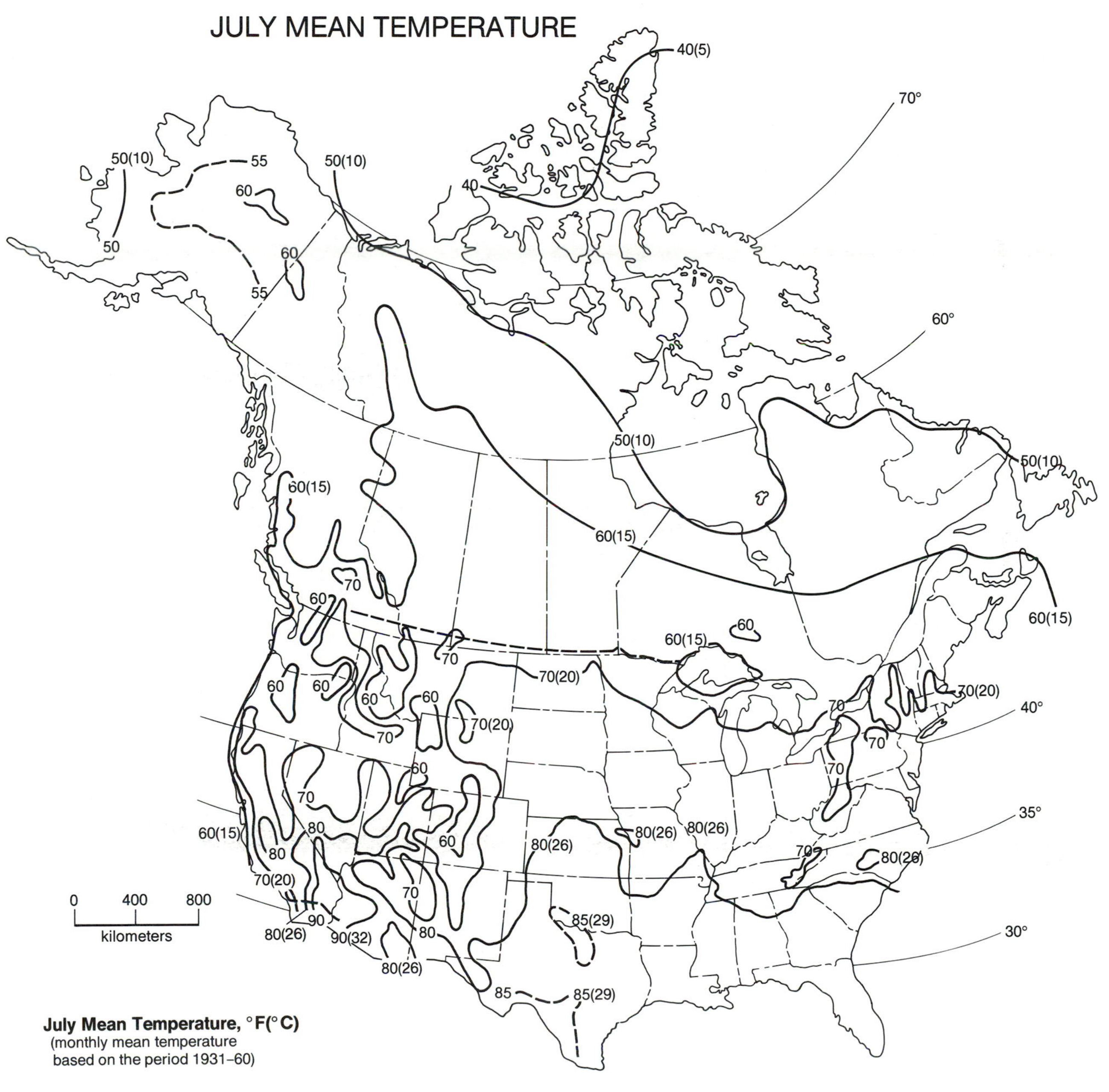

Sources: U.S. National Atlas and Canadian National Atlas

Figure C.1

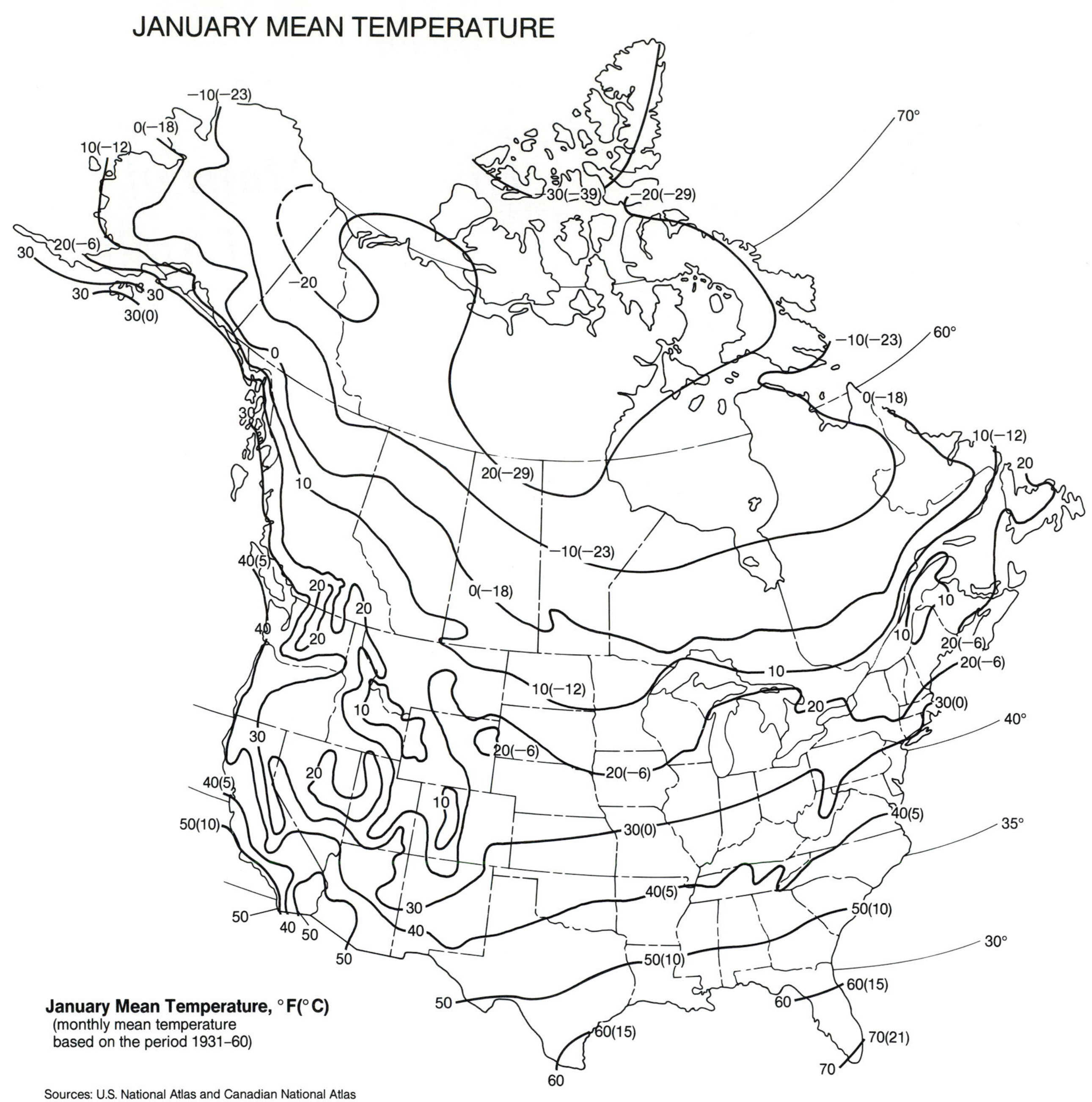
JANUARY MEAN TEMPERATURE
−10(−23)
0(−18)
10(−12)
20(−6)
30
30
30
30(0)
−20
0
30(−39)
−20(−29)
70°
−10(−23)
60°
0(−18)
10(−12)
20
20(−29)
10
−10(−23)
40(5)
20
20
20
0(−18)
40
10
10
20(−6)
20(−6)
10
10(−12)
20
30(0)
40°
10
30
20(−6)
20(−6)
40(5)
20
10
30(0)
40(5)
35°
50(10)
40(5)
50(10)
30
50
40
50
40
50
50(10)
30°
January Mean Temperature, °F(°C)
(monthly mean temperature
based on the period 1931–60)
50
60
60(15)
60(15)
70(21)
60
70
Sources: U.S. National Atlas and Canadian National Atlas

Figure C.2

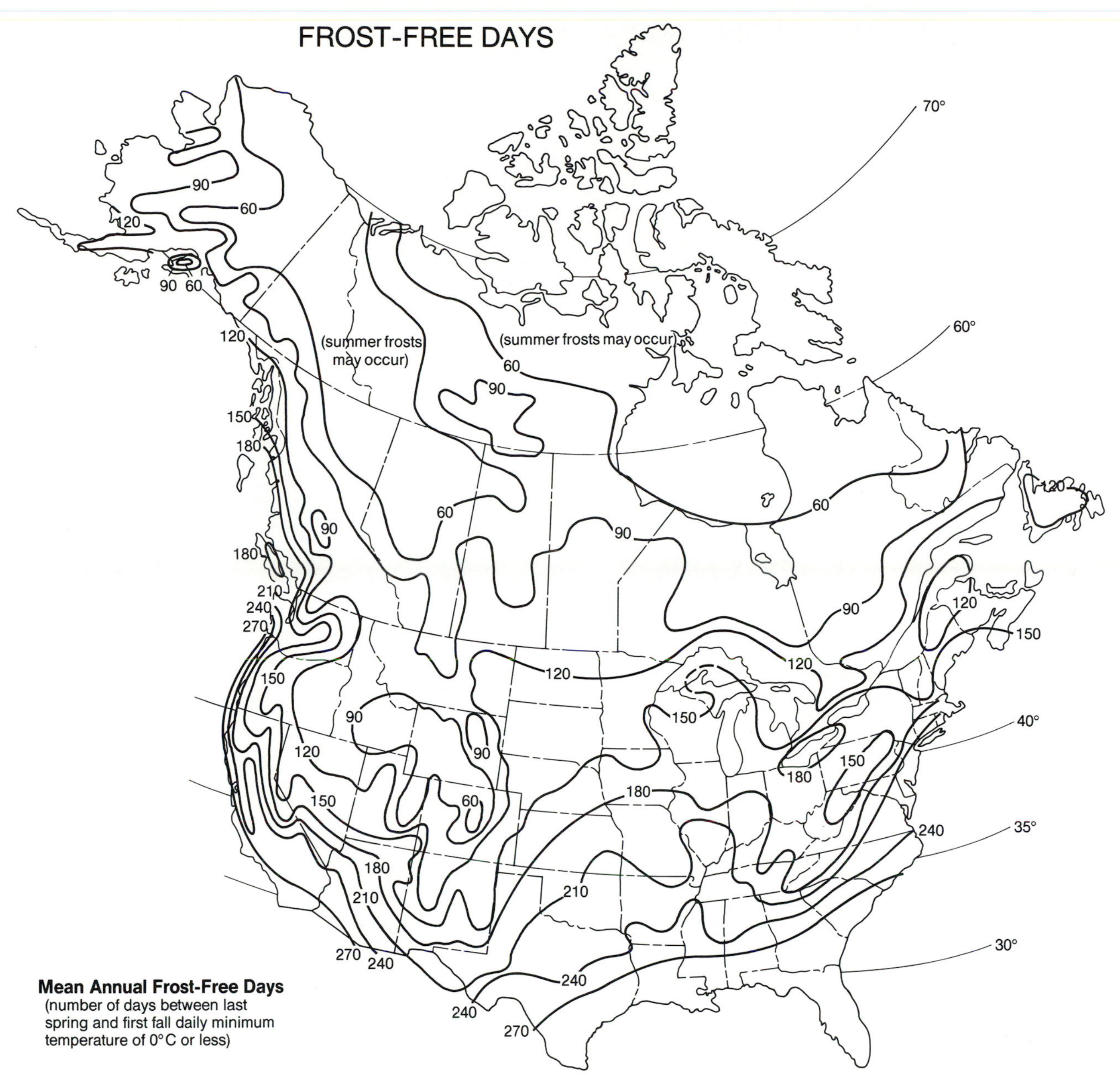
FROST-FREE DAYS
70°
60°
40°
35°
30°
(summer frosts may occur)
(summer frosts may occur)
60
90
120
150
180
210
240
270
Mean Annual Frost-Free Days
(number of days between last spring and first fall daily minimum temperature of 0°C or less)
Sources: U.S. National Atlas and Canadian National Atlas

Figure C.3

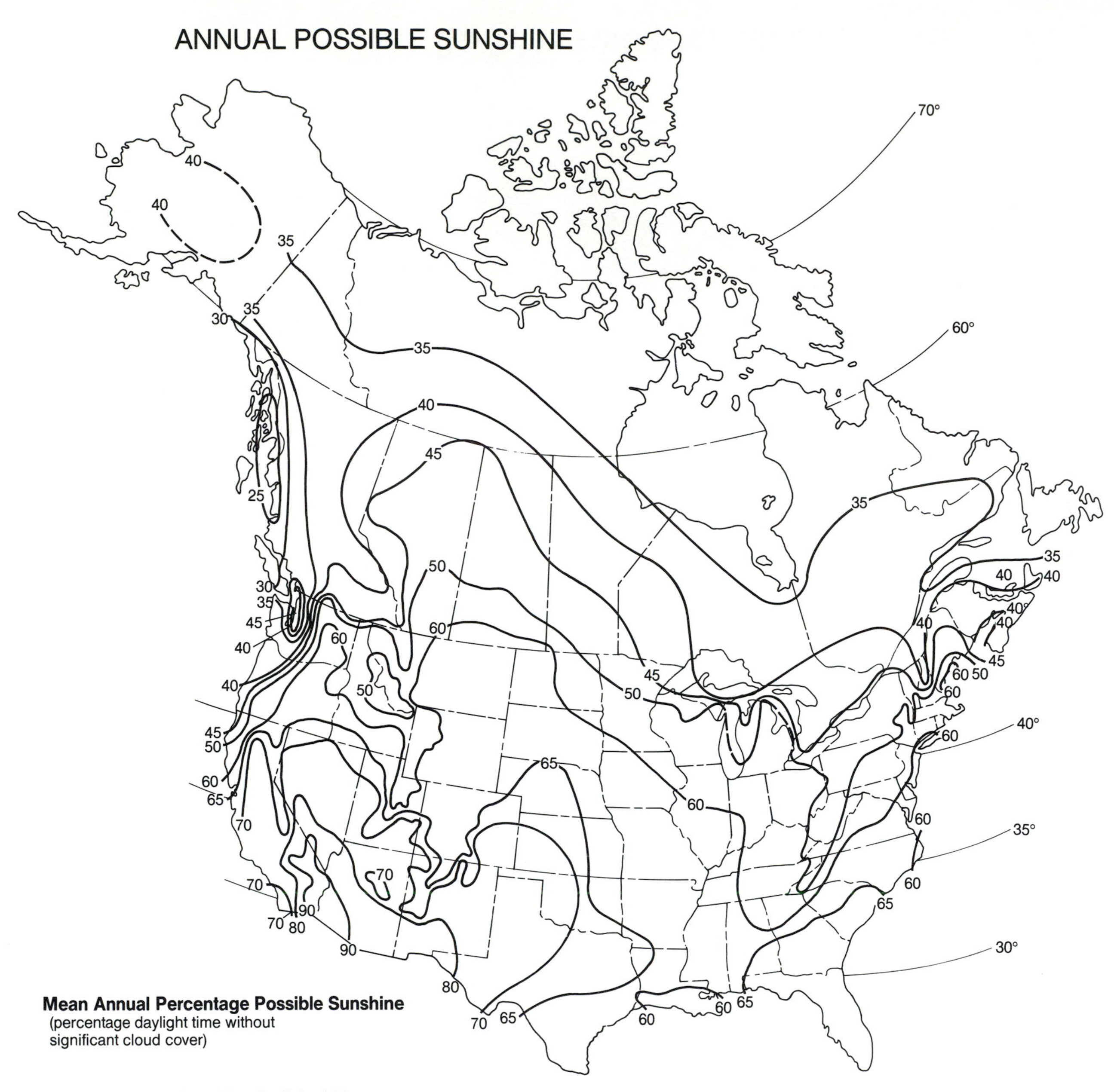
ANNUAL POSSIBLE SUNSHINE
70°
60°
40°
35°
30°
Mean Annual Percentage Possible Sunshine
(percentage daylight time without
significant cloud cover)
Sources: U.S. National Atlas and Canadian National Atlas

Figure C.4

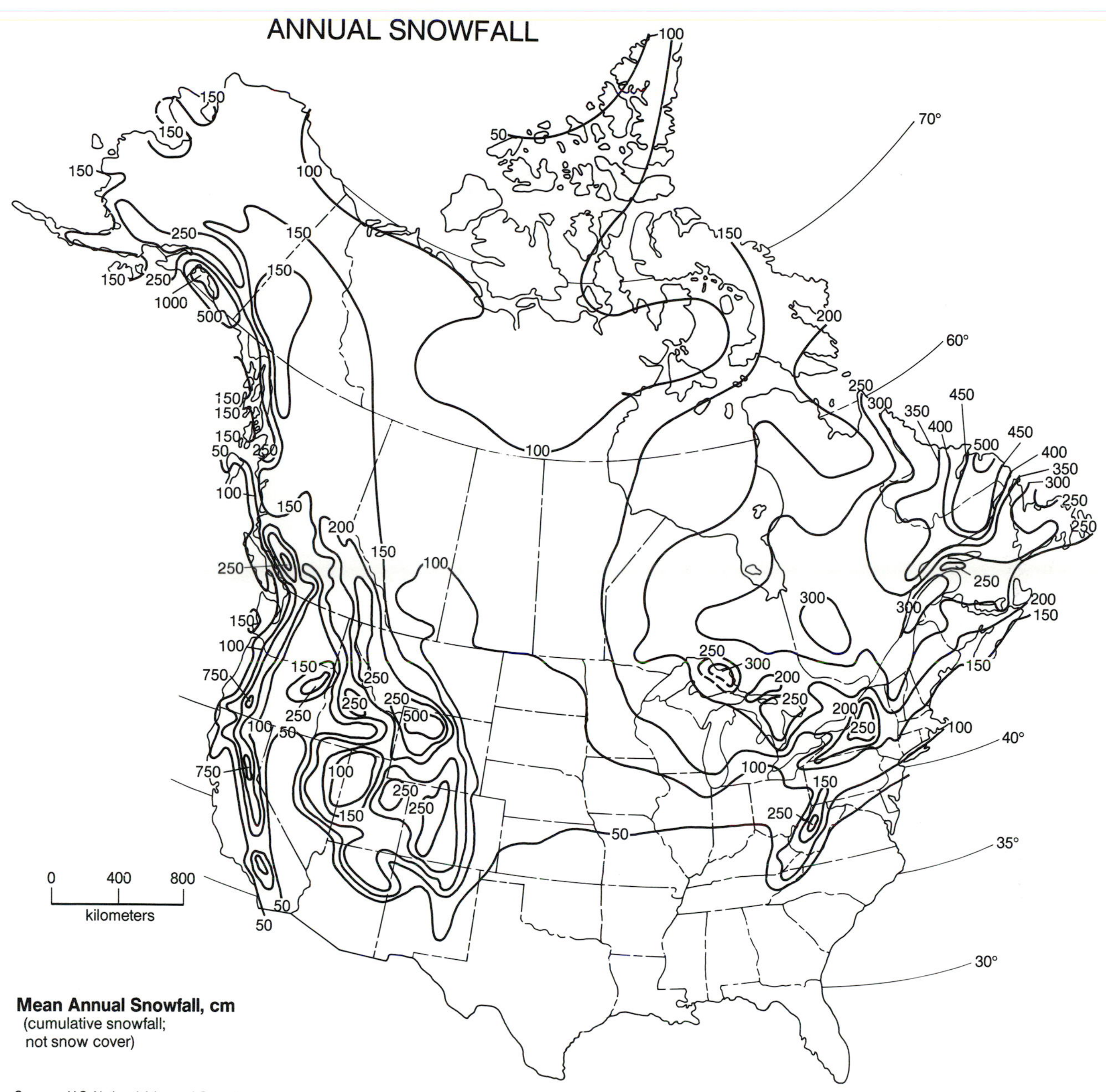
ANNUAL SNOWFALL
Mean Annual Snowfall, cm
(cumulative snowfall;
not snow cover)
0 400 800
kilometers
70°
60°
40°
35°
30°
Sources: U.S. National Atlas and Canadian National Atlas

Figure C.5

Appendix D

Soil Tables

SOIL TEMPERATURE REGIMES

TABLE D-1 DEFINITIONS AND FEATURES OF SOIL TEMPERATURE REGIMES IN THE UNITED STATES

Temperature Regime	Mean Annual Temperature in Root Zone, 5 to 100 centimeters		Characteristics and Some Locations
	C		
Pergelic	<0°	(<32°F)	Permafrost and ice wedges common. Tundra of northern Alaska (and Canada) and high elevations of middle and northern Rocky Mountains
Cryic and frigid[a]	0–8°	(32–47°F)	Cool to cold soils of northern Great Plains of United States (and southern Canada) where spring wheat is the dominant crop. Forested regions of New England
Mesic	8–15°	(47–59°F)	Midwestern and Great Plains regions where corn and winter wheat are common crops
Thermic	15–22°	(59–72°F)	Coastal plain of southeastern United States where temperatures are warm enough for cotton. Central valley of California
Hyperthermic	Over 22°	(Over 72°F)	Citrus areas of Florida peninsula, Rio Grande Valley of Texas, southern California, and low elevations in Puerto Rico and Hawaii. Tropical climates and crops

Source: U.S. Department of Agriculture.

[a]Frigid soils have warmer summers than cryic soils; both have the same mean annual soil temperature. Frigid soils have more than 5°C temperature difference between mean winter and mean summer temperatures at a depth of 50 centimeters (or lithic or paralithic contact, if shallower).

TABLE D-2 SOIL TEMPERATURE CLASSES ACCORDING TO THE CANADIAN SOIL CLASSIFICATION

Extremely Cold

Mean annual soil temperature less than −7°C
Continuous permafrost within 1 m of surface
No significant growing season; less than 15 days above −5°C
Cold to very cool summer with mean summer soil temperature less than −5°C

Very Cold

Mean annual soil temperature between −7°C and −2°C
Discontinuous permafrost may occur below active layer
Short growing season; less than 120 days above 5°C
Moderately cool summer with mean summer soil temperature between 5°C and 8°C

Cold

Mean annual soil temperature between 2°C and 8°C
No permafrost, though seasonal ground frost is the norm
Moderate growing season; 140–220 days above 5°C
Mild summer with mean summer soil temperature between 8°C and 15°C

Cool

Mean annual soil temperature between 5°C and 8°C
Ground frost may or may not occur during dormant season
Moderate growing season; 170–220 days above 5°C
Mild to moderately warm summer with mean summer soil temperature between 15°C and 18°C

Mild

Mean annual soil temperature between 8°C and 15°C
Seasonal ground frost rare
Moderate to nearly continuous growing season; 200–365 days above 5°C
Moderately warm to warm summer with mean summer soil temperature between 15–22°C

Source: The Canadian System of Soil Classification, Ottawa, 1978.

DIAGNOSTIC HORIZONS

TABLE D-3 DIAGNOSTIC SURFACE HORIZONS

Name	Description
Anthropic	Dark colored, thick, high in exchangeable bases, high in phosphate because of long-continued farming
Histic	Thin, high in organic carbon, wet most of the time
Mollic	Dark-colored, thick, high in exchangeable bases, but low in phosphate
Ochric	Too light in color, too thin, or too low in organic carbon to belong to another horizon type
Plaggen	Synthetic, thick, but with textural and chemical characteristics of original natural soil
Umbric	Dark in color but poor in exchangeable bases

Source: U.S. Department of Agriculture.

TABLE D-4 DIAGNOSTIC SUBSURFACE HORIZONS

Name	Description
Albic	Color determined by sand and silt particles, clay and free oxides having been removed
Argillic	Illuvial horizon with significant clay accumulation
Agric	Formed by cultivation, being enriched with clay, humus, or both
Calcic	Enriched with secondary accumulations of calcium carbonate
Cambic	Changed or altered, for instance, by obliteration of the structure of parent material, by liberation of free oxides, or by clay formation
Gypsic	Enriched with calcium sulfate
Natric	Rich in exchangeable sodium; argillic
Oxic	Content of weatherable minerals very low; kaolinitic clay with low exchange capacity
Salic	Enriched with salts more soluble than gypsum
Spodic	With accumulation of free sesquioxides, or organic carbon, or both, but not of clays
Duripan	An horizon cemented by silica or silicates
Fragipan	Partly cemented loamy horizon
Petrocalcic	Strongly cemented by calcium carbonate
Plinthite	Very sesquioxide-rich horizon

Source: U.S. Department of Agriculture.

TABLE D-5 ORDERS AND SUBORDERS, USDA COMPREHENSIVE SOIL CLASSIFICATION SYSTEM

Order	Outline Description	Rapid Recognition Characteristics	Suborders	
Alfisols	Argillic horizon present; base content moderate to high	All mineral horizons present except oxic	With gleying	Aqualfs
			Others in cold climates	Boralfs
			Others in humid climates	Udalfs
			Others in subhumid climates	Ustalfs
			Others in subarid climates	Xeralfs
Aridisols	Semi-desert and desert soils	Ochric or argillic horizon present, no oxic or spodic horizon; usually dry	With argillic horizon	Argids
			Others	Orthids
Entisols	Weakly developed, usually azonal	No diagnostic horizon except ochric, anthropic, albic, or agric	With gleying	Aquents
			With strong artificial disturbance	Arents
			On alluvial deposits	Fluvents
			With sandy or loamy texture	Psamments
			Others	Orthents
Histosols	Developed in organic materials	30% or more organic matter	Rarely saturated, 75% fibric	Folists
			Usually saturated, 75% fibric	Fibrists
			Usually saturated, partly decomposed	Hemists
			Usually saturated, highly decomposed	Saprists
Inceptisols	Moderately developed; not listed elsewhere	Cambic or histic horizon present; no argillic, natric, oxic, or petrocalcic horizon, no plinthite	With gleying	Aquepts
			On volcanic ash	Andepts
			In tropical climates	Tropepts
			With umbric epipedon	Umbrepts
			With plaggen epipedon	Plaggepts
			Others	Ochrepts
Mollisols	With dark A horizon and high base status	Mollic horizon present; no oxic horizon	With albic argillic horizon	Albolls
			With gleying	Aquolls
			On highly calcareous materials	Rendolls
			Others in cold climates	Borolls
			Others in humid climates	Udolls
			Others in subhumid climates	Ustolls
			Others in subarid climates	Xerolls
Oxisols	With oxic horizon	Oxic horizon present	With gleying	Aquox
			With humic A horizon	Humox
			Others in humid climates	Orthox
			Others in drier climates	Ustox
			Usually dry	Torrox

(Continued on Next Page)

TABLE D-5 (*continued*)

Order	Outline Description	Rapid Recognition Characteristics	Suborders	
Spodosols	With spodic horizon	Spodic horizon present	With gleying	Aquods
			With little humus in spodic horizon	Ferrods
			With little iron in spodic horizon	Humids
			With iron and humus	Orthods
Utisols	Argillic horizon present; base status low	Mean annual temperature 8°C or above; soils not listed elsewhere	With gleying	Aquults
			With humic A horizon	Humults
			In humid climates	Udults
			Others in subhumid climates	Ustults
			Others in subarid climates	Xerults
Vertisols	Cracking clay soils	30% or more clay, with gilgai or other signs of up-and-down movement	Usually moist	Uderts
			Dry for short periods	Usterts
			Dry for long periods	Xererts
			Usually dry	Torrerts
Mountain Soils	These vary greatly over short distances; with many steep slopes.			

TABLE D-6 THE UNIFIED SOIL CLASSIFICATION SYSTEM

Letter	Description	Criterion	Further Criteria
G	Gravel and gravelly soils (basically pebble size, larger than 2 mm diameter)	Texture	Based on uniformity of grain size and the presence of smaller materials such as clay and silt:
S	Sand and sandy soils	Texture	W Well graded (uniformly sized grains) and clean (absence of clays, silts, and organic debris) C Well graded with clay fraction, which binds soil together P Poorly graded, fairly clean
M	Very fine sand and salt (inorganic)	Texture; composition	Based on performance criteria of compressibility and plasticity:
C	Clays (inorganic)	Texture; composition	L Low to medium compressibility and low plasticity
O	Organic silts and clays	Texture; composition	H High compressibility and high plasticity
P_t	Peat	Composition	

TABLE D-7 SOIL ORDERS AND GREAT GROUPS OF THE CANADIAN SOIL CLASSIFICATION

Order, Great Group	General Description
Chernozemic Brown Dark brown Black Dark gray	Soils of the semiarid and subhumid grasslands of the Manitoba, Saskatchewan, and Albert. Typically, rich organic accumulation in A horizon and parent material made up principally of silt and clay deposits. Calcium carbonate content is high, and pH is consistently above 7. Soils support the great wheat-growing area of Canada.
Solonetzic Solonetz Solodized solonetz Solod	Soils developed in salinized material mainly within the area of chernozemic soils of the Interior Plains. Very limited geographic coverage (less than 1 percent of Canada) and low agricultural potential owing to the abundant salt.
Luvisolic Gray brown luvisol Gray luvisol	Soils with silicate clay accumulation in the B horizon and fairly strong organic accumulation in A horizon. Found in forested areas and in loamy glacial deposits originally of calcareous composition. In southern Ontario and Quebec, they have been extensively cultivated.
Podzolic Humic podzol Ferro-humic podzol Humo-ferric podzol	Soils with a strong B horizon comprised of iron and aluminum oxides. Characterized by heavy leaching favored by humid climate and sandy parent material. A horizon usually contains significant organic accumulation. Found throughout Canada, covering a total of 15.6 percent of the country.
Brunisolic Melanic brunisol Eutric brunisol Sombric brunisol Dystric brunisol	Soils with a brownish-colored B horizon and a substantial organic accumulation in the A horizon. They form under forest covers in the humid subarctic of northern British Columbia and southern Yukon. Often rocky and thin with a pH around 5–6.
Regosolic Regosol Humic regosol	Soils of recent sandy deposits and active geomorphic environments. Horizons are absent or very weak. They are found with all sorts of climates and vegetation.
Gleysolic Humic gleysol Gleysol Luvic gleysol	Soils of mineral composition that are saturated all or part of the year and characterized by reducing conditions. Gleyed horizons of gray or bluish color are a common trait. Occur in all regions as poorly drained counterpart of other soils.
Organic Fibrisol Mesisol Humisol Folisol	Soils composed of largely organic matter. Found in areas of prolonged saturation and swamp and bog vegetation. Most are underlain by permafrost; located in large area on southern Hudson Bay and prominent within areas of luvisols.
Cryosolic Turbic cryosol Static cryosol Organic cryosol	Soils with permafrost within 1–2 m of surface and mean annual soil temperature under 0°C. Vegetation and texture are highly variable; surface drainage is often poor, and frost action induces mechanical mixing of the active layer. The most extensive soil order in Canada, covering 40 percent of the country.

Source: The Canadian System of Soil Classification, Ottawa, 1978.

TABLE D-8 BEARING CAPACITY VALUES FOR SOIL MATERIALS

Relative Rank	Material	Allowable Bearing Values Tons per ft^2
High	Gravels, sands (well-compacted)	10
	Gravel/sand (compact)	6
	Clay (stiff, dry)	4
	Gravel (loose), coarse sand (compact)	4
	Gravel/sand (loose), fine sand (compact)	3
	Sand (compact)	3
	Fine sand (loose)	2
	Clay (medium-stiff)	2
	Clay (soft)	1
Low	Fill, organic material, silt	(fixed by field test)

Source: Code Manual, New York State Building Code Commission.

Appendix E

Units of Measurements and Conversions

TABLE E-1 CONVERSION FACTORS AND DECIMAL NOTATIONS

Energy, Power, Force, and Pressure

Energy Units and Their Equivalents

joule (abbreviation J): 1 joule = 1 unit of force (a newton) applied over a distance of 1 meter = 0.239 calorie

calorie (abbreviation cal): 1 calorie = heat needed to raise the temperature of 1 gram of water from 14.5°C to 15.5°C = 4.186 joules

British Thermal Unit (abbreviation BTU): 1 BTU = heat needed to raise the temperature of 1 pound of water 1° Fahrenheit from 39.4° to 40.4°F = 252 calories = 1055 joules

Power

watt (abbreviation W): 1 watt = 1 joule per second

horsepower (abbreviation hp): 1 hp = 746 watts

Force and Pressure

newton (abbreviation N): 1 newton = force needed to accelerate a 1-kilogram mass over a distance of 1 meter in 1 second squared

bar (abbreviated b): 1 bar = pressure equivalent to 100,000 newtons on an area of 1 square meter

millibar (abbreviation mb): 1 millibar = one-thousandth ($\frac{1}{1000}$) of a bar

pascal (abbreviation Pa): 1 pascal = force exerted by 1 newton on an area of 1 square meter

atmosphere (abbreviation Atmos.): 1 atmosphere = 14.7 pounds of pressure per square inch = 1013.2 millibars

(*Continued on Next Page*)

TABLE E-1 *(continued)*

Length, Area, and Volume

1 micrometer (μm) = 0.000001 meter = 0.0001 centimeter
1 millimeter (mm) = 0.03937 inch = 0.1 centimeter
1 centimeter (cm) = 0.39 inch = 0.01 meter
1 inch (in.) = 2.54 centimeters = 0.083 foot
1 foot (ft.) = 0.3048 meter = 0.33 yard
1 yard (yd) = 0.9144 meter
1 meter (m) = 3.2808 feet = 1.0936 yards
1 kilometer (km) = 1000 meters = 0.6214 mile (statute) = 3281 feet
1 mile (statute) (mi.) = 5280 feet = 1.6093 kilometers
1 mile (nautical) (mi.) = 6076 feet = 1.8531 kilometers

Area

1 square centimeter (cm^2) = 0.0001 square meter = 0.15550 square inch
1 square inch ($in.^2$) = 0.0069 square foot = 6.452 square centimeters
1 square foot ($ft.^2$) = 144 square inches = 0.0929 square meter
1 square yard (yd^2) = 9 square feet = 0.8361 square meter
1 square meter (m^2) = 1.1960 square yards = 10.764 square feet
1 acre (ac) = 43,560 square feet = 4046.95 square meters
1 hectare (ha) = 10,000 square meters = 2.471 acres
1 square kilometer (km^2) = 1,000,000 square meters = 0.3861 square mile
1 square mile ($mi.^2$) = 640 acres = 2.590 square kilometers

Volume

1 cubic centimeter (cm^3) = 1000 cubic millimeters = 0.0610 cubic inch
1 cubic inch ($in.^3$) = 0.0069 cubic foot = 16.387 cubic centimeters
1 liter (l) = 1000 cubic centimeters = 1.0567 quarts
1 gallon (gal) = 4 quarts = 3.785 liters
1 cubic ft ($ft.^3$) = 28.31 liters = 7.48 gallons = 0.02832 cubic meter
1 cubic yard (yd^3) = 27 cubic feet = 0.7646 cubic meter
1 cubic meter (m^3) = 35.314 cubic feet = 1.3079 cubic yards
1 acre-foot (ac-ft) = 43,560 cubic feet = 1234 cubic meters

Mass and Velocity

Mass (Weight)

1 gram (g) = 0.03527 ounce* = 15.43 grains
1 ounce (oz) = 28.3495 grams = 437.5 grains
1 pound (lb) = 16 ounces = 0.4536 kilogram
1 kilogram (kg) = 1000 grams = 2.205 pounds
1 ton* (ton) = 2000 pounds = 907 kilograms
1 tonne = 1000 kilograms = 2205 pounds

Velocity

1 meter per second (m/sec) = 2.237 miles per hour
1 km per hour (km/hr) = 27.78 centimeters per second
1 mile per hour (mph) = 0.4470 meter per second
1 knot (kt) = 1.151 miles per hour = 0.5144 meter/second

*Avoirdupois, i.e., the customary system of weights and measures in most English-speaking countries.

TABLE E-2 QUANTITIES, DECIMAL EQUIVALENTS, AND SCIENTIFIC NOTATION

Quantity	Decimal Notation	Scientific Notation		Prefix
One trillion (U.S.)	1,000,000,000,000	10^{12}	T	tera-
One billion (U.S.)	1,000,000,000	10^{9}	G	giga-
One million	1,000,000	10^{6}	M	mega-
One thousand	1,000	10^{3}	k	kilo-
One hundred	100	10^{2}	h	hecto-
Ten	10	10	da	deka-
One tenth	0.1	10^{-1}	d	deci-
One hundredth	0.01	10^{-2}	c	centi-
One thousandth	0.001	10^{-3}	m	milli-
One millionth	0.000001	10^{-6}	μ	micro-
One billionth (U.S.)	0.000000001	10^{-9}	n	nano-
One trillionth (U.S.)	0.000000000001	10^{-12}	p	pico-

Glossary

Ablation The wastage of ice or snow by melting and sublimation; in the case of glaciers, it also includes calving.

Abrasion The wearing away of a substance by rasping action; for example, the scouring of bedrock by the boulders carried in the base of a glacier.

Absolute humidity An expression for the water vapor content of air; the mass (weight) of water vapor in a given volume of air (irrespective of the mass of the air); usually expressed in grams of water vapor per cubic meter of air.

Absolute zero The zero point on the Kelvin temperature scale, which represents the state at which there is no molecular vibration in a substance and hence no heat. Corresponds to −273.15 on the Celsius scale.

Absorption The process by which incident radiation is taken up by a substance such as water vapor and converted to other forms of energy.

Abyssal plain The deep-ocean floor; the most extensive part of the ocean basin, which lies between the midoceanic ridges and the trenches, usually 5000 m to 7000 m below sea level.

Acidic soil A soil with a pH less than 7.0.

Acid rain Precipitation whose pH has been significantly lowered by air pollution, from a pH of 5.6 to one as low as 1.5. A serious problem in southeastern Canada, northeastern United States, and northwestern Europe.

Active layer The surface layer in a permafrost environment, which is characterized by freezing and thawing on an annual basis.

Actual evapotranspiration The true amount of moisture given up by the soil over some time period; it is equal to soil-moisture loss plus precipitation in the period of negative moisture balance.

Adaptation A change in an organism that brings it into better harmony with its environment; two types of adaptation are acquired and genetic.

Adaptation strategies A general means of plant adaptation. Three primary strategies are competition, stress toleration, and disturbance tolerance.

Adiabatic cooling A thermodynamic decline in a system, such as a parcel of air, in which there is no transfer of heat or mass from the system. In a rising parcel of air, decompression and expansion result in cooling.

Adsorption A chemical process by which ions are taken up by colloids.

Advection The transfer or exchange of energy in the atmosphere by the lateral movement of air as when cold air gains heat with movement over a warm surface.

Aggradation Filling in of a stream channel with sediment, usually associated with low discharges and/or heavy sediment loads.

Agronomy Agriculture science, much of which is devoted to the study of soil, water, and related phenomena.

Air pressure (see *Sea-level pressure*)

Albedo The percentage of incident radiation reflected by a material. Usage in earth science is usually limited to shortwave radiation and landscape materials.

Alfisols Soils similar to mollisols but associated with forest covers. A whitish layer called an argillic horizon, forms in the B horizon; upper soil is relatively rich in organic matter.

Alkaline soil A soil with a pH greater than 7.0.

Allopatric A theory of speciation in which a new species begins as a small group on the periphery of the parent range.

Alluvial fan A fan-shaped deposit of sediment laid down by a stream at the foot of a slope; very common features in dry regions, where streams deposit their sediment load as they lose discharge downstream.

Alluvium Material deposited by a stream or river.

Alpine meadow A formation of grasses and forbs found in mountains above the treeline; similar to the tundra formation of the arctic.

Andesite An extrusive igneous rock comprised of intermediate amounts of dark minerals; the extrusive counterpart of diorite.

Andesite line The seaward extent of andesite lavas used in some places to delimit the geologic border of the continents.

Angiosperm A flowering, seed-bearing plant; the angiosperms are presently the principal vascular plants on earth.

Angle of repose The maximum angle at which a material can be inclined without failing; in civil engineering the term is used in reference to clayey materials.

Angular momentum A measure of the momentum of an object with respect to its rotation about a point; in the atmosphere the farther poleward a particle of air goes, the closer it gets to the axis of earth rotation and the faster its velocity becomes.

Anion A negatively charged ion.

Anticline A fold characterized by an upward bend, e.g., convex upward in cross-section.

Aphelion The position of the earth in its orbit when it is farthest from the sun—152 million km (94.25 million miles).

Aquiclude An impervious stratum or formation that impedes the movement of groundwater.

Aquifer Any subsurface material that holds a relatively large quantity of groundwater and is able to transmit that water readily.

Archipelago An arc-shaped group of islands, usually of volcanic origin; most archipelagos, e.g., the Japanese Islands, are associated with subduction zones.

Arête The term given to the sharp ridges that separate cirques or faces on a glaciated mountain.

Aridisols Soils of dry environments; poor in organic matter and often heavy in salts.

Artesian flow A pressurized flow of groundwater that reaches an elevation above the water table or even above the ground surface; groundwater that has become sandwiched into an inclined aquifer such that when the lower end is tapped, the water rises under the pressure of the water higher in the aquifer.

Artesian well (see *Artesian flow*)

Asthenosphere The layer immediately under the lithosphere, where the rock appears to be in a plastic state and capable of slow-flowing motion.

Autumnal equinox (see *Equinox*)

Azonal soil A soil order under the traditional USDA soil classification scheme; soils without horizons; those that are usually found in geomorphically active environments such as sand dunes and river valleys.

Backscattering That part of solar radiation directed back into space as a result of diffusion by particles in the atmosphere.

Backshore The zone behind the shore—between the beach berm and the backshore slope.

Backshore slope The bank or bluff landward of the shore that is comprised of *in situ* material.

Backswamps A low, wet area in the floodplain, often located behind a levee.

Backwash The counterpart to swash; the sheet of water that slides back down the beach face.

Bar A unit of force equal to 100,000 newtons per square meter; normal atmospheric pressure at sea level is slightly greater than 1 bar (1.0132 bars).

Barchan A type of sand dune that is crescent-shaped, with its long axis transverse to the wind and its wings tipped downward.

Basalt floods A form of volcanism characterized by massive outflows of lava from long fissures in the crust.

Basaltic rock A general term applied to the rocks that form the ocean basins and the lower crust; relatively high-density, dark-colored igneous rock; basalt and related rocks of the sima layer.

Base In soil chemistry, a mineral that forms cations, e.g., magnesium and potassium.

Base level The lowest elevation to which a river can downcut its channel. For large rivers, this is controlled by sea level; for smaller ones, it may be controlled by a lake, resistant rock, or another river.

Baseflow The portion of streamflow contributed by groundwater; it is a steady flow that is slow to change even during rainless periods.

Basic soil (see *Alkaline soil*)

Basin A rock structure formed by a large downward flexure, often hundreds of km in diameter, e.g., Pasin Basin of France, Michigan Basin of North America.

Batholith A large accumulation of magma in a great chamber; upon cooling, it forms coarse-grained igneous rocks.

Bay-mouth bar A ribbon of sand deposited across the mouth of a bay.

Beam radiation Directed shortwave radiation which passes through the atmosphere without much scattering.

Bed load The stream-carried particles (sediment) that roll along the bottom and are in nearly continuous contact with the streambed.

Bed shear stress The force exerted against a streambed by moving water; it is a function of water density, gravitational acceleration, water depth, and the slope of the channel.

Bergschrund A deep crevasse at the very head of an alpine glacier that opens as the ice pulls away from the mountain side.

Berm A low mound that forms along sandy beaches.

Biochore A major region or division of vegetation defined on the basis of the structural and compositional characteristics of the plant cover; in a general sense, it is the vegetative version of a biome; see also *Biome*.

Biological diversity The number of species of plants, animals, and microorganisms per unit area of land or water.

Biomass The total weight of organic matter per unit area of landscape; also, the total weight of the organic matter in an ecosystem.

Biome A major division (region) of the earth's surface, defined on the basis of the plants and animals inhabiting it; geographically, it may correspond to a climatic zone.

Black body A hypothetical body that is capable of absorbing all radiation incident on it and in turn is the most effective possible emitter of radiation.

Blowout (see *Deflation hollow*)

Boreal forest Subarctic conifer forests of North America and Eurasia; floristically homogeneous forests dominated by fir, spruce, and tamarack; in Russia, it is called *taiga*.

Boundary layer The lower layer of the atmosphere; the lower 300 m of the atmosphere where airflow is influenced by the earth's surface.

Bowen ratio The ratio of sensible-heat flux to latent-heat flux between a surface and the atmosphere.

Braided channel A stream channel characterized by multiple threads or subchannels which appear to weave in and out of one another.

British thermal unit A unit of energy used to measure heat; one BTU is equal to 1054 joules, the amount of heat required to raise the temperature of one pound of water by one degree Fahrenheit.

Brittle rock Rock that ruptures with little or no plastic deformation.

Caatinqa (see *Thornbush*)

Calcic layer A concentration of calcium at or near the surface of soils in arid regions.

Calcification A soil-forming regime of dry environments that results in the accumulation of calcium carbonate and, in grassy areas, a strong organic layer.

Caldera A large, circular depression in a volcano, resulting from an explosion or the loss of magma through a lower vent.

Caliche An accumulation of calcium carbonate at or near the soil surface in an arid environment.

Calorie A unit of energy; the amount of heat required to raise the temperature of water 1°C, from 14.5°C to 15.5°C.

Calving The process by which a glacier loses mass when ice breaks off into the sea.

Canopy The roof of foliage formed by the crowns of trees in a forest.

Capillarity The capacity of a soil to transfer water by capillary action; capillarity is greatest in medium-textured soils.

Capillary fringe The transition between the zone of aeration and the zone of saturation, at the base of which capillary water gives way to groundwater; in fine-textured material the capillary fringe may be several meters thick, whereas in coarse-textured material it is usually only several centimeters thick.

Cation A positively charged ion.

Cation adsorption The process by which cations become attached to colloids.

Cation exchange capacity The total exchangeable cations that a soil can adsorb; the capacity per unit volume of soil increases with finer soil textures.

Cavitation A mechanism of stream and wave erosion brought about by the sudden increase in hydraulic pressure, which causes air bubbles to burst and exert force against rock.

Celsius A temperature scale, also known as centigrade, on which 0° represents the normal freezing point of water, and 100° represents the normal boiling point of water.

Centigrade (see *Celsius*)

Central vent The main passageway by which magma ascends to the surface of a volcano.

Channel precipitation Rain or snow that falls directly on the stream channel and thereby contributes immediately to discharge.

Chapparal A vegetative formation of the Mediterranean climate of California, characterized by shrubs and small trees, often in shrubby thickets.

Chelation A weathering process involving the bonding of mineral ions to a large organic molecule, followed by the removal of the ions with the molecule.

Chemical stability A term referring to the tendency of a mineral to change to another form, such as another mineral; stable minerals are not readily transformed to other states and are usually those, such as quartz, that originated at relatively low temperatures.

Chemical weathering One of the two major types of weathering and generally considered to be the more effective one; it involves the chemical decomposition of rock by a variety of chemical processes, including dissolution, chelation, and hydrolysis.

Chernozem soils One of the great soil groups under the traditional USDA soil classification scheme; soil characterized by a heavy O horizon and calcium carbonate accumulation in the B horizon.

Chinook A dry, often warm, wind that descends the leeward side of a mountain range; in Germany, Austria, and Switzerland such a wind is called a *föhn.*

Circle of illumination The line dividing the illuminated half of the earth from the dark (shadow) half; the alignment of this circle changes with the seasons, thereby changing the daily length of daylight hours at different latitudes.

Climate The representative or general conditions of the atmosphere at a place on earth; it is more than the average conditions of the atmosphere, for climate may also include extreme and infrequent conditions.

Climatology The field of earth science devoted to the study of climate and climatic processes.

Climax community A group of organisms that represents an ecological equilibrium with its environment and therefore is capable of maintaining long-term stability.

Clone A genetically uniform group of plants regenerated by vegetative means (asexual) from a single parent.

Closed forest A forest structure with multiple levels of growth from the ground up; a forest in which undergrowth closes out the area between the canopy and the ground.

Coastal dune A sand dune that forms in coastal areas and is fed by sand from the beach.

Coefficient of runoff A number given to a type of ground surface representing the proportion of a rainfall converted to overland flow; it is a dimensionless number between 0 and 1.0 that varies inversely with the infiltration capacity; impervious surfaces have high coefficients of runoff.

Cold front A contact between a cold air mass and a warm air mass in which the cold air is advancing on the warm air, driving the warm air upward.

Colloid A small clay particle, less than 0.001 mm in diameter, that provides adsorption sites for ions.

Colluvium An unsorted mix of soil and mass-movement debris.

Community A group of organisms that live together in an interdependent fashion.

Community-succession concept A popular concept of vegetation change based on the idea that one plant community succeeds another in occupying a site; succession ends when a climax community is established which represents an equilibrium between vegetation and environment.

Composite volcano A cone-shaped volcano comprised mainly of pyroclastic material, e.g., Fuji-san of Japan and Vesuvius of Italy.

Compressional wave A type of seismic (elastic) wave that generates a back-and-forth motion along the line of energy propagation through a substance; also termed *primary wave*.

Concentration time The time taken for a drop of rain falling on the perimeter of a drainage basin to go through the basin to the outlet.

Condensation The physical process by which water changes from the vapor to the liquid phase.

Condensation nuclei Very small particles of dust or salt suspended in the atmosphere on which condensation takes place to initiate the formation of a precipitation droplet.

Conduction A mechanism of heat transfer involving no external motion or mass transport; instead, energy is transferred through the collision of vibrating molecules.

Cone of ascension The ascent of salt groundwater under a well in response to the development of a cone of depression in the overlying fresh groundwater.

Cone of depression A conical-shaped depression that forms in the water table or an aquifer surface immediately around a well from which groundwater is being rapidly pumped.

Conservation of angular momentum The principle that the angular momentum of a rotating mass, such as the earth or a twirling skater, will not change unless torque (stress) is applied to it; if the radius of rotation decreases (arms in), rotational velocity increases, and vice versa.

Conservation-of-energy principle The principle that energy in an isolated system can be neither destroyed nor created; thus, total energy in the system remains constant.

Conservative zone A type of contact or border along a tectonic plate where lithosphere is neither destroyed nor created; the tectonic movement in conservative zones is often lateral, such as along the San Andreas fault of California.

Constructive zone A type of contact or border along a tectonic plate where new lithosphere is emerging; usually associated with midoceanic ridges.

Continental climate A climate characterized by a large annual temperature range and often relatively dry conditions.

Continental drift The term used to describe the wholesale movement of land masses over great distances on the surface of the earth; the term is generally attributed to Alfred Wegener, who advanced the first coherent theory of continental drift.

Continental shelf The seaward-sloping margin of the continents under water to a depth of 200–300 m; the shoulders of the continents where the rate sediment accumulation and sedimentary rock formation is high.

Continuity of flow A principle that describes the maintenance of flow in a system with changes in flow velocity and system capacity.

Core The innermost of the two major divisions of the solid earth, the other being the mantle; the core includes the *outer core*, which is liquid, and the *inner core*, which is solid; the core is the densest part of the earth.

Coriolis effect The effect of the earth's rotation on the path of airborne objects; winds in the Northern Hemisphere are deflected to the right and those in the Southern Hemisphere are deflected to the left of their original paths; at the equator it is negligible; the Coriolis is an effect apparent only to observers standing on the earth and as such is not a force.

Corrasion Another term for abrasion.

Convection A mechanism of heat transfer which involves mass transport (mixing) of fluid, such as occurs with turbulence in the atmosphere; any mixing motion in a fluid.

Convectional precipitation A type of precipitation resulting from the ascent of unstable air; the instability may be caused by local heating in the landscape or by frontal activity; usually short-term, intensive rainfall.

Convergent precipitation A type of precipitation that takes place when air moves into a low-pressure trough or topographic depression and escapes by moving upward; precipitation in the ITC zone is at least partially convergent.

Cover density The percentage of areal coverage by vegetation of all types in an area.

Critical threshold The point at which shear stress (driving force) equals shear strength (resisting force) and beyond which change, such as slope failure, rock rupture, plant damage, or soil erosion, is imminent.

Crust The outermost zone of the lithosphere, which ranges from 8 km to 65 km (4–40 miles) in thickness and is bounded on the bottom by the Moho discontinuity.

Cumulus A class of clouds characterized by vertical development.

Cuspate foreland A large depositional feature along a coastline, often in the form of a triangular point.

Cyclone A large low-pressure cell characterized by convergent airflow and internal instability; the two main classes of cyclones are midlatitude cyclones and tropical cyclones; in some parts of the United States, tornadoes are also called cyclones.

Cyclonic/frontal precipitation A type of precipitation that results from a large cell of low pressure and the meeting of warm and cold air masses; most of the precipitation is concentrated along the fronts; see *Cold front* and *Warm front*.

Darcy's law The principle that describes the velocity of groundwater flow; velocity is equal to permeability times the hydraulic gradient.

Debris flow A type of mass movement characterized by the downslope flow of a saturated mass of heterogeneous soil material and rock debris.

Declination of the sun The location (latitude) on earth where the sun on any day is directly overhead; declinations range from 23.27° S latitude to 23.27° N latitude.

Deerpark The term that the English use to describe the parkland vegetation of southern England, northwest France, and similar areas; see also *Parkland.*

Deflation hollow A topographic depression caused by wind erosion; also called a *blowout.*

Degradation Scouring and downcutting of a stream channel, usually associated with high discharges.

Denudation A term used to describe the erosion or wearing down of a land mass; also used to describe the process by which a site is stripped of its vegetative cover.

Denude (see *Denudation*)

Deranged drainage Highly irregular drainage patterns in areas of complex geology, such as the Canadian Shield, which have been heavily glaciated.

Desert lag A veneer of coarse particles left on the ground after the fine particles have been eroded away, often by wind.

Desert soils Soils characterized by a weak O horizon and salt accumulation at or near the surface; in the traditional USDA classification scheme, these soils are classed as either Red or Gray desert soils.

Desseminule Any part of a plant from which another plant can be established; seeds, fruits, spores, or vegetative parts.

Destructive zone A type of contact or border along a tectonic plate where lithosphere is destroyed; see also *Subduction.*

Detrital rock Sedimentary rock composed of particles transported to their place of desposition by erosional processes, e.g., sandstone and shale.

Dew point The temperature at which a parcel of air is saturated.

Diagnostic horizon Soil horizons used to define the key conditions and processes of soil formation.

Diapir A conduit in the lithosphere through which magma moves to the surface.

Differential stress Force, directed on a body, that is not equal in all directions.

Diffuse radiation Solar radiation which has been scattered in the atmosphere; light rays moving in random patterns among molecules and particles in the air.

Dike A vertical fissure in the crust that serves as a passageway for magma.

Diorite An intrusive igneous rock that is both darker and of higher density than granite.

Dip (see *Strike and dip*)

Dip-slip fault A fault in which the principal direction of displacement is up or down along the fault plane.

Discharge The rate of water flow in a stream channel; measured as the volume of water passing through a cross-section of a stream per unit of time, commonly expressed as cubic feet (or meters) per second.

Dispersal The process of distribution of a dissemmule from a parent plant to a new location.

Dissociation The process by which water molecules give up hydrogen ions.

Dissolved load Material carried in solution (ionic form) by a stream.

Disturbance Factors, other than the basic requirements, that affect a plant's well-being, e.g., floods, disease, and soil erosion.

Diurnal damping depth The maximum depth in the soil which experiences temperature change over a 24-hour (diurnal) period.

Divine Plan of Nature The idea that the earth, and indeed the entire universe, are designed according to God's great plan.

Doldrums The belt of calm and variable winds in the intertropical convergence zone; in the days of ocean sailing, a source of quiet.

Dolomite A sedimentary rock of chemical origin; it appears to be an altered form of limestone in which some of the calcium is replaced by magnesium.

Dome Large upward flexure of rock often hundreds of km in diameter, e.g., the Ozark Plateaus of Missouri.

Dominant species A plant species that occupies the greatest amount of space in an area as measured by the extent of its foliage or root system.

Dominant stratum A number of plants whose combined foliage makes up the principal layer (story) in a vegetative formation.

Dornveld (see *Thornbush*)

Dowsing The practice of locating groundwater by using a stick, tree branch, or special rod; dowsing is steadfastly practiced by believers, but it has no scientific basis.

Drainage basin The area that contributes runoff to a stream, river, or lake.

Drainage density The number of miles (or km) of stream channels per square mile (or km^2) of land.

Drainage divide The border of a drainage basin or watershed where runoff separates between adjacent areas.

Drainage network A system of stream channels usually connected in a hierarchical fashion (see also *Principle of stream orders*).

Drift (see *Glacial drift* and *Littoral drift*)

Ductile rock Rock that has a relatively large capacity for plastic deformation.

Dust Bowl The name given to the area of severe drought and wind erosion in the Great Plains during the 1930s.

Dwarfism The tendency for a plant to achieve less than full size at maturity because of environmental stress such as inadequate heat or light.

Dynamic equilibrium A term used to describe the behavior of a system, such as a river network, which is continually trending toward a state of equilibrium, but rarely reaches it. The trend may change with changes in the energy available to drive the system.

Earthquake intensity (see *Mercalli scale*)

Earthquake magnitude (see *Richter scale*)

Easterly wave A trough of low pressure in the trade wind belt. They form over water, move westward with the trade winds, and usually produce showers and thunderstorms.

Eccentricity A variation in the shape of the earth's orbit from a more circular to a more elliptical shape over a period of 90,000 to 100,000 years.

Ecological amplitude The variation in tolerance and resource needs from member to member in a plant species.

Ecosystem A group of organisms linked together by a flow energy; also, a community of organisms and their environment.

Ecotone The transition zone between two groups, or zones, of vegetation.

Eddy The term given to a whirling or spiral motion in a fluid.

Edge wave A wave that moves parallel to the shore; usually a secondary wave of complex origins.

Effective frictional surface (see *Roughness length*)

Elastic deformation Change in the shape of a body as a result of differential stress; on the release of the stress, the body returns to its original shape.

Elastic limit The maximum level of elastic deformation of a body, beyond which it ruptures.

Elastic wave An energy wave that causes motion in a material without permanently deforming it.

Electromagnetic spectrum The classification scheme used to describe the array of electromagnetic radiation; the various categories of radiation are distinguished on the basis of wavelength.

Eluviation The removal of colloids and ions from a soil or from one level to another in a soil.

Emissivity The ratio of total radiant energy emitted at a specified wavelength and temperature by a substance to that emitted by a *black body* under the same conditions.

Empirical approach An approach to scientific investigation based on direct observation and measurement.

Energy Generally, the capacity to do work; defined as any quantity that represents force times distance. Joules and calories are energy units commonly used in science.

Energy balance The concept or model that concerns the relationship among energy input, energy storage, work, and energy output of a system such as the atmosphere or oceans.

Energy flux The rate of energy flow into, from, or through a substance; also called radiant flux density and irradiance.

Energy pyramid The attenuation of organic energy in an ecosystem; the decline of energy in an ecosystem as organic matter is passed from one level of organisms to another.

Entisols Soils of recent origins with no horizons or weakly developed ones.

Ephemeral stream A stream without baseflow; one that flows only during or after rainstorms or snowmelt events.

Epiphyte A plant that grows in the superstructure of another plant without rooting in the soil; a nonparasitic aerial plant, e.g., Spanish moss and many orchids.

Equatorial zone The middle belt of latitude, extending 10° or so north and south of the equator.

Equilibrium profile A gently sloping surface extending from the shore through the offshore zone across which wave energy is evenly distributed.

Equinox The dates when the declination of the sun is at the equator, March 20–21 and September 21–22, and the number of hours of dark and daylight in a 24-hour period is the same for all locations on earth. These dates are known as the autumnal equinox and vernal equinox, but which is which depends on the hemisphere.

Erosion The removal of rock debris and soil by an agency such as moving water, wind, or glaciers; generally, the sculpting or wearing down of the land by erosional agents.

Euler's theorem The theorem that describes the movement of a plate on the surface of a sphere; such a plate moves about its own pole, called the Euler pole, along a small-circle path. This theorem helps explain the differences in the rates of sea floor spreading along the midoceanic ridges.

Eutrophication Accelerated biological productivity in a water body as a result of the input of nutrients such as nitrogen and phosphorus.

Evaporite A rock or mineral formed from the evaporation of mineral-rich water, e.g., rock salt and gypsum.

Evapotranspiration The loss of water from the soil through evaporation and transpiration.

Event An episode of a process defined as some quantity of a variable such as a river discharge or wind velocity.

Evolution Biological change over time that results in new or changed relationships between organisms and the environment; irreversible biological change.

Exfoliation A mechanical weathering process involving the breaking off, or "shedding," of slabs of rock in response to the differential expansion of a rock mass.

Exotic river A river, such as the Nile or Colorado, that flows through an arid region after gaining its flow elsewhere; exotic streams usually lose much of their discharge to groundwater recharge and evaporation.

Extrusive rock Igneous rock that forms at the surface and cools quickly.

Fahrenheit A temperature scale on which 32° represents the normal freezing point of water and 212° represents the normal boiling point of water.

Fault A fracture in rock along which there has been displacement of one side relative to the other.

Fault line The linear trend of a fault along the earth's surface as one would see it from the air.

Fault plane The plane representing the fracture surface in a fault.

Fault scarp The part of the fault plane exposed in a fault; in a normal fault it is the upper part of the footwall.

Feedback A return effect of a change; the consequences of a change have a feedback effect if they dampen or amplify the change or the causes of it.

Fell fields Areas of very light plant cover in polar and alpine regions; often rocky areas with scattered lichens, mosses, and small flowering plants.

Ferromagnesian minerals A subgroup of the silicate minerals; rock-forming minerals that are rich in iron and magnesium; dark-colored and relatively high-density minerals, e.g., biotite, hornblende, and olivine.

Fetch The distance of open water in one direction across a water body; it is one of the main controls on wave size.

Firn Névé on a glacier that has survived the entire ablation season; the material that is transformed into glacial ice.

Firn line The lower edge of the accumulation zone on a glacier.

Flank eruption Volcanic eruption that breaks out on the side of a volcano.

Floristic system The principal botanical classification scheme in use today; under this scheme the plant kingdom is made up of divisions, each of which is subdivided into smaller and smaller groups arranged according to the apparent evolutionary relationships among plants.

Fluvioglacial deposits Materials deposited by glacial meltwaters, including outwash plains, kames, and eskers; usually stratified.

Fold A rock structure characterized by a bend in a rock formation.

Föhn (see *Chinook*)

Foot ice An accumulation of grounded ice on and near shore in lakes, oceans, and rivers.

Footwall The lower surface of an inclined fault.

Forest fire (see *Ground fire*)

Formation A structural unit of vegetation that may be considered a subdivision of a biochore; a formation may be made up of several communities. In the traditional terminology, it is called a physiognomic unit; in geology, a major unit of rock.

Free convection Mixing motion in a fluid caused by differences in density. In the atmosphere such differences are usually caused by differential heating of air near the surface.

Free-face A steep slope or cliff formed in bedrock.

Freeze-thaw activity Weathering and mass-movement processes associated with daily and seasonal cycles of freezing and melting.

Frequency The term used to express how often a specified event is equaled or exceeded.

Friable A term used to describe the tendency of a soil to crumble or break up when plowed.

Front (see *Cold front* and *Warm front*)

Frost wedging A mechanical weathering process in which water freezes in a crack and exerts force on the rock, which may result in the breaking of the rock; a very effective weathering process in alpine and polar environments.

Fusion Another word for freezing.

Gabbro A coarse-grained, intrusive igneous rock that is dark and heavy owing to a relatively high percentage of ferromagnesian minerals.

Geodesy The science that measures the geoid and its major features.

Geographic cycle A concept developed by William Morris Davis on the formation of river-eroded landscapes. It describes three stages of landscape development (youth, maturity, and old age) and argues that rejuvenation takes place with uplift of the land, thereby renewing the cycle.

Geoid The term given to the true shape of the earth, which deviates from a perfect sphere because of a slight bulge in the equatorial zone.

Geomechanical regime A soil environment where the formative processes are dominated by mechanical mixing associated with ground frost or wetting and drying of clay.

Geomorphic system A physical system comprised of an assemblage of landforms linked together by the flow of water, air, or ice.

Geomorphology The field of earth science that studies the origin and distribution of landforms, with special emphasis on the nature of erosional processes; traditionally, a field shared by geography and geology.

Geophysics A field of earth science devoted to the study of the earth, including the oceans and the atmosphere, through the application of models and techniques from physics.

Geostrophic winds Winds in the upper troposphere which generally flow parallel to isobars and often reach high velocities.

Geothermal energy Energy in the form of heat that is produced by the earth's interior and flows through the crust mainly by conduction.

Glacial drift A general term applied to all glacial deposits, including moraine and fluvioglacial deposits.

Glacial flour (see *Glacial milk*)

Glacial lake A natural impoundment of meltwater at the edge of a glacier.

Glacial milk Glacial meltwater of a light or cloudy appearance because of clay-sized sediment held in suspension.

Glacial polish Bedrock surfaces that have been made smooth and shiny by glacial abrasion.

Glacial surge A rapid advance of the snout of a glacier.

Glacial uplift Uplift of the crust following isostatic depression under the weight of the continental glaciers.

Gleization A soil-forming regime of poorly drained areas such as bogs and swamps; it results in a heavy organic layer over a layer of blue clay.

Global coordinate system The network of east-west and north-south lines (parallels and meridians) used to measure locations on earth; the system uses degrees, minutes, and seconds as the units of measurement.

Gneiss A metamorphosed form of granite characterized by minerals arranged in bands.

Graben (see *Rift*)

Graded profile The longitudinal profile of a stream representing an equilibrium condition toward which the stream adjusts in response to changes in discharge and sediment load.

Grafting The practice of attaching additional channels to a drainage network; in agricultural areas new channels appear as drainage ditches; in urban areas, as stormsewers.

Granite An intrusive igneous rock comprised mainly of quartz and feldspar; limited in its distribution to the continents.

Granitic rock A general term applied to the rocks that comprise the continental masses; low density, light-colored igneous rock; granite and related rocks of the sial layer.

Granodiorite An intrusive igneous rock that is intermediate in composition between granite and diorite.

Graupel A frozen precipitation particle comprised of a snow crystal and a raindrop frozen together.

Gravity water Subsurface water that responds to the gravitational force (in contrast to capillary water, which responds to molecular forces); the water that percolates through the soil to become groundwater.

Great circle Any circle that circumscribes the full circumference of the earth and the plane of which passes through the center of the earth; the equator is a great circle; the shortest distance between any two points on the globe follows a great-circle route.

Greenbelt A tract of trees and associated vegetation in urbanized areas; it may be a park, nature preserve, or part of a transportation corridor.

Greenwich meridian The zero degree meridian from which east and west longitude are measured; it is named for the town of Greenwich, England, through which the line is drawn.

Groin A wall or barrier built from the beach into the surf zone for the purpose of slowing down longshore transport and holding sand.

Gross sediment transport The total quantity of sediment transported along a shoreline in some time period, usually a year.

Ground fire All-consuming fire that burns trees, ground plants, and topsoil.

Ground frost Frost that penetrates the ground in response to freezing surface temperatures.

Groundwater The mass of gravity water that occupies the subsoil and upper bedrock zone; the water occupying the zone of saturation below the soil-water zone.

Gulf stream A large, warm current in the Atlantic Ocean that originates in and around the Caribbean and flows northwestward to the North Atlantic and northwest Europe.

Gullying Soil erosion characterized by the formation of narrow, steepsided channels etched by rivulets or small streams of water. Gullying can be one of the most serious forms of soil erosion of cropland.

Guyot Volcanic island eroded to sea level and then submerged with subsidence of the ocean floor. Coral reefs may form on them.

Gymnosperm A plant that bears naked seeds; the most common group of gymnosperms are the conifers, needleleaf cone-bearing plants.

Gyre A large circular pattern of ocean currents associated with major systems of pressure and prevailing winds; the largest are the subtropical gyres.

Habitat The surrounding or local environment in which an organism gains its resources.

Habitat versatility (see *Ecological amplitude*)

Hadley cells Large atmospheric circulation cells comprised of rising air near the equator, poleward flow aloft, descending air in the subtropics, and return flow on the surface in the form of trade winds;

concept first proposed by George C. Hadley in 1735.

Hair hygrometer An instrument for measuring atmospheric humidity based on the reaction of human hair to changes in vapor levels.

Hamada (see *Reg*)

Hanging valley A tributary valley that enters the main valley at an elevation well above the valley floor; most common in areas of mountain glaciation, where hanging valleys are often the sites of spectacular waterfalls.

Hanging wall The upper surface of an inclined fault.

Hardpan A hardened soil layer characterized by the accumulation of colloids and ions.

Heat island The area or patch of relatively warm air which develops over urbanized areas.

Heat syndrome A health problem caused by some combination of extreme heat, exposure to solar radiation, physical exertion, and loss of body fluids and salts.

Heat transfer The flow of heat within a substance or the exchange of heat between substances by means of conduction, convection, or radiation.

High latitude The zones poleward of the Arctic and Antarctic circles.

Higher plants Generally, the larger and more advanced plants; the vascular plants; pteridophytes, gymnosperms, and angiosperms.

Hillslope processes The geomorphic processes that erode and shape slopes; mainly mass movements such as soil creep and landslides and runoff processes such as rainwash and gullying.

Histosols Soils dominated by a thick organic layer as a result of hydromorphic soil-forming conditions.

Horizon A layer in the soil that originates from the differentiation of particles and chemicals by moisture movement within the soil column; four major horizons are recognized in a standard soil profile: O, A, B, and C.

Horst A fault characterized by a block displaced upward relative to adjacent rock formations.

Humboldt current A cold current that flows northward along the west coast of South America, contributing to the aridity there.

Humus Organic matter in the soil that has been broken down by physical, chemical, and biological processes into a granular form which is relatively stable.

Hurricane A large tropical cyclone characterized by convergent airflow, ascending air in the interior, and heavy precipitation.

Hydraulic gradient The slope of the water table or any body of groundwater; equal to the difference in the elevation of the water table at two points divided by the distance between them.

Hydraulic pressure (see *Cavitation*)

Hydraulic radius The ratio of the cross-sectional area of a stream to its wetted perimeter.

Hydrograph A streamflow graph which shows the change in discharge over time, usually hours or days; see also *Hydrograph method.*

Hydrograph method A means of forecasting streamflow by constructing a hydrograph that shows the representative response of a drainage basin to a rainstorm; the use of a "normalized" hydrograph for flow forecasting in which the size of the individual storm is filtered out; see also *Hydrograph.*

Hydrologic cycle The planet's water system, described by the movement of water from the oceans to the atmosphere to the continents and back to the sea.

Hydrologic equation The amount of surface runoff (overland flow) from any parcel of ground is proportional to precipitation minus evapotranspiration loss, plus or minus changes in storage water (groundwater and soil water).

Hydrolysis A complex chemical weathering process, or series of processes, involving the reaction of water and an acid on a mineral; it is considered to be the most effective process in the decomposition of granite.

Hydromorphic regime A soil environment where the formative processes are dominated by water, usually a swamp or a bog.

Hydrophyte Water-loving plants; aquatic plants such as water lily and water hyacinth.

Hydrostatic pressure The pressure exerted by elevated groundwater as in artesian flow.

Hygrophyte Water-tolerant plants, such as cattail, which are able to grow in saturated or lightly flooded sites.

Hygroscopic water Molecular water that resides directly on the surface of all materials; it is bound to surfaces under such great pressure that it is immobile and cannot be evaporated or used by plants.

Ice Age (see *Pleistocene Epoch*)

Ice wedging (see *Frost wedging*)

Illuviation The process of accumulation of ions and colloids in a soil.

Inceptisols Soils with horizons in the early phases of development; usually formed on young geomorphic surfaces such as midlatitude glacial deposits.

Individualistic concept A concept of vegetation change contrary to the community-succession concept, especially the idea of a climax community; it argues that the stability of the climax community is a matter of probability because those species in greatest abundance favor regeneration of their own kind.

Infiltration capacity The rate at which a ground material takes in water through the surface; measured in inches or centimeters per minute or hour.

Inflooding Flooding caused by overland flow concentrating in a low area.

Infrared radiation Mainly longwave radiation of wavelengths between 3.0–4.0 and 100 micrometers, but also includes near infrared radiation, which occurs at wavelengths between 0.7 and 3.0–4.0 micrometers.

In situ A term used to indicate that a substance is in place as contrasted with one, such as river sediment, that is in transit.

Instability A physical condition in which a fluid is gravitationally unstable; in the atmosphere it is one in which heavier (denser) air overlies lighter air.

Interception The process by which vegetation intercepts rainfall or snow before it reaches the ground.

Interflow Infiltration water that moves laterally in the soil and seeps into stream channels; in forested areas this water is a major source of stream discharge.

Interglacial A relatively warm and dry period in the Pleistocene Epoch during which most of the continental glaciers are thought to have melted away.

Intermittent A stream with baseflow in all but the dry season when the water table drops below the streambed.

Intertropical convergence zone The belt of convergent airflow and low pressure in the equatorial zone (between the tropics) which is fed by the trade winds.

Intertropical zone The zone between the Tropic of Cancer (23.5°N latitude) and the Tropic of Capricorn (23.5°S latitude).

Intrazonal soils A soil order under the traditional USDA soil classification scheme: soils that form under conditions of impeded drainage.

Intrusive rock Igneous rock that forms within the earth and cools slowly; see also *Plutonic rock.*

Inversion (see *Temperature inversion*)

Ion A minute particle of a dissolved mineral; usually an atom or group of atoms that are electrically charged.

Isostatic depression Large-scale down-warping of the crust in response to an increase in mass (weight) on the surface; in areas of continental glaciation the crust was depressed by the weight of the ice.

Isostatic rebound The uplift or recovery of the earth's crust following isostatic depression; elastic recovery of the crust from large-scale depression; see also *Isostatic depression.*

Jet stream Zone of concentrated geostrophic winds; in the midlatitudes it is called the polar front jet stream because it often coincides in location with the polar front.

Joint line An open crack or fracture in bedrock, usually a result of weathering.

Joule A unit of energy equal to one newton (a unit of force) applied over a one-meter length; in terms of heat, 4186 joules are needed to raise the temperature of 1 kilogram of water 1°C, from 14.5°C to 15.5°C.

Kame A mound- or cone-shaped deposit of sand and gravel laid down by melting water in and around glacial ice.

Kaolinite A type of clay produced in the weathering of granite; it is especially widespread in tropical and subtropical regions.

Kaŕmań vortex street A train of wind eddies downwind from a hill.

Karst topography Irregular topography in areas of carbonate rock, characterized by sinkholes, caverns, and underground drainage channels.

Katabatic wind Any wind blowing down a large incline such as a mountain slope; chinook winds are katabatic winds.

Kelvin A temperature scale based on absolute zero, the temperature at which a substance has no molecular vibration and thus generates no heat. Water freezes at 273.15°K and boils at 373.15°K.

Kilogram A metric unit of mass (weight) equal to 2.208 pounds.

Kinetic energy The energy represented by the motion of a substance; equal to mass times velocity squared, divided by two.

Kettlehole A pit in an outwash plain or moraine left from a buried block of glacial ice which melted away; see also *Pitted topography.*

Laminar flow Flow characterized by one layer of a fluid sliding over another without vertical (turbulent) mixing; the source of flow resistance is limited to intermolecular friction within the fluid.

Laminar sublayer In the atmosphere the layer of essentially calm air immediately adjacent to fixed surfaces such as vegetation and soil; in reality this air is not perfectly calm, but characterized by a faint laminar flow parallel to the surface.

Land cover The materials such as vegetation and concrete that cover the ground; see also *Land use.*

Landscape The composite of natural and human features that characterize the surface of the land at the base of the atmosphere; includes spatial, textural, compositional, and dynamic aspects of the land.

Landslide A type of mass movement characterized by the slippage of a body of material over a rupture plane; often a sudden and rapid movement.

Land use The human activities that characterize an area, e.g., agriculture, industry, and residential.

Latent heat The heat released or absorbed when a substance changes phase as from liquid to gas. For water at 0°C, heat is absorbed or released at a rate of 2.5 million joules per kilogram (597 calories per gram) in the liquid/vapor phase change.

Laterite A layer of iron and aluminum oxide accumulation in tropical soils, mainly the latosols.

Laterization A soil-forming regime of warm, moist environments that produces a strongly leached soil with light topsoil and heavy accumulations of iron and aluminum oxides; also called *ferratillization.*

Latosols One of the great soil groups under the traditional USDA soil classification scheme; soils characterized by a weak O horizon, heavy accumulation of laterite, and a deeply weathered profile.

Lava Molten rock that has reached the surface; see also *Magma.*

Law of plastic deformation A physical principle describing the deformational response of a substance to increasing shear stress.

Leaching The removal of minerals in solution from a soil; the washing out of ions from one level to another in the soil.

Levee A mount of sediment which builds up along a river bank as a result of flood deposition.

Life cycle The biological stages in the complete life of a plant.

Life form The form of individual plants or the form of the individual organs of a plant; in general, the overall structure of the vegetative cover may be thought of as life form as well.

Limb Term applied to the flanks of a fold when viewed in cross-section; in an anticline the limbs slope away from the axis.

Limestone A sedimentary rock of chemical and biological origins; calcium carbonate precipitated from seawater and deposited in the form of the shells and skeletons of sea creatures.

Limiting factors (see *Principle of limiting factors*)

Lithosol An azonal soil comprised of large fragments of bedrock.

Lithosphere The upper layer of the mantle; the unit in which the tectonic plates are defined; it is about 100 km thick and includes the crust.

Little Ice Age A period of climatic change in the Northern Hemisphere generally from the fourteenth through the eighteenth centuries; it was marked by a cooling trend and manifested by glacial advances in the seventeenth and eighteenth centuries.

Littoral drift The material that is moved by waves and currents in coastal areas.

Littoral transport The movement of sediment along a coastline; it is comprised of two components: longshore transport and onshore-offshore transport.

Loess Silt deposits laid down by wind over extensive areas of the midlatitudes during glacial and postglacial times.

Longshore current A current that moves parallel to the shoreline; velocities generally range between 0.25 and 1 m/sec.

Longshore transport The movement of sediment parallel to the coast.

Longwave radiation Radiation at wavelengths greater than 3.0–4.0 micrometers; includes infrared (thermal), radio waves, and microwaves.

Mafic lava Lava with a low quartz content in which silicon dioxide constitutes about 50 percent of the rock; this lava is prevalent in the ocean basins.

Magma Molten rock within the lithosphere which cools to become igneous rock; magma that reaches the surface is called lava.

Magnetic polarization The polarization of magnetized materials such as iron particles in volcanic or sedimentary rock.

Magnitude and frequency The concept concerning the behavior of processes and the resultant changes they produce individually and collectively in the landscape; it involves which events render the greatest change and what kinds of change different-sized events render.

Mantle One of the two major divisions of the solid earth, the other being the core; the mantle includes the lithosphere, asthenosphere, mesosphere (the upper mantle), and the lower mantle; the mantle contains about two-thirds of the earth's mass.

Maquis Shrubby vegetation of the Mediterranean lands of Europe; apparently a response to climate and long-term disturbance by various land uses; also called *macchia* and *garique:* see also *Chapparal.*

Maritime climate A climate characterized by a small annual temperature range and often high rainfall.

Mass balance The relative balance in a system, based on the input and output of material such as sediment or water; the state of equilibrium between the input and output of mass in a system.

Mass budget (see *Mass balance*)

Mass movement A type of hillslope process characterized by the downslope movement of rock debris under the force of gravity; it includes soil creep, rock fall, landslides, and mudflows; also termed *mass wasting.*

Meander A bend or loop in a stream channel.

Meander belt The width of the train of active meanders in a river valley.

Mechanical weathering One of the two major types of weathering; it produces physical fragmentation of rock by ice wedging, rock expansion, and a variety of other mechanisms.

Megastorm A great cluster of thunderstorms covering an area of 10,000 mi^2 or more, and lasting 12 to 18 hours.

Mercalli scale A scale for rating the intensity of an earthquake; intensity is a measure of an earthquake's destructive effect in the landscape.

Meridians The north-south-running lines of the global coordinate system; meridians converge at the North and South poles; the Prime Meridian marks 0° longitude.

Mesa A flat-topped mass of bedrock that rises sharply above the surrounding terrain; it is usually capped by a resistant formation of rock and has the general aspect of a broad table.

Mesophyll The inner tissue of a leaf where moisture is stored and from which it is released in transpiration.

Mesophyte Plants with intermediate water requirements, usually found in sites with well-drained soils but adequate soil moisture in most months.

Mesosphere A subdivision of the atmosphere that lies above the stratosphere, extending from 50 km to 90 km altitude.

Metastable state In chemical weathering the condition of a mineral when it is intermediate between stability and instability.

Meteorology The field of earth science that studies the weather, with emphasis on forecasting short-term changes and events.

Microflora Minute plant life in the soil, mainly bacteria, algae, and fungi, that consume vegetal matter and in turn help produce humus; the most effective consumers of the organic matter deposited on the soil by plants.

Middle latitude Generally, the zone between the pole and the equator in both hemispheres; usually given as 35°–55° latitude.

Midoceanic ridge The volcanic mountain chain located in the interior of an ocean basin along a zone of seafloor spreading.

Migration The successful growth and establishment of a plant in a new location.

Millibar A unit of force (or pressure) equal to one-thousandth of a bar; normal atmospheric pressure at sea level is 1013.2 millibars (mb); see also *Bar.*

Mineral A naturally occurring inorganic substance with a characteristic crystal (molecular) structure that is fundamentally the same in all samples.

Mixing ratio An expression for the vapor content of air; the weight (mass) of water vapor relative to the weight of the dry air occupying the same space; usually measured in grams per kilogram.

Moho discontinuity The lower boundary of the crust, where seismic wave velocities show an appreciable increase; the exact nature of the Moho and its significance in the lithosphere is not known; also called the *M discontinuity.*

Moisture deficit A term in the soil-moisture balance; the difference between actual and potential evapotranspiration; the difference between the demand for and availability of soil moisture for evapotranspiration.

Moisture index The difference between precipitation and evapotranspiration; used in the Thornthwaite System of climate classification to distinguish different climatic zones.

Mollisols Soils of the grasslands, usually in semiarid zones. Rich in organic matter with calcium carbonate nodules at depth.

Monsoon A seasonal wind system in South Asia which blows from sea to land in summer, bringing moisture to the continent, and from land to sea in winter, bringing dry conditions to India and neighboring lands.

Montmorillonite A type of clay that is notable for its capacity to shrink and expand with wetting and drying.

Moraine The material deposited directly by a glacier; also, the material (load) carried in or on a glacier; as landforms, moraines usually have hilly or rolling topography.

Mudflow A type of mass movement characterized by the downslope flow of a saturated mass of clayey material.

Neat tide (see *Tide*)

Near-deserts A term sometimes used to describe deserts with heavier than average vegetative covers; specifically, the American deserts with diverse plants, including large ones such as the sahuaro cactus.

Net photosynthesis The energy balance of a plant; the balance between the energy produced in photosynthesis and that used in respiration.

Net sediment transport The balance between the quantites of sediment moved in two (opposite) directions along a shoreline.

Névé Partially melted and compacted snow; it generally has a density of at least 500 kg/m^3.

Newton A measure of force; the force necessary to accelerate a 1-kilogram mass 1 meter in 1 second squared.

Nitrogen fixation The process by which gaseous nitrogen is converted by microorganisms living in association with certain plants to a form that can be stored in the soil and utilized by plants.

Nonferromagnesian minerals A subgroup of the silicate minerals; light-colored, low-density, rock-forming minerals, e.g., quartz and orthoclase feldspar; also called *Aluminosilicate* minerals.

Nonparallel slope retreat A mode of slope retreat in which the slope angle grows smaller as the slope is eroded back; see also *Parallel slope retreat.*

Nonsedentary A settlement type, such as nomadism, characterized by groups shifting about an area.

Normal fault A fault in which the hanging wall is displaced downward relative to the footwall.

Nuée ardente A dense, "glowing cloud" of hot volcanic gas and ash which moves downhill at high speeds, scorching the landscape.

Oblique-slip fault A fault that combines both strike-slip and dip-slip displacements.

Obliquity A variation in the angle of the earth's axis from 21.8 to 24.4 degrees with a period of 40,000 years.

Occluded front A frontal condition in a midlatitude cyclone in which the cold and warm fronts have merged, forcing the warm-air sector upward.

Ocean trench A great trough in the ocean floor, between 7500 m and 11,000 m below sea level; trenches are associated with subduction and lie along island arcs or orogenic belts.

Open forest A forest structure with a strong upper one or two stories and limited undergrowth; a forest that is largely open at ground level.

Open system A system characterized by a throughflow of material and/or energy; a system to which energy or material is added and released over time.

Orogenic belt A major chain of mountains on the continents; one of the major geologic subdivisions of the continents.

Orographic precipitation A type of precipitation that results when moist air is forced to rise when passing over a mountain range; most areas of exceptionally heavy rainfall are areas of orographic precipitation.

Oscillatory wave A wave in which there is no mass transport of water; the motion of the wave is circular; thus, water particles return to their original position with the passage of each wave.

Outflooding Flooding caused by a stream or river overflowing its banks.

Outwash plain A fluvioglacial deposit comprised of sand and gravel with a flat or gentle sloping surface; usually found in close association with moraines.

Overland flow Runoff from surfaces on which the intensity of precipitation or snowmelt exceeds the infiltration capacity; also called Horton overland flow, for hydrologist Robert E. Horton.

Oxbow A crescent-shaped lake or pond formed in an abandoned segment of river channel.

Oxisols Soils of tropical, moist environments which have undergone intense weathering; weak in organic matter, but heavy in oxides of iron and aluminum.

Ozone One of the minor gases of the atmosphere; a pungent, irritating form of oxygen that performs the important function of absorbing ultraviolet radiation.

Paleosol A soil exhibiting relict features, the result of some past conditions and processes.

Pangaea An ancient supercontinent that was comprised of the world's major land masses packed together around Africa several hundred million years ago; the breakup of Pangaea led to the formation of today's continents.

Parallels The east-west-running lines of the global coordinate system; the equator, the Arctic Circle, and the Antarctic Circle are parallels; all parallels run parallel to one another.

Parallel slope retreat A mode of slope retreat in which the slope angle remains essentially constant as the slope is eroded back; see also *Nonparallel slope retreat.*

Parent material The particulate material in which a soil forms; the two types of parent material are *residual* and *transported.*

Parkland A savanna formation of the midlatitudes characterized by prairie or meadows, with patches and ribbons of broadleaf trees.

Patterned ground Ground in which vegetation, water features, or stones are arranged in a geometric pattern, e.g., circles or polygons; it is widespread in cold environments.

Peak annual flow The largest discharge produced by a stream or river in a given year.

Pedalfers A general class of soil characterized by accumulations of iron and aluminum; soils in areas that receive at least 60 cm of precipitation annually.

Pediment Long, gentle slope at the foot of a cliff or free-face; it is usually composed of bedrock with a light covering of rock debris; common in dry regions.

Pedocals A general class of soil characterized by calcium accumulations and found in areas that receive less than 60 cm of precipitation annually.

Pedon The smallest geographic unit of soil defined by soil scientists of the U.S. Department of Agriculture.

Percolation test A soil-permeability test performed in the field to determine the suitability of a material for wastewater disposal; the test most commonly used by sanitarians and planners to size soil-absorption systems.

Perennial stream A stream that receives inflow of groundwater all year; a stream that has a permanent baseflow.

Peridoite A coarse-grained igneous rock found at depth in the lithosphere; a very dark, high-density rock.

Periglacial environment An area where frost-related processes are a major force in shaping the landscape.

Permafrost A ground-heat condition in which the soil or subsoil is permanently frozen; long-term frozen ground in periglacial environments.

Permeability The rate at which soil or rock transmits groundwater (or gravity water in the area above the water table); measured in cubic feet (or meters) of water transmitted through a specified cross-sectional area when under a hydraulic gradient of 1 foot per 1 foot (or 1 m per 1 m).

pH (see *Soil pH*)

Phase change Reorganization of a substance at the atomic or molecular level resulting in a change in physical state as from liquid to vapor; also called *phase transition.*

Phase transition (see *Phase change*)

Phenological adaptation A form of plant adaptation in which the stages in the life cycle (e.g., flowering, pollination, seed germination) are in phase (adjusted to) the seasons and the periodicity of certain events in the year.

Phloem (see *Xylem and phloem*)

Photoperiod The duration of the daily light period when photochemical activity can take place in a plant.

Photosynthesis The process by which green plants synthesize water and carbon dioxide and, with the energy from absorbed light, convert it into plant materials in the form of sugar and carbohydrates.

Piedmont glacier A large glacier usually formed from the merger of many alpine glaciers.

Piezometric surface The theoretical elevation (datum) to which groundwater would adjust if released from the differential pressure under which it normally exists.

Pioneer One of the communities of the community-succession concept; the first community of plants to occupy a new site.

Piping The formation of horizontal tunnels in a soil due to sapping, i.e., erosion by seepage water; piping often occurs in areas where gullying is or was active and is limited to soils resistant to cave-in.

Pitted topography Glacial terrain characterized by a pocked surface; the pits result from buried ice blocks that have melted away.

Plane of the ecliptic The plane defined by one complete revolution of the earth around the sun.

Plant production The rate of output of organic material by a plant; the total amount of organic matter added to the landscape over some period of time, usually measured in grams per square meter per day or year.

Plant stress Limitations placed on photosynthesis by too much or too little of the basic requirements, namely, light, heat, water, carbon dioxide, and certain minerals.

Plastic deformation Irreversible change in the shape of a body without rupturing.

Plate tectonics The geophysical theory or model in which the lithosphere is partitioned into great plates which move laterally on the surface of the earth; plate tectonics emerged as a serious scientific proposition with the articulation of the theory of continental drift.

Playa A dry lake bed in the desert.

Pleistocene Epoch The present Ice Age which began 1–2 million years ago.

Plucking The process by which a glacier removes blocks of rock from the bedrock; an erosional process associated with melting and refreezing at the base of glacial ice.

Plunging fold A fold whose axis is inclined rather than horizontal to the earth's surface.

Plutonic rock Deep intrusive rock; igneous rock that forms in large chambers well within the crust.

Podzolization A soil-forming regime of cool, moist environments that produces a strongly leached soil with a distinctive hardpan layer.

Podzols One of the great soil groups under the traditional USDA soil-classification scheme; soils characterized by a strong O horizon, a leached A horizon, and a B horizon containing oxides of iron and aluminum.

Point bar Deposit in a stream channel on the inside of a meander or bend.

Polar front The zone or line of contact in the mid-latitudes between polar/arctic air and tropical air; it often coincides with the polar front jet stream.

Polar glacier A glacier characterized by frozen conditions throughout and no meltwater. Common in the Antarctic, these glaciers are frozen fast to the underlying ground.

Polar zone The upper high latitudes, 75°–90° latitude.

Polypedon A group of pedons having similar characteristics; also called a *soil body*.

Pools and riffles Features of stream channels; pools are quiescent places separated by riffles, or reaches of rapid flow.

Pore water pressure The pressure exerted by groundwater against the particles through which it is flowing.

Porosity The total volume of pore (void) space in a given volume of rock or soil; expressed as the percentage of void volume to the total volume of the soil or rock sample.

Potential energy The energy represented by the elevation of mass above a critical datum plane (elevation), e.g., the elevation of rainwater above sea level.

Potential evapotranspiration The projected or calculated loss of soil water in evaporation and transpiration over some time period given an inexhaustible supply of soil water.

Precession A variation in the earth's orbit, which has a period of 21,000 to 23,000 years and results in the equinox date.

Precipitable water vapor The total amount of water in the atmosphere; the average depth of water added to the earth's surface if all the moisture in the atmosphere were to condense and fall to earth.

Precipitation The term used for all moisture—solid and liquid—that falls from the atmosphere.

Pressure cell A body of air, usually covering a large area, which is defined on the basis of air pressure, either high pressure or low pressure.

Pressure gradient The change in pressure over distance between two points; on weather maps the pressure gradient is measured along a line drawn at right angles to the isobars.

Primary consumer An organism that eats plants as its sole source of substance; it may be either a plant (e.g., bacteria or algae) or an animal (e.g., deer or buffalo).

Prime meridian (see *Greenwich Meridian*)

Principle of limiting factors The biological principle that the maximum obtainable rate of photosynthesis is limited by whichever basic resource of plant growth is in least supply.

Principle of stream orders The relationship between stream order and the number of streams per order; the relationship for most drainage nets is an inverse one, characterized by many low-order streams and fewer and fewer streams with increasingly higher orders; see also *Stream order.*

Productivity (see *Plant production*)

Progradation A term used to describe a shoreline that builds seaward.

Pruning In hydrology the cutting back of a drainage net by diverting or burying streams; usually associated with urbanization or agricultural development.

Psychrometer An instrument for measuring atmospheric humidity, comprised of two thermometers—a wet bulb and a dry bulb; humidity is measured by the difference in readings between the two thermometers.

Pteridophyte A low, nonwoody plant that reproduces via spores rather than seeds; the largest group of pteridophytes is the ferns.

Pyroclastic materials Fragments of volcanic rock thrown out in a volcanic explosion.

Quickflow (see *Stormflow*)

Quicksand Sand that is incapable of supporting overburden (added weight) because of high pore water pressure.

Radiation The process by which radiant (electromagnetic) energy is transmitted through free space; the term used to describe electromagnetic energy, as in infrared radiation or shortwave radiation.

Radiation beam The column of solar radiation flowing into or through the atmosphere.

Rainfall intensity The rate of rainfall measured in inches or centimeters of water deposited on the surface per hour or minute.

Rainforest A forest formation dominated by a heavy cover of evergreen trees, with abundant secondary vegetation in the form of epiphytes and lianas; in addition to the equatorial and tropical rainforests, the conifer forests of the very humid portions of the marine west coast are often classed as rainforest.

Rainshadow The dry zone on the leeward side of a mountain range of orographic precipitation.

Rainsplash Soil erosion from the impact of raindrops.

Rainwash Soil erosion by overland flow; erosion by sheets of water running over a surface; usually occurs in association with rainsplash; also called *wash.*

Range The geographic area occupied by a species, genus, or family of organisms.

Rating curve A graph that shows the relationship between the discharge and stage of various flow events on a river; once this relationship is established it may be used to approximate discharge using stage data alone.

Rational method A method for computing the discharge from a small drainage basin in response to a given rainstorm; computation is based on the coefficient of runoff, rainfall intensity, and basin area.

Reach A stretch or segment of stream channel.

Recharge The replenishment of groundwater with water from the surface.

Recurrence The number of years on the average that separate events of a specific magnitude, e.g., the average number of years separating river discharges of a given magnitude or greater.

Reflected wave A wave that rebounds off the shore or an obstacle and is redirected seaward through shore-bound incident waves.

Reg A rocky desert landscape; also called a *hamada.*

Regolith The weathered material overlying the bedrock; usually coarse, unsorted.

Relative humidity An expression for the water vapor content of air at a given temperature; the vapor content of a body of air expressed as a percentage of the amount of vapor held by a parcel of air when it is saturated.

Relief The range of topographic elevation within a prescribed area.

Residual soil Soil formed in parent material derived from the underlying bedrock, i.e., from *in situ* material.

Respiration The internal cellular processes of a plant by which energy is used for biological maintenance.

Reverse fault A fault in which the hanging wall is displaced upward relative to the footwall.

Revolution The motion of a planet in its orbital path around the sun.

Ria coast A heavily indented coast marked by prominent headlands and deep reentrants.

Richter scale A scale for rating the magnitude of an earthquake, i.e., the amount of energy released at the focus of the earthquake. The Richter scale is a logarithmic scale; for each unit, magnitude increases about 32 times.

Riffles (see *Pools and riffles*)

Rift (graben) A fault in a zone of tensional stress characterized by a block displaced downward relative to adjacent rock formations.

Rift valley A valley formed by a rift fault, e.g., the valleys of the large lakes of East Africa.

Rimed A term used to describe snow crystals on which condensation has taken place as they fall to earth.

Rip current A relatively narrow jet of water that flows seaward through the breaking waves; it serves as a release for water that builds up near shore.

Riprap Rubble such as broken concrete and rock placed on a surface to stabilize it and reduce erosion.

Roche moutonnées An erosional feature sculpted from a rock knob or dome by a glacier; it is smooth on the side of ice advance and jagged on the downice side; the term is derived from *moutonnée*, a French word for wavy wigs of the eighteenth century that were pomaded with mutton tallow.

Rockfall A type of mass movement involving the fall of rock fragments from a cliff or slope face.

Root wedging A mechanical weathering process in which a root grows inside a crack, placing stress on the rock and widening the crack.

Rotation The spinning motion of a sphere, such as that of the earth about its axis.

Roughness length The height of the zone or envelope of calm air over a surface which marks the base of the zone of turbulent airflow.

Ruderal Plants that have evolved special means to tolerate disturbance.

Runoff In the broadest sense runoff refers to the flow of water from the land as both surface and subsurface discharge; the more restricted and common use, however, refers to runoff as surface discharge in the form of overland flow and channel flow.

Rupture Deformation of a substance by fracturing.

Salination Salt saturation of soil as a result of a rise in the water table due to irrigation.

Saltation The principal mode of transport by wind; it is characterized by particles "hopping" over the ground.

Sand bypassing A means of artificially feeding sand to the beach downshore from a barrier such as a breakwater across a bay-mouth.

Sand sea A huge field of sand dunes such as the Empty Quarter of Saudi Arabia.

Sapping An erosional process that usually accompanies gullying in which soil particles are eroded by water seeping from a bank.

Saturated adiabatic lapse rate The rate of decline in the temperature of a rising parcel of air after it has reached saturation; it is variable but averages $-0.6°C/100$ m; this rate is less than the dry adiabatic lapse rate ($0.98°C/100$ m), because of the heat released in condensation.

Saturation absolute humidity The maximum mass of water vapor that can be held in a cubic meter of air at a given temperature; see also *Absolute humidity*.

Saturation mixing ratio The maximum mass of water vapor that can be held in a given mass of dry air at a given temperature and pressure.

Saturation vapor pressure The maximum value that vapor pressure can attain in air at a given temperature; see also *Vapor pressure*.

Savanna A biochore characterized by trees and shrubs scattered among a cover of grasses and forbs; the tropical savanna is the most extensive savanna formation and is found in the areas of the tropical wet/dry climate.

Scattering The process by which minute particles suspended in the atmosphere diffuse incoming solar radiation.

Sclerophyll forest Forest of the Mediterranean climate, characterized by small, widely spaced evergreen hardwood trees; generally considered the least prominent of the world's forest formations.

Scouring A mechanism of erosion by streams and glaciers in which particles carried at the bed abrade underlying rock.

Sea A term used to describe the choppy sort of waves in an area of wave generation.

Sea breeze A local wind that blows from sea to land as a result of the differential heating of land and water in the coastal zone; usually a daily occurrence.

Sea-level pressure The pressure exerted by the atmosphere on the earth's surface at sea level; measured by the height of a column of mercury in a mercurial barometer; normal (average) sea-level pressure is 29.92 inches, or 76 cm of mercury; 14.7 pounds of pressure per square inch; or 1013.2 millibars of force per square meter.

Seamount A volcanic mountain in an ocean basin whose origin is not connected with a midoceanic ridge or a subduction zone; volcanic rises in the abyssal plains such as Bermuda and Hawaii.

Secondary consumer An animal that preys on primary consumers; in the soil moles are secondary consumers.

Sediment sink A coastal environment, such as a baymouth, where massive amounts of sediment are deposited.

Seepage The process by which groundwater or interflow water seeps from the ground.

Seepage lake A lake that gains its water principally from the seepage of groundwater into its basin.

Seif A large sand dune that is elongated in the general direction of the formative winds; a dune formed by winds from more than one direction.

Seismology A branch of geophysics devoted to the study of earthquakes and the interpretation of seismic waves.

Semipermanent pressure cells Large pressure cells, such as the subtropical highs, which exist most of the time in all seasons within a zone of latitude.

Sensible heat Heat that raises the temperature of a substance and thus can be sensed with a thermometer. In contrast to latent heat, it is sometimes called the heat of dry air.

Septic system Specifically, a sewage system that relies on a septic tank to store and/or treat wastewater; generally, an on-site (small-scale) sewage-disposal system that depends on the soil to dispose of wastewater.

7th approximation The modern soil-classification system of the U.S. Department of Agriculture; it uses six levels of classification, beginning with orders—Entisols, Inceptisols, Aridosols, Mollisols, Spodosols, Alfisols, Ultisols, Vertisols, Oxisols, and Histosols.

Shear stress Differential stress acting on a body in which the forces are directed at angles to one another.

Shear wave A type of seismic (elastic) wave that produces motion transverse (perpendicular) to the direction of seismic energy propagation; also termed *secondary wave.*

Shield A major geologic subdivision of the continents; the relatively low elevation interior of a geologically stable continent. The term is derived from the shape of a battle shield placed handle side down.

Shield volcano A volcano comprised mainly of lava, with the overall shape of a shield or dome. The Hawaiian Islands are shield volcanoes.

Shifting agriculture An agricultural practice in tropical and equatorial areas characterized by the movement of farmers from plot to plot as soil becomes exhausted under cultivation.

Sial layer The upper part of the crust; the part that forms the continents and is comprised of relatively light, granitic rocks. The term is a contraction of *si*licon and *al*uminum.

Silicate minerals The principal rock-forming group of minerals; minerals composed of a basic ion of silicon and oxygen, called the silicon oxygen tetrahedron, combined with one or more additional elements.

Silicic lava Lava with a high quartz content in which silicon dioxide constitutes 70 percent or more of the rock; this lava is prevalent on the continents.

Sima layer The lower part of the crust; the part that forms the ocean basin and is comprised of relatively heavy, basaltic rock. The term is a contraction of *si*licon and *ma*gnesium.

Single-thread channel A stream channel characterized by a single course; it may be straight or meandering.

Sinkhole A pitlike depression in areas of karst topography, caused by the removal of limestone or dolomite by underground drainage; also called a sink or doline.

Slip face The lee side of a sand dune where windblown sand accumulates and slides downslope.

Slope failure A slope that is unable to maintain itself and fails by mass movement such as a landslide, slump, or similar movement.

Slope form The configuration of a slope, e.g., convex, concave, or straight.

Sluiceway A large drainage channel or spillway for glacier meltwater.

Slump A type of mass movement characterized by a back rotational motion along a rupture plane.

Small circle Any circle drawn on the globe that represents less than the full circumference of the earth; thus, the plane of a small circle does not pass through the center of the earth. All parallels except the equator are small circles.

Soil-absorption systems The term applied to sewage-disposal systems that rely on the soil to absorb wastewater; see also *Septic system.*

Soil body (see *Polypedon*)

Soil Conservation Service An agency of the U.S. Department of Agriculture that is responsible for soil mapping and analysis in the United States.

Soil creep A type of mass movement characterized by a very slow downslope displacement of soil, generally without fracturing of the soil mass; the mechanisms of soil creep include freeze-thaw activity and wetting and drying cycles.

Soil-forming factors The major factors responsible for the formation of a soil: climate, parent material, vegetation, topography, and drainage.

Soil-heat flux The rate of heat flow into, from, or through the soil.

Soil mantle A traditional term used to describe the composite mass of soil material above the bedrock.

Soil material Any rock or organic debris in which soil formation takes place.

Soil-moisture balance A model that describes the changes in the availability of soil moisture as a product of precipitation, evapotranspiration, and storage water in the soil.

Soil-moisture deficiency The amount of water needed to raise the moisture content of a soil to field capacity.

Soil-moisture recharge A term in the soil-moisture balance; the replenishment of soil moisture following a period of soil-moisture loss, i.e., following a period of negative moisture balance.

Soil-nutrient system The system defined by the flow of nutrients between the soil and the plant cover.

Soil order A major level of soil classification in the traditional USDA scheme; the three orders defined are zonal, azonal, and intrazonal.

Soil pH The degree of alkalinity or acidity of a soil; the ratio of hydrogen ions to hydroxyl ions. On the pH scale 7.0 is neutral.

Soil profile The sequence of horizons, or layers, of a soil.

Soil regime A particular combination of soil-forming conditions, generally related to climate, that gives rise to certain soil processes and in turn a distinctive soil profile.

Soil structure The term given to the shape of the aggregates of particles that form in a soil; four main structures are recognized; blockly, platy, granular, and prismatic.

Soil texture The cumulative sizes of particles in a soil sample; defined as the percentage by weight of sand, silt, and clay-sized particles in a soil.

Soil-water balance (see *Soil-moisture balance*)

Solar constant The rate at which solar radiation is received on a surface (perpendicular to the radiation) at the edge of the atmosphere. Average strength is 1353 joules/$m^2 \cdot$ sec, which can also be stated as 1.94 cal/$cm^2 \cdot$ min.

Solifluction A type of mass movement in periglacial environments, characterized by the slow flowage of soil material and the formation of lobe-shaped features; prevalent in tundra and alpine landscapes.

Solstice The dates when the declination of the sun is at 23.27° N latitude (the Tropic of Cancer) and 23.27° S (the Tropic of Capricorn)—June 21–22 and December 21–22, respectively. These dates are known as the winter and summer solstices, but which is which depends on the hemisphere.

Solum That part of soil material capable of supporting life; the true soil according to the agronomist; the upper part of the soil mass, including the topsoil and soil horizons.

Solution weathering A type of weathering in which a mineral dissolves on contact with water carrying a solvent such as carbonic acid.

Speciation The process by which new species originate.

Species A group, or taxon, of individuals able to freely interbreed among themselves, but unable to breed with other groups; the smallest taxon of the floristic system of plant classification.

Specific heat The relative increase in the temperature of a substance with the absorption of energy.

Specific humidity An expression for the vapor content of air similar to the *mixing ratio*; the weight (mass) of water vapor relative to the weight of the moist air (vapor plus dry air) to which it belongs; measured in grams per kilogram.

Spheroidal weathering A form of chemical weathering in which a boulder sheds thin plates of rock debris.

Spodosols Soils with pronounced zones of illuviation characterized by accumulations of iron and aluminum oxides. Formed in moist, cool climates.

Spore A reproductive cell in plants; generally any nonsexual reproductive cell; among the vascular plants, the pteridophytes are spore-bearing.

Spring tide (see *Tide*)

Squall line The narrow zone of intensive turbulence and rainfall along a cold front.

Stage The elevation of the water surface in a river channel.

Stand A floristically uniform growth of vegetation, often of similar size and age, that dominates an area.

Stefan-Boltzmann equation The intensity of energy radiation from a body increases with the fourth power of its temperature times a constant.

Stemflow Precipitation water that reaches the ground by running from the vegetation canopy down the trunk of a tree, shrub, or the stem of grass; see also *Interception.*

Steppe Short-grass prairie of the semiarid climatic zones; widespread in Eurasia and North America.

Stomata The openings in the foliage of a plant through which moisture is released during transpiration; the stomata open and close in response to air temperature, humidity, and other factors.

Stormflow The portion of streamflow that reaches the stream relatively quickly after a rainstorm, adding a surcharge of water to baseflow.

Story A layer or level of tree crowns in a forest.

Stratosphere The subdivision of the atmosphere that lies above the troposphere; it is characterized by stability and temperature that increases with altitude.

Stream order The relative position, or rank, of a stream in a drainage network. Streams without tributaries, usually the small ones, are first-order; streams with two or more first-order tributaries are second-order, and so on.

Stress A force acting on a body or substance; see also *Plant stress* or *Shear stress.*

Stress-threshold concept A concept of vegetation change based on the magnitude-and-frequency concept; vegetation changes according to the magnitude of various stresses and disturbances in the form of floods, drought, disease, and fire, for example, in the plant environment.

Striation Scour line etched into bedrock by the rock debris on the base of a glacier.

Strike and dip Directional properties of a geologic structure such as a fault. Strike is the directional trend of a formation along the surface; dip is the angle of incline of the formation measured at a right angle to strike.

Strike-slip fault A fault in which the main direction of displacement is lateral, or along the fault line.

Subarctic zone The belt of latitude between 55° and the Arctic and Antarctic circles.

Subduction The process by which a tectonic plate is consumed or destroyed as it slides into the earth along the contact with an adjacent plate; subduction zones are places of frequent earthquakes and volcanism and are usually marked on the surface by ocean trenches.

Subhumid In the Thornthwaite System of climate classification, the climatic zone transitional between humid and semiarid climates.

Sublimation A physical process by which a solid is changed directly to a gas (or vice versa) without passing through the liquid phase.

Subtropical high-pressure cells Large cells of high pressure, centered at 25°–30° latitude in both hemispheres, which are fed by air descending from aloft; these cells are the main cause of aridity in tropics and subtropics.

Subtropical zone The zone of latitude near the tropics in both hemispheres; between 23.5° and 35°.

Succulent habit A form of plant adaptation to arid conditions characterized by fleshy bodies and/or foliage with the capacity to store large amounts of water.

Summer solstice (see *Solstice*)

Sun angle The angle formed between the beam of incoming solar radiation and a plane at the earth's surface or a plane of the same attitude anywhere in the atmosphere.

Surge A large and often destructive wave caused by intensive atmospheric pressure and strong winds.

Suspended load The particles (sediment) carried aloft in a stream of wind by turbulent flow; usually clay- and silt-sized particles.

Swash The thin sheet of water that slides up the beach face after a wave breaks.

Swell A wave with a relatively smooth form, usually found at some distance from the area of wave generation.

Syncline A fold characterized by a downward bend, i.e., concave downward in cross-section.

Synoptic scale The scale of geographic coverage most commonly used on daily weather charts depicting air masses, winds, and storm cells.

System An interconnected set of objects or things; two or more components such as organisms, cities, or streams linked together in some fashion; e.g., energy systems, ecosystems, road systems.

Système international (*S.I.*) The preferred system of units according to an international consensus of scientists; the basic units of energy, time, and space are the joule, second, and square meter, respectively.

Taiga (see *Boreal forest*)

Talus An accumulation of rock debris at the foot of a slope as a result of rockfall.

Taxon Any unit (category) of classification of organisms.

Tectonic plate A large sheet of lithosphere that moves as a discrete entity on the surface of the earth; the lithosphere is subdivided into seven major plates and many smaller ones; each major plate, with the exception of the Pacific plate, contains a continent.

Temperate forest A forest of the midlatitude regions that could be described as climatically temperate, e.g., broadleaf deciduous forests of Europe and North America, comprised of beeches, maples, and oaks.

Temperate glacier A glacier characterized by abundant meltwater and an internal temperature slightly below the freezing threshold.

Temperature inversion An atmospheric condition in which the cold air underlies warm air; inversions are highly stable conditions and thus not conducive to atmospheric mixing.

Temperature profile The change in temperature along a line or transect through an environment, usually expressed in a graphical format.

Terrace A surface formed by wave erosion or river processes and elevated above the existing level of the ocean or floodplain.

Tertiary consumer A carnivore that preys on secondary as well as primary consumers; animals, such as birds of prey, that are near the ends of the food chains.

Theory of continental drift (see *Continental drift*)

Thermal conductivity A thermal property of a substance describing its capacity to transmit heat given a thermal gradient of 1°K per meter (or 1°C/m).

Thermal diffusivity A thermal property of a substance that describes the rate at which a given temperature, represented, for example, by an isotherm, passes through a substance. It is defined as the ratio of thermal conductivity to volumetric heat capacity.

Thermal gradient The change in temperature over distance in a substance; usually expressed in degrees Celsius per centimeter or meter.

Thermal regime The annual or seasonal pattern of temperatures for a place or region; usually used in climatology.

Thornbush A vegetative formation of the tropical savanna regions, characterized by short, thorny trees and shrubs; called *caatinqa* in northeastern Brazil and *dornveld* in South Africa.

Threshold The level of magnitude of a process at which sudden or rapid change is initiated.

Thrust fault A fault in which the hanging wall is driven laterally over the footwall.

Thunderstorm An intensive convectional storm that produces heavy precipitation, strong local winds, as well as thunder and lightning.

Tide A large wave caused by bulges in the sea in response to the lunar and solar gravitational forces. The largest tides are *spring tides*, which occur when the moon and sun are aligned with the earth; the smallest are *neap tides*, which occur when the moon and sun are positioned at a right angle relative to the earth.

Tolerance The range of stress or disturbance a plant is able to withstand without damage or death.

Tombolo A depositional feature along some shorelines which forms a neck of land between an island and the mainland.

Topographic relief (see *Relief*)

Topsoil The uppermost layer of the soil, characterized by a high organic content; the organic layer of the soil.

Township and range A system of land subdivision in the United States which uses a grid to classify land units; standard subdivisions include townships and sections.

Traction A mode of sediment transport by wind in which particles move in contact with the ground; also called *creep*.

Trade winds The system of prevailing easterly winds, which flow from the subtropical highs to the intertropical convergence zone (ITC) in both hemispheres; also called the *tropical easterlies*.

Transform fault A strike-slip fault; in particular, the term is applied to the faults that run transverse to the midoceanic ridges.

Transmission The lateral flow of groundwater through an aquifer; measured in terms of cubic feet (or meters) transmitted through a given cross-sectional area per hour or day.

Transpiration The flow of water through the tissue of a plant and into the atmosphere via stomatal openings in the foliage.

Transported soil Soil formed in parent material comprised of deposits laid down by water, wind, or glaciers.

Travel time (see *Concentration time*)

Tree line The upper limit of tree growth on a mountain where forest often gives way to alpine meadow.

Tropical cyclone (see *Hurricane*)

Tropics Correctly used, this term refers to the Tropic of Capricorn and the Tropic of Cancer; however, it is often used to refer to areas equatorward of the tropics.

Troposphere The lowermost subdivision of the atmosphere; the layer that contains the bulk of the atmosphere's mass and is characterized by convectional mixing and temperature that decreases with altitude.

Tsunami A large and often destructive wave caused by tectonic activity such as faulting on the ocean floor.

Tundra Landscape of cold regions, characterized by a light cover of herbaceous plants and underlain by permafrost.

Turbulent flow Flow characterized by mixing motion in which the primary source of flow resistance is the mixing action between slow-moving and faster-moving molecules in a fluid.

Turgor A term used to describe the status of water pressure in plant foliage; when a plant wilts, it loses its turgor, and the leaves become puckered and limp.

Typhoon (see *Hurricane*)

Ultisols Soils in an advanced state of development in warm, moist climates. Pronounced eluviation; often poor in bases.

Ultraviolet radiation Electromagnetic radiation of wavelengths shorter than visible, but longer than X-rays.

Unified system A soil-classification scheme used in civil engineering, based on soil performance when it is placed under stress.

Uniformitarianism A concept attributed to sixteenth-century geologist James Hutton; the types of processes operating on the earth today are the same ones that were active in the geologic past; often condensed to "the present is the key to the past."

U.S. Geological Survey An agency of the U.S. Department of Interior that is responsible for mapping and analyzing rock types, minerals, earthquakes, river flow, and related phenomena.

Urban climate The climate in and around urban areas; it is usually somewhat warmer, foggier, and less well lighted than the climate of the surrounding region.

Urbanization The term used to describe the process of urban development, including suburban residential and commercial development.

Valley wall The side slope of a river valley where the floodplain gives way to upland surfaces.

Vapor pressure An expression for the water vapor content of air; it is the pressure exerted by the weight of the water vapor molecules independent of the weight of the other gases in the air; expressed in millibars or newtons per square meter.

Vascular plants Plants in which cells are arranged into a pipelike system of conducting, or vascular, tissue; xylem and phloem are the two main types of vascular tissue.

Vegetative regeneration Asexual regeneration by plants in which some part of the plant, such as a root or a special organ such as a rhizone, is able to propagate new stems.

Vein Igneous rock or a deposit of minerals that form in a joint line or fracture.

Velocity The rate of movement in one direction, expressed as distance over time, e.g., m/sec, km/hr, or mph.

Vernal equinox (see *Equinox*)

Vertisols Soils dominated by geomechanical mixing in areas of montmorillonite clay.

Viscosity A measure of the resistance to flow in a fluid due to intermolecular friction. At a temperature of 10°C, molasses has a higher viscosity than water does.

Visible light Electromagnetic radiation at wavelengths between 0.4 and 0.7 micrometer; the radiation that comprises the bulk of the energy emitted by the sun.

Warm front A contact between a cold air mass and a warm air mass in which the warm air is moving against the cold air, sliding upward along the contact.

Water table The upper boundary of the zone of groundwater; in fine-textured materials it is usually a transition zone rather than a boundary line. The configuration of the water table often approximates that of the overlying terrain.

Water witching (see *Dowsing*)

Watt A unit of power that is often used as an energy expression; equal to 1 joule per second.

Wave period The time it takes a wave to travel the distance of one wavelength.

Wave refraction The bending of a wave, which results in an approach angle more perpendicular to the shoreline.

Wave of translation A wave that produces a mass transport of water; in coastal areas it is often a breaking wave; see *Oscillatory wave.*

Weathering The breakdown and decay of earth materials, especially rock; see also *Chemical weathering.*

Westerlies The prevailing eastward flow of air over land and water in the midlatitudes of both hemispheres; also called the *prevailing westerlies.*

West-wind drift Ocean current, or drift current, which flows eastward in the midlatitudes and subarctic, driven by the prevailing westerly winds.

Wetland A term generally applied to an area where the ground is permanently wet or wet most of the year and is occupied by water-loving (or tolerant) vegetation such as cattails, mangrove, or cypress.

Wetted perimeter The distance from one side of a stream to the other, measured along the bottom.

Wien's law A physical law stating that the wavelength of maximum-intensity radiation grows longer as the absolute temperature of the radiating body decreases; also called *Wien's Displacement Law.*

Wind chill Heat loss from the skin as a function of both air temperature and wind velocity.

Wind power The power generated by wind; proportional to the cube of speed or velocity.

Wind wave A wave generated by the transfer of momentum from wind to a water surface.

Winter solstice (see *Solstice*)

Work A concept closely related to energy, work is the product of force and distance and is accomplished when the application of force yields movement of an object in the direction of the force.

Xerophyte Plants capable of surviving prolonged periods of soil drought, e.g., the cacti.

Xylem and phloem Conducting tissue in vascular plants through which the plant fluids are transmitted.

Zenith For any location on earth, the point that is directly overhead to an observer. The zenith position of the sun is the one directly overhead.

Zenith angle The angle formed between a line perpendicular to the earth's surface (at any location) and the beam of incoming solar radiation (on any date).

Zonal soil A soil order under the traditional USDA soil-classification scheme; soils with well-developed horizons that reflect the climate conditions of the region in which they are found.

Zone of eluviation The level, or zone, in a soil losing materials in the form of colloids and ions; the zone of removal.

Zone of illuviation The level, or zone, in a soil where colloids and ions accumulate.

Zone of saturation (see *Groundwater*)

Credits

CHAPTER 1

Figures 1.1, 1.2, 1.3, 1.4, 1.5, 1.11: From Marsh, W.M., and Dozier, J., *Landscape: An Introduction to Physical Geography,* Addison-Wesley, 1981. Figure 1.6: From Spencer, E., *Physical Geology,* Addison-Wesley, 1983. Figure 1.7: Data from U.S. Nautical Almanac, 1976. Figure 1.10: Courtesy of U.S. Weather Service. Figure 1.12: Data from U.S. Air Force Reference Atmosphere.

CHAPTER 2

Figures 2.1, 2.2, 2.3: From Marsh, W.M., and Dozier, J., *Landscape: An Introduction to Physical Geography,* Addison-Wesley, 1981. Figure 2.5: Courtesy of NOAA/US Dept. of Commerce.

CHAPTER 3

Figures 3.1, 3.3, 3.4, 3.8, 3.9, 3.10, 3.11, 3.12, 3.17: From Marsh, W.M., and Dozier, J., *Landscape: An Introduction to Physical Geography,* Addison-Wesley, 1981. Figures 3.2, 3.13: Prepared by Jeff Dozier. Figure 3.5: From Marsh, W.M., *Landscape Planning: Environmental Applications,* Addison-Wesley, 1983. Figure 3.6: Courtesy of Woo and Williams, Inc. Used by permission. Figure 3.7: Data from U.S. National Almanac, 1976. Figure 3.14: Based on data from Budyko, M.I., *Climate and Life,* Academic Press, 1974. Used by permission. Figure 3.16: After T.R. Oke, *Boundary Layer Climates,* Methuen, 1978. Table 3.1: From Landsberg, H.E., *Physical Climatology,* 2nd ed., Gray Printing Co., 1960.

CHAPTER 4

Figures 4.1, 4.2, 4.3, 4.4, 4.5, 4.6, 4.9, 4.10, 4.12, 4.14: From Marsh, W.M., and Dozier, J., *Landscape: An Introduction to Physical Geography,* Addison-Wesley, 1981. Figure 4.7: From Marsh, W.M., *Landscape Planning: Environmental Applications,* Addison-Wesley, 1983. Figure 4.8: Compiled by Troy L. Péwé from numerous sources. Used by permission. AIT 4.1: Photograph by Laurence Pringle/Photo Researchers.

CHAPTER 5

Figures 5.1, 5.2, 5.8, 5.9, 5.12, 5.13, 5.14, 5.15, 5.17: From Marsh, W.M., and Dozier, J., *Landscape: An Introduction to Physical Geography,* Addison-Wesley, 1981. Figure 5.16: Adapted from Perry, A.H., and Walker, J.M., *The Ocean-Atmosphere System,* Longman, 1977. Used by permission. Figure 5.18: G.F. Wieczorek, U.S. Geological Survey.

CHAPTER 6

Figures 6.1, 6.2, 6.3, 6.13, o.14, 6.16: From Marsh, W.M., and Dozier, J., *Landscape: An Introduction to Physical Geography*, Addison-Wesley, 1981. AIT Figure 6.3: Bill Males. Figures 6.7*a*, 6.9, 6.12, 6.15, 6.20: Courtesy of NOAA. Figure 6.10: From Marsh, W.M. *Landscape Planning: Environmental Applications*, Addison-Wesley, 1983. Figure 6.17: From *Climates of the United States*, Washington, D.C., U.S. Department of Commerce, 1974. Figure 6.18: From NASA-Marshall Space Flight Center, courtesy of J. Michael Fritsch. Figure 6.19: From NOAA/NESS Geostationary GOES Satellite, courtesy of Jeff Dozier. 6.20: United Press International.

CHAPTER 7

Figure 7.4: From Perry, A.H., and Walker, J.M., *The Ocean-Atmosphere System*, Longman, 1977. Used by permission. Figure 7.8: Data from the Smithsonian Institution, Washington, D.C. Figure 7.9: Photography by W.M. Marsh. Figure 7.12: Reprinted from the *Geographical Review*, 38 (1948). Figure 7.14: From Schachori, A., et al., "Effect of Mediterranean Vegetation on the Moisture Regime," in *Forest Hydrology*, Pergamon, 1967. Used by permission. Figure 7.16: From Marsh, W.M., and Dozier, J., *Landscape: An Introduction to Physical Geography*, Addison-Wesley, 1981. AIT Figure 7.1: Photography by William Marsh. AIT 7.3: Photograph by Leonard Hungerford.

CHAPTER 8

Figures 8.2, 8.6, 8.7: From Marsh, W.M., and Dozier, J., *Landscape: An Introduction to Physical Geography*, Addison-Wesley, 1981. Note 8.A: Photography by William Marsh. Note 8.B: George Holton/Photo Researchers. Figure 8.9: From U.S. Environment Data Service. Figure 8.10: Photograph by Nash Uebelhart. Figure 8.11: Barbara Rios. Figure 8.12: Photography by U.S. Department of Agriculture.

CHAPTER 9

Figure 9.1: From *Understanding Climate Change*, U.S. National Academy of Sciences, 1975. Figure 9.2: From Spencer, E., *Physical Geology*, Addison-Wesley, 1983. Figure 9.3: Photograph by Bruce D. Marsh. Figure 9.4: From Marsh, B.J., "Towards the Understanding and Prediction of Climate Variations," *Quarterly Journal of the Royal Meteorological Society* 102, 433. Reprinted by permission. Figure 9.5: From Clawson, Marion, *America's Land and Its Uses*, The Johns Hopkins Press and Resources For the Future, Inc. Reprinted by permission. Figure 9.6: From Marsh, W.M., and Dozier, J., *Landscape: An Introduction to Physical Geography*, Addison-Wesley, 1981. Figure 9.7: After Oke, T.R., *Boundary Layer Climates*, Methuen, 1976. Figure 9.10: Adapted in part from Davenport, 1965. Figure 9.11: Adapted from data and maps by the Southeast Michigan Council of Governments, Detroit, Michigan, 1976. Figure 9.12: From U.S.-Canada Memorandum of Intent on Transboundary Air Pollution and the U.S. Department of Energy, 1983. Tables 9.1, 9.2: From Marsh, W.M., and Dozier, J., *Landscape: An Introduction to Physical Geography*, Addison-Wesley, 1981. AIT 9.1: From Spencer, E., *Physical Geology*, Addison-Wesley, 1983. Note 9: Maps compiled by Tim Ball.

CHAPTER 10

Figures 10.2, 10.3, 10.4, 10.17: From Marsh, W.M., and Dozier, J., *Landscape: An Introduction to Physical Geography*, Addison-Wesley, 1981. Figure 10.6: Adapted from Küchler, 1967, and Polunin, 1967. AIT 10.2: Blue Ridge Ariel Survey. AIT 10.3: Charlie Oti/National Audobon Society. AIT 10.4: Photograph by U.S. Geological Survey. AIT 10.5: Dr. E.R. Degginger. AIT 10.6: Photography by Josef Muench. Used by permission. AIT 10.7: Photograph by W.M. Marsh.

CHAPTER 11

Figure 11.2: From Gates, D.M., "Leaf Temperature and Energy Exchange," *Archiv für Metiorologie, Geophysik und Bioklimatologie*, B12 (1963): 321–336. Used by permission of the author. Figures 11.3, 11.4, 11.6, 11.9, Table 11.1: From Marsh, W.M., and Dozier, J., *Landscape: An Introduction to Physical Geography*, Addison-Wesley, 1981. Figure 11.4: Photograph by Michael Treshow (frost damage). Figure 11.7: Adapted from Jen-Hu Chang. Figure 11.8: Adapted from Ryther, J.H., "Photosynthesis in the Ocean as a Function of Light Intensity," *Limnology and Oceanography*, 1 (1956): 61–70. Figure 11.12: Adapted from Hutchinson, B.A., and Matt, D.R., "The Distribution of Solar Radiation Within a Deciduous Forest," *Ecological Monographs*, 47 (1977): 185–207. Figure 11.14: Data from H.T. Odum, 1957.

CHAPTER 12

Figures 12.1, 12.6, 12.8, 12.9, 12.11, 12.12, 12.13: From Marsh, W.M., and Dozier, J., *Landscape: An Introduction to Physical Geography*, Addison-Wesley, 1981. Figure 12.2: Photograph courtesy of Harvard Forest, Harvard University. Figure 12.3: Photograph by Charles Schlinger. Figure 12.4: Photographs by Deborah Coffey, Patrick W. Hassett, and William M. Marsh. Figure 12.5: From Thornthwaite, C.W., and Mather, J.P., "The Water Balance," *Publications in Climatology* 8, 1, Drexel Institute of Technology, 1955. Used by permission of C.W. Thornthwaite Associates, Elmer, N.J. Figure 12.7: Photograph by U.S. Soil Conservation Service. Figure 12.10: Kent and Donna Dannen/Photo Researchers. Figure 12.14: From Redfield, Alfred C., "Development of a New England Salt Marsh," *Ecological Monographs* 42:2, 1972. Copyright 1972 by the Ecological Society of America. Used by permission. AIT 12.1: Chas Cunningham, U.S. Geological Survey. AIT 12.2: Photography by William M. Marsh. Note 12: Adapted from Clausen, J., and Hiesey, W.M., "Experimental Studies on the Nature of Species IV, Genetic Structure of Ecological Races," Carnegie Institute, Washington Publication No. 615, 1958. Used by permission.

CHAPTER 13
Figure 13.2: Original illustration by Peter Van Dusen. Figures 13.3, 13.4, 13.5, 13.7, 13.8, 13.9, 13.11, 13.13, 13.14, 13.15, 13.16: From Marsh, W.M., and Dozier, J., *Landscape: An Introduction to Physical Geography,* Addison-Wesley, 1981. Figure 13.6: From Marsh, W., *Landscape Planning: Environmental Applications,* Addison-Wesley, 1983. Note 13: Transmission electron micrographs by Kenneth M. Towe, Smithsonian Institution.

CHAPTER 14
Figures 14.2, 14.5, 14.6, 14.7, 14.11, 14.17: From Marsh, W.M., and Dozier, J., *Landscape: An Introduction to Physical Geography,* Addison-Wesley, 1983. Figure 14.12: From Soil Geography Unit, Soil Conservation Service, U.S. Department of Agriculture, 1972. Figure 14.13: From U.S. Department of Agriculture. Figure 14.14: From Land Resource Research Institute, Agriculture Canada, Ottawa. Figure 14.15: Photograph by Troy W. Péwé. Figure 14.16: From *Cropland Erosion,* Soil Conservation Service, U.S. Department of Agriculture, 1977. Tables 14.1, 14.2, 14.4: From Marsh, W.M., and Dozier, J., *Landscape: An Introduction to Physical Geography,* Addison-Wesley, 1981. Table 14.3: From U.S. Department of Agriculture. AIT 14.1, 14.2, 14.3: Courtesy of U.S. Soil Conservation Service. AIT 14.4: David Moon/ Black Star. AIT 14.5, 14.6: Photographs by William M. Marsh.

CHAPTER 15
Figure 15.1: J.K. Hillers, U.S. Geological Survey. Figures 15.2, 15.3, 15.5, 15.6, 15.9, 15.13, 15.16, 15.17, 15.18, 15.19, 15.20: From Marsh, W.M., and Dozier, J., *Landscape: An Introduction to Physical Geography,* Addison-Wesley, 1981. Figure 15.4: Compiled from multiple sources. Figure 15.7: Adapted in part from Phillips, O.M., *The Heart of the Earth,* Freeman Cooper, 1968. Figure 15.8: Adapted in part from Spencer, E., *Physical Geology,* Addison-Wesley, 1983. Figure 15.10*a*: Photograph by U.S. Geological Survey. Figures 15.11, 15.12: From U.S. Geological Survey. Figure 15.15: Adapted by permission from Strahler, A.N., *Physical Geography,* 2nd ed., Wiley, 1960.

CHAPTER 16
Figures 16.1, 16.2, 16.5, 16.6, 16.9, 16.10, 16.11, 16.12, 16.13, 16.15, 16.16, 16.17, 16.18, 16.19, 16.22, 16.23, 16.24, 16.27, 16.29: From Marsh, W.M., and Dozier, J., *Landscape: An Introduction to Physical Geography,* Addison-Wesley, 1981. Figures 16.2, 16.3, 16.4: From U.S. Geodynamics Committee, National Academy of Science. Figure 16.7: Original drawing by Peter Van Dusen after Marsh, B.D., 1979; and Marsh, B.D., and Carmichael, I., 1974. Figure 16.8: Compiled from a variety of sources. Figure 16.14: Map used by permission from Chapman, D.S., and Pollack, N.N., "Global Heat Flow: A New Look," *Earth and Planetary Science,* Letter 23, 1975. Used by permission. Figure 16.20: Photograph by E.L. Fitzgerald. Figure 16.21: From Spencer, E., *Physical Geology,* Addison-Wesley, 1983. Figure 16.26: Adapted from McConnell, R.B., "Rift and Shield Structure in East Africa," *Report* to International Geological Congress of London, 1948:18, 1951. Figure 16.28: Illustration after Stearns and MacDonald, 1946; photographs by B.D. Marsh. Figure 16.30: Photograph courtesy of U.S. Geological Survey. Figure 16.31: Photograph by W.M. Marsh.

CHAPTER 18
Figure 18.18*a*: U.S. Geological Survey. Figure 18.18*b*: Robert Perron.

CHAPTER 20
Figure 20.9: Courtesy NASA. Figure 20.12*a*: Courtesy of Royal Canadian Air Force. Figure 20.13: Courtesy of U.S. Army Corps of Engineers. Figure 20.15: Courtesy of U.S. Air Force. AIT 20.1: From Virginia Department of Transportation.

CHAPTER 21
Figures 21.1, 21.10: From Marsh, W.M., *Landscape Planning: Environmental Applications,* Addison-Wesley, 1983. Figures 21.4, 21.5, 21.6, 21.9, 21.11, 21.12, 21.17, 21.18, 21.21: From Marsh, W.M., and Dozier, J., *Landscape: An Introduction to Physical Geography,* Addison-Wesley, 1981. Figures 21.13, 21.16: From Spencer, E., *Physical Geology,* Addison-Wesley, 1983. Figure 21.14: Photograph by Jeff Dozier. Figure 21.15 and Note 21: From U.S. Army Corps of Engineers. AIT 21.1: Photograph by W.M. Marsh.

CHAPTER 22
Figures 22.1, 22.19: From *Encyclopedia of Geomorphology,* edited by R.W. Fairbridge, copyright 1968 Dowdan, Hutchinson and Ross, Inc., Stroudsburg, Pa. Used by permission. Figures 22.2, 22.3, 22.4, 22.5, 22.7, 22.11, 22.12, 22.15, 22.21, 22.22, 22.23, 22.24: From Marsh, W.M., and Dozier, J., *Landscape: An Introduction to Physical Geography,* Addison-Wesley, 1981. Figure 22.6: From Hausser, C.L., and Marcus, M.G., "Historical Variations of Lemon Creek Glacier, Alaska, and Their Relationship to the Climatic Record," *Journal of Glaciology,* 1964. Used by permission of the International Glaciology Society. Figure 22.8*a*: Photograph by U.S. Navy. Figure 22.8*b*: Map from U.S. Central Intelligence Agency. Figure 22.9. Photograph by J. Dozier. Figures 22.10, 22.13: Photographs by W. Marsh. Figures 22.14, 22.16, 22.19, 22.20: Photographs by U.S. Geological Survey. Note 22*a* & *b*: Photographs by A.L. Washburn.

CHAPTER 23
Figures 23.1, 23.2, 23.3, 23.5, 23.6, 23.8, 23.12, 23.14, 23.16, 23.22: From Marsh, W., and Dozier, J., *Landscape: An Introduction to Physical Geography,* Addison-Wesley, 1981. Figure 23.4: From Environmental Data Service, NOAA. Figures 23.7, 23.10: Photographs by W. Marsh. Figure 23.9: From Bagnold, R.A., *Physics of Blown Sand and Desert Dunes,* The Royal Society, Proceedings Series

A, Volume 167. Used by permission of the author and The Royal Society. Figure 23.15: Photograph courtesy of Ralpho/Photo Researchers.

CHAPTER 24
Figure 24.6: Photograph courtesy of U.S. Forest Service. Figure 24.7: Photograph courtesy of Royal Canadian Air Force. Figures 24.9, 24.10: Photographs by Troy L. Péwé. Used by permission.

APPENDIXES
Figure A.5: Drawing by Joe Smigiel. Figure A.6: From U.S. Geological Survey. Figure A.7: From Marsh, W.M., *Landscape Planning: Environmental Applications*, Addison-Wesley, 1983. Figure B.2: Image courtesy of the Environmental Research Institute of Michigan. Figure B.3: From NOAA.

Index